Springer Series in Statistics

More information about this series at http://www.springer.com/series/692

Thomas W. Yee

Vector Generalized Linear and Additive Models

With an Implementation in R

 Springer

Thomas W. Yee
Department of Statistics
University of Auckland
Auckland, New Zealand

ISSN 0172-7397   ISSN 2197-568X   (electronic)
Springer Series in Statistics
ISBN 978-1-4939-2817-0   ISBN 978-1-4939-2818-7   (eBook)
DOI 10.1007/978-1-4939-2818-7

Library of Congress Control Number: 2015945942

Springer New York Heidelberg Dordrecht London

Printed on acid-free paper

Springer Science+Business Media LLC New York is part of Springer Science+Business Media (www.springer.com)

To my parents and Selina and Annie

Preface

Beauty will result from the form and correspondence of the whole, with respect to the several parts, of the parts with regard to each other, and of these again to the whole; that the structure may appear an entire and compleat body, wherein each member agrees with the other, and all necessary to compose what you intend to form.
—The First Book of Andrea Palladio's Architecture, 1570

In the early 1970s, the *generalized linear model* (GLM) class of statistical models was proposed by Nelder and Wedderburn (1972), providing a unified framework for several important regression models. They showed that the linear model, Poisson regression, logistic regression and probit analysis, and others could be treated as special cases of GLMs, and that one algorithm could be used to estimate them all. The unified GLM framework also provides an elegant overriding theoretical structure resulting in inference, diagnostics, software interface, etc. that applies to all of them. Prior to GLMs, these methods were largely treated as unrelated. Since then, GLMs have gained universal acceptance in the statistical world, and as Senn (2004, p. 7) writes, "Nelder and Wedderburn was a paper that changed the statistical landscape for ever and it is simply impossible now to envisage the modelling world without it."

Although GLMs were a great advance, they are largely confined to one-parameter distributions from an exponential family. There are many situations where practical data analysis and regression modelling demands much greater flexibility than this. To this end, this book describes a much larger and more flexible statistical framework for fixed-effects regression modelling that greatly extends GLMs. It comprises about half-a-dozen major classes of statistical models, and having at its heart two classes, called *vector generalized linear models* (VGLMs) and *vector generalized additive models* (VGAMs). (Other classes are listed below.) Each class is related to each other in a natural way within this framework, e.g., VGAMs are a smooth or data-driven version of VGLMs.

The purpose of this book is to introduce the framework and each major subclass of models. VGLMs might be thought of loosely as multivariate GLMs, which are not confined to the exponential family. VGAMs replace the linear functions of VGLMs by smooths. The remaining classes can be thought of as extensions of VGLMs and VGAMs. As a software implementation of our approach, the R package **VGAM** (and its companion **VGAMdata**) are used. Following the grand tradition of GLMs, we demonstrate its usefulness for data analysis over a wide range of problems and application areas. It is hoped that, in a way similar to what was pioneered by GLMs, the overall framework described in this book will provide a natural vehicle for thinking about regression modelling, and performing much more of applied statistics as a coherent whole.

What advantages does the VGLM/VGAM framework confer? It provides the same benefits that GLMs gave, but on a much larger scale. For example, the theory and methodology of this book unifies areas such as univariate distributions, categorical data analysis, aspects of quantile regression, and extremes. And the advantages which generalized additive models add to GLM-type analyses in terms of smoothing, VGAMs do for VGLMs. It has long been my belief that what should be 'ordinary' and 'routine' regression modelling in applied statistics has been hampered by a lack of common framework. For VGLMs/VGAMs, the user has the freedom to easily vary model elements within a large flexible framework, e.g., currently **VGAM** implements over 150 family functions.

The underlying conception is to treat almost all distributions and classical models as generalized regression models. By 'classical', we mean models and distributions which are amenable to estimation by first and second derivative methods, particularly Fisher scoring. For years there has been a need to broaden the scope of GLMs and GAMs, and to fortify it with some necessary infrastructure to make them more fully operable. The end is that the framework instils structure in many classical regression models.

Audience

The book was written with three types of readers in mind. The first are existing users of the **VGAM** R package needing a primary source for the methodology, and having sufficient examples and details to be useful. The second are people interested in a new and very general modelling framework. The third are students/teachers in courses on general regression following a basic regression course on GLMs. For these, some notes for instructors are given below.

Some assumptions have been made about the background of the reader. Firstly is a basic working knowledge of R. There are now many books that provide this, e.g., Venables and Ripley (2002), Dalgaard (2008), Maindonald and Braun (2010), Cohen and Cohen (2008), Zuur et al. (2009), de Vries and Meys (2012). Further references are listed at the end of Chap. 8. Second is a mid-undergraduate level knowledge of statistical theory and practice. Thirdly, some chapters assume a basic familiarity with linear algebra and calculus up to mid-undergraduate level.

Some of the applied chapters (in Part II) would benefit from some prior exposure to those subject areas, e.g., econometrics, plant ecology, etc. because only a minimal attempt can be made to establish the background, motivation, and notation there.

How This Book is Organized

There are two parts. Part I describes the general theory and computational details behind each major class of models. These classes are displayed in the flowchart of Fig. 1.2. Readers familiar with LMs, GLMs, and GAMs might jump directly into the VGLM and VGAM chapters (Chap. 2 is a summary of these topics), otherwise they could go through sequentially from the beginning. However, Chap. 1 (especially Sect. 1.3) is crucial, because it sets the scene and gives a brief overview of the entire framework. The other major classes of models are various generalizations of reduced-rank methods. The main motivation for these is dimension reduction, and they consequently operate on latent variables. They are:

- RR-VGLMs: these *reduced-rank VGLMs* (Chap. 5) are based on linear combinations of the explanatory variables, and should be useful for many readers.
- QRR-VGLMs: these *quadratic RR-VGLMs* (Chap. 6) being relevant to mainly ecologists could be generally skipped on first reading.
- RR-VGAMs: these allow for *constrained additive ordination* (CAO; Chap. 7), which is a smooth version of RR-VGLMs.

Chapter 8 on the VGAM package should be read by all software users.

Part II explores some major application areas. Practitioners of certain topics in Part II will usually not need to concern themselves with the other parts, because they are mainly separate. However, readers will be able to at least browse through the range of other applications there. The breadth of coverage should demonstrate the versatility of the framework developed in Part I; I feel this is a major attractive feature of the book. Chapter 18 is a specialist chapter addressed more to R programmers wishing to extend the capabilities of the current software, e.g., by writing new VGAM family functions.

Most acronyms are listed on pp. xxiii–xxiv, and the appendices, notation, and glossary are located at the back of the book. R packages are denoted in sans-serif fonts, e.g., stats, splines; and R commands in typewriter font and functions ending in parentheses, e.g., coef(). All logarithms are to base e unless otherwise specified.

The scope of this book is broad. To keep things to a manageable size, many topics which I would like to have included have had to be omitted. Rather than citing hundreds of journal articles, I have largely cited books and review papers. Other references and sources of data, etc. can be found in the packages' online help files. I apologize beforehand for many peoples' work that should probably have been cited but are not, due to space limitations or ignorance. A slightly unfortunate consequence of the book's breadth is that some of the notation had to be recycled, however, despite some symbols having multiple meanings, this should not raise excessive confusion because the topics where they are used are far enough apart.

Topics considered too specialized or technical for most readers have been marked with a dagger (†); these may be safely skipped without losing the overall discourse.

Book Website and Software

Resources are available at the book's webpage, starting at

$$\texttt{http://www.stat.auckland.ac.nz/~yee}$$

Amongst other things, these should include R scripts, complements, errata, and any latest information. Readers wishing to run the software examples will need access to R (see `http://www.R-project.org`), and the VGAM and VGAMdata packages installed along side. The latter comprises some data used here. Version 1.0-0 or later for both packages are needed for compatibility with this book.

Note for Instructors

Most of the contents are pitched at the level of a senior undergraduate or first-year postgraduate in statistics. It might be considered a follow-on course after a first course in GLMs. Students are assumed to have a reasonable working R background already, because only a few important R topics that are crucial are covered (at a cursory level too) in the first chapter. The first few chapters of Fox and Weisberg (2011) would be useful preparation, as well as those listed above.

For a course based on this book, it is recommended that about half of the time be devoted to Part I, so that the overall framework can be seen. The remaining time might be used on a selection of topics drawn from Part II depending on interest and need. Each Part II chapter is at an introductory level, and the references at the end of each chapter can be pursued for a deeper coverage.

Some of the exercises might be suitable for homework or tutorial purposes. Overall, they are a blend of short analyses involving real and simulated data, mathematical statistics problems, and statistical computing (R programming) tasks. While most problems are short, some are less straightforward and require more advanced R programming to obtain an elegant solution. The most demanding or time-consuming problems are marked with a dagger (†) and should be considered optional or for a sizeable project.

The book is primarily focused on estimation. Theoretical rigour that was unnecessary to the main messages has been omitted, and the presentation is chiefly driven by a pragmatic approach. Most of the material is self-contained, and I have tried to strike a balance between theory, computational details and practice for an applied statistician.

Acknowledgements

This is a synthesis of work that started while I was a PhD student at the University of Auckland, and it has spanned the two decades since. Not surprisingly, there are many people to thank, both directly and indirectly. I am indebted to my PhD supervisors Chris Wild and Neil Mitchell who had a very positive effect during the formative years of my academic life. Ross Ihaka and Trevor Hastie pointed me in the right direction soon after in those very early days.

Many graduate students have contributed to this body of work in one way or another over the years, especially James Gray, Kai Huang, Víctor Miranda, and Alvin Sou. Successive heads of departments have helped arrange my teaching to allow for a faster progression of this work. Additionally, I wish to acknowledge the hospitality of many (mostly statistics) departments all across this terrestrial ball, which I have had the good fortune to visit over the years.

Many thanks are extended to Garry Tee for proofreading the book in its entirety. Also to many others who have given valuable feedback on portions thereof; those are Michel Dedual, Stéphane Guindon, Lian Heng, Richard Huggins, Murray Jorgensen, Ioannis Kosmidis, Louise McMillan, Alastair Scott, Alec Stephenson, Jakub Stoklosa, Berwin Turlach, Yong Wang, Alan Welsh, and Chris Wild. All errors, however, are my own. Thanks to John Young for his translation of several quotes from the ancients used at the beginning and ending of several chapters. For help with the typesetting (especially LaTeX and knitr), my thanks to Trevor Hastie, Alan Izenman, Andrew Robinson, and Yihui Xie. Also, the R-core members and the R community at large have been of great direct and indirect help; these include the countless users who have sent in bug reports and suggestions over the years. Feedback from workshop attendees has been helpful too. I would also like to thank Hannah Bracken at Springer.

Last but not least, thanks to my wife Selina who has been very understanding during the intensive preparations.

Future Directions

This book is a snapshot from a project that is a work in progress, and it will probably never be finished. Large tracts of theory and methodology presently lie undeveloped, and many existing parts require refinement. It is hoped that, over time, bugs will be attended to and further additions made to the software. Consequently, a warning to the reader: low-level internal changes to the software are ongoing, however high-level usage will change less. (Good coding practice such as the use of the generic functions listed in Chap. 8, as well as the suggestions of Sect. 8.5.1, should help minimize such problems). The NEWS file in the packages lists successive changes between versions.

Over time, I hope to further develop the theory and continue to add to the functionality of VGAM. Due to the volume of emails received (which exceeds the number I can reply to), please send bug reports only. And my apologies if I am unable to reply at all!

Auckland, New Zealand   Thomas W. Yee
May 2015

It is a good thing for an uneducated man to read books of quotations.
—Winston Churchill, *My Early Life*

Contents

Acronyms

ALD	Asymmetric Laplace distribution (Sect. 15.3.2)
ALS	Asymmetric least squares (Sect. 15.4.1)
AML	Asymmetric maximum likelihood (Sect. 15.4.2)
BPM	Bivariate probit model (Sect. 10.3.3)
CAO	Constrained additive ordination (Chap. 7, Fig. 1.2)
CDF	Cumulative distribution function, $P(Y \leq y)$ or $P(Y_1 \leq y_1, Y_2 \leq y_2, \ldots)$ (Sect. 11.1.1)
CQO	Constrained quadratic ordination, cf. QRR-VGLM (Chap. 6, Fig. 1.2)
CV	Cross-validation (Sect. 2.4.7.6)
EIM	Expected (Fisher) information matrix (3.11)
ENDF	Effective nonlinear degrees of freedom (Sects. 2.4.7.4, 7.2.1)
EVT	Extreme value theory (Chap. 16)
flop	Floating point operation
GAM	Generalized additive model (Sect. 2.5)
GEV	Generalized extreme value (Sect. 16.2)
GLM	Generalized linear model (Sect. 2.3)
GLS	Generalized least squares (Sect. 2.2.2)
GPD	Generalized Pareto distribution (Sect. 16.3)
GRC	Goodman's RC(R) model (Sect. 5.7)
HWE	Hardy-Weinberg equilibrium (Sect. 14.5)
iff	If and only if, $\iff$
i.i.d.	Independent and identically distributed
IRLS	Iteratively reweighted least squares (Sect. 3.2)
LHS	Left hand side
LM	Linear model (Sect. 2.2)
LRT	Likelihood ratio test (Sect. A.1.4.1)
LS	Least squares (Sect. 2.2)
LSD	Least significant difference (Sect. 5.7.3.4)
MGF	Moment generating function
MLE	Maximum likelihood estimation or estimator
MLM	Multinomial logit model (Sects. 1.2.4, 14.2)
MPV	Median predicted value (Sect. 16.2.2)
MVUE	Minimum variance unbiased estimator (Sect. A.1.3.1)
NBD	Negative binomial distribution (Sects. 1.2.2, 11.3)

NCS   Natural cubic spline ((2.47), Sects. 2.4.3.2, 4.2.1)
NEF   Natural exponential family (Sect. 11.2)
NR   Newton-Raphson (Sects. 3.2, A.1.2.4)
OIM   Observed information matrix ((3.15), Sect. A.1.2.4)
OLS   Ordinary least squares (Sect. 2.2.2)
PDF   Probability density function, $f(y)$ or $f(\boldsymbol{y})$
PGF   Probability generating function (Sect. 11.5)
PMF   Probability mass function, $P(Y = y)$ or $P(Y_1 = y_1, Y_2 = y_2, \ldots)$
               (Sect. 11.1)
QLP   Quasi-likelihood Poisson (Sect. 11.3.4.1)
QRR-VGLM   Quadratic reduced-rank vector generalized linear model (Chap. 6,
               Fig. 1.2)
QSE   Quasi-standard error (Sect. 5.7.3)
RCIM   Row–column interaction model (Sect. 5.7, Fig. 1.2)
ResSS   Residual sum of squares (2.4)
RHS   Right hand side
RKHS   Reproducing kernel Hilbert space (Sect. 4.2.1.7)
RR   Reduced-rank (Sect. 1.3.3, Chap. 5)
RRR   Reduced-rank regression (Sect. 1.3.3, Chap. 5)
SE   Standard error (an estimated standard deviation)
SFS   Simulated Fisher scoring (Sect. 9.2.2)
SUR   Seemingly unrelated regressions (Sects. 4.2.2, 10.2.3)
SVD   Singular value decomposition (Sect. A.3.4)
UQO   Unconstrained quadratic ordination (Sect. 6.7)
VCM   Varying coefficient model (Sect. 10.2.1)
VGAM   Vector generalized additive model (Chap. 4, Fig. 1.2)
VGLM   Vector generalized linear model (Chap. 3, Fig. 1.2)
VLM   Vector linear model (3.22), Fig. 1.2
VMM   Vector measurement model (4.2)
WLS   Weighted least squares (Sect. 2.2.2)
ZA   Zero-altered (Sect. 17.1)
ZI   Zero-inflated (Sect. 17.1)

Part I
General Theory

Chapter 1
Introduction

We hope that the approach developed in this paper will prove to be a useful way of unifying what are often presented as unrelated statistical procedures, and that this unification will simplify the teaching of the subject to both specialists and non-specialists.
—Nelder and Wedderburn (1972)

1.1 Introduction

In 1972 a class of statistical models called *generalized linear models* (GLMs) was proposed which provided a unified framework for several important regression models. The paper, Nelder and Wedderburn (1972), showed that the linear model (LM), Poisson regression, logistic regression plus others were special cases of GLMs, and that an algorithm called *iteratively reweighted least squares* (IRLS) could be used for their estimation. Prior to the advent of GLMs these methods were largely treated as disparate. Since then GLMs have become accepted universally by applied and theoretical statisticians. The unified framework results in streamlining inference, diagnostics, computation, software interface, etc.

Unfortunately, GLMs are mainly restricted to one-parameter distributions (or two with an unknown scale parameter) within the classical exponential family. Routine regression modelling and data analysis requires much greater flexibility than afforded by GLMs. This book describes a very large class of models called *vector generalized linear models* (VGLMs) which are purposely general so as to confer greater usefulness. This class can be extended naturally in various directions so as to offer a suite of variants, e.g., *vector generalized additive models* (VGAMs) for smoothing, reduced-rank VGLMs (RR-VGLMs) for dimension reduction, etc. The purpose of this book is to describe all these classes for regression modelling and how they are interrelated. As a software implementation of the framework, the R package VGAM and its companion VGAMdata are also described. Altogether, it is hoped that this work will provide a broader and unified framework to perform generalized regression modelling.

© Thomas Yee 2015
T.W. Yee, *Vector Generalized Linear and Additive Models*,
Springer Series in Statistics, DOI 10.1007/978-1-4939-2818-7_1

1.1.1 Outline of This Book and a Quick Start

Users familiar with GLMs and GAMs who are wishing to get into VGLMs and VGAMs as quickly as possible might look at the following sections, in decreasing order of cost-benefit.

- Sections 1.2–1.4 to obtain an overview.
- Chapter 8—to obtain an overall picture of the **VGAM** package.
- Part II can be perused, depending on interest, for some basic theory and examples of fitting some models.
- Chapters 3 and 4—to find out what VGLMs and VGAMs are in detail.
- The rest of Part I. The classes RR-VGLMs, QRR-VGLMs, etc. are important variants but some users may not need them.
- The rest of Part II.

This chapter exposes the reader to the main ideas of the VGLM/VGAM framework, so that users can start immediately. Much of the core/essential details are touched upon here, and when read in conjunction with Chap. 8, should provide a minimal amount of information needed to conduct the simplest type of **VGAM** analysis. The rest of the book describes these and other advanced features, in addition to the technical details behind them.

VGAM implements several large classes of regression models, of which vector generalized linear and additive models are most commonly used (Table 1.1). In a nutshell the overall key ideas or 'features' are:

- parameter link functions $g_j(\theta_j)$ applied to all parameters,
- multivariate responses, and sometimes multiple responses too,
- linear predictors $\eta_j = \boldsymbol{\beta}_j^T \boldsymbol{x}$ and additive predictors $\eta_j = \sum_{k=1}^{d} f_{(j)k}(x_k)$,
- constraints on the functions $(\mathbf{H}_1, \ldots, \mathbf{H}_p)$,
- η_j-specific covariates (i.e., $\eta_j(\boldsymbol{x}_{ij})$) via the `xij` facility,
- reduced-rank regression (RRR), latent variables $\boldsymbol{\nu} = \mathbf{C}^T \boldsymbol{x}_2$, ordination,
- Fisher scoring, IRLS, maximum likelihood estimation, vector smoothing,
- the **VGAM** package. It presently fits over 150 models and distributions.

Altogether the entirety represents a broad and unified framework. Section 1.3 gives a brief overview of the general framework and this book.

The scope of **VGAM** is potentially very broad, and Part II of this book tries to convey this by a few select application areas.

1.2 Six Illustrative Models

To motivate VGLMs/VGAMs and their variants, let's consider a few specific models from applied statistics. They serve as concrete examples to illustrate structures described throughout the whole book, e.g., constraints such as parallelism and exchangeability, η_j-specific covariates, and intercept-only parameters. We also provide some initial information on how some of these models may be fitted by **VGAM**. Before doing so, we give some basic notation and a description of VGLMs and VGAMs. The glossary (pp.561–565) gives most of the notation of this book.

The data is written $(\boldsymbol{x}_i, \boldsymbol{y}_i)$ for $i = 1, \ldots, n$, where $\boldsymbol{x}_i$ is a vector of explanatory variables, $\boldsymbol{y}_i$ is a (possibly vector) response, for n independent observations. Sometimes we drop the i and write $\boldsymbol{x} = (x_1, \ldots, x_p)^T$ to focus only on the variables, with $x_1 = 1$ denoting an intercept if there is one.

We wish to fit a regression model involving parameters θ_j. Basically, VGLMs model each parameter, transformed if necessary, as a linear combination of the explanatory variables. That is,

$$g_j(\theta_j) \;=\; \eta_j \;=\; \boldsymbol{\beta}_j^T \boldsymbol{x} \;=\; \beta_{(j)1}\, x_1 + \cdots + \beta_{(j)p}\, x_p, \quad j = 1, \ldots, M, \qquad (1.1)$$

where g_j is a parameter link function such as a logarithm or logit (1.17). Note that potentially *every* parameter is modelled using *all* explanatory variables x_k, and the parameters need not be a mean such as for GLMs.

VGAMs extend (1.1) to

$$g_j(\theta_j) \;=\; \eta_j \;=\; \sum_{k=1}^{d} f_{(j)k}(x_k), \quad j = 1, \ldots, M, \qquad (1.2)$$

i.e., an additive model for each parameter. The functions $f_{(j)k}$ are merely assumed to be smooth and are estimated by smoothers such as splines, therefore the whole approach is data-driven rather than model-driven. In (1.2) $f_{(j)1}(x_1) = \beta_{(j)1}$ is simply an intercept.

The distinction between p in (1.1) and d in (1.2) is made more apparent in Sects. 1.5.2.2 and 1.5.3. For now, p is the number of parameters looking at one η_j at a time, and d is the number of terms in an S formula.

1.2.1 The Linear Model

For a response Y assumed to be distributed as $N(\mu, \sigma^2)$ with

$$\mu \;=\; \eta_1 \;=\; \boldsymbol{\beta}_1^T \boldsymbol{x}, \qquad (1.3)$$

the *linear model* (LM) is statistically ubiquitous. The standard theory of GLMs treats σ as a scale parameter, because the classical exponential family is restricted to only one parameter. However, in the VGLM/VGAM framework it is more natural to couple (1.3) with

$$\log \sigma \;=\; \eta_2 \;=\; \boldsymbol{\beta}_2^T \boldsymbol{x}. \qquad (1.4)$$

A log link is generally suitable because $\sigma > 0$. Incidentally, a consequence is that some mean-variance relationships can be modelled together, such as in Sect. 5.5.1.

Modelling η_2 as *intercept-only* means that the typical assumption of constant variance (homoscedasticity) is made:

$$\log \sigma \;=\; \eta_2 \;=\; \beta_{(2)1}. \qquad (1.5)$$

If there were a covariate x_2, then we might test the null hypothesis $H_0 : \beta_{(2)2} = 0$ in

$$\log \sigma \;=\; \eta_2 \;=\; \beta_{(2)1} + \beta_{(2)2}\, x_2 \qquad (1.6)$$

as a test for no heteroscedasticity.

The VGAM family $\texttt{uninormal()}$ implements (1.3)–(1.4), the response being an univariate normal. The typical call is of the form

```
vglm(y ~ x2 + x3 + x4, family = uninormal, data = udata)
```

where $x_1 = 1$ is implicitly the intercept term. The function $\texttt{uninormal()}$ assigned to the $\texttt{family}$ argument is known as a VGAM *family function*. It makes the usual LM assumptions (Sect. 2.2): independence and normality of the errors $y_i - \mu_i$, linearity (1.3), and constant variance (1.5). It is good style for the data frame $\texttt{udata}$ to hold all the variables, viz. $\texttt{y, x2, x3}$ and $\texttt{x4}$.

1.2.2 Poisson and Negative Binomial Regression

The Poisson distribution is as fundamental to the analysis of count data as the normal (Gaussian) is to continuous responses. It has the simple probability mass function (PMF)

$$P(Y = y; \mu) \;=\; \frac{e^{-\mu}\,\mu^y}{y!}\,, \qquad y = 0, 1, 2, \ldots, \qquad \mu > 0, \qquad (1.7)$$

resulting in $E(Y) = \mu = \mathrm{Var}(Y)$. As μ is positive,

$$\eta \;=\; \log \mu \qquad (1.8)$$

is a very natural recommendation. This loglinear association is highly interpretable: an increase of the kth variable by one unit, keeping other variables fixed, implies

$$\log \mu(x_1, \ldots, x_{k-1}, x_k + 1, x_{k+1}, \ldots, x_p) - \log \mu(x_1, \ldots, x_p) \;=\; \beta_k. \qquad (1.9)$$

That is,

$$\mu(x_1, \ldots, x_{k-1}, x_k + 1, x_{k+1}, \ldots, x_p) \;=\; \mu(\boldsymbol{x} + \boldsymbol{e}_k) \;=\; \mu(\boldsymbol{x}) \cdot e^{\beta_k} \qquad (1.10)$$

so that e^{β_k} is the multiplicative effect on $\mu(\boldsymbol{x})$ of increasing x_k by one unit. Hence a positive/negative value of β_k corresponds to an increasing/decreasing effect, respectively.

Counts sometimes arise from an underlying rate, e.g., if λ is the mean rate per unit time, then $\mu = \lambda t$ is the mean number of events during a period of time t. For example, if λ is the mean number of earthquakes per annum in a specified geographical region which exceed a certain magnitude, then $\mu = \lambda t$ is the expected number of earthquakes during a t-year period. For such, (1.8) means that

$$\log \mu \;\equiv\; \eta \;=\; (\log t) + \log \lambda. \qquad (1.11)$$

When we want to model the rate, adjusting for time t, and provided that t is known, then the Poisson regression (1.11) involves (known) *offsets*—see Sect. 3.3. If all time periods are equal, then the offsets $\log t_i$ can be omitted.

The same type of calculation holds with rates arising from a population of N individuals, say. Then $\mu = \lambda N$ where λ is the rate per unit size of population (e.g., 100,000 people) and

$$\eta = (\log N) + \log \lambda \qquad (1.12)$$

also involves an offset. Of course, (1.11) and (1.12) can be combined for $\mu_i = \lambda_i t_i N_i$, e.g., populations of various sizes with different follow-up times.

In practice, the property that the mean and variance coincide is often not realized with real data. This may be indicated by the sample variance exceeding the sample mean. This feature is called *overdispersion* with respect to the Poisson distribution, and there are several common causes of it (Sect. 11.3). The most common remedy is to allow

$$\mathrm{Var}(Y) = \phi \cdot \mu \qquad (1.13)$$

in the standard Poisson regression (1.8), where ϕ is estimated by the method of moments; then $\hat{\phi} > 1$ indicates overdispersion relative to a Poisson distribution. The quasi-Poisson estimate $\hat{\boldsymbol{\beta}}$ coincides with the usual maximum likelihood estimate (MLE).

A better method of handling overdispersed data is to perform negative binomial (NB) regression. An NB random variable Y has a probability function that can be written as

$$P(Y = y; \mu, k) = \binom{y + k - 1}{y} \left(\frac{\mu}{\mu + k}\right)^y \left(\frac{k}{k + \mu}\right)^k, \quad y = 0, 1, 2, \ldots, \quad (1.14)$$

with positive parameters μ ($= E(Y)$) and k. The quantity k^{-1} is known as the dispersion or ancillary parameter, and the Poisson distribution is the limit as $k \to \infty$. The NB has $\mathrm{Var}(Y) = \mu + \mu^2/k \geq \mu$, so that overdispersion relative to the Poisson is accommodated—however, underdispersion ($\phi < 1$) isn't. The NB can be motivated in various ways; one is as a mixture where the μ parameter of a Poisson distribution is gamma distributed (Sect. 11.3).

Many software implementations are restricted to an intercept-only estimate of k (called the NB-2 by some authors), e.g., one cannot fit $\log k = \beta_{(2)1} + \beta_{(2)2}\, x_2$. In contrast, the VGLM/VGAM framework can naturally fit

$$\log \mu = \eta_1 = \boldsymbol{\beta}_1^T \boldsymbol{x}, \qquad (1.15)$$
$$\log k = \eta_2 = \boldsymbol{\beta}_2^T \boldsymbol{x}, \qquad (1.16)$$

which is known as an NB-H. In **VGAM** this is achieved by a call of the form

```
vglm(y ~ x2 + x3 + x4, family = negbinomial(zero = NULL), data = ndata)
```

Implicit in (1.16) is the notion that any positive parameter would be better having the log link as the default than an identity link $g_j(\theta_j) = \theta_j$.

One feature that many **VGAM** family functions possess is the ability to handle *multiple responses*. For example,

```
vglm(cbind(y1, y2) ~ x2 + x3 + x4, family = negbinomial(zero = NULL), data = ndata)
```

Table 1.1 A simplified summary of **VGAM** and most of its framework. The latent variables $\boldsymbol{\nu} = \mathbf{C}^T \boldsymbol{x}_2$, or $\nu = \boldsymbol{c}^T \boldsymbol{x}_2$ if rank $R = 1$. Here, $\boldsymbol{x}^T = (\boldsymbol{x}_1^T, \boldsymbol{x}_2^T)$. Abbreviations: A = additive, C = constrained, I = interaction, Q = quadratic, RC = row–column, RR = reduced-rank, VGLM = vector generalized linear model. See also Fig. 1.2 and Table 5.1.

$\boldsymbol{\eta} = (\eta_1, \ldots, \eta_M)^T$	Model	Modelling function	Reference
$\mathbf{B}_1^T \boldsymbol{x}_1 + \mathbf{B}_2^T \boldsymbol{x}_2 \; (= \mathbf{B}^T \boldsymbol{x})$	VGLM	`vglm()`	Yee and Hastie (2003)
$\mathbf{B}_1^T \boldsymbol{x}_1 + \sum_{k=p_1+1}^{p_1+p_2} \mathbf{H}_k \, \boldsymbol{f}_k^*(x_k)$	VGAM	`vgam()`	Yee and Wild (1996)
$\mathbf{B}_1^T \boldsymbol{x}_1 + \mathbf{A}\,\boldsymbol{\nu}$	RR-VGLM	`rrvglm()`	Yee and Hastie (2003)
$\mathbf{B}_1^T \boldsymbol{x}_1 + \mathbf{A}\,\boldsymbol{\nu} + \begin{pmatrix} \boldsymbol{\nu}^T \mathbf{D}_1 \boldsymbol{\nu} \\ \vdots \\ \boldsymbol{\nu}^T \mathbf{D}_M \boldsymbol{\nu} \end{pmatrix}$	QRR-VGLM	`cqo()`	Yee (2004a)
$\mathbf{B}_1^T \boldsymbol{x}_1 + \sum_{r=1}^{R} \boldsymbol{f}_r(\nu_r)$	RR-VGAM	`cao()`	Yee (2006)
$(\beta_0 + \alpha_i)\,\mathbf{1} + \boldsymbol{\gamma} + \mathbf{A}\boldsymbol{\nu}_i$	RCIM	`rcim()`	Yee and Hadi (2014)

regresses two responses simultaneously. The responses are treated independently. An important use is that some regression coefficients are allowed to common to all responses, e.g., the effect of a variable such as x3 might allowed to be the same for all the responses' means. In the above example we have $\boldsymbol{\eta} = (\eta_1, \eta_2, \eta_3, \eta_4)^T = (\log \mu_1, \log k_1, \log \mu_2, \log k_2)^T$.

There are other variants of the NB distribution within the VGLM/VGAM framework and implemented in **VGAM**, e.g., NB-1, the zero-inflated and zero-altered versions (ZINB and ZANB), as well as an interesting reduced-rank variant known as the RR-NB (also referred to as the NB-P); see Sect. 5.5.2.3. Other aspects of the NB distribution are expounded in Sects. 11.3 and 17.1.

1.2.3 Bivariate Odds Ratio Model

Logistic regression, where we have a single binary response Y, is one of the most well-known techniques in the statistician's toolbox. It is customary to denote $Y = 1$ and 0 as "success" and "failure", respectively, so that $E(Y)$ equals the probability of success, $P(Y = 1) = p$, say. Then the logistic regression model can be written

$$\text{logit } p(\boldsymbol{x}) \;\equiv\; \log \frac{p(\boldsymbol{x})}{1 - p(\boldsymbol{x})} \;=\; \eta(\boldsymbol{x}), \qquad (1.17)$$

where the quantity $p/(1 - p)$ is known as the *odds* of event $Y = 1$. Odds are very interpretable, e.g., a value of 3 means that the event is 3 times more likely to occur than not occur. Also, if $\beta_{(1)k}$ is the coefficient of x_k in (1.17), then $\beta_{(1)k}$ is the log odds ratio of $Y = 1$ for an observation with $x_k + 1$ versus an observation with x_k (keeping all other variables in $\boldsymbol{x}$ fixed at their values). That is, the odds of $Y = 1$ for an observation with $x_k + \Delta$ is $\exp\{\beta_{(1)k}\,\Delta\}$ multiplied by the odds of $Y = 1$ for an observation with x_k, keeping all other variables in $\boldsymbol{x}$ fixed.

In some applications it is natural to measure *two* binary responses, Y_1 and Y_2, say, e.g., measurements of deafness in both ears, the presence/absence of cataracts in elderly patients' eyes, the presence/absence of two plant species at sites in a very large forest region, etc. Then a natural regression model for such is to couple two logistic regressions together with an equation for the *odds ratio*. (The responses are often dependent, and the odds ratio is a natural measure for the association between two binary variables). Specifically, the *bivariate odds ratio model* (also known as a *bivariate logistic model*) is

$$\text{logit } p_j(\boldsymbol{x}) = \eta_j(\boldsymbol{x}), \qquad j = 1, 2, \tag{1.18}$$

$$\log \psi(\boldsymbol{x}) = \eta_3(\boldsymbol{x}). \tag{1.19}$$

The joint probability $p_{11}(\boldsymbol{x})$ can be obtained from the two marginals $p_j(\boldsymbol{x}) = P(Y_j = 1|\boldsymbol{x})$ and the non-negative odds ratio

$$\psi(\boldsymbol{x}) = \frac{p_{00}(\boldsymbol{x})\,p_{11}(\boldsymbol{x})}{p_{01}(\boldsymbol{x})\,p_{10}(\boldsymbol{x})} = \frac{P(Y_1 = 0, Y_2 = 0|\boldsymbol{x})\,P(Y_1 = 1, Y_2 = 1|\boldsymbol{x})}{P(Y_1 = 0, Y_2 = 1|\boldsymbol{x})\,P(Y_1 = 1, Y_2 = 0|\boldsymbol{x})}. \tag{1.20}$$

Then Y_1 and Y_2 are independent if and only if $\psi = 1$.

As its name suggests, ψ is the ratio of two odds:

$$\psi(\boldsymbol{x}) = \frac{\text{odds}(Y_1 = 1|Y_2 = 1, \boldsymbol{x})}{\text{odds}(Y_1 = 1|Y_2 = 0, \boldsymbol{x})}, \tag{1.21}$$

i.e., ψ is the odds ratio of event $Y_1 = 1$ for an observation with covariates $(Y_2 = 1, \boldsymbol{x})$ relative to an observation with covariates $(Y_2 = 0, \boldsymbol{x})$. Equation (1.21) is quite interpretable, and this can be seen especially when the joint distribution (Y_1, Y_2) is presented as a 2×2 table.

In the VGLM/VGAM framework there is no reason why other link functions could not be used for the marginal probabilities (1.18)—such as the probit $[\Phi^{-1}(p)]$ or complementary log–log $[\log(-\log(1 - p))]$—indeed, the `binom2.or()` VGAM family function accommodates this. In fact, a different link for each linear predictor is viable. A summary of some link functions in VGAM is given in Table 1.2. For users wanting to write their own link function, some details are in Chap. 18.

The typical call to fit the model in VGAM is of the form

```
vglm(cbind(y00, y01, y10, y11) ~ x2 + x3, family = binom2.or, data = bdata)
```

where the LHS matrix contains the joint frequencies, e.g., $\text{y01} = (Y_1 = 0, Y_2 = 1)$. The argument name `family` is for upward compatibility and simplicity: it mimics the same argument in `glm()` which is for the half-dozen exponential family members only. For such, the concept of an error distribution can be made rigorous. In VGAM a `family` is viewed loosely, and usually it refers to some full-likelihood specified statistical model worth fitting in its own right.

Table 1.2 Some **VGAM** link functions currently available. They are grouped approximately according to their domains. As with the entire book, all logarithms are natural: to base e.

Function	Link $g_j(\theta_j)$	Domain of θ_j	Link name
`cauchit()`	$\tan(\pi(\theta - \frac{1}{2}))$	$(0,1)$	Cauchit
`cloglog()`	$\log\{-\log(1-\theta)\}$	$(0,1)$	Complementary log–log
`foldsqrt()`	$\sqrt{2\theta} - \sqrt{2(1-\theta)}$	$(0,1)$	Folded square root
`logit()`	$\log \dfrac{\theta}{1-\theta}$	$(0,1)$	Logit
`multilogit()`	$\log \dfrac{\theta_j}{\theta_{M+1}}$	$(0,1)^M$	Multi-logit; $\sum\limits_{j=1}^{M+1} \theta_j = 1$
`probit()`	$\Phi^{-1}(\theta)$	$(0,1)$	Probit (for "probability unit")
`fisherz()`	$\frac{1}{2} \log \dfrac{1+\theta}{1-\theta}$	$(-1,1)$	Fisher's Z
`rhobit()`	$\log \dfrac{1+\theta}{1-\theta}$	$(-1,1)$	Rhobit
`loge()`	$\log \theta$	$(0,\infty)$	Log (logarithmic)
`logneg()`	$\log(-\theta)$	$(-\infty,0)$	Log-negative
`negloge()`	$-\log(\theta)$	$(0,\infty)$	Negative-log
`reciprocal()`	θ^{-1}	$(0,\infty)$	Reciprocal
`nbcanlink()`	$\log(\theta/(\theta+k))$	$(0,\infty)$	NB canonical link (Sect. 11.3.3)
`extlogit()`	$\log \dfrac{\theta - A}{B - \theta}$	(A,B)	Extended logit
`explink()`	e^θ	$(-\infty,\infty)$	Exponential
`identitylink()`	θ	$(-\infty,\infty)$	Identity
`negidentity()`	$-\theta$	$(-\infty,\infty)$	Negative-identity
`logc()`	$\log(1-\theta)$	$(-\infty,1)$	Log-complement
`loglog()`	$\log\log(\theta)$	$(1,\infty)$	Log–log
`logoff(`θ`, offset = `A`)`	$\log(\theta + A)$	$(-A,\infty)$	Log with offset

With certain types of data, sometimes one wishes to fit the model subject to the constraint $p_1(\boldsymbol{x}) = p_2(\boldsymbol{x})$, e.g., if the Y_j are the presence/absence of deafness in the LHS and RHS ears. This corresponds to an *exchangeable* error structure, and constraining $\eta_1 = \eta_2$ can be handled with the constraints-on-the-functions framework described in Sect. 3.3. As well, it is not uncommon to constrain ψ to be intercept-only, because of numerical problems that may arise when ψ is modelled too flexibly. This too can be achieved with constraints on the functions.

1.2.4 Proportional Odds and Multinomial Logit Models

Suppose that response Y is *ordinal* (an ordered categorical or grouping variable or factor), e.g., $Y = 1 =$ 'low', $Y = 2 =$ 'medium', $Y = 3 =$ 'high'. McCullagh and Nelder (1989) give strong reasons why ordinal responses are more naturally modelled in terms of the cumulative probabilities $P(Y \leq j|\boldsymbol{x})$ rather than $P(Y = j|\boldsymbol{x})$ directly. To this end, the *proportional odds model* for a general ordinal Y taking levels $\{1, 2, \ldots, M + 1\}$ may be written

$$\text{logit } P(Y \leq j|\boldsymbol{x}) \ = \ \eta_j(\boldsymbol{x}), \tag{1.22}$$

subject to the constraint that

$$\eta_j(\boldsymbol{x}) \ = \ \beta^*_{(j)1} + \boldsymbol{x}^T_{[-1]} \boldsymbol{\beta}^*_{[-(1:M)]}, \qquad j = 1, \ldots, M. \tag{1.23}$$

Here, $\boldsymbol{x}_{[-1]}$ is $\boldsymbol{x}$ with the first element (the intercept) deleted. The superscript "$*$" denotes regression coefficients that are to be estimated. Equation (1.23) describes M parallel surfaces in $(p-1)$-dimensional space. The VGAM family functions `cumulative()` and `propodds()` fit this model and variants thereof.

Here are some further comments.

(i)   *Selecting different link functions*

Let $\gamma_j(\boldsymbol{x}) = P(Y \leq j|\boldsymbol{x})$. The proportional odds model is also known as the *cumulative logit model*; there are M simultaneous logistic regressions applied to the γ_j. If we replace the logit link in (1.22) by a probit link say, then this may be referred to as a *cumulative probit model*. For this, the VGAM family function `cumulative()` has an argument `link` that can be assigned `probit`. More generally, models for (1.23) with any link are termed *cumulative link models*.

(ii) *Non-proportional odds model*

In (1.23) the linear predictors are parallel on the logit scale because the estimable regression coefficients $\boldsymbol{\beta}^*_{[-(1:M)]}$ in (1.23) are common for all j. Consequently they do not intersect, therefore the probabilities $P(Y = j|\boldsymbol{x})$ do not end up negative or greater than unity for some $\boldsymbol{x}$. This is known as the so-called *parallelism* or *proportional odds* assumption. An assumption that ought to be checked in practice, it gives rise to the property that the odds of $Y \leq j$ given $\boldsymbol{x}_1$, relative to $\boldsymbol{x}_2$, say, does not depend on j, hence its name. This means that the thresholds in the latent variable motivation (Sect. 14.4) do not affect the regression parameters of interest.

(iii) *Partial proportional odds model*

An intermediary between the proportional odds and the fully non-proportional odds models is to have some explanatory variables parallel and others not. Some authors call this a *partial proportional odds model*. As an example, suppose $p = 4$, $M = 2$ and

$$\eta_1 \ = \ \beta^*_{(1)1} + \beta^*_{(1)2}\, x_2 + \beta^*_{(1)3}\, x_3 + \beta^*_{(1)4}\, x_4,$$
$$\eta_2 \ = \ \beta^*_{(2)1} + \beta^*_{(1)2}\, x_2 + \beta^*_{(2)3}\, x_3 + \beta^*_{(1)4}\, x_4.$$

The parallelism assumption applies to x_2 and x_4 only. This may be fitted by

```
vglm(ymatrix ~ x2 + x3 + x4, cumulative(parallel = TRUE ~ x2 + x4 - 1), cdata)
```

or equivalently,

```
vglm(ymatrix ~ x2 + x3 + x4, cumulative(parallel = FALSE ~ x3), data = cdata)
```

There are several other extensions that may easily be handled by the constraint matrices methodology described in Sect. 3.3.

(iv) *Common* VGAM *family function arguments*

Rather than (1.22) many authors define the proportional odds model as

$$\text{logit } P(Y \geq j + 1|\boldsymbol{x}) \;=\; \eta_j(\boldsymbol{x}), \qquad j = 1, \ldots, M, \tag{1.24}$$

because $M = 1$ coincides with logistic regression. Many VGAM categorical family functions share a number of common arguments and one such argument is `reverse`. Here, setting `reverse = TRUE` will fit (1.24). Other common arguments include `link` (and its variants which usually start with "l"), `parallel`, `zero`, and initial values of parameters (which usually start with "i"). These are described in Sect. 8.1.2.

When a factor Y is unordered (*nominal*) it is customary to fit a *multinomial logit model*

$$\log \frac{P(Y = j|\boldsymbol{x})}{P(Y = M + 1|\boldsymbol{x})} \;=\; \eta_j(\boldsymbol{x}), \qquad j = 1, \ldots, M. \tag{1.25}$$

The model is particularly useful for exploring how the relative chances of falling into the response categories depend upon the covariates because

$$\frac{P(Y = j|\boldsymbol{x})}{P(Y = s|\boldsymbol{x})} \;=\; \exp\left\{\eta_j(\boldsymbol{x}) - \eta_s(\boldsymbol{x})\right\} \;=\; \exp\left\{\boldsymbol{x}^T \left(\boldsymbol{\beta}_j - \boldsymbol{\beta}_s\right)\right\}. \tag{1.26}$$

This can be interpreted as the relative risk of response j relative to response s, given $\boldsymbol{x}$. The interpretation of the coefficient $\beta_{(j)k}$ is based on increasing the kth variable by one unit, keeping other variables fixed:

$$\beta_{(j)k} \;=\; \log \frac{P(Y = j|x_1, \ldots, x_{k-1}, x_k + 1, x_{k+1}, \ldots, x_p)}{P(Y = j|x_1, \ldots, x_{k-1}, x_k, x_{k+1}, \ldots, x_p)}. \tag{1.27}$$

From (1.26) it is easy to show that

$$P(Y = j|\boldsymbol{x}) \;=\; \frac{\exp\{\eta_j(\boldsymbol{x})\}}{\sum\limits_{s=1}^{M+1} \exp\{\eta_s(\boldsymbol{x})\}}, \qquad j = 1, \ldots, M. \tag{1.28}$$

Clearly the numerator is positive and the $M + 1$ probabilities add to unity, which are the essential properties of this regression model. Identifiability constraints such as $\eta_{M+1}(\boldsymbol{x}) \equiv 0$ are required because (1.28) may be multiplied by $1 = e^c/e^c$ for any constant c. The VGAM family function `multinomial()` chooses the last level of Y to be the *baseline* or *reference group* by default, but there is an argument called `refLevel` to allow the choice of some other. It is most common to select the first or last levels to be baseline, and if the first then this coincides with logistic regression.

Chapter 14 gives details about the proportional odds model and other models for categorical data.

1.3 General Framework

In this section we describe the general framework and notation by briefly sketching each major class of models. Fuller details can be found in the relevant Part I chapter and the references in Table 1.1. To realize its full potential, it is stressed that it is more important to see the forest than a few individual trees, i.e., grasping the overriding VGLM/VGAM framework is paramount over concentrating on a few special cases.

1.3.1 Vector Generalized Linear Models

Suppose the observed response $\boldsymbol{y}$ is a Q-dimensional vector. VGLMs are models for which the conditional distribution of $\boldsymbol{Y}$ given explanatory $\boldsymbol{x}$ is of the form

$$f(\boldsymbol{y}|\boldsymbol{x};\mathbf{B}) \;=\; f(\boldsymbol{y},\eta_1,\ldots,\eta_M) \tag{1.29}$$

for some known function $f(\cdot)$, where $\mathbf{B} = (\boldsymbol{\beta}_1\,\boldsymbol{\beta}_2\,\cdots\,\boldsymbol{\beta}_M)$ is a $p \times M$ matrix of regression coefficients, and the jth linear predictor is

$$\eta_j \;=\; \boldsymbol{\beta}_j^T\boldsymbol{x} \;=\; \sum_{k=1}^{p}\beta_{(j)k}\,x_k, \quad j = 1,\ldots,M, \tag{1.30}$$

where $\boldsymbol{x} = (x_1,\ldots,x_p)^T$ with $x_1 = 1$ if there is an intercept. VGLMs are thus like GLMs but allow for multiple linear predictors, and they encompass models outside the limited confines of the classical exponential family. In (1.30) some of the $\beta_{(j)k}$ may be, for example, set to zero using quantities known as constraint matrices which are described later.

The η_j of VGLMs may be applied directly to parameters of a distribution rather than just to μ_js as for GLMs. A simple example is a univariate distribution with a location parameter ξ and a scale parameter $\sigma > 0$, where we may take $\eta_1 = \xi$ and $\eta_2 = \log \sigma$. In general, $\eta_j = g_j(\theta_j)$ for some parameter link function g_j and parameter θ_j. In the formulation *all* the explanatory variables can potentially be used to model *each* parameter. In **VGAM**, there are currently over a dozen links to choose from (Table 1.2).

There is no particular relationship between Q and M in general: it depends specifically on the model or distribution to be fitted. Often M is the number of independent parameters. Table 1.3 lists the values for the 6 illustrative models.

Table 1.3 Values of Q and M for the six illustrative models of Sect. 1.2. The categorical models have a $(M + 1)$-level factor response.

Model	Q	M
Normal	1	2
Poisson	1	1
Negative binomial	1	2
Proportional odds model	$M + 1$	M
Multinomial logit model	$M + 1$	M
Bivariate odds ratio model	4	3

VGLMs are estimated by IRLS. This algorithm is remarkably adaptable, and it allows for many different enhancements. Its scope is wide, and this book and accompanying software is a reflection of some of its flexibility. For data $(\boldsymbol{x}_i, \boldsymbol{y}_i)$, $i = 1, \ldots, n$, most models that can be fitted have a log-likelihood

$$\ell \; = \; \sum_{i=1}^{n} w_i \, \ell_i, \tag{1.31}$$

and this will be assumed here. The w_i are known and fixed positive prior weights. Let $\boldsymbol{x}_i$ denote the explanatory vector for the ith observation. Then one can write

$$\boldsymbol{\eta}_i \; = \; \begin{pmatrix} \eta_1(\boldsymbol{x}_i) \\ \vdots \\ \eta_M(\boldsymbol{x}_i) \end{pmatrix} \; = \; \mathbf{B}^T \boldsymbol{x}_i \; = \; \begin{pmatrix} \boldsymbol{\beta}_1^T \boldsymbol{x}_i \\ \vdots \\ \boldsymbol{\beta}_M^T \boldsymbol{x}_i \end{pmatrix}. \tag{1.32}$$

VGLMs are covered in Chap. 3.

1.3.2 Vector Generalized Additive Models

VGAMs provide additive-model extensions to VGLMs, i.e., (1.30) becomes

$$\eta_j(\boldsymbol{x}) \; = \; \beta_{(j)1} + \sum_{k=2}^{d} f_{(j)k}(x_k), \quad j = 1, \ldots, M, \tag{1.33}$$

a sum of smooth functions of the individual covariates, just as with ordinary GAMs (Hastie and Tibshirani, 1990). The component functions comprising $\boldsymbol{f}_k = (f_{(1)k}(x_k), \ldots, f_{(M)k}(x_k))^T$ are centred for uniqueness, and they are estimated *simultaneously* using *vector smoothers*. VGAMs are thus a visual data-driven method that is well-suited for exploring data. They retain the simplicity of interpretation that GAMs possess because each x_k has an additive effect in (1.33), but the linearity assumption imposed by (1.1) is relaxed.

In practice we may wish to constrain the effect of a covariate to be the same for some of the η_j and to have no effect for others, e.g., for VGAMs,

$$\eta_1 = \beta_{(1)1} + f_{(1)2}(x_2) + f_{(1)3}(x_3),$$
$$\eta_2 = \beta_{(2)1} + f_{(1)2}(x_2),$$

so that $f_{(1)2} \equiv f_{(2)2}$ and $f_{(2)3} \equiv 0$. We can achieve this using "constraints-on-the-functions" which are used to enforce relationships between the $\beta_{(j)k}$ of VGLMs, etc. For VGAMs, we can represent these models using

$$\boldsymbol{\eta}(\boldsymbol{x}) = \boldsymbol{\beta}_{(1)} + \sum_{k=2}^{d} \boldsymbol{f}_k(x_k) \; = \; \mathbf{H}_1 \boldsymbol{\beta}_{(1)}^* + \sum_{k=2}^{d} \mathbf{H}_k \boldsymbol{f}_k^*(x_k) \tag{1.34}$$

where $\mathbf{H}_1, \mathbf{H}_2, \ldots, \mathbf{H}_d$ are known and fixed full column-rank *constraint matrices*, $\boldsymbol{f}_k^*$ is a vector containing a possibly reduced set of component functions, and $\boldsymbol{\beta}_{(1)}^*$

is a vector of unknown intercepts. With no constraints at all, $\mathbf{H}_1 = \mathbf{H}_2 = \cdots = \mathbf{H}_d = \mathbf{I}_M$ and $\boldsymbol{\beta}^*_{(1)} = \boldsymbol{\beta}_{(1)}$. Like the $\boldsymbol{f}_k$, the $\boldsymbol{f}^*_k$ are centred for uniqueness. For VGLMs, the $\boldsymbol{f}_k$ are linear so that

$$\mathbf{B}^T = \left(\mathbf{H}_1 \boldsymbol{\beta}^*_{(1)} \quad \mathbf{H}_2 \boldsymbol{\beta}^*_{(2)} \quad \cdots \quad \mathbf{H}_p \boldsymbol{\beta}^*_{(p)} \right). \tag{1.35}$$

VGAMs are covered in Chap. 4.

1.3.3 RR-VGLMs

Reduced-rank VGLMs are a surprisingly useful and interesting class of models. One of its primary aims is for dimension reduction. Partition $\boldsymbol{x}$ into $(\boldsymbol{x}_1^T, \boldsymbol{x}_2^T)^T$ and $\mathbf{B} = (\mathbf{B}_1^T \ \mathbf{B}_2^T)^T$. In general, $\mathbf{B}$ is a dense matrix of full rank, i.e., $\min(M, p)$. Thus there are $M \times p$ regression coefficients to estimate, and even when M and p are moderate, for some data sets this is too large so that the model overfits.

One solution is based on a simple and elegant idea: replace $\mathbf{B}_2$ by a reduced-rank regression (RRR). This dimension reduction technique is generally attributed to Anderson (1951) but it obtained its generally used name from Izenman (1975). Essentially RRR operates by determining a low-rank matrix which is an optimal approximation to a full rank matrix. The low rank matrix is expressed as a product of two 'thin' matrices $\mathbf{A}$ and $\mathbf{C}$, i.e., $\mathbf{B}_2 = \mathbf{A}\,\mathbf{C}^T$ where $\mathbf{A}$ is $M \times R$ and $\mathbf{C}$ is $p_2 \times R$, and where the *rank* R is the reduced dimension and $p_2 = \dim(\boldsymbol{x}_2)$. The resulting $\boldsymbol{\eta}$ given in Table 1.1 can be written in terms of an R-vector of latent variables $\boldsymbol{\nu} = \mathbf{C}^T \boldsymbol{x}_2$. The concept of a latent variable is very important in many fields such as economics, medicine, biology (especially ecology) and the social sciences.

RRR can reduce the number of regression coefficients enormously, if the rank R is kept low relative to large p_2 and M. Ideally, the problem can be reduced down to one or two dimensions and therefore plotted. The RRR is applied to $\mathbf{B}_2$ because we want to make provision for some variables that define $\boldsymbol{x}_1$ which we want to leave alone, e.g., the intercept. In practice, other variables usually chosen to be part of $\boldsymbol{x}_1$ are covariates such as age and sex that the regression is adjusting for. Variables which can be thought of as playing some role in an underlying latent variable or gradient should belong to $\boldsymbol{x}_2$, e.g., temperature, solar radiation and rainfall in an ecological application to represent 'climate'.

It transpires that *reduced-rank VGLMs* (RR-VGLMs) are simply VGLMs where the constraint matrices corresponding to $\boldsymbol{x}_2$ are equal and unknown, so need to be estimated. As a consequence, the modelling function `rrvglm()` calls an alternating algorithm which toggles between estimating $\mathbf{A}$ and $\mathbf{C}$.

Incidentally, special cases of RR-VGLMs have appeared in the literature. For example, an RR-multinomial logit model, is known as the *stereotype* model (Anderson, 1984). Another is Goodman (1981)'s RC model which is an RRR applied to several Poisson regressions—in fact it is a sub-variant called a *row–column interaction model* (RCIM) which is applied to a $\mathbf{Y}$ (no $\mathbf{X}$) only. Other useful RR-VGLMs are the RR-NB and RR zero-inflated Poisson (RR-ZIP; Sect. 5.5.2.2). RR-VGLMs are the subject of Chap. 5.

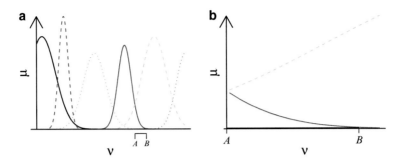

Fig. 1.1   Rank-1 constrained quadratic ordination (CQO). (**a**) The mean abundance is $\mu_s(\nu) = E(Y_s|\nu)$ for $s = 1, \ldots, S$ species, and $\nu = \boldsymbol{c}^T \boldsymbol{x}_2$ is a latent variable. (**b**) Zooming in on the subinterval $[A, B]$ in (**a**). This is approximately linearly on the η scale, meaning an RR-VGLM would be suitable.

1.3.4 QRR-VGLMs and Constrained Ordination

RR-VGLMs form an optimal linear combination of the explanatory variables $\boldsymbol{x}_2$ and then fit a VGLM to these and $\boldsymbol{x}_1$. The $\boldsymbol{\eta}$ is *linear* with respect to $\boldsymbol{\nu}$, and because the link functions are monotonic, the response is *monotonic* with respect to $\boldsymbol{\nu}$.

In contrast, suppose a biologist has an $n \times S$ response matrix $\mathbf{Y}$ collected at n sites upon S species; e.g., y_{ij} is the abundance of species j at site i. In biology, species' responses can be thought of as *unimodal* such as in Fig. 1.1a (e.g., animals and plants die when the temperature is too hot or cold). Another situation where unimodal responses occur is in psychology, where increasing the dosage of some stimulus initially raises the level of some response but beyond a certain dosage the response starts falling.

These examples suggest fitting a *quadratic* in $\boldsymbol{\nu}$ on the $\boldsymbol{\eta}$ scale. This gives rise to the class of *quadratic RR-VGLMs* (QRR-VGLMs), so as to perform *constrained quadratic ordination* (CQO). The results are bell-shaped curves/surfaces on axes defined in terms of latent variables such as environmental gradients. This is illustrated in Fig. 1.1a. In contrast, RR-VGLMs perform a constrained linear ordination—this is illustrated in Fig. 1.1b.

As a specific example, a simple model for Poisson counts of S species is

$$\log \mu_j(\nu_i) = \eta_j(\nu_i) = \beta_{(j)1} + \beta_{(j)2}\,\nu_i + \beta_{(j)3}\,\nu_i^2, \quad j=1,\ldots,S, \quad (1.36)$$

subject to $\beta_{(j)3} < 0$ to ensure bell-shaped response curves. The latent variable $\nu = \boldsymbol{c}^T \boldsymbol{x}_2$ is an optimal linear combination of the environmental variables and is interpreted as a 'gradient'. Usually the interpretation is based on the magnitudes of the values of $\boldsymbol{c}$, called the constrained coefficients or loadings.

CQO is detailed in Chap. 6.

1.3.5 RR-VGAMs and Constrained Ordination

RR-VGLMs and QRR-VGLMs are model-driven. A data-driven version of these two classes of models is available in the form of *reduced-rank VGAMs* (RR-VGAMs). These are VGAMs which smooth latent variables as well as those variables in $\boldsymbol{x}_1$. For example, each quadratic in (1.36) might be replaced by a smooth function

$$\log \mu_j(\nu_i) \;=\; \eta_j(\nu) \;=\; \beta_{(j)1} + f_{(j)1}(\nu_{i1}), \quad j = 1, \ldots, S, \qquad (1.37)$$

and estimated by a smoother such as a spline. RR-VGAMs have potential use in community ecology, when one wants to see what the data-driven response curve of each species looks like when plotted against an optimally estimated gradient. More details are given in Chap. 7.

1.3.6 RCIMs

Row–column interaction models operate on a matrix $\mathbf{Y}$ only; there is no explicit $\mathbf{X}$. They apply some link function to a parameter (such as the cell mean) to equal a row effect plus a column effect plus an optional interaction modelled as a reduced-rank regression, i.e., $\mathbf{AC}^T$ as with RR-VGLMs. Technically, RCIMs are RR-VGLMs with special constraint matrices and indicator variables, set up to handle the row and column positions of each cell y_{ij}. They fit several useful models, e.g., Goodman's RC association model for Poisson counts, median polish, simple Rasch models and quasi-variances. RCIMs are described in Sect. 5.7.

1.4 An Overview of VGAM

Figure 1.2 gives an overview of all the major classes of models in the framework. Recall we have data $(\boldsymbol{x}_i, \boldsymbol{y}_i)$, for $i = 1, \ldots, n$ independent observations. Starting from the LM $\boldsymbol{\mu} = \mathbf{X}\boldsymbol{\beta}$, or equivalently,

$$\mu_i \;=\; \sum_{k=1}^{p} \beta_k \, x_{ik}, \qquad (1.38)$$

one extends this mainly in three directions: toward

(i) *generalized responses*: that is, from the normal distribution to the classical exponential family and beyond. This includes rates, quantiles, proportions, counts, directions, survival times and positive data.
(ii) *nonparametric models*: that is, from linear modelling to additive models involving smoothing.
(iii) *multivariate responses*: that is, handle a response that is vector-valued. Additionally, sometimes several multivariate responses can be handled and the response vectors are treated as being independent of each other, e.g.,

```
fit1 <- vglm(cbind(y1, y2) ~ x2 + x3 + x4, poissonff, data = pdata)
fit2 <- vglm(cbind(y1, y2) ~ x2 + x3 + x4, binom2.or, data = bdata)
fit3 <- vglm(cbind(y1, y2, y3, y4) ~ x2 + x3 + x4, binormalcop, data= Bdata)
```

Here, `fit1` has multiple responses `y1` and `y2` that are treated independently. whereas for `fit2` the `y1` and `y2` are treated as correlated. For `fit3` the first bivariate response is `y1` and `y2` which is treated independently of the second bivariate response `y3` and `y4`,

The first extension which handles a weighted multivariate response is the *vector linear model* (VLM) class, which has

$$\boldsymbol{y}_i = \sum_{k=1}^{p} \mathbf{H}_k \, \boldsymbol{\beta}^*_{(k)} \, x_{ik} + \boldsymbol{\varepsilon}_i, \qquad \boldsymbol{\varepsilon}_i \sim (\mathbf{0}, \mathbf{W}_i) \quad \text{independently}, \qquad (1.39)$$

where the $\mathbf{W}_i$ are *known* positive-definite symmetric matrices and the constraint matrices $\mathbf{H}_k$ are *known* and of full column-rank. This implies

$$\boldsymbol{\mu}_i = \sum_{k=1}^{p} \mathbf{H}_k \, \boldsymbol{\beta}^*_{(k)} \, x_{ik}. \qquad (1.40)$$

In practice, the VLM has few direct applications, but it serves as the computational building block of the VGLM class.

As an extension of VLMs, VGLMs have

$$\begin{pmatrix} g_1(\theta_1) \\ \vdots \\ g_M(\theta_M) \end{pmatrix} = \boldsymbol{\eta} = \sum_{k=1}^{p} \mathbf{H}_k \, \boldsymbol{\beta}^*_{(k)} \, x_k = \mathbf{B}^T \boldsymbol{x} = \left(\mathbf{B}_1^T \ \mathbf{B}_2^T \right) \begin{pmatrix} \boldsymbol{x}_1 \\ \boldsymbol{x}_2 \end{pmatrix} \qquad (1.41)$$

where the g_j are link functions and θ_j are the parameters in the model or distribution. If a set of the $\mathbf{H}_k$ are equal, *unknown* and to be estimated, then this produces the "reduced-rank" class, termed RR-VGLMs.

Written in S4 (Chambers, 1998), the **VGAM** modelling functions are used in a similar manner as `glm()`, and `gam()` in **gam**. Given a `vglm()`/`vgam()` object, standard generic functions such as `coef()`, `fitted()`, `predict()`, `summary()`, `vcov()` are available (Tables 8.5, 8.6, 8.7). The typical usage is like

```
vglm(yvector ~ x2 + x3 + x4, family = aVGAMfamilyFunction, data = adata)
vgam(ymatrix ~ s(x2) + x3,    family = aVGAMfamilyFunction, data = adata)
```

Many models have a multivariate response and/or **VGAM** family functions which handle multiple (independent) responses, therefore the LHS of the formula can be a matrix (otherwise a vector).

1.5 Some Background Topics

For completeness, the following background material that the reader is assumed to be familiar with is summarized. Also, this section revises several sub-themes that are woven throughout this book.

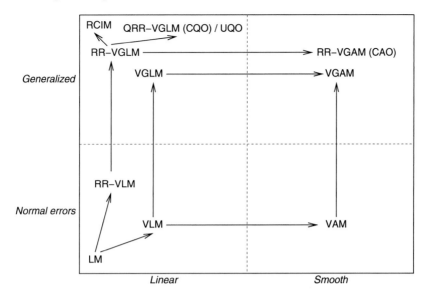

Fig. 1.2 Flowchart for different classes of models. Legend: LM = linear model, V = vector, G = generalized, A = additive, O = ordination, Q = quadratic, U = unconstrained, RCIM = row–column interaction model. See also Table 1.1. Apart from the LM, the models of the bottom half are more to be viewed as computational building blocks.

1.5.1 The Penalty Function Approach

In mathematical modelling it is common to estimate the parameters by balancing two opposing quantities. For example, a smoother fitted to a scatter plot data set (x_i, y_i), $i = 1, \ldots, n$, will closely go through the data cloud if the residual sum of squares $\sum_i (y_i - \hat{y}_i)^2$ is small, e.g., Fig. 4.3. Choosing a curve that makes this quantity approach 0 means setting the residuals to 0, implying that the curve interpolates the data. The resulting curve will almost be certainly be too wiggly— a case of *overfitting*. So, a technique to obtain a more moderate and generalizable curve is to include a penalty for how wiggly it is.

In general, the minimization problem is

$$\min_{\boldsymbol{\theta}} \quad A + \lambda B \tag{1.42}$$

where $\boldsymbol{\theta}(\lambda)$ is the vector of parameters and λ (≥ 0) is the balancing or trade-off parameter. The smaller quantity A is, the closer the fit is with the data. Simply minimizing A would result in an extreme fit that would not generalize well for future data. But if we add a quantity B to the objective function that increases as A decreases then we can regularize the fit.

As $\lambda \to 0^+$ the fit will become complicated because B becomes negligible. As $\lambda \to \infty$ the fit becomes simpler because λB is forced to remain small relative to A, i.e., the penalty B is forced to decrease quicker relative to the increase in A.

In the subject of statistics, the penalty approach (1.42) is adopted commonly. The following are a few instances, and in this book we see the approach appear a number of times.

AIC, BIC   The Akaike information criterion and Bayesian information criterion are commonly used to compare models. They balance goodness of fit by the number of parameters. These information criteria have known λ and are traditionally used on multiple models that are not nested for the purpose of model selection. See Sect. 9.3.

Smoothing splines   The objective function (4.3) for a vector smoothing spline minimizes a residual sum of squares plus a measure of the wiggliness of all the component functions. There are techniques such as cross-validation which are used to try and obtain a reasonable value for λ for a given data set (Sect. 2.4.7.6). See Sect. 4.2.1.

LASSO   Although not covered in this book, this method has generated much interest and research activity in recent times. Proposed by Tibshirani (1996), the method estimates the β_k of an LM by minimizing

$$\sum_{i=1}^{n}\left(y_i - \beta_1 - \sum_{k=2}^{p} x_{ik}\,\beta_k\right)^2 + \lambda \sum_{k=2}^{p} |\beta_k|, \qquad (1.43)$$

and called the 'least absolute shrinkage and selection operator' (LASSO). The penalty is an l_1 norm, and with increasing λ, the shrinking is such that $\beta_k = 0$ for values of k belong to some set of variables, and thus x_k is no longer selected in the regression. For λ sufficiently large, all the coefficients become 0 (except the intercept term which is unpenalized). This can be seen in Fig. 1.3, where the paths of the LASSO coefficients based on an LM fitted to the **azpro** data frame described in Sect. 11.3.4.3 are traced. The first plot has λ on a log-scale as its x-axis, and the second plot has the l_1 norm of $(\beta_2, \ldots, \beta_p)^T$.

Trees   Also not covered in this book, in the topic of classification and regression trees, a popular algorithm for choosing a tree of reasonable size is to contrast the number of leaves (the penalty term B) with some measure of impurity, such as the Gini index or deviance. The basics of trees are described in, e.g., James et al. (2013).

P-splines   These are similar to smoothing splines; see Sect. 2.4.5.

1.5.2 Snippets of the S Language

This book describes the **VGAM R** package as a software implementation of VGLMs/VGAMs, etc. (Chap. 8). R is largely based on the S language, therefore we summarize some aspects of the language here, especially those pertaining to general regression modelling. This includes S formulas to describe the response

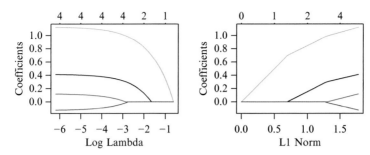

Fig. 1.3 Paths of the estimated LASSO coefficients in an LM fit. The response `log(los)` is regressed against the variables `admit` (*black*), `age75` (*red*), `procedure` (*green*) and `sex` (*blue*) and intercept, in the data frame `azpro` from COUNT. The first plot has $\log \lambda$ as its x-axis, whereas second plot has the quantity $\sum_{k=2}^{p} |\beta_k|$ in (1.43). The upper numbers are the number of variables in the model. Package glmnet is used here.

and explanatory variables, data frames to store the variables in, object-oriented features used to streamline analyses, and generic functions to extract quantities and perform actions on objects, etc.

The full language is actually quite deep, because it can redefine itself. But we will use the language at the level of an applied statistician doing straightforward analyses.

A basic question from the outset is: why R? There are many compelling reasons such as it being free, open-source, fully featured (including 6500+ packages between early and mid-2015), and is based on the S language (Ihaka and Gentleman, 1996) which is quite elegant and was award-winning (1999 ACM Software System Award). R is very powerful and has superb graphics. It runs on the most common platforms including Windows, Macintosh and Linux/Unix systems. It was developed by Ross Ihaka and Robert Gentleman in the early 1990s at the Statistics Department of the University of Auckland. Being based on S, it inherited a base of many S-PLUS users who switched over during the early 2000s. R tends to be first when it comes to the implementation of new statistical methodology. Being a programmable language, it offers unrivalled flexibility and functionality. It is not really GUI-based, hence it is a stumbling block to some novices who are unwilling to ascend the rather steep initial learning curve. Probably its weakest point is its difficulties handling huge data sets due to it being memory-bound. The software can be downloaded from the Comprehensive R Archive Network (CRAN; `http://CRAN.R-project.org`).

1.5.2.1 Arguments in `lm()`

The function `lm()` is a prototype of other modelling functions such as `glm()` and `vglm()`. It has arguments

```
> args(lm)

  function (formula, data, subset, weights, na.action, method = "qr",
      model = TRUE, x = FALSE, y = FALSE, qr = TRUE, singular.ok = TRUE,
      contrasts = NULL, offset, ...)
  NULL
```

Some of the arguments are described a little below and in Table 1.4, however, here are some brief comments about some of the other arguments.

- The `weights` argument in general can be assigned a vector with positive values. They are called *prior weights*, and are the w_i in (1.31). The default is usually a vector of ones. A trick is sometimes to assign a very small value such as 10^{-10} so that observation effectively is not present during estimation, however its fitted value can be returned as an ordinary observation. (This is another way of performing prediction; the `predict()` generic is much more conventional.)
- The `data` argument can be assigned a data frame or list that contains all the variables. This is good practice because having the variables scattered about as separate vectors in the environment from which `lm()` is called is dangerous. This is particularly important when predictions are made from the fitted model.

1.5.2.2 S Formulas

The first argument of modelling functions is typically `formula`. It is assigned an S formula which is of the form response $\sim$ expression. The distinguishing feature of a formula is the "$\sim$" character which means "is modelled as a function of." A simple formula has the form

```
y ~ x2 + x3 + ... + xp
```

For us, the LHS is a vector or a matrix response, and the RHS are the explanatory variables $x_1 = 1, x_2, \ldots, x_p$ which forms the model matrix known as $\mathbf{X}_{\mathrm{LM}}$ (see later). The "+" joining the terms, in most cases, means the terms are additive, e.g., $\cdots + \beta_k x_k + \beta_{k+1} x_{k+1} + \cdots$. By default a `1` is included in a formula, however, intercepts may be suppressed by adding the term `-1` or `0`. It is good style to have all the variables in the formula residing in a data frame.

More generally we can write a formula as a sum of d terms. The (LM) model matrix has $p \equiv p_{\mathrm{LM}}$ columns made up of d subsets of columns. For example,

```
y ~ x2 + poly(x3, 2) + bs(x4, 3) + x5
```

produces $1 + 1 + 2 + 3 + 1 = 8 = p$ regression coefficients to be estimated but there are $d = 5$ terms. Confusingly, sometimes $\boldsymbol{x} = (x_1, \ldots, x_d)^T$ is meant while sometimes $\boldsymbol{x} = (x_1, \ldots, x_p)^T$ is meant. The use of d is more natural when dealing with additive models such as (1.2), while $\boldsymbol{x} = (x_1, \ldots, x_p)^T$ is more natural dealing with conventional linear/parametric models.

The terms of an object `fit` may be obtained by using the `terms()` generic, and the `"term.labels"` attribute gives the term labels, omitting the intercept. For example,

```
> d.AD <- data.frame(counts    = c(18, 17, 15, 20, 10, 20, 25, 13, 12),
                     outcome   = gl(3, 1, 9),
                     treatment = gl(3, 3))
> glm.D93 <- glm(counts ~ outcome + treatment, poisson, data = d.AD)
> attr(terms(glm.D93), "term.labels")

  [1] "outcome"    "treatment"
```

Also, the `"assign"` attribute of the model matrix maps each term to specified columns of the LM matrix. For example,

```
> attr(model.matrix(glm.D93), "assign")

 [1] 0 1 1 2 2
```

shows that the intercept occupies the first column, `outcome` takes up the next two columns, and `treatment` the last two columns.

S model formulas, which are based on the Wilkinson and Rogers (1973) notation, allow operators such as interactions and nesting. As an example consider

```
y ~ -1 + x1 + x2 + x3 + f1:f2 + f1*x1 + f2/f3 + f3:f4:f5 + (f6 + f7)^2
```

where variables beginning with an x are numeric and those with an f are factors. An *interaction* `f1*f2` is expanded to `1 + f1 + f2 + f1:f2` where the terms `f1` and `f2` are *main effects*. An interaction between two factors can be expressed using `factor:factor`: γ_{jk}. There are other types of interactions, e.g., between a factor and numeric, `factor:numeric`, produce $\beta_j x_k$. Interactions between two numerics, `numeric:numeric`, produce a cross-product term such as $\beta x_2 x_3$.

The term `(f6 + f7)^2` expands to `f6 + f7 + f6:f7`. A term `(f6 + f7 + f8)^2 - f7:f8` would expand to all main effects and all second-order interactions except for `f7:f8`.

Nesting is achieved by the operator "/", e.g., `f2/f3` is shorthand for `1 + f2 + f3:f2`, or equivalently,

```
1 + f2 + f3 %in% f2
```

An example of nesting is `f2` = state and `f3` = county. Another example is `mother/foetus` for teratological experiments (Sect. 11.4) with individual-level covariates.

From Table 1.4, there are times when the mathematical meaning of operators such as `+`, `-`, `*`, `/`, etc. is desired. Then the *identity function* `I()` is needed, e.g.,

```
vglm(y ~ -1 + offset(o) + x1 + I(x2 - 1) + poly(x3, 2, raw = TRUE) + I(x4^2),
     uninormal, data = udata)
```

fits the LM

$$y_i = o_i + \beta_1 x_{i1} + \beta_2 (x_{i2} - 1) + \beta_3 x_{i3} + \beta_4 x_{i3}^2 + \beta_5 x_{i4}^2 + \varepsilon_i, \quad (1.44)$$
$$\varepsilon_i \sim \text{ i.i.d. } N(0, \sigma^2), \quad i = 1, \ldots, n, \quad (1.45)$$

where o is a vector containing the (known and fixed) o_i. Alternatively, offsets may be inputted using the `offset` argument (see Table 1.4). The acronym "i.i.d." stands for 'independent and identically distributed', and the statement of (1.45) holds various assumptions that should be checked (Sect. 2.2.1.1).

1.5.2.3 More on S Formulas

The S formula language endowed with the operators of Table 1.4 is flexible enough to cater for a variety of simple functional forms as building blocks. For example, the interaction operator allows for step functions in one or more dimensions, as well

Table 1.4 Upper table: S formula operators. Sources: Chambers and Hastie (1991), Chambers (1998) and Chambers (2008). Mid-table: logical operators. Lower table: some commonly used arguments in modelling functions such as `glm()` and `vglm()`.

Operator/function	Comment
+	Addition of a term
1	Intercept (present by default)
-	Omit the following term, e.g., `-1` suppresses an intercept
.	All variables in a data frame except for the response
0	No intercept (alternative method)
:	Interaction (tensor product) between two terms
*	Interaction (expansion), e.g., `A * B = A + B + A:B`
/	Nesting, same as `%in%`, e.g., `A / B = A + B:A`
^	Higher-order 'expansion', e.g., `(A + B)^2 = A + B + A:B`
~	"is modelled as a function of", defines a S formula
offset()	Offset, a vector or matrix of fixed and known values, e.g., `offset(log.time)`
I()	Identity or insulate, allows standard arithmetic operations to have their usual meaning, e.g., `I((x2 - x3)^2)` for the variable $(x_2 - x_3)^2$

Operator	Comment
&	Vector operator: and
|	Vector operator: or
!	Vector operator: not

Argument	Comment
contrasts	Handling of factor contrasts. See Sect. 1.5.2.4, e.g., `contrasts = c("contr.sum", "contr.poly")`
na.action	Handling of missing values. See Sect. 1.5.2.6, e.g., `na.action = na.pass`
offset	Offset, an alternative to `offset()` in the `formula` argument, e.g., `offset = log(followup.time)`
subset	Subset selection, e.g., `subset = 20 < age & sex == "M"`
weight	Prior weights, known and fixed, w_i in (1.31)

as piecewise-linear functions that are continuous or discontinuous. Figure 1.4 gives some examples. The fits are of the form of an LM regressing a response against a variable x, or x_2 and x_3. The x, x_2 and x_3 are random samples from the unit interval. The threshold value x_0, called `x0`, was assigned an arbitrary value of 0.4. The property that

```
> as.numeric(c(TRUE, FALSE))

 [1] 1 0
```

is useful in S formulas, e.g., Fig. 1.4a, is of the form

```
lm(y ~ I(x - x0 < 0), data = ldata)
```

and the first colour image, Fig. 1.4k, is

```
lm(y ~ I((x2 - x0 > 0) * (x3 - x0 > 0)), data = ldata)
```

Figure 1.4a has the same form of a stump in a regression tree with univariate threshold splits, i.e., $x_k < x_0$ for some variable x_k and threshold x_0. Figure 1.4f is a constrained version of Fig. 1.4b,c because the slopes on either side of x_0 have opposite signs. Figure 1.4l has the same form as a regression tree in two variables; it has four leaves and three layers (including the root) but, of course, x_0 need not be the same as the threshold for each of the two variables.

Finally, to illustrate the "." shortcut:

```
lm(y ~ . , data = ldata)
```

means **y** is regressed against all the other variables in data frame **ldata**. One can use, e.g.,

```
lm(y ~ . - x5, data = ldata)
```

to exclude **x5** as a regressor.

Model matrices are obtained first by combining the formula with the data frame to produce a model frame, from which the model matrix is constructed.

1.5.2.4 Factors

Factors in regression are used for grouping variables and are represented by dummy or indicator variables whose values equal 0 or 1. If there are L levels, then there are $L - 1$ such variables. Table 1.5 gives three examples.

The **contrast** argument of **lm()** may be assigned a list whose entries are values (numeric matrices or character strings naming functions) to be used as replacement values for the **contrasts()** replacement function, and whose names are the names of columns of **data** containing **factors**. For example,

```
lm(y ~ f1 + f2, data = ldata, contrasts = list(f1 = "contr.sum"))
lm(y ~ f1 + f2, data = ldata, contrasts = list(f1 = "contr.sum", f2 = "contr.poly"))
```

Here, assigning it to the **contrast** argument will have an effect for that model only. One can change them globally, for example, by

```
options(contrasts = c("contr.SAS", "contr.poly"))
```

The R default is

```
> options()$contrasts

          unordered           ordered
    "contr.treatment"     "contr.poly"
```

Here are some notes.

1. By default, **contr.treatment()** is used so that each coefficient compares that level with level 1 (omitting level 1 itself). That is, the *first* level is baseline.
2. Function **contr.SAS()** makes the last level of the factor the baseline level.

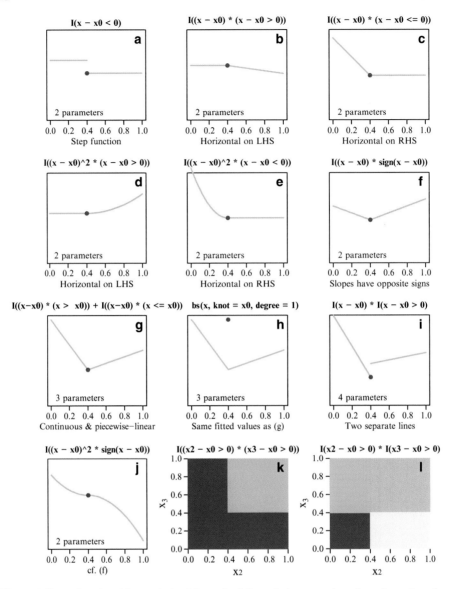

Fig. 1.4 Some functional forms derived from the S formula language based on elementary functions and operators. The formula heads each plot, followed by footnotes. The number of parameters in the regression is given. An intercept term is assumed in all—and the blue ● point indicates its value at the location $x = x_0$. Plots (k)–(l) are contour images with various colours denoting different fitted values (blue for the intercept). The value $\mathtt{x0} = x_0 = 0.4$ here, and variables x, x_2 and x_3 are defined on the unit interval. Function $\mathtt{bs()}$ resides in the **splines** package.

3. Function `contr.sum()` constrains the coefficients to sum to zero. For example, the one-way analysis of variance model

```
lm(y ~ f1, data = ldata, contrasts = list(f1 = "contr.sum"))
```

Table 1.5 Dummy variables D_1, D_2, D_3, from `contr.treatment()` (LHS; the default in R) when applied to the nominal ethnicity variable of `xs.nz`; the Europeans are the reference group. Middle: the same with `contr.SAS()`; the Others are the reference group. RHS: the same with `contr.sum()`.

	`contr.treatment()`			`contr.SAS()`			`contr.sum()`		
Ethnicity	D_1	D_2	D_3	D_1	D_2	D_3	D_1	D_2	D_3
European	0	0	0	1	0	0	1	0	0
Maori	1	0	0	0	1	0	0	1	0
Polynesian	0	1	0	0	0	1	0	0	1
Other	0	0	1	0	0	0	−1	−1	−1

with `f1` taking on 3 levels has $y_{ij} = \alpha + \beta_i + \varepsilon_{ij}$ with $\beta_1 + \beta_2 + \beta_3 = 0$, for $j = 1, \ldots, n_i$. Then β_3 is omitted from the regression and $\hat{\alpha} + \hat{\beta}_i = \overline{y}_{i\bullet}$ for $i = 1, 2, 3$ with $\hat{\beta}_3 = -\hat{\beta}_1 - \hat{\beta}_2$.

4. Function `contr.poly()` is used for equally spaced, equally replicated orthogonal polynomial contrasts.

Other useful functions for manipulating factors are `factor()`, `as.factor()`, `ordered()`, `levels()`, `options()`.

For `cumulative()` and `propodds()` the response should represent an ordered factor—if `is.ordered()` applied to a factor response is `FALSE` then a warning is issued. Similarly, if the response for a `multinomial()` fit is an ordered factor, then a warning is also issued.

1.5.2.5 Centring Variables

For linear predictors η_j the intercept may have little to no meaning if $x_2 = x_3 = \cdots = 0$ makes no physical sense, i.e., if $\boldsymbol{x}_{[-1]} = \boldsymbol{0}$ is unrealizable. For example, suppose in human data, $\boldsymbol{x} = (1, \texttt{age}, \texttt{sex}, \texttt{height}, \texttt{weight})^T$ is measured in years, female/male $= 0/1$, metres and kilograms. The fitted intercept is largely uninterpretable because males just born with zero height and weight are hard to find! However, some of the x_k can be centred so that the intercept can refer to some representative person, e.g.,

```
hdata <- transform(hdata, age30    = age    - 30,
                          height1.7 = height - 1.7,
                          weight90  = weight - 90)
```

means that a regression performed with these new variables might have fitted intercepts corresponding to a 30-year-old female American of height 1.7 m and weight 90 kg. In short, centring is an optional technique that can enhance the interpretability of a statistical model.

Scaling the x_k can also help, e.g., by dividing its standard deviation. If the variable is approximately normally distributed, then ± 2 units of the standardized variable will provide a range that covers about 95% of that variable.

Standardizing by centring and scaling may be achieved by `scale()` or its smart version `sm.scale()` (Sects. 8.2.5 and 18.6).

Table 1.6 Standard functions that argument **na.action** can be assigned, for modelling functions such as **lm()** and **vglm()**. The model frame comprises the explanatory and response variables specified by **formula**.

na.omit()	Any row with an NA in the model frame is deleted. This is the system default (options("na.action"))
na.fail()	Issues an error message if there are any NAs in the model frame
na.pass()	Any NA is treated the same as an ordinary value so that the model frame remains unchanged. Some VGAM family functions, such as for the block Gumbel model (Sect. 16.2.2) for extreme value data, allow NAs in the response in order to handle ragged arrays, therefore this option is sometimes preferred
na.exclude()	Similar to na.omit() but its class is "exclude", whereas na.omit() has the class "omit". Some functions such as naresid() and napredict() need to distinguish between these two types, e.g., NAs are used to pad cases omitted by na.exclude()

1.5.2.6 Missing Values

Missing values are represented by **NA** for "not available" or "not applicable". Modelling functions usually come with an **na.action** argument that can be assigned one of the functions listed in Table 1.6. Note that the model frame is constructed from the data frame specified by argument **data** (otherwise from **.GlobalEnv**), based on the variables in the formula. Other variables in the data frame that are not part of the formula are ignored so that, e.g., rows from a data frame with missing values in unneeded variables will not be deleted by **na.omit()** in the fitted object. In VGAM, the model matrix $\mathbf{X}_{\mathrm{LM}}$ is then constructed from model frame. In general, **na.action** must be assigned a function that takes a data frame as argument and returns another data frame as result with the same variables.

If any data frame has any **NAs**, then the data frame returned by the functions has an attribute called **"na.action"** which has a certain **class**. For example,

```
> ldata <- data.frame(x2 = c(1, 2,  3), y = c(0, 10, NA))
> fit <- lm(y ~ x2, data = ldata, model = TRUE, na.action = na.exclude)
> attr(fit$model, "na.action")

 3
 3
attr(,"class")
[1] "exclude"

> class(attr(fit$model, "na.action"))

[1] "exclude"
```

1.5.2.7 Recycling and Storage of Elements

Two-dimensional data structures such as matrices and data frames store their elements so that the left-most index varies fastest. e.g.,

```
> c(matrix(1:10, 5, 2))

 [1]  1  2  3  4  5  6  7  8  9 10
```

This is also true for arrays of any dimension, and the second left-most index varies the second fastest, etc.

Recycling of values is an important feature of the S language and is used a lot in VGAM.

1.5.3 More on the Number of Parameters and Terms

We saw from the Sect. 1.5.2.2 that a term in an S formula can generate multiple columns of the model matrix, e.g, `poly(x3, 2, raw = TRUE)` in (1.44) creates two columns. Something to bear in mind is that sometimes it is necessary to distinguish between the number of parameters being estimated and the number of 'raw' or original covariates. This is because each raw covariate x_k is commonly expanded using basis functions. In general, p denotes the number of regression coefficients to be estimated (in some form or another because there are at least two of them defined—p_{LM} and p_{VLM}), and d is the number of raw covariates. Often, d is the number of terms in a simple additive model because each term is a function of only one of the original explanatory variables (no interactions assumed). We tend to express additive models (1.2) in terms of d.

Here is another simple example:

```
> y ~ poly(x2, 2, raw = TRUE) + poly(x3, 2, raw = TRUE) + x4
```

The corresponding (LM) model matrix has 6 columns $\{1, x_2, x_2^2, x_3, x_3^2, x_4\}$ which are mapped to $\boldsymbol{x} = (x_1, x_2, \ldots, x_p)^T$. But one might sometimes refer to $(1, x_2, x_3, x_4)^T$ as the original $\boldsymbol{x}$. An even simpler term that creates the need for duality in notation is when x_k as a factor; then, it expands into $L - 1$ columns of the model matrix, where L is the number of levels.

When there is no need to distinguish between d and p usually the latter is used, and often $d = p$ and they are used loosely and interchangeably.

1.6 Summary

The purpose of this chapter is to provide a brief overview of the statistical framework including a short description of each major class of models, and to provide some preliminary background material for such.

Being a large project, VGAM is still in development. The project is largely driven by practical expediencies, and is designed to be hopefully general. With a potentially large capability, it is structured by its large and unified framework. It is hoped that the package will be useful to many statistical practitioners.

Bibliographic Notes

The S3 language was defined in Chambers and Hastie (1991), and S4 in Chambers
(1998); see also Chambers (2008). Many of VGAM's features are similar to `glm()`,
therefore readers unfamiliar with these functions are referred to Chambers and
Hastie (1991). Readers unfamiliar with R and aspects of the S language are also
directed to more user-friendly account such as Zuur et al. (2009) and de Vries and
Meys (2012), as well as the statistical modelling books listed in the bibliographic
notes of Chap. 8.

References for categorical and count data analyses are given at the end of
Chaps. 14 and 11, respectively. Keogh and Cox (2014) describe the odds ratio
in the context of case-control studies. Imai et al. (2008) describe a common framework for statistical analysis and development, and implement the ideas in `zelig`.
The subject of this book fits well under the umbrella of data analysis, and an
overview of this topic is Huber (2011).

Exercises

Ex. 1.1.   Show for the proportional odds model (1.22)–(1.23) that the odds
of $Y \leq j$ given $\boldsymbol{x}_1$, relative to $\boldsymbol{x}_2$, does not depend on j.   [McCullagh and Nelder
(1989)]

Ex. 1.2.   Suppose $Y \sim \text{Poisson}(\mu)$ with $\eta \equiv \log \mu$, and $Y^* = I(Y > 0)$ is
binary. Show that a complementary log–log link applied to a binary regression
of Y^* results in the same η, i.e., $\texttt{cloglog}(P[Y^* = 1]) = \eta$.

Ex. 1.3.   Suppose Y has a standard Cauchy distribution, i.e., $f(y) = [\pi(1 +$
$y^2)]^{-1}$ for all real y. Show that $F^{-1}(p) = \tan(\pi(p - \frac{1}{2}))$.

Ex. 1.4.   Page et al. (2007) investigate the suicide rate in England and Wales
during heat waves that occurred in 1995 and 2003. The following data were adapted
from their Table 2 and other sources.

Run the following code and then fit a basic Poisson regression in order to estimate the death rate per 100,000 persons per year (taken as $365\frac{1}{4}$ days) during the
heat waves. The variables are: `popn` is an estimate of the population of England
and Wales during that year, `days` is the duration of each heat wave in days, `dates`
gives the time period of each heat wave (formatted as yyyymmdd), and `deaths` is
the number of suicides (estimated).

```
sdata <- data.frame(popn = c(51395000L, 52721000L, 52721000L),
                    deaths = c(104, 57, 119),
                    days = c(6, 5, 10),
                    dates = c("19950729-19950803", "20030713-20030717",
                              "20030804-20030813"))
```

Ex. 1.5.   One example where Poisson regression with offsets is commonly used
is the estimation of *catch per unit effort* (CPUE) in fisheries studies. Consider
the `lake0` trout data. Although the units of CPUE is usually the number of fish
caught per hour of fishing, we will use the number of fish caught per visit here and
call it CPUE_2.

(a) For the usual CPUE definition why is it necessary to assume that, over the years, the mean fishing duration per visit to the lake was the same? Do you think this is realistic?

(b) Plot the rainbow trout $CPUE_2$ versus year, the brown trout $CPUE_2$ versus year, and the raw counts over time. Comment.

(c) Fit a simple intercept-only Poisson regression with offsets to estimate the rainbow trout $CPUE_2$. Do the same for the brown trout $CPUE_2$.

(d) Repeat (c) but add **year** as a covariate. Are these **year** terms justified?

(e) Fit a logistic regression to the proportion of brown trout with **year** as a covariate. Is there any evidence of a change over time? If so, does this mean that the proportion of brown trout in the lake is changing over time? Justify your answer by giving some possible explanations.

Ex. 1.6. Consider the logistic regression model (1.17). Some software represent a binary explanatory variable with $x_k^\ddagger = 1$ and -1 for true and false respectively, e.g., when a factor with 2 levels is converted into a numerical variable. Suppose logit $P(Y = 1) = \eta^\ddagger = \sum \beta_k^\ddagger x_k^\ddagger$ is fitted, where $Y = 1$ or 0 is the usual response. Express $\beta_k^\ddagger$ in terms of the usual β_k, and obtain an expression for the standard error $SE(\hat{\beta}_k^\ddagger)$ in terms of $SE(\hat{\beta}_k)$.

Ex. 1.7. Consider the bivariate odds ratio model (1.18)–(1.19).

(a) Show that the joint probability $P(Y_1 = 1, Y_2 = 1)$ can be expressed as

$$
p_{11} = \begin{cases} \frac{1}{2}(\psi - 1)^{-1}\{a - \sqrt{a^2 + b}\}, & \psi \neq 1; \\ p_1\, p_2, & \psi = 1, \end{cases} \tag{1.46}
$$

where $a = 1 + (p_1 + p_2)(\psi - 1)$ and $b = -4\psi(\psi - 1)p_1 p_2$.

(b) Show that (1.21) leads to (1.20).

Ex. 1.8. Consider Fig. 1.4a. Write three different S formula terms that have the effect of fitting a 3-step function with known breakpoints, i.e., suppose $0 < a < b < 1$ for some known a and b. Then the fitted values of the LM are

$$
\hat{\mu}_i = \begin{cases} \hat{\beta}_1, & 0 < x_i \le a; \\ \hat{\beta}_2, & a < x_i \le b; \\ \hat{\beta}_3, & b < x_i \le 1. \end{cases}
$$

Ex. 1.9. Show that a multinomial logit model (1.28) with constraints $\sum_{t=1}^{M} \boldsymbol{\beta}_t^* = \mathbf{0}$ implies that the "median" response can be viewed as the reference category. That is, for $s = 1, \ldots, M$,

$$
\log\left(P(Y = s|\boldsymbol{x}) \div \left[\prod_{t=1}^{M} P(Y = t|\boldsymbol{x}) \right]^{1/M} \right) = \boldsymbol{\beta}_s^*. \tag{1.47}
$$

The denominator can be regarded as a geometric mean. [Tutz (2012)]

Ex. 1.10.   Negative Binomial Distribution

(a) Suppose a negative binomial distribution (1.14) is fitted to Y with the constraint $k \propto \mu$. What then is $Var(Y)$?

(b) Given log links for both μ and k parameters, what constraint matrices need to be chosen to achieve (a)?

Ex. 1.11. If $A = 1 - e^{-\lambda}$ and $B = 2\,e^{-\lambda}/\lambda^2$ in (1.42), find the optimal λ and the value of the penalty (objective function) at the solution.

Ex. 1.12. Complete-Case Analysis
The function `na.omit()` performs what is known as *complete-case analysis* whereby any row of an $n \times p$ matrix is deleted if there is a missing value in any of the variables of that row. Suppose that $\mathbf{X}$ is $n \times p$ with missing values in positions that are independent and random with probabilities that are uniform over all elements. Assume both n and p are greater than the number of missing values. Let N be the number of rows in a complete-case analysis of $\mathbf{X}$.

(a) Suppose there is 1 missing value. Explain why $E(N) = n - 1$.
(b) Suppose there are 2 missing values. Show that $E(N) = (n-1)\left[1 - p/(np-1)\right]$.
(c) Suppose there are 3 missing values. Obtain the probability function of N.

Ex. 1.13. Consider the ethnicity variable of `xs.nz`.

(a) Apply `contr.SAS()` to a simple vector equivalent of it, and obtain output of the same format as Table 1.5. Explain it to show your understanding of it.
(b) Repeat the same with `contr.sum()`.

Ex. 1.14. Given a data frame `pdata` with variables y and x2, show how to fit Poisson regressions subject to the following scenarios. That is, how might one call `glm()`? Assume the values of x2 are positive.

(a) We want $\mu \propto x_2^{\beta_2}$ for some regression coefficient β_2.
(b) We want $\mu \propto x_2$. List two calls to `glm()` to achieve this. What dangers are there in this model? Which call is to be preferred, if any?

Ex. 1.15. Given the Poisson probability function (1.7), show that

(a) $E(Y) = \sum_{y=0}^{\infty} y f(y) = \mu$,
(b) $\mathrm{Var}(Y) \equiv E[(Y - \mu)^2] = \mu$.

Ex. 1.16. Some S Formulas

(a) For Fig. 1.4h, express $\lim_{x \to x_0^-}$ of the fitted values as a function of the parameters.
(b) For Fig. 1.4i, which is discontinuous at x_0, express $\lim_{x \to x_0^+}$ of the fitted values as a function of the parameters.

R is a ratty, old, cobbled-together piece of crap (compared with what we now know is possible).
—Ross Ihaka, 2012-08

Chapter 2
LMs, GLMs and GAMs

...I have long been a proponent of the following unified field theory for statistics: "Almost all of statistics is linear regression, and most of what is left over is non-linear regression."
—R. I. Jennrich (discussion of Green (1984))

2.1 Introduction

This chapter reviews a few details about two building blocks for this book: LMs and smoothing. These topics naturally draw in two others, viz. GLMs and GAMs, whose inclusion poses the risk of distracting the reader from the main thrust of this book by having to include some material that is not quite necessary. Some justification and details, such as computational, are deferred to the next two chapters, where they are described under more general conditions.

2.2 LMs

LMs operate on data $(\boldsymbol{x}_i, y_i)$, $i = 1, \ldots, n$, where y_i is the ith response, $\boldsymbol{x}_i$ is explanatory, and $p \leq n$ (classically, it is usually not considered good practice if n is not much more than p). The data is assumed to follow the model

$$Y_i \;=\; \beta_1\, x_{i1} + \cdots + \beta_p\, x_{ip} + \varepsilon_i, \quad \varepsilon_i \sim N(0,\,\sigma^2) \text{ independently.} \qquad (2.1)$$

Note that R has the first column of the model matrix $\mathbf{X}$ equal to ones if there is an intercept, i.e., $x_{i1} = 1$. Often, people call (2.1) a *multiple linear regression*. In the wide literature, the ε_i are known under various names, such as the (statistical) error, white noise, innovations, random noise. The mean of Y_i, given $\boldsymbol{x}_i$, is

$$E(Y_i|\boldsymbol{x}_i) \;=\; \mu_i(\boldsymbol{x}_i) \;=\; \boldsymbol{\beta}^T \boldsymbol{x}_i \;=\; \sum_{k=1}^{p} \beta_k\, x_{ik}\ . \qquad (2.2)$$

T.W. Yee, *Vector Generalized Linear and Additive Models*,
Springer Series in Statistics, DOI 10.1007/978-1-4939-2818-7_2

Confusingly, some people call the LM a "general linear model", which has the same acronym as a "generalized linear model"(GLM). The general linear model includes generalized least squares, whereas the LM usually refers to ordinary least squares.

It is more compact to write the LM using vectors and matrices:

$$\boldsymbol{Y} = \mathbf{X}\boldsymbol{\beta} + \boldsymbol{\varepsilon}, \quad \varepsilon_i \sim N_n(\mathbf{0}, \sigma^2 \mathbf{I}_n), \tag{2.3}$$

where $\boldsymbol{Y} = (Y_1, \ldots, Y_n)^T$ and $x_{ik} = (\mathbf{X})_{ik}$. Later, we will sometimes write $\mathbf{X}$ as $\mathbf{X}_{\mathrm{LM}}$ to distinguish it from another type of *model matrix* (*design matrix*).

The LM makes strong assumptions. Embedded in the above equations, these are the following.

(i) *Independence of the errors*: data such as time series often violate this assumption because the errors are correlated with each other, in particular, with observations from the past. Independence of the ε_i is not explicitly stated in (2.3) because of a property of the multivariate normal distribution that $\mathrm{Cov}(\varepsilon_i, \varepsilon_j) = 0$ iff ε_i and ε_j are independent.

(ii) *Linearity*: the mean function defines a hyperplane in p-dimensional space. Each covariate x_k is usually modelled having a linear effect on the mean response, keeping the other variables in $\boldsymbol{x}$ fixed. Actually, LMs are called LMs because the mean is linear with respect to the parameters, so that a polynomial in x_k is a LM.

(iii) *Errors that are normal with mean zero and constant variance*: for example, a common form of *heteroscedasticity* is when $\mathrm{Var}(\varepsilon_i)$ increases with increasing mean μ_i.

In practice, all these assumptions should be checked as part of the model-building process; see Sect. 2.3.2. Relaxing the linearity assumption is the main motivation of *additive models*, e.g., see Sect. 2.5 and Chap. 4, and is a major theme of this book.

To estimate $\boldsymbol{\beta}$, it is most common to minimize the residual sum of squares

$$\mathrm{ResSS}(\boldsymbol{\beta}) = \sum_{i=1}^{n} \varepsilon_i^2 = (\boldsymbol{y} - \mathbf{X}\boldsymbol{\beta})^T(\boldsymbol{y} - \mathbf{X}\boldsymbol{\beta}) \tag{2.4}$$

as a function of $\boldsymbol{\beta}$. Of course, this is why the name 'least squares' (LS) is used, and the estimator corresponds to the MLE because the errors are assumed to be normally distributed. Setting the derivative of ResSS with respect to $\boldsymbol{\beta}$ as $\mathbf{0}$ gives the *normal equations*

$$\mathbf{X}^T\mathbf{X}\boldsymbol{\beta} = \mathbf{X}^T\boldsymbol{Y}. \tag{2.5}$$

Provided that $\mathbf{X}$ is of full column-rank, then

$$\widehat{\boldsymbol{\beta}} = \left(\mathbf{X}^T\mathbf{X}\right)^{-1}\mathbf{X}^T\boldsymbol{y} \tag{2.6}$$

is the LS estimate, and so

$$\widehat{\boldsymbol{y}} = \mathbf{X}\widehat{\boldsymbol{\beta}} = \mathbf{X}\left(\mathbf{X}^T\mathbf{X}\right)^{-1}\mathbf{X}^T\boldsymbol{y} = \boldsymbol{\mathcal{H}}\boldsymbol{y}, \quad \text{say}, \tag{2.7}$$

where $\mathcal{H}$ is known as the *hat matrix*. Then $\widehat{y}_i = \boldsymbol{x}_i^T \widehat{\boldsymbol{\beta}}$ are the *fitted values*, and $r_i = y_i - \widehat{y}_i$ the *residuals*. It then follows that $\mathrm{Var}(\widehat{\boldsymbol{\beta}}) = \sigma^2 (\mathbf{X}^T \mathbf{X})^{-1}$, and

$$\widehat{\boldsymbol{\beta}} \sim N_p \left(\boldsymbol{\beta}, \ \sigma^2 \left(\mathbf{X}^T \mathbf{X} \right)^{-1} \right). \tag{2.8}$$

Provided that an estimate of σ can be obtained, (2.8) can be used for inference, e.g., to calculate the standard errors for the $\widehat{\beta}_k$, and construct confidence intervals. We shall see below that the quantity S^2, defined as $\sum_{i=1}^n (Y_i - \widehat{Y}_i)^2 / (n-p)$, is unbiased for σ^2, therefore it is natural to use $\widehat{\sigma} = s$.

Suppose we wish to test the null hypothesis $H_0 : \mathbf{A}\boldsymbol{\beta} = \boldsymbol{c}$ for some $q \times p$ matrix of known constants $\mathbf{A}$ having rank-q, and $\boldsymbol{c}$ is a vector of known constants. For example, this can be used to test that a subset of the β_k are all 0. Fitting the LM under this constrained (smaller) model results in a residual sum of squares which can be denoted by ResSS_0. Likewise, the unconstrained (larger or full) model has a ResSS_1 which will be generally lower. Then the test statistic F_0 is distributed as

$$F_0 = \frac{(\mathrm{ResSS}_0 - \mathrm{ResSS}_1)/q}{\mathrm{ResSS}_1/(n-p)} \sim F_{q, n-p}. \tag{2.9}$$

The null hypothesis is rejected at the α-significance level if $F_0 > F_{q,n-p}(1-\alpha)$. The function `linearHypothesis()` in `car` may be used to test hypotheses of this form. The special cases of testing $H_0 : \beta_k = 0$ versus $H_1 : \beta_k \neq 0$, for all k one-at-a-time, is printed out in the `summary(lmObject)` output. For these, the test statistics are labelled "`t values`" because $T^2 \sim F_{1,\nu}$ for $T \sim t_\nu$, and have the form $t_{0k} = (\widehat{\beta}_k - 0)/\mathrm{SE}(\widehat{\beta}_k)$. The 2-sided p-values are printed as `Pr(>|t|)`, and any controversial 'significance stars' adjacent to them.

A $100(1-\alpha)\%$ confidence ellipsoid for $\boldsymbol{\beta}$ comprises values of $\boldsymbol{\beta}$ such that

$$\left(\widehat{\boldsymbol{\beta}} - \boldsymbol{\beta} \right)^T \mathbf{X}^T \mathbf{X} \left(\widehat{\boldsymbol{\beta}} - \boldsymbol{\beta} \right) \leq p \widehat{\sigma}^2 F_{p, n-p}(1-\alpha).$$

Often we wish to look at one coefficient of $\boldsymbol{\beta}$ at a time, then a $100(1-\alpha)\%$ confidence interval for β_k is $\widehat{\beta}_k \pm t_{n-p}(1-\alpha/2)\,\mathrm{SE}(\widehat{\beta}_k)$, where the standard error is $s \cdot [(\mathbf{X}^T \mathbf{X})^{-1}]_{kk}$. A call of the form `confint(lmObject)` returns such intervals.

For prediction at a value $\boldsymbol{x}_0$, say, a $100(1-\alpha)\%$ prediction interval for $y(\boldsymbol{x}_0)$ is

$$\boldsymbol{x}_0^T \widehat{\boldsymbol{\beta}} \pm t_{n-p}(1-\alpha/2)\,\widehat{\sigma} \sqrt{1 + \boldsymbol{x}_0^T \left(\mathbf{X}^T \mathbf{X} \right)^{-1} \boldsymbol{x}_0}.$$

Similarly, a $100(1-\alpha)\%$ confidence interval for $\mu(\boldsymbol{x}_0)$ is

$$\boldsymbol{x}_0^T \widehat{\boldsymbol{\beta}} \pm t_{n-p}(1-\alpha/2)\,\widehat{\sigma} \sqrt{\boldsymbol{x}_0^T \left(\mathbf{X}^T \mathbf{X} \right)^{-1} \boldsymbol{x}_0}.$$

Prediction intervals focus on a future (random) value of y and are consequently wider than a confidence interval for the (fixed) conditional mean $E(Y|\boldsymbol{x}_0)$. This is intuitively so because confidence intervals simply need to account for the uncertainty in $\widehat{\boldsymbol{\beta}}$, whereas prediction intervals have the additional randomness due to $\mathrm{Var}(\varepsilon_i)$ as well.

2.2.1 The Hat Matrix

From (2.7), we have the $n \times n$ matrix

$$\mathcal{H} = \mathbf{X} \left(\mathbf{X}^T \mathbf{X}\right)^{-1} \mathbf{X}^T, \tag{2.10}$$

which is referred to as the 'hat' matrix because it adds a 'hat' to $\boldsymbol{y}$ (i.e., $\widehat{\boldsymbol{y}}$) when $\boldsymbol{y}$ is premultiplied by it. Assuming that $\mathbf{X}$ is of rank-p, the hat matrix has the following properties.

(i) $\mathcal{H} = \mathcal{H}^T$, i.e., symmetric.
(ii) $\mathcal{H}^2 = \mathcal{H}$, i.e., idempotent.
(iii) $\mathcal{H}\mathbf{1} = \mathbf{1}$ if the LM has an intercept, i.e., all rows sum to unity.
(iv) rank($\mathcal{H}$) = trace($\mathcal{H}$) = p.
(v) $\mathcal{H}$ has p unit eigenvalues and $n - p$ zero eigenvalues.
(vi) $0 \leq h_{ii} \leq 1$, where $h_{ij} = (\mathcal{H})_{ij}$ is the (i,j)-element of $\mathcal{H}$. If the LM has an intercept, then $n^{-1} \leq h_{ii} \leq 1$.
(vii) $\mathrm{Var}(r_i) = \sigma^2(1 - h_{ii})$. This serves as motivation for (2.12).

The proofs are not difficult and are left as an exercise (Ex. 2.1). The hat matrix $\mathcal{H}$ is also known as a *projection matrix* because it is the orthogonal projection of $\boldsymbol{Y}$ onto the column (range) space of $\mathbf{X}$; it has the two properties of being symmetric and idempotent.

To show that S^2 is unbiased for σ^2, $E[\mathrm{ResSS}] = E[(\boldsymbol{Y} - \widehat{\boldsymbol{Y}})^T(\boldsymbol{Y} - \widehat{\boldsymbol{Y}})] = E[\boldsymbol{Y}^T(\mathbf{I} - \mathcal{H})^T(\mathbf{I} - \mathcal{H})\boldsymbol{Y}] = E[\boldsymbol{Y}^T(\mathbf{I} - \mathcal{H})\boldsymbol{Y}] = \mathrm{trace}((\mathbf{I} - \mathcal{H})\,\sigma^2\mathbf{I}) + \boldsymbol{\mu}(\mathbf{I} - \mathcal{H})\boldsymbol{\mu} = \sigma^2(n - p) + 0 = \sigma^2(n - p)$, by formulas given in Sect. A.2.5. These results are generalized later for linear smoothers in, e.g., Sect. 2.4.7.4.

The importance of the hat matrix, especially for diagnostic checking, is due to its interpretation as containing the weights of all the observations in obtaining the fitted value at a particular point. This can be seen by focusing on the ith row of (2.7):

$$\widehat{y}_i = \sum_{j=1}^{n} h_{ij}\, y_j, \tag{2.11}$$

so that h_{ij} can be interpreted as the weight associated with datum (x_j, y_j) to give the fitted value for datum (x_i, y_i). For the $p = 2$ case, plotting the h_{ij} versus x_{j2} for $j = 1, \ldots, n$, is analogous to the equivalent kernels for smoothers considered in the next chapter (e.g., Sect. 2.4.7.3).

However, it is usually the diagonal elements of $\mathcal{H}$ that are of greatest relevance to people fitting LMs. Element h_{ii} measures how much impact y_i has on $\widehat{y}_i$, and consequently it quantifies the amount of influence that observation i has on the fit. Indeed, the h_{ii} are called *leverages* or *leverage scores*, and they measure how far $\boldsymbol{x}_i$ is away from the centre of all the data ($\overline{\boldsymbol{x}}$). Intuitively, as ResSS is being minimized, observations $\boldsymbol{x}_i$ that are isolated from the rest cause the regression plane $\mu(\boldsymbol{x})$ to be 'pulled' unduly towards them. The leverages can be defined as $\partial \widehat{y}_i / \partial y_i$. Since the sum of the h_{ii} is p, any individual diagonal element that is substantially higher than the mean value can be considered influential. It is common to use the rule-of-thumb: if $h_{ii} > 2p/n$, say, then observation i is influential or has high leverage. Others use $h_{ii} > 3p/n$ instead. As high-leverage points may 'pull' the regression line or plane towards them, they do not necessarily have a large residual.

2.2.1.1 LM Residuals and Diagnostics

A major component in checking the underlying LM assumptions is the examination of the residuals. Indeed, the use of diagnostic plots and other tools to check the adequacy of a fitted regression model separates competent practitioners from amateurs. From above, the ordinary residuals do not have equal variances, therefore

$$r_i^{\text{std}} \;=\; \frac{y_i - \widehat{y}_i}{s\sqrt{1 - h_{ii}}} \tag{2.12}$$

are used commonly, called *standardized residuals* or *(internally) Studentized residuals*. Here, s follows from the result that S^2 is an unbiased estimator for σ^2. However, if $(\boldsymbol{x}_i, y_i)$ is an outlier, then s may be affected, therefore it is safer to use the *(externally) Studentized residuals* (or simply *Studentized residuals*)

$$r_i^{\text{stu}} \;=\; \frac{y_i - \widehat{y}_i}{s^{[-i]}\sqrt{1 - h_{ii}}} \tag{2.13}$$

where $s^{[-i]}$ is the estimate of σ by deleting observation i. The function `hatvalues()` returns diag($\boldsymbol{\mathcal{H}}$), and `rstandard()` and `rstudent()` return (2.12) and (2.13), respectively.

LM diagnostics are quite a large subject and beyond the scope of this book. Here is a small listing of popular diagnostic plots for detecting violations in some of the underlying LM assumptions. Here, 'residuals' are best standardized or Studentized residuals, although ordinary residuals are often used.

1. When the residuals are plotted against their fitted values, ideally one would want a patternless horizontal band. A common form of departure from this is a 'funnel-effect', where there is less spread at lower fitted values—this suggests non-constant variance of the errors.
2. Plot the *partial residuals* against each x_k, e.g., $r_i + \widehat{\beta}_k x_{ik}$ versus x_{ik}. These types of residuals attempt to remove the effect of x_k from the regression temporarily. If indeed x_k has a linear effect on the response, then removing $\widehat{\beta}_k x_{ik}$ from the residual should leave white-noise. Any nonlinear trend would suggest that the $\beta_k x_k$ term might be generalized to some smooth function $f_k(x_k)$, i.e., an additive model. Not surprising, partial residuals are central to the backfitting algorithm for fitting VGAMs (Sect. 4.3.2).
3. Check the normality of the errors by a normal Q-Q plot of the residuals or a histogram, e.g., with `qqnorm()` and `hist()`.
4. As a check of the independence of the errors, plotting $\widehat{\varepsilon}_{i-1}$ versus $\widehat{\varepsilon}_i$ and obtaining a pattern with a nonzero slope suggests a violation of the independence assumption. For this, the Durbin-Watson test is popular, and there are implementations for this in, e.g., `car` and `lmtest`. More fundamentally, how the data was generated and collected needs to be considered in the broader context of the analysis.

Some of the above are produced by a call of the form `plot(lmObject)`. Common remedies to violations of the LM assumptions include transforming the response or the x_k, adding polynomial terms in x_k, using weighted least squares (WLS) or generalized least squares (GLS), and fitting additive models.

From an inference point-of-view, it is generally thought that independence of the errors is the most crucial assumption, followed by constant variance of the errors and linearity with respect to each x_k. Normality of the errors is the least important assumption, and the Shapiro-Wilk test can be used for this (`shapiro.test()`). Of course, gross features such as outliers and high-leverage points must be attended to. Other potential problems include

- multicollinearity (sets of x_k which are highly correlated or almost linearly dependent, e.g., $x_2 = \log$ `width` and $x_3 = \log$ `breadth` would be highly correlated with $x_4 = \log$ `area` if regions were approximately rectangular in shape),
- interactions (e.g., the effect of x_s on Y changes with the value of another explanatory variable x_t); these can be complicated and hard to deal with,
- variable selection, e.g., trying to determine which variables should be included in the model. This problem is exacerbated in an age of Big Data, where many variables are collected routinely.

Consequently, good linear modelling requires skill and diligence; it is an art as well as a science.

Regarding variable selection, a simple technique for moderate p that can be performed manually, called *backward elimination*, involves fitting a model with all the explanatory variables in, and then removing the least significant variable (i.e., the one with the largest p-value) and refitting a new model. This procedure is repeated until all remaining variables are 'significant'. The criterion for a 'significant' variable might be one whose p-value less than 5% or 10%; it should be decided upon beforehand.

For `lm()` fits, `influence.measures()` is the primary high-level function for diagnostics. It returns a number of quantities such as DFBETAS, DFFITS, Cook's distances, and diag($\mathcal{H}$). These are 'leave-one-out' diagnostics because they measure the effect of removing one observation at a time. In particular, DFBETA are the quantities $\widehat{\beta} - \widehat{\beta}_{[-i]}$, the **diff**erence in the **beta** vector of coefficients. DFBETAS is the scaled version of DFBETA, DFFITS is a scaled version of DFFIT (which is $\widehat{y}_i - \boldsymbol{x}_i^T \widehat{\beta}_{[-i]} = \boldsymbol{x}_i^T (\widehat{\beta} - \widehat{\beta}_{[-i]})$). These quantities plus more are defined in Belsley et al. (1980, Chap.2). Cook's distances measure the effect of the ith observation on $\widehat{\beta}$, in a way that picks up high-leverage points and observations with large Studentized residuals—`cooks.distance()` can return these quantities.

2.2.2 WLS and GLS

The assumption in (2.3) that $\text{Var}(\boldsymbol{\varepsilon}) = \sigma^2 \mathbf{I}_n$ is commonly unrealistic in real data analysis, as the previous section indicated. A generalization that relaxes homoscedastic errors is to allow $\text{Var}(\boldsymbol{\varepsilon}) = \sigma^2 \text{diag}(w_1, \ldots, w_n)^{-1}$ for positive known weights w_i. Then fitting an LM by minimizing $\sum_i w_i (y_i - \widehat{y}_i)^2$ is known as *weighted least squares* (WLS). Note that the ε_i are still independent, as in *ordinary least squares* (OLS).

It is necessary to generalize WLS further. If $\text{Var}(\boldsymbol{\varepsilon}) = \sigma^2 \boldsymbol{\Sigma}$ for any known positive-definite matrix $\boldsymbol{\Sigma}$, then estimating $\boldsymbol{\beta}$ by minimizing $(\boldsymbol{y} - \mathbf{X}\boldsymbol{\beta})^T \boldsymbol{\Sigma}^{-1} (\boldsymbol{y} - \mathbf{X}\boldsymbol{\beta})$ is called *generalized least squares* (GLS). GLS is needed when considering VGLMs and VGAMs, whereas WLS is sufficient for fitting GLMs. GLS allows the errors ε_i to be correlated.

The following formulas hold for GLS, and it is left to the reader to supply their proofs (Ex. 2.5):

$$\widehat{\boldsymbol{\beta}} = \left(\mathbf{X}^T \boldsymbol{\Sigma}^{-1} \mathbf{X}\right)^{-1} \mathbf{X}^T \boldsymbol{\Sigma}^{-1} \boldsymbol{y}, \qquad (2.14)$$

$$\mathrm{Var}(\widehat{\boldsymbol{\beta}}) = \sigma^2 \left(\mathbf{X}^T \boldsymbol{\Sigma}^{-1} \mathbf{X}\right)^{-1}, \qquad (2.15)$$

$$\frac{(\boldsymbol{y} - \mathbf{X}\widehat{\boldsymbol{\beta}})^T \boldsymbol{\Sigma}^{-1} (\boldsymbol{y} - \mathbf{X}\widehat{\boldsymbol{\beta}})}{\sigma^2} \sim \chi^2_{n-p}. \qquad (2.16)$$

Also, its hat matrix (defined as $\mathcal{H}$ such that $\widehat{\boldsymbol{y}} = \mathcal{H}\boldsymbol{y}$), is idempotent, but not symmetric in general. Equation (2.16) implies that $\widehat{\sigma}^2 = \mathrm{ResSS}/(n-p)$ is an unbiased estimator of σ^2.

2.2.3 Fitting LMs in R

The fitting function `lm()` is used to fit LMs. It has arguments

```
> args(lm)

  function (formula, data, subset, weights, na.action, method = "qr",
      model = TRUE, x = FALSE, y = FALSE, qr = TRUE, singular.ok = TRUE,
      contrasts = NULL, offset, ...)
  NULL
```

Arguments `formula` and `data` should be used all the time. Arguments `subset`, `weights` and `na.action` can be very useful at times, and `offset` can be incorporated into the formula instead by addition of the term `offset(<value>)`.

LMs may be fitted using `glm()`, albeit with slightly less efficiency. The normal distribution being known as the *Gaussian* distribution, `gaussian()` is the default for its `family` argument.

2.3 GLM Basics

This section gives a bare-bones and incomplete overview of GLMs. Such is not really required for an understanding of the VGLMs described in Chap. 3, however it does provide some background for such. Hence this section is given more for completion than for necessity.

GLMs as proposed by Nelder and Wedderburn (1972) provide a unifying framework for a number of important models in the exponential family. In particular, these include the normal (Gaussian), binomial and Poisson distributions. One consequence is that a single algorithm (IRLS) can be used to fit them all.

As with LMs, we have independent sample data $(\boldsymbol{x}_i, y_i)$, $i = 1, \ldots, n$, where y_i is the response (more general now), n is the sample size, and $\boldsymbol{x}_i = (x_{i1}, \ldots, x_{ip})^T$ is a vector of p explanatory variables ($x_{i1} = 1$ is the intercept if there is one). A GLM is composed of three parts:

(i)   a *random component* $f(y; \mu)$ specifying the distribution of Y,
(ii) a *systematic component* $\eta = \boldsymbol{\beta}^T \boldsymbol{x}$ specifying the variation in Y accounted for by known covariates, and
(iii) a *link function* $g(\mu) = \boldsymbol{\beta}^T \boldsymbol{x}$ that ties the two together. Often, g is simply called the *link*.

The η is known as the *linear predictor*, and the random component $f(y; \mu)$ is typically an exponential family distribution with $E(Y|\boldsymbol{x}) = \mu(\boldsymbol{x})$ being the *mean function*. GLMs thus fit

$$g(\mu(\boldsymbol{x}_i)) \;=\; \eta_i \;=\; \boldsymbol{\beta}^T \boldsymbol{x}_i \;=\; \beta_1 \, x_{i1} + \cdots + \beta_p \, x_{ip}, \tag{2.17}$$

where g is known. The required properties of g are strict monotonicity and being twice-differentiable in the range of μ. The main purpose of g is to transform the mean, which is usually bounded, into an unbounded parameter space where the optimization problem is unfettered and therefore simpler. Another purpose is that it often aids interpretability. In the VGLM framework described in Chap. 3, we write (2.17) as

$$g_1(\mu(\boldsymbol{x}_i)) \;=\; \eta_{i1} \;=\; \boldsymbol{\beta}_1^T \boldsymbol{x}_i \;=\; \beta_{(1)1} \, x_{i1} + \cdots + \beta_{(1)p} \, x_{ip} \tag{2.18}$$

to allow for more than one linear predictor in a model. The linear predictor (2.18) is central to this book. It takes the form of a weighted average of the covariate values $x_{i1}, \ldots, x_{ip}$ for object i—it is a plane in p-dimensional space.

For one observation, the probability density or mass function (PDF or PMF) of the exponential family can be written

$$f(y; \theta, \phi) \;=\; \exp\left\{ \frac{y\,\theta - b(\theta)}{\phi} + c\,(y, \phi) \right\}, \tag{2.19}$$

where θ is called the *natural parameter* or *canonical parameter*, ϕ is a possibly-known *dispersion parameter* (or *scale parameter*), and b and c are known functions. When ϕ is known, the distribution of Y is a one-parameter canonical exponential family member. When ϕ is unknown, it is often a nuisance parameter and then it is estimated by the method of moments. In most of GLM theory, the role of ϕ is curious and unfortunate: it is often treated as an unknown constant but not as a parameter. The VGLM framework views this as a deficiency, because of the original framework's inability to handle more than one parameter gracefully. The VGLM framework tends to estimate all parameters by full maximum likelihood estimation. This makes life easier in general, and estimation and inference is simpler.

At this stage, it is a good idea to handle known prior weights, A_i say, which may be entered into modelling functions such as `glm()` and `vglm()` via the `weights` argument. We will write $\phi_i = \phi/A_i$, where usually $A_i = 1$. In the case of the Y_i being binomial proportions, $N_i Y_i \sim \text{Binomial}(N_i, \mu_i)$ where the N_i can be assimilated into the A_i by $A_i = N_i$. Another reason is because we want to maximize a log-likelihood of the form $\sum_{i=1}^n A_i \, \ell_i$; most of this book dwells on maximizing the log-likelihood (3.7) so that the A_i here can be absorbed into the prior weights w_i there (this is largely a change of notation).

Noting that $\ell(\mu; y) = \log f(y; \mu)$, the log-likelihood is

$$\ell(\boldsymbol{\theta}, \phi; y) \;=\; \sum_{i=1}^{n} \frac{y_i\,\theta_i - b(\theta_i)}{\phi/A_i} + c\left(y_i, \frac{\phi}{A_i}\right), \qquad (2.20)$$

and the score function is

$$U(\boldsymbol{\theta}) \;=\; \frac{\partial \ell}{\partial \boldsymbol{\beta}} \;=\; \sum_{i=1}^{n} \frac{y_i - b'(\theta_i)}{\phi/A_i}. \qquad (2.21)$$

Then, using (A.17) and (A.18),

$$E(Y_i) \;=\; b'(\theta_i) \;\; \text{and} \;\; \mathrm{Var}(Y_i) \;=\; \frac{\phi}{A_i}\,b''(\theta_i). \qquad (2.22)$$

The *variance function* is $V(\mu) = b''(\theta(\mu))$, i.e., the variance of Y as a function of the mean.

It is noted at this stage that the MLE $\widehat{\boldsymbol{\beta}}$ is obtained by solving the estimating equation

$$\boldsymbol{U_\beta} \;=\; \sum_{i=1}^{n} \frac{\partial \mu_i}{\partial \boldsymbol{\beta}}\,\mathrm{Var}(Y_i)^{-1}\,(y_i - \mu_i) \;=\; \boldsymbol{0}. \qquad (2.23)$$

To see this,

$$
\begin{aligned}
\boldsymbol{0} \;=\;& \frac{\partial \ell}{\partial \boldsymbol{\beta}} \;=\; \sum_{i=1}^{n} \frac{\partial \ell_i}{\partial \theta_i} \frac{\partial \theta_i}{\partial \boldsymbol{\beta}} \;=\; \sum_{i=1}^{n} U(\theta_i) \frac{\partial \theta_i}{\partial \boldsymbol{\beta}} \\
\;=\;& \sum_{i=1}^{n} \frac{Y_i - b'(\theta_i)}{\phi/A_i} \frac{1}{(\partial \mu_i/\partial \theta_i)} \frac{\partial \mu_i}{\partial \theta_i} \frac{\partial \theta_i}{\partial \boldsymbol{\beta}} \\
\;=\;& \sum_{i=1}^{n} \frac{(Y_i - \mu_i)}{\phi/A_i} \frac{1}{b''(\theta_i)} \frac{\partial \mu_i}{\partial \boldsymbol{\beta}} \\
\;=\;& \sum_{i=1}^{n} \left\{\frac{Y_i - \mu_i}{\mathrm{Var}(Y_i)}\right\} \frac{\partial \mu_i}{\partial \boldsymbol{\beta}}, \qquad (2.24)
\end{aligned}
$$

which is the LHS of (2.23). This equation serves as the motivation for quasi-likelihood models described later because it only depends on the first two moments of Y.

Table 2.3 lists the most common members of the exponential family and how they are fitted in **VGAM**. For GLMs, the canonical parameter is a link function applied to the mean. Consequently, this particular link is known as the *canonical link*. With this link and given ϕ, $\mathbf{X}^T \boldsymbol{y}$ are a set of sufficient statistics for $\boldsymbol{\beta}$.

Iteratively reweighted least squares (IRLS) forms the core algorithm behind GLMs and GAMs. It is described under more general conditions in Sect. 3.2. For GLMs, it involves regressing at each iteration (the ath iteration, say) the *adjusted dependent variable* (or *modified dependent variable* or *working dependent variate* or *pseudo-response*) $z_i^{(a-1)}$ against $\boldsymbol{x}_i$ with (working) weights $w_i^{(a-1)}$; these are given by

$$z_i^{(a-1)} \;=\; \eta_i^{(a-1)} + \frac{y_i - \mu_i^{(a-1)}}{d\mu_i^{(a-1)}/d\eta_i} \;\; \text{and} \;\; w_i^{(a-1)} \;=\; \frac{A_i}{V\left(\mu_i^{(a-1)}\right)} \left(\frac{d\mu_i^{(a-1)}}{d\eta_i}\right)^2.$$

$$(2.25)$$

It can be achieved by WLS: $\mathbf{X}$ is the model matrix, and $\mathbf{W}^{(a-1)}$ is $\mathrm{diag}(w_1^{(a-1)}, \ldots, w_n^{(a-1)})$.

The above weights are actually obtained by the *expected* negative Hessian (rather than the *observed*), so this is Fisher scoring. For some models this equates to Newton-Raphson. With the new $\boldsymbol{\beta}^{(a)}$, a new $\boldsymbol{\eta}^{(a)}$ and $\boldsymbol{\mu}^{(a)}$ are computed, and the cycle is continued till convergence. When close to the solution, the convergence rate is quadratic for GLMs with a canonical link, meaning that the number of correct decimal places doubles (asymptotically) at each iteration.

At convergence, we have

$$\widehat{\mathrm{Var}}(\widehat{\boldsymbol{\beta}}) \;=\; \widehat{\phi} \left(\mathbf{X}^T \mathbf{W}^{(a)} \mathbf{X}\right)^{-1},$$

$$(2.26)$$

which is returned by the function `vcov()`.

Convergence problems with GLMs can occur in practice, although they are not particularly common. Section 3.5.4 gives an example.

2.3.1 Inference

There is an elegant body of theory for inference pertaining to GLM models that naturally extend that of ordinary linear theory. Under certain conditions (e.g., grouped binary data), the *deviance* $D(\boldsymbol{y}; \boldsymbol{\mu})$ can be used to measure goodness-of-fit of a model. The *scaled deviance* satisfies

$$\frac{D(\boldsymbol{y}; \boldsymbol{\mu})}{\phi} \;=\; 2\left\{\ell(\boldsymbol{y}; \boldsymbol{y}) - \ell(\boldsymbol{\mu}; \boldsymbol{y})\right\}$$

$$(2.27)$$

which is non-negative. The term $\ell(\boldsymbol{y}; \boldsymbol{y})$ corresponds to a *saturated model*—one where $\boldsymbol{\mu}$ maximizes ℓ over $\boldsymbol{\mu}$ unconstrained. For GLMs, a saturated model has $b'(\widehat{\theta}_i) = \widehat{\mu}_i = y_i$. The opposite extreme is a *null model*, which is what we call intercept-only, i.e., the R formula is of the form y $\sim$ 1. If $\phi = 1$ then $D =$

$$2\sum_{i=1}^{n} A_i \left[\{y_i\, \theta(y_i) - b(\theta(y_i))\} - \{y_i\, \widehat{\theta}_i - b(\widehat{\theta}_i)\}\right] \;\left(= \sum_{i=1}^{n} A_i\, d_i, \text{say}\right), \quad (2.28)$$

where $\widehat{\theta}$ is the MLE. This is called the deviance of a model even when the scale parameter is unknown, or is known to have a value other than one. The scaled deviance is sometimes called the *residual deviance*.

Smaller values of D indicate a better fit. The deviance, which is a generalization of the residual sum of squares for the LM, is a function of the data and of the fitted values, and when divided by a dispersion parameter ϕ, it is sometimes asymptotically χ^2 (e.g., as $n \to \infty$, or as the number of binomial trials $N_i \to \infty$). More generally, to test if a smaller model (i.e., one with fewer variables) is applicable given a larger model, it is only necessary to examine the increase in the deviance, and to compare it to a χ^2 distribution with degrees of freedom equal to the difference in the numbers of independent parameters in the two models (as each parameter has 1 degree of freedom). In the model-building process, this enables a test to be carried out as to which variables can be deleted to form a smaller model, or which variables need to be added to form a larger model.

For a Gaussian family with identity link, ϕ is the variance σ^2, and D is the residual sum of squares, i.e.,

$$D = \sum_{i=1}^{n} A_i (y_i - \mu_i)^2.$$

Hence

$$D/\phi \sim \chi^2_{n-p},$$

leading to the unbiased estimator

$$\widehat{\phi} = D/(n - p) \tag{2.29}$$

because, with an abuse of notation, $E(\chi^2_\nu) = \nu$.

An alternative estimator is the sum of squares of the standardized residuals divided by the residual degrees of freedom:

$$\tilde{\phi} = \frac{1}{n - p} \sum_{i=1}^{n} \frac{(y_i - \widehat{\mu}_i)^2}{V(\widehat{\mu}_i)/A_i}. \tag{2.30}$$

This formula may be used to estimate the scale parameter ϕ. It has much less bias than (2.29). For the Gaussian errors, $\tilde{\phi} = \widehat{\phi}$.

If $\mathcal{M}_0$ is a submodel within a model $\mathcal{M}$ (that is, nested) with $q < p$ parameters, and if ϕ is known, then

$$\frac{D_{\mathcal{M}_0} - D_{\mathcal{M}}}{\phi} \dot\sim \chi^2_{p-q}.$$

If ϕ is unknown, then

$$\frac{D_{\mathcal{M}_0} - D_{\mathcal{M}}}{\tilde{\phi}(p - q)} \dot\sim F_{p-q,n-p}.$$

2.3.2 GLM Residuals and Diagnostics

As with LMs, diagnostics are available for GLMs to help check the underlying assumptions. These tend to be based on residuals, however, GLM residuals are more difficult to utilize compared to LMs because they are harder to interpret.

There are several residual-types that may be defined for GLMs. The `resid()` (or `residuals()`) method function returns one of five types of residuals for `glm()`

objects, depending on the `type` argument. For simplicity we set $A_i = 1$ here, but more general formulas are given in Sect. 3.7 for VGLMs. The five residual types are:

(i) *Deviance residuals* are

$$r_i^D = \text{sign}(y_i - \widehat{\mu}_i)\sqrt{d_i}, \quad \text{where} \quad D = \sum_{i=1}^{n} d_i \qquad (2.31)$$

(cf. (2.28)). This residual type is the default, and is the most useful for diagnostic purposes.

(ii) *Pearson residuals* are related to the working residuals, and are

$$r_i^P = \frac{y_i - \widehat{\mu}_i}{\sqrt{V(\widehat{\mu}_i)}}. \quad \text{Note that} \quad X^2 = \sum_{i=1}^{n} \left(r_i^P\right)^2 = \sum_{i=1}^{n} \frac{(y_i - \widehat{\mu}_i)^2}{V(\widehat{\mu}_i)}$$

is the Pearson chi-squared statistic.

(iii) *Working residuals* are

$$r_i^W = (y_i - \widehat{\mu}_i)\frac{\partial \widehat{\eta}_i}{\partial \widehat{\mu}_i}. \qquad (2.32)$$

They arise from the final IRLS iteration (cf. (2.25)).

(iv) *Response residuals* are simply $r_i^R = y_i - \widehat{\mu}_i$.

(v) *Partial residuals* are used for enhancing plots of the component functions of GAMs. For $\eta_i = \beta_1 + \beta_2\, x_{i2} + \cdots + \beta_p\, x_{ip}$, these are $\beta_k(x_{ik} - \overline{x}_k) + r_i^W$. For GAMs having η_i of the form $\beta_1 + f_2(x_{i2}) + \cdots + f_p(x_{ip})$, these are $\tilde{f}_k(x_{ik}) + r_i^W$ for $k = 2, \ldots, p$ and where the $\tilde{f}_k$ are centred f_ks.

The first four types of residuals coincide for the Gaussian family. For the types of plots listed in Sect. 2.2.1.1, some people maintain that deviance residuals are the most informative for GLMs. Figure 2.21 is an example of the first four types of residuals.

2.3.3 Estimation of ϕ

In the VGLM/VGAM framework, it is usually preferable to estimate the dispersion parameter by full maximum likelihood because it is simpler, inference is simplified too, and the models can be more flexible. However, for some GLM families such as the gamma, there are problems such as bias, and extreme sensitivity in very small values (McCullagh and Nelder, 1989, Chap. 8).

The most common GLM estimation method for ϕ is to use (2.30) which is based on the method of moments. It is unbiased for the LM, and is generally consistent for GLMs.

2.3.4 Fitting GLMs in R

GLMs are well-served by the modelling function `glm()`. It has arguments

```
> args(glm)

    function (formula, family = gaussian, data, weights, subset,
        na.action, start = NULL, etastart, mustart, offset, control = list(...),
        model = TRUE, method = "glm.fit", x = FALSE, y = TRUE, contrasts = NULL,
        ...)
    NULL
```

They may be also fitted by `vglm()` with family functions having the same name as `glm()` but with an "`ff`" appended (they must be different because they are incompatible), e.g., `gaussianff()`.

2.3.5 Quasi-Likelihood Models

It is not uncommon for one to be unsure about the full distribution (2.19) of the response, e.g., when the variance of the data is much greater than the model suggests (*overdispersion*). Wedderburn (1974) proposed the use of the quasi-likelihood to help alleviate this problem. Specifically, it replaces the assumptions tied in with (2.19) by the weaker variance assumption that

$$\mathrm{Var}(Y) \; = \; \frac{\phi}{A}\, V(\mu), \tag{2.33}$$

where ϕ is assumed constant across samples. This can be very useful in applied work when the data are limited and information on the distribution of Y is lacking. However, one may have enough prior knowledge to specify, or data to reliably estimate, a relationship between the first two moments, as required by the quasi-likelihood model. In the absence of a likelihood function, one may estimate $\boldsymbol{\beta}$ by solving (2.23) because it only depends on the first two moments. What is the justification for using this? We have already seen that it yields the MLE of $\boldsymbol{\beta}$ for families belonging to the exponential family (2.19).

Now the term in braces in (2.24) is an expression for $\partial \ell_i / \partial \mu_i$. Coupled with (2.33), this suggests that

$$q(\mu; y) \; = \; \int_y^\mu \frac{y - t}{(\phi/A)\, V(t)}\, dt \tag{2.34}$$

might behave like a log-likelihood function for μ. Indeed, if

$$U_i \; = \; \frac{Y_i - \mu_i}{(\phi/A_i)\, V(\mu_i)}$$

then $E(U_i) = 0$ and $\mathrm{Var}(U_i) = 1/\{(\phi/A_i)\, V(\mu_i)\} = -E(\partial U_i / \partial \mu_i)$. These are the same properties that a log-likelihood derivative has (see (A.17)–(A.18)). So the overall conclusion is to solve the estimating equation

Table 2.1 Some quasi-likelihood functions. **VGAM** families are `quasibinomialff()` and `quasipoissonff()` (adapted from McCullagh and Nelder, 1989, Table 9.1).

Distribution	$V(\mu)$	$\phi \, q(\mu; y)$	`glm()` family
Gaussian	1	$-(y - \mu)^2/2$	
Binomial	$\mu(1 - \mu)$	$y \operatorname{logit} \mu + \log(1 - \mu)$	`quasibinomial()`
Poisson	μ	$y \log \mu - \mu$	`quasipoisson()`
Gamma	μ^2	$-y/\mu - \log \mu$	
Inverse Gaussian	μ^3	$-y/(2\mu^2) + 1/\mu$	

$$\sum_{i=1}^{n} \frac{\partial \, q(\mu_i; y_i)}{\partial \, \beta} = \mathbf{0}. \tag{2.35}$$

A list of some quasi-likelihood functions is given in Table 2.1 and the `glm()` family functions for estimating them. In R, the `glm()` family functions `quasibinomial()` and `quasipoisson()` solve for β in (2.35) for the binary and Poisson cases, and $\hat{\phi}$ in (2.30) is printed out in the `summary()` as the 'Dispersion parameter'. For `vglm()` in **VGAM**, the family functions are called `quasibinomialff()` and `quasipoissonff()`.

2.3.6 Binary Responses

We now dwell a little on one specific GLM, viz. the binomial family. In contrast, the Gaussian family is well-served in multitudes of books, and count data is described in Sect. 11.3 via negative binomial regression.

It is noted that two popular models for handling overdispersion with respect to the Poisson and binomial distributions are the negative binomial (Sect. 11.3) and beta-binomial (Sect. 11.4) distributions. Also, the variants positive-binomial, zero-inflated binomial and zero-altered binomial (Chap. 17) are available.

2.3.6.1 Links

Figure 2.1a,d plots four of the most commonly used link functions. The default is the logit link, which is not only very interpretable in terms of log-odds, it is usually indistinguishable in practice from the probit link unless n is large. Of these, only the complementary log–log link is asymmetric; the other three satisfy $p(\frac{1}{2} - c) = p(\frac{1}{2} + c)$ for $0 < c < \frac{1}{2}$. The cloglog link ties in with Poisson regression very simply because if $Y \sim \text{Poisson}(\mu)$ with $\eta = \log \mu$, then cloglog $P[Y > 0] = \log(-\log P[Y = 0]) = \log(-\log e^{-\mu}) = \eta$.

The links correspond to the CDFs of some standardized distributions (Table 12.3): the logit link of a logistic distribution, the probit of a normal, the cauchit of a Cauchy, and the cloglog of a extreme-value (log-Weibull) distribution (Chap. 16).

Figure 2.1b are plots of the reciprocal first derivatives, $1/g_j'(p)$, which are relevant to (2.25) because of the term $d\mu/d\eta$. For grouped binomial data

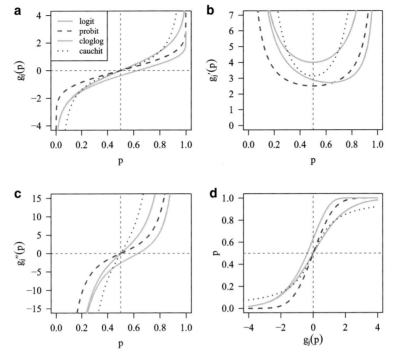

Fig. 2.1 Properties of some common link functions g_j suitable for a probability. (**a**) $g_j(p)$; (**b**) $g'_j(p)$; (**c**) $g''_j(p)$; (**d**) $g_j^{-1}(p)$. The legend in (**a**) is common for all plots. The calls to (**a**)–(**c**) are of the form `link.function(p, deriv = d)` for `d = 0, 1` and `2` (Table 1.2).

where Y_i represents the proportion of successes out of N_i trials, we have $N_i Y_i \sim$ Binomial(N_i, μ_i), so that (2.25) with a logit link becomes

$$z_i \;=\; \boldsymbol{\beta}^T \boldsymbol{x}_i + \frac{y_i - \mu_i}{\mu_i(1-\mu_i)}, \quad \text{with working weights} \quad w_i \;=\; N_i \mu_i(1 - \mu_i).$$

2.3.6.2 The Hauck-Donner Phenomenon

This effect, which was first observed by Hauck and Donner (1977), gives one reason why the likelihood ratio test is to be preferred over the Wald test (these tests are described in Sect. A.1.4.2). They used the following example. Suppose that $n = 200$ for a logistic regression involving two groups of 100 observations each. The observed proportion of 1s in group k is p_k, and let $x_2 = 0$ and 1, denote groups 1 and 2 respectively. Then the coefficient of x_2 is the log odds ratio, and we wish to test equality of the population proportions in the two groups via $H_0 : \beta_{(1)2} = 0$ for the model logit $P(Y = 1) = \beta_{(1)1} + \beta_{(1)2} x_2$.

For two illustrative values of p_1, and allowing p_2 to vary as 0.01(0.01)0.99 (i.e., 0.01 to 0.99 in steps of 0.01), plotted in Fig. 2.2 is the square of the usual Wald statistic for x_2, which is χ_1^2 under H_0. The LRT statistic $-2\log\lambda$ has the same asymptotic distribution. The glaring feature is the quadratic shape about p_1 of both test statistics. Another feature is that the LRT increases as $|p_2 - p_1|$ increases.

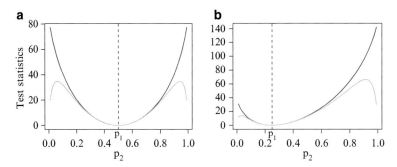

Fig. 2.2 Wald (*orange*) and likelihood ratio test (*blue*) statistics plotted against p_2, for: (**a**) $p_1 = 0.5$, (**b**) $p_1 = 0.25$ (*vertical dashed lines*). Actually, the Wald statistic here is the square of the usual Wald statistic, and $p_2 = 0.01(0.01)0.99$ is discrete. The data follows Hauck and Donner (1977).

This monotonicity is a good thing: there is increasing evidence against the null hypothesis the more the two sample proportions differ. However, the Wald statistic initially increases but then decreases near the boundaries. Thus Wald tests of binomial (and Poisson GLMs) can be highly unreliable, because a moderate p-value indicates no effect or a very large effect. Not only does it show aberrant behaviour, it is also less powerful than a LRT.

2.3.6.3 Problems Due to Complete Separation

It is well-known that multiple maximums of the log-likelihood function cannot occur with logistic regression because ℓ is globally concave, meaning that the function can have at most one maximum (Amemiya, 1985). However, it is possible for the likelihood function to have *no* maximum, in which case the MLE is said to not exist. The problem occurs when there is *complete separation* (Albert and Anderson, 1984). For example, the data used by Allison (2004) is

```
> cs.data <- data.frame(y = rep(0:1, each = 5), x2 = c((-5):(-1), 1:5))
```

These data are plotted in Fig. 2.3 as solid blue circles. Suppose that we fit $\text{logit} P(Y = 1|x_2) = \beta_{(1)1} + \beta_{(1)2} x_2$. Then it may easily be shown that the log-likelihood function increases as a function of $\beta_{(1)2}$ and that it flattens out as $\beta_{(1)2} \longrightarrow \infty$, i.e., the MLE does not exist. Complete separation occurs when there exists some vector $\boldsymbol{\beta}$ such at $y_i = 1$ whenever $\boldsymbol{\beta}^T \boldsymbol{x}_i > 0$, and $y_i = 0$ whenever $\boldsymbol{\beta}^T \boldsymbol{x}_i < 0$.

There is a related problem called *quasi-complete separation*. This occurs if there exists a $\boldsymbol{\beta}$ such that $\boldsymbol{\beta}^T \boldsymbol{x}_i \geq 0$ whenever $y_i = 1$ and $\boldsymbol{\beta}^T \boldsymbol{x}_i \leq 0$ whenever $y_i = 0$, and when equality holds for at least one observation in each category of the response variable. Adding $(0, 0)$ and $(0, 1)$ to the previous data set will result in quasi-complete separation (Fig. 2.3). Once again, the MLE does not exist.

In practice, quasi-complete separation is far more common than complete separation. It most often occurs when an explanatory variable x_k is a dummy variable, and for one value of x_k, either every observation has $y = 1$ or $y = 0$. In general, consider the 2×2 table of y versus every dichotomous explanatory variable. If there is a zero in any cell, then the MLE will not exist. Convergence failure in

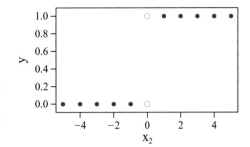

Fig. 2.3 Completely separable data (*blue circles*). Adding the two orange hollow points results in quasi-completely separable data. The logistic regression estimate of the slope will tend to infinity in both cases.

logistic regression is most commonly caused by this. It occurs more often when the sample size is small, however it is certainly possible in large data sets too.

Allison (2004) gives practical advice about a course of action to take if there is complete separation or quasi-complete separation. Actually, the two cases are best considered separately. One possibility is to use bias-reduction. Heinze and Schemper (2002) have shown that this method always yields finite estimates of parameters under complete or quasi-complete separation. However, the result that bias-reduction of β gives a finite answer as a by-product is not a sufficient reason for the blind application of the technique.

Of course, the problem can occur regardless of the link function chosen. And it can also occur for the multinomial logit and cumulative link models described in Chap. 14. Bias-reducing techniques have been developed for GLMs, which have the effect that each coefficient in $\hat{\beta}$ is finite for binary responses; see Sect. 9.4 for an appetizer.

Complete separation is the subject of Altman et al. (2004, Chap.10).

2.4 Univariate Smoothing Methods

2.4.1 The Classical Smoothing Problem

Smoothing is a powerful tool for exploratory data analysis, and it allows a *data-driven* approach to the more *model-driven* LM modus operandi. The simplest form of smoothing operates on scatter plot data. As a preliminary (and imperfect) example, consider Fig. 2.5a which is a scatter plot of the proportion of rainbow trout caught from Lake Otamangakau (`lake0`) between the years 1974 to 1989 by a certain angler. A smoother called a regression spline has been fitted (Fig. 2.5b). The main feature of the data has been picked up, namely it is flat on the LHS, and then is followed by a decrease over the years. However, the smooth $\hat{f}(x)$ is probably *too* flexible here and it overfits. (Another weakness with this example is that smoothing proportions directly is not a good idea, because the smooth can sometimes assume negative values or values greater than 1. A logistic regression GAM would be recommended instead.)

Four broad categories of smoothers are:

[1.] Regression or series smoothers (polynomial regression, regression splines, P-splines, Fourier regression, filtering, wavelets),

[2.] Smoothing splines (with roughness penalties, e.g., cubic smoothing splines, O-splines, P-splines),

[3.] Local regression (Nadaraya-Watson estimator, kernel smoothers, Lowess, Loess, it generalizes to local likelihood),

[4.] Nearest-neighbour smoothers (running means, running lines, running medians).

We will only be concerned with a small selection of these for two reasons: not all so naturally generalize to the vector-y case, and **VGAM** currently implements only two methods (regression splines and smoothing O-splines). However, we also consider two other methods, local regression and P-splines, because of their contribution to our understanding of vector smoothing as a whole, and because they have favourable properties, respectively. P-splines ("P" for "penalized") can be considered a hybrid method, therefore they appear twice in the list as they share similarities with regression splines and smoothing splines. Section 4.1.1 gives an overview of how smoothing relates to VGLMs and VGAMs.

Smoothing has many general uses, e.g., data visualization and exploratory data analysis, prediction, derivative estimation (e.g., growth curves, acceleration), and it is used as a building block for many modern statistical techniques, such as in Chap. 4. Examples of each of these uses can be found throughout this book.

For our purposes, the *classical smoothing problem* is to estimate an arbitrary smooth function f based on the model

$$y_i = f(x_i) + \varepsilon_i, \quad \varepsilon_i \sim (0, \sigma_i^2) \text{ independently,} \tag{2.36}$$

for data $(x_i, y_i, w_i = 1)$, $i = 1, \ldots, n$. If there is no *a priori* function form for f, then it may be estimated by a smoother. They do not impose any particular form on the function apart from being smooth. Since the errors have mean 0, we are modelling the conditional mean $E(Y|x) = f(x)$. Ordinarily, it is usually assumed that all the σ_i are equal. For this, the data sets of Figs. 2.4a and 2.5 appear to violate this assumption. This section describes conventional smoothing methods for the univariate problem (2.36). Indeed, hundreds of journal articles have addressed this problem and its direct extensions.

It is needful to generalize (2.36) to the weighted case so that $\text{Var}(\varepsilon_i) = \sigma_i^2 = \sigma^2 w_i^{-1}$, where the w_i are known and positive, and σ is unknown and may be estimated. This can be written $\text{Var}(\varepsilon) = \sigma^2 \mathbf{W}^{-1}$ where $\mathbf{W} = \text{diag}(w_1, \ldots, w_n) = \mathbf{\Sigma}^{-1}$. For example, in the `lake0` example, one might assign $w_i = \{y_i(1-y_i)/N_i\}^{-1}$ where $N_i = $ `total.fish`$_i$ because y_i is a sample proportion (however, in this case, some $y_i = 1$, which is problematic).

Without loss of generality, let the data be ordered so that $x_1 < x_2 < \cdots < x_n$. Consider the unweighted case of $w_i = 1$ for all i. The fundamental idea behind all smoothing methods is the concept of a *neighbourhood*. Here, only observations whose x_i values are 'close' (are neighbours) to a *target point* x_0, say, are used to estimate f at x_0, and the more closer x_i is to x_0, the more influence or weight that (x_i, y_i) has for $\widehat{f}(x_0)$. As an extreme case, observation (x_1, y_1) has the least effect on $\widehat{f}(x_n)$ because x_1 is the furthest away from x_n. This concept is especially easy to see for local regression, e.g., in Fig. 2.13 the shaded region denotes the effective neighbourhood, and the kernel function at the bottom provides the relative weights given to the (x_i, y_i). Rather than 'neighbourhood', many writers use the word *window* to describe the *localness* idea because observations lying outside the window are effectively ignored. This hypothetical window glides along the x-axis to estimate the entire f. The window may or may not have distinct sides depending on whether the weights are strictly zero past a certain distance away from the target point. In Table 2.2, all but one kernel function vanishes beyond a certain distance away from its centre, hence such 'windows' have distinct sides. The window of Fig. 2.13 has blurry edges because the Gaussian kernel function is strictly positive.

We shall see later that the size of the window or neighbourhood about a target point x_0 is crucial because it controls how smooth or wiggly the smooth is. It is fundamentally related to bias and variance. For example, a very large neighbourhood that effectively includes all the data corresponds to little or no smoothing at all, and this has relatively little variance but much bias.

While kernel function smoothing methods are probably the easiest to understand, other smoothing methods such as splines are motivated completely differently, however its asymptotic properties can be shown to be based on the neighbourhood idea, e.g., Sect. 2.4.7.3 shows that, under certain conditions, a cubic smoothing spline operates like a kernel function smoother.

This section describes a few common methods for fitting the classical smoothing problem (2.36). The purpose is to lay a foundation for methods that apply to the vector y case (Sect. 4.2). Some books covering this large topic can be found in the bibliographic notes.

The reader should be aware that this section largely adopts commonly used notation from the smoothing literature, and consequently there is some recyling of notation, e.g., K denotes the number of knots for splines, as well as the kernel function for local smoothers. This however should not present any severe problems because these topics are quite separate and their context is easily grasped.

2.4.2 Polynomial Regression

Polynomial regression is a common technique that involves fitting polynomial functions of each x_k in order to provide more flexibility than the usual linear $\beta_k x_k$ term. One reason for its widespread use is that polynomials are easy to work with in just about every way—mathematically, computationally and they have high interpretability for low degrees. For (2.36), one may use an S formula having the term `poly(x2, degree)` for $degree = 1, 2, \ldots$, else explicit terms such as `I(x2^2)` (or `poly(x2, degree, raw = TRUE)`). The former has the advantage of using orthogonal polynomials which are numerically stable, but at the expense of having coefficients that are not so interpretable.

The Weierstrass approximation theorem, which states that every continuous function on a closed interval $[a, b]$ can be uniformly approximated as closely as desired by a polynomial, might lead us to believe that fitting a sufficiently high order polynomial will be a good idea for estimating most fs in general, however this is not the case. The reasons include the following.

- Polynomials are not very local but have a global nature. Their derivatives are continuous functions for all orders. For example, a small perturbation on the LHS of the curve may result in a large change on the RHS. Consequently the concept of a local neighbourhood does not really exist. Polynomials are thus flexible but not flexible enough. This can be seen in Fig. 2.4a; polynomials of degree 1–4 are unable to conform to the trend (which is admittedly complex).

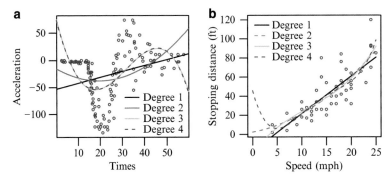

Fig. 2.4 Polynomials of degree 1–4 fitted to two data sets. (a) mcycles from MASS. (b) cars from datasets.

The cubic and quartics are almost indistinguishable. It is left to the reader to confirm that polynomials up to the 10th degree do not offer much improvement (Ex. 2.17).

- They usually have edge effects: they do not model the boundaries well, especially if the degree of the polynomial is high. This results in significant bias in regions of the x-space. Figure 2.4b illustrates this. The trend should be monotonically increasing, and if a polynomial of degree that is too high is fitted then often the boundaries are not modelled well. Here, while the linear and quadratic functions appear well-behaved over the range of x, the quartic at the LHS corner curls upwards, and therefore would be dangerous for prediction.
- They are sensitive to outliers and high-leverage points.
- The polynomial degree is discrete rather than continuous.

It is probably a safe general recommendation that one should avoid fitting quartics or higher, and even fitting a cubic should be done cautiously and with trepidation.

2.4.3 Regression Splines

Usually a better alternative to polynomial regression is the use of *regression splines*. These are *piecewise* polynomials, hence the neighbourhood concept is built in directly. Each piece, usually of low degree, is defined on an x-region that is delimited by *knots* (or *breakpoints*). The (x, y) positions where each pair of *segments* join are called *joints*. The more knots, the more flexible the family of curves become. It is customary to force the piecewise-polynomials to join smoothly at the knots, e.g., a popular choice called *cubic splines* are piecewise-cubic polynomials with continuous zeroth, first and second derivatives. By forcing the first few derivatives to be continuous at the knots, the entire curve has the appearance of one nice smooth curve. Using splines of degree > 3 seldom confers any additional benefit. Figures 2.5b and 2.6 are examples of cubic regression splines. The coloured segments of Fig. 2.6 join up at the joints, and the vertical lines mark the knots.

The word 'spline' comes from a thin flexible strip used by engineers and architects in the pre-computer days to construct ship hulls and the aerofoils of wings. Splines were attached to important positions on a 2-dimensional plan (e.g., floor of a design loft or on enlarged graph paper) using lead weights called "ducks"

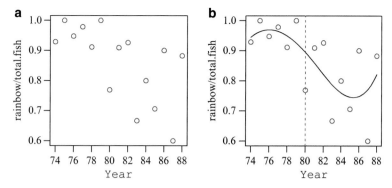

Fig. 2.5 (**a**) Proportion of fish caught that are rainbow trout from Lake Otamangakau (`lake0`) caught by an angler who frequented the spot. The variable `Year` is `year-1900`. (**b**) Smoothing the same data with a cubic regression spline (truncated power basis) with one knot located at the year 1980. A boundary effect on the RHS is evident.

and then released. The resting shapes assumed by the splines would minimize the strain energy according to some calculus of variations criterion. Splines were used by ancient Greek mathematicians (including Diocles) for drawing curves in diagrams (e.g., conic sections). In more modern times, I. J. Schoenberg is attributed to be the first to use 'splines' in the mathematical literature, and is known as the father of splines. The physical meaning of splines is especially relevant to the smoothing spline (Sect. 2.4.4), where it is related to curvature and Hooke's Law for elastic bodies such as springs.

Regression splines are one example of multiple regression on a family of *basis vectors*. A simpler example is polynomial regression where the set $\mathcal{S} = \{1, x, x^2, \ldots, x^r\}$ form the usual basis of polynomials of degree r. We say $\mathcal{S}$ *spans* this function space. There are two common bases for cubic splines:

1. *Truncated power series*   These are easy to understand, but are not recommended in practice because they may be ill-conditioned. Some details are in Sect. 2.4.3.1.
2. *B-splines*   These are more complex but are used in practice because they are numerically more stable. The functions `bs()` and `ns()` are two implementations of B-splines. Some details are in Sect. 2.4.3.2. Called "B-splines" by Schoenberg, "B" is usually taken to stand for "basic" or "basis", and for some others, "beautiful".

From the glossary, a function $f \in \mathcal{C}^k[a,b]$ if derivatives $f', f'', \ldots, f^{(k)}$ all exist and are continuous in an interval $[a,b]$. For example, ordinary polynomials $\in \mathcal{C}^k[a,b]$ for all k, a and b, but $|x| \notin \mathcal{C}^1[a,b]$ if $0 \in [a,b]$. Then a *spline of degree r* (some given positive integer) with knots $\xi_1, \ldots, \xi_K$ (such that $a < \xi_1 < \xi_2 < \cdots < \xi_K < b$) is a function $f(x)$ defined over an interval (a,b) if it satisfies the following properties:

(i) for any subinterval (ξ_j, ξ_{j+1}), $f(x)$ is a polynomial of degree r (*order $r+1$*);
(ii) $f \in \mathcal{C}^{r-1}(a,b)$, i.e., $f(x), f'(x), \ldots, f^{(r-1)}(x)$ are continuous;
(iii) the rth derivative of $f(x)$ is a step function with jumps at $\xi_1, \ldots, \xi_K$.

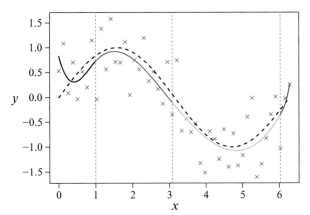

Fig. 2.6 Smoothing some data with a regression spline (B-spline). Each segment of the spline is coloured differently. The term is effectively `bs(x, knots = c(1, 3.08, 6.03))`. The true function is the sine function (*dashed*) and $n = 50$.

The name *cubic spline* is used for the curve arising from the special case of $r = 3$. An example of a plot of the first few derivatives of a cubic spline fit is Fig. 2.11—a step function can be seen in the latter plot.

Regression splines have at least two advantages: they are computationally and statistically simple, and standard parametric inferences are available. The first advantage is partly because they are an LM representation of a smooth function. An example of the second advantage is testing whether a particular knot can be removed and the same polynomial equation used to explain two adjacent segments (by using $H_0 : \beta_j = 0$ in (2.40)). In R, this corresponds to one of the t-test statistics printed by `summary()`. However, they do have drawbacks such as the difficulty choosing the number of knots and their locations, and their smoothness cannot be controlled continuously as a function of a single smoothing parameter.

Regarding the first drawback, in a paper reflecting a lot of experience fitting regression splines, Wold (1974) made the following recommendations for cubic splines:

(i) They should have as few knots as possible, ensuring that a minimum of 4 or 5 observations lie between knot points.
(ii) No more than one stationary point and one inflexion point should fall between two knots (because a cubic is not flexible enough to allow for too many of such points).
(iii) Stationary points should be centred in intervals, and inflexion points should be located near knot points.

These recommendations might be actioned using the `knots` argument of `bs()` and `ns()`. As an illustration, consider Fig. 2.6. Overall, the fit is alright except at the left-hand side boundary. This example illustrates how regression splines may be poor at the boundaries, especially if the knot placement is careless. How `bs()` and `ns()` choose their knots by default is described on p.58.

2.4.3.1 Truncated Power Series

Following the notation of Green and Silverman (1994), a cubic spline may be written as

$$f(x) = d_s (x - \xi_s)^3 + c_s (x - \xi_s)^2 + b_s (x - \xi_s) + a_s, \qquad \xi_s \leq x \leq \xi_{s+1}, \qquad (2.37)$$

for $s = 0, \ldots, K$. For given constants a_s, b_s, c_s and d_s, we define $\xi_0 = a$ and $\xi_{K+1} = b$. The coefficients are interrelated because of the various continuity conditions, e.g., f is continuous at ξ_{s+1} implies

$$d_s \left(\xi_{s+1} - \xi_s\right)^3 + c_s \left(\xi_{s+1} - \xi_s\right)^2 + b_s \left(\xi_{s+1} - \xi_s\right) + a_s = a_{s+1} \tag{2.38}$$

for $s = 0, \ldots, K - 1$. Thus there are $4(K + 1) - 3K = K + 4$ parameters. The *truncated power series basis* for a cubic spline with K knots is

$$\left\{1, \, x, \, x^2, \, x^3, \, (x - \xi_1)_+^3, \ldots, (x - \xi_K)_+^3\right\} \tag{2.39}$$

where $u_+ = \max(0, u)$ is the positive part of u. Figure 2.7 gives an example of these functions with $\xi_k = k$ for $k = 1, \ldots, 5$. When x is large the curves $(x - \xi_k)_+^3$ become almost vertical and parallel, therefore ill-conditioning occurs.

With (2.39) we can express the spline as

$$f(x) = \beta_1 + \beta_2\, x + \beta_3\, x^2 + \beta_4\, x^3 + \sum_{s=1}^{K} \beta_{4+s}\, (x - \xi_s)_+^3 \, . \tag{2.40}$$

As an example, here is the essential code behind the `lake0` example of Fig. 2.5:

```
> Pos <- function(x) pmax(x, 0)    # Same as ifelse(x > 0, x, 0)
> lake0 <- transform(lake0, Year = year - 1900)   # Because of ill-conditioning
> knot <- 80   # For the year 1980; a prespecified knot
> fit.trout <- lm(rainbow / total.fish ~ Year + I(Year^2) + I(Year^3) +
                  I(Pos(Year-knot)^3), data = lake0)
> model.matrix(fit.trout)

     (Intercept) Year I(Year^2) I(Year^3) I(Pos(Year - knot)^3)
1              1   74      5476    405224                     0
2              1   75      5625    421875                     0
3              1   76      5776    438976                     0
4              1   77      5929    456533                     0
5              1   78      6084    474552                     0
6              1   79      6241    493039                     0
7              1   80      6400    512000                     0
8              1   81      6561    531441                     1
9              1   82      6724    551368                     8
10             1   83      6889    571787                    27
11             1   84      7056    592704                    64
12             1   85      7225    614125                   125
13             1   86      7396    636056                   216
14             1   87      7569    658503                   343
15             1   88      7744    681472                   512
attr(,"assign")
[1] 0 1 2 3 4
```

The variable `Year = year-1900` has been used to ameliorate the ill-conditioning, e.g., 1974^3 and 1988^3 are both large numbers, and are treated almost as having the same value since the computations are performed using finite arithmetic. The example reflects the recommendation that the truncated power series is unsuitable for general use because the model matrices may be ill-conditioned (the columns almost linearly dependent). In general, a B-spline basis is superior.

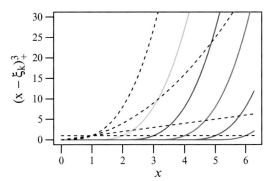

Fig. 2.7 Truncated power series basis for cubic splines (2.39). The *black dashed lines* are $1, x, x^2, x^3$. The *coloured solid lines* are $(x - \xi_k)^3_+$ for knots $\xi_1 = 1, \xi_2 = 2, \xi_3 = 3, \xi_4 = 4$ and $\xi_5 = 5$.

2.4.3.2 B-Splines

The reason why B-splines are used is mainly computational: their very good numerical properties arise from the fact that B-splines have minimal support. That is, $B_{s,q}(x) > 0$ for x belonging to the smallest overall region defined by the knots. With minimal support, the concept of a neighbourhood is made computationally more stable because the coefficient of one B-spline is related to the fewest number of coefficients associated with the other B-splines, i.e., the amount of overlap is minimal. This can be seen in Fig. 2.8: a B-spline of order q consists of q segments only. Hence a B-spline on the LHS of a scatter plot is only concerned with data around the neighbourhood there. In contrast, the value $x = 6$ in Fig. 2.7 is associated with all the truncated power series basis functions.

It is convenient to consider splines of a general order, Q say. Some special cases are as follows.

$Q = 1$:   These are similar to a shifted unit rectangle function or boxcar function.
$Q = 2$:   *Linear spline* which has continuous derivatives up to order $Q - 2 = 0$ at the knots—i.e., the function is continuous and is piecewise-linear. In fact, it is a scaled density of a triangle distribution (Table 12.10).
$Q = 3$:   *Quadratic spline* (*parabolic spline*) which has continuous derivatives up to order $Q - 2 = 1$ at the knots.
$Q = 4$:   *Cubic spline*, these are very popular, and have been described as the lowest-order spline for which the discontinuities in the $f^{(Q-1)}(\xi_s)$ are imperceptible (Hastie et al., 2009).

Let ξ_s for $s = 1, \ldots, K$, be K *interior knots*, and let ξ_0 and ξ_{K+1} be the two *boundary knots*. Then we can augment these knots with $2Q$ others to obtain a vector $\boldsymbol{\tau} = (\tau_1, \ldots, \tau_{K+2Q})^T$ satisfying the following inequality:

$$\tau_1 \le \tau_2 \le \cdots \le \tau_Q \quad \le \xi_0 \tag{2.41}$$
$$< \xi_1 \le \cdots \le \xi_K \tag{2.42}$$
$$< \xi_{K+1} \le \tau_{K+Q+1} \le \cdots \le \tau_{K+2Q}, \tag{2.43}$$

where $\tau_{Q+s} = \xi_s$ for $s = 1, \ldots, K$. Usually $\tau_1 = \cdots = \tau_Q = \xi_0$ and $\tau_{K+Q+1} = \cdots = \tau_{K+2Q} = \xi_{K+1}$ is chosen.

Denote by $B_{s,q}(x)$ the sth B-spline basis function of order q (degree $q - 1$) for the knot sequence $\boldsymbol{\tau}$ for $q = 1, \ldots, Q$. They are defined recursively as follows (de Boor, 2001):

(1) For $s = 1, \ldots, K + 2Q - 1$,

$$B_{s,1}(x) = \begin{cases} 1, & \tau_s \leq x < \tau_{s+1}, \\ 0, & \text{otherwise.} \end{cases} \qquad (2.44)$$

(2) Then for $s = 1, \ldots, K + 2Q - q$ and $q > 1$,

$$\begin{aligned} B_{s,q}(x) &= \omega_{s,q} \, B_{s,q-1}(x) + (1 - \omega_{s+1,q}) \, B_{s+1,q-1}(x) & (2.45) \\ &= \frac{x - \tau_s}{\tau_{s+q-1} - \tau_s} B_{s,q-1}(x) + \frac{\tau_{s+q} - x}{\tau_{s+q} - \tau_{s+1}} B_{s+1,q-1}(x), & (2.46) \end{aligned}$$

where $\omega_{s,q} \equiv (x - \tau_s)/(\tau_{s+q-1} - \tau_s)$ for $\tau_{s+q-1} > \tau_s$, while $\omega_{s,q} \equiv 0$ if $\tau_{s+q-1} = \tau_s$. These may be computed using stable and efficient recursive algorithms. Note that $B_{s,q}$ only depends on the $q+1$ knots $\tau_s, \ldots, \tau_{s+q}$, and vanishes outside the interval $[\tau_s, \tau_{s+q})$ and is positive in its interior. If $\tau_s = \tau_{s+q}$, then $B_{s,q} = 0$.

Thus with $Q = 4$, $B_{s,4}$ (for $s = 1, \ldots, K + 4$) are the $K + 4$ cubic B-spline basis functions for $\boldsymbol{\tau}$.

Some B-splines of orders 1 to 4 are plotted in Fig. 2.8. Essentially, the code is

```
knots <- c(1:3, 5, 7, 8, 10)   # Interior knots
x.vector <- seq(0, 11, by = 0.01)
for (ord in 1:4) {
  B.matrix <- bs(x = x.vector, degree = ord-1, knots = knots, intercept = TRUE)
  matplot(x.vector, B.matrix, type = "l")
}
```

The significance of the argument `intercept` is due to the $B_{s,Q}$ in (2.46) including the intercept because $\sum_{s=1}^{K+Q} B_{s,Q}(x) = 1$ for $x \in [\xi_0, \xi_{K+1}]$. Function `bs()` has `intercept = FALSE` as the default, because usually it is called within an S formula that has an intercept by default. Figure 2.10 shows B-spline basis functions corresponding to a regression spline LM fitted without an intercept term and with a `bs(x, intercept = TRUE)`-type term. Also, note however that `bs()` presently does not accept a value of `degree = 0`, hence the first case `ord = 1` might be computed as follows (it is assumed that `intercept = TRUE`).

```
allknots <- sort(c(Boundary.knots, knots))
B1.matrix <- matrix(0, length(x), length(knots) + intercept)
for (s in 1:(length(allknots)-1))
  B1.matrix[, s] <- as.numeric(allknots[s] <= x & x < allknots[s+1])   # 0 or 1
```

Here are some additional notes.

1. `> args(bs)`

   ```
   function (x, df = NULL, knots = NULL, degree = 3, intercept = FALSE,
       Boundary.knots = range(x))
   NULL
   ```

   If argument `knots` is supplied, then the function returns a matrix of dimension `c(length(x), df = length(knots) + degree + intercept)`. Alternatively, the second dimension may be inputted directly by the `df` argument.

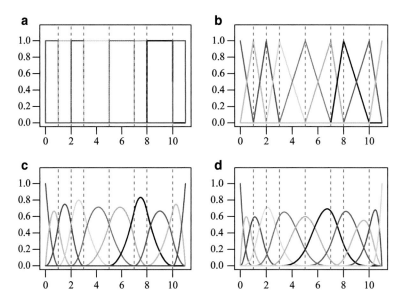

Fig. 2.8 B-splines of order 1–4 ((**a**)–(**d**)), where the interior knots are denoted by *vertical lines*. The boundary knots are at 0 and 11. The basis functions have been plotted from left to right.

It can be seen from the argument `Boundary.knots` that $\xi_0 = \min(x_i)$ and $\xi_{K+1} = \max(x_i)$. By default, the internal knots selected by `bs()` and `ns()` are of the form `quantile(x.inside, probs)` with equally spaced `probs` values. In fact, if `nIknots` is the number of internal knots, then `probs = (1:nIknots)/(nIknots + 1)`, and `x.inside` are the x values inside the interval described by the 2-vector `Boundary.knots`.

Note that predicting `bs()` outside the boundary knots is not recommended, because $B_{s,Q}(x)$ is not well defined outside of $[\xi_0, \xi_{K+1}]$. In fact a warning is issued if this is attempted.

2. A well-known type of cubic spline on $[\xi_0, \xi_{K+1}]$ called a *natural cubic spline* (NCS) has second and third derivatives, which are zero at ξ_0 and ξ_{K+1}:

$$f''(\xi_0) = f'''(\xi_0) = f''(\xi_{K+1}) = f'''(\xi_{K+1}) = 0. \qquad (2.47)$$

These are called the *natural boundary conditions*. NCSs are implemented by `ns()`, which has defaults

```
> args(ns)

function (x, df = NULL, knots = NULL, intercept = FALSE,
    Boundary.knots = range(x))
NULL
```

The lack of a `degree` argument is due to only *cubic* NCSs being implemented. Given knots $\xi_1, \ldots, \xi_K$, an NCS is linear on $(-\infty, \xi_0]$ and $[\xi_{K+1}, \infty)$. Function `ns()` has $K + 2$ parameters including the intercept because $K + 2 = (K + 4) - 2 \times 2 + 2$: each boundary constraint in (2.47) deducts one parameter from the total number, and there are two extra knots at ξ_0 and ξ_{K+1}.

In practice, often there is not a huge difference between `bs()` and `ns()` terms, when they are calibrated to be as similar to each other as possible. However,

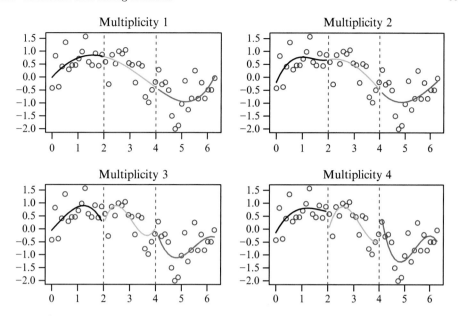

Fig. 2.9 Smoothing some data with cubic regression splines with knots of varying multiplicities. The knots are at $x = 2$ and 4.

usually the natural boundary constraints means `ns()` behaves better at the edges than `bs()`. The `ns()` function is also a double-edged sword compared to `bs()`, because it is defined for all x, but as with all smoothers, prediction beyond the range of the data poses more potential danger.

3. It is possible to move adjacent knots closer and closer together until they coincide. We then say the set of distinct knots has varying multiplicities. Then we need to define $B_{s,1} \equiv 0$ if $\tau_s = \tau_{s+1}$, and use the maxim *anything times zero is zero* in (2.46) to avoid division by 0. It transpires that if a knot is duplicated then it loses one continuous derivative for each new knot there. It can be shown that *the number of continuity conditions at a knot ξ_s, plus the multiplicity of knots at ξ_s, equals Q*. The effect of this important formula can be seen in Fig. 2.9: the cubic spline ($Q = 4$) with a knot of multiplicity m means that only $f^{(0)}, f^{(1)}, \ldots, f^{(Q-m-1)}$ exist there.

4. Although *safe prediction* in R will be sufficient for most users, it will not handle nested expressions involving data-dependent functions such as `I(bs(x))` and `poly(scale(x), 2)`. In such cases, *smart prediction* will work; see Sect. 18.6 for details.

2.4.4 Smoothing Splines

For regression splines, the user typically controls the flexibility of the smoother by selecting a small number of basis functions, e.g., by assigning the `df` argument an appropriate value that is less than 10, say. In contrast, the regularization approach to smoothing is to start off with many basis functions (e.g., n of them) and penalize some characteristic of these basis functions in order to control the flexibility of the fit. A popular example of this approach is the *cubic smoothing spline*. These are defined as minimizers of the objective function

$$S(f) = \sum_{i=1}^{n} (y_i - f(x_i))^2 + \lambda \int_a^b \{f''(x)\}^2 \, dx, \qquad (2.48)$$

over a space of "smooth" functions. In fact, it is an infinite-dimensional space of functions known as $\mathcal{W}_2^2[a, b]$ (a *Sobolev space* of order 2 on $[a, b]$ described in Table A.3). Here, $a < x_1 < \cdots < x_n < b$ for some a and b is again assumed, and the *smoothing parameter* satisfies $\lambda \geq 0$.

The first term of $S(f)$ penalizes lack-of-fit since it is a residual sum of squares. The second term penalizes the wiggliness or lack of smoothness, e.g., the integral equals zero for constant and linear functions. These two opposing quantities are balanced with each other by λ. Larger values of λ produce more smooth curves, indeed, as $\lambda \to \infty$, $f''(x) \to 0$ and the solution becomes the least squares line. The other extreme is as $\lambda \to 0^+$, and the solution tends to a twice-differentiable function that interpolates the data (x_i, y_i). These two extremes are often unacceptable as a solution, so it is surmised that there is some λ value which balances the two adequately. The quantity (2.48) fits into the *"penalty function"* approach described in Sect. 1.5.1, and is expounded by Green and Silverman (1994) specifically for splines.

Let $\boldsymbol{\Sigma} = \mathbf{W}^{-1} = \text{diag}(w_1^{-1}, \ldots, w_n^{-1})^T$ to handle known prior weights as in the weighted classical smoothing problem (2.36). Then the *penalized least squares* criterion can be written

$$S(f) = (\boldsymbol{y} - \boldsymbol{f})^T \boldsymbol{\Sigma}^{-1} (\boldsymbol{y} - \boldsymbol{f}) + \lambda \boldsymbol{f}^T \mathbf{K} \boldsymbol{f} \qquad (2.49)$$

where $\mathbf{K}$ is a roughness penalty matrix described below. Setting its derivative with respect to $\boldsymbol{f}$ to $\mathbf{0}$ yields the solution

$$\widehat{\boldsymbol{f}} = \mathbf{S}(\lambda) \, \boldsymbol{y} \qquad (2.50)$$

where $\mathbf{S}(\lambda) = (\mathbf{I}_n + \lambda \boldsymbol{\Sigma} \mathbf{K})^{-1}$ is known as the *influence* or *smoother matrix*. We shall see that it has properties similar to the LM hat matrix $\mathcal{H}$ (2.10).

Here are some notes.

1. One can select λ by trial-and-error such as by eye, however, more objective methods such as cross-validation are described below.
2. Following on from the description of a spline as a thin wooden strip in Sect. 2.4.3, one justification for the penalty term of $S(f)$ is that the energy to bend it is proportional to $\int_a^b$ curvature^2 with[1] respect to arc length, which is approximately proportional to $\int_a^b f''(t)^2 \, dt$. From Hooke's Law, springs exert an energy that is proportional to $\sum_{i=1}^n (y_i - f(x_i))^2$. Hence (2.48) does have a real physical meaning.

---

[1] The *curvature* of a curve $y = f(x)$ is $|f''(x)| \left\{ 1 + [f'(x)]^2 \right\}^{-3/2}$. If the $\{ \}$ term is dropped (because the assumption $|f'(x)| \ll 1$ is almost always made in physics and engineering), then $|f''(x)|$ is left as an approximation to the curvature. In natural cubic spline interpolation, we are finding a curve with minimal (approximate) curvature over an interval, for the quantity $\int [f''(x)]^2 \, dx$ is being minimized.

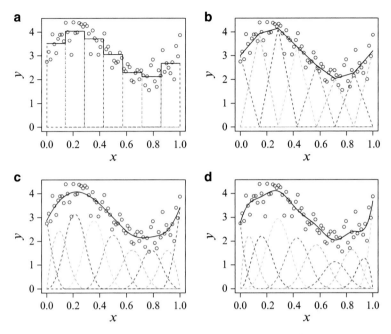

Fig. 2.10 (a)–(d) Linear combinations of B-splines of degrees 0–3 fitted to some scatter plot data; the formula is similar to (2.55). The knots are equally spaced on the unit interval.

3. Importantly, Reinsch (1967) used the calculus of variations to show that the solution of (2.48) is a cubic spline with knots at the distinct values of the x_i (provided that $n \geq 3$ and $\lambda > 0$). Another important result is that if NCSs interpolate the data (x_i, y_i) then they uniquely minimize $\int \{f''\}^2$ over all functions in $\mathcal{W}_2^2[a, b]$ (for $a < x_1 < \cdots < x_n < b$). Both these results are stated in, e.g., Green and Silverman (1994, Thms 2.3, 2.4).

4. Actually, the optimization problem (2.48) derives from minimizing

$$\int_a^b \{f''(x)\}^2 \, dx \quad \text{subject to the constraint} \quad \sum_{i=1}^{n} \{y_i - f(x_i)\}^2 \leq A$$

for some A. Then (2.48) arises from applying the Lagrange multipliers technique to this (Reinsch, 1967).

5. As $n \to \infty$, λ should become smaller, consequently some authors replace λ in (2.48) by λ/n.

6. There are alternative regularizations to the penalty of (2.48), e.g.,

$$\int_a^b f'(x)^2 \, dx, \tag{2.51}$$

whose solution is a linear spline. In general, using $\int_a^b [f^{(\nu)}(x)]^2 \, dx$ produces a spline solution of degree $2\nu - 1$.

7. More generally, a Sobolev space of order m on $[a, b]$ is written $\mathcal{W}_2^m[a, b]$. The case of $m = 2$ corresponds to cubic splines. Absolutely continuity is a stronger

condition than uniform continuity, given the definition of $\mathcal{C}^k[a, b]$ on p.53, it can be shown that

$$\mathcal{C}^2[a, b] \subset \mathcal{W}_2^2[a, b]. \tag{2.52}$$

As an example, Fig. 2.11a is a plot of a cubic smoothing spline fitted to the `lake0` data. Only a little nonlinearity is afforded to it. It suggests a gradual decline in the proportion of rainbow trout caught there over time. The first 3 derivatives are also shown in Fig. 2.11b–d, and these become increasingly more jagged. The third derivative is a step function, and the second derivative is a piecewise-linear function.

2.4.4.1 Computation by the Reinsch Algorithm

Cubic smoothing splines may be computed in several ways. All of the following methods except for the first can be efficiently computed in $O(n)$ operations.

1. Direct method (2.50). *Not* recommended because it involves $O(n^3)$ operations due to an order-n matrix inversion.
2. B-splines—this numerically stable method is probably the most commonly used algorithm nowadays, and is implemented in R by splines.
3. Reinsch algorithm—using clever linear algebra, one can transform the problem into a banded system that can be efficiently solved. Green and Silverman (1994, Sect.2.3.3) gives a succinct description and this is summarized even more below. It forms the basis of the Fessler (1991) algorithm for vector splines (Sect. 4.2.1).
4. State-space approach—this is based on Kalman filter computations in time series analysis (Wecker and Ansley, 1983; Kohn and Ansley, 1987).

Elements of the Reinsch (1967) algorithm are as follows. Firstly, it may be shown that the roughness penalty matrix can be expressed[2] as $\mathbf{K} = \lambda \mathbf{Q}\mathbf{T}^{-1}\mathbf{Q}^T$, where $\mathbf{Q}$ is a banded $n \times (n-2)$ matrix and $\mathbf{T}$ is symmetric tridiagonal of order $n-2$. Also, it may be shown that $\mathbf{Q}^T \boldsymbol{f} = \mathbf{T}\boldsymbol{\gamma}$ for some vector $\boldsymbol{\gamma}$. Secondly, starting at (2.50),

$$\boldsymbol{f} \;=\; \left(\mathbf{I}_n + \lambda \mathbf{W}^{-1}\mathbf{K}\right)^{-1} \boldsymbol{y} \;=\; (\mathbf{W} + \lambda \mathbf{K})^{-1}\,\mathbf{W}\,\boldsymbol{y}. \tag{2.53}$$

Then $\boldsymbol{f} = \boldsymbol{y} - \lambda \mathbf{W}^{-1}\mathbf{Q}\mathbf{T}^{-1}\mathbf{Q}^T \boldsymbol{f}$ and hence $\boldsymbol{f} = \boldsymbol{y} - \lambda \mathbf{W}^{-1}\mathbf{Q}\boldsymbol{\gamma}$. Premultiply both sides by $\mathbf{Q}^T$ and substitute $\mathbf{Q}^T \boldsymbol{f} = \mathbf{T}\boldsymbol{\gamma}$ to give

$$\left(\mathbf{T} + \lambda \mathbf{Q}^T \mathbf{W}^{-1}\mathbf{Q}\right)\boldsymbol{\gamma} \;=\; \mathbf{Q}^T \boldsymbol{y}. \tag{2.54}$$

This is the key equation. The LHS is a symmetric positive-definite band matrix with bandwidth 5 (half-bandwidth 3). One can decompose this into the rational Cholesky decomposition $\mathbf{L}\mathbf{D}\mathbf{L}^T$, where $\mathbf{L}$ is a unit lower diagonal band matrix and $\mathbf{D}$ is a diagonal matrix with positive diagonal elements. The matrices $\mathbf{Q}$ and $\mathbf{T}$ can be found in $O(n)$ operations, and hence $\mathbf{L}$ and $\mathbf{D}$ require only linear time for their computation.

---

[2] Explicitly, letting $h_i = x_{i+1} - x_i$ for $i = 1, \ldots, n-1$, their nonzero elements are: $(\mathbf{T})_{ii} = (h_i + h_{i+1})/3$, $(\mathbf{T})_{i,i-1} = (\mathbf{T})_{i,i+1} = h_i/6$, $(\mathbf{Q})_{ii} = h_i^{-1}$, $(\mathbf{Q})_{i+1,i} = -(h_i^{-1} + h_{i+1}^{-1})$ and $(\mathbf{Q})_{i+2,i} = h_{i+1}^{-1}$.

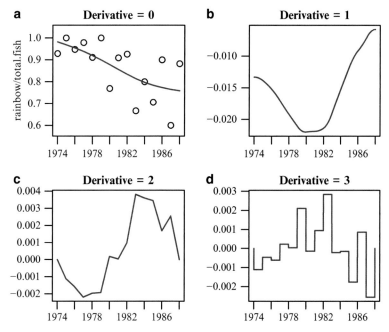

Fig. 2.11 (a) Cubic smoothing spline fitted to the proportion of fish caught that are rainbow trout from `lake0`. The x-axis is `year`. The smoother has 1 nonlinear degrees of freedom. (b)–(d) Derivatives of the smooth of orders 1–3. In contrast, Fig. 2.15 fits a local linear regression to these data.

So the steps are:

(i) compute $\mathbf{Q}^T \boldsymbol{y}$,

(ii) find the non-zero bands of $\left(\mathbf{T} + \lambda \mathbf{Q}^T \mathbf{W}^{-1} \mathbf{Q}\right)$ and hence its rational Cholesky decomposition factors $\mathbf{L}$ and $\mathbf{D}$,

(iii) solve $\mathbf{L}\mathbf{D}\mathbf{L}^T \boldsymbol{\gamma} = \mathbf{Q}^T \boldsymbol{y}$,

(iv) compute $\widehat{\boldsymbol{f}} = \boldsymbol{y} - \lambda \mathbf{W}^{-1} \mathbf{Q} \boldsymbol{\gamma}$.

Unfortunately the Reinsch algorithm becomes numerically unstable as n gets very large and/or if the x_i are very unequally spaced. Like regression splines, a more numerically stable algorithm can be devised, based on B-splines.

2.4.4.2 B-Splines

Here, we express $\widehat{f}$ as a linear combination of B-splines like Sect. 2.4.3.2 and Fig. 2.10:

$$\widehat{f}(x) = \sum_{s=1}^{K+Q} \beta_s B_{s,Q}(x) \qquad (2.55)$$

so that the elements of the roughness penalty matrix from (2.49) are $(\mathbf{K})_{st} = \int_a^b B''_{s,Q}(x) \, B''_{t,Q}(x) \, dx$. These integrals are not difficult to compute because the integrands are merely quadratics.

2.4.4.3 O-Splines

As stated above, an important property of cubic smoothing splines is that the knots are the (distinct) x_i. However, for large n, having so many knots is overkill. Consequently, O-splines are used to reduced the computational cost by choosing an 'effective' number of knots ($K \ll n$, say) that hopefully results in a fitted curve that does not differ appreciably from the *full-knot* solution. The result has been called a *low-rank* spline smoother (e.g., Ruppert et al., 2003) or *reduced-knot* smoother.

How might the K knots be chosen? Ideally, they should 'mimic' the x_is, hence one technique is to take a simple random sample of them. Another suggestion is to place relatively more knots in regions where f is wiggly as opposed to simple. A good strategy would be to choose quantile-based knots, and another to use equally spaced knots. For these two, it is possible to construct f and distributions of the x_i that cause the other strategy to perform poorly. O-splines use quantile-based knots, whereas P-splines (Sect. 2.4.5) choose equally spaced knots. The former is implemented in the R function `smooth.spline()` and also in VGAM as a whole.

As for the value of K itself, the upper function of Fig. 2.12 is a plot of K versus n used by `smooth.spline()`. As $n \to \infty$, $K = 200 + (n - 3200)^{1/5}$ grows very slowly. To 'fill the space' of the x_is, the software selects the sth knot to be approximately the $s/(K + 1)$th sample quantiles of the unique x_is. (In contrast, P-splines choose equally spaced knots). But $K = n$ for $n \leq 50$ because of the light computational cost. It should be noted that O-splines use the natural boundary constraints (2.47) so that the solution is linear beyond the range of the data. Some more details are given in Wand and Ormerod (2008).

The function `vsmooth.spline()` described in Sect. 4.4.2 also follows a similar idea. However, it reduces K with greater severity because $M > 1$ increases the computational cost quickly as M grows. Currently,

$$K = \begin{cases} n, & n \leq 40, \\ \lfloor 40 + (n - 400)^{1/4} \rfloor, & n > 40, \end{cases} \qquad (2.56)$$

which is the lower function of Fig. 2.12.

Incidentally, the "O" in "O-splines" is due to F. O'Sullivan, the author of a software implementation of the above, named BART, which was written in the mid-1980s. It forms the innards of `smooth.spline()`. By default, this function will implement O-splines, but if argument `all.knots = TRUE` then the full-knot solution to (2.48) will be returned. The early S-PLUS `gam()` function was built on BART, as is `gam()` in gam presently. More details about the O-spline algorithm are given in Sect. 4.2.1.3 for the general M case.

2.4.5 P-Splines

Rather than using smoothing splines, it is more convenient to smooth using the "penalized B-splines" of Eilers and Marx (1996), also known as "P-splines". They are another example of a low-rank smoother and have several compelling advantages. Their solution can be conveniently computed because it involves straightforward linear algebra computations, therefore estimation can proceed in a similar

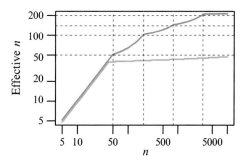

Fig. 2.12 O-splines: number of knots K selected from n unique x_i, for smooth. spline() is the top function. The lower function is (2.56) for vsmooth.spline(). Both axes are on a logarithmic scale. The top function intersects with the *dashed lines* at $(50, 50)$, $(200, 100)$, $(800, 140)$ and $(3200, 200)$; logarithmic interpolation is used for other n values.

manner to GLMs. There is no need for backfitting (Sect. 4.3) and all functions are estimated simultaneously. Furthermore, inference is straightforward, and automatic smoothing parameter selection is a less daunting problem. The R package mgcv conveys P-splines to the common GAM class plus about a dozen more distributions.

P-splines extend regression splines by penalizing the coefficients of adjacent B-splines. Suppose that $x_i \in [a, b]$ for some simple scatter plot data. Let the regression spline be

$$f(x) = \sum_{s=1}^{K+Q-1} \beta_s \, B_{s,q}(x), \qquad (2.57)$$

where there are $K + 1$ equidistant knots $\xi_s = a + s(b-a)/K$ (for $s = 0, 1, \ldots, K$) in $[a, b]$ (i.e., $K - 1$ internal knots). We can write (2.57) as $\boldsymbol{f} = \mathbf{X}\boldsymbol{\beta}$ where $(\mathbf{X})_{ij} = B_{j,Q}(x_i)$. Then $\boldsymbol{\beta}$ can be estimated by minimizing

$$S(\boldsymbol{\beta}) = (\boldsymbol{y} - \mathbf{X}\boldsymbol{\beta})^T \mathbf{W}(\boldsymbol{y} - \mathbf{X}\boldsymbol{\beta}) + \lambda \boldsymbol{\beta}^T \, \mathbf{D}_{[d]}^T \, \mathbf{D}_{[d]} \boldsymbol{\beta} \qquad (2.58)$$

where $\lambda > 0$ is the smoothing parameter, and $\mathbf{D}_{[d]}$ $((K + Q - 1 - d) \times (K + Q - 1))$ is the matrix representation of the dth-order differencing operator Δ^d, e.g., $\Delta^1 \beta_s = \beta_s - \beta_{s-1}$ and $\Delta^2 \beta_s = \Delta(\Delta\beta_s) = \Delta\beta_s - \Delta\beta_{s-1} = \beta_s - \beta_{s-1} - (\beta_{s-1} - \beta_{s-2}) = \beta_s - 2\beta_{s-1} + \beta_{s-2}$. In practice, the values $d = 2$ and 3 are common. In general, the roughness penalty term in (2.58) is $\lambda \sum_{s=d+1}^{K+Q-1} \left(\Delta^d \beta_s\right)^2$, and this penalty may not make sense with non-equidistant knots. The form the $\mathbf{D}_{[d]}^T$ takes on is similar to:

```
> (D_1 <- diff(diag(4)))

     [,1] [,2] [,3] [,4]
[1,]   -1    1    0    0
[2,]    0   -1    1    0
[3,]    0    0   -1    1

> (D_2 <- diff(diff(diag(4))))   # Same as diff(diag(4), diff = 2)

     [,1] [,2] [,3] [,4]
[1,]    1   -2    1    0
[2,]    0    1   -2    1
```

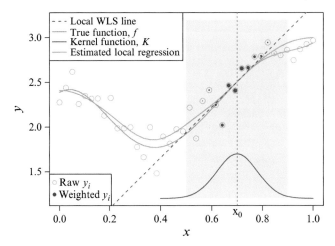

Fig. 2.13 Local linear regression with $n = 40$ points. The kernel weights have been divided by 10 for scaling purposes. The *vertical line* is at the target point $x_0 = 0.7$ so that three *curves/lines* intersect at $(x_0, \widehat{f}(x_0))$. The *shaded region* is effectively the window for computing $\widehat{f}(x_0)$.

It may be seen in (2.58) that $\boldsymbol{\beta}$ appears in both terms. This means that the coefficients controls both the amount of goodness-of-fit and the wiggliness. The solution obtained by setting $\partial S / \partial \boldsymbol{\beta} = \mathbf{0}$ is

$$\widehat{\boldsymbol{\beta}} \;=\; \left(\mathbf{X}^T \mathbf{W} \mathbf{X} + \lambda \, \mathbf{D}_{[d]}^T \mathbf{D}_{[d]} \right)^{-1} \mathbf{X}^T \mathbf{W} \boldsymbol{y}. \tag{2.59}$$

Then the variance-covariance matrix of $\widehat{\boldsymbol{\beta}}$ is easy: $\mathrm{Var}(\widehat{\boldsymbol{\beta}}) =$

$$\sigma^2 \left(\mathbf{X}^T \mathbf{W} \mathbf{X} + \lambda \, \mathbf{D}_{[d]}^T \mathbf{D}_{[d]} \right)^{-1} \mathbf{X}^T \mathbf{W} \mathbf{X} \left(\mathbf{X}^T \mathbf{W} \mathbf{X} + \lambda \, \mathbf{D}_{[d]}^T \mathbf{D}_{[d]} \right)^{-1}, \tag{2.60}$$

and so

$$\mathrm{Var}(\widehat{\boldsymbol{y}}) \;=\; \mathrm{Var}(\mathbf{X}\widehat{\boldsymbol{\beta}}) \;=$$
$$\sigma^2 \, \mathbf{X} \left(\mathbf{X}^T \mathbf{W} \mathbf{X} + \lambda \, \mathbf{D}_{[d]}^T \mathbf{D}_{[d]} \right)^{-1} \mathbf{X}^T \mathbf{W} \mathbf{X} \left(\mathbf{X}^T \mathbf{W} \mathbf{X} + \lambda \, \mathbf{D}_{[d]}^T \mathbf{D}_{[d]} \right)^{-1} \mathbf{X}^T. \tag{2.61}$$

2.4.6 Local Regression

Local regression refers to a major class of smoothers that includes the Nadaraya-Watson smoother (2.63), local polynomial kernel estimators (2.65), and variants such as Loess and Lowess. No local regression smoother is currently implemented in **VGAM**, so we describe it here mainly for completeness and for preparation of some theoretical properties in the vector case (Sect. 4.2.2.1). We give scant attention to any practical aspects, and only briefly mention that a popular smoother is loess() (Cleveland et al., 1991) and its older variant lowess()—see Sect. 2.4.6.5.

Consider the classical smoothing problem (2.36) with $\sigma_i^2 = \sigma^2$ and $w_i = 1$ as related to Fig. 2.13. To estimate $f(x_0)$, one computes a WLS fit to the (x_i, y_i) with weights determined by the distance x_0 is from the x_i. In fact, these weights are $K_h(x_0 - x_i)$ where K is some *kernel function*, h is the positive smoothing parameter known as the *bandwidth*, and

$$K_h(u) = h^{-1} \cdot K\left(\frac{u}{h}\right) \tag{2.62}$$

is a scaled version of K that integrates to unity for all h. The bandwidth scales the distance by adjusting the window size in a similar manner that the standard deviation does to a normal distribution. Small/large values of h mean a small/large effective window size about x_0. An h that is too low results in too few observations, therefore is prone to overfit. As $h \to \infty$, the solution becomes an essentially unweighted LS fit (because all weights are equal) to all the data, e.g., $\widehat{f}(x) = \bar{y}$ for all x if a polynomial of degree $r = 0$ is fitted.

Some popular kernel functions are given in Table 2.2 and are plotted in Fig. 2.14. For convenience they possess the following properties:

(i) symmetric,
(ii) have unit area,
(iii) centred at the origin,
(iv) nonincreasing going away from the origin.

Regarding the latter property, apart from the uniform kernel which assigns an equal weight to observations within the window, other kernel functions strictly decrease as the distance from the origin increases. This is called the unimodal property. The significance of the Epanechnikov kernel is that it minimizes the asymptotic mean integrated squared error (2.84).

Given the kernel weights, the WLS fit is a polynomial of degree r (Fig. 2.13), hence the name *local polynomial kernel estimator* is sometimes used. The case $r = 1$ is known as a *local linear regression* or *local linear kernel* smoother. The case $r = 0$ gives the *local constant* or simple *Nadaraya-Watson estimator*

$$\widehat{f}_{nw}(x_0) = \frac{\sum\limits_{i=1}^{n} K\left(\frac{x_0 - x_i}{h}\right) y_i}{\sum\limits_{i=1}^{n} K\left(\frac{x_0 - x_i}{h}\right)} = \sum_{i=1}^{n}\left\{\frac{K_h(x_0 - x_i)}{\sum\limits_{t=1}^{n} K_h(x_0 - x_t)}\right\} y_i. \tag{2.63}$$

Clearly, it takes a weighted average of the y_i, and more weight is assigned to those x_i that are closer to x_0. The quantities in braces are normalized kernel weights.

More generally, an explicit expression for the rth-degree local polynomial kernel estimator can be obtained as follows. At a target point x, the estimator $\widehat{f}(x; r, h)$ is obtained by fitting the polynomial $\beta_1 + \beta_2(\cdot - x) + \cdots + \beta_{r+1}(\cdot - x)^r$ to the (x_i, y_i) using weighted least squares with kernel weights $K_h(x_i - x)$. The value of $\widehat{f}(x; r, h)$ is the intercept $\widehat{\beta}_1$ because of the centring, where $\widehat{\boldsymbol{\beta}} = (\widehat{\beta}_1, \ldots, \widehat{\beta}_{r+1})^T$ minimizes the WLS criterion

$$\sum_{i=1}^{n} \left\{ y_i - \beta_1 - \beta_2(x_i - x) - \cdots - \beta_{r+1}(x_i - x)^r \right\}^2 K_h(x_i - x). \tag{2.64}$$

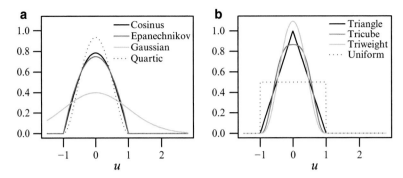

Fig. 2.14 Kernel functions from Table 2.2. All but one has compact support.

Table 2.2 Some popular kernel functions for local regression. All but one is defined on $[-1, 1]$, and the quartic is also known as the biweight. They are graphed in Fig. 2.14.

Kernel	$K(u)$	Kernel	$K(u)$
Cosinus	$\frac{\pi}{4} \cos(\pi u/2) \cdot I(|u| \leq 1)$	Triangle	$(1 - |u|) \cdot I(|u| \leq 1)$
Epanechnikov	$\frac{3}{4}(1 - u^2) \cdot I(|u| \leq 1)$	Tricube	$\frac{70}{81}(1 - |u|^3)^3 \cdot I(|u| \leq 1)$
Gaussian	$\phi(u) = \exp(-\frac{1}{2}u^2)/\sqrt{2\pi}$	Triweight	$\frac{35}{32}(1 - u^2)^3 \cdot I(|u| \leq 1)$
Quartic	$\frac{15}{16}(1 - u^2)^2 \cdot I(|u| \leq 1)$	Uniform	$\frac{1}{2} \cdot I(|u| \leq 1)$

The centring about x is for mathematical convenience. The solution is

$$\widehat{\boldsymbol{\beta}}_x = \left(\mathbf{X}_x^T \mathbf{W}_x \mathbf{X}_x\right)^{-1} \mathbf{X}_x^T \mathbf{W}_x \boldsymbol{y} \qquad (2.65)$$

where $\boldsymbol{y} = (y_1, \ldots, y_n)^T$, $\mathbf{W}_x = \text{diag}(K_h(x_1 - x), \ldots, K_h(x_n - x))$ and the model matrix specific to x is

$$\mathbf{X}_x = \begin{pmatrix} 1 & (x_1 - x) & \ldots & (x_1 - x)^r \\ \vdots & \vdots & & \vdots \\ 1 & (x_n - x) & \ldots & (x_n - x)^r \end{pmatrix}, \qquad (2.66)$$

which is $n \times (r + 1)$. Since the estimator of $f(x)$ is the intercept, we have

$$\widehat{f}(x; r, h) = e_1^T \widehat{\boldsymbol{\beta}}_x = e_1^T \left(\mathbf{X}_x^T \mathbf{W}_x \mathbf{X}_x\right)^{-1} \mathbf{X}_x^T \mathbf{W}_x \boldsymbol{y}. \qquad (2.67)$$

Substituting $r = 0$ into this yields the Nadaraya-Watson estimator (2.63). Similarly, the local linear estimator ($r = 1$) can be written as

$$\widehat{f}(x; 1, h) = n^{-1} \sum_{i=1}^{n} \frac{\{\widehat{s}_2(x; h) - \widehat{s}_1(x; h) \cdot (x_i - x)\} K_h(x_i - x) y_i}{\widehat{s}_2(x; h) \widehat{s}_0(x; h) - \widehat{s}_1(x; h)^2} \qquad (2.68)$$

where

$$\widehat{s}_r(x; h) = n^{-1} \sum_{i=1}^{n} (x_i - x)^r K_h(x_i - x). \qquad (2.69)$$

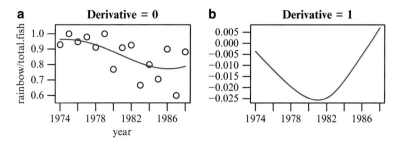

Fig. 2.15 (a) Local linear regression fitted to the proportion of fish caught that are rainbow trout from lake0. The smoother has a bandwidth of $h = 2.5$ and uses the Gaussian kernel function. (b) $\widehat{f}'(x)$. In contrast, Fig. 2.11 fits a cubic smoothing spline to these data.

2.4.6.1 Derivative Estimation

Sometimes the first or second derivative of f is of more interest than f itself. For example, in the study of human growth curves of height as a function of age, the "speed" and "acceleration" of growth have important biological significance.

The νth derivative of f is easily estimated from above. As

$$\frac{d^\nu}{du^\nu} \beta_{p+1} \cdot (u-x)^p \bigg|_{u=x} = \begin{cases} p!\,\beta_{p+1}, & \nu = p, \\ 0, & \text{otherwise}, \end{cases}$$

we simply extract the $(\nu+1)$th coefficient of $\boldsymbol{\beta}_x$ to give the estimate

$$\widehat{f}^{(\nu)}(x; r, h) = \nu!\, e_{\nu+1}^T \left(\mathbf{X}_x^T \mathbf{W}_x \mathbf{X}_x \right)^{-1} \mathbf{X}_x^T \mathbf{W}_x \, \boldsymbol{y} \tag{2.70}$$

for all $\nu = 0, \ldots, r$. Of course, (2.67) is a special case of this. Note that $\widehat{f}^{(\nu)}(x; r, h)$ is not in general equal to the νth derivative of $\widehat{f}(x; r, h)$.

As an example, Fig. 2.15 fits a local linear regression to the lake0 data. In contrast, Fig. 2.11 fits a cubic smoothing spline to these data. Both estimates of $f(x)$ and $f'(x)$ are similar for the two smoothers, which is not surprising for this small simple scatter plot. As $r = 1$, $\widehat{f''}(x)$ is not available through (2.70), however it might be estimated using a local quadratic polynomial kernel estimator.

2.4.6.2 Bias and Variance †

Compared to some other smoothers, the large-sample properties of local polynomial kernel estimators are readily derived. This section demonstrates some of this ability. From (2.65), write

$$\widehat{\boldsymbol{\beta}}_x = \boldsymbol{\Delta}_x^{-1} \boldsymbol{\theta}_x \tag{2.71}$$

where $\boldsymbol{\Delta}_x = n^{-1}\mathbf{X}_x^T\mathbf{W}_x\mathbf{X}_x$ and $\boldsymbol{\theta}_x = n^{-1}\mathbf{X}_x^T\mathbf{W}_x\boldsymbol{y}$, so that their means are finite. If $\mathrm{Var}(y_i) = \sigma^2$ then it follows from (2.65) that

$$E\left[\widehat{\boldsymbol{\beta}}_x\right] \;=\; \left(\mathbf{X}_x^T\mathbf{W}_x\mathbf{X}_x\right)^{-1}\mathbf{X}_x^T\mathbf{W}_x\,\boldsymbol{f}, \tag{2.72}$$

$$\mathrm{Var}\left[\widehat{\boldsymbol{\beta}}_x\right] \;=\; \sigma^2\left(\mathbf{X}_x^T\mathbf{W}_x\mathbf{X}_x\right)^{-1}\mathbf{X}_x^T\mathbf{W}_x^2\mathbf{X}_x\left(\mathbf{X}_x^T\mathbf{W}_x\mathbf{X}_x\right)^{-1}, \tag{2.73}$$

where $\widehat{\boldsymbol{\beta}}_x = (\widehat{f}(x), \ldots, \widehat{f}^{(r)}(x)/r!)^T$ and $\boldsymbol{f} = (f(x_1), \ldots, f(x_n))^T$.

Suppose the *design points* x_i are a random sample from some distribution with density function g (this is called a *random design*). Define

$$\mu_q \;=\; \int_{-\infty}^{\infty} u^q\,K(u)\,du \quad\text{and}\quad \nu_q \;=\; \int_{-\infty}^{\infty} u^q\,K^2(u)\,du, \tag{2.74}$$

so that $\mu_0 = 1$, and $\mu_q = \nu_q = 0$ for odd $q \geq 1$. We assume that $\nu_2 < \infty$ and $\mu_4 < \infty$. For the purposes of Sect. 4.2.2.1, we shall mainly consider the $r = 1$ case. Then it can be shown, subject to regularity conditions (e.g., Sect. 4.2.2.1), that when the x_i are uniformly distributed and x_0 is away from the boundaries, then asymptotically

$$\mathrm{Bias}[\widehat{f}(x_0)] \;\sim\; \frac{h^2}{2}\,\mu_2\,f''(x_0), \tag{2.75}$$

$$\mathrm{Bias}[\widehat{f}'(x_0)] \;\sim\; \frac{h^2}{3!\,\mu_2}\mu_4\,f'''(x_0), \tag{2.76}$$

$$\mathrm{Var}\left[\widehat{f}(x_0)\right] \;\sim\; \frac{\nu_0\,\sigma^2}{n\,h\,g(x_0)}, \tag{2.77}$$

$$\mathrm{Var}\left[\widehat{f}'(x_0)\right] \;\sim\; \frac{\nu_2\,\sigma^2}{n\,h^3\,\mu_2^2\,g(x_0)}. \tag{2.78}$$

In these formulas, the bias-variance trade-off can be seen immediately, e.g., as h decreases, the biases decrease and the variances increase. Another observation is that in order for the estimator of $f(x_0)$ to be consistent, it is necessary for $h \to 0$ and $nh \to \infty$ as $n \to \infty$. In fact, it can be shown that to minimize the asymptotic mean integrated squared error (2.84), the optimum rate is $h = O(n^{-1/5})$. Additionally, it can be shown that the asymptotic bias of $\widehat{f}(x_0)$ from a local polynomial regression of degree r is $O(h^{r+1})$ for odd r, and $O(h^{r+2})$ for even r. This suggests that a higher degree r should be chosen for large samples if f is very wiggly.

To verify (2.75)–(2.76), one needs to show that, for example,

$$n^{-1}\sum_{i=1}^{n} \alpha(x_i)\,(x_i - x)\,K_h(x_i - x) \;\sim\; h^2\,\{\alpha'(x)\,g(x) + \alpha(x)\,g'(x)\}\,\mu_2$$

for $h > 0$ and some smooth function $\alpha(x)$. The following standard argument is used to obtain the asymptotic mean of the LHS. Call the LHS I_1, say, and let $z = (x_i - x)/h$. Then apply two Taylor series about x:

$$
\begin{aligned}
I_1 \;\sim\;& \int_{-\infty}^{\infty} \left[\alpha(x) + \alpha'(x)\,(x_i - x) + \alpha''(x)\,\frac{(x_i - x)^2}{2} + \cdots \right] (x_i - x) \cdot \\
& \frac{1}{h}\, K\!\left(\frac{x_i - x}{h}\right) \left[g(x) + g'(x)\,(x_i - x) + g''(x)\,\frac{(x_i - x)^2}{2} + \cdots \right] dx_i \\
=\;& \frac{1}{h} \int_{-\infty}^{\infty} (zh)\, K(z) \left[\alpha(x) + \alpha'(x)\, zh + \frac{1}{2}\,\alpha''(x)\,(zh)^2 + \cdots \right] \cdot \\
& \left[g(x) + g'(x)\, zh + \frac{1}{2}\,g''(x)\,(zh)^2 + \cdots \right] h\, dz \\
\sim\;& h \int_{-\infty}^{\infty} z\, K(z)\, \{\alpha'(x)\, g(x)\, zh + \alpha(x)\, g'(x)\, zh\}\, dz \\
=\;& h^2\, \{\alpha'(x)\, g(x) + \alpha(x)\, g'(x)\} \int_{-\infty}^{\infty} z^2\, K(z)\, dz.
\end{aligned}
$$

A similar argument to the above can be used to show the following:

$$
n^{-1} \sum_{i=1}^{n} \alpha(x_i)\, K_h(x_i - x)\,(x_i - x)^t \;\sim
$$

$$
\begin{cases}
\alpha(x)\, g(x) + \\
\quad h^2\, \mu_2 \left\{ \frac{1}{2}\,\alpha(x)\, g''(x) + \alpha'(x)\, g'(x) + \frac{1}{2}\,\alpha''(x)\, g(x) \right\}, & t = 0, \\[4pt]
h^2\, \mu_2 \left\{ \alpha(x)\, g'(x) + \alpha'(x)\, g(x) \right\} + \\
\quad h^4\, \mu_4 \left\{ \frac{1}{6}\,\alpha\, g''' + \frac{1}{2}\,\alpha'\, g'' + \frac{1}{2}\,\alpha''\, g' + \frac{1}{6}\,\alpha'''\, g \right\}, & t = 1, \\[4pt]
h^2\, \mu_2\, \alpha(x)\, g(x) + \\
\quad h^4\, \mu_4 \left\{ \frac{1}{2}\,\alpha(x)\, g''(x) + \alpha'(x)\, g'(x) + \frac{1}{2}\,\alpha''(x)\, g(x) \right\}, & t = 2.
\end{cases}
\tag{2.79}
$$

One makes good use of the above when working out the elements of $\boldsymbol{\Delta}_x$ and $E[\boldsymbol{\theta}_x]$ (the latter uses $\alpha = f$). For the local linear case,

$$
\boldsymbol{\Delta}_x^{-1} \;\sim\; \frac{1}{g(x)} \begin{pmatrix} 1 & -g'(x)/g(x) \\ -g'(x)/g(x) & 1/\{h^2\, \mu_2\} \end{pmatrix},
\tag{2.80}
$$

which can be used to premultiply the 2-vector $E[\boldsymbol{\theta}_x]$. This gives the first element

$$
f(x) + \frac{h^2\, \mu_2}{g^2(x)} \left\{ \frac{1}{2}\, f(x)\, g(x)\, g''(x) + \frac{1}{2}\, f''(x)\, g^2(x) - f(x)\, [g'(x)]^2 \right\}.
$$

Subtracting $f(x)$ from this gives the asymptotic bias. If the x_i are uniformly distributed (a *fixed design*), then $g(x)$ is a constant, leading to the bias term (2.75). Similarly, deriving the second element gives $\mathrm{Bias}[\widehat{f}'(x_0)] \sim$

$$
\begin{aligned}
\frac{h^2}{g(x_0)} &\left[\frac{\mu_4}{\mu_2} \left\{ \frac{1}{6}\, f(x_0)\, g'''(x_0) + \frac{1}{2}\, f'(x_0)\, g''(x_0) + \frac{1}{2}\, f''(x_0)\, g'(x_0) + \frac{1}{6}\, f'''(x_0)\, g(x_0) \right\} - \right. \\
& \left. \mu_2\, \frac{g'(x_0)}{g(x_0)} \left\{ \frac{1}{2}\, f(x_0)\, g''(x_0) + f'(x_0)\, g'(x_0) + \frac{1}{2}\, f''(x_0)\, g(x_0) \right\} \right].
\end{aligned}
$$

Then uniformly distributed x_i implies (2.76).

The variance terms (2.77)–(2.78) follow from a similar standard argument that shows $n^{-1} \sum_{i=1}^{n} \alpha(x_i) \, K_h^2(x_i - x) \, (x_i - x)^t \sim$

$$
\begin{cases}
h^{-1} \nu_0 \, \alpha(x) \, g(x) + \\
h \, \nu_2 \left\{ \frac{1}{2} \alpha(x) \, g''(x) + \alpha'(x) \, g'(x) + \frac{1}{2} \alpha''(x) \, g(x) \right\}, & t = 0, \\
h \, \nu_2 \left\{ \alpha(x) \, g'(x) + \alpha'(x) \, g(x) \right\} + \\
h^3 \, \nu_4 \left\{ \frac{1}{6} \alpha \, g''' + \frac{1}{2} \alpha' \, g'' + \frac{1}{2} \alpha'' \, g' + \frac{1}{6} \alpha''' \, g \right\}, & t = 1, \\
h \, \nu_2 \, \alpha(x) \, g(x) + \\
h^3 \, \nu_4 \left\{ \frac{1}{2} \alpha(x) \, g''(x) + \alpha'(x) \, g'(x) + \frac{1}{2} \alpha''(x) \, g(x) \right\}, & t = 2.
\end{cases}
\tag{2.81}
$$

These are used in the $n^{-1} \left(\mathbf{X}_x^T \mathbf{W}_x^2 \, \mathbf{X}_x \right)$ part of the formula of (2.73). Multiplying $\boldsymbol{\Delta}_x^{-1} \mathbf{X}_x^T \mathbf{W}_x^2 \, \mathbf{X}_x \boldsymbol{\Delta}_x^{-1}$ together and setting $g' = 0$ gives the required results.

Of further interest, the equivalent kernel (Sect. 2.4.7.3) of a smoother are the weights assigned to y_i in order to obtain $\widehat{f}(x)$. That is, $\widehat{f}(x) = \sum_{i=1}^{n} \omega_i^* \, y_i$ where the ω_i^* are known as the *equivalent kernel* of $\widehat{f}(x)$. For local linear regression, the equivalent kernels are easily found by

$$
\begin{pmatrix} \widehat{f}(x) \\ \widehat{f}'(x) \end{pmatrix} = \boldsymbol{\Delta}_x^{-1} \, \boldsymbol{\theta}_x = n^{-1} \sum_{i=1}^{n} K_h(x_i - x) \, \boldsymbol{\Delta}_x^{-1} \begin{pmatrix} 1 \\ x_i - x \end{pmatrix} y_i,
$$

therefore the ith vector of this sum which multiplies y_i is

$$
n^{-1} \, K_h(x_i - x) \begin{pmatrix} \dfrac{1}{g(x)} - \dfrac{g'(x)}{g^2(x)} (x_i - x) \\[2ex] -\dfrac{g'(x)}{g^2(x)} + \dfrac{x_i - x}{h^2 \, g(x) \, \mu_2} \end{pmatrix}.
\tag{2.82}
$$

The first element is the asymptotic equivalent kernel for $\widehat{f}(x)$ [cf. (4.44)]. For uniformly distributed x_i, this is proportional to $K_h(x_i - x)$, which makes intuitive sense.

The second element of (2.82) is the asymptotic equivalent kernel for $\widehat{f}'(x)$. For a simple example of $n = 101$ equally spaced points on $[0, 1]$ with $h = 0.2$ and a Gaussian kernel, Fig. 2.16 is a plot of these for three values of x_0. The weights for the y_i are positive to the immediate RHS of x_0, and negative on the LHS; this makes sense given the central finite-difference formula in Sect. 9.2.5: $f'(x) \approx [f(x + h/2) - f(x - h/2)]/h$, whose error is $O(h^2)$.

The above argument may be simplified for the $r = 0$ case to show that the Nadaraya-Watson estimator also has $O(h^2)$ bias in the interior. But it can also be shown that the bias at the boundaries is $O(h)$, which may be quite severe. This can be seen quite simply by smoothing data of the form $y_i = \alpha + \beta \, x_i$ where the x_i are not equally spaced: the Nadaraya-Watson estimate will be nonlinear! If $f(x)$ is quite flat, then the Nadaraya-Watson estimator can perform better than local linear regression, but if $f(x)$ is steep and curved, then local linear regression should be the better choice.

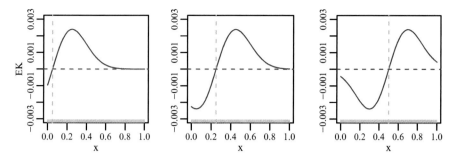

Fig. 2.16 Asymptotic equivalent kernel for $\widehat{f'}(x)$ from (2.82). The 101 x_i are equally spaced on $[0, 1]$ (as shown by the rugplot). The x_0 values are 0.05, 0.25, 0.5 (*vertical dashed lines*), and the bandwidth is 0.2. The kernel function $K = \phi(\cdot)$.

2.4.6.3 On Choosing r, h and K

In the early 1990s, Fan and co-workers showed that, for the νth derivative of f, the case of even $r - \nu$ had the same amount of variability as the next odd $r - \nu$ value. It was therefore recommended that the lowest odd degree $r = \nu + 1$ be chosen, and occasionally, $r = \nu + 3$. Thus, for most applications where f is of primary interest, a local linear regression was suggested. Ruppert et al. (2003, pp.85–6) suggest $r = 1$ when f is monotonically increasing, otherwise $r = 2$ is a good choice (partly supported by simulations). In conclusion, probably $r = 1$ and/or $r = 2$ are a good choice for many data sets, and occasionally $r = 3$.

However, in practice, the choice of the bandwidth h is the most crucial. Much research has been directed towards this very difficult problem, and ideas such as variable bandwidths have been investigated. Some packages reflecting bandwidth selection are **bbefkr**, **KernSmooth**, **lokern**, and **np**. The choice of the kernel function has long been known to be less important than bandwidth selection.

2.4.6.4 Further Comments

Practically, a major drawback of local regression as described above is the sparse data problem: if some of the (sorted) x_i have big gaps between them and the bandwidth is too small, then there may not be any observations at all within the window. For $r = 0$, this results in the denominator of (2.63) being 0 or nearly so, hence the estimate is unstable or undefined. This problem does not occur so much with splines. Hence the standard local regression formulation needs modification for coal-face general practice. One such modification is to use a nearest neighbourhood such as (2.86).

Since $\widehat{f}(x_0)$ is essentially some WLS fit, many properties of local regression estimators are consequently naturally defined for vector responses. For example, the equivalent kernel, influential measures, bias and variance, degrees of freedom, etc. Some of these are considered for the vector case in Sect. 4.2.2.

2.4.6.5 Lowess and Loess

In passing, it mentioned that two popular methods based on local regression are Lowess (Cleveland, 1979) and Loess (Cleveland and Devlin, 1988). Lowess stands for *lo*cally *we*ighted *s*catterplot *s*moother, and it robustifies the locally WLS method described above. Loess is the more modern of the two (Cleveland et al., 1991) and it can perform multivariate smoothing for x.

The basic idea of Loess is to fit a polynomial of degree r locally (the window sizes of which are determined by a nearest-neighbours scheme) and obtain the fitted values. Then the residuals are assigned weights: larger/smaller residuals receive small/large weights respectively. Another local polynomial of degree r (with weights given by the product of the initial weight and new weight) is fitted. Thus observations showing large residuals at the initial fit are *downweighted* in the second fit. The above process is repeated a few times. Cleveland (1979) recommended 3 iterations and $r = 1$, which are the software defaults.

Loess can be invoked simply, e.g.,

```
fit.lo <- loess(y ~ x, data = ldata)
plot(y ~ x, data = ldata)
lines(predict(fit.lo) ~ x, data = ldata)   # The variable x is assumed sorted here
```

and for additive models, it is implemented in **gam**, e.g.,

```
gam(y ~ lo(x2) + lo(x3), binomial, data = bdata)
```

Both Lowess and Loess measure the size of a neighbourhood using the 'span'; the larger the value, the larger the neighbourhood.

2.4.7 Some General Theory

In this subsection, a sprinkling of general theory relating to scatter plot smoothing is provided. Here, there is a fundamental trade-off between the bias and variance of the estimator, and this phenomenon is governed by the smoothing parameter. One criterion that compares the two quantities directly at a value x is the (pointwise) *mean squared error* (MSE; Sect. A.1.3.1)

$$\mathsf{MSE}(\widehat{f}(x)) = E\left[\left(\widehat{f}(x) - f(x)\right)^2\right] = \mathrm{Var}\left(\widehat{f}(x)\right) + \left(E\,\widehat{f}(x) - f(x)\right)^2. \quad (2.83)$$

A similar quantity to the above, known as the *mean integrated squared error* (MISE), is

$$\mathsf{MISE}(\widehat{f}(\cdot)) = \int_{-\infty}^{\infty} \mathsf{MSE}(\widehat{f}(x))\, g(x)\, W(x)\, dx, \quad (2.84)$$

which is a global measure of precision. Here, $W(x)$ is a weighting function that might be needed for the integral to exist; it is assumed that $W(x) = 1$ unless otherwise stated. This MISE weighs the MSE of $\widehat{f}$ by the density of the design points g. For some smoothers, this criterion can be minimized with respect to the smoothing parameter.

2.4.7.1 Linear Smoothers

A smoother is said to be *linear* if

$$\widehat{\boldsymbol{y}} \ = \ \mathbf{S}\boldsymbol{y}, \tag{2.85}$$

where the influence matrix $\mathbf{S}$ can depend on $\boldsymbol{x}$ but *not* on $\boldsymbol{y}$. The rank of $\mathbf{S}$ might be n or much less than n—called *full-rank* and *low-rank smoothers*, respectively.

All four smoothers described in this section (regression splines, cubic smoothing splines, P-splines and local polynomial kernel smoothers) are linear smoothers, provided that the smoothing parameters are fixed. Strictly speaking, the use of automatic smoothing parameter selection procedures makes a smoother nonlinear because then $\mathbf{S}$ does depend on $\boldsymbol{y}$. However, as Ruppert et al. (2003) confess, we commonly pretend the smoothing parameter is fixed and, as an approximation, treat the smoother as linear. Other linear smoothers not discussed here include bin smoothers, running-mean smoothers and running-line smoothers.

One can define a symmetric nearest neighbourhood of x_i as the set of indices around about i as:

$$\mathcal{N}_i \ = \ \left\{ \min\left(i - \frac{\lfloor sn \rfloor - 1}{2}, 1 \right), \ldots, i - 1, i, i + 1, \ldots, \right.$$
$$\left. \max\left(i + \frac{\lfloor sn \rfloor - 1}{2}, n \right) \right\}, \tag{2.86}$$

(Buja et al., 1989) where $0 < s < 1$ is known as the *span*. For $j \in N_i$, one computes the mean of observations (x_j, y_j) to get $\widehat{f}(x_i)$ for the running mean smoother.

An example of a nonlinear smoother is the running median smoother. To see this, suppose that n is large and the size of the symmetric nearest neighbourhood in the interior is 3 observations (these are x_{i-1}, x_i and x_{i+1}). In the absence of ties, the tridiagonal part of the smoother matrix will have two 0s and one 1 in order to pick off the median of three y_i observations. The position of the 1 can only be determined by looking at the y_i, therefore the influence matrix does not depend on $\boldsymbol{x}$ alone.

The theory for linear smoothers is much simpler than for nonlinear smoothers, and this is probably the reason why they are used much more commonly—their properties are well-understood. In probably all respects, linear smoothers generalize all the properties of simple linear regression.

2.4.7.2 Eigenvalues

Many properties of smoothers can be seen by examining the eigenvalues and eigenvectors of $\mathbf{S}$. For example, a cubic smoothing spline has all eigenvalues of $\mathbf{S}(\lambda)$ in $(0, 1]$, with exactly two unit eigenvalues with corresponding eigenvectors $\mathbf{1}$ and $\boldsymbol{x}$, i.e.,

$$\mathbf{S}\,\mathbf{1} \ = \ 1\,\mathbf{1} \quad \text{and} \quad \mathbf{S}\,\boldsymbol{x} \ = \ 1\,\boldsymbol{x}. \tag{2.87}$$

These correspond to constant and linear functions: smoothing y_i that are constant or lie on a line with respect to x_i results in fitted values equal to y_i because such

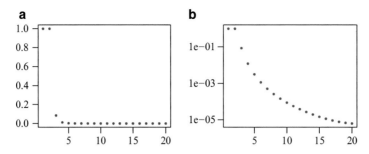

Fig. 2.17 (a) Eigenvalues of the smoother matrix of a cubic smoothing spline. Here, $n = 20$ and the x_i are equidistant on $[0, 1]$. (b) The same on a logarithmic scale. The two unit eigenvalues correspond to constant and linear functions. The corresponding eigenvectors are plotted in Fig. 2.18.

functions are not penalized by the roughness penalty criterion (2.48). Later, we shall see that such functions which are not penalized belong to the null space $\mathcal{H}_0$ in the RKHS framework of Sect. 4.2.1.7.

If $\mathbf{S}$ is symmetric, then we can write

$$\mathbf{S}\, \boldsymbol{v}_i \;=\; \theta_i\, \boldsymbol{v}_i, \quad i = 1, \ldots, n, \tag{2.88}$$

where θ_i are real eigenvalues. Smoothers with some $0 < \theta_i < 1$ are called *shrinking smoothers*. If all the θ_i are 0 or 1, then the smoother is called a *regression smoother*. For cubic smoothing splines, the $\boldsymbol{v}_i$ are approximately orthogonal polynomials of increasing order, and

$$\theta_i \;=\; 1/(1 + \lambda \rho_i)$$

where $\rho_1 \leq \rho_2 \leq \cdots \leq \rho_n$ so that $\theta_1 \geq \theta_2 \geq \cdots \geq \theta_n$. Figure 2.17 illustrates how quickly these eigenvalues can decay. Now $\{\boldsymbol{v}_1, \ldots, \boldsymbol{v}_n\}$ forms an orthonormal basis for $\mathbb{R}^n$, and the spectral decomposition of $\mathbf{S}$ is

$$\mathbf{S} \;=\; \mathbf{V} \operatorname{Diag}(\theta_1, \ldots, \theta_n) \mathbf{V}^T \;=\; \sum_{i=1}^{n} \theta_i\, \boldsymbol{v}_i\, \boldsymbol{v}_i^T \;\approx\; \sum_{i=1}^{n^*} \theta_i\, \boldsymbol{v}_i\, \boldsymbol{v}_i^T.$$

The approximation thereof holds for some appropriate $n^* \ll n$ because $\theta_i \approx 0$ for $i > n^*$. The predicted values then are

$$\widehat{\boldsymbol{y}} \;=\; \mathbf{S}\boldsymbol{y} \;\approx\; \sum_{i=1}^{n^*} \theta_i \cdot (\boldsymbol{v}_i^T \boldsymbol{y}) \cdot \boldsymbol{v}_i.$$

This shows that the fitted values are largely determined by the first few eigenvalues and eigenvectors. The high-frequency eigenvectors (Fig. 2.18) are not very important because their effect is dampened by those almost-zero eigenvalues. This suggests a low-rank approximation (e.g., Hastie, 1996) whereby a few of the largest eigenvalues are retained and the remainder set to zero. This idea can be used to motivate P-splines (Sect. 2.4.5).

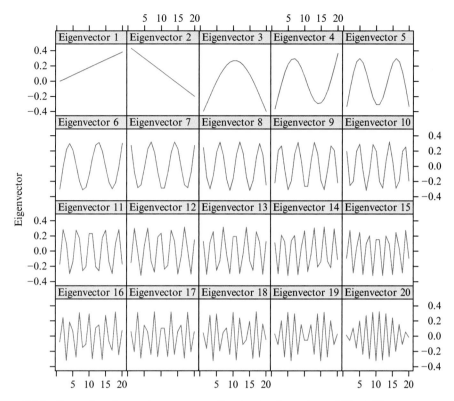

Fig. 2.18 Successive eigenvectors corresponding to the eigenvalues of Fig. 2.17.

2.4.7.3 Equivalent Kernels

Some properties of a smoother may be seen by considering its so-called *equivalent kernel*, e.g., these may be used to compare different types of linear smoothers (e.g., Buja et al., 1989). For a typical linear smoother, plotting a row of the influence matrix $\mathbf{S}$ (see (2.85)) against the x_i values gives the form of neighbourhood used and the weighting function.

For some smoothers, it is possible to derive analytical expressions for their equivalent kernel as $n \to \infty$. We saw this was the case for local linear smoothers in Sect. 2.4.6.2. This is also the case for the cubic smoothing spline: consider the weighted cubic smoothing problem

$$S(f) \;=\; \sum_{i=1}^{n} w_i \left\{ y_i - f(x_i) \right\}^2 \;+\; \lambda \int_a^b \left\{ f''(x) \right\}^2 dx, \qquad (2.89)$$

where $w_i > 0$ are known and they sum to unity. Silverman (1984) showed that

$$\widehat{f}(t) \;=\; \sum_{i=1}^{n} F(t, x_i)\, w_i\, y_i \qquad (2.90)$$

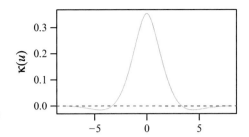

Fig. 2.19 Equivalent kernel of a cubic spline, $\kappa(u)$ (Eq. (2.92)).

asymptotically, with weighting function

$$F(t,x) \approx \frac{1}{g(x)\,h(x)}\, \kappa\!\left(\frac{t-x}{h(x)}\right) \tag{2.91}$$

with $h(x) = (\lambda/g(x))^{\frac{1}{4}}$, and kernel

$$\kappa(u) = \frac{1}{2}\, e^{-|u|/\sqrt{2}}\, \sin\!\left(\frac{|u|}{\sqrt{2}} + \frac{\pi}{4}\right). \tag{2.92}$$

The latter function is plotted in Fig. 2.19. Although κ is an even function that integrates to unity, its values are not all positive everywhere. The elements of the smoother matrix are given by $(\mathbf{S})_{ij} = w_j\, F(x_i, x_j)$.

Equation (2.90) holds for large n, small λ and x_i not too close to the boundary. That $g^{-1/4}(x)$ is bounded by $g^0(x)$ and $g^{-1}(x)$ indicates that the behaviour of the smoothing spline is between fixed-kernel smoothing and smoothing based on an average of a fixed number of neighbouring values.

2.4.7.4 Effective Degrees of Freedom

All smoothers allow the user to vary the amount of smoothing via the smoothing parameter, e.g., bandwidth h, λ, etc. However, it would be useful to have some measure of the amount of smoothing done that applies to *all* linear smoothers. One such measure is the *effective degrees of freedom* (EDF) of a smooth. It is useful for a number of reasons, e.g., comparing different types of smoothers while keeping the amount of smoothing roughly equal.

Using some basic results pertaining to the hat matrix of the linear model (Sect. 2.2.1) by replacing $\mathcal{H}$ by $\mathbf{S}$, these results suggest the following three definitions for the effective degrees of freedom of a smooth:

$$df = \text{trace}(\mathbf{S}), \tag{2.93}$$
$$df^{\text{var}} = \text{trace}(\mathbf{S}\mathbf{S}^T), \text{ and} \tag{2.94}$$
$$df^{\text{err}} = n - \text{trace}(2\mathbf{S} - \mathbf{S}^T\mathbf{S}). \tag{2.95}$$

More generally, with weights $\mathbf{W}$, these are

$$df = \text{trace}(\mathbf{S}), \tag{2.96}$$
$$df^{\text{var}} = \text{trace}(\mathbf{W}\mathbf{S}\mathbf{W}^{-1}\mathbf{S}^T), \text{ and} \tag{2.97}$$
$$df^{\text{err}} = n - \text{trace}(2\mathbf{S} - \mathbf{S}^T\mathbf{W}\mathbf{S}\mathbf{W}^{-1}). \tag{2.98}$$

It can be shown that if $\mathbf{S}$ is a (symmetric) projection matrix then trace($\mathbf{S}$), trace($2\mathbf{S} - \mathbf{SS}^T$) and trace($\mathbf{SS}^T$) coincide. For cubic smoothing splines, it can be shown that

$$\text{trace}(\mathbf{SS}^T) \leq \text{trace}(\mathbf{S}) \leq \text{trace}(2\mathbf{S} - \mathbf{SS}^T), \tag{2.99}$$

and that all three of these functions are decreasing in λ.

Computationally, df is the most popular and the cheapest because only the diagonal elements of the smoother matrix are needed. Its cost is $O(n)$ for most smoothers.

Practically, the EDF lies in the interval $[2, n]$, where a linear fit corresponds to the smallest value and an interpolating function to the largest value. As the EDF increases, the fit becomes more wiggly. Very crudely, a smooth with an EDF of about 3 or 3.5 might have about the same flexibility as a quadratic, say. A value of 4 or 5 degrees of freedom is often used as the default value in software, as this can accommodate a reasonable amount of nonlinearity but without being excessive—it should handle f having one or two stationary points.

Unfortunately, there is scope for confusion when citing the EDF because some authors do not include the constant function because the function has already been centred. For example, `smooth.spline()` and `vsmooth.spline()` have a `df` argument that corresponds to the EDF above: the value 2 means a linear LS fit, etc. However, the `df` argument of `s()` in gam's `gam()`, and `vgam()`, is such that the value 1 corresponds to a linear function. Its default is

```
> args(s)

  function (x, df = 4, spar = 0, ...)
  NULL
```

There may be less opportunity for confusion if the *effective nonlinear degrees of freedom* (ENDF) is cited, e.g., it has value 0 for a linear function.

Zhang (2003) examines calibration issues with regard to their EDF relating to local regression and spline smoothers.

2.4.7.5 Standard Errors

Plots of smooths are commonly supplemented with ± 2 pointwise standard error bands in order to prevent the over-interpretation of the estimated function. For example, Fig. 17.3 shows that the `weight` smooth has its widest pointwise standard errors at the boundaries. Such plots give the viewer some idea about how much to trust $\widehat{f}$, and which parts of the smooth have greater certainty.

From (2.85) it immediately follows that

$$\text{Var}(\widehat{\boldsymbol{f}}) = \sigma^2 \mathbf{SS}^T, \tag{2.100}$$

and so its diagonal elements may be extracted. However, this becomes impractical with large n because the entire $\mathbf{S}$ is needed. For cubic splines, the approximation $\sigma^2 \mathbf{S}$ has been used instead (and justified by a Bayesian argument, e.g., Wahba (1990); Silverman (1985)). Its cost is $O(n)$, and the approximation has been found to work well in practice.

2.4.7.6 Automatic Smoothing Parameter Selection

Choosing the smoothing parameter is arguably the most important decision for a
specified method. Ideally, we want an automated way of choosing the 'right' value.
In this section, we restrict ourselves to linear smoothers.

Occasionally it is possible to estimate λ by maximum likelihood, e.g., Wecker
and Ansley (1983) for smoothing splines. However, a more general and popular method is the *cross-validation* (CV) technique. The idea is to leave out
point (x_i, y_i) one at a time, and estimate the smooth at x_i based on the remaining $n-1$ points. Then λ_{CV} can be chosen to minimize the *cross-validation sum of
squares*

$$\mathsf{CV}(\lambda) \;=\; \frac{1}{n} \sum_{i=1}^{n} \left\{ y_i - \widehat{f}_\lambda^{[-i]}(x_i) \right\}^2, \tag{2.101}$$

where $\widehat{f}_\lambda^{[-i]}(x_i)$ is the fitted value at x_i, computed by leaving out (x_i, y_i).

While one could compute (2.101) naïvely, a more efficient way is to set the
weight of the ith observation to zero and increasing the remaining weights so that
they sum to unity. Then

$$\widehat{f}_\lambda^{(-i)}(x_i) \;=\; \sum_{\substack{j=1 \\ j \neq i}}^{n} \frac{s_{ij}}{1 - s_{ii}} \, y_j \,. \tag{2.102}$$

From this,

$$\widehat{f}_\lambda^{(-i)}(x_i) \;=\; \sum_{j=1, j \neq i}^{n} s_{ij} \, y_j + s_{ii} \, \widehat{f}_\lambda^{(-i)}(x_i)$$

and

$$y_i - \widehat{f}_\lambda^{(-i)}(x_i) \;=\; \frac{y_i - \widehat{f}_\lambda(x_i)}{1 - s_{ii}}.$$

Thus, $\mathsf{CV}(\lambda)$ can be written

$$\mathsf{CV}(\lambda) \;=\; \frac{1}{n} \sum_{i=1}^{n} \left\{ \frac{y_i - \widehat{f}_\lambda(x_i)}{1 - s_{ii}(\lambda)} \right\}^2. \tag{2.103}$$

This only requires the addition of the diagonal elements of the smoother matrix.

In practice, CV sometimes gives questionable performance. A popular alternative is the *generalized cross validation* (GCV) technique, where

$$\mathsf{GCV}(\lambda) \;=\; \frac{n^{-1} \| (\mathbf{I} - \mathbf{S}(\lambda) \, \boldsymbol{y}) \|^2}{[n^{-1} \, \mathrm{trace}(\mathbf{I} - \mathbf{S}(\lambda))]^2} \tag{2.104}$$

is minimized. The rationale for this expression is to replace s_{ii} by its average value,
$\mathrm{trace}(\mathbf{S})/n$, which is easier to compute:

$$\mathrm{GCV}(\lambda) \;=\; \frac{1}{n} \sum_{i=1}^{n} \left\{ \frac{y_i - \widehat{f}_\lambda(x_i)}{1 - \mathrm{trace}(\mathbf{S})/n} \right\}^2.$$

GCV enjoys several asymptotic optimality properties. However, neither method can be trusted always, especially with small n, e.g., an interpolating spline ($\lambda = 0$) has some positive probability of arising for a given data set.

Both CV and GCV are used when σ^2 is unknown. They are related to other criterion, such as Mallow's C_p (unbiased risk estimator; UBRE). When σ^2 is known, minimizing the UBRE is a popular choice. Another popular criterion is AIC.

2.4.7.7 Testing for Nonlinearity

Suppose we wish to compare two smooths $\widehat{f}_1 = \mathbf{S}_1 \mathbf{y}$ and $\widehat{f}_2 = \mathbf{S}_2 \mathbf{y}$, e.g., $\widehat{f}_2$ might be less smooth than $\widehat{f}_1$, and we wish to test if it picks up any significant bias. A standard case that often arises is when $\widehat{f}_1$ is linear, in which case we want to test if the linearity is real. We must assume that $\widehat{f}_2$ is unbiased, and that $\widehat{f}_1$ is unbiased under H_0. Letting ResSS_j be the residual sum of squares for the jth smooth, and γ_j be $\mathrm{trace}(2\mathbf{S} - \mathbf{S}^T\mathbf{S})$, then

$$\frac{(\mathrm{ResSS}_1 - \mathrm{ResSS}_2)/(\gamma_2 - \gamma_1)}{\mathrm{ResSS}_2/(n - \gamma_1)} \;\dot\sim\; F_{\gamma_2 - \gamma_1, n - \gamma_1} \qquad (2.105)$$

approximately, which follows from a standard F test applied to a LM (2.9).

An approximate score test for VGAMs, given in Sect. 4.3.4, tests for the linearity of component functions.

2.5 Generalized Additive Models

GAMs are a nonparametric extension of GLMs, and they provide a powerful data-driven class of models for exploratory data analysis. GAMs extend (2.17) to

$$g(\mu(\boldsymbol{x}_i)) \;=\; \eta_i \;=\; \beta_1 + f_2(x_{i2}) + \cdots + f_p(x_{ip}), \qquad (2.106)$$

a sum of smooth functions of the individual covariates. As usual with these types of models, an intercept is included because the f_k are centred for identifiability. GAMs loosen the linearity assumption of GLMs; this is very useful as it allows the data to 'speak for themselves'. Smoothers are used to estimate the f_k. They still assume additivity of the effects of the covariates, although interaction terms may be accommodated.

We will see later that the VGAM framework writes (2.106) as

$$g_1(\mu(\boldsymbol{x}_i)) \;=\; \eta_1(\boldsymbol{x}_i) \;=\; \beta_{(1)1} + f_{(1)2}(x_{i2}) + \cdots + f_{(1)d}(x_{id}), \qquad (2.107)$$

to have provision for handling multiple additive predictors η_j. For VGAM's `vgam()` function, the `s()` function represents the smooths $f_{(j)k}(x_k)$, and it has arguments `df` and `spar` to regulate the amount of smoothness. However, `df` ≥ 1 only is allowed, with a value of unity corresponding to a linear fit.

2.5.1 Why Additive Models?

One of the reasons additive models are popular is that they do not suffer from the *curse of dimensionality* (Bellman, 1961) because all the smoothing is done univariately: $\eta_j(\boldsymbol{x}) = \sum_{k=1}^{d} f_{(j)k}(x_k)$. In one dimension, the concept of a neigbourhood poses the least problems because the x_i are spread out in only one dimension. However, as the dimension of $\boldsymbol{x}$ increases, the volume of the space increases so fast that the data rapidly becomes more and more isolated in d-space. Smoothers then require a larger neighbourhood to find enough data points, hence the estimate becomes less localized and can be severely biased. Theoretically, the sparsity problem might be overcome by a sample size that grows exponentially with the dimensionality, however, this is impractical in most applications.

Modelling the $\eta_j(\boldsymbol{x})$ additively has another advantage: they have simple interpretation. Each covariate has an additive effect, therefore each effect can be determined by keeping the other x_k fixed (although this may be unrealistic in the presence of multicollinearity). The fitted functions are easily plotted separately and examined. However, this simplicity comes at a cost, e.g., interactions are not so readily handled.

One family of models which hold additive models as a special case is based on classical analysis of variance (ANOVA) and called *smoothing spline ANOVA* (SS-ANOVA). Here, functions replace the usual parameters, e.g., one-way SS-ANOVA corresponds to an additive model. A simple example of a two-way SS-ANOVA with covariates x_2 and x_3 is

$$\mu = \beta_{(1)1} + f_2(x_2) + f_3(x_3) + f_{23}(x_2, x_3),$$

where f_k represents the main effects for x_k, and f_{23} is a second-order interaction between x_2 and x_3. More generally, the unique SS-ANOVA decomposition of a multivariable function f is

$$f(x_2, \ldots, x_d) = \beta_{(1)1} + \sum_{k=1}^{d} f_k(x_k) + \sum_{s<t} f_{st}(x_s, x_t) + \cdots \quad (2.108)$$

with the constraints $E_k(f_k) = 0$, $E_s E_t(f_{st}) = 0$, etc. where the E_k are averaging operators. In practice, it is necessary to drop high-order interactions from the model space in order to avoid the curse of dimensionality. Additive models and models with second-order interactions are the most commonly used.

It should be noted that even bivariate smoothing of the form $f(x_s, x_t)$ raises difficulties: although possibly suffering from a mild case of the curse of dimensionality, plotting the functions meaningfully can require some effort, and their interpretation may be difficult.

SS-ANOVA has been extended to generalized SS-ANOVA (GSS-ANOVA), i.e., to η in the classical exponential family. It would be natural then to define the Vector SS-ANOVA class as those VGLM/VGAM families having an ANOVA decomposition (2.108) applied to each η_j.

Now just to show the simplest of GAMs, we now fit a nonparametric logistic regression with one covariate, albeit with a grain of salt.

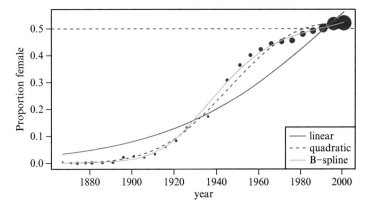

Fig. 2.20 Fitted values from some logistic regression models applied to `chinese.nz`. The response is the proportion of New Zealand Chinese who are female. The terms are `year`, `poly(year, 2)`, `bs(year, 4)`. Area sizes of the points are proportional to the number of people.

2.5.2 Binomial 'example'

In the following, we (controversially) illustrate how ordinary logistic regression can potentially misfit data. To do this, we do something that is not strictly correct, in order to make a point.

The *simple linear logistic regression* model logit $P(Y = 1) = \beta_{(1)1} + \beta_{(1)2}\, x_2$ for a single covariate x_2 results in a sigmoid curve that slopes upward or downward depending on the sign of $\beta_{(1)2}$. The limitation of sigmoid curves seems largely unappreciated by many practitioners. To illustrate the potential inadequacies of this model, consider the `chinese.nz` data frame, which gives the proportion of females in the Chinese population of New Zealand from the mid-1800s to the start of this century. These data are of historical interest, and the number of individuals involved is large enough for the sample proportions to be clearly seen.

Figure 2.20 plots the fitted values of the basic model, as well as some alternatives. Specifically, the underlying code are the first three models of:

```
vglm(cbind(female, male) ~ year,           binomialff, data = chinese.nz)
vglm(cbind(female, male) ~ poly(year, 2),  binomialff, data = chinese.nz)
vglm(cbind(female, male) ~ bs(year, 4),    binomialff, data = chinese.nz)
vgam(cbind(female, male) ~ s(year, df = 3), binomialff, data = chinese.nz)
```

It can be clearly seen that there is underfitting in the 'ordinary' logistic regression. Any predictions based on this model would have severe biases. Applying a quadratic yields a large improvement, and the regression spline is even more flexible.

The above is unwarranted for any formal inference because the data are longitudinal: most people appearing in one year will appear in adjacent years, hence the binomial independence assumption does not hold. Thus the plot should be used little more than for descriptive purposes. It is left to the reader to confirm that the last two models are very similar (Ex. 2.11, 2.23).

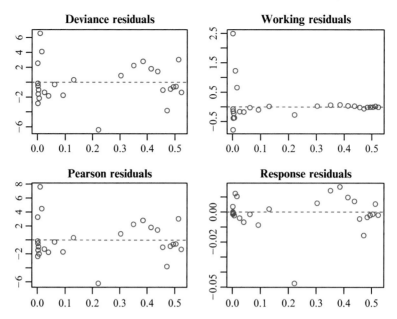

Fig. 2.21 Four residual types for the regression spline fit of Fig. 2.20. The fitted values are plotted on the x-axis.

To give some idea about what residuals can look like, Fig. 2.21 plots four types of residuals versus fitted values for the regression spline fit of Fig. 2.20. It shows that the response residuals are very different from the others, and there is much similarity between the Pearson and deviance residuals here.

Bibliographic Notes

Many books cover the theory of LMs, e.g., Rao (1973), Seber and Lee (2003), Rencher and Schaalje (2008), Christensen (2011). Books for linear modelling based on R include Fox and Weisberg (2011) and Faraway (2015). LM diagnostics are the focus of Belsley et al. (1980) and Cook and Weisberg (1982).

GLMs too are well-served in the literature. McCullagh and Nelder (1989) remains the standard text, and some other descriptions are Firth (1991), Azzalini (1996), Lindsey (1997), Fahrmeir and Tutz (2001), Venables and Ripley (2002), Faraway (2006), Dobson and Barnett (2008), Aitkin et al. (2009), Myers et al. (2010), Agresti (2015). Much of Hastie and Pregibon (1991) is still relevant to glm(). Some R package written to overcome situations where glm() may fail include biglm (to handle extra large data sets), and glm2 (which addressed possible convergence problems). The brglm package implements bias-reduction, which gives a finite solution to completely separated data (Sects. 2.3.6.3 and 9.4).

The smoothing literature is very large, due partly to it being an active research area during the 1980s and 1990s especially. General texts with accessible material on smoothing include Hastie and Tibshirani (1990), Ruppert et al. (2003) (especially P-splines), Hastie et al. (2009), James et al. (2013, Chap.7). Schimek (2000) surveys many topics related to smoothing, especially computational.

An additive model with at least one smooth term might be called a semiparametric regression model, and Harezlak et al. (2015) is a recent introductory book to such. Ruppert et al. (2009) reviews semiparametric regression from 2002–7. The class of partially linear models, as defined by $Y_i = \boldsymbol{x}_i^T \boldsymbol{\beta} + g(\boldsymbol{t}_i) + \varepsilon_i$, is the subject of Härdle et al. (2000).

B-splines are well-covered in the mathematical literature. Of these, some starters include de Boor (2001) and Schumaker (2007). For smoothing splines and their extensions specifically, a good introductory book is Green and Silverman (1994). More intermediate treatments include Eubank (1999), Ruppert et al. (2003). Advanced treatments based on RKHS theory (Sect. 4.2.1.7) include Wahba (1990), Wang (2011), Gu (2013). The two latter references cover the subject of SS-ANOVA models; they are implemented by **ssanova()** in **gss**. A book specifically on RKHS is Berlinet and Thomas-Agnan (2004), however, Nosedal-Sanchez et al. (2012) is the simplest introduction to RKHS and is specifically focused on smoothing splines.

For kernel smoothing, local regression and likelihood, see e.g., Härdle (1987), Härdle (1990), Wand and Jones (1995), Fan and Gijbels (1996), Loader (1999). Loess was first described within S3 by Cleveland et al. (1991).

The two most comprehensive references on GAMs are Hastie and Tibshirani (1990) and Wood (2006). The current approach of **VGAM** is much more similar to the former, with respect to the theory and its software (**gam**). The latter is more focused on automatic smoothing parameter selection based on P-splines and GCV, as implemented by **mgcv**. Another GAM book is Ruppert et al. (2003). An elementary applied GAM book for novice users only is Zuur (2012). Härdle et al. (2004) gives more examples and theory on a number of topics considered in this chapter, as does Gentle et al. (2012). An accessible overview of some of the ideas behind **mgcv** is Marra and Radice (2010). The utility of GAMs was recognized quite quickly and introduced into many fields during the 1990s, e.g., Yee and Mitchell (1991) into plant ecology.

Linear algebra and matrices for statisticians are presented at a moderate level by Banerjee and Roy (2014), and at a more advanced level by Harville (1997) and Seber (2008). Yanai et al. (2011) is an accessible introduction to projection matrices, generalized inverses and SVD.

Exercises

Ex. 2.1.   Prove all the properties of the hat matrix $\boldsymbol{\mathcal{H}}$ listed in Sect. 2.2.1.

Ex. 2.2.   **Hat Matrices**

(a) Prove that $\boldsymbol{\mathcal{H}}$ projects $\boldsymbol{Y}$ orthogonally onto the column (range) space of $\mathbf{X}$.
(b) Obtain an expression for h_{ii} for an LM through the origin: $y_i = \beta_1 x_i + \varepsilon_i$, for $i = 1, \ldots, n$.
(c) Repeat (b) for the simple linear regression model $y_i = \beta_1 + \beta_2 x_{i2} + \varepsilon_i$.

Ex. 2.3.   Explain why $\boldsymbol{\mathcal{H}}\mathbf{1} = \mathbf{1}$ is a good idea on p.36. Why would $\boldsymbol{\mathcal{H}}\boldsymbol{x} = \boldsymbol{x}$ also be a good property?

Ex. 2.4.   The degrees of freedom for an LM can be defined as $\sum_{i=1}^{n} \mathrm{Cov}(\widehat{y}_i, y_i)/\sigma^2$. Show that this equals p.

Ex. 2.5.   GLS
Prove the results (2.14)–(2.16), as well as its hat matrix being idempotent but not symmetric in general.

Ex. 2.6.   Show that the score function (2.21) leads to

$$U_{\boldsymbol{\beta}} = \sum_{i=1}^{n} A_i \, \boldsymbol{x}_i \left(y_i - \boldsymbol{x}_i^T \boldsymbol{\beta}\right) = \mathbf{0}$$

for multiple linear regression, and

$$U_{\boldsymbol{\beta}} = \sum_{i=1}^{n} A_i \, \boldsymbol{x}_i \left(y_i - \frac{\exp\{\boldsymbol{x}_i^T \boldsymbol{\beta}\}}{1 + \exp\{\boldsymbol{x}_i^T \boldsymbol{\beta}\}}\right) = \mathbf{0}$$

for logistic regression ($A_i Y_i \sim \text{Binomial}(A_i, \mu_i)$ with $\eta = \text{logit}\,\mu$).

Ex. 2.7.   Given the exponential family (2.19), verify all the columns of Table 2.3 from θ to $b''(\theta)$. What are the $c(y, \phi)$ functions?

Ex. 2.8.   Using (2.19) with $\phi_i = \phi/w_i$ and where the diagonal elements of $\mathbf{W} =$ $\text{diag}(w_1, \ldots, w_n)$ are known prior weights, show that $\mathbf{X}^T \mathbf{W} \boldsymbol{y}$ are a set of sufficient statistics for $\boldsymbol{\beta}$ for a GLM having a canonical link and known ϕ. Hint: use (A.4).

Ex. 2.9.   The moment generating function (MGF) of a random variable Y is defined as $M_Y(t) = E(e^{tY})$ for real t, wherever this expectation exists.

(a) Show that $E(Y) = M_Y'(0)$, $E(Y^2) = M_Y''(0)$, and deduce that $E(Y^k) = M_Y^{(k)}(0)$ for $k = 0, 1, 2, \ldots$.
(b) Obtain an expression for $M_Y(t)$ for Y belong to the exponential family (2.19).
(c) Apply (b) to the Poisson distribution to verify that $E(Y) = \text{Var}(Y) = \mu$.

Ex. 2.10.   LRT, Score and Wald Tests

(a) Generate 5 observations from Poisson($\mu = 3$) with the random number seed initialized to some value. Then compute the MLE.
(b) Suppose we wish to test $H_0 : \mu = 3$ versus $H_1 : \mu \neq 3$. Compute the p-values from Wald, score and likelihood ratio tests for this. Comment.
(c) Reset the random number generator and generate 5 observations from Poisson($\mu = 30$). Test $H_0 : \mu = 30$ versus $H_1 : \mu \neq 30$. Repeat in a similar way to (b). Comment.

Ex. 2.11.   Nonparametric Logistic Regressions
Fit the four models given in Sect. 2.5.2 to the `chinese.nz` data frame. Obtain the same as Fig. 2.20, and then add the fitted values of the `vgam()` model—hence, show its similarity with the regression spline fit.

Ex. 2.12.   Suppose somebody wrote a `log10link()` function so that $\eta = \log_{10} \mu$ could be fitted somewhat like a Poisson regression. How would its estimate of $\boldsymbol{\beta}$ be related to the usual MLE $\widehat{\boldsymbol{\beta}}$?

Ex. 2.13.   IRLS Initial Values

(a) Show that one iteration of IRLS starting from any value of $\boldsymbol{\mu}^{(0)}$ will give the LS solution for the normal case, e.g., `gaussianff()`.

(b) Write down expressions for z_i and the working weights w_i for the Poisson and binomial models (each with canonical link functions).

Ex. 2.14.   GLM Residuals

(a) Show that the first four residual types in Sect. 2.3.2 simplify to $y_i - \widehat{\mu}_i$ for the Gaussian case.
(b) Obtain a formula for the deviance residuals of a standard Poisson regression.
(c) Obtain a formula for the deviance residuals of a standard logistic regression.

Ex. 2.15.   Exponential Family Members
Consider models in the exponential family (2.19).

(a) Show that the two-parameter NB distribution (1.14) is not a standard member of the exponential family. Show that it is a member if k is a known.
(b) Find the canonical parameter θ of the NB, and show that the canonical link is $\log(\mu/(\mu + k))$.
(c) Consider a Pareto distribution (Table 12.8) where the scale parameter b is known. Is this distribution a member of the exponential family? If so, what is its canonical link?

Ex. 2.16.   For $p \neq 0$, 1, 2, and $V(\mu) = \mu^p$, show that $q(\mu; y) = y\,\mu^{1-p}/(1-p) - \mu^{2-p}/(2-p)$ where $\mu > 0$. What is the canonical parameter?   [McCullagh and Nelder (1989)]

Ex. 2.17.   Polynomial Regression
Fit polynomials up to the 10th degree to `mcycles` in MASS, and add them to a scatter plot. Comment.

Ex. 2.18.   Running-Mean Smoother

(a) Find $\mathbf{S}$ such that $\widehat{\boldsymbol{y}} = \mathbf{S}\,\boldsymbol{y}$ for a running-mean smoother with $n = 10$, and span $= 0.5$ as defined by (2.86).
(b) Compute the eigenvalues of $\mathbf{S}$ and show that a few are negative. How many unit eigenvalues are there? Comment.
(c) What range of span values would allow for a maximum of 3 values in a symmetric nearest neighbourhood?   [Buja et al. (1989)]

Ex. 2.19.

(a) Determine the coefficients a, b and c so that $f(x)$ is a cubic spline, where

$$f(x) = \begin{cases} x^3, & x \in [0,2], \\ \frac{1}{3}(x-2)^3 + a\,(x-2)^2 + b\,(x-2) + c, & x \in [2,5]. \end{cases}$$

(b) Do the same for coefficients a–d, where

$$f(x) = \begin{cases} \frac{1}{3}x^3 + x - 1, & x \in [-3,0], \\ a\,x^3 + b\,x^2 + c\,x + d, & x \in [0,1], \end{cases}$$

with $f(1) = 7$.

(c) Do the same for coefficients a–h so that $f(x)$ is a natural cubic spline satisfying $f(-1) = 1$, $f(0) = 2$ and $f(2) = 2$, where

$$f(x) = \begin{cases} a\,x^3 + b\,x^2 + c\,x + d, & x \in [-1, 0], \\ e\,x^3 + f\,x^2 + g\,x + h, & x \in [0, 2]. \end{cases}$$

(d) Use R to plot the solution of (c) on $[-2, 3]$. What are $f(-2)$ and $f(3)$?

Ex. 2.20.   Express the order-3 Bspline $B_{s,3}$ as a linear combination of order-1 B-splines. The coefficients should be in terms of $\omega_{s,3}$, etc.   [de Boor (2001)]

Ex. 2.21.   Parabolic B-Splines
Plot the 5 parabolic B-spline basis functions whose support is in $[0, 6]$ for the knot sequence $\{0, 1, 1, 3, 4, 6, 6, 6\}$. Comment.   [de Boor (2001)]

Ex. 2.22.   Regression Splines and `lake0`

(a) Modify the `lake0` code that produces Fig. 2.5b to 'work' for the raw variable `year` (taking on values 1974,...,1988). Verify that the LM fails to give finite regression estimates due to the ill-conditioning.
(b) Figure 2.5b is fitted with a *cubic* regression spline. Modify the code to use a *quadratic* regression spline—keep the same knot. Does the fitted curve change appreciably?

Ex. 2.23.   Derivative Estimation for GAMs
Consider the `chinese.nz` data set, and let $x_0 = 1936$.

(a) Create a subset of the data frame *without* the year x_0. Fit 3 separate logistic regressions to the subset; the $\eta(x)$ should be (i) linear, (ii) quadratic, and (iii) a smooth function, of x = `year`. Use `vgam()` for (iii). The response is the proportion that are female. Call the fitted curves $\widehat{p}_j(x)$ for $j = 1, 2, 3$.
(b) Predict the values for $p_j(x_0)$.
(c) Plot the sample proportions versus year. Predict the $p_j(x)$ along a fine grid of time, and add the fitted curves to your plot.
(d) For model (iii) compute $\hat{p}'_3(x_0)$, the first derivative of $\widehat{p}_3(x)$ evaluated at x_0. Plot the sample proportions versus year again, then add $\widehat{p}_3(x)$ and the tangent line at $\widehat{p}_3(x_0)$ to the plot.

Ex. 2.24.   Nadaraya-Watson Estimator: Bias and Variance

(a) Verify each entry of (2.79).
(b) Verify each entry of (2.81).
(c) Work out the expressions for the bias and variance, as in (2.72)–(2.73), for the Nadaraya-Watson estimator ($r = 0$).

Ex. 2.25.   Variance Estimates in Local Linear Regression
Given (2.81), derive the results (2.77)–(2.78).

Ex. 2.26.   Equivalent Degrees of Freedom—Projection Matrix
Show that the 3 definitions of the EDF of a smooth, based on the trace in (2.93)–(2.95), coincide when $\mathbf{S}$ is a projection matrix.

Ex. 2.27.   Nadaraya-Watson Estimate

(a) Generate scatter plot data coming from $x_i = 0((n-1)^{-1})1$, $y_i = e^{-2x_i}\sin(5x_i) + \varepsilon_i$, where $\varepsilon_i \sim N(0, \sigma^2 = 0.01)$ i.i.d., for $n = 101$, i.e., the design points are equally spaced on the unit interval. Use `set.seed()` for reproducibility.

(b) By eye, determine a reasonable value for the bandwidth h so that your Nadaraya-Watson estimate fits reasonably well.

(c) For your bandwidth, what is the ENDF value? Comment on the how the smoother handles the boundaries.

(d) Determine the values of h so that ENDF ≈ 4, 5, and 6.

Ex. 2.28.   GCV and CV

Show that $\mathsf{GCV}(\lambda)$ can be written as a weighted version of $\mathsf{CV}(\lambda)$, i.e., $\mathsf{CV}(\lambda) = n^{-1}\sum_i w_i^* \{\ldots\}^2$ starting from (2.101). Obtain an expression for the weights w_i^*.

We believe that the generalized linear models here developed could form a useful basis for courses in statistics.
—Nelder and Wedderburn (1972)

Table 2.3 Summary of GLMs supported by VGAM. A is a known prior weight. The "ff" stands for "family function", to avoid clashes with some functions associated with glm(). All families have dpqr-type functions supplied in standard R. The ϕ and $b(\theta)$ columns relate to (2.19), where θ is the natural parameter.

Distribution	PMF/PDF $f(y;\theta)$	Support	θ	Range of θ	$\mu(\theta)$	$\mathrm{Var}(Y)$	ϕ	$b(\theta)$	VGAM family
Gaussian	$(2\pi\sigma^2)^{-\frac{1}{2}}\cdot$ $\exp\left\{-\dfrac{1}{2}\left(\dfrac{y-\mu}{\sigma}\right)^2\right\}$	$(-\infty,\infty)$	μ	$\mu\in\mathbb{R},\,0<\sigma$	θ	$\dfrac{\sigma^2}{A}$	σ^2	$\theta^2/2$	gaussianff()
Binomial	$\dbinom{A}{Ay}\mu^{Ay}(1-\mu)^{A(1-y)}$	$0\left(\dfrac{1}{A}\right)1$	$\mathrm{logit}\,\mu$	$0<\mu<1$	$\dfrac{e^\theta}{1+e^\theta}$	$\dfrac{\mu(1-\mu)}{A}$	1	$\log\left(1+e^\theta\right)$	binomialff()
Poisson	$\dfrac{e^{-\mu}\mu^y}{y!}$	$0(1)\infty$	$\log\mu$	$0<\mu$	e^θ	$\dfrac{\mu}{A}$	1	e^θ	poissonff()
Gamma	$\dfrac{(k/\mu)^k\,y^{k-1}}{\Gamma(k)}\,\exp\{-ky/\mu\}$	$(0,\infty)$	$-1/\mu$	$0<\mu,\,0<k$	$-1/\theta$	$\dfrac{\mu^2}{Ak}$	$\dfrac{1}{k}$	$-\log(-\theta)$	gammaff()
Inverse Gaussian	$\left(\dfrac{\lambda}{2\pi y^3}\right)^{\frac{1}{2}}\cdot$ $\exp\left\{-\dfrac{\lambda}{2\mu^2}\dfrac{(y-\mu)^2}{y}\right\}$	$(0,\infty)$	$\dfrac{-1}{2\mu^2}$	$0<\mu,\,0<\lambda$	$\dfrac{1}{\sqrt{-2\theta}}$	$\dfrac{\mu^3}{\lambda A}$	$\dfrac{1}{\lambda}$	$-\sqrt{-2\theta}$	inverse.gaussianff()

Chapter 3
VGLMs

Life is really simple, but we insist on making it complicated.
—Confucius

3.1 Introduction

VGLMs lie at the heart of the overall statistical framework, being the linear or parametric class. This chapter describes them somewhat incrementally. Firstly, Sect. 3.2 describes them as a general method for maximum likelihood estimation based on iteratively reweighted least squares. Section 3.3 adds to this the very important feature of constraint matrices—these allow the η_j to share relationships with each other, etc. To do this, we take the liberty to use VGAMs to aid their definition. Then Sect. 3.4 describes a second important feature, which is called the `xij` facility: allowing a covariate x_k to have different values for different η_j—meaning that it is a function of $\boldsymbol{x}_{ij}^T\boldsymbol{\beta}$. The remaining sections describe allied topics such as inference and diagnostics. Relating many aspects of VGLMs with the VGAM software is deferred to Chap. 8. The nonparametric version of VGLMs, called VGAMs, is described in the next chapter.

One might loosely think of VGLMs as multivariate GLMs, however, this is only partly true, because GLMs are intertwined with the exponential family. One might then consider a multivariate generalization of (2.19),

$$f(\boldsymbol{y}_i; \boldsymbol{\theta}_i, \boldsymbol{\phi}) \;=\; \exp\left\{ \frac{\boldsymbol{y}_i^T\boldsymbol{\theta}_i - b(\boldsymbol{\theta}_i)}{a(\boldsymbol{\phi})} + c(\boldsymbol{y}_i, \boldsymbol{\phi}) \right\}, \tag{3.1}$$

for which models based on, e.g., the multinomial distribution are accommodated ($\boldsymbol{\theta}_i$ are the natural parameters, and $\boldsymbol{\phi}$ the dispersion parameters). However, the VGLM/VGAM framework is wider still, the main reason being that unnecessary restrictions limit their scope and practical usage.

© Thomas Yee 2015
T.W. Yee, *Vector Generalized Linear and Additive Models*,
Springer Series in Statistics, DOI 10.1007/978-1-4939-2818-7_3

3.2 Iteratively Reweighted Least Squares

Suppose that the observed response $\boldsymbol{y}$ is a Q-dimensional vector. VGLMs are defined as a model for which the conditional distribution of $\boldsymbol{Y}$ given explanatory variable(s) $\boldsymbol{x}$ is of the form

$$f(\boldsymbol{y}|\boldsymbol{x};\mathbf{B},\boldsymbol{\phi}) \;=\; h(\boldsymbol{y},\eta_1,\ldots,\eta_M,\boldsymbol{\phi}) \tag{3.2}$$

for some known function $h(\cdot)$, where $\mathbf{B} = (\boldsymbol{\beta}_1\,\boldsymbol{\beta}_2\,\cdots\,\boldsymbol{\beta}_M)$ is a $p \times M$ matrix of unknown regression coefficients, and the jth linear predictor is

$$\eta_j \;=\; \eta_j(\boldsymbol{x}) \;=\; \boldsymbol{\beta}_j^T\boldsymbol{x} \;=\; \sum_{k=1}^{p} \beta_{(j)k}\,x_k, \quad j=1,\ldots,M. \tag{3.3}$$

Here, $\boldsymbol{x} = (x_1,\ldots,x_p)^T$ with $x_1 = 1$ if there is an intercept. Note that (3.3) means that *all* the parameters may be potentially modelled as functions of $\boldsymbol{x}$. It can be seen that VGLMs are like GLMs but allow for multiple linear predictors, and they encompass models outside the small confines of the exponential family. Although multivariate exponential family densities have been defined by many, such as (3.1), VGLMs are not confined to these, so as to make them more general.

In (3.2), the quantity $\boldsymbol{\phi}$ is an optional vector of scaling parameters which is included for backward compatibility with GLMs. While GLMs are encumbered by a scale factor, the VGLM philosophy is generally to estimate all parameters by MLE, although it is by no means restricted to such. This tends to make inference simpler.

We saw in Chap. 1 that in the S language, terms such as poly() and bs() will usually generate more than one column of the model matrix $\mathbf{X}$, and hence they represent more than one k value in (3.3). This can be notationally confusing. A simple example is a polynomial term of x2 such as poly(x2, 2, raw = TRUE), which defines variables x_2 and $x_3 = x_2^2$ for the second and third columns of $\mathbf{X}$. Consequently, it should be understood that η_j is called a linear predictor because it is linear with respect to the coefficients $\boldsymbol{\beta}_j$, and not necessarily linear with respect to the x_k.

Let $\boldsymbol{x}_i$ denote the explanatory vector for the ith observation, for $i = 1,\ldots,n$. Then one can write

$$\boldsymbol{\eta}_i \;=\; \boldsymbol{\eta}(\boldsymbol{x}_i) \;=\; \begin{pmatrix} \eta_1(\boldsymbol{x}_i) \\ \vdots \\ \eta_M(\boldsymbol{x}_i) \end{pmatrix} \;=\; \mathbf{B}^T\boldsymbol{x}_i \;=\; \begin{pmatrix} \boldsymbol{\beta}_1^T\boldsymbol{x}_i \\ \vdots \\ \boldsymbol{\beta}_M^T\boldsymbol{x}_i \end{pmatrix} \tag{3.4}$$

$$=\; \begin{pmatrix} \beta_{(1)1} & \cdots & \beta_{(1)p} \\ & \vdots & \\ \beta_{(M)1} & \cdots & \beta_{(M)p} \end{pmatrix} \boldsymbol{x}_i \;=\; \begin{pmatrix} \boldsymbol{\beta}_{(1)} & \cdots & \boldsymbol{\beta}_{(p)} \end{pmatrix} \boldsymbol{x}_i. \tag{3.5}$$

Let

$$\mathbf{X} \;\equiv\; \mathbf{X}_{\mathrm{LM}} \;=\; (\boldsymbol{x}_1,\ldots,\boldsymbol{x}_n)^T \tag{3.6}$$

be the usual $n \times p$ (LM) model matrix, or simply LM matrix, constructed from the (first) argument formula of vglm().

For models of the form (3.2), the log-likelihood can be expressed in the form

$$\ell(\boldsymbol{\beta}) = \sum_{i=1}^{n} w_i \, \ell_i \{\eta_1(\boldsymbol{x}_i), \ldots, \eta_M(\boldsymbol{x}_i)\}, \tag{3.7}$$

where

$$\eta_j = \eta_j(\boldsymbol{x}_i) = \boldsymbol{\beta}_j^T \boldsymbol{x}_i, \quad \boldsymbol{\beta} = (\boldsymbol{\beta}_{(1)}^T, \ldots, \boldsymbol{\beta}_{(p)}^T)^T \quad \text{and} \quad \boldsymbol{\beta}^{\dagger} = (\boldsymbol{\beta}_1^T, \ldots, \boldsymbol{\beta}_M^T)^T. \tag{3.8}$$

The w_i are known fixed positive prior weights.

Applying a Newton-like algorithm for maximizing the likelihood,

$$\boldsymbol{\beta}^{(a)} = \boldsymbol{\beta}^{(a-1)} + \boldsymbol{\mathcal{I}}\left(\boldsymbol{\beta}^{(a-1)}\right)^{-1} \boldsymbol{U}\left(\boldsymbol{\beta}^{(a-1)}\right) \tag{3.9}$$

at iteration a, which can be written in IRLS form as

$$\boldsymbol{\beta}^{(a)} = \left(\sum_{i=1}^{n} \mathbf{X}_i^T \mathbf{W}_i^{(a-1)} \mathbf{X}_i\right)^{-1} \left(\sum_{i=1}^{n} \mathbf{X}_i^T \mathbf{W}_i^{(a-1)} \boldsymbol{z}_i^{(a-1)}\right). \tag{3.10}$$

Here, $\mathbf{X}_i$ is the $M \times Mp$ matrix $\mathbf{X}_i = \boldsymbol{x}_i^T \otimes \mathbf{I}_M$, and $\mathbf{W}_i$ is an $M \times M$ matrix with (j,k)th element

$$(\mathbf{W}_i)_{jk} = -E\left(\frac{\partial^2 \ell_i}{\partial \eta_j \, \partial \eta_k}\right), \tag{3.11}$$

and $\boldsymbol{z}_i$ are the *working responses* (cf. (2.25)) is an M-vector given by

$$\boldsymbol{z}_i^{(a-1)} = \mathbf{X}_i \boldsymbol{\beta}^{(a-1)} + \mathbf{W}_i^{-1(a-1)} \boldsymbol{u}_i^{(a-1)} \tag{3.12}$$

$$= \boldsymbol{\eta}_i^{(a-1)} + \boldsymbol{r}_i^{W(a-1)}, \tag{3.13}$$

[cf. (3.58)], where the *score vector* $\boldsymbol{u}_i$ has jth element

$$(\boldsymbol{u}_i)_j = \frac{\partial \ell_i}{\partial \eta_j}. \tag{3.14}$$

In practice, the *expected information matrices* (EIMs; (3.11)) are used because *all* the working weight matrices need to be positive-definite, not just their sum. This generally holds in a much larger parameter space than for the *observed information matrices* (OIMs)

$$(\mathbf{W}_i)_{jk} = \frac{-\partial^2 \ell_i}{\partial \eta_j \, \partial \eta_k}. \tag{3.15}$$

Using the EIM and OIM corresponds to Fisher scoring and the Newton-Raphson algorithms, respectively. Hence the latter is seldom used within the VGLM framework except when the algorithms coincide. An important step when implementing VGLMs, therefore, is to derive the EIM where possible. In cases where they are intractable, one can try approximating them numerically (Sect. 9.2).

In (3.10), $\boldsymbol{\beta}^{(a)}$ is the solution to the GLS problem

$$z^{(a-1)} \; = \; \mathbf{X}_{\mathrm{VLM}} \, \boldsymbol{\beta}^{(a)} + \boldsymbol{\varepsilon}^{(a-1)}, \tag{3.16}$$

where (suppressing superscript (a) for simplicity)

$$z \; = \; \left(z_1^T, \dots, z_n^T\right)^T, \tag{3.17}$$

$$\mathbf{X}_{\mathrm{VLM}} \; \equiv \; \mathbf{X}_{\mathrm{LM}} \otimes \mathbf{I}_M \; = \; \left(\mathbf{X}_1^T, \dots, \mathbf{X}_n^T\right)^T, \tag{3.18}$$

$$\mathrm{Var}(\boldsymbol{\varepsilon}) \; = \; \mathrm{diag}\left(w_1^{-1}\mathbf{W}_1^{-1}, \dots, w_n^{-1}\mathbf{W}_n^{-1}\right) \; = \; \mathbf{W}^{-1}. \tag{3.19}$$

The VLM matrix, $\mathbf{X}_{\mathrm{VLM}}$, is known informally as the 'big' model matrix, whereas the LM matrix, $\mathbf{X}_{\mathrm{LM}}$, is known as the 'small' one. Tables 8.5, 8.6 describe how these and other quantities may be properly extracted from a fitted VGLM object.

Actually, (3.18) holds only when there are *trivial constraints* (the *constraint matrices* are $\mathbf{H}_k = \mathbf{I}_M$ for all k). With more generality,

$$\mathbf{X}_{\mathrm{VLM}} \; = \; \left((\mathbf{X}\,\boldsymbol{e}_1) \otimes \mathbf{H}_1 \;\middle|\; (\mathbf{X}\,\boldsymbol{e}_2) \otimes \mathbf{H}_2 \;\middle|\; \cdots \;\middle|\; (\mathbf{X}\,\boldsymbol{e}_p) \otimes \mathbf{H}_p\right) \tag{3.20}$$

for any valid constraint matrices $\mathbf{H}_k$ (see Sect. 3.3), so that $\mathbf{X}_{\mathrm{VLM}}$ is $(nM) \times p_{\mathrm{VLM}}$ where $p_{\mathrm{VLM}} = \sum_{k=1}^p \texttt{ncol}(\mathbf{H}_k)$. Lastly, with even more generality, when the $\texttt{xij}$ argument is specified, $\mathbf{X}_{\mathrm{VLM}}$ becomes (3.41) as described in Sect. 3.4. Notationally, sometimes we let $\mathcal{R}_k = \texttt{ncol}(\mathbf{H}_k)$ be the rank of $\mathbf{H}_k$.

At convergence, the estimated variance-covariance matrix is

$$\widehat{\mathrm{Var}}\left(\widehat{\boldsymbol{\beta}^*}\right) \; = \; \widehat{\phi}\left(\mathbf{X}_{\mathrm{VLM}}^T \, \mathbf{W}^{(a)} \, \mathbf{X}_{\mathrm{VLM}}\right)^{-1}, \tag{3.21}$$

where $\boldsymbol{\beta}^*$, enumerated as in (3.30), are all the regression coefficients to be estimated. The square roots of its diagonal elements are the standard errors, etc. For most VGLMs, $\phi \equiv 1$ so it need not be estimated.

3.2.1 Computation †

The quantity minimized at the heart of the VLM is the weighted sum of squares

$$\mathrm{ResSS} \; = \; \sum_{i=1}^n w_i \left\{z_i^{(a-1)} - \boldsymbol{\eta}_i^{(a-1)}\right\}^T \mathbf{W}_i^{(a-1)} \left\{z_i^{(a-1)} - \boldsymbol{\eta}_i^{(a-1)}\right\}. \tag{3.22}$$

Sometimes referred to as a 'residual sum of squares', this quantity is minimized at each IRLS iteration.

How is (3.16) computed? The GLS system of equations is converted to OLS by the standard method of premultiplying both sides by the Cholesky decompositions of the $w_i \mathbf{W}_i$. This has the effect of standardizing the errors and removing the correlation between them. Specifically, let $\mathbf{W} = \mathbf{U}^T \mathbf{U} = \mathrm{diag}(\mathbf{U}_1^T \mathbf{U}_1, \dots, \mathbf{U}_n^T \mathbf{U}_n)$ be such. Then premultiplying (3.16) by $\mathbf{U}^{(a-1)}$ gives

$$z^{**(a-1)} \; = \; \mathbf{X}_{\mathrm{VLM}}^{**(a-1)} \, \boldsymbol{\beta}^{(a)} + \boldsymbol{\varepsilon}^{**(a-1)}, \tag{3.23}$$

Algorithm 3.1 `vglm.fit()`

---

1. Construct $\mathbf{X}_{\mathrm{VLM}}$ from $\mathbf{X}_{\mathrm{LM}}$ and $\mathbf{H}_1, \ldots, \mathbf{H}_p$.
2. Initialize: $a = 1$, `converged = FALSE`, `one.more = TRUE`, and assign $\boldsymbol{\eta}_i^{(0)}$ e.g., from $\boldsymbol{\beta}^{(0)}$ or $\boldsymbol{\mu}^{(0)}$.
3. If necessary, compute $\boldsymbol{\mu}_i^{(0)}$, $\ell^{(0)}$, define $\boldsymbol{\beta}^{(0)}$, etc.
4. Compute the first and second derivatives: $\boldsymbol{u}_i^{(0)}$ and $\mathbf{W}_i^{(0)}$.
5. Compute $\boldsymbol{z}^{(0)} = (\boldsymbol{z}_1^{(0)T}, \ldots, \boldsymbol{z}_n^{(0)T})^T$ where $\boldsymbol{z}_i^{(0)} = \boldsymbol{\eta}_i^{(0)} + \left(\mathbf{W}_i^{(0)}\right)^{-1} \boldsymbol{u}_i^{(0)}$.

6. **While** `one.more`

   (a) Regress $\boldsymbol{z}^{(a-1)}$ against $\mathbf{X}_{\mathrm{VLM}}$ with weights $\mathbf{W}^{(a-1)} = \mathrm{diag}(w_1 \mathbf{W}_1^{(a-1)}, \ldots, w_n \mathbf{W}_n^{(a-1)})$ to obtain $\boldsymbol{\beta}^{(a)}$. Use $\mathbf{X}_{\mathrm{VLM}}^{**(a-1)}$ in (3.23) to do this.

   (b) Assign $\boldsymbol{\eta}^{(a)} = \mathbf{X}_{\mathrm{VLM}} \, \boldsymbol{\beta}^{(a)}$.

   (c) Compute $\boldsymbol{\mu}^{(a)}$, $\ell^{(a)}$, etc. from $\boldsymbol{\eta}^{(a)}$.

   (d) Test for convergence, e.g., `converged` = if $\ell^{(a)} - \ell^{(a-1)} < \varepsilon$ or $\|\boldsymbol{\beta}^{(a)} - \boldsymbol{\beta}^{(a-1)}\|_\infty < \varepsilon$.

   (e) Let `one.more` = $a < $ `maxit` and not `converged`. If `one.more` then $a = a + 1$ and then compute $\boldsymbol{u}_i^{(a-1)}$, $\mathbf{W}_i^{(a-1)}$, and $\boldsymbol{z}_i^{(a-1)} = \boldsymbol{\eta}_i^{(a-1)} + \left(\mathbf{W}_i^{(a-1)}\right)^{-1} \boldsymbol{u}_i^{(a-1)}$.

7. Save quantities such as a, $\boldsymbol{\eta}^{(a)}$, $\boldsymbol{\mu}^{(a)}$, $\boldsymbol{\beta}^{(a)}$ on the object.

---

say, where $\mathrm{Var}(\boldsymbol{\varepsilon}^{**(a-1)}) = \sigma_{**}^2 \, \mathbf{I}_{nM}$. The linearly transformed response $\boldsymbol{z}^{**(a-1)}$ has OLS applied to it, therefore the Gauss-Markov theorem applies and so the GLS estimate is the best linear unbiased estimator (BLUE) for $\boldsymbol{\beta}^{(a)}$.

The OLS normal equations (3.23) are currently computed using orthogonal methods, that are less prone to numerical problems due to ill-conditioned design matrices. Specifically, the QR algorithm is invoked using modified LINPACK subroutines to give stable ordering and rank estimation. Applied to a large $n \times m$ matrix, the QR decomposition costs approximately $2nm^2$ floating point operations (flops), therefore a major component of fitting a VGLM with trivial constraints involves about $2nM(Mp)^2 = 2nM^3p^2$ flops at each IRLS iteration for $\mathbf{X}_{\mathrm{VLM}}^{**(a-1)}$. Also, if $\mathbf{X}_{\mathrm{VLM}}$ is $nM \times Mp$ in dimension, then the storage requirements involves nM^2p doubles. It can be seen that both storage and time costs for fitting a VGLM grows most rapidly with respect to M, followed by p and then n. One way to reduce the storage is to reduce the number of parameters, e.g., through imposing constraints such as intercept-only for some of the parameters θ_j. Another way is to perform a reduced-rank regression (Chap. 5) in the form of an RR-VGLM.

A simplified description of the algorithm, as implemented by `vglm.fit()` (which is called by `vglm()`), is given as Algorithm 3.1. Here are some accompanying miscellaneous notes.

- Here, $\boldsymbol{\mu}$ generically stands for the fitted values of the model, as returned by `fitted()`.
- The initialization step of $\boldsymbol{\eta}$ can be first brought about by an initial $\boldsymbol{\beta}^{(0)}$ (in which case $\boldsymbol{\eta}^{(0)} = \mathbf{X}_{\mathrm{VLM}} \, \boldsymbol{\beta}^{(0)}$), or an initial $\boldsymbol{\mu}^{(0)}$ (in which case $\boldsymbol{\eta}^{(0)} = g(\boldsymbol{\mu}^{(0)})$). These alternatives are reflected in the arguments `coefstart` and `mustart`, respectively. The preferred choice is `etastart` for $\boldsymbol{\eta}^{(0)}$ because only a few VGAM family functions support `mustart` (notably, those for GLMs and a few other $M = 1$ families). More information about these arguments is given in Sect. 8.3.1.
- Regarding the inversion of the working weight matrices, this involves computing the Cholesky decomposition $w_i \mathbf{W}_i = \mathbf{U}_i^T \mathbf{U}_i$ to each working weight matrix,

where $\mathbf{U}_i$ is upper-triangular. Details are given in Sect. A.3.1. The total cost of the Cholesky decompositions is about $nM^3/3$ flops, and the total cost for the substitutions is about $2nM^2$ flops.

- In testing for convergence, if $\ell^{(a)} < \ell^{(a-1)}$, then half-stepping may be performed to obtain an improvement for that step; details are at Sect. 3.5.4.
- The number of iterations required for convergence is a after the algorithm has run, and it appears at the bottom of the summary() output.

3.2.2 Advantages and Disadvantages of IRLS

It is quite amazing that linearly regressing a pseudo-response with working weights that both change from iteration to iteration can lead to the solution of a myriad of nonlinear problems.

IRLS has its merits and demerits. One advantage is that it can be built on existing LS software in a programming environment. Such exists in *every* statistical package, because LS computations are fundamental to statistical regression. Another is that generalized additive extensions can be naturally built on the IRLS algorithm (Chap. 4). Another advantage is that the estimated variance-covariance matrix of the estimate $\widehat{\boldsymbol{\theta}}$ is a natural by-product—this is not so readily available in some others such as the Expectation-Maximization (EM) algorithm. IRLS is conceptually simple and a versatile algorithm that is easy to implement. Also, it would be easy to switch to a robust version by simply reducing large weights. Altogether, there is a natural simplicity behind the entire scheme.

Some disadvantages of IRLS include the relatively large memory requirement to store $\mathbf{X}_{\text{VLM}}$, and at least the first derivatives of the likelihood function are needed. Actually these can be relaxed, for example, algorithms exist for piecewise block computations (e.g., biglm), and derivatives can be approximated by finite differences and/or simulation (see Sect. 9.2.2). Almost all **VGAM** family functions implement Fisher scoring. Regularity conditions (Sects. 3.6.1 and A.1.2.2) must be met, in general, by the assumed model.

3.3 Constraints on the Component Functions

In (3.3), the effect of each x_k on η_j is linear. One can generalize this to

$$\eta_j = \eta_j(\boldsymbol{x}) = \sum_{k=1}^{p} f_{(j)k}(x_k), \quad j = 1, \ldots, M. \tag{3.24}$$

where the *component functions* $f_{(j)k}$ are some (usually smooth) functions to be estimated from the data. Equation (3.24) gives rise to the class of *vector generalized additive models* (VGAMs) and these are detailed in Chap. 4. VGAMs are data-driven rather than model-driven, i.e., they allow the data to speak for themselves rather than having an a priori model imposed on the data. For identifiability, each component function is centred, e.g., informally written $E[f_{(j)k}] = 0$. VGLMs are special cases of VGAMs, with all component functions constrained to be linear: $f_{(j)k}(x_k) = \beta_{(j)k} x_k$.

Additionally, a very important facility is to allow the $f_{(j)k}$ to be constrained, e.g., often one wishes to force some of them to be equal or zero. For example, with a bivariate odds ratio model applied to eye data, one should constrain $\eta_1 = \eta_2$ because of symmetry; then, we say the error structure is exchangeable. Another example is the proportional odds model, where there is a parallelism (or proportional odds) assumption $\beta_{(1)k} = \cdots = \beta_{(M)k}$, for $k = 2, \ldots, p$. Another example is when $g_j(\theta_j) = \beta_{(j)1}$, a constraint known as *intercept-only* for θ_j; this models θ_j as simply as possible by using a scalar parameter.

Writing (3.24) in vector form, with the allowance of a vector of known *offsets*, o, gives linear/additive predictors of the form

$$\begin{aligned}\boldsymbol{\eta} = (\eta_1, \ldots, \eta_M)^T &= \boldsymbol{o} + \boldsymbol{\beta}_{(1)} + \boldsymbol{f}_2(x_2) + \cdots + \boldsymbol{f}_p(x_p) \\ &= \boldsymbol{o} + \mathbf{H}_1 \boldsymbol{\beta}_{(1)}^* + \mathbf{H}_2 \boldsymbol{f}_2^*(x_2) + \cdots + \mathbf{H}_p \boldsymbol{f}_p^*(x_p), \quad (3.25)\end{aligned}$$

where $\boldsymbol{f}_k^* = \left(f_{(1)k}^*(x_k), \ldots, f_{(\mathcal{R}_k)k}^*(x_k)\right)^T$ is a vector containing a (possibly reduced) set of component functions, $\boldsymbol{f}_1^* = \boldsymbol{\beta}_{(1)}^*$ is a vector of unknown intercepts, and the $\mathbf{H}_1, \ldots, \mathbf{H}_p$ are *constraint matrices*. Valid matrices $\mathbf{H}_k$ have the following properties:

 (i) they are $M \times \mathcal{R}_k$ for $1 \leq \mathcal{R}_k \leq M$,
 (ii) they have full column-rank, i.e., $\left(\mathbf{H}_k^T \mathbf{H}_k\right)^{-1}$ exists,
(iii) they are known,
(iv) they are fixed.

Usually the elements of $\mathbf{H}_k$ are 0s and 1s. It will be seen in Chap. 5 that RR-VGLMs relax property (iii) so that some of the $\mathbf{H}_k$ are estimated. Note that starred quantities in (3.25) are unknown and need to be estimated. As with the $\boldsymbol{f}_k$, the $\boldsymbol{f}_k^*$ are centred. The default value of the $M \times 1$ offset vector $\boldsymbol{o}$ is $\boldsymbol{0}$, effectively meaning no offsets at all.

Returning back to VGLMs, note that

$$\boldsymbol{\eta}_i = \boldsymbol{o}_i + \mathbf{B}^T \boldsymbol{x}_i = \boldsymbol{o}_i + \begin{pmatrix} \boldsymbol{\beta}_1^T \boldsymbol{x}_i \\ \vdots \\ \boldsymbol{\beta}_M^T \boldsymbol{x}_i \end{pmatrix} \quad (3.26)$$

where $\boldsymbol{o}_i$ is the optional offset vector (ith row of the offset matrix). With constraints on the functions, we have

$$\boldsymbol{\eta}_i = \boldsymbol{o}_i + x_{i1} \mathbf{H}_1 \boldsymbol{\beta}_{(1)}^* + x_{i2} \mathbf{H}_2 \boldsymbol{\beta}_{(2)}^* + \cdots + x_{ip} \mathbf{H}_p \boldsymbol{\beta}_{(p)}^* \quad (3.27)$$

$$= \boldsymbol{o}_i + \sum_{k=1}^p x_{ik} \mathbf{H}_k \begin{pmatrix} \boldsymbol{\beta}_{(1)k}^* \\ \vdots \\ \boldsymbol{\beta}_{(\mathcal{R}_k)k}^* \end{pmatrix} \quad (3.28)$$

so that

$$\mathbf{B}^T = \left(\mathbf{H}_1 \boldsymbol{\beta}_{(1)}^* \,\middle|\, \mathbf{H}_2 \boldsymbol{\beta}_{(2)}^* \,\middle|\, \cdots \,\middle|\, \mathbf{H}_p \boldsymbol{\beta}_{(p)}^* \right) \quad (3.29)$$

for vector

$$\boldsymbol{\beta}^* = \left(\boldsymbol{\beta}_{(1)}^{*T}, \ldots, \boldsymbol{\beta}_{(p)}^{*T}\right)^T \tag{3.30}$$

to be estimated. Equation (3.29) focuses on the rows of $\mathbf{B}$, whereas (3.5) focuses on the columns.

Here are some simple concrete examples of (3.25), with no offsets for simplicity.

(1) *Exchangeability in a bivariate odds ratio model (1.18)–(1.19):* $\eta_1 = \eta_2$

$$
\begin{aligned}
\eta_1 &= \beta_{(1)1}^* + f_{(1)2}^*(x_2) + f_{(1)3}^*(x_3),\\
\eta_2 &= \beta_{(1)1}^* + f_{(1)2}^*(x_2) + f_{(1)3}^*(x_3),\\
\eta_3 &= \beta_{(2)1}^* + f_{(2)2}^*(x_2) + f_{(2)3}^*(x_3).
\end{aligned}
$$

Then

$$\mathbf{H}_1 = \mathbf{H}_2 = \mathbf{H}_3 = \begin{pmatrix} 1 & 0 \\ 1 & 0 \\ 0 & 1 \end{pmatrix}.$$

(2) *Odds ratio is intercept-only in an exchangeable bivariate odds ratio model*

$$
\begin{aligned}
\eta_1 &= \beta_{(1)1}^* + f_{(1)2}^*(x_2) + f_{(1)3}^*(x_3),\\
\eta_2 &= \beta_{(1)1}^* + f_{(1)2}^*(x_2) + f_{(1)3}^*(x_3),\\
\eta_3 &= \beta_{(2)1}^*.
\end{aligned}
$$

Then

$$\mathbf{H}_1 = \begin{pmatrix} 1 & 0 \\ 1 & 0 \\ 0 & 1 \end{pmatrix}, \qquad \mathbf{H}_2 = \mathbf{H}_3 = \begin{pmatrix} 1 \\ 1 \\ 0 \end{pmatrix}. \tag{3.31}$$

(3) *The η_j differ by scalars; e.g., parallelism in the proportional odds model (1.23)*

$$
\begin{aligned}
\eta_1 &= \beta_{(1)1}^* + f_{(1)2}^*(x_2) + f_{(1)3}^*(x_3),\\
\eta_2 &= \beta_{(2)1}^* + f_{(1)2}^*(x_2) + f_{(1)3}^*(x_3),\\
\eta_3 &= \beta_{(3)1}^* + f_{(1)2}^*(x_2) + f_{(1)3}^*(x_3).
\end{aligned}
$$

Then

$$\mathbf{H}_1 = \mathbf{I}_3, \qquad \mathbf{H}_2 = \mathbf{H}_3 = \begin{pmatrix} 1 \\ 1 \\ 1 \end{pmatrix}.$$

(4) *Negative binomial mimicking a quasi-Poisson model ((1.13), (1.15), (1.16))*
Consider an NB regression with $\eta_1 = \log\mu$ and $\eta_2 = \log k$. Stipulating that $\mathbf{H}_1 = \mathbf{I}_2$ and $\mathbf{H}_2 = \cdots = \mathbf{H}_p = (1,1)^T$ means that $\mathrm{Var}(Y) = \mu(1+\mu/k) = \mu(1 + \exp\{\beta_{(1)1}^* - \beta_{(2)1}^*\})$ $(= \phi\,\mu$, say), where ϕ is estimated by MLE, and it has associated confidence intervals and SEs. This is known as an NB-1 model. Other NB variants are described in Sect. 11.3.

3.3.1 Fitting Constrained Models in VGAM

There are two ways of fitting VGLMs/VGAMs with constraints in VGAM. The first is through family function-specific arguments like `parallel`, `exchangeable` and `zero`—this is the most convenient method, but it provides only limited flexibility. The second is through the `constraints` argument of `vglm()`/`vgam()`, which gives full flexibility but at the expense of having to set up each individual $\mathbf{H}_k$. Casual users of VGAM should be able to manage with the first method. Most family functions with $M > 1$ η_js have a `zero` argument, and many others have `parallel` and `exchangeable`, etc.; these arguments provide a convenient way of enforcing a useful type of constraint for that particular model.

Arguments such as `parallel`, `exchangeable` and `zero` are described below, and readers are directed to Chap. 8 for further details. Users wishing to constrain component functions in more elaborate ways need to understand the underlying details given in Sect. 3.3.1.3, and they should note that arguments such as `parallel`, `exchangeable` and `zero` merely provide a convenient short-cut for the method described in that section.

The default value of the arguments `parallel` and `exchangeable` is the logical `FALSE`, and if set to `TRUE`, then it may or may not apply to the intercepts—that depends on that particular family. For example, setting `cumulative(parallel = TRUE)` never affects the intercepts in a cumulative logit model, whereas setting `binom2.or(exchangeable = TRUE)` does.

3.3.1.1 The `zero` Argument

The `zero` argument specifies which linear predictors are to be modelled with an intercept term only ("intercept-only"), for example,

```
zfit <- vglm(cbind(y1, y2) ~ x2,
             binom2.or(zero = 3, exchangeable = TRUE), data = bdata)
```

fits the bivariate odds ratio model

$$
\begin{aligned}
\text{logit } P(Y_1 = 1) &= \eta_1 &= \beta^*_{(1)1} + \beta^*_{(1)2}\, x_2, \\
\text{logit } P(Y_2 = 1) &= \eta_2 &= \beta^*_{(1)1} + \beta^*_{(1)2}\, x_2, \\
\log \psi &= \eta_3 &= \beta^*_{(2)1},
\end{aligned}
\tag{3.32}
$$

i.e., with an intercept-only log odds ratio. Thus the estimated odds ratio is simply the scalar $\widehat{\psi} = \exp\{\widehat{\beta}^*_{(2)1}\}$.

The argument `zero` is often applied to models where at least one of the parameters has good reason to be treated as an unknown scalar than to be modelled with respect to the covariates. Parameters such as shape parameters, scale parameters, odds ratios, correlation coefficients or powers are common examples. Indeed, most of the family functions described in Chaps. 11–13 implement this idea as the default. In general, `zero` can be assigned a vector whose (integer) values lie between 1 and M inclusive. Assigning negative values are also allowed for multiple responses (Sect. 3.5.1).

For families which are prone to numerical problems, judicious use of the `zero` argument is recommended, at least in the early phases of the model-building process, and at least for initial-value generation for more complex models (see the stepping-stone model of Sect. 8.5).

A `NULL` value assigned to `zero` indicates that no η_j are intercept-only. This value should be set if the `constraints` argument is used. If `zero` has a value j, then it means that the jth row of all the $\mathbf{H}_k$s are all 0s, except for $\mathbf{H}_1$. Consequently, $\eta_j = \beta_{(j)1}$ is intercept-only.

3.3.1.2 Arguments Such as `parallel` and `exchangeable`

The `parallel` and `exchangeable` arguments differ from `zero` in that they may be assigned `TRUE` or `FALSE`, or else a basic[1] S formula consisting of a logical as response and additive simple terms as explanatory variables. These arguments always have `FALSE` as the default, and their syntax is best illustrated by examples. Recall that, by default, an intercept term is included in an S formula, and that it can be dropped by adding a `0` or a `-1` term.

- `parallel = TRUE` means the parallelism assumption is applied to all terms (or equivalently, to all columns of the model matrix), except possibly the intercept. Each **VGAM** family function decides whether the constraint is applied to the intercept or not—and for some families it is very clear whether it is or not.
- `parallel = TRUE ~ x2 + x5` means that the parallelism assumption is only applied to x2, x5 and the intercept.
- `parallel = TRUE ~ -1` and `parallel = TRUE ~ 0` both means that the parallelism assumption is applied to *no* variables at all. Similarly, `parallel = FALSE ~ -1` and `parallel = FALSE ~ 0` means that the parallelism assumption is applied to *all* the variables, including the intercept.
- `exchangeable = FALSE ~ x2 - 1` applies the exchangeability constraint to all terms (including the intercept) except for x2.
- `exchangeable = TRUE ~ s(x2, df = 2) - 1` applies the exchangeability constraint only to the smooth term `s(x2, df = 2)`. Note that one doesn't need to worry about white spaces, etc. when typing in the terms to match (but you *do* have to worry about white spaces when using the `constraints` argument; Sect. 3.3.1.3); nevertheless, the term must be syntactically identical. Typing `exchangeable = TRUE ~ s(x2) - 1` would have no effect.

It can be seen therefore that special care must be taken regarding the intercept term when a constraints argument is assigned a formula. The user in such a case must specify explicitly whether or not the constraint applies to the intercept term.

Note there is room for contradiction between arguments and within arguments, e.g., `binom2.or(zero = 2:3, exchangeable = TRUE)` is ambiguous with covariates. Currently, there is very little safety by way of warnings to safeguard against

---

[1] In being simple, the formula cannot handle terms such as `.` and `-x2`, nor interactions and nested terms, etc.

this, and the results are unpredictable. Users should try to keep things simple and/or use the `constraints` argument (see Sect. 3.3.1.3). Upon fitting, users can use the `constraints()` extractor function to double-check their results (Table 8.5).

3.3.1.3 The `constraints` Argument

The constraint matrices $\mathbf{H}_1, \ldots, \mathbf{H}_p$ may be inputted into `vglm()`/`vgam()` using the `constraints` argument. It is assigned a `list` containing *all* the $\mathbf{H}_k$, or functions that create them. Each of these must be carefully named with the variable name or term name. For example, the equivalent of `zfit` above can be obtained from

```
H1 <- matrix(c(1, 1, 0, 0, 0, 1), 3, 2)
H2 <- H1[, 1, drop = FALSE]
zfit2 <- vglm(cbind(y1, y2) ~ x2, binom2.or(zero = NULL), data = bdata,
              constraints = list("(Intercept)" = H1,
                                 x2           = H2))
```

If the family function has a `zero` argument, then it needs to be set to `NULL` because otherwise there is a high likelihood of contradiction between the arguments. If any explanatory variable is a factor, then only the name of the factor (not any of its levels) needs to appear in `constraints`.

When using the `constraints` argument with functions such as `s()`, `bs()` etc., one must be very careful to get any white spaces right. For example,

```
Hlist <- list("(Intercept)"  = diag(3),
              "poly(x2,3)"    = matrix(1, 3, 1),  # Incorrect spacings!
              "bs(x3,df = 3)" = matrix(1, 3, 1))  # Incorrect spacings!
vglm(y ~ poly(x2,3) + bs(x3,df = 3), ..., constraints = Hlist)
```

will fail, but it will succeed with

```
Hlist <- list("(Intercept)"   = diag(3),
              "poly(x2, 3)"    = matrix(1, 3, 1),  # Correct spacings!
              "bs(x3, df = 3)" = matrix(1, 3, 1))  # Correct spacings!
```

If unsure, apply something like

```
> as.character( ~ poly(x2,3))

  [1] "~"              "poly(x2, 3)"

> as.character( ~ bs(x3, df=3))

  [1] "~"              "bs(x3, df = 3)"
```

to each term separately to determine how white spaces are apportioned. The output might then be copy-and-pasted. At present, it is necessary to get it exactly right, because the names of the `constraints` list are matched with the character representation of each term in the formula.

3.3.1.4 Simple Examples from the Normal Distribution

Given a random sample $Y_i \sim N(\mu = \theta, \ \sigma^2 = \theta^2)$, how might θ be estimated? One way is to use `uninormal()`, which has default $\eta_1 = \mu$ and $\eta_2 = \log \sigma$, and then use the identity link function for σ and set $\mathbf{H}_1 = (1,1)^T$. Suppose $\theta = 10$ and $n = 100$. Then

```
> theta <- 10; n <- 100; set.seed(123)
> udata <- data.frame(y1 = rnorm(n, mean = theta, sd = theta))
> Hlist <- list("(Intercept)" = rbind(1, 1))
> fit1a <- vglm(y1 ~ 1, uninormal(lsd = "identitylink"), udata, constraints = Hlist)
> fit1b <- vglm(y1 ~ 1, data = udata,
              uninormal(lsd = "identitylink", parallel = TRUE ~ 1))
> coef(fit1a, matrix = TRUE)

                mean     sd
  (Intercept) 9.7504 9.7504
```

are two equivalent models.

Consider a similar problem but from $N(\mu = \theta, \ \sigma^2 = \theta)$. Now $\log \mu = \log \theta = 2 \log \sigma$ so that

```
> udata <- data.frame(y2 = rnorm(n, mean = theta, sd = sqrt(theta)))
> Hlist2 <- list("(Intercept)" = rbind(1, 0.5))
> fit2a <- vglm(y2 ~ 1, uninormal(lmean = "loge"), udata, constraints = Hlist2)
> fit2b <- vglm(y2 ~ 1, uninormal(var = TRUE, lvar = "identitylink",
                        parallel = TRUE ~ 1), data = udata)
> (cfit2a <- coef(fit2a, matrix = TRUE))

               loge(mean) loge(sd)
  (Intercept)     2.2844   1.1422

> loge(cfit2a[1, "loge(mean)"], inverse = TRUE)  # Estimated mean

  [1] 9.8197
```

are two ways of fitting this model. Here, `fit2a` is based on $\theta = \sigma$, while `fit2b` estimates $\theta = \sigma^2$. It is left to the reader as an exercise to estimate θ from a random sample from $N(\mu = \theta, \ \sigma^2 = \sqrt{\theta})$ (Ex. 3.8).

3.3.1.5 The s() Term and Constraints

Jumping ahead a little to fit VGAMs, the specification of arguments `df` and `spar` in `s()` under `zero`, `parallel` and `exchangeable`, etc. is simple: the successive values of `df` correspond to successive columns of the constraint matrix of that variable. If the length of `df` is less than the number of columns of the constraint matrix, then values are recycled. This recycling also applies similarly to `spar`, e.g.,

```
vgam(cbind(y1, y2) ~ s(x2, df = c(4, 1)),
     binom2.or(exchangeable = TRUE, zero = NULL), data = bdata)
```

fits the model

$$\text{logit } P(Y_j = 1) = \beta^*_{(1)1} + f^*_{(1)2}(x_2), \qquad j = 1, 2,$$
$$\log \psi(\boldsymbol{x}) = \beta^*_{(2)1} + \beta^*_{(2)2} \, x_2,$$

where $f^*_{(1)2}$ has 4 degrees of freedom (1 degree of freedom denotes a linear fit).

For VGAMs fitted with regression splines, it is noted that `bs()` and `ns()` cannot be assigned a vector of `df` values, hence one term is needed for each component function having a different degree of freedom. For example,

```
vglm(y ~ ns(x2, df=4) + ns(x2, df=3), negbinomial(zero = NULL), data = ndata,
     constraints = list("(Intercept)"    = diag(2),
                        "ns(x2, df = 4)" = rbind(1, 0),
                        "ns(x2, df = 3)" = rbind(0, 1)))
```

fits the negative binomial additive model

$$\log \mu = \beta^*_{(1)1} + f^*_{(1)2}(x_2),$$
$$\log k = \beta^*_{(2)1} + f^*_{(2)2}(x_2),$$

where $f^*_{(1)2}$ has 4 degrees of freedom, and $f^*_{(2)2}$ has 3 degrees of freedom.

To round-up this section, as it can be seen, the `constraints` argument is powerful and flexible, albeit a little cumbersome. Users are encouraged to use `zero/parallel/exchangeable` etc. where possible, but to be aware of possible conflicts.

3.4 The `xij` Argument

Initially, we had (3.3), viz.

$$\eta_j(\boldsymbol{x}_i) = \boldsymbol{\beta}_j^T \boldsymbol{x}_i = \sum_{k=1}^p x_{ik} \beta_{(j)k}, \quad j = 1, \ldots, M, \tag{3.33}$$

as the jth linear predictor. Importantly, this can be generalized to

$$\eta_j(\boldsymbol{x}_{ij}) = \boldsymbol{\beta}_j^T \boldsymbol{x}_{ij} = \sum_{k=1}^p \beta_{(j)k} x_{ikj}, \tag{3.34}$$

where x_{ikj} is the 'jth value' of variable x_k for η_j and observation i. Writing this another way,

$$\eta_j(\boldsymbol{x}_i^*, \boldsymbol{x}_{ij}^*) = \boldsymbol{\beta}_j^{*T} \boldsymbol{x}_i^* + \boldsymbol{\beta}_j^{**T} \boldsymbol{x}_{ij}^*. \tag{3.35}$$

Usually $\boldsymbol{\beta}_j^{**} = \boldsymbol{\beta}^{**}$, say. In (3.35), the variables in $\boldsymbol{x}_i^*$ each have the same value for all η_j, and the variables in $\boldsymbol{x}_{ij}^*$ have different values for differing η_j. This allows for covariate values that are specific to each η_j, a facility which is very important in many applications. Here are a few simple examples.

1. Suppose that two binary responses, $Y_j = 1$ or 0, measure the presence/absence of a disease in the jth eye, where $j = 1, 2$ for the left and right eye, respectively. There is a single covariate, called *intraocular ocular pressure* (IOP), which measures the internal fluid pressure within each eye. With data from n people, it would be natural to fit an exchangeable bivariate odds ratio model:

$$\text{logit } P(Y_{ij} = 1) = \eta_j = \beta^*_{(1)1} + \beta^*_{(1)2} x_{i2j}, \quad j = 1, 2; \quad i = 1, \ldots, n;$$
$$\log \psi = \eta_3 = \beta^*_{(2)1}. \tag{3.36}$$

Note that the regression coefficient for x_{i21} and x_{i22} is the same, and $x_{i21} \neq x_{i22}$ in general, because each person's eyes may have different intraocular pressures. The constraint matrices $\mathbf{H}_1$ and $\mathbf{H}_2$ are the same as (3.31) and (3.32).

2. The $\mathcal{M}_{tbh}$ capture-recapture models fitted in Sects. 17.2.4 and 17.2.6 include a time-varying covariate. For example, to model the effect on a capture at time j from memory effects resulting from previous captures, a capture history variable can be fed into the model in the form (3.34), because the η_j are largely functions of the jth sampling occasion.

3. Suppose that an econometrician is interested in peoples' choice of transport between two cities, and that there are four choices: $Y = 1$ for "bus", $Y = 2$ "car", $Y = 3$ "train" and $Y = 4$ for "plane". Assume that people only choose one of these means. Suppose that there are three covariates: $X_2 = $ cost, $X_3 = $ journey time, and $X_4 = $ the person's income. Of the covariates, only X_4 (and the intercept X_1) are the same for all transport choices; the cost and journey time differ according to the means chosen. Suppose that a random sample of n people is collected from some population, and that each person has access to all these transport modes.[2] For such data, a natural regression model would be a multinomial logit model with $M = 3$: for $j = 1, \ldots, M$,

$$
\begin{aligned}
\eta_j &= \log \frac{P(Y = j)}{P(Y = M + 1)} \\
&= \beta_{(j)1}^* + \beta_{(1)2}^* \left(x_{i2j} - x_{i24} \right) + \beta_{(1)3}^* \left(x_{i3j} - x_{i34} \right) + \beta_{(1)4}^* \, x_{i4}, \quad (3.37)
\end{aligned}
$$

where, for the ith person, x_{i2j} is the cost for the jth transport means, and x_{i3j} is the journey time of the jth transport means. The income variable is x_{i4}; it has the same value regardless of the transport means.

Equation (3.37) implies $\mathbf{H}_1 = \mathbf{I}_3$ and $\mathbf{H}_2 = \mathbf{H}_3 = \mathbf{H}_4 = \mathbf{1}_3$. Note also that if the last response category is used as the baseline or reference group (the default of `multinomial()`), then $x_{ik,M+1}$ can be subtracted from x_{ikj} for $j = 1, \ldots, M$—this is the natural way that $x_{ik,M+1}$ enters into the model. An example based on this scenario is given in Sect. 14.2.1.

The use of the `xij` argument with the **VGAM** family function `multinomial()` has a very important application in economics with *consumer choice* or *discrete choice* modelling. In that field, the term "multinomial logit model" includes a variety of models such as the "generalized logit model" where (3.33) holds, the "conditional logit model" where (3.34) holds, and the "mixed logit model", which is a combination of the two, where (3.35) holds. The generalized logit model focuses on the individual as the unit of analysis, and it uses individual characteristics as explanatory variables, e.g., age of the person in the transport example. The conditional logit model assumes different values for each alternative, and the impact of a unit of x_k is assumed to be constant across alternatives, e.g., journey time in the choice of transport mode. The conditional logit model was proposed for econometrics by McFadden (1974), and it has been used in many fields including biomedical research to estimate relative risks in matched case-control studies. Unfortunately, there is confusion in the literature for the terminology of the models.

---

[2] If not, then this is known as a "varying choice set" in the discrete-choice model literature. This presently is outside the VGLM/VGAM framework.

Some authors call `multinomial()` with (3.33) the "generalized logit model", while others call the mixed logit model the "multinomial logit model", and they view the generalized logit and conditional logit models as special cases. In VGAM terminology, there is no need to give different names to all these slightly-different special cases. They can all be called multinomial logit models, although it may be added that some have covariate-specific linear/additive predictors. The important thing is that the framework accommodates $\boldsymbol{x}_{ij}$, so one tries to avoid making life unnecessarily complicated. And `xij` can apply in theory to any VGLM and not just to the multinomial logit model.

3.4.1 The Central Formula

VGAM handles an explanatory variable taking different values for each η_j, (3.35), using the `xij` argument. It is assigned an S formula, or a list of S formulas. Each formula, which must have M *different* terms on the RHS, forms the diagonal of a diagonal matrix that premultiplies a constraint matrix. In detail, recall that (3.33) can accommodate constraint matrices, and be written

$$\boldsymbol{\eta}(\boldsymbol{x}_i) \; = \; \mathbf{B}^T \boldsymbol{x}_i \; = \; \sum_{k=1}^{p} \mathbf{H}_k \, \boldsymbol{\beta}_{(k)}^* \, x_{ik}, \tag{3.38}$$

where the $\boldsymbol{\beta}_{(k)}^* = (\beta_{(1)k}^*, \ldots, \beta_{(\mathcal{R}_k)k}^*)^T$ are to be estimated. This may be written [cf. (3.27)]

$$\boldsymbol{\eta}(\boldsymbol{x}_i) \; = \; \sum_{k=1}^{p} \operatorname{diag}(x_{ik}, \ldots, x_{ik}) \, \mathbf{H}_k \, \boldsymbol{\beta}_{(k)}^*. \tag{3.39}$$

To handle (3.34)–(3.35), we can generalize (3.39) to

$$\boldsymbol{\eta}_i \; = \; \boldsymbol{o}_i + \sum_{k=1}^{p} \operatorname{diag}(x_{ik1}, \ldots, x_{ikM}) \, \mathbf{H}_k \, \boldsymbol{\beta}_{(k)}^* \tag{3.40}$$

$$= \; \boldsymbol{o}_i + \sum_{k=1}^{p} \mathbf{X}_{(ik)}^{\#} \, \mathbf{H}_k \, \boldsymbol{\beta}_{(k)}^*, \; \text{say},$$

with provision for offsets $\boldsymbol{o}_i$. This is the central formula for the `xij` facility and the most general for VGLMs. Then the big model matrix has the block form (cf. (3.20))

$$\mathbf{X}_{\text{VLM}} \; = \; \begin{pmatrix} \mathbf{X}_{(11)}^{\#} \mathbf{H}_1 & \cdots & \mathbf{X}_{(1p)}^{\#} \mathbf{H}_p \\ \vdots & & \vdots \\ \mathbf{X}_{(n1)}^{\#} \mathbf{H}_1 & \cdots & \mathbf{X}_{(np)}^{\#} \mathbf{H}_p \end{pmatrix}. \tag{3.41}$$

To summarize, for VGLMs, the fundamental computation at each IRLS iteration is the fitting of a VLM to the $\boldsymbol{z}_i$ upon $\mathbf{X}_{\text{VLM}}$ with working weights $\mathbf{W}_i$. The model matrix $\mathbf{X}_{\text{VLM}}$ incorporates the constraint matrices $\mathbf{H}_k$ and $\boldsymbol{x}_{ij}$. The VLM for GLMs is a simple weighted least squares regression.

3.4.2 Using the xij *Argument in* VGAM

To fit such models in **VGAM**, one needs to use the `xij` and `form2` arguments. Each component of the argument `xij` list is a formula having M *different* terms (ignoring the intercept) which specifies the *successive* diagonal elements of the matrix $\mathbf{X}^{\#}_{(ik)}$ in (3.40). By "*different*", this guarantees there will be M terms. The constraint matrices themselves are not affected by the `xij` argument.

Here are two examples revisited.

1. ```
   fit.eyes <- vglm(formula = cbind(leye, reye) ~ iop,
                    data = eyesData,
                    family = binom2.or(exchangeable = TRUE, zero = 3),
                    xij = list(iop ~ liop + riop + fill1(liop)),
                    form2 = ~ iop + liop + riop + fill1(liop))
   ```

   Here, `liop` and `riop` are the intraocular pressures of the left and right eyes. The specific values of the vector `iop` are not needed (unless plotted—see Sect. 3.4.4.1) because they are overwritten by `liop` and `riop` when forming $\mathbf{X}_{\mathrm{VLM}}$. One could call `iop` a "dummy" vector since its purpose is only for labelling; however, it is not a dummy variable! The function `fill1()` makes the number of terms equal to 3 ($= M$ for `binom2.or()`), and the value it returns is a structure of 0s the same dimension as `liop`—here it is just a vector. Each response term in the formulas in `xij` connects with the same term in `formula`, so essentially it is for labelling purposes only (here, "iop"). One can see this labelling by using `model.matrix(fit.eyes, type = "vlm")` to get $\mathbf{X}_{\mathrm{VLM}}$; there will be a column called "iop". The argument `form2` should contain *all* terms used in arguments `formula` and `xij`, except possibly the response; it creates an all-encompassing LM matrix from which columns are extracted (this matrix is called $\mathbf{X}_{\mathbf{form2}}$).

   The desired model is

   $$\begin{aligned}
   \mathrm{logit}\, P(Y_{i1} = 1) &= \beta^*_{(1)1} + \beta^*_{(1)2}\, x_{i21}, \\
   \mathrm{logit}\, P(Y_{i2} = 1) &= \beta^*_{(1)1} + \beta^*_{(1)2}\, x_{i22}, \\
   \log \psi &= \beta^*_{(2)1}.
   \end{aligned}$$

   By the way,

   ```
   bad.eyes1 <- vglm(cbind(leye, reye) ~ liop + riop, data = eyesData,
                     binom2.or(exchangeable = TRUE, zero = 3))
   ```

   would result in the model

   $$\begin{aligned}
   \mathrm{logit}\, P(Y_{i1} = 1) &= \beta^*_{(1)1} + \beta^*_{(1)2}\, x_{i21} + \beta^*_{(1)3}\, x_{i22}, \\
   \mathrm{logit}\, P(Y_{i2} = 1) &= \beta^*_{(1)1} + \beta^*_{(1)2}\, x_{i21} + \beta^*_{(1)3}\, x_{i22}, \\
   \log \psi &= \beta^*_{(2)1},
   \end{aligned} \tag{3.42}$$
   ```

which is incorrect. Another similar model is

$$
\begin{aligned}
\operatorname{logit} P(Y_{i1} = 1) &= \beta^*_{(1)1} + \beta^*_{(1)2}\, x_{i21}, \\
\operatorname{logit} P(Y_{i2} = 1) &= \beta^*_{(1)1} + \beta^*_{(1)3}\, x_{i22}, \\
\log \psi &= \beta^*_{(2)1},
\end{aligned}
$$

which can be fitted with

```
bad.eyes2 <- vglm(cbind(leye, reye) ~ liop + riop, binom2.or, data = eyesData,
             constraints = list("(Intercept)" = matrix(c(1,1,0, 0,0,1), 3, 2)
                          liop = rbind(1, 0, 0),
                          riop = rbind(0, 1, 0)))
```

This differs from (3.36) because it allows for a different regression coefficient for each eye, i.e., the effect of intraocular pressure on each eye is different. Such a model is not exchangeable. For this reason, it too is incorrect.

2. Let's fit (3.37). Suppose the journey cost and time variables have had the cost and time by plane subtracted from them. Then, using ".trn" to denote train,

```
gfit <- vglm(cbind(bus, car, train, plane) ~ Cost + Time + Income,
         family = multinomial(parallel = TRUE ~ Cost + Time + Income - 1),
         xij = list(Cost ~ Cost.bus + Cost.car + Cost.trn,
                 Time ~ Time.bus + Time.car + Time.trn),
         form2 = ~ Cost.bus + Cost.car + Cost.trn +
                 Time.bus + Time.car + Time.trn +
                 Cost + Time + Income, data = gotowork)
```

should do the job. It has $\mathbf{H}_1 = \mathbf{I}_3$ and $\mathbf{H}_2 = \mathbf{H}_3 = \mathbf{H}_4 = \mathbf{1}_3$ because the lack of parallelism only applies to the intercept. However, unless `Cost` is the same as `Cost.bus` and `Time` is the same as `Time.bus`, this model should not be plotted with the `"vgam"` methods function for `plot()`; see Sect. 3.4.4.1 for details.

Incidentally, it can be argued that $\beta^*_{(1)4}$ in (3.37) is better replaced by $\beta^*_{(j)4}$. Then the above code, but with one line replaced by

```
family = multinomial(parallel = FALSE ~ 1 + Income),
```

should fit this model. Equivalently,

```
family = multinomial(parallel = TRUE ~ Cost + Time - 1),
```

3.4.3 More Complicated Examples

The `xij` facility allows a lot of flexibility among all the regression coefficients with respect to linear constraints. The following are some further examples. To exploit the full flexibility, sometimes some manipulation is required. We assume below that the data frame `adata` contains the variables `x1` and `x2` from the outset. Recall that the crucial equation is (3.40).

Example 1 Suppose that

$$
\eta = \beta_1 x_1 + \beta_2 x_2.
$$

How can one fit this subject to $\beta_1 + \beta_2 = \delta$, say, where δ is known? The answer is to let

$$
\begin{aligned}
\eta &= \beta_1 x_1 + (\delta - \beta_1) \, x_2 \\
&= \delta x_2 + \beta_1 (x_1 - x_2),
\end{aligned}
$$

i.e., use an offset δx_2 and regress on x_3, where $x_3 = x_1 - x_2$. Use code of the form

```
adata <- transform(adata, x3 = x1 - x2)
vglm(y ~ offset(delta * x2) - 1 + x3, aVGAMfamilyFunction, data = adata)
```

or alternatively,

```
vglm(y ~ 0 + I(x1 - x2), aVGAMfamilyFunction, offset = delta * x2, data = adata)
```

Example 2 Suppose that

$$
\begin{aligned}
\eta_1 &= \beta_{(1)1} \, x_1 + \beta_{(1)2} \, x_2, & (3.43) \\
\eta_2 &= \beta_{(2)1} \, x_1 + \beta_{(2)2} \, x_2. & (3.44)
\end{aligned}
$$

How can one fit (3.43)–(3.44) subject to $\beta_{(1)1} + \beta_{(1)2} = \beta_{(2)1} + \beta_{(2)2}$? The answer is to let

$$
\begin{aligned}
\eta_2 &= \beta_{(2)1} \, x_1 + \left(\beta_{(1)1} + \beta_{(1)2} - \beta_{(2)1} \right) x_2 \\
&= \beta_{(1)1} \, x_2 + \beta_{(1)2} \, x_2 + \beta_{(2)1} (x_1 - x_2).
\end{aligned}
$$

Letting $x_3 = x_1 - x_2$, then

$$
\begin{aligned}
\eta &= \begin{pmatrix} x_1 & 0 \\ 0 & x_2 \end{pmatrix} \begin{pmatrix} 1 \\ 1 \end{pmatrix} \beta_{(1)1} + x_2 \begin{pmatrix} 1 \\ 1 \end{pmatrix} \beta_{(1)2} + x_3 \begin{pmatrix} 0 \\ 1 \end{pmatrix} \beta_{(2)1} \\
&= \begin{pmatrix} x_1 & 0 \\ 0 & x_2 \end{pmatrix} \begin{pmatrix} 1 \\ 1 \end{pmatrix} \beta_{(1)1}^* + (x_2 \, \mathbf{I}_2) \begin{pmatrix} 1 \\ 1 \end{pmatrix} \beta_{(1)2}^* + (x_3 \, \mathbf{I}_2) \begin{pmatrix} 0 \\ 1 \end{pmatrix} \beta_{(1)3}^*.
\end{aligned}
$$

Use code of the form

```
adata <- transform(adata, X1 = x1, x3 = x1 - x2)
Hlist <- list(X1 = rbind(1, 1), x2 = rbind(1, 1), x3 = rbind(0, 1))
fit2 <- vglm(y ~ -1 + X1 + x2 + x3, aVGAMfamilyFunction, data = adata,
             constraints = Hlist, xij = list(X1 ~ -1 + x1 + x2),
             form2 = ~ -1 + x1 + x2 + x3 + X1)
```

Here, the value of X1 is ignored, therefore it could be assigned any numerical vector of the appropriate length. Alternatively, one could use `form2 = Select(adata, TRUE, as.formula = TRUE)`, which chooses every variable in `adata` to be on the RHS of an S formula—this is overkill but it would work.

Example 3 How can one fit (3.43)–(3.44) subject to $\beta_{(1)1} + \beta_{(2)2} = \beta_{(2)1} + \beta_{(1)2}$? The answer is to let

$$
\begin{aligned}
\eta_2 &= \beta_{(2)1} \, x_1 + \left(\beta_{(2)1} + \beta_{(1)2} - \beta_{(1)1} \right) x_2 \\
&= \beta_{(2)1} (x_1 + x_2) + \beta_{(1)2} \, x_2 - \beta_{(1)1} x_2.
\end{aligned}
$$

Write this as

$$
\boldsymbol{\eta} = \begin{pmatrix} x_1 & 0 \\ 0 & x_2 \end{pmatrix} \begin{pmatrix} 1 \\ -1 \end{pmatrix} \beta_{(1)1} + x_2 \begin{pmatrix} 1 \\ 1 \end{pmatrix} \beta_{(1)2} + x_3 \begin{pmatrix} 0 \\ 1 \end{pmatrix} \beta_{(2)1}
$$

$$
= \begin{pmatrix} x_1 & 0 \\ 0 & x_2 \end{pmatrix} \begin{pmatrix} 1 \\ -1 \end{pmatrix} \beta_{(1)1}^* + (x_2\,\mathbf{I}_2) \begin{pmatrix} 1 \\ 1 \end{pmatrix} \beta_{(1)2}^* + (x_3\,\mathbf{I}_2) \begin{pmatrix} 0 \\ 1 \end{pmatrix} \beta_{(1)3}^*
$$

where $x_3 = x_1 + x_2$, say. This can be fitted by code of the form

```
adata <- transform(adata, X1 = x1, x3 = x1 + x2)
Hlist <- list(X1 = rbind(1, -1), x2 = rbind(1, 1), x3 = rbind(0, 1))
fit3 <- vglm(y ~ -1 + X1 + x2 + x3, aVGAMfamilyFunction, data = adata,
             xij = list(X1 ~ -1 + x1 + x2), constraints = Hlist,
             form2 = ~ -1 + x1 + x2 + x3 + X1)
```

Example 4   Suppose that

$$
\eta_1 = \beta_{(1)1}\,x_1 + \beta_{(1)2}\,x_2, \tag{3.45}
$$
$$
\eta_2 = \beta_{(2)1}\,x_1 + \beta_{(2)2}\,x_2, \tag{3.46}
$$
$$
\eta_3 = \beta_{(3)1}\,x_1 + \beta_{(3)2}\,x_2. \tag{3.47}
$$

How can one fit this subject to $\beta_{(1)1} + \beta_{(2)1} + \beta_{(3)1} = \delta$, where δ is known? The answer is to let

$$
\eta_3 = \left(\delta - \beta_{(1)1} - \beta_{(2)1} \right) x_1 + \beta_{(3)2}\,x_2
$$
$$
= \delta\,x_1 + \beta_{(1)1}\,(-x_1) + \beta_{(2)1}\,(-x_1) + \beta_{(3)2}\,x_2.
$$

Thus

$$
\boldsymbol{\eta} = \begin{pmatrix} 0 \\ 0 \\ \delta\,x_1 \end{pmatrix} + x_1 \begin{pmatrix} 1 & 0 \\ 0 & 1 \\ -1 & -1 \end{pmatrix} \begin{pmatrix} \beta_{(1)1} \\ \beta_{(2)1} \end{pmatrix} + x_2\,\mathbf{I}_3 \begin{pmatrix} \beta_{(1)2} \\ \beta_{(2)2} \\ \beta_{(3)2} \end{pmatrix}.
$$

This can be fitted by code of the form

```
fit4 <- vglm(y ~ offset(cbind(0, 0, delta * x1)) + x1 + x2 - 1,
             aVGAMfamilyFunction, data = adata,
             constraints = list(x1 = matrix(c(1, 0, -1, 0, 1, -1), 3, 2),
                                 x2 = diag(3)))
```

Example 5   How can one fit (3.45)–(3.47) subject to $\beta_{(1)1} + \beta_{(2)1} + \beta_{(3)1} = \beta_{(1)2} + \beta_{(2)2} + \beta_{(3)2}$? The answer is to let

$$
\eta_3 = \beta_{(3)1}\,x_1 + \left(\beta_{(1)1} + \beta_{(2)1} + \beta_{(3)1} - \beta_{(1)2} - \beta_{(2)2} \right) x_2.
$$

Usually (trivial constraints)

$$
\boldsymbol{\eta} = x_1\,\mathbf{I}_3 \begin{pmatrix} \beta_{(1)1} \\ \beta_{(2)1} \\ \beta_{(3)1} \end{pmatrix} + x_2\,\mathbf{I}_3 \begin{pmatrix} \beta_{(1)2} \\ \beta_{(2)2} \\ \beta_{(3)2} \end{pmatrix}
$$

but here,

$$\eta_3 = \beta_{(3)1}(x_1 + x_2) + \beta_{(1)1}x_2 + \beta_{(2)1}x_2 + \beta_{(1)2}(-x_2) + \beta_{(2)2}(-x_2).$$

So

$$\eta = \begin{pmatrix} x_1 & 0 & 0 \\ 0 & x_1 & 0 \\ x_2 & x_2 & x_1 + x_2 \end{pmatrix} \begin{pmatrix} \beta_{(1)1} \\ \beta_{(2)1} \\ \beta_{(3)1} \end{pmatrix} + \begin{pmatrix} x_2 & 0 \\ 0 & x_2 \\ -x_2 & -x_2 \end{pmatrix} \begin{pmatrix} \beta_{(1)2} \\ \beta_{(2)2} \end{pmatrix}.$$

The second term is easy:

$$x_2 \begin{pmatrix} 1 & 0 \\ 0 & 1 \\ -1 & -1 \end{pmatrix} \begin{pmatrix} \beta_{(1)2} \\ \beta_{(2)2} \end{pmatrix}.$$

The first term is problematic if dealt with wholly. Instead, one needs to break it up by columns:

$$\begin{pmatrix} x_1 \\ 0 \\ x_2 \end{pmatrix} \beta_{(1)1} + \begin{pmatrix} 0 \\ x_1 \\ x_2 \end{pmatrix} \beta_{(2)1} + \begin{pmatrix} 0 \\ 0 \\ x_1 + x_2 \end{pmatrix} \beta_{(3)1}$$

which equals

$$\begin{pmatrix} x_1 & 0 & 0 \\ 0 & a_1 & 0 \\ 0 & 0 & x_2 \end{pmatrix} \begin{pmatrix} 1 \\ 0 \\ 1 \end{pmatrix} \beta_{(1)1} + \begin{pmatrix} a_2 & 0 & 0 \\ 0 & x_1 & 0 \\ 0 & 0 & x_2 \end{pmatrix} \begin{pmatrix} 0 \\ 1 \\ 1 \end{pmatrix} \beta_{(2)1} + \begin{pmatrix} a_3 & 0 & 0 \\ 0 & a_4 & 0 \\ 0 & 0 & x_3 \end{pmatrix} \begin{pmatrix} 0 \\ 0 \\ 1 \end{pmatrix} \beta_{(3)1}$$

where $x_3 = x_1 + x_2$, and a_j has any value. Thus $p = 4$. The model can be fitted using code of the form

```
adata <- transform(adata, X1 = x1, X2 = x1, X3 = x1, x3 = x1 + x2,
                   a1 = 0 * x1, a2 = 0 * x1, a3 = 0 * x1, a4 = 0 * x1)
fit5 <- vglm(y ~ X1 + X2 + X3 + x2 - 1, aVGAMfamilyFunction, data = adata,
             constraints = list(X1 = rbind(1, 0, 1),
                                X2 = rbind(0, 1, 1),
                                X3 = rbind(0, 0, 1),
                                x2 = matrix(c(1, 0, -1, 0, 1, -1), 3, 2)),
             form2 = Select(adata, TRUE, as.formula = TRUE),
             xij = list(X1 ~ -1 + x1 + a1 + x2,
                        X2 ~ -1 + a2 + x1 + x2,
                        X3 ~ -1 + a3 + a4 + x3))
```

Once again, the values assigned to X1, X2 and X3 do not matter.

3.4.4 Smoothing

The examples in Sect. 3.4.2 are reasonably straightforward because the variables are entered linearly. Things become more tricky if data-dependent functions are used in any xij terms, e.g., bs() or poly() (see Sect. 18.6). In particular, regression splines such as bs() can be used to estimate a general smooth function $f(x_{ij})$, which is very useful for exploratory data analysis.

For the `eyesData` example, the code

```
fit.wrong <- vglm(cbind(leye, reye) ~ bs(iop), data = eyesData,
            family = binom2.or(exchangeable = TRUE, zero = 3),
            xij = list(bs(iop) ~ bs(liop) + bs(riop) + fill(bs(liop))),
            form2 = ~ bs(iop) + bs(liop) + bs(riop) + fill(bs(liop)))
```

is incorrect because the basis functions for `bs(liop)` and `bs(riop)` are not identical since the knots differ. Consequently, they represent two different functions despite having common regression coefficients.

Fortunately, it is possible to force the two `bs()` terms to have identical basis functions by using a trick: combine the vectors temporarily. To do this, one can use

```
BS <- function(x, ..., df = 3)
  bs(c(x, ...), df = df)[1:length(x), , drop = FALSE]
```

This computes a B-spline evaluated at `x`, but using other arguments as well, to form an overall vector from which to obtain the (common) knots. Then the usage of `BS()` can be something like

```
fit.BS <- vglm(cbind(leye, reye) ~ BS(iop), data = eyesData,
          binom2.or(exchangeable = TRUE, zero = 3),
          xij = list(BS(iop) ~ BS(liop, riop) +
                             BS(riop, liop) +
                             fill1(BS(liop, riop))),
          form2 = ~  BS(iop) + BS(liop, riop) + BS(riop, liop) +
                     fill1(BS(liop, riop)) + iop + liop + riop)
```

So `BS(liop, riop)` is the smooth term for `liop`, and `BS(riop, liop)` is the smooth term for `riop`.

The generic `predict()` should work as usual with `vglm()` models that utilize the `xij` argument, provided that the argument `newdata` is assigned a data frame with *all* the variables, i.e., those needed by the argument `form2`. However, `Select()` should not be assigned to `form2` when there are `BS()` or `NS()` terms.

Plotting the terms of `fit.BS` correctly, however, requires finesse, and this is explained in the next section.

3.4.4.1 Plotting †

Plotting each term of a VGLM can be achieved by coercing the object into a `"vgam"` object and calling the corresponding methods function; the call is of the form `plot(as(vglmObject, "vgam"))`. If `vglmObject` uses the `xij` argument, then some finesse is required. The details are as follows.

The important rules for a valid plot are:

(i) terms in RHS of `formula` should match both

    (a) the LHS term (response) of each formula in the `xij` list, and
    (b) the first term of the RHS of each formula in the `xij` list.

For example, `term1` and `term2` in

```
tfit <- vglm(response ~ term1 + term2 + term3 + ..., aVGAMfamilyfunction, adata,
             xij = list(term1 ~ term1  + term1a + term1b + ...,
                        term2 ~ term2  + term2a + term3b + ...),
             form2 = ~  term1  + term2  + term3  + ... +
                        term1a + term1b + ... +
                        term2a + term2b + ... )
plot(as(tfit, "vgam"), se = TRUE)
```

Here, the first component functions of `term1` and `term2` should plot correctly against their first (inner) arguments, e.g., if `term1` is `myfun(x5, x6, df = 4)` then its first inner argument is `x5`.

(ii) the `varxij`th (inner) argument of each such term is used for the plotting. The default is `varxij = 1`, meaning the first. For example, if `term1` was `NS(dum1, dum2)`, then it has two variables `dum1` and `dum2`, and so its component functions would be plotted against `dum1`. If `term2` was `NS(dum3, dum4)`, then its component functions would be plotted against `dum3` by default.

The above rules arise because the default for `plot()` applied to a `"vgam"` object is `raw = TRUE`, meaning that if a constraint matrix $\mathbf{H}_k$ has $\mathcal{R}_k$ columns then $\mathbf{H}_k$ is temporarily replaced by

$$\mathbf{H}_k^* = \begin{pmatrix} \mathbf{I}_{\mathcal{R}_k} \\ \mathbf{O} \end{pmatrix}, \tag{3.48}$$

so that

$$\mathbf{H}_k^* \, \boldsymbol{f}_k^*(x_k) = \begin{pmatrix} \boldsymbol{f}_k^*(x_k) \\ \mathbf{0} \end{pmatrix}.$$

Since only the first $\mathcal{R}_k$ component functions of x_k are plotted, these are then just the $\widehat{\boldsymbol{f}}_k^*(x_k)$ plotted against x_k.

The call `plot(as(VGLMobject, "vgam"))` uses the `formula` argument of `VGLMobject` to obtain the x_k variable for which the plots of the component functions $\widehat{\boldsymbol{f}}_k^*(x_k)$ are produced. Since x_k may vary for each η_j, this means that only one of them is potentially correct. By default, VGAM chooses the *first* argument (more generally the `varxij`th one) of the *first* term of the RHS of each formula in the `xij` list. For example, the term `NS(dum1, dum2)` has two variables `dum1` and `dum2`, and so all the raw component functions for that term are plotted against `dum1`.

With terms affected by `xij`, the default values of some other arguments need to be changed to give a more accurate representation. For example, something like `xlab = "dum1"` should be assigned in `plot(as(., "vgam"))` because `NS(dum1, dum2)` produces an `xlab` equalling `dum1` and `dum2` written on two lines. Also, it may be necessary to use the `which.term` and `which.cf` arguments to select only 'correct' component functions.

The above method of how `plot(as(., "vgam"))` works means that essentially only one-column constraint matrices are handled. If necessary, use the `which.cf` argument to select the component function. Note that although the `xij` argument is not restricted to one-column constraint matrices, plotting essentially is.

Example 1
The call

```
Fit1 <-
  vglm(cbind(leye, reye) ~ BS(liop, riop),
       binom2.or(exchangeable = TRUE, zero = 3), data = eyesData,
       xij = list(BS(liop, riop) ~ BS(liop, riop) + BS(riop, liop) + fill1(BS(liop))),
       form2 = ~ liop + riop + BS(liop, riop) + BS(riop, liop) + fill1(BS(liop)))
plot(as(Fit1, "vgam"), se = TRUE)
```

plots the estimated smooth component function against `liop`. To plot the (same) component function against `riop`, try

```
plot(as(Fit1, "vgam"), varxij = 2, se = TRUE)
```

Example 2
We jump ahead to Chap. 14 where a discrete-choice model similar to the one described in this section is fitted. Following from the analysis of `TravelMode` in AER in Sect. 14.2.1, we will allow regression spline smoothing of the cost variable(s).

```
> NS <- function(x, ..., df = 4)
    ns(c(x, ...), df = df)[1:length(x), , drop = FALSE]
> tfit2 <-
    vglm(mode ~ NS(gcost.air, gcost.trn, gcost.bus) + wait + income,
         multinomial(parallel = FALSE ~ 1), data = TravelMode2,
         xij = list(NS(gcost.air, gcost.trn, gcost.bus) ~
                    NS(gcost.air, gcost.trn, gcost.bus) +
                    NS(gcost.trn, gcost.bus, gcost.air) +
                    NS(gcost.bus, gcost.air, gcost.trn),
                    wait   ~  wait.air + wait.trn + wait.bus,
                    income ~   inc.air +  inc.trn +  inc.bus),
         form2 = ~  NS(gcost.air, gcost.trn, gcost.bus) +
                    NS(gcost.trn, gcost.bus, gcost.air) +
                    NS(gcost.bus, gcost.air, gcost.trn) +
                    wait + income +
                      inc.air +   inc.trn +   inc.bus +
                    gcost.air + gcost.trn + gcost.bus +
                      wait.air +  wait.trn +  wait.bus)
> plot(as(tfit2, "vgam"), se = TRUE, lcol = "orange", scol = "blue",
       which.term = 1, xlab = "gcost", ylab = "Fitted smooth", noxmean = TRUE)
```

This gives Fig. 3.1. The fitted function appears linear where the bulk of the data are. An approximate p-value for testing linearity of the function can be obtained from `lrtest(tfit2, fit.travel)`, and has the value 0.082, which indicates weak evidence against linearity. The first few rows of $\mathbf{X}_{VLM}$, which is (3.41), can be obtained by

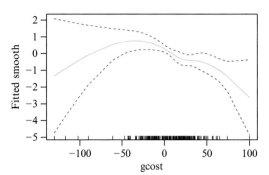

Fig. 3.1 An extension of the fitted model **fit.travel** from Sect. 14.2.1, where the cost variable is smoothed with regression splines. The variable **gcost** stands for 'generalized cost'.

```
> head(model.matrix(tfit2, type = "vlm"), 4)

     (Intercept):1 (Intercept):2 (Intercept):3
1:1              1             0             0
1:2              0             1             0
1:3              0             0             1
2:1              1             0             0
     NS(gcost.air, gcost.trn, gcost.bus)1 NS(gcost.air, gcost.trn, gcost.bus)2
1:1                            0.53076                              0.377890
1:2                            0.51110                              0.390758
1:3                            0.53076                              0.377890
2:1                            0.88402                              0.012843
     NS(gcost.air, gcost.trn, gcost.bus)3 NS(gcost.air, gcost.trn, gcost.bus)4
1:1                           0.144633                             -0.053282
1:2                           0.150849                             -0.052711
1:3                           0.144633                             -0.053282
2:1                           0.066754                             -0.036113
     wait income
1:1    69     35
1:2    34      0
1:3    35      0
2:1    64     30
```

The stacking of η_j-specific values can be seen, which result from the parallelism constraints, cf. (3.41).

3.4.5 Last Word

The **xij** argument operates *after* the ordinary $\mathbf{X}_{\mathrm{VLM}}$ matrix is created. Then selected columns of $\mathbf{X}_{\mathrm{VLM}}$ are modified from information in the constraint matrices, **xij** and **form2** arguments, i.e., from $\mathbf{X}_{\mathrm{form2}}$ and $\mathbf{H}_k$. This whole operation is possible because $\mathbf{X}_{\mathrm{VLM}}$ remains structurally the same. The crucial equation is (3.40).

Some examples of the **xij** argument found elsewhere include a pooled SUR model (Sect. 10.2.3.1), a positive-Bernoulli $\mathcal{M}_{tbh}$ model (Sect. 17.2.4), and a discrete-choice model (Sect. 14.2.1). Other **xij** examples are given in the online help of **fill1()** and **vglm.control()**.

3.5 Other Topics

3.5.1 Multiple Responses

A feature that some **VGAM** family functions possess is the ability to handle multiple responses. For example,

```
nfit <- vglm(cbind(y1, y2, y3) ~ x2 + x3, negbinomial, data = ndata)
```

might be $S = 3$ species' counts regressed upon some environmental variables for data collected at n sites. This can be achieved by invoking **vglm()** thrice, so why then offer this capability? One reason is because some of the regression coefficients might be constrained to be equal over all the responses. In fact, the $\mathbf{H}_k$ allow linear constraints over all variables and responses. Another reason is that the ordination methods described in Chaps. 6–7 need to handle multiple responses naturally. In this example, the enumeration of the η_js are

$$\boldsymbol{\eta} = (\eta_1, \ldots, \eta_6)^T = (\log \mu_1, \, \log k_1, \, \log \mu_2, \, \log k_2, \, \log \mu_3, \, \log k_3)^T \quad (3.49)$$

so that Species 1 specifies the first two η_js, then Species 2 for the next two, etc. One generally writes the overall M-vector as

$$\boldsymbol{\eta} = \left(\boldsymbol{\eta}_{(1)}^T, \ldots, \boldsymbol{\eta}_{(S)}^T\right)^T = (\eta_1, \ldots, \eta_M)^T \quad (3.50)$$

say, for S multiple responses (i.e., the $\boldsymbol{\eta}_{(s)}$ is the usual set of linear predictors for the sth response).

Some care is needed to distinguish between *multiple* responses and *multivariate* responses. Multiple responses are treated independently of each other, and are to be inputted side-by-side on the LHS of the formula using **cbind()**. Each response may be a vector, of dimension Q_1. Multivariate responses correspond to $Q_1 > 1$, and are handled by some full-likelihood model that takes into account their joint distribution. The quantity Q_1 is not to be confused with the length of each column of the response matrix, which is n. It is possible to have multiple multivariate responses, e.g.,

```
mfit <- vglm(cbind(y1, y2, y3, y4, y5, y6) ~ x2, binormal, data = bdata)
```

treats the response **cbind(y1, y2)** as one bivariate normal response, **cbind(y3, y4)** as another set of N_2-distributed random vectors, and **cbind (y5, y6)** as another. Thus one might say there are $S = 3$ (multiple) responses, and each response is bivariate. For **binormal()**, $M_1 = 5$ because the standard bivariate normal distribution can be considered as having 5 parameters to estimate. Hence, $M = 5 \times 3 = 15$ is the total number of linear/additive predictors η_j. More details about **binormal()** are in Sect. 13.2.1.

Here are some basic results. For 'one' response, the number of linear/additive predictors is denoted by M_1. The quantity M is always the total number of linear/additive predictors η_j, hence $M = M_1 \cdot S$, and with multiple responses, $\dim(\boldsymbol{y}_i) = SQ_1 = Q$. Table A.4 summarizes the notation.

With multiple responses, the overall log-likelihood can be written

$$\ell(\boldsymbol{\beta}^*) = \sum_{i=1}^{n} \sum_{s=1}^{S} w_{is} \, \ell_{is}\{\eta_1(\boldsymbol{x}_i), \ldots, \eta_M(\boldsymbol{x}_i)\}, \quad (3.51)$$

and all the derivations naturally extend to cover this case. Importantly, the ordering of $\boldsymbol{\eta}$ is as (3.50). One can input the prior weights of each response side-by-side
as in the response, e.g.,

```
vglm(cbind(y1, y2) ~ x2 + x3, negbinomial, data = ndata, weights = cbind(w1, w2))
```

If the inputted prior weights are a vector or one-column matrix, then recycling is
used, e.g.,

```
vglm(cbind(y1, y2) ~ x2 + x3, negbinomial, data = ndata, weights = w1)
```

recycles `w1` for both responses. By default, all prior weights are unity.

Returning back to the `zero` argument of Sect. 3.3.1.1, it is noted that this
argument may be assigned negative values in order to handle multiple responses
more easily. A negative j value means that each $\eta_{|j|}$ for each response is intercept-
only, e.g.,

```
vglm(cbind(y1, y2, y3) ~ x2 + x3, negbinomial(zero = -2), data = ndata)
```

models all three k_j parameters in (3.49) as scalars.

3.5.2 Deviance

The generic function `deviance()` returns the deviance for **VGAM** models. However,
it mainly applies to GLMs because they are readily defined for such. A few details
are as follows. The deviance D is defined by

$$\frac{D(\boldsymbol{y}; \boldsymbol{\mu})}{\phi} \;=\; 2\left\{\ell(\boldsymbol{y}; \boldsymbol{y}) - \ell(\boldsymbol{\mu}; \boldsymbol{y})\right\}, \tag{3.52}$$

since $\widehat{\mu}_i = y_i$ when the log-likelihood is maximized over μ_i unconstrained, i.e.,
when the number of parameters is equal to the number of observations. This is
called the *saturated* model. (That $\widehat{\mu}_i = y_i$ is easily shown for the normal, binomial
and Poisson families). The LHS of (3.52) is known as the *scaled deviance*. When
the GLM is an LM, then $\phi = \sigma^2$ is the variance, the numerator is the residual sum
of squares $\sum_{i=1}^{n} w_i(y_i - \mu_i)^2$, and the scaled deviance is χ^2_{n-p} distributed. For the
standard binomial and Poisson families, $\phi = 1$.

For GLMs, if ϕ is unknown, then it is most commonly estimated by the method
of moments. This results in a $\widehat{\phi}$ that is generally consistent, and in the case of
an ordinary LM, it is unbiased. Contrary to popular opinion, the deviance is *not*
asymptotically χ^2_{n-p} in general. Indeed, the use of χ^2_{n-p} for the scaled deviance is
only justified in some cases by 'small-dispersion asymptotics'. While the deviance
itself may have some complicated distribution, differences in the deviance between
two nested models is well-approximated by a $\chi^2_{p_1-p_2}$ distribution, where p_1 and p_2
are the number of parameters in the complex and simpler models, respectively.
Further details can be found, e.g., in Firth (1991).

For the few VGLMs that have a definable deviance, it follows the usual formula,
viz.,

$$D \;=\; 2\left(\ell_{\max} - \ell\right), \tag{3.53}$$

where $\ell_{\max}$ is the maximum achievable log-likelihood. There usually is no scaling
parameter, because this is estimated along with the other parameters in a full-
likelihood model.

The deviance is not defined for most models, and if so then `deviance()` will return a `NULL`. If a VGAM family function has a `@deviance` slot programmed in, then this will usually mean that function will be used for testing convergence while the model is estimated, otherwise usually the log-likelihood is used instead (almost all VGAM family functions have the log-likelihood function programmed in the `@loglikelihood` slot).

Quantities such as the deviance and log-likelihood should be retrieved from a fitted object using the appropriate accessor function (Tables 8.5, 8.6, 8.7), e.g., `logLik()` for ℓ.

3.5.3 Convergence

Section A.1.2.4 describes Newton-like algorithms that are relevant to the VGLM/VGAM framework. Although Fisher scoring does not converge as fast as the quadratically convergent Newton-Raphson algorithm in general, it is much faster than the EM algorithm that is so popular in many fields. It has the additional advantage over the EM algorithm in that the standard errors are automatically available as a by-product. As Dempster et al. (1977, p.35) write, "...it is important to remember that Newton-Raphson or Fisher-scoring algorithms can be used in place of EM. The Newton-Raphson algorithm is clearly superior from the point of view of rate of convergence near a maximum, since it converges quadratically."

Note that there are two approximations in the estimation process. Firstly, there is the quadratic approximation to ℓ, and secondly, there is the OIM approximated by the EIM. Regarding the quadratic approximation, this is illustrated with the V1 data set in Fig. 3.2. In World War II, some of the grids (each $\frac{1}{4}$ of a square km) about south London were heavily hit by V1 flying bombs, and we define a 'success' here as 3 or more hits. A simple intercept-only logistic regression gives $\hat{\mu} = 0.075$ as the probability of being heavily hit. Figure 3.2a plots ℓ as a function of μ, and Fig. 3.2b as a function of $\eta = \text{logit}\,\mu$. It may be seen that the logit link increases the symmetry about $\hat{\mu}$. Hence the quadratic approximation to ℓ about its MLE is slightly better under the logit transformation than under the raw parameter μ. Consequently, it is to be expected that its SEs will be more accurate. For completion, applying the delta method (A.29) to these data, approximate 95% confidence intervals for μ from both methods are

```
> V1 <- transform(V1, hit3 = hits >= 3)
> fit.logit <- vglm(hit3 ~ 1, binomialff, weights = ofreq, data = V1)
> fit.ident <- vglm(hit3 ~ 1, binomialff(link = "identitylink"), weights = ofreq, V1)
> logit(predict(fit.logit)[1] + 1.96 * c(-1, 1) * sqrt(vcov(fit.logit)), inv = TRUE)

  [1] 0.055828 0.099159

> fitted(fit.ident)[1] + 1.96 * c(-1, 1) * sqrt(vcov(fit.ident))

  [1] 0.053188 0.096117
```

Here, they do not appear to be markedly different. Incidentally, since the Poisson model fits the data well, if the response was whether the square grids were hit versus not hit, then a complementary log–log link would be the most natural choice.

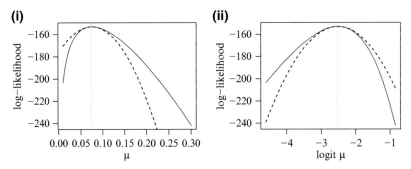

Fig. 3.2 Log-likelihood ℓ as a function of (i) μ and (ii) $\eta = \text{logit}\,\mu$, for the V1 data frame. The model is a logistic regression with 3 or more hits defining 'success'. The *vertical dashed lines* are at the MLE $\hat{\mu}$. The *dashed curves* are the quadratic approximation to ℓ at $\hat{\mu}$.

3.5.4 Half-Stepping

The Newton-Raphson algorithm unfortunately does not guarantee convergence to a maximum of the likelihood, even when the $\mathbf{W}_i$ are positive-definite. VGAM uses two enhancements to aid convergence: parameter link functions and half-stepping.

The use of parameter link functions is mainly for two reasons.

1. To transform the parameter to $(-\infty, \infty)$, hence does not suffer from parameter estimates that go outside its boundary.
2. To make the log-likelihood better approximated by a quadratic.

Sometimes, it may aid interpretation too.

Half-stepping is the other enhancement. Sometimes an iteration may step too far and reduce the likelihood. This problem, which can be due to a log-likelihood that is far from locally quadratic, occurs when $\ell^{(a)} < \ell^{(a-1)}$. It can be rectified by repeatedly halving the step until the likelihood is increased, because the score vector points in the direction of greatest ascent in ℓ at the current iteration. In more detail (cf. (3.9)),

$$\boldsymbol{\beta}^{(a)} = \boldsymbol{\beta}^{(a-1)} + \left(\mathbf{X}^{*T}\mathbf{W}^{(a-1)}\mathbf{X}^*\right)^{-1}\mathbf{X}^{*T}\mathbf{W}^{(a-1)}\boldsymbol{z}^{(a)} = \boldsymbol{\beta}^{(a-1)} + \boldsymbol{h}^{(a-1)} \quad (3.54)$$

where $\boldsymbol{h}^{(a-1)}$ is referred to as the *step*. Half-stepping involves replacing $\boldsymbol{h}^{(a-1)}$ by $\frac{1}{2}\boldsymbol{h}^{(a-1)}$, $\frac{1}{4}\boldsymbol{h}^{(a-1)}$, $\frac{1}{8}\boldsymbol{h}^{(a-1)}$, ..., if $\ell^{(a)} < \ell^{(a-1)}$ until $\ell^{(a)} > \ell^{(a-1)}$. It entails computing ℓ at

$$\boldsymbol{\beta}^{(a-1)} + \alpha\left(\boldsymbol{\beta}^{(a)} - \boldsymbol{\beta}^{(a-1)}\right) = (1-\alpha)\boldsymbol{\beta}^{(a-1)} + \alpha\,\boldsymbol{\beta}^{(a)}, \quad a = 1, 2, \ldots, \quad (3.55)$$

for $\alpha = \frac{1}{2}, \frac{1}{4}, \frac{1}{8}, \ldots$ until an improvement occurs. Note that half-stepping requires both $\boldsymbol{\beta}^{(a)}$ and $\boldsymbol{\beta}^{(a-1)}$, therefore it can be implemented at the first iteration only if the initial value of the coefficient vector $\boldsymbol{\beta}^{(0)}$ is known. This is only so if `coefstart` is inputted.

Currently, half-stepsizing is presently only implemented in `vglm()`. Furthermore, the value of α in (3.55) is specified in `vglm.control()$stepsize`, and defaults to unity. The value provides an upper bound on the size of any step during the Newton-type iterations. At any particular iteration, the stepsize is the minimum of this value and twice the previous stepsize. Half-stepping can be turned off by setting `vglm.control(half.stepsizing = FALSE)` because it is `TRUE` by default.

To illustrate that plain Fisher scoring may fail occasionally without half-stepping, consider the toy data set of Ridout (1990) involving a dilution series method to estimate the density of organisms (this sort of experiment is described in, e.g., McCullagh and Nelder, 1989, Sect.1.2.4).

```
> ridout <- data.frame(v = 10^(3:1), r = c(4, 3, 3), n = c(5, 5, 5))
> (ridout <- transform(ridout, logv = log(v)))

       v r n   logv
1 1000 4 5 6.9078
2  100 3 5 4.6052
3   10 3 5 2.3026
```

It is left to the reader (Ex. 3.11) to run the following:

```
glm.fail <- glm(r/n ~ offset(logv) + 1, weight = n,
                binomial(link = cloglog), data = ridout, trace = TRUE, maxit = 25)
```

and confirm that the iterations oscillate between two local solutions. With **VGAM**, the half-stepping feature results in a slight improvement:

```
vglm.okay <- vglm(cbind(r, n-r) ~ offset(logv) + 1,
                  binomialff(link = "cloglog"), data = ridout, trace = TRUE)
```

Then

```
> c(deviance(glm.fail), deviance(vglm.okay))

 [1] 17.866 17.437
```

According to Ridout (1990), ℓ has its maximum at $\widehat{\theta} = -5.4007$, and is nearly quadratic in the vicinity of the MLE.

The morale of this example is that it is a good idea to monitor convergence all the time, e.g., by setting `trace = TRUE`, especially as the VGLM/VGAM framework fits many types of statistical models, some of which are naturally ill-conditioned and/or require greater skill levels to fit safely. Indeed, the manner a model converges (or doesn't) provides important insights into the underlying problem modelling (Osborne, 1992). Slow convergence, or a lack of it, should raise alarm bells to the statistical modeller, and generate caution on any use of it. In the case of this data set, the model provides a very poor fit to the observed data, e.g., the residual deviance is 17.44, and the Pearson statistic is 54.32 (both with 2 degrees of freedom; the large discrepancy between the two statistics arises because, of the three fitted probabilities, one is close to 0 and another is close to 1) (Ridout, 1990).

For handling working weight matrices that are not positive-definite, see Sect. 9.2.1. How **VGAM** tests for convergence and some practicalities are described in Sect. 8.2.4.

3.6 Inference

3.6.1 Regularity Conditions

Section A.1 summarizes some classical likelihood theory relevant to the VGLM class. In particular, regularity conditions that must be satisfied in order for Fisher scoring to be expected to work properly, and for correct inferences to be made.

3.6.2 Parameter Link Functions

As mentioned above, parameter link functions are recommended for several reasons. They are the first port of call for handling range restrictions, e.g., $\theta_j \in (0, \infty)$ is naturally handled by $\eta_j = \log(\theta_j)$. Secondly, and partly as a consequence, the log-likelihood is usually more quadratic-shaped about the solution when an appropriate transformation is chosen, hence convergence will be faster and inferences more accurate. Intuitively, it is due to the Taylor series approximation to ℓ about the solution being more accurate. This can be seen in Fig. A.2 where a simple negative binomial model is fitted to the `machinists` data set. This intercept-only model has $\widehat{\mu} = 0.4831$ and $\widehat{k} = 0.4743$, and the parameter of interest is $\theta = k$. In Fig. A.2a there is pronounced asymmetry about $\widehat{\theta}$, whereas Fig. A.2b is clearly more quadratic-shaped as a function of $\eta = \log k$.

A third reason is that the initial-value parameter space where successful convergence will occur tends to be larger when the θ_j have been transformed suitably; see, e.g., Green (1984, Figs. 1–2).

3.6.3 Hypothesis Testing

Hypothesis testing under classical likelihood theory is described in Sect. A.1.4.2. The function `linearHypothesis()` in car may be used to conduct Wald tests for linear combinations of the fitted coefficients, given only $\widehat{\boldsymbol{\beta}}$ and $\widehat{\mathrm{Var}}(\widehat{\boldsymbol{\beta}})$ (as accessed by the extractor functions `coef()` and `vcov()`). An example, applied to the nonproportional odds model, is in Sect. 14.4.3.

3.6.4 Residual Degrees of Freedom

In the LM, the quantity $n_{\mathrm{LM}} - p_{\mathrm{LM}}$ is known as the *(total) (residual) degrees of freedom*. It can be seen from (3.23) that the VLM-degrees of freedom can be defined as

$$n_{\mathrm{VLM}} - p_{\mathrm{VLM}} = nM - p_{\mathrm{VLM}}. \tag{3.56}$$

The call `df.residual(vglmObject, type = "vlm")` returns this quantity.

In some applications, it is useful to know the residual degrees of freedom corresponding to a specific η_j. For $\mathbf{H}_k = \mathbf{I}_M$, this is

$$n_{\text{LM}} - p_{(j)\text{LM}} \;=\; n - p_{(j)\text{VLM}}, \qquad j = 1, \ldots, M, \tag{3.57}$$

where $p_{(j)\text{VLM}} = \dim(\boldsymbol{\beta}_j^*) + \dim(\boldsymbol{\beta}_j^{**})$ in (3.35). The M values are returned by df.residual(vglmObject, type = "lm"). Of course, taking the sum of the M values will not necessarily equal the degrees of freedom of the VLM because of the constraints on the functions, e.g., one coefficient may appear in several η_js, such as in a parallelism constraint.

As a simple example, consider fit1a from Sect. 3.3.1.4.

```
> df.residual(fit1a, type = "lm")

  mean    sd
    99    99

> df.residual(fit1a, type = "vlm")

 [1] 199
```

3.7 Residuals and Diagnostics

As with GLMs, there are several types of residuals that can be defined for VGLMs. However, for some specific VGLMs, only a subset of these residual types are properly defined. For example, working and Pearson residuals are always defined according to (3.12), but deviance and response residuals may not be. Currently, the VGAM default are working residuals, cf. deviance residuals for glm(). In general, residuals in the VGLM framework are best seen in the context of the IRLS algorithm of Sect. 3.2. To reduce clutter, the superscript (a) denoting the iteration number is omitted, and quantities at the final IRLS iteration are assumed.

In practice, residuals should be extracted by the extractor function resid() (or residuals()). These have a type argument that can be used to select from the following residual types.

3.7.1 Working Residuals

As in the univariate case, these are the differences between the *working* responses and the linear predictors at the final IRLS iteration. From (3.12), they are defined as

$$\boldsymbol{r}_i^W \;=\; \boldsymbol{z}_i - \boldsymbol{\eta}_i \;=\; \mathbf{W}_i^{-1}\, \boldsymbol{u}_i. \tag{3.58}$$

Of course, they are on the $\boldsymbol{\eta}$-scale. For GLMs,

$$\frac{\partial \ell_i}{\partial \eta} \;=\; \frac{y_i - \widehat{\mu}_i}{\operatorname{Var}(Y_i)} \frac{\partial \widehat{\mu}_i}{\partial \widehat{\eta}_i} \quad \text{and} \quad \mathbf{W}_i \;=\; \left(\frac{\partial \mu_i}{\partial \eta_i}\right)^2 \frac{1}{\operatorname{Var}(Y_i)}, \; \text{so that} \; r_i^W = (y_i - \widehat{\mu}_i)\, \frac{\partial \widehat{\eta}_i}{\partial \widehat{\mu}_i}.$$

If `fit` is a `vglm()` or `vgam()` object then

```
predict(fit) + resid(fit, type = "working")
```

is an $n \times M$ matrix of adjusted dependent variables, i.e., the ith row is $\boldsymbol{z}_i^T$.

Working residuals are used in partial residual plots—see Chap. 4. An example of them is Fig. 10.1.

3.7.2 Pearson Residuals

Pearson residuals, $\boldsymbol{r}_i^P$, are related to the working residuals and defined by

$$\boldsymbol{r}_i^P = \sqrt{w_i}\,\mathbf{W}_i^{1/2}\,\boldsymbol{r}_i^W = \sqrt{w_i}\,\mathbf{W}_i^{-1/2}\,\boldsymbol{u}_i. \tag{3.59}$$

The justification for this is that each IRLS step minimizes (3.22), and at convergence, (3.22) reduces to

$$\sum_{i=1}^{n} w_i\,\boldsymbol{u}_i^T\mathbf{W}_i^{-1}\boldsymbol{u}_i = \sum_{i=1}^{n} \left(\boldsymbol{r}_i^P\right)^T \boldsymbol{r}_i^P = \sum_{i=1}^{n}\sum_{j=1}^{M} \left(r_{ij}^P\right)^2.$$

For GLMs,

$$r_i^P = \sqrt{w_i}\,\frac{y_i - \widehat{\mu}_i}{\sqrt{V(\widehat{\mu}_i)}} \quad\text{and}\quad X^2 = \sum_{i=1}^{n} \left(r_i^P\right)^2 = \sum_{i=1}^{n} w_i\,\frac{(y_i - \widehat{\mu}_i)^2}{V(\widehat{\mu}_i)} \tag{3.60}$$

is the Pearson chi-squared statistic.

In the above, $\mathbf{W}_i^{1/2}$ must be obtained by the spectral decomposition of $\mathbf{W}_i$ (Sect. A.3.4) and not its Cholesky decomposition, therefore they are not cheap to compute. Note also that when $w_i = 1$, $\boldsymbol{r}_i^P \overset{\cdot}{\sim} (\boldsymbol{0}, \mathbf{I}_M)$ as $n_i \to \infty$.

3.7.3 Response Residuals

These are simply

$$y_{ij} - \widehat{\mu}_{ij}, \tag{3.61}$$

obtained from the components of $\boldsymbol{y}_i - \widehat{\boldsymbol{\mu}}_i$. Note that $\widehat{\boldsymbol{\mu}}_i$ is not necessarily the mean, but is generically the 'fitted value' as returned by `fitted()`. For some models, the fitted values are selected quantiles, for others such as the Cauchy distribution, they are an estimated location parameter.

This residual type is not always defined, e.g., the dimensions of `depvar(fit)` and `fitted(fit)` must be the conformable so that their difference can be computed. Of course, for GLMs, they are well-defined.

3.7.4 Deviance Residuals

Chambers and Hastie (1991, Sect.6.2.1, 6.4.1) define *deviance residuals* for GLMs and GAMs as

$$r_i^D = \sqrt{w_i}\,\text{sign}(y_i - \widehat{\mu}_i)\,\sqrt{d_i}, \quad \text{where } D = \sum_{i=1}^{n} \left(r_i^D\right)^2 = \sum_{i=1}^{n} w_i\,d_i \quad (3.62)$$

is the deviance. Deviance residuals in VGAM are as in (3.62), but are only defined for some regression models, e.g., requiring a univariate response Y_i, a well-defined μ_i and ℓ_i.

The deviance is defined generally by (3.53).

3.7.5 Hat Matrix

The *hat* or *projection matrix* for VGLMs might be defined as

$$\boldsymbol{\mathcal{H}} = \mathbf{U}\,\mathbf{X}_{\text{VLM}} \left(\mathbf{X}_{\text{VLM}}^T \mathbf{W}\,\mathbf{X}_{\text{VLM}}\right)^{-1} \mathbf{X}_{\text{VLM}}^T \mathbf{U}^T \quad (3.63)$$

computed at the final IRLS iteration, where it satisfies

$$\widehat{z}^{**(a)} = \boldsymbol{\mathcal{H}}\,z^{**(a)}. \quad (3.64)$$

As with LMs and GLMs, and based on (3.16), it is useful for regression diagnostics such as influence and residuals, where the diagonal elements play the most important role. The projection matrix $\boldsymbol{\mathcal{H}}$ retains the properties of symmetry and being idempotent, i.e., $\boldsymbol{\mathcal{H}} = \boldsymbol{\mathcal{H}}^2$. Since $\mathbf{U}\mathbf{X}_{\text{VLM}} = \mathbf{QR} = \mathbf{X}_{\text{VLM}}^{**}$ in (3.23) is its QR decomposition, we have $\boldsymbol{\mathcal{H}} = \mathbf{Q}\mathbf{Q}^T$ whose diagonal is easily computed as the row sums of $\mathbf{Q} \circ \mathbf{Q}$, the Hadamard product of $\mathbf{Q}$ with itself. The methods function `hatvalues(vglmObject)` returns an $n \times M$ matrix arrangement of the diagonal elements of $\boldsymbol{\mathcal{H}}$. Then trace$(\boldsymbol{\mathcal{H}}) = p_{\text{VLM}}$, which is equal to the rank of $\mathbf{Q}$, and has p_{VLM} and $n_{\text{VLM}} - p_{\text{VLM}}$ unit and zero eigenvalues respectively.

The hat matrix is also useful for bias-reduction (see Sect. 9.4), a method for removing the $O(n^{-1})$ bias from a maximum likelihood estimate. For a substantial class of models including GLMs, bias-reduction can be formulated in terms of a minor adjustment of the score vector within an IRLS algorithm. One by-product, for logistic regression, is that while the MLE can be infinite, the adjustment leads to estimates that are always finite.

The generic function `hatplot(vglmObject)` produces an index plot of the leverages, with horizontal dashed lines at 2 and 3 multiple of $p_{\text{VLM}}n/M$, which is the 'average' value per plot. With small samples, usually 3 is used as the rough rule-of-thumb, otherwise 2 is commonly used. Observations above these thresholds might be interpreted as being influential.

It can be shown that

$$\text{Var}(\widehat{\boldsymbol{\eta}}_i) = \mathbf{U}_i^{-1}\,\boldsymbol{\mathcal{H}}_{ii}\,\mathbf{U}_i^{-T}, \quad (3.65)$$

where the order-M matrices $\boldsymbol{\mathcal{H}}_{ii}$ are the ith central block matrices of $\boldsymbol{\mathcal{H}}$, i.e., $\boldsymbol{\mathcal{H}}$ is made up of blocks $\boldsymbol{\mathcal{H}}_{ij}$ for $i,j = 1,\ldots,n$. In VGAM, the square roots of the diagonal elements of (3.65) are returned by `predict(vglmObject, se = TRUE)`.

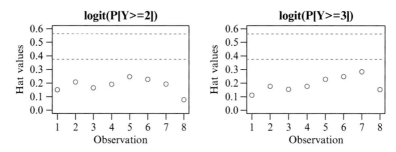

Fig. 3.3 Hat values from a proportional odds model fitted to the severity of pneumoconiosis data set in coalminers, called **pneumo**.

To see how the MLE is affected by removing one observation at a time, `dfbeta(vglmObject)` can return an $n \times p_{\text{VLM}}$ matrix of differences in the regression coefficients due to deleting observation i. Notationally, the ith row of `dfbeta(vglmObject)` might be written as the transpose of $\widehat{\boldsymbol{\beta}}^* - \widehat{\boldsymbol{\beta}}^*_{[-i]}$. As a simple example, for the proportional odds model fitted to the **pneumo** data of Sect. 14.4.2 and Table 14.2,

```
> pneumo <- transform(pneumo, let = log(exposure.time))
> fit.pom <- vglm(cbind(normal, mild, severe) ~ let, propodds, data = pneumo)
> hatplot(fit.pom, multiplier = 2:3, ylim = c(0, 0.6), col = "blue")
```

This produces Fig. 3.3. All the 8 exposure times lie below the thresholds, suggesting than none of them individually are influential to the extent of raising concern.

Bibliographic Notes

IRLS is a very powerful and general technique for the numerical maximum likelihood estimation. The method is generally attributed to Beaton and Tukey (1974). Some references include Green (1984), McCullagh and Nelder (1989), del Pino (1989), Jorgensen (2001), Rubin (2006). Dempster et al. (1980) give general properties of IRLS, and convergence results can be found in Byrd and Pyne (1979) and Birch (1980).

The term VGLM was coined *after* VGAMs, since the latter made use of vector splines (Sect. 4.2.1). VGLMs are similar to the multivariate GLMs of Fahrmeir and Tutz (2001). VGAMs were proposed by Yee and Wild (1996), and VGLMs were developed in Yee and Hastie (2003). The function `vglm()` is largely Green (1984, Eq.(7)), while `vgam()` is largely Yee and Wild (1996). Green's method, which is central to **VGAM**, has been extended by Jørgensen (1984).

Davison and Snell (1991) describe several types of residuals for GLMs. Fox and Weisberg (2011, Chap.6) provide R examples of regression diagnostics for LMs and GLMs, as does Faraway (2006, Sect.6.4).

Exercises

Ex. 3.1.   The constraints Argument I

(a) How can one fit (3.45)–(3.47) subject to $\beta_{(3)2} = 0$, $\beta_{(1)2} + \beta_{(2)2} = 0$ and $\beta_{(1)1} = \beta_{(2)1}$? Write down the equations in the form of (3.27).
(b) Write down the form of the call to vglm() in order to fit such a model (as in Sect. 3.4.3).

Ex. 3.2.   The constraints Argument II
Repeat Ex. 3.1 but fitting (3.45)–(3.47) subject to $\beta_{(1)1} + 2\beta_{(2)1} + \beta_{(3)1} = 1$ and $\beta_{(1)2} = 3\beta_{(2)2} = \beta_{(3)2}$.

Ex. 3.3.   The xij Argument I

(a) How can one fit (3.45)–(3.47) subject to $\beta_{(1)1} + \beta_{(1)2} = \beta_{(2)2} + \beta_{(3)2}$? Write down the equations in the form of (3.40).
(b) Write down the form of the call to vglm() in order to fit such a model.

Ex. 3.4.   The xij Argument II
Repeat Ex. 3.3 but fitting (3.45)–(3.47) subject to $\beta_{(1)1} + \beta_{(1)2} = \beta_{(2)1} + \beta_{(3)2}$.

Ex. 3.5.   The xij Argument III
Repeat Ex. 3.3 but fitting (3.45)–(3.47) subject to $\beta_{(1)1} + \beta_{(2)1} = \beta_{(3)2}$.

Ex. 3.6.   The xij Argument IV
Repeat Ex. 3.3 but fitting (3.45)–(3.47) subject to $\beta_{(1)1} + \beta_{(1)2} = \beta_{(2)1} + \beta_{(2)2}$.

Ex. 3.7.   The xij Argument V
Repeat Ex. 3.3 but fitting (3.45)–(3.47) subject to $\beta_{(2)1} + \beta_{(3)1} = \beta_{(1)2}$ and $\beta_{(1)1} + \beta_{(2)2} = 0$.

Ex. 3.8.   Simple Constraints—Normal Distribution

(a) Generate a random sample of 100 observations from a $N(\mu = \theta,\ \sigma^2 = \sqrt{\theta})$ distribution with $\theta = 10$. Then estimate θ using uninormal().
(b) Repeat (a) using a $N(\mu = \theta,\ \sigma^2 = 1 + \sqrt{\theta})$ distribution.
(c) Repeat (a) using a $N(\mu = 2\theta,\ \sigma^2 = 1 + \sqrt{\theta})$ distribution.

Ex. 3.9.   Simple Constraints—Beta Distribution

(a) Generate a random sample of 100 observations from a beta distribution with shape parameters $s_1 = e^1$ and $s_2 = e^2$. Then estimate the s_j using betaR().
(b) Repeat (a) but estimate s_1 knowing that $s_1 + s_2 = e^1 + e^2$.
(c) Repeat (a) but estimate s_1 knowing that $s_2 = e^1 s_1$.
(d) Comment on your results, e.g., which of (a)–(c) seems to give the best estimate?

Ex. 3.10.   Fit a Poisson regression to the V1 data. Display some evidence that the convergence rate is quadratic.

Ex. 3.11.   Dilution Series Data
Run the analyses on the ridout data frame in Sect. 3.5.4. Confirm that the log-likelihood is nearly quadratic in the vicinity of the MLE.

Ex. 3.12.   **Deviance**

(a) Show that the deviance of an unweighted Poisson regression is $2\sum_{i=1}^{n}\{y_i$
$\log(y_i/\widehat{\mu}_i) + \widehat{\mu}_i - y_i\}$.
(b) Derive the deviance function for a logistic regression fitted to proportions y_i,
where $N_i Y_i \sim \text{Binomial}(N_i, \ \mu_i)$, for $i = 1, \ldots, n$.

Ex. 3.13.   Show that (3.22) at the final IRLS iteration is $\widehat{\text{ResSS}}$ =
$\sum_{i=1}^{n} \boldsymbol{u}_i^T \mathbf{W}_i^{-1} \boldsymbol{u}_i$ evaluated at $\widehat{\boldsymbol{\beta}}^*$.

Ex. 3.14.   Prove (3.65).

Ex. 3.15.   **Link Functions for Binomial Regression**
Using the `V1` data, obtain a plot similar to Fig. 3.2b but with a complementary
log–log link.

Ex. 3.16.   **The `constraints` and `zero` Arguments**

(a) Run the following code to generate some artificial data.

```
set.seed(1)
n <- 200
ndata <- data.frame(x2 = sort(runif(n)))
ndata <- transform(ndata, y1 = rnbinom(n, mu = exp(1), size = exp(2 - x2)),
                          y2 = rnbinom(n, mu = exp(2 + x2), size = exp(3)))
```

Knowing that x_2 affects only some parameters, simultaneously fit two negative binomial regressions to these data, in two ways, as follows. The first call
to `vglm()` should only use the `zero` argument. The second call to `vglm()`
should use the `constraints` argument to input the constraint matrices.
(b) Apply `df.residual()` to one of your fits, to obtain the LM-type residual
degrees of freedom. Repeat for the VLM-type residual degrees of freedom.
(c) Fit an NB regression to `y2` alone using `x2` as the covariate, and then plot
several residual types against `x2`. Comment.
(d) Plot the hat values of your fit in (c). Comment.

In our experience VGAM *is a useful package, but at times is confusing to
work with.*
—Zuur (2012)

Chapter 4
VGAMs

As I am sure the editor and authors will agree, this is a fun area to work in.
—G. Wahba, Foreword to Schimek (2000)

4.1 Introduction

VGAMs are VGLMs based on smoothing. As with GAMs, they are particularly useful for exploratory data analysis to allow the data to "speak for themselves". Rather than restricting the linear predictors η_j to be linear in $\boldsymbol{x}$ as with VGLMs, one allows for flexible curves determined by the data so that the η_j are referred to as *additive predictors*. The method used in this chapter to estimate these curves is smoothing. VGAMs thus extend the GAM class outside the small confines of the exponential family and allow multiple η_j.

4.1.1 Smoothing for VGLMs and VGAMs

"Vector GAMs" were coined because of their use of vector smoothers, and in particular, the cubic vector smoothing spline described in Sect. 4.2.1. Vector smoothers should not be confused with multivariate smoothers. There are four types of smoothing:

(1) y-scalar, x-univariate,
(2) y-scalar, $\boldsymbol{x}$-multivariate,
(3) $\boldsymbol{y}$-vector, x-univariate,
(4) $\boldsymbol{y}$-vector, $\boldsymbol{x}$-multivariate

(see, e.g., Miller and Wegman, 1987); vector smoothers for us are case (3), and are the subject of Sect. 4.2. Case (1) was covered in Sect. 2.4. Case (4) is the least common, and it is not considered here at all. Multivariate smoothers, where $\boldsymbol{x}$ is a vector, suffer from the curse of dimensionality when $\dim(\boldsymbol{x})$ is large, e.g., $d > 2$ not including the intercept (see Sect. 2.5.1). Vector $\boldsymbol{x}$ can be computationally handled and interpreted more easily by considering one x_k at a time—these are two reasons why additive models are popular (Sect. 2.5.1).

T.W. Yee, *Vector Generalized Linear and Additive Models*,
Springer Series in Statistics, DOI 10.1007/978-1-4939-2818-7_4

Recall for VGAMs (1.34) that

$$\boldsymbol{\eta}(\boldsymbol{x}) \;=\; \boldsymbol{\beta}_{(1)} + \sum_{k=2}^{d} \boldsymbol{f}_k(x_k) \;=\; \mathbf{H}_1\,\boldsymbol{\beta}_{(1)}^* + \sum_{k=2}^{d} \mathbf{H}_k\,\boldsymbol{f}_k^*(x_k) \qquad (4.1)$$

where $\boldsymbol{f}_k^*(x_k) = (f_{(1)k}^*(x_k), \ldots, f_{(\mathcal{R}_k)k}^*(x_k))^T$. Currently there are two practical approaches for estimating the $\boldsymbol{f}_k^*$:

(i) *Regression splines.* These is a parametric way to represent a nonparametric function. They typically involve the use of `bs()` and `ns()` from **splines**, in conjunction with `vglm()`, e.g.,

```
vglm(y ~ bs(x2) + ns(x3, df = 4), VGAMfamily, data = vdata)
```

Of course, regression splines may be used for LMs, GLMs, etc. Section 2.4.3 described this powerful technology.

(ii) *Vector smoothing splines.* These fit into the classical Hastie and Tibshirani (1990) framework and involve a procedure called *vector backfitting* (Sect. 4.3). An example of their use is of the familiar form

```
vgam(y ~ s(x2) + s(x3, df = 2), VGAMfamily, data = vdata)
```

Vector splines fit nicely into the penalty function approach of Sect. 1.5.1—see Sect. 4.3.1 for details.

(iii) *P-splines.* This more modern approach is more amenable to automatic smoothing parameter selection and inference: the wiggliness of the curves may be automatically chosen by some objective criterion. These low-rank smoothers were sketched in Sect. 2.4.5, and they are currently under development for VGAMs.

Much of this chapter is aimed towards (ii). We draw upon results from ordinary univariate smoothing in Sect. 2.4, as they form the building blocks of GAMs and we shall see that their ideas naturally extend to the vector $\boldsymbol{y}$ case.

Like GLMs, GAMs have historically been restricted to the exponential family. They were developed from the mid-1980s both theoretically (Hastie and Tibshirani, 1990) and in software (e.g., **S-PLUS**' `gam()`), and have since become an invaluable tool in the statistician's toolbox. The original GAM software required prespecified smoothing parameters—this is also the case in the present **R** implementation **gam** with its `gam()` function. The treatment of VGAMs in this chapter largely follows from this.

The use of regression splines is sprinkled throughout this book, and kernel smoothers are only covered (Sect. 4.2.2.1) to bolster our understanding of vector smoothing as a whole—traditionally their asymptotic properties are the most mathematically tractable. But smoothing splines are described in relatively more detail (Sect. 4.2.1), because they are the only 'real' smoother currently implemented in **VGAM**.

A major enhancement to GAM methodology was the development of automatic smoothing parameter selection (Wood, 2006) and its **R** implementation, the package **mgcv**. It is hoped that in the not too distant future, the development of P-spline VGAMs will be completed and confer automatic smoothing parameter selection to **VGAM**.

4.1.2 The Vector Smoothing Problem

The vector smoothing problem involves a vector response $\boldsymbol{y}_i = (y_{i1}, \ldots, y_{iM})^T$ measured at each x_i. For example, one might regress $Y_1 =$ diastolic and $Y_2 =$ systolic blood pressures (DBP/SBP) versus $X =$ age with data from n people in a cross sectional study. This is illustrated in Fig. 4.1 for a random sample of $n = 100$ participants from `xs.nz`. The plots suggest that both types of blood pressure increase over the 20–50 year age range.

In general, for scatter plot data $(x_i, \boldsymbol{y}_i, \boldsymbol{\Sigma}_i)$, $i = 1, \ldots, n$, the *vector measurement model* (VMM), which generalizes the classical smoothing problem (2.36), is

$$\boldsymbol{y}_i = \boldsymbol{f}(x_i) + \boldsymbol{\varepsilon}_i, \qquad \boldsymbol{\varepsilon}_i \sim (\boldsymbol{0}, \boldsymbol{\Sigma}_i) \text{ independently,} \qquad (4.2)$$

where $\boldsymbol{y}_i \in \mathbb{R}^M$, $\boldsymbol{f}(x_i) = (f_1(x_i), \ldots, f_M(x_i))^T$ is a vector of M smooth *component functions* to be estimated. Sometimes a vector smoothing problem may be simplified to M separate ordinary smoothing problems. In the case of vector smoothing splines, Sect. 4.2.1 lists some conditions that suffice for this.

In this book we are not primarily interested in (4.2) to vector smooth data directly. Instead, its main role is to serve as the building block for estimating VGAMs. In a nutshell, the response vector $\boldsymbol{y}_i$ in (4.2) becomes the $\boldsymbol{z}_i$ in the IRLS algorithm (3.12), and the $\boldsymbol{\Sigma}_i$ are treated as known because they are the inverse of the working weight matrices $\mathbf{W}_i$. Further details are given in Sect. 4.3. The vector $\boldsymbol{x}$ case is handled one x_k at a time by a procedure known as modified (vector) backfitting.

In practice, tied observations $(x, \boldsymbol{y}_j, \boldsymbol{\Sigma}_j)$ for $j = 1, \ldots, T$, may be replaced by a single observation $(x, \boldsymbol{y}^*, \boldsymbol{\Sigma}^*)$, where

$$\boldsymbol{y}^* = \left(\sum_{j=1}^T \boldsymbol{\Sigma}_j^{-1} \right)^{-1} \left(\sum_{j=1}^T \boldsymbol{\Sigma}_j^{-1} \boldsymbol{y}_j \right) \text{ and } \boldsymbol{\Sigma}^* = \left(\sum_{j=1}^T \boldsymbol{\Sigma}_j^{-1} \right)^{-1}.$$

The assumption $x_1 < x_2 < \cdots < x_n$ is still made.

Returning to the blood pressures example, if we assume that the $\boldsymbol{\Sigma}_i$ are equal then the sample variance of the residuals gives

$$\widehat{\boldsymbol{\Sigma}}_i = \begin{pmatrix} 68.87 & 43.12 \\ 43.12 & 101.13 \end{pmatrix}.$$

Not surprisingly, this gives a positive sample correlation coefficient (0.64 actually), e.g., elevated SBP levels tend to be associated with elevated DBPs.

4.2 Vector Smoothing Methods

This section describes two methods for fitting the VMM (4.2). This problem is important because it forms the 'backbone' of the estimating algorithm of the VGAM class, viz. vector backfitting.

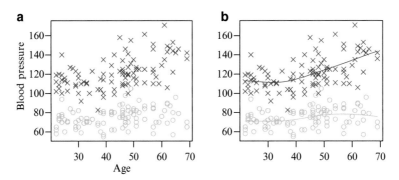

Fig. 4.1 (**a**) Scatter plot of diastolic ($\circ$) and systolic ($\times$) blood pressures (mm Hg) versus age, for a random sample of 100 European-type females from `xs.nz`. (**b**) A vector smoothing spline fit overlaid on the same. Each component function has 2 effective nonlinear degrees of freedom (ENDF).

4.2.1 Vector Splines

Fessler (1991) proposed the *vector (smoothing) spline* for fitting the VMM. It minimizes $S(f_1, \ldots, f_M) =$

$$\sum_{i=1}^{n} \{\boldsymbol{y}_i - \boldsymbol{f}(x_i)\}^T \, \mathbf{W}_i \, \{\boldsymbol{y}_i - \boldsymbol{f}(x_i)\} \; + \; \sum_{m=1}^{M} \lambda_m \int_a^b \{f_m''(x)\}^2 \, dx \qquad (4.3)$$

over $f_j \in \mathcal{W}_2^2[a, b]$. Here, the $\mathbf{W}_i = \boldsymbol{\Sigma}_i^{-1}$ are known. The quantity S is a natural extension of the objective function of an ordinary (weighted) cubic smoothing spline, (2.48), and it fits naturally in the roughness penalty approach described in Green and Silverman (1994). The smoothing parameters $\lambda_1, \ldots, \lambda_M$ are nonnegative, and the design points satisfy $a \le x_1 < \cdots < x_n \le b$.

A fundamental result is that the solution to (4.3) consists of M natural cubic splines (NCSs). The argument follows that of the $M = 1$ case and is as follows. Suppose that each f_m is any curve that is not a NCS with knots at the x_i. Let $\widetilde{\boldsymbol{f}}(x_i)$ be a vector of NCS interpolants to the values $\boldsymbol{f}(x_i)$; since by definition, $\widetilde{f}_m(x_i) = f_m(x_i)$ for all i and m, it follows readily that

$$\sum_{i=1}^{n} \{\boldsymbol{y}_i - \widetilde{\boldsymbol{f}}(x_i)\}^T \mathbf{W}_i \{\boldsymbol{y}_i - \widetilde{\boldsymbol{f}}(x_i)\} \; = \; \sum_{i=1}^{n} \{\boldsymbol{y}_i - \boldsymbol{f}(x_i)\}^T \mathbf{W}_i \{\boldsymbol{y}_i - \boldsymbol{f}(x_i)\}.$$

Because of the optimality property of the NCS interpolant, $\int \widetilde{f}_m''^2 \, dx < \int f_m''^2 \, dx$ for all m, and hence (since $\lambda_m > 0$) we can conclude that $S(\widetilde{\boldsymbol{f}}) < S(\boldsymbol{f})$. This means that, unless each f_m is an NCS, we can find an NCS for each component function which attains a smaller value of the penalized sum of squares (4.3); then, it follows that the minimizers of S must each be an NCS. Thus, each f_m is a piecewise-cubic polynomial with continuous first and second derivatives on $[a, b]$. This important result has practical consequences, e.g., each component function can be written as a linear combination of B-splines. We exploit this property in Sect. 4.2.1.3.

Miller and Wegman (1987) identify a number of special cases when the vector spline can be estimated by M separate ordinary smoothing splines:

(1) When all the $\boldsymbol{\Sigma}_i$ are diagonal.
(2) When all the $\boldsymbol{\Sigma}_i$ are equal. If its spectral decomposition is $\mathbf{P}\boldsymbol{\Lambda}\mathbf{P}^T$, say, then it is possible to univariately smooth each component of $\mathbf{P}^T\boldsymbol{y}_i$ since this response has a diagonal covariance matrix. An example of this case is given in Sect. 4.2.1.2.
(3) More generally, if the $\boldsymbol{\Sigma}_i$ are *simultaneously diagonalizable* ($\boldsymbol{\Sigma}_i = \mathbf{P}\boldsymbol{\Lambda}_i\mathbf{P}^T$), then separate weighted smoothing is applicable.

More generally, Fessler (1991) proposed an algorithm based on the Reinsch method described in Sect. 2.4.4.1, and implemented his technique in a C program called VSPLINE. The cost of estimating the $f_m(\cdot)$ is approximately $n(\frac{27}{2}M^3 + O(M^2))$ flops. Unfortunately, it can be numerically unstable as in the $M = 1$ case, and therefore a B-spline-based algorithm has been developed (Sect. 4.2.1.3). It is currently invoked by `s()` within `vgam()`, and `vsmooth.spline()`.

Equation (4.3) can be written as

$$(\boldsymbol{y} - \boldsymbol{f})^T \boldsymbol{\Sigma}^{-1}(\boldsymbol{y} - \boldsymbol{f}) + \boldsymbol{f}^T \mathbf{K}\boldsymbol{f}, \tag{4.4}$$

where $\boldsymbol{f} = (f_1(x_1), \ldots, f_M(x_1), \ldots, f_1(x_n), \ldots, f_M(x_n))^T$, $\boldsymbol{y} = (\boldsymbol{y}_1^T, \ldots, \boldsymbol{y}_n^T)^T$ and $\boldsymbol{\Sigma} = \mathrm{diag}(\boldsymbol{\Sigma}_1, \ldots, \boldsymbol{\Sigma}_n)$. The roughness penalty matrix now equals $\mathbf{K} = (\mathbf{Q}\mathbf{T}^{-1}\mathbf{Q}^T) \otimes \mathrm{diag}(\lambda_1, \ldots, \lambda_M)$ where $\mathbf{Q}$ and $\mathbf{T}$ are defined in Sect. 2.4.4.1. Differentiating (4.4) with respect to $\boldsymbol{f}$ shows that the minimization occurs when

$$\widehat{\boldsymbol{f}} = \mathbf{A}(\boldsymbol{\lambda})\,\boldsymbol{y} \quad \text{where} \quad \mathbf{A}(\boldsymbol{\lambda}) = (\mathbf{I}_{nM} + \boldsymbol{\Sigma}\mathbf{K})^{-1}. \tag{4.5}$$

This expression for the smoother matrix reveals that it does not depend on $\boldsymbol{y}$, therefore vector splines are linear smoothers, provided that $\boldsymbol{\lambda}$ is known and fixed. Many of its properties presented below derive from this fact.

Some basic properties of vector splines are as follows.

1. They have $2M$ unit eigenvalues corresponding to constant and linear functions for each component function f_m.
2. Large values of λ_m produce smoother curves for $f_m(\cdot)$, and $\lambda_m = 0$ corresponds with no smoothing at all: $\widehat{f}_m(x_i) = y_{im}$ is an interpolation spline. As $\lambda_m \to \infty$, $\widehat{f}_m(x_2) \to \widehat{\beta}_{(m)1} + \widehat{\beta}_{(m)2}\,x_2$ is the LS solution.

4.2.1.1 Other 'vector splines'

Compared with the literature on the ordinary cubic spline, there is a paucity of literature on vector splines. However, the vector splines here differ from others bearing the same name in the literature, e.g., Wegman (1981), Wahba (1982) considered vector splines on the sphere, and Amodei and Benbourhim (1991). The vector spline of Wang (2011, Sect.8.2.3), which has the closest similarity with the one considered here, is based on a large framework called semiparametric linear regression models, that includes the projection pursuit regression model $y_i = \beta_1 + \sum_{k=1}^{R} f_k(\boldsymbol{\beta}_k^T \boldsymbol{x}_i) + \varepsilon_i$, and the Gaussian varying-coefficient model (Sect. 10.2.1), as special cases.

4.2.1.2 Equivalent Kernels

The same principle for ordinary smoothers may be applied to linear vector smoothers, except that there are M 'curves'—one for each response. Some of the effects of joint smoothing can be seen by considering a simple example of the simultaneously diagonalizable $\boldsymbol{\Sigma}_i$ case.

Figure 4.2 displays the equivalent kernel for $\widehat{f}_j(0)$ from a vector spline fitted to a data set generated from

$$\boldsymbol{y}_i = \begin{pmatrix} \sin x_i \\ \cos x_i \end{pmatrix} + \boldsymbol{\varepsilon}_i, \qquad \boldsymbol{\varepsilon}_i \ \sim \ N_2\left(\mathbf{0}, \ \boldsymbol{\Sigma}_i = \frac{1}{100}\begin{pmatrix} 1 & \rho \\ \rho & 1 \end{pmatrix}\right) \tag{4.6}$$

independently, where the x_i are equidistant on $[-\pi, \pi]$ and $n = 25$. Data were generated at 5 different values of ρ. For simplicity, only the 25th row of $\mathbf{A}(\boldsymbol{\lambda})$ is plotted, with $\boldsymbol{\lambda}$ chosen to give the same amount of smoothing ($\lambda_1 = \lambda_2$, and with about 8 degrees of freedom ($= df_{(m)}$)). The 25th row corresponds to a fitted value of the first component function evaluated at $x = 0$ which is located in the central interior.

Several features can be seen. The equivalent kernel for the first component function is very similar to the theoretical EK an ordinary cubic spline ((2.92), as plotted in Fig. 2.19). As expected, when $\rho = 0$, $\widehat{f}_1(0)$ is a weighted average of the y_{i1} only because the errors between the component functions are independent. The weighting of y_{i1} for the first component function does not differ much with respect to ρ, whereas the weighting of y_{i2} for the first function component increases in magnitude with increasing $|\rho|$, but its sign is the opposite of ρ.

4.2.1.3 Computation by B-Splines

VGAM implements the following algorithm based on B-splines for computing vector splines. It is to be expected to be more numerically stable than the Reinsch-Fessler algorithm. The vector smoothing spline solution consists of component functions f_m that are spline functions; the solution to (4.3) consists of M NCSs, therefore we can write

$$f_m(x) \;=\; \sum_{j=1}^{n^*+2} \theta_{jm}\, B_j(x), \quad m = 1, \dots, M, \tag{4.7}$$

where n^* is the 'effective' n as for O-splines (Sect. 2.4.4.3), θ_{jm} are coefficients, and $B_j(x)$ are B-spline basis functions. Defining the $n \times (n^* + 2)$ matrix $\mathbf{B}$ and the $(n^* + 2) \times (n^* + 2)$ penalty matrix $\boldsymbol{\Omega}$ by

$$\left[(\mathbf{B})_{ij}\right] \;=\; B_j(x_i) \quad \text{and} \quad \left[(\boldsymbol{\Omega})_{ij}\right] \;=\; \int_a^b B_i''(x)\, B_j''(x)\, dx,$$

we can rewrite (4.3) as

$$(\boldsymbol{y} - \mathbf{B}_*\boldsymbol{\theta})^T \mathbf{W}(\boldsymbol{y} - \mathbf{B}_*\boldsymbol{\theta}) + \boldsymbol{\theta}^T \boldsymbol{\Omega}_* \boldsymbol{\theta}, \tag{4.8}$$

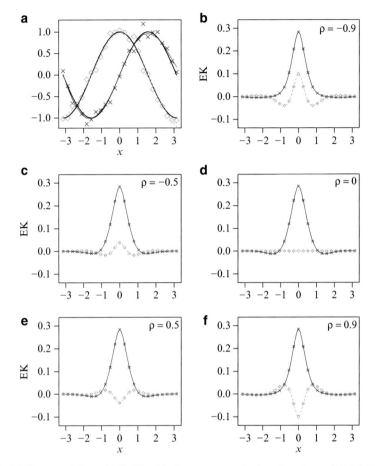

Fig. 4.2 (**a**) Data used from (4.6). The *black curves* are the true functions; (**b**)–(**f**) Equivalent kernels based on (4.6) for various values of ρ. The row of the influence matrix corresponds to $\widehat{f}_1(x = 0)$, and is midway between the boundaries.

where $\boldsymbol{y} = (\boldsymbol{y}_1^T, \dots, \boldsymbol{y}_n^T)^T$, $\mathbf{W} = \mathrm{Diag}(\mathbf{W}_1, \dots, \mathbf{W}_n)$, $\mathbf{B}_* = \mathbf{B} \otimes \mathbf{I}_M$, $\boldsymbol{\Omega}_* = \boldsymbol{\Omega} \otimes \mathrm{Diag}(\boldsymbol{\lambda})$, and $\boldsymbol{\lambda} = (\lambda_1, \dots, \lambda_M)^T$. The columns of $\mathbf{B}$ are the B-spline basis functions evaluated at the sorted values of x_i, therefore $\mathbf{B}$ has bandwidth 4 and has $O(n^{1/4}) \ll n$ columns when n is large. One can think of (4.8) as generalized ridge regression. Note that $\boldsymbol{\Omega}$ is symmetric and has half-bandwidth 4 so that $\boldsymbol{\Omega}_*$ has half-bandwidth $3M + 1$.

Let $\boldsymbol{\theta} = (\theta_{11}, \dots, \theta_{1M}, \dots, \theta_{(n^*+2)1}, \dots, \theta_{(n^*+2)M})^T$ be all the parameters that are estimated. Setting the derivative of (4.8) with respect to $\boldsymbol{\theta}$ as $\mathbf{0}$ gives the solution

$$\left(\mathbf{B}_*^T \mathbf{W} \mathbf{B}_* + \boldsymbol{\Omega}_*\right) \widehat{\boldsymbol{\theta}} = \mathbf{B}_*^T \mathbf{W} \boldsymbol{y}. \tag{4.9}$$

The matrix $\mathbf{M} = \mathbf{B}_*^T \mathbf{W} \mathbf{B}_* + \boldsymbol{\Omega}_*$ is symmetric and positive-definite and has half-bandwidth $4M$, and therefore $\widehat{\boldsymbol{\theta}}$ can be solved for by computing its Cholesky decomposition. The fitted values are $\widehat{\boldsymbol{f}}_m = \mathbf{B}\,\widehat{\boldsymbol{\theta}}^{(m)}$ where $\boldsymbol{\theta} = \mathrm{vec}\left(\widehat{\boldsymbol{\theta}}^{(1)}, \ldots, \widehat{\boldsymbol{\theta}}^{(M)}\right)$. The influence matrix is

$$\mathbf{A}(\boldsymbol{\lambda}) \;=\; \mathbf{B}_* \left(\mathbf{B}_*^T \mathbf{W} \mathbf{B}_* + \boldsymbol{\Omega}_*\right)^{-1} \mathbf{B}_*^T \mathbf{W}. \tag{4.10}$$

The diagonal elements of this smoother matrix (for df calculations) may be obtained by firstly applying the Hutchinson and de Hoog (1985) algorithm (Sect. A.3.1) to $\mathbf{M}$ to get the $6M+1$ central bands of its (symmetric) inverse. Then the M central bands of $\mathbf{B}_* \mathbf{M}^{-1} \mathbf{B}_*^T$ are computed by performing block-quadratic-form-type calculations. The diagonal of this matrix gives the pointwise Bayesian variances. Finally, the diagonal elements of this matrix post-multiplied by $\mathbf{W}$ are computed; this is $\mathrm{Diag}(\mathbf{A})$.

4.2.1.4 Linear Constraints on the Functions

Actually, the estimation of vector splines is based on fitting a linear model first and then a vector spline is fitted to the residuals. This modification leads to better numerical properties for additive models where there are multiple x_k, especially when they are highly correlated. In such cases, the algorithm used is called *modified vector backfitting* which is described in Sect. 4.3.2.2. For this, we adapt the notation here to match Chap. 3 by writing $x_1 = 1$ and $x_2 = x$.

Consider linear constraints on the component functions:

$$\boldsymbol{f}(x_2) \;=\; \mathbf{H}_1\,\boldsymbol{\beta}_{(1)}^* + \mathbf{H}_2\,\boldsymbol{f}_2^*(x_2), \tag{4.11}$$

where $\boldsymbol{f}_2^*(x_2) = (f_{(1)2}^*(x_2), \ldots, f_{(\mathcal{R}_2)2}^*(x_2))^T$ is a possibly reduced vector of unknown component functions, and $\boldsymbol{\beta}_{(1)}^*$ is a vector of unknown intercepts. Computationally, vector splines should be decomposed into a linear and nonlinear component

$$\boldsymbol{f}(x_2) \;=\; \left\{\mathbf{H}_1\,\boldsymbol{\beta}_{(1)}^* + x_2\,\mathbf{H}_2\,\boldsymbol{\beta}_{(2)}^*\right\} + \mathbf{H}_2\,\boldsymbol{r}_{(2)}^*(x_2). \tag{4.12}$$

The projection step involves minimizing the VLM residual sum of squares

$$\sum_{i=1}^{n} \left(\boldsymbol{y}_i - \sum_{k=1}^{2} \mathbf{H}_k\,\boldsymbol{\beta}_{(k)}^*\,x_{ik}\right)^T \mathbf{W}_i \left(\boldsymbol{y}_i - \sum_{k=1}^{2} \mathbf{H}_k\,\boldsymbol{\beta}_{(k)}^*\,x_{ik}\right).$$

Next, the nonlinear fit $\boldsymbol{r}_{(2)}^*$ is obtained by vector spline smoothing the scatter plot data

$$\left(x_{i2},\; \left(\mathbf{H}_2^T \mathbf{W}_i \mathbf{H}_2\right)^{-1} \mathbf{H}_2^T \mathbf{W}_i \left(\boldsymbol{y}_i - \sum_{k=1}^{2} \mathbf{H}_k\,\widehat{\boldsymbol{\beta}}_{(k)}^*\,x_{ik}\right),\; \left(\mathbf{H}_2^T \mathbf{W}_i \mathbf{H}_2\right)^{-1}\right). \tag{4.13}$$

Figure 4.3 illustrates the decomposition (4.12) for a single component function. The fitted LS line is evident, and then a linear combination of B-splines makes up the nonlinear component.

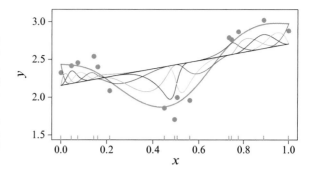

Fig. 4.3 O-smoothing spline (*blue curve*) fitted to scatter plot data •, which is equal to a least squares fit (*black line*) plus the sum of individual B-spline basis functions; cf. Fig. 2.8. It corresponds to the decomposition (4.12). The rugplot denotes the position of the x_i.

The function `vsmooth.spline()`, which fits a vector spline by this method, is described in Sect. 4.4.2.

4.2.1.5 An Iterative Vector Spline Solution

An iterative solution to the vector spline problem can be seen by partitioning $\boldsymbol{f}(x_i)$ in (4.3) into $(f_1(x_i), \boldsymbol{f}_2(x_i)^T)^T$, and supposing an estimate of $\boldsymbol{f}_2(x_i)$ is available. Then (4.3) can be written

$$
\sum_{i=1}^{n} \begin{pmatrix} y_{i1} - f_1(x_i) \\ \boldsymbol{y}_{i2} - \boldsymbol{f}_2(x_i) \end{pmatrix}^T \begin{pmatrix} w_{i11} & \boldsymbol{w}_{i21}^T \\ \boldsymbol{w}_{i21} & \mathbf{W}_{i22} \end{pmatrix} \begin{pmatrix} y_{i1} - f_1(x_i) \\ \boldsymbol{y}_{i2} - \boldsymbol{f}_2(x_i) \end{pmatrix} + \sum_{j=1}^{M} \lambda_j \int_a^b \{f_j''(t)\}^2 \, dt \;\; =
$$

$$
C_1 + \sum_{i=1}^{n} w_{i11} \left[\left\{ y_{i1} + \frac{\boldsymbol{w}_{i21}^T(\boldsymbol{y}_{i2} - \boldsymbol{f}_2(x_i))}{w_{i11}} \right\} - f_1(x_i) \right]^2 + \lambda_1 \int_a^b \{f_1''(t)\}^2 \, dt
$$

for some constant C_1. That is, f_1 can be estimated by applying a cubic smoothing spline to a modified response with weights w_{i11}. An iterative solution is therefore to update each component function f_j from the others in this manner. Initial values for the f_j could be obtained by smoothing (x_i, y_{ij}) with weights $(\mathbf{W}_i)_{jj}$.

The above can be adapted for the additive model so that there is a nested loop in the computation. While the method is not implemented in **VGAM** presently, the idea allows VGAMs to be fitted using GAM-type software, i.e., it bypasses the need for a vector smoother.

4.2.1.6 Choosing Vector Smoothing Parameters

Fessler (1991) suggested several criteria for vector spline smoothing parameter selection. These can be regarded as applying more generally to all linear vector smoothers. He suggested

$$
\mathsf{MSE}(\boldsymbol{\lambda}) \;=\; \frac{1}{n} \sum_{i=1}^{n} \left(\widehat{\boldsymbol{f}}(x_i) - \boldsymbol{f}(x_i) \right)^T \left(\widehat{\boldsymbol{f}}(x_i) - \boldsymbol{f}(x_i) \right),
$$

$$\text{UR}(\boldsymbol{\lambda}) \;=\; \frac{1}{n}\left\{\boldsymbol{y}^T(\mathbf{I}_{nM}-\mathbf{A})^T(\mathbf{I}_{nM}-\mathbf{A})\boldsymbol{y}\right\} - \frac{2}{n}\,\text{trace}\,(\boldsymbol{\Sigma}(\mathbf{I}_{nM}-\mathbf{A})) + \frac{1}{n}\,\text{trace}\,(\boldsymbol{\Sigma})\,,$$

and $\text{CV}(\boldsymbol{\lambda}) =$

$$\frac{1}{n}\sum_{i=1}^{n}\left(\boldsymbol{y}_i-\widehat{\boldsymbol{f}}(x_i)\right)^T \left(\mathbf{I}_M-\mathbf{A}_{(nn)}(\boldsymbol{\lambda})^T\right)^{-1}\boldsymbol{\Sigma}_i^{-1}\left(\mathbf{I}_M-\mathbf{A}_{(nn)}(\boldsymbol{\lambda})\right)^{-1}\left(\boldsymbol{y}_i-\widehat{\boldsymbol{f}}(x_i)\right),$$

where $\mathbf{A}_{(nn)}$ is the nth $M \times M$ block diagonal submatrix of $\mathbf{A}(\boldsymbol{\lambda})$.

His software **VSPLINE** supports span selection by unbiased risk, CV and GCV. For the latter, $\boldsymbol{\lambda}^{\text{GCV}}$ is chosen by minimizing the generalized cross validation (GCV) score: $\text{GCV}(\boldsymbol{\lambda}) =$

$$\frac{1}{n}\sum_{i=1}^{n}\left(\boldsymbol{y}_i-\widehat{\boldsymbol{f}}(x_i)\right)^T \boldsymbol{\Sigma}_i^{-1}\left(\boldsymbol{y}_i-\widehat{\boldsymbol{f}}(x_i)\right) \Big/ \left(\frac{1}{n}\,\text{trace}\{\mathbf{I}_{nM}-\mathbf{A}(\boldsymbol{\lambda})\}\right)^2. \quad (4.14)$$

This may be computed cheaply by the Hutchinson and de Hoog (1985) algorithm.

4.2.1.7 RKHS †

Reproducing kernel Hilbert space (RKHS) theory provides an elegant mathematical framework for handling penalized regression problems, including the vector spline. Indeed, one major reason for its use is its generality. For $M = 1$, the objective function (4.3) can be written in the form

$$n^{-1}\sum_{i=1}^{n}\left[y_i-f(x_i)\right]^2 + \lambda\,J(f), \quad (4.15)$$

where the penalty $J(f)$ can be expressed as an inner product $\langle f, f \rangle$, with

$$\langle f, g \rangle \;=\; \int_a^b f''(t)\,g''(t)\,dt. \quad (4.16)$$

The ideas behind RKHS are based on vector spaces, Banach spaces and Hilbert spaces. The latter allows concepts from finite-dimensional linear algebra to be applied to infinite-dimensional spaces of functions. As the area is specialized, only some sketchy details are given here, and the reader is directed to the references in the bibliography of Chap. 2 for full details. The description here is based on Wang (2011) and Nosedal-Sanchez et al. (2012). Some terms used in this section are defined in the Glossary.

A vector space $\mathcal{V}$ is a set[1] whose elements (generically called "vectors", but can be an entity such as a function, matrix or sequence) fundamentally satisfy 8 axioms, e.g., involving operators called 'addition' and 'multiplication'. A vector space endowed with a real-valued function called a norm (denoted by $\|\cdot\|$) is known as a normed vector space. The norm generalizes the concept of a length, and must satisfy 4 basic mathematical properties, e.g., $\|\boldsymbol{u}\| \geq 0$. A complete normed space is known as a Banach space—by 'complete', it is meant that every

---

[1] More strictly, $\mathcal{V}$ is defined over a field, and the operators called 'addition' and 'multiplication', and there is an element called $\mathbf{0}$.

Cauchy sequence in the normed space converges to an element of that normed space. Now an inner product is a mapping $\mathcal{V} \times \mathcal{V} \to \mathbb{R}$, and it generalizes the dot product $\boldsymbol{u}^T \boldsymbol{v}$ of two vectors. The squared norm can be defined using an inner product by $\| \cdot \|^2 = \langle \cdot, \cdot \rangle$. Then put simply, a Hilbert space is a complete inner product space. The inner product supplies structure to the space by allowing not only length to be measured but also angles and projections, e.g., two vectors $\boldsymbol{u}$ and $\boldsymbol{v}$ in a Hilbert space $\mathcal{H}$ are orthogonal (perpendicular) if $\langle \boldsymbol{u}, \boldsymbol{v} \rangle = 0$.

Hilbert spaces can be finite-dimensional or infinite-dimensional, e.g., $\mathcal{V} = \mathbb{R}^n = \{\boldsymbol{x} = (x_1, \ldots, x_n)^T : x_i \in \mathbb{R}\}$ with inner product $\langle \boldsymbol{u}, \boldsymbol{v} \rangle = \boldsymbol{u}^T \mathbf{A} \boldsymbol{v}$ (for any positive-definite matrix $\mathbf{A}$) is n-dimensional, whereas for positive integer m, the Sobolev space $\mathcal{V} = \{f : f^{(\nu)}(0) = 0 \text{ and } f^{(\nu)} \text{ are absolutely continuous on } \mathbb{R}, \text{ for } \nu = 0, \ldots, m-1, \text{ and } f^{(m)} \in \mathcal{L}_2\}$ with inner product $\langle f, g \rangle = \int_{-\infty}^{\infty} f^{(m)}(s) \, g^{(m)}(s) \, ds$ is infinite-dimensional. In this section, a "Sobolev space" is used loosely to mean a function space with a norm involving derivatives.

A Hilbert space $\mathcal{H}$ of real-valued functions defined on some domain Ω is an RKHS if there exists a function $K(s,t)$ (called the *reproducing kernel*) defined on $\Omega \times \Omega$ such that

(i) for every s, $K(s,t)$ as a function of t is an element of $\mathcal{H}$; and
(ii) for every $t \in \Omega$ and every $f \in \mathcal{H}$,

$$f(t) = \langle f(s), \ K(s,t) \rangle_s. \tag{4.17}$$

The subscript s on the inner product indicates that the inner product applies to functions of s. This equation is known as the *reproducing property*.

RKHSs possess a number of properties, including the following (Aronszajn, 1950).

1. *Uniqueness.* If an RK K exists, then it is unique.
2. *Existence.* An RK K exists iff for every $t \in \Omega$, $f(t)$ is a continuous functional for f running through $\mathcal{H}$. Continuous functionals are defined below.
3. $K(\cdot, \cdot)$ *is a non-negative definite function.* That is, for all real $c_1, \ldots, c_n$,

$$\sum_{i=1}^{n} \sum_{j=1}^{n} c_i \, c_j \, K(t_i, t_j) \geq 0$$

for all $t_1, \ldots, t_n$ in Ω.
4. *Converse.* To every non-negative definite function $K(\cdot, \cdot)$ there corresponds a uniquely determined quadratic form forming a Hilbert space and admitting $K(\cdot, \cdot)$ as a reproducing kernel. This is known as the Moore–Aronszajn theorem.
5. *RK sums.* If $\mathcal{H}$ has RK K, and $\mathcal{H}_0$ and $\mathcal{H}_1$ are two subspaces of $\mathcal{H}$ (such that $\mathcal{H}_0 \oplus \mathcal{H}_1 = \mathcal{H}$ with $\mathcal{H}_0 \cap \mathcal{H}_1 = \{0\}$) and having RKs K_0 and K_1, respectively, then

$$K(\cdot, t) = K_0(\cdot, t) + K_1(\cdot, t). \tag{4.18}$$

Soon, it will be seen that $K(\cdot, t_i)$ provides a basis for representing the solution.

An example of an RKHS is a Sobolev space defined on $[0,1]$: $\mathcal{H} = \{f : f \text{ is absolutely continuous on } [0,1], \ f(0) = f(1) = 0, \ f' \in \mathcal{L}_2[0,1]\}$ with $\langle f, g \rangle = \int_0^1 f'(t) \, g'(t) \, dt$. Then it can be shown that $K(s,t) = (1-t)s$ for $s \leq t$, and $K(s,t) = (1 - s)t$ for $t \leq s$. In contrast, the space of squared integrable functions on the closed unit interval, $\mathcal{L}_2[0,1]$, is a Hilbert space but not an RKHS, because the boundedness property (4.19) does not hold.

A function $\phi : \mathcal{V} \to \mathbb{R}$ is known as a *functional*, e.g., $\phi(f) = \int_a^b [f''(t)]^2 \, dt$, where a and b are fixed. An important type of functional is the *evaluation functional*: let $\mathcal{V}$ be a vector space of functions $f : \Omega \to \mathbb{R}$ for some bounded domain $\Omega \in \mathbb{R}^d$, then e is an evaluation functional if $e(f) = f(t)$ for $t \in \Omega$ and $f \in \mathcal{V}$. It is commonly written $e_t(f)$ to emphasize that the function f is evaluated at t. Observe that $e_t(\cdot)$ is a linear functional because $e_t(c_1 \, f + c_2 \, g) = c_1 \, f(t) + c_2 \, g(t) = c_1 \, e_t(f) + c_2 \, e_t(g)$, for constants c_i. Then an alternative definition of an RKHS is that it is a Hilbert space $\mathcal{H}$ of real-valued functions defined on a set Ω having evaluation functionals that are bounded, i.e., there exists some constant C such that

$$|\phi(t)| \leq C \|f\|, \qquad \forall f \in \mathcal{H}. \tag{4.19}$$

Additionally, it may be shown that boundedness here is also equivalent to the evaluation functionals being continuous, i.e., $\lim_{n\to\infty} f_n = f$ implies $\lim_{n\to\infty} e_t(f_n) = e_t(f)$. An important result known as the *Reisz representation theorem* states that for a Hilbert space $\mathcal{H}$ and a linear continuous functional ϕ, there exists a unique vector $g \in \mathcal{H}$ such that

$$\phi(f) = \langle f, g \rangle, \quad \forall f \in \mathcal{H}. \tag{4.20}$$

Then g is called the *representation* or *representer* of ϕ. Applying this result specifically to the evaluation functional and an RKHS, we can see from (4.17) that $g = K$. We shall see soon that instead of expressing the estimate of the vector spline component function f_j in terms of basis functions only, representers (in a subspace called $\mathcal{H}_1$) will be used instead.

Note that, from the reproducing property (4.17),

$$K(s,t) = \langle K(\cdot, s), \, K(\cdot, t) \rangle. \tag{4.21}$$

Example 1: Linear Interpolating Splines

A simple example developed in Nosedal-Sanchez et al. (2012) concerns scatter plot data (t_i, y_i), $i = 0, \ldots, n$, where $y_0 = 0$ and $0 = t_0 < t_1 < \cdots < t_n = 1$. The problem is to find the smoothest function f that interpolates these data, subject to the constraint that $f \in \mathcal{H}$ where $\mathcal{H} = \{f : f$ is absolutely continuous on $[0,1]$, $f(0) = 0$, $\int_0^1 (f'(t))^2 \, dt < \infty\}$. Define $\langle f, g \rangle = \int_0^1 f'(t) \, g'(t) \, dt$ on $\mathcal{H}$, so that the squared norm can be used for measuring smoothness. It can be shown that if $K(s,t) = \min(s,t)$ on $[0,1] \times [0,1]$ then (4.17) holds:

$$\begin{aligned} \langle f(s), \, K(s,t) \rangle_s &= \int_0^1 f'(s) \cdot \frac{\partial K(s,t)}{\partial s} \, ds \\ &= \int_0^t f'(s) \cdot 1 \, ds + \int_t^1 f'(s) \cdot 0 \, ds = f(t) - f(0) = f(t). \end{aligned}$$

Thus $K(\cdot, \cdot)$ is the reproducing kernel, and $\mathcal{H}$ is an RKHS. Writing $K_i(s) = \min(s, t_i)$ for $i = 1, \ldots, n$, then $\langle f, K_i \rangle = f(t_i)$, and

$$\widehat{f}(t) = \sum_{i=1}^{n} c_i \, K_i(t) \tag{4.22}$$

$$= \begin{cases} (c_1 + \cdots + c_n) \, t, & t \leq t_1; \\ c_1 \, t_1 + (c_2 + \cdots + c_n) \, t, & t_1 \leq t \leq t_2; \\ c_1 \, t_1 + c_2 \, t_2 + (c_3 + \cdots + c_n) \, t, & t_2 \leq t \leq t_3; \\ \quad \vdots & \quad \vdots \\ c_1 \, t_1 + \cdots + c_{n-1} \, t_{n-1} + c_n \, t, & t_{n-1} \leq t \leq t_n, \end{cases}$$

where the $c_i \in \mathbb{R}$ are coefficients that can be estimated. Note that

$$\langle K_i, \; K_j \rangle = K_j(t_i) = K_i(t_j) = \min(t_i, t_j)$$

by (4.21).

Example 2: Vector Splines

For vector splines, we want to minimize a penalized least squares criterion, subject to each component function f_j belonging to the RKHS of an ordinary cubic smoothing spline. We start off by considering the ordinary cubic smoothing spline case by solving the penalized least squares criterion (4.15), written equivalently as

$$\min_{f \in \mathcal{H}} \; n^{-1} \sum_{i=1}^{n} [y_i - f(t_i)]^2 + \lambda \, \|P_1 f\|^2 \tag{4.23}$$

where P_1 is the projection of f onto some RKHS (called $\mathcal{H}_1$ below). We write $t_i = x_i$ in this example.

Following Wang (2011), a real-valued function f with continuous derivatives up to $f^{(m-1)}$ at a, and $f^{(m)} \in \mathcal{L}_2[a, b]$, has Taylor expansion

$$f(t) = \sum_{\nu=0}^{m-1} \frac{(t-a)^\nu \, f^{(\nu)}(a)}{\nu!} + \int_a^b \frac{(t-s)_+^{m-1}}{(m-1)!} \, f^{(m)}(s) \, ds, \tag{4.24}$$

where the integral is called the remainder. For cubic smoothing splines, we have $m = 2$. Also, let $\mathcal{W}_2^m[a, b]$ be

$$\{f : f, f', \ldots, f^{(m-1)} \text{ are absolutely continuous on } [a, b], \; f^{(m)} \in \mathcal{L}_2[a, b]\},$$

and let the inner product be

$$\langle f, g \rangle = \sum_{\nu=0}^{m-1} f^{(\nu)}(a) \cdot g^{(\nu)}(a) + \int_a^b f^{(m)}(s) \cdot g^{(m)}(s) \, ds, \tag{4.25}$$

which is an RKHS. The integral vanishes for polynomials f and g of degree $m - 1$ or less. As these functions are not penalized, they span the *null space*, which is called $\mathcal{H}_0$ in the following. Decompose

$$\mathcal{W}_2^m[a, b] = \mathcal{H}_0 \oplus \mathcal{H}_1 \tag{4.26}$$

so that $\mathcal{H}_1 = \{f : f^{(\nu)}(a) = 0 \text{ for } \nu = 0, \ldots, m-1; \ f^{(m)} \in \mathcal{L}_2[a,b]\}$. Then we could write (4.25) as $\langle f, g \rangle_{\mathcal{H}} = \langle f, g \rangle_{\mathcal{H}_0} + \langle f, g \rangle_{\mathcal{H}_1}$. From the Taylor series,

$$\mathcal{H}_0 = \text{span}\left\{1, (t-a), \ldots, \frac{(t-a)^{m-1}}{(m-1)!}\right\}, \quad (= \{\phi_1(t), \ldots, \phi_m(t)\}, \text{ say}),$$

and it can be shown that these basis functions are orthonormal, i.e., $\langle \phi_j, \phi_k \rangle = 1$ if $j = k$, otherwise 0.

Because of (4.26), the RK of $\mathcal{H}$ can be expressed as the sum of two RKs, as in (4.18). For the first,

$$K_0(s,t) = \sum_{\nu=1}^{m} \phi_\nu(s)\,\phi_\nu(t) = \sum_{\nu=1}^{m} \frac{(s-a)^{\nu-1}}{(\nu-1)!}\frac{(t-a)^{\nu-1}}{(\nu-1)!}.$$

Now, for $\mathcal{H}_1$, it has RK

$$K_1(s,t) = \int_a^b G_m(t,u)\,G_m(s,u)\,du, \quad s,t \in [a,b], \tag{4.27}$$

where

$$G_m(t,s) = \frac{(t-s)_+^{m-1}}{(m-1)!} \tag{4.28}$$

is known as a *Green's function*. This can be seen by reconciling the Taylor series remainder term with the Reisz representation theorem under $\langle f, g \rangle_{\mathcal{H}_1}$: then it is seen that the representer $g = K_1$. Thus

$$f(t) = \int_a^b G_m(t,s)\,f^{(m)}(s)\,ds$$

for $f \in \mathcal{H}_1$.

By the Kimeldorf–Wahba representer theorem (see, e.g., Wahba, 1990, Thm 1.3.1), the solution is of the form

$$\widehat{f}_\lambda(t) = \sum_{\nu=1}^{m} d_\nu\,\phi_\nu(t) + \sum_{i=1}^{n} c_i\,K_1(t_i,t). \tag{4.29}$$

This means that a minimization problem over a possibly infinite-dimensional Hilbert space simplifies to an optimization problem in $\mathbb{R}^n$. Compared to our previous description of splines based on the more familiar subject of B-splines, the quantities $K_1(t_i,t)$ dictate the knot placement of the space of natural cubic splines.

From (4.29), the penalized least squares criterion (4.23) can be written as

$$n^{-1}\left\|\boldsymbol{y} - \mathbf{T}\boldsymbol{d} - \boldsymbol{\Xi}\boldsymbol{c}\right\|^2 + \lambda\,\boldsymbol{c}^T\boldsymbol{\Xi}\boldsymbol{c},$$

where $\boldsymbol{c} = (c_1, \ldots, c_n)^T$, $\boldsymbol{d} = (d_1, \ldots, d_m)^T$, $(\mathbf{T})_{ij} = \phi_j(t_i)$, and the Gram matrix $\boldsymbol{\Xi}$ has elements $(\boldsymbol{\Xi})_{ij} = \langle K_1(t,t_i),\ K_1(t,t_j)\rangle$. The solution is

$$\boldsymbol{c} = \mathbf{M}^{-1}\left[\mathbf{I}_n - \mathbf{T}\left(\mathbf{T}^T\mathbf{M}^{-1}\mathbf{T}\right)^{-1}\mathbf{T}^T\mathbf{M}^{-1}\right]\boldsymbol{y},$$

$$\boldsymbol{d} = (\mathbf{T}^T\mathbf{M}^{-1}\mathbf{T})^{-1}\mathbf{T}^T\mathbf{M}^{-1}\boldsymbol{y},$$

where $\mathbf{M} = \boldsymbol{\Xi} + n\lambda\mathbf{I}_n$, however, this is not practical computationally. Instead, setting the derivatives with respect to $\boldsymbol{c}$ and $\boldsymbol{d}$ to $\mathbf{0}$ yields the system of equations

$$\begin{aligned} \mathbf{M}\boldsymbol{\Xi}\boldsymbol{c} + \boldsymbol{\Xi}\mathbf{T}\boldsymbol{d} &= \boldsymbol{\Xi}\boldsymbol{y}, \\ \mathbf{T}^T\boldsymbol{\Xi}\boldsymbol{c} + \mathbf{T}^T\mathbf{T}\boldsymbol{d} &= \mathbf{0}, \end{aligned}$$

which can then be simplified to other sets of equations, one of which is

$$\begin{aligned} \mathbf{M}\boldsymbol{c} + \mathbf{T}\boldsymbol{d} &= \boldsymbol{y}, \\ \mathbf{T}^T\boldsymbol{c} &= \mathbf{0}. \end{aligned}$$

This set is amenable to the QR method.

Now for a (cubic) vector (smoothing) spline, we have each component function $f_j \in \mathcal{H}_j = \mathcal{H}_{0j} \oplus \mathcal{H}_{1j}$ for $j = 1, \ldots, M$, where $\mathcal{H}_j = \mathcal{W}_2^m[a, b]$ is Sobolev space defined earlier—it is an RKHS, and $\mathcal{H}_{0j}$ is an m-dimensional subspace that is not penalized. Once again, $m = 2$. The objective function to be minimized is

$$\sum_{i=1}^{n} \{\boldsymbol{y}_i - \boldsymbol{f}(x_i)\}^T \boldsymbol{\Sigma}_i^{-1} \{\boldsymbol{y}_i - \boldsymbol{f}(x_i)\} + \sum_{j=1}^{M} \lambda_j \|P_j f_j\|^2, \tag{4.30}$$

where P_j is the projection of f_j onto $\mathcal{H}_{0j}$.

By the Kimeldorf–Wahba representation theorem, the minimizer has the form

$$f_j(x) = \sum_{\nu=1}^{m} d_{\nu j} \, \phi_\nu(x) + \sum_{i=1}^{n} c_{ij} \, K_{1j}(x_i, x), \tag{4.31}$$

where $\{\phi_1, \ldots, \phi_m\} = \{1, x - a\}$ is the orthonormal basis of $\mathcal{H}_{0j}$, and $K_{1j}(\cdot, \cdot)$ is the RK of $\mathcal{H}_{1j}$. The objective function can be written in matrix form:

$$\left(\boldsymbol{y} - \widetilde{\mathbf{T}}\boldsymbol{d} - \widetilde{\boldsymbol{\Xi}}\boldsymbol{c}\right)^T \boldsymbol{\Sigma}^{-1} \left(\boldsymbol{y} - \widetilde{\mathbf{T}}\boldsymbol{d} - \widetilde{\boldsymbol{\Xi}}\boldsymbol{c}\right) + \boldsymbol{c}^T \widetilde{\boldsymbol{\Xi}}\boldsymbol{c}, \tag{4.32}$$

where $\boldsymbol{y} = (\boldsymbol{y}_1^T, \ldots, \boldsymbol{y}_n^T)^T$, $\widetilde{\mathbf{T}} = \mathbf{T} \otimes \mathbf{I}_M$, $\boldsymbol{c} = (c_{11}, \ldots, c_{1M}, c_{21}, \ldots, c_{2M}, \ldots, c_{nM})^T$, $\boldsymbol{d} = (d_{11}, \ldots, d_{1M}, d_{21}, \ldots, d_{2M}, \ldots, d_{mM})^T$, $\widetilde{\boldsymbol{\Xi}} = \boldsymbol{\Xi} \otimes \mathrm{diag}(\lambda_1, \ldots, \lambda_M)$, and $\boldsymbol{\Sigma} = \mathrm{diag}(\boldsymbol{\Sigma}_1, \ldots, \boldsymbol{\Sigma}_n)$. Note here that it is more convenient to absorb the λ_j into the $\widetilde{\boldsymbol{\Xi}}$. The solution could be computed by solving

$$\begin{pmatrix} \mathbf{I}_{nM} + \widetilde{\boldsymbol{\Xi}}\boldsymbol{\Sigma}^{-1} & \widetilde{\boldsymbol{\Xi}}\boldsymbol{\Sigma}^{-1}\widetilde{\mathbf{T}} \\ \widetilde{\mathbf{T}}^T\boldsymbol{\Sigma}^{-1} & \widetilde{\mathbf{T}}^T\boldsymbol{\Sigma}^{-1}\widetilde{\mathbf{T}} \end{pmatrix} \begin{pmatrix} \widetilde{\boldsymbol{\Xi}}\boldsymbol{c} \\ \boldsymbol{d} \end{pmatrix} = \begin{pmatrix} \widetilde{\boldsymbol{\Xi}}\boldsymbol{\Sigma}^{-1}\boldsymbol{y} \\ \widetilde{\mathbf{T}}^T\boldsymbol{\Sigma}^{-1}\boldsymbol{y} \end{pmatrix}.$$

4.2.2 Local Regression for Vector Responses †

The estimation of VGAMs in Sect. 4.3 by vector backfitting and vector splines within an IRLS algorithm engenders questions about its properties, such as convergence and statistical efficiency. To this end, in this section we extend the results of local regression for the $M = 1$ case (Sect. 2.4.6) to the $M = 2$ VMM case. We can gain insight into the advantages of vector smoothing and its asymptotic properties. This section is based on Welsh and Yee (2006), who investigated local linear regression under the more general conditions of seemingly unrelated regressions (SUR; Sect. 10.2.3). We will summarize some results from that article which are more relevant to the context here.

In this vein we summarize in this section some results of vector smoothing based on local regression so that light may be shed on the gain in efficiency that vector smoothing may bring in general (not just for vector smoothing splines), as opposed to smoothing M component functions separately. The starting point is to realize that, at each IRLS iteration, we are effectively solving the VMM as a building block. In particular, the working weight matrices $\mathbf{W}_i$ differ by i and are treated as known in the smoothing.

The material presented here applies to a univariate normal likelihood, since it maximizes a sum of squares. More generally, local regression can be applied to log-likelihoods from the exponential family, say, and this is the topic of *local likelihood estimation.*

Without loss of generality, we let each component function be locally approximated by an r-degree polynomial. Let $\mathbf{X}_x$ be defined as in (2.66), $\mathbf{X}_x^* = \mathbf{X}_x \otimes \mathbf{I}_M$, and $\mathbf{W}_x^* = \mathrm{Diag}(K_h(x_1 - x)\mathbf{W}_1, \ldots, K_h(x_n - x)\mathbf{W}_n)$. For the VMM (4.2), an r-degree *vector local polynomial kernel estimator* is

$$\widehat{\boldsymbol{f}}(x; r, h) \;=\; \widehat{\boldsymbol{\beta}}_x^{[1]}, \tag{4.33}$$

where $\widehat{\boldsymbol{\beta}}_x = \left(\widehat{\boldsymbol{\beta}}_x^{[1]\,T}, \ldots, \widehat{\boldsymbol{\beta}}_x^{[r+1]\,T} \right)^T$ minimizes the GLS criterion

$$\sum_{i=1}^{n} \left\| \boldsymbol{y}_i - \sum_{s=0}^{r} \boldsymbol{\beta}_x^{[s+1]}(x_i - x)^s \right\|_{K_h(x_i - x)\,\mathbf{W}_i}^2 . \tag{4.34}$$

One can see that the weight matrices $\mathbf{W}_i$ are downweighted the further x_i is away from the target point x, and the $M = 1$ case corresponds to the ordinary local polynomial kernel estimator (2.67).

The solution to (4.34) is

$$\widehat{\boldsymbol{\beta}}_x \;=\; \left(\mathbf{X}_x^{*\,T} \mathbf{W}_x^* \mathbf{X}_x^* \right)^{-1} \mathbf{X}_x^{*\,T} \mathbf{W}_x^* \boldsymbol{y}$$

[cf. (2.65)], so that (2.67) becomes

$$\widehat{\boldsymbol{f}}(x; r, h) \;=\; \left(\boldsymbol{e}_1^T \otimes \mathbf{I}_M \right) \widehat{\boldsymbol{\beta}}_x.$$

Thus, to handle vector responses, the essential difference is that the usual weighted multiple linear regression is supplanted by a weighted multivariate regression.

The estimation of the νth derivatives is straightforward: for $\nu = 0, \ldots, r$,

$$\widehat{\boldsymbol{f}}^{(\nu)}(x; r, h) \;=\; \nu! \left(\boldsymbol{e}_{\nu+1}^T \otimes \mathbf{I}_M \right) \left(\mathbf{X}_x^{*\,T} \mathbf{W}_x^* \mathbf{X}_x^* \right)^{-1} \mathbf{X}_x^{*\,T} \mathbf{W}_x^* \boldsymbol{y} \;=\; \nu! \, \widehat{\boldsymbol{\beta}}_x^{[\nu+1]}$$

[cf. (2.70)].

4.2.2.1 Local Regression for $M = 2$ †

Traditionally, local regression methods have been used by mathematical statisticians, because their asymptotic properties are relatively easy to obtain compared with splines. In this spirit, the vector smoothing problem (4.3) was considered from a local regression viewpoint by Welsh and Yee (2006). That paper, which considered a more general setting involving seemingly unrelated regressions (SUR; Sect. 10.2.3), has results relating to the vector smoothing problem extracted and presented here. They mainly considered the $M = 2$ component function case, and showed that the placement of the kernel weights in WLS estimators is very important in the SUR problem (to ensure that the estimator is asymptotically unbiased) but not in the vector measurement model (4.2). While the component estimators are asymptotically uncorrelated in the SUR model, they are asymptotically correlated in the VMM.

We consider the vector regression model in which observations $y_i^{(j)}$ are related to known, univariate explanatory variables $x_i^{(j)}$, $i = 1, \ldots, n$, $j = 1, \ldots, M$, by

$$
\begin{aligned}
y_i^{(1)} &= f_1(x_i^{(1)}) + \varepsilon_i^{(1)} \\
&\vdots \\
y_i^{(M)} &= f_M(x_i^{(M)}) + \varepsilon_i^{(M)},
\end{aligned}
\tag{4.35}
$$

where $f_1, \ldots, f_M$ are unknown regression functions and $\boldsymbol{\varepsilon}_i = (\varepsilon_i^{(1)}, \ldots, \varepsilon_i^{(M)})^T$ are independent random vectors with $E(\boldsymbol{\varepsilon}_i) = \mathbf{0}$.

Welsh and Yee (2006) considered the following two cases for $\mathrm{Var}(\boldsymbol{\varepsilon}_i)$, because the estimators have different asymptotic properties.

| **Case A**: $\mathrm{Var}(\boldsymbol{\varepsilon}_i) = \boldsymbol{\Sigma}_i$ | Here, the elements of $\boldsymbol{\Sigma}_i$ are treated as unknown constants. Write the diagonal elements as σ_{ji}^2, and the off-diagonal elements as $\sigma_{ji}\sigma_{ki}\rho_{jki}$. There are $nM(M+1)/2$ nuisance parameters so they cannot all be unknown. |

| **Case B**: $\mathrm{Var}(\boldsymbol{\varepsilon}_i) = \boldsymbol{\Sigma}(\boldsymbol{x}_i)$ | Here, $\boldsymbol{x}_i = (x_i^{(1)}, \ldots, x_i^{(M)})^T$. |

Note that the vector smoothing fitted at each IRLS iteration more closely corresponds to Case A but with effectively known working weights $\boldsymbol{\Sigma}_i^{-1}$. Hence we do not concern ourselves so much with Case B.

We can write (4.35) explicitly as M separate but dependent regression models. Let $\boldsymbol{y}^{(j)} = \left(y_1^{(j)}, \ldots, y_n^{(j)}\right)^T$, $\boldsymbol{x}^{(j)} = \left(x_1^{(j)}, \ldots, x_n^{(j)}\right)^T$, $\boldsymbol{\varepsilon}^{(j)} = \left(\varepsilon_1^{(j)}, \ldots, \varepsilon_n^{(j)}\right)^T$, and $\boldsymbol{f}^{(j)}(\boldsymbol{x}^{(j)}) = \left(f_j(x_1^{(j)}), \ldots, f_j(x_n^{(j)})\right)^T$ so we can write each model as

$$
\boldsymbol{y}^{(j)} = \boldsymbol{f}^{(j)}(\boldsymbol{x}^{(j)}) + \boldsymbol{\varepsilon}^{(j)}, \qquad j = 1, \ldots, M,
$$

where $E(\boldsymbol{\varepsilon}^{(j)}) = \mathbf{0}$, $\mathrm{Var}(\boldsymbol{\varepsilon}^{(j)}) = \mathbf{D}_{jj} = \mathrm{diag}(\sigma_{j1}^2, \ldots, \sigma_{jn}^2)$ and $\mathrm{Cov}(\boldsymbol{\varepsilon}^{(j)}, \boldsymbol{\varepsilon}^{(k)}) = \mathbf{D}_{jk} = \mathrm{diag}(\sigma_{j1}\sigma_{k1}\rho_{jk1}, \ldots, \sigma_{jn}\sigma_{kn}\rho_{jkn})$.

In the following, it is more convenient to stack vectors in a different order than what is usually used. For this, the superscript "$(*)$" is used to reflect this change. If we let $\boldsymbol{y}^{(*)} = \left(\boldsymbol{y}^{(1)T}, \ldots, \boldsymbol{y}^{(M)T}\right)^T$, $\boldsymbol{x} = \left(\boldsymbol{x}^{(1)T}, \ldots, \boldsymbol{x}^{(M)T}\right)^T$, $\boldsymbol{f}^{(*)}(\boldsymbol{x}) = \left(\boldsymbol{f}_1(\boldsymbol{x}^{(1)})^T, \ldots, \boldsymbol{f}_M(\boldsymbol{x}^{(M)})^T\right)^T$ and $\boldsymbol{\varepsilon} = \left(\boldsymbol{\varepsilon}^{(1)T}, \ldots, \boldsymbol{\varepsilon}^{(M)T}\right)^T$, then we can stack the component models so that

$$\boldsymbol{y}^{(*)} = \boldsymbol{f}^{(*)}(\boldsymbol{x}) + \boldsymbol{\varepsilon}^{(*)},$$

where $E(\boldsymbol{\varepsilon}^{(*)}) = \mathbf{0}$ and $\mathbf{V} = \mathrm{Var}(\boldsymbol{\varepsilon}^{(*)}) = (\mathbf{D}_{jk})$ is an $nM \times nM$ matrix made up as an $M \times M$ array of $n \times n$ diagonal matrices.

If we have $x_i^{(j)} = x_i^{(k)}$, then (4.35) is a nonparametric VMM (4.2) which may be useful for modelling multiple measurements made on the same unit. For example, relating systolic and diastolic blood pressure to covariates such as body mass index (BMI), age, measures of stress, etc., on n independent subjects. If, in addition, $f_1 = f_2 = \cdots = f_M$, then the VMM corresponds to a marginal model for clustered data with a cluster-level covariate. All of these models can be viewed as generalizations of the classical univariate smoothing model ($M = 1$).

When the mean functions in (4.35) are linear, $f_j(x) = \beta_{(j)1} + \beta_{(j)2} x$ so if we put $\boldsymbol{\beta}^{(*)} = (\beta_{(1)1}, \beta_{(1)2}, \ldots, \beta_{(M)1}, \beta_{(M)2})^T$ and $\mathbf{X}^{(*)} = \mathrm{diag}(\mathbf{X}^{(1)}, \ldots, \mathbf{X}^{(M)})$, where $\mathbf{X}^{(j)} = \left(\mathbf{1}, \boldsymbol{x}^{(j)}\right)$, then the LS estimator of $\boldsymbol{\beta}^{(*)}$ is the solution to

$$\mathbf{0} = \mathbf{X}^{(*)T}\left(\boldsymbol{y}^{(*)} - \mathbf{X}^{(*)}\boldsymbol{\beta}^{(*)}\right) = \begin{pmatrix} \mathbf{X}^{(1)T}\left(\boldsymbol{y}^{(1)} - \mathbf{X}^{(1)}\boldsymbol{\beta}^{(1)}\right) \\ \vdots \\ \mathbf{X}^{(M)T}\left(\boldsymbol{y}^{(M)} - \mathbf{X}^{(M)}\boldsymbol{\beta}^{(M)}\right) \end{pmatrix}, \quad (4.36)$$

where $\boldsymbol{\beta}^{(j)} = (\beta_{(j)1}, \beta_{(j)2})^T$. The least squares estimator effectively fits each component model separately (i.e., marginal fitting). We can often obtain a more efficient estimator by fitting the combined component models appropriately. In particular, the WLS estimator which, when $\mathbf{V}$ is known, is the solution to

$$\mathbf{0} = \mathbf{X}^{(*)T}\mathbf{V}^{-1}\left(\boldsymbol{y}^{(*)} - \mathbf{X}^{(*)}\boldsymbol{\beta}^{(*)}\right), \quad (4.37)$$

is at least as efficient as, and often more efficient than, the least squares estimator. When $\mathbf{V}$ is unknown, we replace it in (4.37) by a consistent estimator $\widehat{\mathbf{V}}$ without affecting the asymptotic efficiency of the estimator.

In the nonparametric case, let x_j denote the point at which we want to estimate f_j, $\mathbf{X}_{x_j}^{(j)} = \left(\mathbf{1}, \boldsymbol{x}^{(j)} - x_j\mathbf{1}, \ldots, (\boldsymbol{x}^{(j)} - x_j\mathbf{1})^r\right)$ be an $n \times (r+1)$ matrix, $\boldsymbol{x} = (x_1, \ldots, x_M)^T$, and $\mathbf{X}_{\boldsymbol{x}}^{(*)} = $ block diagonal $(\mathbf{X}_{x_1}^{(1)}, \ldots, \mathbf{X}_{x_M}^{(M)})$ be an $nM \times (r+1)M$ matrix. The local regression approach approximates the model (4.35) by making a Taylor expansion of the mean function to obtain a multivariate linear model with $(r+1)M$-vector regression parameter

$$\boldsymbol{\delta}_{\boldsymbol{x}} = \left(f_1(x_1), f_1'(x_1), \ldots, f_1^{(r)}(x_1), \ldots, f_M(x_M), f_M'(x_M), \ldots, f_M^{(r)}(x_M)\right)^T$$

and then estimates $\boldsymbol{\delta_x}$ by fitting the approximate model locally. Let $\mathbf{K}_j = \text{diag}(K_{h_j}(x_1^{(j)} - x_j), \ldots, K_{h_j}(x_n^{(j)} - x_j))$, $j = 1, \ldots, M$, denote the kernel weights, and put $\mathbf{K}^{(*)} = \text{block diagonal}(\mathbf{K}_1, \ldots, \mathbf{K}_M)$. Then the local polynomial estimator derived from the LS estimator (4.36) is the solution to $\mathbf{0} =$

$$\mathbf{X}_{\boldsymbol{x}}^{(*)T} \mathbf{K}^{(*)} \left(\boldsymbol{y}^{(*)} - \mathbf{X}_{\boldsymbol{x}}^{(*)} \boldsymbol{\delta_x} \right) = \begin{pmatrix} \mathbf{X}_{x_1}^{(1)T} \mathbf{K}_1 \left(\boldsymbol{y}^{(1)} - \mathbf{X}_{x_1}^{(1)} \boldsymbol{\delta}_{x_1}^{(1)} \right) \\ \vdots \\ \mathbf{X}_{x_M}^{(M)T} \mathbf{K}_M \left(\boldsymbol{y}^{(M)} - \mathbf{X}_{x_M}^{(M)} \boldsymbol{\delta}_{x_M}^{(M)} \right) \end{pmatrix}, \quad (4.38)$$

where $\boldsymbol{\delta}_{x_j}^{(j)} = (f_j(x_j), f_j'(x_j), \ldots, f_j^{(r)}(x_j))^T$. As in the linear case, this estimator fits each component model separately, and the question arises as to whether we can improve on such marginal smoothing. An obvious approach is to try to use the WLS estimator (4.37) rather than the LS estimator (4.36), to fit the approximate polynomial model locally. However, it is much less clear how to introduce the local kernel weights $\mathbf{K}$ into (4.37). Consider the four possibilities

$$\mathbf{0} = \begin{cases} \mathbf{X}_{\boldsymbol{x}}^{*T} \mathbf{V}^{-1} \mathbf{K}^{(*)} \left(\boldsymbol{y}^{(*)} - \mathbf{X}_{\boldsymbol{x}}^{(*)} \boldsymbol{\delta_x} \right), \\ \mathbf{X}_{\boldsymbol{x}}^{(*)T} \mathbf{K}^{(*)1/2} \mathbf{V}^{-1} \mathbf{K}^{(*)1/2} \left(\boldsymbol{y}^{(*)} - \mathbf{X}_{\boldsymbol{x}}^{(*)} \boldsymbol{\delta_x} \right), \\ \mathbf{X}_{\boldsymbol{x}}^{(*)T} \mathbf{K}^{(*)} \mathbf{V}^{-1} \left(\boldsymbol{y}^{(*)} - \mathbf{X}_{\boldsymbol{x}}^{(*)} \boldsymbol{\delta_x} \right), \\ \mathbf{X}_{\boldsymbol{x}}^{(*)T} \mathbf{V}^{-T/2} \mathbf{K}^{(*)} \mathbf{V}^{-1/2} \left(\boldsymbol{y}^{(*)} - \mathbf{X}_{\boldsymbol{x}}^{(*)} \boldsymbol{\delta_x} \right), \end{cases} \quad (4.39)$$

where $\mathbf{A}^{-T/2} = (\mathbf{A}^{-1/2})^T$. These estimators are all the same when $\mathbf{V}$ is diagonal but not otherwise. We compare the asymptotic efficiencies of these estimators to that of (4.38). Again, when $\mathbf{V}$ is unknown, we replace it in (4.39) by a consistent estimator.

Welsh and Yee (2006) show that, for the SUR model ($x_i^{(j)} \neq x_i^{(k)}$ for some $j \neq k$), incorporating the correlation structure into local polynomial estimators can gain or lose asymptotic efficiency, but for the VMM ($x_i^{(j)} = x_i^{(k)}$) leads to gains in asymptotic efficiency for Case A but not for Case B. Derivative estimators behave very differently in both the bias and the variance under cases A and B, and this is also true of the covariance between the estimators of the mean functions and their derivatives in the VMM. In addition, the placement of the kernel weights in WLS estimators is very important in the SUR problem (to ensure that the estimator is asymptotically unbiased) but not in the VMM, and the component estimators are asymptotically uncorrelated in the SUR model but asymptotically correlated in the VMM. These interesting results add to our understanding of the problem of smoothing dependent data.

4.2.2.2 Bivariate Local Linear Estimators

The issues are the same for any $M \geq 2$, so there is no real loss of generality in restricting attention to the bivariate ($M = 2$) case. We can simplify $\mathbf{X}_{x_j}^{(j)}$ to $\mathbf{X}_{x_j}^{(j)} = \mathbf{1}$ to obtain a vector version of the Nadaraya-Watson estimator, but local polynomial estimators with $r \geq 1$ are well-known to enjoy advantages over the Nadaraya-Watson estimator (design adaption, automatic edge-effect properties, derivative estimation, etc.) so it makes sense to consider at least local linear estimators. Including higher powers of $\boldsymbol{x}^{(j)} - x_j \boldsymbol{1}$ in a general local polynomial estimator can reduce the asymptotic bias and enable us to estimate higher-order derivatives of f_1 and f_2. However, to obtain theoretical results for local polynomial estimators of order r, we have to analytically invert order-$M(r+1)$ matrices and this quickly becomes intractable. In general, we do not have to use the same order polynomial for both components, and we could allow both parametric and nonparametric components. For simplicity, we restrict attention to the simplest case in which both components are nonparametric and both polynomials are linear $(r = 1)$.

For the bivariate local linear case, $\boldsymbol{y}^{(*)} = \left(\boldsymbol{y}^{(1)T}, \boldsymbol{y}^{(2)T}\right)^T$, $\mathbf{X}_{x_j}^{(j)} = \left(\boldsymbol{1}, \boldsymbol{x}^{(j)} - x_j \boldsymbol{1}\right)$ is an $n \times 2$ matrix, $\boldsymbol{x} = (x_1, x_2)^T$ is a 2-vector, $\mathbf{X}_{\boldsymbol{x}}^{(*)} = \mathrm{diag}\left(\mathbf{X}_{x_1}^{(1)}, \mathbf{X}_{x_2}^{(2)}\right)$ is a $2n \times 4$ matrix and

$$\boldsymbol{\Sigma}_i = \begin{pmatrix} \sigma_{1i}^2 & \sigma_{1i}\,\sigma_{2i}\,\rho_i \\ \sigma_{1i}\,\sigma_{2i}\,\rho_i & \sigma_{2i}^2 \end{pmatrix}. \tag{4.40}$$

Let $\mathbf{W}_i = \boldsymbol{\Sigma}_i^{-1}$ denote the weight matrix and define $\overline{\mathbf{W}} = n^{-1} \sum_{i=1}^n \boldsymbol{\Sigma}_i^{-1}$. Let

$$\alpha_i = 1/\{\sigma_{1i}^2(1-\rho_i^2)\}, \quad \beta_i = -\rho_i/\{\sigma_{1i}\sigma_{2i}(1-\rho_i^2)\}, \quad \gamma_i = 1/\{\sigma_{2i}^2(1-\rho_i^2)\},$$

so that

$$\mathbf{W}_i = \boldsymbol{\Sigma}_i^{-1} = \begin{pmatrix} \alpha_i & \beta_i \\ \beta_i & \gamma_i \end{pmatrix} \quad \text{and} \quad \overline{\mathbf{W}}^{-1} = \frac{1}{\overline{\alpha}\,\overline{\gamma} - \overline{\beta}^2} \begin{pmatrix} \overline{\gamma} & -\overline{\beta} \\ -\overline{\beta} & \overline{\alpha} \end{pmatrix}.$$

It is also useful to note that

$$\mathbf{V}^{-1} = \begin{pmatrix} \mathrm{diag}(\alpha_1,\ldots,\alpha_n) & \mathrm{diag}(\beta_1,\ldots,\beta_n) \\ \mathrm{diag}(\beta_1,\ldots,\beta_n) & \mathrm{diag}(\gamma_1,\ldots,\gamma_n) \end{pmatrix}.$$

The solutions to the first three sets of estimating equations in (4.39) can be written compactly as

$$\widehat{\boldsymbol{\delta}}_{\boldsymbol{x}}^{(*)}(a) = \left\{ \mathbf{X}_{\boldsymbol{x}}^{(*)T} \mathbf{K}^{(*)a} \mathbf{V}^{-1} \mathbf{K}^{(*)(1-a)} \mathbf{X}_{\boldsymbol{x}}^{(*)} \right\}^{-1} \mathbf{X}_{\boldsymbol{x}}^{(*)T} \mathbf{K}^{(*)a}\, \mathbf{V}^{-1} \mathbf{K}^{(*)(1-a)} \boldsymbol{y}^{(*)}$$
$$= \boldsymbol{\Delta}_n^{(*)}(a)^{-1}\, \boldsymbol{\theta}_n^{(*)}(a), \tag{4.41}$$

where $a \in \{0, 0.5, 1\}$. Multiplying out the terms in (4.41), we see that the 4×4 matrix $n\, \boldsymbol{\Delta}_n^{(*)}(a)$ equals

$$\sum_{i=1}^n \begin{pmatrix} \alpha_i\,K_{h_1}(x_i^{(1)} - x_1)\,\boldsymbol{\Gamma}_{i(11)} & \begin{array}{l} \beta_i\,K_{h_1}(x_i^{(1)} - x_1)^a \times \\ K_{h_2}(x_i^{(2)} - x_2)^{1-a}\,\boldsymbol{\Gamma}_{i(12)} \end{array} \\ \begin{array}{l} \beta_i\,K_{h_1}(x_i^{(1)} - x_1)^{1-a} \times \\ K_{h_2}(x_i^{(2)} - x_2)^a\,\boldsymbol{\Gamma}_{i(21)} \end{array} & \gamma_i\,K_{h_2}(x_i^{(2)} - x_2)\,\boldsymbol{\Gamma}_{i(22)} \end{pmatrix},$$

where

$$\boldsymbol{\Gamma}_{i(jk)} = \begin{pmatrix} 1 & x_i^{(k)} - x_k \\ x_i^{(j)} - x_j & (x_i^{(j)} - x_j)(x_i^{(k)} - x_k) \end{pmatrix}$$

and, similarly, $n\,\boldsymbol{\theta}_n^{(*)}(a)$ is

$$
\sum_{i=1}^{n}
\begin{pmatrix}
\alpha_i\,K_{h_1}(x_i^{(1)}-x_1)\,y_i^{(1)} + \beta_i\,K_{h_1}(x_i^{(1)}-x_1)^a K_{h_2}(x_i^{(2)}-x_2)^{1-a}\,y_i^{(2)} \\[2mm]
\begin{aligned}
&\alpha_i\,K_{h_1}(x_i^{(1)}-x_1)(x_i^{(1)}-x_1)\,y_i^{(1)} + \\
&\beta_i\,K_{h_1}(x_i^{(1)}-x_1)^a K_{h_2}(x_i^{(2)}-x_2)^{1-a}(x_i^{(1)}-x_1)\,y_i^{(2)}
\end{aligned} \\[2mm]
\beta_i\,K_{h_1}(x_i^{(1)}-x_1)^{1-a}K_{h_2}(x_i^{(2)}-x_2)^a\,y_i^{(1)} + \gamma_i\,K_{h_2}(x_i^{(2)}-x_2)\,y_i^{(2)} \\[2mm]
\begin{aligned}
&\beta_i\,K_{h_1}(x_i^{(1)}-x_1)^{1-a}K_{h_2}(x_i^{(2)}-x_2)^a(x_i^{(2)}-x_2)\,y_i^{(1)} + \\
&\gamma_i\,K_{h_2}(x_i^{(2)}-x_2)(x_i^{(2)}-x_2)\,y_i^{(2)}
\end{aligned}
\end{pmatrix}.
$$

The solution to the fourth estimating equation in (4.39) is

$$
\begin{aligned}
\widehat{\boldsymbol{\delta}}_{\boldsymbol{x}}^{(*)} &= \left\{ \mathbf{X}_{\boldsymbol{x}}^{(*)T}\,\mathbf{V}^{-T/2}\,\mathbf{K}^{(*)}\mathbf{V}^{-1/2}\,\mathbf{X}_{\boldsymbol{x}}^{(*)} \right\}^{-1} \mathbf{X}_{\boldsymbol{x}}^{(*)T}\,\mathbf{V}^{-T/2}\,\mathbf{K}^{(*)}\,\mathbf{V}^{-1/2}\,\boldsymbol{y}^{(*)} \\[2mm]
&= \left[\boldsymbol{\Delta}_n^{(*)} \right]^{-1} \boldsymbol{\theta}_n^{(*)}.
\end{aligned}
\tag{4.42}
$$

It is generally difficult to obtain $\mathbf{V}^{-1/2}$ explicitly. However, if we use the Cholesky decomposition $\mathbf{V}^{-1} = \mathbf{U}^T\mathbf{U}$, where $\mathbf{U}$ is the upper triangular Cholesky factor of $\mathbf{V}^{-1}$, and define $\mathbf{V}^{-1/2} = \mathbf{U}$, then we can show that

$$
\mathbf{V}^{-1/2} = \begin{pmatrix}
\mathrm{diag}(\alpha_1^{1/2},\ldots,\alpha_n^{1/2}) & \mathrm{diag}(\beta_1/\alpha_1^{1/2},\ldots,\beta_n/\alpha_n^{1/2}) \\[2mm]
\mathbf{O} & \mathrm{diag}(\{(\alpha_1\gamma_1-\beta_1^2)/\alpha_1\}^{1/2},\ldots)
\end{pmatrix}.
$$

It follows that $\widehat{\boldsymbol{\delta}}_{\boldsymbol{x}}^{(*)}$ in (4.42) is the product of the inverse of $n\,\boldsymbol{\Delta}_n^{(*)}$, which is

$$
\sum_{i=1}^{n}
\begin{pmatrix}
\alpha_i\,K_{h_1}(x_i^{(1)}-x_1)\,\boldsymbol{\Gamma}_{i(11)} & \beta_i\,K_{h_1}(x_i^{(1)}-x_1)\,\boldsymbol{\Gamma}_{i(12)} \\[2mm]
\beta_i\,K_{h_1}(x_i^{(1)}-x_1)\,\boldsymbol{\Gamma}_{i(21)} & G_i\,\boldsymbol{\Gamma}_{i(22)}
\end{pmatrix}
$$

(where $G_i = (\beta_i^2\,K_{h_1}(x_i^{(1)}-x_1))/\alpha_i + ((\gamma_i\alpha_i-\beta_i^2)\,K_{h_2}(x_i^{(2)}-x_2))/\alpha_i)$, and $n\,\boldsymbol{\theta}_n^{(*)}$ which is

$$
\sum_{i=1}^{n}
\begin{pmatrix}
\left\{\alpha_i\,y_i^{(1)} + \beta_i\,y_i^{(2)}\right\} K_{h_1}(x_i^{(1)}-x_1) \\[3mm]
\left\{\alpha_i\,y_i^{(1)} + \beta_i\,y_i^{(2)}\right\} K_{h_1}(x_i^{(1)}-x_1)(x_i^{(1)}-x_1) \\[3mm]
\begin{aligned}
&\beta_i K_{h_1}(x_i^{(1)}-x_1)\,y_i^{(1)} + \frac{\beta_i^2}{\alpha_i} K_{h_1}(x_i^{(1)}-x_1)\,y_i^{(2)} + \\
&\frac{\gamma_i\alpha_i-\beta_i^2}{\alpha_i} K_{h_2}(x_i^{(2)}-x_2)\,y_i^{(2)}
\end{aligned} \\[3mm]
\begin{aligned}
&\left\{\beta_i K_{h_1}(x_i^{(1)}-x_1)\,y_i^{(1)} + \frac{\beta_i^2}{\alpha_i} K_{h_1}(x_i^{(1)}-x_1)\,y_i^{(2)} + \right. \\
&\left. \frac{\gamma_i\alpha_i-\beta_i^2}{\alpha_i} K_{h_2}(x_i^{(2)}-x_2)\,y_i^{(2)}\right\}(x_i^{(2)}-x_2)
\end{aligned}
\end{pmatrix}.
$$

This is clearly of the same basic form as $\widehat{\boldsymbol{\delta}}_{\boldsymbol{x}}^{(*)}(a)$ but with different weights.

4.2.2.3 Conditions

Here, we will treat $\mathbf{V}$ as known. First, the estimators with known $\mathbf{V}$ give a lower bound on the asymptotic efficiency of the estimators with an estimated $\mathbf{V}$. Unless the estimators with known $\mathbf{V}$ are asymptotically more efficient than the unweighted estimator, there is no point in exploring the complexity of the general case. Second, our results are relevant to the general case. In the homoscedastic case in which $\boldsymbol{\Sigma}_i = \boldsymbol{\Sigma}$, the terms in $\boldsymbol{\Sigma}$ can be estimated at the parametric $n^{-1/2}$ rate, even though the mean is only estimated at a nonparametric rate. Obviously in this case, there is no loss of generality in treating the variance as known.

The asymptotic results are quite different for the cases $x^{(1)} \neq x^{(2)}$ (SUR) and $x^{(1)} = x^{(2)}$ (VMM), so we treat these cases separately. For the SUR model ($x^{(1)} \neq x^{(2)}$) results see Welsh and Yee (2006). For the VMM ($x^{(1)} = x^{(2)}$), we state the conditions for our main theorems with a single kernel function K and common bandwidths $h = h_1 = h_2$. We may use a single kernel function but different bandwidths (to allow for different curvature in f_1 and f_2, say)—this case is considered briefly in Sect. 4.2.2.4.

Our results require the following conditions.

VMM Conditions

  (i) The random variables x_i are independent and identically distributed with common density function $g(x)$. The derivative $g'(x)$ of $g(x)$ exists and is continuous.
 (ii) The point x where estimation is taking place is an interior point of the support of g, and $g(x) > 0$.
(iii) The third derivatives f_j''' of the component functions f_j are continuous on an open interval about x.
 (iv) The kernel function K is a density function with compact support which is symmetric about zero.
  (v) The bandwidth $h \to 0$ such that $nh^3 \to \infty$ as $n \to \infty$.
 (vi) For Case A, the matrices $\boldsymbol{\Sigma}_i$ and $\overline{\mathbf{W}}$ are bounded and non-singular.

The above conditions ensure that we can estimate both the mean functions and their derivatives together. If we consider the estimators of the mean functions and ignore the estimators of their derivatives, then we can discard conditions (vi) and then weaken the conditions by replacing (iii) and (v) by

(iii′) The second derivatives f_j'' of the component functions f_j are continuous on an open interval about x.
 (v′) The bandwidth $h \to 0$ such that $nh \to \infty$ as $n \to \infty$.

4.2.2.4 The Vector Measurement Model

Now suppose that $\boldsymbol{x}^{(1)} = \boldsymbol{x}^{(2)}$. We take $x_1 = x_2 = x$, say, as some terms diverge as $n \to \infty$ when $x_1 \neq x_2$. Then $\mathbf{X}_x^{(1)} = \mathbf{X}_x^{(2)} = \mathbf{X}_x$ say, and we can write $\mathbf{X}_x^{(*)} = \mathbf{I}_2 \otimes \mathbf{X}_x$. This structure allows more flexibility in the positioning of the kernel weights in the local estimator in the sense that all four estimators in (4.39) are consistent.

Generally, the choice of bandwidth depends on the estimand and the distribution of the covariate. Since we have the same covariate in each component, we may choose to use different bandwidths h_1 and h_2 or the same bandwidth $h_1 = h_2$ in our estimators according to whether f_1 and f_2 are different or similar. We consider both of these cases.

For the case $h_1 \neq h_2$, it can be shown that, under the VMM conditions,

(a) the bias of $\widehat{\boldsymbol{\delta}}_x^{(*)}(a)$ or $\widehat{\boldsymbol{\delta}}_x^{(*)}$ does not depend on the correlation so it is the same as the asymptotic bias from separate marginal fits to the two components,
(b) the diagonal blocks in the asymptotic variance are of the same order as in the SUR case,
(c) the off-diagonal blocks are of the same order as the diagonal blocks (so the component estimators are asymptotically correlated when ρ_i or $\rho(\boldsymbol{x}) \neq 0$), and
(d) the constants in the asymptotic variance depend on both components of the model.

When $h_1 = h_2$ and $\mathbf{K}_1 = \mathbf{K}_2$, all four weighted estimators are identical and the results are simpler, so that we can obtain explicit results which give insight into the general asymptotic behaviour of the estimators $\widehat{\boldsymbol{\delta}}_x^{(*)}$. The estimator

$$
\widetilde{\boldsymbol{\delta}}_x^{(*)} = \begin{pmatrix} \left(\mathbf{X}_x^T \mathbf{K}_1 \mathbf{X}_x\right)^{-1} \mathbf{X}_x^T \mathbf{K}_1 \, \boldsymbol{y}^{(1)} \\ \left(\mathbf{X}_x^T \mathbf{K}_2 \mathbf{X}_x\right)^{-1} \mathbf{X}_x^T \mathbf{K}_2 \, \boldsymbol{y}^{(2)} \end{pmatrix}.
$$

has asymptotic bias

$$
\text{bias}\left(\widetilde{\boldsymbol{\delta}}_x^{(*)}\right) \sim h^2 \begin{pmatrix} \mu_2 \, f_1''(x)/2 \\ \dfrac{(\mu_4 - \mu_2^2) g'(x) \, f_1''(x)}{2 \, \mu_2 \, g(x)} + \dfrac{\mu_4 \, f_1'''(x)}{3! \, \mu_2} \\ \mu_2 \, f_2''(x)/2 \\ \dfrac{(\mu_4 - \mu_2^2) \, g'(x) \, f_2''(x)}{2 \, \mu_2 \, g(x)} + \dfrac{\mu_4 \, f_2'''(x)}{3! \, \mu_2} \end{pmatrix}. \tag{4.43}
$$

Its asymptotic variance is

$$
\overline{\boldsymbol{\Sigma}} \otimes \frac{1}{n \, h \, g(x)} \begin{pmatrix} \nu_0 & -\nu_0 \, g'(x)/g(x) \\ -\nu_0 \, g'(x)/g(x) & \nu_2/(h\mu_2)^2 \end{pmatrix}
$$

for Case A. And for both cases A and B, we obtain

$$
\text{Var}(\widetilde{\boldsymbol{\delta}}_x^{(*)}) \sim \boldsymbol{\Sigma} \otimes \frac{1}{n \, h \, g(x)} \begin{pmatrix} \nu_0 & -\nu_0 \, g'(x)/g(x) \\ -\nu_0 \, g'(x)/g(x) & \nu_2/(h\mu_2)^2 \end{pmatrix}.
$$

For the weighted estimators, we have the following results.

Theorem 1. Suppose that $h_1 = h_2$, $\mathbf{K}_1 = \mathbf{K}_2$ and the VMM conditions (i), (ii), (iii$'$), (iv) and (v$'$) hold. Then, for all four weighted estimators (4.39), the asymptotic bias of $(\widehat{f}_1(x), \widehat{f}_2(x))^T$ is

$$\frac{1}{2}\,h^2\,\mu_2\begin{pmatrix} f_1''(x) \\ f_2''(x) \end{pmatrix},$$

and the asymptotic variance of $(\widehat{f}_1(x), \widehat{f}_2(x))^T$ is $\nu_0\,\{n\,h\,g(x)\}^{-1}\overline{\mathbf{W}}^{-1}$ for Case A.

Theorem 2. Suppose that $h_1 = h_2$, $\mathbf{K}_1 = \mathbf{K}_2$ and the VMM conditions (i)–(vi) hold. Then, for all four weighted estimators (4.39), for Case A, the bias satisfies (4.43), and the asymptotic variance is

$$\overline{\mathbf{W}}^{-1} \otimes \frac{1}{n\,h\,g(x)}\begin{pmatrix} \nu_0 & -\nu_0\,g'(x)/g(x) \\ -\nu_0\,g'(x)/g(x) & \nu_2/(h\mu_2)^2 \end{pmatrix}.$$

Here are some remarks.

1. The off-diagonal blocks in the variance matrix are of the same order as the diagonal blocks, therefore the estimates of (f_1, f_1') and (f_2, f_2') are asymptotically correlated.
2. The asymptotic variance is the same as in the unweighted case under homoscedasticity $\boldsymbol{\Sigma}_i = \boldsymbol{\Sigma}$ or $\boldsymbol{\Sigma}(x) = \boldsymbol{\Sigma}$. That is, there is no gain in smoothing jointly compared to smoothing marginally. The covariance between the 2×2 matrices $\mathrm{Var}(\widehat{f}_1(x), \widehat{f}_1'(x))$ and $\mathrm{Var}(\widehat{f}_2(x), \widehat{f}_2'(x))$ depends on ρ. (This outcome is analogous to the parametric case where the same thing occurs.) This means that, as in the SUR model, the weighted and unweighted estimators have the same asymptotic variance. However, whether we use a weighted or an unweighted estimator, in the VMM, the off-diagonal blocks in the asymptotic variance matrix depend on ρ and ignoring this correlation can result in a large loss of efficiency in some applications. For example, suppose that we want to estimate $\theta = f_1(x) - f_2(x)$ or $\theta = f_1'(x) - f_2'(x)$ corresponding to the mean difference (or difference in change) between measurements at x. Then ignoring the asymptotic correlation between the estimates can lead to a substantial loss in efficiency, as reflected in unnecessarily wide confidence intervals.
3. In the heteroscedastic case, and for the VMM, under Case A, the weighted estimators have smaller asymptotic variance for estimating $f_j(x)$ and $f_j'(x)$ than the unweighted estimator. However, for Case B, the weighted estimators have the same asymptotic variance for estimating $f_j(x)$ and $f_j'(x)$ as the unweighted estimator. The results under Case A are therefore quite different from those under Case B.

Further insight into the nature of the weighted estimators can be obtained by examining their equivalent kernels. Equivalent kernels for estimating any single component of the model by a linear vector smoother are obtained in the same way as equivalent kernels for linear smoothers, except that there are M 'curves'— one for each $y_i^{(j)}$. The proofs of Theorems 1–2 allow us to derive formulas for the asymptotic equivalent kernels; we present results only for Case A. In the case $K_1 = K_2$ and $h_1 = h_2$, the asymptotic equivalent kernel for $\widehat{f}_1(x)$ weights $y_i^{(1)}$ by

$$\frac{1}{n}\, K_h(x_i - x)\, \left(\frac{\alpha_i\, \overline{\gamma} - \beta_i\, \overline{\beta}}{\overline{\alpha}\, \overline{\gamma} - \overline{\beta}^2}\right)\left(\frac{1}{g(x)} - \frac{g'(x)}{g^2(x)}\,(x_i - x)\right),\qquad (4.44)$$

and $y_i^{(2)}$ by

$$\frac{1}{n}\, K_h(x_i - x)\, \left(\frac{\beta_i\, \overline{\gamma} - \gamma_i\, \overline{\beta}}{\overline{\alpha}\, \overline{\gamma} - \overline{\beta}^2}\right)\left(\frac{1}{g(x)} - \frac{g'(x)}{g^2(x)}\,(x_i - x)\right).\qquad (4.45)$$

Similarly, the asymptotic equivalent kernel can be derived for $\widehat{f}'_1(x)$ with weights with respect to $y_i^{(1)}$ and $y_i^{(2)}$. In the homoscedastic case, the terms involving the components of $\mathbf{W}_i$ equal one in (4.44) and for $\widehat{f}'_1(x)$ with respect to $y_i^{(1)}$, and zero for (4.45) and for $\widehat{f}'_1(x)$ with respect to $y_i^{(2)}$, so in this case, the weighted estimator is effectively smoothing marginally. In the heteroscedastic case, numerical exploration shows that even when the correlation is quite large, relatively little weight is given to $y_i^{(2)}$ for estimating f_1 and f'_1.

4.2.2.5 Conclusions

We have considered the problem of estimating the regression functions and their first derivatives in the vector regression model (4.35) under the assumption that the errors are correlated. Specifically, we have explored whether we can construct linear local polynomial estimators which are more efficient than unweighted local polynomial estimators which ignore the correlation structure and estimate each component regression function and its derivative separately. Our results depend on the covariate structure and the form of the variance model. In particular, the results are different for the SUR model, in which $x_i^{(j)} \neq x_i^{(k)}$ for some $j \neq k$, and the VMM in which $x_i^{(j)} = x_i^{(k)}$. They also depend on whether the variance is modelled as constants (Case A) or as smooth functions of the covariates (Case B). In the univariate case, the SUR model reduces to a marginal model, and the VMM to a (marginal) cluster-level model so this distinction is natural.

Under the SUR model, only the second estimator in (4.39) is consistent when the correlation is nonzero, while under the VMM, when the kernels and bandwidths are equal, all four estimators in (4.39) are always identical. For estimating the regression functions, we find that the asymptotic bias of the (consistent) weighted and unweighted estimators is the same. The components of the consistent weighted estimator under the SUR model are asymptotically uncorrelated while under the VMM, the components of the weighted estimator are asymptotically correlated. Table 4.1 shows the estimator of the regression function with the smaller asymptotic variance in the different cases.

The results for the derivative estimators are more complicated, because the asymptotic biases of the (consistent) weighted estimators are the same as those of the unweighted estimators under Case A but differ under Case B. The asymptotic

Table 4.1 The estimator of the regression or derivative function with the smaller asymptotic variance in the different cases. Here, "Either" means the weighted estimator will sometimes be more efficient, and will sometimes be less efficient.

Case		SUR	VMM
		Not all covariates are equal	Common covariates
A	Variance is unrelated to covariates	$\rho_i = \rho$ Weighted	Weighted
		$\rho_i \neq \rho$ Either	
B	Variance is a function of covariates	$\rho(x) = \rho$ Equal	Equal
		$\rho(x) \neq \rho$ Either	

variances of the derivative estimators follow the same pattern as the regression estimators in Table 4.1. However, note that the asymptotic correlation between the weighted regression and derivative estimators of a component of the model bears the same relation to that of the unweighted estimators as the asymptotic variances of the weighted to the unweighted estimators in Case A, but they include different terms in Case B.

Finally, simulation results show that the asymptotic results apply in finite samples, and hence it is important when selecting an estimator for vector regression to distinguish between Cases A and B.

4.2.3 On Linear Vector Smoothers

The general theory relating to linear smoothers described in Sect. 2.4.7 quite naturally extends to the vector smoothing situation. Some results of practical importance are summarized below.

4.2.3.1 Degrees of Freedom for Linear Vector Smoothers

Following on from Sect. 2.4.7.4, for vector smoothers with $\mathrm{Var}(\boldsymbol{y}) = \boldsymbol{\Sigma}$, there are at least three definitions for the degrees of freedom of a linear vector smoother: they are $df = \mathrm{trace}(\mathbf{A})$, $df^{\mathrm{var}} = \mathrm{trace}(\boldsymbol{\Sigma}^{-1}\mathbf{A}\boldsymbol{\Sigma}\mathbf{A}^T)$ and $df^{\mathrm{err}} = tr(\mathbf{I}_{nM} - 2\mathbf{A} + \mathbf{A}^T\boldsymbol{\Sigma}^{-1}\mathbf{A}\boldsymbol{\Sigma})$, respectively. These are the degrees of freedom for the overall smooth $\boldsymbol{f}$. However, we need to define the degrees of freedom for each of the M component functions of $\boldsymbol{f}$ as well. Intuitively, one would want the latter to be the sum of the former. Correspondingly, we define the degrees of freedom of the jth component function $f_j(\cdot)$ as the sum of those diagonal elements corresponding to the jth component function. For example, for vector splines, the degrees of freedom of the jth smooth is the sum of elements (j,j), $(M{+}j,M{+}j)$, ..., $((n-1)M{+}j,(n-1)M{+}j)$ of $\mathbf{A}(\boldsymbol{\lambda})$, $\boldsymbol{\Sigma}^{-1}\mathbf{A}(\boldsymbol{\lambda})\boldsymbol{\Sigma}\mathbf{A}(\boldsymbol{\lambda})^T$ and $\mathbf{I}_{nM} - 2\mathbf{A}(\boldsymbol{\lambda}) + \mathbf{A}(\boldsymbol{\lambda})^T\boldsymbol{\Sigma}^{-1}\mathbf{A}(\boldsymbol{\lambda})\boldsymbol{\Sigma}$ for $df_{(j)}$, $df_{(j)}^{\mathrm{var}}$ and $df_{(j)}^{\mathrm{err}}$, respectively. They can be written compactly, e.g.,

$$df_{(m)} = \mathrm{trace}\left\{\mathrm{diag}(\mathbf{1}_n \otimes \boldsymbol{e}_m)\, \mathbf{A}(\boldsymbol{\lambda})\, \mathrm{diag}(\mathbf{1}_n \otimes \boldsymbol{e}_m)\right\}, \quad m = 1, \ldots, M.$$

Hastie and Tibshirani (1990, App.B) describe an approximation for df^{err} for cubic splines as

$$2 \operatorname{trace} \mathbf{S}(\lambda) - \operatorname{trace}(\mathbf{S}(\lambda)\,\mathbf{S}(\lambda)^T) \;\approx\; \frac{5}{4}\operatorname{trace}\mathbf{S}(\lambda) - \frac{1}{2}. \qquad (4.46)$$

In our work with vector splines, this approximation also works well for each component function, i.e., $df^{\mathrm{err}}_{(m)} \approx 1.25\, df_{(m)} - 0.5$, no matter what the correlations are.

4.2.3.2 Standard Errors

Following on from Sect. 2.4.7.5, the variance-covariance matrix of a linear vector smoother fit $\widehat{\boldsymbol{f}}$ is $\mathbf{A}(\boldsymbol{\lambda})\,\boldsymbol{\Sigma}\,\mathbf{A}(\boldsymbol{\lambda})^T$. In theory, this can be used to form pointwise standard error bands for each of the M component functions of $\boldsymbol{f}$. Unfortunately, it is impractical if nM is large, as the complete influence matrix must be computed. However, for vector splines, the Bayesian-based alternative $\mathbf{A}(\boldsymbol{\lambda})\,\boldsymbol{\Sigma}$ is used. This is computationally quite cheap, because only the central $2M-1$ bands of $\mathbf{A}(\boldsymbol{\lambda})$ are required and these may be efficiently computed using the Hutchinson and de Hoog (1985) algorithm, and gives similar results to $\mathbf{A}(\boldsymbol{\lambda})\,\boldsymbol{\Sigma}\,\mathbf{A}(\boldsymbol{\lambda})^T$.

4.3 The Vector Additive Model and VGAM Estimation

Recall from Sect. 3.2 that the IRLS algorithm used to estimate VGLMs. Simply put, VGAMs are estimated by fitting a vector additive model at each iteration of the IRLS algorithm to the pseudo-response $\boldsymbol{z}_i$ with explanatory variables $\boldsymbol{x}_i$ and working weight matrices $\mathbf{W}_i$. Many of the results pertaining to the univariate additive models generalize naturally to the vector case. Consequently, this section draws upon Hastie and Tibshirani (1990, Chap.5).

4.3.1 Penalized Likelihood

A heuristic argument is given to show that using vector smoothing splines can be justified by a penalized likelihood argument. Consider penalizing ℓ for lack of smoothness by maximizing

$$\ell\{\eta_1(\boldsymbol{x}_i),\ldots,\eta_M(\boldsymbol{x}_i)\} - \sum_{k=2}^{d}\sum_{j=1}^{\mathcal{R}_k} \lambda_{(j)k}\int \{f^{*''}_{(j)k}(x_k)\}^2\, dx_k\,. \qquad (4.47)$$

We saw from Sect. 3.2 that maximizing a log-likelihood ℓ (3.7) by Fisher scoring amounted to minimizing a GLS criterion (3.22) at each IRLS iteration. This entailed fitting a weighted multivariate regression to working responses $\boldsymbol{z}_i$ to some $\mathbf{X}_{\mathrm{VLM}}$ with working weights $\mathbf{W}_i$. Hence, with trivial constraints, the solution minimizes

$$\sum_{i=1}^{n}\{\boldsymbol{z}_i - \boldsymbol{\eta}_i\}^T\,\boldsymbol{\Sigma}_i^{-1}\,\{\boldsymbol{z}_i - \boldsymbol{\eta}_i\} + \sum_{k=2}^{d}\sum_{j=1}^{M}\lambda_{(j)k}\int\{f''_{(j)k}(x_k)\}^2\, dx_k,$$

and when $d = 2$, this reduces to the vector spline problem (4.3).

4.3.2 Vector Backfitting

The (weighted) *vector additive model* (VAM) for a response M-vector $\boldsymbol{y}_i$ is

$$\boldsymbol{y}_i = \mathbf{H}_1\boldsymbol{\beta}_{(1)}^* + \sum_{k=2}^{d} \mathbf{H}_k\boldsymbol{f}_k^*(x_{ik}) + \boldsymbol{\varepsilon}_i, \quad \boldsymbol{\varepsilon}_i \sim (\mathbf{0}, \boldsymbol{\Sigma}_i), \tag{4.48}$$

where $\boldsymbol{f}_k^*(x_{ik}) = (f_{(1)k}^*(x_{ik}), \ldots, f_{(\mathcal{R}_k)k}^*(x_{ik}))^T$, and $E(\boldsymbol{f}_k^*) = \mathbf{0}$ for all k. It is applied to observations $(\boldsymbol{x}_i, \boldsymbol{y}_i, \boldsymbol{\Sigma}_i)$, $i = 1, \ldots, n$. The VAM is clearly a natural extension of the univariate additive model to vector or multivariate responses. As with the ordinary additive model, the component functions of the kth variable, $f_{(m)k}^*(\cdot)$, are arbitrary smooth functions.

For VGAMs, the VAM is fitted to the adjusted dependent vectors $\boldsymbol{z}_i$ with working weights $\mathbf{W}_i = \boldsymbol{\Sigma}^{-1}$ described in Sect. 3.2. This can be done by a *vector backfitting algorithm* using vector smoothers as the basic building block. The essential idea of backfitting is to smooth *partial residuals* against one covariate at a time, having adjusted for all other covariates. This algorithmic procedure is iterated until convergence.

As with the univariate additive model, conditional expectations provide a simple motivation. Under the VAM (4.48),

$$E\left[\boldsymbol{y}_i - \mathbf{H}_1\boldsymbol{\beta}_{(1)}^* - \sum_{s \neq k}\mathbf{H}_s\boldsymbol{f}_s^*(x_{is})\,\bigg|\,x_k\right] = \mathbf{H}_k\boldsymbol{f}_k^*(x_{ik}) \tag{4.49}$$

for all k. The vectors under the expectation are the partial residuals. Here, the effects of all covariates except for the kth are 'removed' from $\boldsymbol{y}_i$ for the updating of $\boldsymbol{f}_k^*$. The above formula motivates the vector backfitting algorithm (Algorithm 4.1), which is presented with trivial constraints for simplicity.

There are actually two common updating techniques depending on whether the latest updates are used. Algorithm 4.1 is of the block-Jacobi type. The block Gauss-Seidel type involves replacing $\sum_{s \neq k}\boldsymbol{f}_s^{(a-1)}(x_{is})$ in (a) by $\sum_{s < k}\boldsymbol{f}_s^{(a)}(x_{is}) + \sum_{s > k}\boldsymbol{f}_s^{(a-1)}(x_{is})$, i.e., the most recent estimate of the component functions are used. Also, it is noted that any vector norm would do for testing convergence. The algorithm is justified more fully below.

It is possible to apply the iterative vector spline solution idea of Sect. 4.2.1.5 by looping over the k and the j; Yee (1998) considers this procedure. In the backfitting with vector smoothing here, we only have to loop over the k. Vector backfitting has advantages such as handling high correlations better (especially when the algorithm is modified by a projection step) and it is faster, because it treats up to M component functions at a time as one computational block.

To justify the vector backfitting algorithm as a technique for estimating the vector additive model, note that when there are d (> 1) covariates,

$$\left(\boldsymbol{y} - \sum_{k=1}^{d}\boldsymbol{f}_j\right)^T\boldsymbol{\Sigma}^{-1}\left(\boldsymbol{y} - \sum_{k=1}^{d}\boldsymbol{f}_j\right) + \sum_{k=1}^{d}\boldsymbol{f}_k^T\mathbf{K}_j\boldsymbol{f}_k,$$

Algorithm 4.1 *The vector backfitting algorithm (with trivial constraints).*

Initialize: $\boldsymbol{\beta}_{(1)} = \text{average}\{\boldsymbol{z}_1, \ldots, \boldsymbol{z}_n\}$, and $\boldsymbol{f}_2^{(0)} \equiv \boldsymbol{f}_d^{(0)} \cdots \equiv \boldsymbol{0}$.
Iterate: For $a = 1, 2, \ldots$

(i) Iterate: For $k = 2, \ldots, d$:

   (a) Compute the function $\widetilde{\boldsymbol{f}}$, as the weighted vector smooth of

$$\text{observations} \left(x_{ik}, \quad \boldsymbol{z}_i - \boldsymbol{\beta}_{(1)} - \sum_{s \neq k} \boldsymbol{f}_s^{(a-1)}(x_{is}), \quad \boldsymbol{\Sigma}_i \right).$$

   (b) Adjust the intercept: $\boldsymbol{\beta}_{(1)} = \boldsymbol{\beta}_{(1)} + \text{average}\{\widetilde{\boldsymbol{f}}(x_{1k}), \ldots, \widetilde{\boldsymbol{f}}(x_{nk})\}$.

   (c) Compute: $\boldsymbol{f}_k^{(a)}(x_{ik}) = \widetilde{\boldsymbol{f}}(x_{ik}) - \text{average}\left\{\widetilde{\boldsymbol{f}}(x_{1k}), \ldots, \widetilde{\boldsymbol{f}}(x_{nk})\right\}$.

(ii) Test for convergence: If $\|\boldsymbol{f}_k^{(a)} - \boldsymbol{f}_k^{(a-1)}\| < \varepsilon \ \forall k$ and positive $\varepsilon \approx 0$ then stop.

---

where lack-of-smoothness penalties $\boldsymbol{f}_k^T \mathbf{K}_k \boldsymbol{f}_k$ are imposed on each function $\boldsymbol{f}_k$. The quantities in this formula are defined as in (4.4). By differentiating with respect to $\boldsymbol{f}_k$, this is minimized when

$$\widehat{\boldsymbol{f}}_k = (\mathbf{I}_{nM} + \boldsymbol{\Sigma}\mathbf{K}_k)^{-1} \left(\boldsymbol{y} - \sum_{s \neq k} \widehat{\boldsymbol{f}}_s \right), \qquad k = 2, \ldots, d,$$

which corresponds to a weighted backfitting algorithm using vector splines.

4.3.2.1 Consistency and Convergence

Properties such as the convergence of the vector backfitting algorithm are examined by considering the $nMd \times nMd$ system of estimating equations

$$\begin{pmatrix} \mathbf{I}_{nM} & \mathbf{A}_1 & \mathbf{A}_1 & \cdots & \mathbf{A}_1 \\ \mathbf{A}_2 & \mathbf{I}_{nM} & \mathbf{A}_2 & \cdots & \mathbf{A}_2 \\ \vdots & & \ddots & & \vdots \\ \mathbf{A}_{d-1} & \cdots & \mathbf{A}_{d-1} & \mathbf{I}_{nM} & \mathbf{A}_{d-1} \\ \mathbf{A}_d & \cdots & \mathbf{A}_d & \mathbf{A}_d & \mathbf{I}_{nM} \end{pmatrix} \begin{pmatrix} \boldsymbol{f}_1 \\ \boldsymbol{f}_2 \\ \vdots \\ \boldsymbol{f}_{d-1} \\ \boldsymbol{f}_d \end{pmatrix} = \begin{pmatrix} \mathbf{A}_1 \boldsymbol{y} \\ \mathbf{A}_2 \boldsymbol{y} \\ \vdots \\ \mathbf{A}_{d-1} \boldsymbol{y} \\ \mathbf{A}_d \boldsymbol{y} \end{pmatrix}, \qquad (4.50)$$

or

$$\widehat{\mathbf{P}}\boldsymbol{f} = \widehat{\mathbf{Q}}\boldsymbol{y}, \text{ say.}$$

This is the usual estimating equations with $\mathbf{A}_j$ replacing $\mathbf{S}_j$. The following theorem gives the consistency and convergence result for the vector backfitting algorithm using vector splines.

Theorem With vector splines, the estimating equations (4.50) are consistent for every $\boldsymbol{y}$, and the solution is unique unless there exists a $\boldsymbol{g} \neq \boldsymbol{0}$ such that $\widehat{\mathbf{P}}\boldsymbol{g} = \boldsymbol{0}$ (concurvity). Furthermore, a vector backfitting algorithm using vector splines converges to some solution of the estimating equations.

Proof These are consequences of Hastie and Tibshirani (1990, Sect.5.3.7).   □

It may be noted that Hastie and Tibshirani (1990, Sect.5.3.6) recognize that their convergence results also hold for some non-univariate smoothers, however, they do not mention vector smoothers explicitly. Also note that it can be shown that if all the smoothers are vector splines, then exact concurvity only exists if the covariates are exactly collinear—a result which is known for ordinary cubic splines.

When the $\boldsymbol{\Sigma}_i$ are not all equal, the theorem may be generalized to deal with the non-symmetric case. This is important, because $\mathbf{A}_k(\boldsymbol{\lambda}_k)$ will not be symmetric in most applications. It is done by premultiplying $\boldsymbol{y}$ by $\boldsymbol{\Sigma}^{-\frac{1}{2}}$. The resulting influence matrix is symmetric with eigenvalues in $[0, 1]$, and unit eigenvalues corresponding to linear component functions of the kth variable. The transformation in fact leads to M separate unweighted additive models. Hence, all results on existence, uniqueness and convergence of algorithms apply.

4.3.2.2 Modified Vector Backfitting

The vector backfitting algorithm described above is actually not implemented entirely as stated. Instead, a VLM is first fitted to estimate the linear parts of the component functions,

$$f_{(j)k}^*(x_k) \;=\; \beta_{(j)k}^* \, x_k + g_{(j)k}^*(x_k), \tag{4.51}$$

and then the vector of residuals

$$\boldsymbol{r}_i \;=\; \boldsymbol{z}_i - \sum_{k=1}^{d} \mathbf{H}_k \, \widehat{\boldsymbol{\beta}}_{(k)}^* \, x_{ik}$$

is formed and ordinary vector backfitting is performed on the $\boldsymbol{r}_i$ to estimate the $g_{(j)k}^*(x_k)$. Clearly, this is an extension of (4.12). The reason for this modification is that the first step is a projection onto the space of linear fits, and the second step relates to a nonprojection component (they correspond to $\mathcal{H}_{0j}$ and $\mathcal{H}_{1j}$ as related to (4.30)). The result is that it generally leads to fewer vector backfitting iterations required, especially when the x_k are correlated.

4.3.3 Degrees of Freedom and Standard Errors

As a further approximation in the case of more than one covariate, the three forms of the degrees of freedom are defined additively, e.g., $df^{\,\mathrm{err}}$ becomes $nM - \left(M + \sum_{k=2}^{d} [\mathrm{trace}(\mathbf{A}_k) - M]\right)$. For more than a single component function, the breakdown by successive diagonal elements applies as before.

With more than one covariate, one can express $\widehat{\boldsymbol{f}}_j = \mathbf{R}_j \boldsymbol{y}$ for some $nM \times nM$ matrix $\mathbf{R}_j$ at convergence. Direct computation of $\mathbf{R}_j$ is too costly in practice. Instead, *very* approximate Bayesian standard errors (SEs) are available and can be plotted in **VGAM**. These pointwise ± 2 standard error bands are useful to give an indication about which parts of the curve are more variable.

For covariate x_k, VGAM computes SEs as the square root of

$$\sigma_{kk} \left(x_k - \overline{x}_k\right)^2 + \operatorname{diag}(\mathbf{A}_k \mathbf{W}^{-1}), \tag{4.52}$$

being the sum of the variances of the linear and nonlinear components of (4.51). The latter term is not quite right—the true variance of $\widehat{g}^*_{(j)k}(x_k)$ is complicated and involves a large hat matrix. In (4.52) $\mathbf{A}_k$ is the influence matrix of the vector smooth of the kth covariate and $\mathbf{W}$ the weight matrix at the final local scoring iteration. However, the pointwise ± 2 SE bands plotted for linear terms is correct because the latter term is zero for such. For nonlinear terms the pointwise ± 2 SE bands are slightly too wide.

Currently, it is not possible for VGAM to compute SEs at new data, nor to handle non-trivial constraint matrices.

4.3.4 Score Tests for Linearity †

Suppose

$$\eta_{ij} \;=\; \sum_{k=1}^{d} f_{(j)k}(x_{ik}) \;=\; \boldsymbol{\beta}_j^T \boldsymbol{x}_i + \sum_{k=1}^{d} s_{(j)k}(x_{ik})$$

is an additive predictor using modified backfitting for a VGAM. The $s_{(j)1}, \ldots, s_{(j)p}$ are nonlinear part of the smooth component functions $f_{(j)1}, \ldots, f_{(j)d}$ where $j = 1, \ldots, M$. We want to test $H_0 : f_{(j)t}$ is linear versus $H_1 : f_{(j)t}$ is nonlinear, for some $t \in \{1, \ldots, d\}$ and $j \in \{1, \ldots, M\}$, i.e., $H_0 : s_{(j)t} = 0$ versus $H_1 : s_{(j)t} \neq 0$. The following approximate score test is applied. It follows Chambers and Hastie (1991, p.306) by adjusting only for the linear parts while keeping the other nonlinear parts fixed.

Define the Pearson chi-squared statistic for VGAMs to be the same as with VGLMs (Sect. 3.7.2). Then $X^2 \,\dot\sim\, \chi^2_{nM}$ because $\boldsymbol{z}_i \,\dot\sim\, (\boldsymbol{\eta}_i, w_i^{-1} \mathbf{W}_i^{-1})$. Allowing for constraint matrices $\mathbf{H}_k$ ($M \times \operatorname{ncol}(\mathbf{H}_k)$), at convergence under H_1,

$$\boldsymbol{z}_i - \sum_{k=1}^{d} \mathbf{H}_k \, \widehat{\boldsymbol{s}}_k(x_{ik}) \;\;\sim\;\; \widehat{\mathbf{B}}^T \boldsymbol{x}_i.$$

The notation "$\sim$" here means when the LHS is regressed upon $\boldsymbol{x}_i$ it yields $\widehat{\mathbf{B}}$—but the regression adjusts for weights $w_i \mathbf{W}_i$ and constraint matrices $\mathbf{H}_k$ in the design matrix. Write the columns of $\mathbf{H}_k$ as $(\boldsymbol{h}_{1k}, \ldots, \boldsymbol{h}_{\operatorname{ncol}(\mathbf{H}_k)k})$. Suppose

$$\boldsymbol{h}_{jt}\, \widehat{s}_{(j)t}(x_{it}) \;\;\sim\;\; \boldsymbol{h}_{jt} \widehat{\boldsymbol{\gamma}}^T_{(j)t} \boldsymbol{x}_i, \qquad j = 1, \ldots, \operatorname{ncol}(\mathbf{H}_t).$$

Then

$$\boldsymbol{z}_i - \sum_{k \neq t} \mathbf{H}_k\, \widehat{\boldsymbol{s}}_k(x_{ik}) - \sum_{m \neq j} \boldsymbol{h}_{mt}\, \widehat{s}_{(m)t}(x_{it}) \;\;\sim\;\; \left(\widehat{\mathbf{B}}^T + \boldsymbol{h}_{jt} \widehat{\boldsymbol{\gamma}}^T_{(j)t}\right) \boldsymbol{x}_i,$$

and so

$$
\begin{aligned}
\boldsymbol{z}_i - \boldsymbol{\eta}_i &= \left(\widehat{\mathbf{B}}^T \boldsymbol{x}_i + \sum_{k=1}^d \mathbf{H}_k \, \widehat{\boldsymbol{s}}_k(x_{ik}) + \boldsymbol{r}_i^W \right) - \\
&\quad \left(\left(\widehat{\mathbf{B}}^T + \boldsymbol{h}_{jt} \, \widehat{\boldsymbol{\gamma}}_{(j)t}^T \right) \boldsymbol{x}_i + \sum_{k \neq t} \mathbf{H}_k \, \widehat{\boldsymbol{s}}_k(x_{ik}) + \sum_{m \neq j} \boldsymbol{h}_{mt} \, \widehat{\boldsymbol{s}}_{(m)t}(x_{it}) \right) \\
&= \boldsymbol{h}_{jt} \left\{ \widehat{\boldsymbol{s}}_{(j)t}(x_{it}) - \widehat{\boldsymbol{\gamma}}_{(j)t}^T \boldsymbol{x}_i \right\} + \boldsymbol{r}_i^W .
\end{aligned}
$$

Let

$$
\widehat{\boldsymbol{\delta}}_{itj} = \left\{ \widehat{s}_{(j)t}(x_{it}) - \widehat{\boldsymbol{\gamma}}_{(j)t}^T \boldsymbol{x}_i \right\} \boldsymbol{h}_{jt} \quad (= \widehat{\delta}_{itj} \, \boldsymbol{h}_{jt}, \text{ say}).
$$

[With no constraints, $\boldsymbol{h}_{jt} = \boldsymbol{e}_j$]. Then the difference in the Pearson chi-squared statistic between H_0 and H_1 is

$$
\begin{aligned}
X_{0t}^2 - X_{1t}^2 &= \sum_{i=1}^n w_i \left(\boldsymbol{r}_i^W + \widehat{\boldsymbol{\delta}}_{itj} \right)^T \mathbf{W}_i \left(\boldsymbol{r}_i^W + \widehat{\boldsymbol{\delta}}_{itj} \right) - \sum_{i=1}^n w_i \, \boldsymbol{r}_i^{WT} \mathbf{W}_i \, \boldsymbol{r}_i^W \\
&= 2 \sum_{i=1}^n w_i \, \boldsymbol{r}_i^{WT} \mathbf{W}_i \, \widehat{\boldsymbol{\delta}}_{itj} + \sum_{i=1}^n w_i \, \widehat{\boldsymbol{\delta}}_{itj}^T \mathbf{W}_i \, \widehat{\boldsymbol{\delta}}_{itj} \quad (4.53) \\
&= 2 \sum_{i=1}^n w_i \, \boldsymbol{u}_i^T \, \widehat{\boldsymbol{\delta}}_{itj} + \sum_{i=1}^n w_i \, \widehat{\boldsymbol{\delta}}_{itj}^T \mathbf{W}_i \, \widehat{\boldsymbol{\delta}}_{itj} . \quad (4.54)
\end{aligned}
$$

With no constraints, this simplifies to $w_i \boldsymbol{u}$, and

$$
X_{0t}^2 - X_{1t}^2 = 2 \sum_{i=1}^n w_i \, \widehat{\delta}_{itj} \, u_{ij} + \sum_{i=1}^n w_i \, \widehat{\delta}_{itj}^2 \, (\mathbf{W}_i)_{jj} .
$$

The quantity (4.54) is returned by `vgam.nlchisq()` and is printed by the `summary()` generic. One can test $H_0 : \boldsymbol{s}_t = \boldsymbol{0}$ versus $H_1 : \boldsymbol{s}_t \neq \boldsymbol{0}$ using (4.53), and dropping the "j",

$$
\widehat{\boldsymbol{\delta}}_{it} = \mathbf{H}_t \left\{ \widehat{\boldsymbol{s}}_t(x_{it}) - \begin{pmatrix} \widehat{\boldsymbol{\gamma}}_{(1)t}^T \\ \vdots \\ \widehat{\boldsymbol{\gamma}}_{(\texttt{ncol}(\mathbf{H}_t))t}^T \end{pmatrix} \boldsymbol{x}_i \right\} .
$$

From practical point-of-view, some comments on the bias of the above score test are made in Sect. 4.4.

It is noted that several workers have developed alternative methods for conducting tests within a spline-additive model context. For example, Cantoni and Hastie (2002) propose a likelihood-ratio-type test statistic for smoothing splines based on mixed-effects models which enables tests of linearity to be conducted. Their test can be applied to additive models too, and in contrast with the test presented here, the exact distribution of their test statistic is derived. A second example is Fan and Jiang (2005) who extend the generalized likelihood ratio (GLR) tests to additive models using the backfitting estimator and show that, under the null models, the newly proposed GLR statistics follow asymptotically rescaled chi-squared distributions.

4.4 On More Practical Aspects

4.4.1 Using the Software

Here are some notes about using **VGAM** to fit VGAMs.

1. The call is typically of the form

```
vgam(y ~ s(x2) + s(x3, df = 2) + x4 + s(x5, df = c(1, 4)),
     VGAMfamily, data = vdata)
```

Now **s** is symbolic for a vector smoothing spline, and it defaults to 4 degrees of freedom (1 means linear) for *each* component function, therefore its ENDF is 3. So here, the smooths with respect to **x3** have about the flexibility of a curve somewhat between a line and a quadratic. Values of **df** are recycled if necessary, and the above call implies $M \geq 2$—if not, then a warning message is issued.

The function **s()** should only be used with **vgam()**—not with **vglm()**. If all terms in the formula are parametric, then **vglm()** should be called instead.

Variables **x2**, **x3** and **x5** should be numeric with at least 7 unique values; this is partly because smoothing with $n < 7$ is rather nonsensical.

2. It is usual to input argument **df** rather than **spar**. The latter denotes scaled versions of the smoothing parameters $\lambda_{(j)k}$ in (4.47). The present implementation details are to have **lambda** $\propto 256^{3 \cdot \mathbf{spar} - 1}$ where **lambda** is $\lambda_{(j)k}$ after scaling (4.3) from $[a, b]$ to $[0, 1]$. When **df** is specified, a root-finding procedure is applied to **spar** on an interval $[-\frac{3}{2}, \frac{3}{2}]$ (the default).

3. The p-values printed by the **summary()** generic are based on the approximate score test described in Sect. 4.3.4. Because this test is sub-optimal, the χ^2-statistics for nonlinearity tend to be biased upward (too large), hence the associated p-values for testing linearity are generally anticonservative (too small).

4.4.2 vsmooth.spline()

The stand-alone function **vsmooth.spline()** fits vector smoothing splines. It currently has arguments

```
> args(vsmooth.spline)

  function (x, y, w = NULL, df = rep(5, M), spar = NULL, i.constraint = diag(M),
      x.constraint = diag(M), constraints = list('(Intercepts)' = i.constraint,
          x = x.constraint), all.knots = FALSE, var.arg = FALSE,
      scale.w = TRUE, nk = NULL, control.spar = list())
  NULL
```

and operates in a similar spirit to **smooth.spline()**. By default, O-splines are fitted (Sect. 2.4.4.3) so that if n is large then $n^* \ll n$ unless **all.knots = TRUE**. At present, **VGAM** uses a piecewise function to choose the 'effective' number of design points, n^*, as plotted in Fig. 2.12. For small data sets ($n < 50$), a knot is placed at every distinct data point x_i, so $n^* = n$. For larger data sets, the number of knots is $O(n^{1/5})$ when `all.knots = FALSE` (the default). Given n^*, the number of B-spline coefficients is $(n^* + 2)M$ and the number of knots is $n^* + 6$. Given $x_1 < x_2 < \cdots < x_n$, three boundary knots are chosen at both $\min(x_i)$ and $\max(x_i)$, and n^* knots at $x_{\lfloor 1+(j-1)(n-1)/(n^*-1)\rfloor}$ for $j = 1, \ldots, n^*$. The result are four knots each at $\min(x_i)$ and $\max(x_i)$, and the spline is linear beyond the boundaries.

As described in Sect. 4.2.1.4, `vsmooth.spline()` decomposes each component function into a linear and nonlinear part. When the generic `coef()` is applied to such a fit, the coefficients belonging to the linear and nonlinear parts are returned in two separate lists.

Like its univariate counterpart `smooth.spline()`, there is a `predict()` methods function that allows different order derivatives to be computed. Note that the `df` value for `vsmooth.spline()` is one higher than the `df` value for `s()`; this is because `s()`'s corresponds to number of nonlinear degrees of freedom and the function is centred. For example, `df = 2` for `vsmooth.spline()` and `df = 1` for `s()` are linear functions.

Constraints can be inputted in two ways in `vsmooth.spline()`. The first is through the arguments `i.constraint` and `x.constraint`, which stand for intercept-constraint and x-constraint, respectively. These may be assigned a constraint matrix, or a function returning one. The second way is through the argument `constraints`, which is a list with one or more constraint matrices. When none of these arguments are assigned, $\mathbf{I}_M$ is assumed (trivial constraints) for both intercept and x. The `constraints` slots of `vgam()` and `vsmooth.spline()` objects have the same form.

As with `vglm()`, the weight matrices $\mathbf{W}_i = \boldsymbol{\Sigma}_i^{-1}$ may be inputted in matrix-band format (Sect. 18.3.5).

4.4.3 Example: Cats and Dogs

To illustrate some basic aspects of VGAM fitting, let's fit a bivariate odds ratio model to examine how the probability of having a household cat and dog varies as a function of peoples' ages, in the `xs.nz` data frame. We choose a more homogeneous subset by restricting the analysis to European women, and we regress the two binary responses nonparametrically against age. We create this subframe first because we want to make use of it later, and also take the liberty to remove any missing values from those variables at this early stage. The initial model allows all 3 parameters to be smooth functions of `age`, however, the marginal probabilities are afforded more flexible than the odds ratio on their respective η_j scale.

```
> f.euro <- subset(xs.nz, sex == "F" & ethnicity == "European")
> f.euro.cd <- subset(f.euro, !is.na(age) & !is.na(cat) & !is.na(dog))
> fit1.cd <- vgam(cbind(cat, dog) ~ s(age, df = c(4, 4, 2)),
                  binom2.or(zero = NULL), data = f.euro.cd)
```

The data frame `f.euro.cd` has 2569 rows so it is not a small data set. A plot of the component functions with SEs, and under a common y-axis scale to make the component functions more comparable, can be obtained as follows (Fig. 4.4).

```
> plot(fit1.cd, se = TRUE, scol = "limegreen", lcol = "blue", scale = 4)
```

The first two plots suggest that an exchangeable error structure might be present. Testing this by

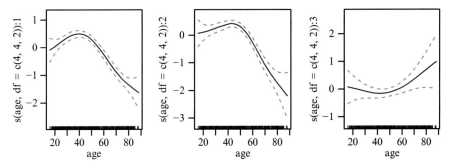

Fig. 4.4 A bivariate odds ratio model fitted as a VGAM to a subset of female Europeans from **xs.nz**, with household cat and dog ownership as the responses. The plots are the $\widehat{f}_{(j)2}(x_2)$.

```
> fit2.cd <- vgam(cbind(cat, dog) ~ s(age, df = c(4, 2)),
                  binom2.or(zero = NULL, exchangeable = TRUE), data = f.euro.cd)
> lrtest(fit1.cd, fit2.cd)
```

rejects the hypothesis strongly.

Whether the component functions are nonlinear can be answered by

```
> summary(fit1.cd)


Call:
vgam(formula = cbind(cat, dog) ~ s(age, df = c(4, 4, 2)), family = binom2.or(zero = NULL),
    data = f.euro.cd)

Number of linear predictors:     3

Names of linear predictors: logit(mu1), logit(mu2), loge(oratio)

Dispersion Parameter for binom2.or family:    1

Residual deviance:  6160.2 on 7694.3 degrees of freedom

Log-likelihood: -3080.1 on 7694.3 degrees of freedom

Number of iterations:  6

DF for Terms and Approximate Chi-squares for Nonparametric Effects

                         Df Npar Df Npar Chisq  P(Chi)
(Intercept):1             1
(Intercept):2             1
(Intercept):3             1
s(age, df = c(4, 4, 2)):1  1     3.0        77.2 0.00000
s(age, df = c(4, 4, 2)):2  1     2.8        68.8 0.00000
s(age, df = c(4, 4, 2)):3  1     0.9         5.8 0.01267
```

These results suggest that all three functions are nonlinear. A plot of the four fitted joint probabilities can be obtained by

```
> ooo <- with(f.euro.cd, order(age))   # Used to join by age
> mycol <- c("orange", "green", "blue", "purple"); mylty <- c(1, 1, 2, 2)
> with(f.euro.cd,
       matplot(age[ooo], fitted(fit1.cd)[ooo, ], type = "l", col = mycol, las = 1,
               xlab = "Age", lty = mylty, ylab = "Fitted joint probabilities"))
> legend("topleft", c("No cat or dog", "Dog only", "Cat only", "Cat and dog"),
         lty = mylty, col = mycol)
```

which gives Fig. 4.5.

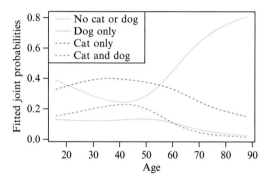

Fig. 4.5 Joint probabilities from `fit.cd1`.

This plot suggests that the probability of having both types of pets in the home attains its maximum for women who are middle-aged. Possibly, this might correspond mainly to mothers of families with children. Not surprisingly, pet ownership declines with age.

One use of VGAMs is to replace nonparametric component functions by parametric ones. Let's try the following parametric replacement.

```
> Hlist <- list("(Intercept)" = diag(3),
               "bs(age, degree = 1, knot = 40)" = rbind(1, 0, 0),
               "bs(age, degree = 1, knot = 50)" = rbind(0, 1, 0),
               "poly(age, 2)" = rbind(0, 0, 1))   # Correct white space needed!
> fit3.cd <- vglm(cbind(cat, dog) ~ bs(age, degree = 1, knot = 40) +
                                    bs(age, degree = 1, knot = 50) +
                                    poly(age, 2),
                  binom2.or(zero = NULL), data = f.euro.cd, constraints = Hlist)
```

One might then compare them separately by

```
> plot(fit1.cd, se = TRUE, scol = "orange", lcol = "blue", scale = 4)
> plot(as(fit3.cd, "vgam"), se = TRUE, scol = "orange", lcol = "blue", scale = 4)
```

however, it's sometimes more effective to overlay them (Fig. 4.6):

```
> for (cf in 1:3) {   # Loop over the component functions
    plot(fit1.cd, which.cf = cf,
         se = TRUE, scol = "blue", lcol = "blue", scale = 3.5)
    plot(as(fit3.cd, "vgam"), which.term = cf, raw = TRUE, add = TRUE, overlay = TRUE,
         se = TRUE, scol = "orange", lcol = "orange", scale = 3.5)
  }
```

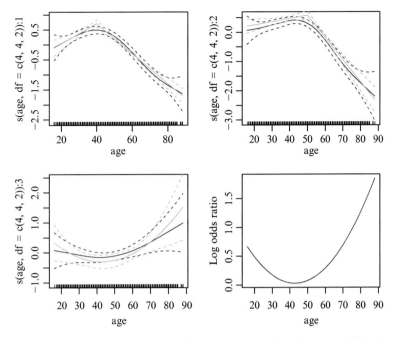

Fig. 4.6 The first three plots are the fitted component functions overlaid; the models are `fit1.cd` are `fit3.cd`. The fourth plot is the estimated log odds ratios of `fit3.cd`.

A natural question is to ask whether the parametric fit is acceptable. One might wish to use either of

```
> lrtest(fit3.cd, fit1.cd)
> lrtest(fit1.cd, fit3.cd)
```

however, comparing a model with an `s()` term with another with a `bs()` term is treacherous. It is safer to perform the test manually, using something like

```
> pchisq(2 * (logLik(fit1.cd) - logLik(fit3.cd)),
         df = df.residual(fit3.cd) - df.residual(fit1.cd), lower.tail = FALSE)

 [1] 0.10755
```

which is very heuristic and approximate. This suggests that the parametric model `fit3.cd` is reasonable. This is in accord with the close resemblance of the functions in Fig. 4.6. Now

```
> coef(fit3.cd, matrix = TRUE)
```

	logit(mu1)	logit(mu2)	loge(oratio)
(Intercept)	-0.11032	-0.90412	0.33448
bs(age, degree = 1, knot = 40)1	0.81614	0.00000	0.00000
bs(age, degree = 1, knot = 40)2	-1.59353	0.00000	0.00000
bs(age, degree = 1, knot = 50)1	0.00000	0.37173	0.00000
bs(age, degree = 1, knot = 50)2	0.00000	-2.51084	0.00000
poly(age, 2)1	0.00000	0.00000	13.21548
poly(age, 2)2	0.00000	0.00000	14.40974

which the reader can show (Ex. 4.1) that the estimated odds ratio is greater than unity for all ages represented in the data set. It is not surprising that these two types of pets have a positive association in terms of household ownership. Another observation is that the point estimates of the odds ratio are closest to unity at around 43 years, which suggests that choice of the type of pet is made almost independently at around that age.

Bibliographic Notes

All of the references on GAMs listed in the bibliographic notes of Chap. 2 have a direct bearing on VGAMs. Further to the references on Hilbert space theory in Chap. 2, of relevance to statisticians are Small and McLeish (1994) and Berlinet and Thomas-Agnan (2004). The iterative vector spline solution described in Sect. 4.2.1.5 is based on Yee (1998). The algorithm of Sect. 4.2.1.3 is due to Yee (2000).

The package mgcv can fit a few models with $M > 1$ additive predictors using automatic smoothing parameter selection, e.g., the Cox model, the multinomial logit model, and zero-inflated Poisson model.

Exercises

Ex. 4.1.   Confirm that the estimated odds ratio in the model fit3.cd of Sect. 4.4.3 is greater than unity for all ages represented in the data set. Show that the odds ratio is closest to unity at around 43 years.

Ex. 4.2.   Consider the xs.nz data frame, with acne as the response and age as the explanatory variable. Restrict your analyses to European-type people.

(a) Fit two GAMs, one for each gender. Plot the fitted values for each gender on a single plot. Comment and discuss the interpretation of the curves, e.g.,

  (i) Does it mean that older people tend to have less of an acne problem?
  (ii) Has the prevalence of acne changed over time?
  (iii) Has there been availability of treatment for acne only in more modern times?
  (iv) Are younger people these days more concerned about their looks and health?

(b) Do the curves coincide?—conduct an approximate hypothesis test.

Ex. 4.3.   Fit a VGAM bivariate odds ratio model to the male Europeans from the xs.nz data frame with worry and worrier as the responses. Use age as the covariate, and model the log odds ratio as a function of age too. Comment on the estimated component functions and fitted values.

Ex. 4.4.   **London Tube Patronage**
Fit two simultaneous GAMs to the tube10 data for the CharingCross and SouthKensington stations—use each quarter-hour time block as values of the explanatory variable. Using Poisson or negative binomial regression, describe quantitatively and qualitatively the differences between the two stations.

Ex. 4.5.   VGAMs Fitted to Two Tree Species
Consider the species Tawa and Rewarewa from the `hunua` data frame.

(a) Use `vgam()` to fit simultaneous smooth logistic regressions to the two species
    against `altitude`. Overlay the (centred) component functions, along with their
    standard error bands. Comment.
(b) On one plot, overlay the fitted values of both species against altitude. Com-
    ment.
(c) † Repeat (a) but plot the *uncentred* component functions. Repeat (b) but
    include the SE bands.

Ex. 4.6.   Kauri Trees in Two Forests
Consider the species *Agathis australis* in the `hunua` and `waitakere` data frames.

(a) Use `vgam()` to fit smooth logistic regressions to the species against `altitude`—
    one for each data set. Overlay the (centred) component functions, along with
    their standard error bands, on a single plot. Comment.
(b) Are the species' distributions the same in both forests? i.e., are the two com-
    ponent functions the same? Conduct some approximate hypothesis test. Com-
    ment.

Ex. 4.7.   RKHS for Cubic Smoothing Splines
Consider the cubic smoothing spline as a special case of a vector spline with $M = 1$.

(a) Show that $\{1, t - a\}$ is an orthonormal basis with respect to the inner prod-
    uct (4.25) for a cubic smoothing spline defined on (a, b).
(b) For $\Omega = [0, b]$ show that the RK for $\mathcal{H}_1$ is

$$K_1(s, t) = \frac{(\min(s, t))^2}{6} \{3 \max(s, t) - \min(s, t)\}, \qquad s, t \in [0, b].$$

Ex. 4.8.   Linear Interpolating Splines
For the linear interpolating spline example of Sect. 4.2.1.7, express $\widehat{f}(t)$ in the
form $\alpha_m + \beta_m t$ for each segment, given data $f(0) = 0$, $f(0.1) = 1$, $f(0.3) = 0.5$,
$f(0.6) = 0.9$ and $f(1) = 2$.

**Ex. 4.9.   Equivalent Kernels for Vector Local Linear Regression
for $M = 2$** Following from (4.44) and (4.45) obtain asymptotic equivalent kernel for $\widehat{f}_1'(x)$ with respect to $y_i^{(1)}$ and $y_i^{(2)}$.

Ex. 4.10. p-Values for Testing Linearity in VGAMs

(a) Generate values of a variable x_2 such that $x_{i2} \sim \text{Unif}(0, 1)$ independently,
    for $i = 1, \ldots, n = 100$. Then do the following 1000 times. Generate Pois-
    son counts with $\eta_i = 1 + x_{i2}$. Using calls of the form `vgam(y ~ s(x2),
    poissonff, ...)`, obtain the p-value for testing linearity of the component
    function. Examine the 1000 p-values and comment.
(b) Repeat (a) but with $n = 30$. Do the results change much?
(c) Repeat (a) but with $n = 1000$. Do the results change much?

**Ex. 4.11. p-Values for Testing Linearity in VGAMs (Regression
Splines)**
Repeat Ex. 4.10(a)–(c) but use calls of the form `vglm(y ~ bs(x2), poissonff,
...)`. Obtain the $p$-values using `lrtest()`.

Ex. 4.12. **Sleep in `xs.nz`**

Consider the response variable `sleep` in the data frame `xs.nz`.

(a) Fit an additive model to these data, using age, sex and ethnicity as explanatory variables. Delete all the missing values from these variables first. Use `uninormal()` and allow both η_j to be additive in the variables.

(b) Plot the estimated component functions of your model and interpret them.

(c) Some of the values of `sleep` are unrealistic. Delete those values which you think are not 'true', and repeat your analyses above. Does the deletion of these values make much of a difference to the results?

(d) Repeat (a)–(b), but fit a Poisson regression instead. What are the advantages of each type of regression model?

A councillor ought not to sleep the whole night through, a man to whom the populace is entrusted, and who has many responsibilities.
—Homer

Chapter 5
Reduced-Rank VGLMs

True love yields not to high rank.
—Sextus Aurelius Propertius

5.1 Introduction

This chapter describes the first of three major classes of models that spring as
extensions of the VGLM and VGAM classes. These main variants are summarized in Table 5.1, and their primary defining characteristic is that they operate
on R *latent variables* which can be written $\boldsymbol{\nu} = (\nu_1, \ldots, \nu_R)^T$. The phrase "latent
variable" has several shades of meaning in statistics, such as a random variable
which cannot be measured directly, or an unobserved or latent trait. For example,
a person's quality of life might be ascertained using measurable variables such as
wealth, health, employment, environment, education, leisure time and social belonging. Other examples of the use of latent variables are to measure happiness,
business confidence and emotional intelligence. The concept of latent variables is
used in many fields such as psychology, economics, medicine, biology (especially
ecology) and the social sciences.

Here, a latent variable is defined as a linear combination of some (measured)
explanatory variables $\boldsymbol{x}_2$, i.e.,

$$\boldsymbol{\nu} = \mathbf{C}^T \boldsymbol{x}_2 \qquad (5.1)$$

for some $p_2 \times R$ matrix of coefficients $\mathbf{C}$. Although the *rank* $R \leq p_2$, usually R
is very low such as 0, 1 or 2, hence $\mathbf{C}$ is described as a 'thin' matrix. The role of
reduced-rank regression (RRR) is to 'replace' a large vector $\boldsymbol{x}_2$ by a smaller vector $\boldsymbol{\nu}$
as explanatory variables. The matrix $\mathbf{C}$ contains the *constrained* (or *canonical*)
coefficients, which are highly interpretable. They are sometimes called the *weights*
or *loadings*.

The idea of taking a linear combination of $\boldsymbol{x}$ pervades much of statistics. One
reason is its simple interpretation as a weighted average of $\boldsymbol{x}$ values. Another is

© Thomas Yee 2015

T.W. Yee, *Vector Generalized Linear and Additive Models*,
Springer Series in Statistics, DOI 10.1007/978-1-4939-2818-7_5

Table 5.1 A summary of the models described in Chaps. 5–7 (upper table). The latent variables are $\boldsymbol{\nu} = \mathbf{C}^T \boldsymbol{x}_2$, or $\nu = \boldsymbol{c}^T \boldsymbol{x}_2$ if $R = 1$. These models compare with VGLMs where $\boldsymbol{\eta} = \mathbf{B}_1^T \boldsymbol{x}_1 + \mathbf{B}_2^T \boldsymbol{x}_2$. Abbreviations: A = additive, C = constrained, L = linear, O = ordination, Q = quadratic, RR = reduced-rank. The lower table are subvariants: Goodman's row–column association model and row–column interaction models. See also Table 1.1.

η_i	Model	Purpose	R function	See also
$\mathbf{B}_1^T \boldsymbol{x}_{1i} + \mathbf{A}\,\boldsymbol{\nu}_i$	RR-VGLM	CLO	`rrvglm()`	Chap. 5
$\mathbf{B}_1^T \boldsymbol{x}_{1i} + \mathbf{A}\,\boldsymbol{\nu}_i +$ $\sum_{j=1}^{M} (\boldsymbol{\nu}_i^T \mathbf{D}_j \boldsymbol{\nu}_i)\, \boldsymbol{e}_j$	QRR-VGLM	CQO	`cqo()`	Chap. 6
$\mathbf{B}_1^T \boldsymbol{x}_{1i} + \sum_{r=1}^{R} \boldsymbol{f}_r(\nu_{ir})$	RR-VGAM	CAO	`cao()`	Chap. 7
$(\mu + \alpha_i)\mathbf{1}_M + \boldsymbol{\gamma} + \mathbf{A}\boldsymbol{\nu}_i$	GRC	GRC	`grc()`	Eq. (5.25)
$(\mu + \alpha_i)\mathbf{1}_M + \boldsymbol{\gamma} + \mathbf{A}\boldsymbol{\nu}_i$	RCIM	RCIM	`rcim()`	Sect. 5.7

computational: $\boldsymbol{\beta}^T \boldsymbol{x}$ is easily differentiated with respect to $\boldsymbol{\beta}$. The quantity $\boldsymbol{c}^T \boldsymbol{x}$ is the mainstay of many multivariate techniques, such as principal components analysis and linear discriminant analysis.

Notationally, recall that the data comprises $n \times Q$ $\mathbf{Y}$ and $n \times p$ $\mathbf{X}$. In the next few chapters, it is common for each column of $\mathbf{Y}$ to pertain to a different species of plant or animal—and there are S of them. Furthermore, we often fit a one-parameter model to each species, such as a binomial for presence/absence, or a Poisson regression for counts. Hence we often have $Q = M = S$ and use $s = 1, \ldots, S$ to index across species. In statistical ecology, the $\mathbf{Y}$ matrix is sometimes referred to as "species-by-site" data. This might be better called "site-by-species" data since $\mathbf{Y}$ is $n \times S$ instead of $S \times n$. Since one application of fitting regression models to $\mathbf{Y}$ as a function of $\mathbf{X}$ is to arrange the species in a meaningful order, we have R axes in the ordination (Chap. 6).

5.2 What Are RR-VGLMs?

Partition $\boldsymbol{x}$ into $(\boldsymbol{x}_1^T, \boldsymbol{x}_2^T)^T$ and $\mathbf{B} = (\mathbf{B}_1^T\ \mathbf{B}_2^T)^T$, cf. (3.4). In general, $\mathbf{B}$ is a dense matrix of full rank, i.e., the rank is $\min(M, p)$. Thus there are $M \times p$ regression coefficients to estimate. For some data sets, both M and p are "too" large for the given n, and that results in overfitting. For example, in a classification problem of the letters "A" to "Z" and "0" to "9" using a 16×16 grey-scale character digitization, we have $M + 1 = 26 + 10 = 36$ and $p = 1 + 16^2 = 257$ in a multinomial logit model. Then there would be $M \times p = 9252$ coefficients! Unless the data set were huge, this would mean that the standard errors of each coefficient would be very large and some simplification would be necessary. Additionally, fitting the model in the first case would require a large amount of computer memory.

One solution is based on a simple and elegant idea: replace $\mathbf{B}_2$ by a reduced-rank regression. This will cut down the number of regression coefficients enormously if the rank R is kept low. Ideally, the problem can be reduced down to one or two dimensions—a successful application of dimension reduction—and therefore can be plotted. The RRR is applied to $\mathbf{B}_2$ because we want to make provision for some variables $\boldsymbol{x}_1$ that are to be left alone, meaning $\mathbf{B}_1$ remains unchanged. In practice, we often let $\boldsymbol{x}_1 = 1$ for the intercept term only (the software default).

"Reduced-rank regression" was coined by Izenman (1975) and the idea goes back at least to Anderson (1951). Since then, most applications of RRR have been to continuous responses (normal errors), and unfortunately this is one reason why RRR has never become as popular as it should be. It has been mainly confined to the statistical and econometrics literature.

The RR-VGLM class arises by endowing the VGLM class with RRR capabilities. It has

$$\boldsymbol{\eta} = \mathbf{B}_1^T \boldsymbol{x}_1 + \mathbf{A}\,\mathbf{C}^T \boldsymbol{x}_2 = \mathbf{B}_1^T \boldsymbol{x}_1 + \mathbf{A}\,\boldsymbol{\nu} \qquad (5.2)$$

where

$$\mathbf{C} = \begin{pmatrix} \boldsymbol{c}_{(1)} & \boldsymbol{c}_{(2)} & \cdots & \boldsymbol{c}_{(R)} \end{pmatrix} = \begin{pmatrix} \boldsymbol{c}_1, \ldots, \boldsymbol{c}_{p_2} \end{pmatrix}^T \text{ is } p_2 \times R, \qquad (5.3)$$

$$\mathbf{A} = \begin{pmatrix} \boldsymbol{a}_{(1)} & \boldsymbol{a}_{(2)} & \cdots & \boldsymbol{a}_{(R)} \end{pmatrix} = \begin{pmatrix} \boldsymbol{a}_1, \ldots, \boldsymbol{a}_M \end{pmatrix}^T \text{ is } M \times R. \qquad (5.4)$$

Both $\mathbf{A}$ and $\mathbf{C}$ are of full column-rank. The effect is that the dense matrix $\mathbf{B}_2$ is approximated by the product of two thin matrices $\mathbf{C}$ and $\mathbf{A}^T$. Of course, $R \leq \min(M, p_2)$, but ideally we want $R \ll \min(M, p_2)$. One can think of (5.2) as an RRR of the coefficients of $\boldsymbol{x}_2$ after having adjusted for the variables in $\boldsymbol{x}_1$. Equation (5.2) can also be thought of as a two-phase regression where the first phase is to choose a few optimal directions in the form of the best $\mathbf{C}$, and the second phase is to regress upon the new variables obtained by projecting the original $\boldsymbol{x}_2$ variables onto the new axes. Thus $\boldsymbol{\nu}$ of an RR-VGLM assumes the role of $\boldsymbol{x}_2$ in an ordinary VGLM.

Strictly speaking, we call models given by (5.2) *partial RR-VGLMs*, because only a subset of the regressors have a reduced-rank representation. We drop the "partial" for convenience. Since $\dim(\boldsymbol{x}_1) = p_1$ and $\dim(\boldsymbol{x}_2) = p_2$ then $p_1 + p_2 = p$. To distinguish between variables belonging to $\boldsymbol{x}_1$ and $\boldsymbol{x}_2$, the argument `noRRR` of `rrvglm()` receives a formula with terms that are left untouched by RRR, i.e., those for $\boldsymbol{x}_1$. By default, `noRRR` $\sim$ `1` so that $\boldsymbol{x}_1 = 1$. Here's an example:

```
rrvglm(y ~ x2 + x3 + x4 + x5, multinomial(parallel = TRUE ~ x2 - 1),
       data = mdata, noRRR = ~ x2 + x3)
```

This means that only variables `x4` and `x5` are subject to an RRR, i.e., $\boldsymbol{x}_2 = (x_4, x_5)^T$, and $\boldsymbol{x}_1 = (1, x_2, x_3)^T$ are left alone. The variable x_2 receives a parallelism constraint so that $\mathbf{H}_2 = \mathbf{I}_M$. The function `rrvglm()` has `Rank = 1` as default.

5.2.1 Why Reduced-Rank Regression?

Here are a few reasons why RRR can be useful.

1. They can be readily interpretable. One can think of $\boldsymbol{\nu}$ as a vector of R latent variables—linear combinations of the original predictor variables that give more explanatory power. They often can be thought of as a proxy for some underlying variable behind the mechanism of the process generating the data. For some models, such as the cumulative logit model, this argument is natural and well-known (Sect. 14.4.1.1). Indeed, this chapter overlaps with several models for categorical data, because such regression models naturally produce a large number of parameters. Therefore they are good candidates for reduced-rank modelling. In fields such as plant ecology, the idea is an important one—see Sect. 6.1.

2. If $R \ll \min(M, p_2)$, then a more parsimonious model can result. The resulting number of parameters is often much less than the full model. This difference is $(M - R)(p_2 - R)$, which is substantial when $R \ll \min(M, p_2)$.
3. It allows for a flexible nonparametric generalization—the RR-VGAM class—the subject of Chap. 7.
4. The reduced-rank approximation to $\mathbf{B}_2$ provides a vehicle for a low-dimensional view of the data, e.g., the biplot. This is illustrated later.

Unfortunately, much of the standard theory of RRR (see, e.g., Reinsel and Velu (1998)) and its ramifications are not directly transferable to RR-VGLMs in general, in a similar way that LM theory is not to GLMs. Another complication in inference of RR-VGLMs is that the solution to a lower-rank problem is not nested within a higher-rank problem.

5.2.2 Normalizations

The factorization (5.2) is not unique, because

$$\boldsymbol{\eta} \; = \; \mathbf{B}_1^T \boldsymbol{x} + \mathbf{A}\,\mathbf{M}\,\mathbf{M}^{-1} \boldsymbol{\nu}$$

for any nonsingular matrix $\mathbf{M}$. The following lists some common uniqueness constraints.

1. Restrict $\mathbf{A}$ to the form

$$\mathbf{A} \; = \; \begin{pmatrix} \mathbf{I}_R \\ \widetilde{\mathbf{A}} \end{pmatrix}, \quad \text{say.} \tag{5.5}$$

   This is referred to as a *corner constraint*, and it corresponds to `Corner = TRUE` in `rrvglm.control()`. The argument `Index.corner = 1:Rank` specifies the rows of (5.5) which $\mathbf{I}_R$ resides, because it may be necessary or more convenient to store $\mathbf{I}_R$ in rows other than the first R. Of course, only the elements of $\widetilde{\mathbf{A}}$ are to be estimated.
2. Another normalization of $\mathbf{A}$, which makes direct comparisons with other statistical methods possible, is based on the singular value decomposition (SVD; Sect. A.3.4)

$$\mathbf{A}\mathbf{C}^T \; = \; (\mathbf{U}\mathbf{D}^{\alpha}) \left(\mathbf{D}^{1-\alpha}\, \mathbf{V}^T \right) \tag{5.6}$$

   for some specified $0 \le \alpha \le 1$. For the alternating method of estimation described below, $\alpha = \frac{1}{2}$ is the default as it scales both sides symmetrically. The parameter α is `Alpha = 0.5` in `rrvglm.control()`.
3. Sometimes we want to choose $\mathbf{M}$ so that the latent variables are uncorrelated, i.e., $\widehat{\mathrm{Var}}(\widehat{\boldsymbol{\nu}}_i)$ is diagonal. Furthermore, we can scale $\mathbf{M}$ so that $\widehat{\mathrm{Var}}(\widehat{\boldsymbol{\nu}}_i) = \mathbf{I}_R$, i.e., unit variances.
4. For the stereotype model described in Sect. 5.2.3, we could choose $\mathbf{M}$ so that the columns of $\mathbf{C}$ are orthogonal with respect to the within-group covariance matrix—this type of normalization is similar to linear discriminant analysis.

5.2.3 The Stereotype Model

When RRR is applied to the *multinomial logit model* (MLM; see Sects. 1.2.4 and 14.2) the result is known as a *stereotype model* (Anderson, 1984). He defined the rank-1 stereotype model to have the additional constraint of ordered elements of $\mathbf{A}$, however, VGAM treats all the elements of $\widetilde{\mathbf{A}}$ as unconstrained. A rank-2 RR-MLM might be fitted like

```
rr.mlm2 <- rrvglm(factor.with.many.levels ~ x2 + x3 + x4 + x5,
                  multinomial, data = mdata, noRRR = ~ 1 + x2, Rank = 2)
```

5.3 A Few Details

It transpires that RR-VGLMs are VGLMs where some of the constraint matrices (those corresponding to $\boldsymbol{x}_2$) are equal, unknown and estimated rather than being known and fixed. This important characterization can be seen by considering

$$\boldsymbol{\eta} \; = \; \mathbf{B}_1^T \boldsymbol{x}_1 + \mathbf{A}\,\boldsymbol{\nu} \; = \; \mathbf{B}_1^T \boldsymbol{x}_1 + \sum_{k=1}^{p_2} \mathbf{A} \begin{pmatrix} c_{k1} \\ \vdots \\ c_{kR} \end{pmatrix} x_{2k} \qquad (5.7)$$

where $(\mathbf{C})_{ij} = c_{ij}$. A comparison of (5.7) with (3.25) shows that $\mathbf{A}$ matches with $\mathbf{H}_k$ for all variables in $\boldsymbol{x}_2$, and the kth row of $\mathbf{C}$ matches with $\boldsymbol{\beta}_{(k)}^{*T}$. This characterization naturally leads to the algorithm that is used to estimate RR-VGLMs, described in the next section.

5.3.1 Alternating Algorithm

Like VGLMs, RR-VGLMs are estimated by the IRLS algorithm. At iteration a, one can minimize a residual sum of squares

$$\sum_{i=1}^{n} \left(\boldsymbol{z}_i^{(a)} - \widehat{\mathbf{B}}_1^T \boldsymbol{x}_{1i} - \widehat{\mathbf{A}}\,\widehat{\mathbf{C}}^T \boldsymbol{x}_{2i} \right)^T \mathbf{W}_i \left(\boldsymbol{z}_i^{(a)} - \widehat{\mathbf{B}}_1^T \boldsymbol{x}_{1i} - \widehat{\mathbf{A}}\,\widehat{\mathbf{C}}^T \boldsymbol{x}_{2i} \right) \qquad (5.8)$$

by fixing $\mathbf{A}$ and solving for $\boldsymbol{\nu} = \mathbf{C}^T \boldsymbol{x}_2$ and $\mathbf{B}_1$, and then keeping $\boldsymbol{\nu}$ fixed, solving for $\mathbf{A}$ and $\mathbf{B}_1$. Equation (5.8), which is a specific case of (3.22), is called an RR-VLM (see Fig. 1.2).

This alternating algorithm has been called by various names, e.g., the criss-cross method by Gabriel and Zamir (1979). Each alternation is a full minimization, therefore it is a 'zigzag' method in the terminology of Smyth (1996). Given $\boldsymbol{\nu}$, solving for $\mathbf{A}$ and $\mathbf{B}_1$ is easily obtained by using $\boldsymbol{x}_1$ and $\boldsymbol{\nu}$ as covariates in a VLM. Given $\mathbf{A}$, the quantities $\mathbf{C}$ and $\mathbf{B}_1$ can be solved from (5.7) by recognizing that this falls within the constraints-on-the-functions framework (Sect. 3.3) with $\mathbf{H}_{p_1+1} = \mathbf{H}_{p_1+2} = \cdots = \mathbf{H}_{p_1+p_2} = \mathbf{A}$ assumed to be known, and $\boldsymbol{\beta}_{(p_1+k)}^* = (c_{k1}, \ldots, c_{kR})^T$ to be estimated. It entails fitting a VLM to the $\boldsymbol{z}_i^{(a)}$, with each of the p_2 constraint matrices of $\boldsymbol{x}_2$ equalling $\mathbf{A}$.

The alternating algorithm can sometimes experience large speed gains by performing a line search with respect to the elements of $\mathbf{C}$ (Sect. 5.4.2).

5.3.2 SEs of RR-VGLMs

With corner constraints (5.5), let $\boldsymbol{\theta} = (\boldsymbol{\theta}_1^T, \boldsymbol{\theta}_2^T, \boldsymbol{\theta}_3^T)^T$ be the vector of all coefficients to be estimated, where $\boldsymbol{\theta}_1 = \mathrm{vec}(\widetilde{\mathbf{A}}) = (\widetilde{\boldsymbol{a}}_{(1)}^T, \ldots, \widetilde{\boldsymbol{a}}_{(R)}^T)^T$, $\boldsymbol{\theta}_2 = (\boldsymbol{\beta}_1^{*T}, \ldots, \boldsymbol{\beta}_{p_1}^{*T})^T$ and $\boldsymbol{\theta}_3 = \mathrm{vec}(\mathbf{C}) = (\boldsymbol{c}_{(1)}^T, \ldots, \boldsymbol{c}_{(R)}^T)^T$. Partition

$$-E\left(\frac{\partial^2 \ell}{\partial \boldsymbol{\theta}\,\partial \boldsymbol{\theta}^T}\right) \;=\; -E\left(\ddot{\ell}\right) \;=\; -E\begin{pmatrix}\ddot{\ell}_{11} & \ddot{\ell}_{12} & \ddot{\ell}_{13} \\ \ddot{\ell}_{21} & \ddot{\ell}_{22} & \ddot{\ell}_{23} \\ \ddot{\ell}_{31} & \ddot{\ell}_{32} & \ddot{\ell}_{33}\end{pmatrix} \qquad (5.9)$$

where $\ddot{\ell}_{jk} = \partial^2 \ell/(\partial \boldsymbol{\theta}_j\,\partial \boldsymbol{\theta}_k^T)$.

Standard errors of $\widehat{\boldsymbol{\theta}}$ can be obtained by computing the complete $-E[\partial^2 \ell/(\partial \boldsymbol{\theta}\,\partial \boldsymbol{\theta}^T)]$ matrix (5.9) evaluated at $\widehat{\boldsymbol{\theta}}$ and inverting it. The block matrices

$$-E\begin{pmatrix}\ddot{\ell}_{11} & \ddot{\ell}_{12} \\ \ddot{\ell}_{21} & \ddot{\ell}_{22}\end{pmatrix} \quad \text{and} \quad -E\begin{pmatrix}\ddot{\ell}_{22} & \ddot{\ell}_{23} \\ \ddot{\ell}_{32} & \ddot{\ell}_{33}\end{pmatrix} \qquad (5.10)$$

may be obtained by fixing $\boldsymbol{\theta}_3$ and $\boldsymbol{\theta}_1$, respectively, and are easily computed using the description presented in Sect. 5.3.1. The most difficult part is $-\ddot{\ell}_{13}$, which may be calculated using profile likelihoods (Richards (1961); see, e.g., (18.6), Seber and Wild (1989, Eq.(2.70))):

$$-E\left[\frac{\partial^2 \ell}{\partial \boldsymbol{\theta}_1\,\partial \boldsymbol{\theta}_3^T}\right] \;=\; -E\left[\frac{\partial \boldsymbol{\theta}_3^T(\boldsymbol{\theta}_1)}{\partial \boldsymbol{\theta}_1}\left(-\frac{\partial^2 \ell}{\partial \boldsymbol{\theta}_3\,\partial \boldsymbol{\theta}_3^T}\right)\right]. \qquad (5.11)$$

The matrix $\partial \boldsymbol{\theta}_3^T(\boldsymbol{\theta}_1)/\partial \boldsymbol{\theta}_1$ is given in Yee and Hastie (2003, App.B). This method also works for Fisher scoring. Equation (5.11) is computed by the RR-VGLM `summary()` methods function.

5.4 Other RR-VGLM Topics

5.4.1 Summary of RR-VGLM Software

The function `rrvglm()`, which operates very much like `vglm()`, should operate on all VGAM family functions with $M \geq 2$, although RRR does not make sense for many of them. The special case of models with $M = 2$ (Sect. 5.5) has a simple tractable formula for the coupling of two parameters, and it furnishes several interesting regression models such as RR-zero inflated Poisson and RR-NB (NB-P) models. The function `rrvglm()` returns an object of class `"rrvglm"` which, not surprisingly, inherits much of a `"vglm"` object.

Table 5.2 lists generic functions which are also applicable to RR-VGLMs.

Table 5.2 A summary of **VGAM** functions and generic functions for RR-VGLMs. The $\boldsymbol{x}$ variables have been assumed to be in the order $(\boldsymbol{x}_1^T, \boldsymbol{x}_2^T)^T$. See also Table 8.7.

Function	Purpose
`rrvglm()`	Fits RR-VGLMs
`grc()`	Goodman's RC association model (for two-way tables), Eq. (5.25)
`rcim()`	Row–column interaction models (for two-way tables), Eq. (5.24)
`rrar()`	RR-autoregressive time series model family function (for `vglm()` only)
`biplot()`	Biplot for RR-VGLMs ($R = 2$ only)
`coef(fit)`	$(\widehat{\boldsymbol{\beta}}_{(1)}^{*T}, \ldots, \widehat{\boldsymbol{\beta}}_{(p_1)}^{*T}, \mathrm{vec}(\widehat{\mathbf{C}}^T)^T)^T$
`coef(fit, matrix = TRUE)`	$\widehat{\mathbf{B}} = \left(\widehat{\mathbf{B}}_1^T \| \widehat{\mathbf{A}}\widehat{\mathbf{C}}^T \right)^T$
`Coef()`	Various coefficients, e.g., $\widehat{\mathbf{A}}$, $\widehat{\mathbf{B}}_1$, $\widehat{\mathbf{C}}$
`concoef()`	Constrained coefficients $\widehat{\mathbf{C}}$
`constraints(fit)`	$\mathbf{H}_1, \ldots, \mathbf{H}_{p_1}$ and p_2 repetitions of $\widehat{\mathbf{A}}$
`latvar()`	Matrix of latent variable values $(\widehat{\boldsymbol{\nu}}_1, \ldots, \widehat{\boldsymbol{\nu}}_n)^T$
`lvplot()`	Currently the same as `biplot()`
`summary()`	Summary function for 'all' RR-VGLMs

5.4.2 Convergence

The alternating algorithm can be very slow at converging, however, **VGAM** allows a line search to be performed which may improve the speed substantially. It is invoked by `Linesearch = TRUE`. The slowness is due to high correlations between the elements of $\widetilde{\mathbf{A}}$ and $\mathbf{C}$, and sometimes the fix is to choose different rows for storing $\mathbf{I}_R$ in the corner constraints (argument `Index.corner`).

During initialization, the unknown elements in $\mathbf{C}$ are chosen randomly using `rnorm()`. One should use `set.seed()` before invoking `rrvglm()` if reproducibility is required.

5.4.3 Latent Variable Plots and Biplots

Biplots are available for RR-VGLMs, and they provide a graphical summary of the rank-2 approximation to $\mathbf{B}_2$. These are based on (5.2), and show that the k-j element of $\mathbf{B}_2$ is the inner-product of the kth row of $\mathbf{C}$ and the jth row of $\mathbf{A}$. The rows of $\mathbf{C}$ are usually represented by arrows, and the rows of $\mathbf{A}$ by points. Currently, `lvplot(fit)` is equivalent to `biplot(fit)`.

If `fit2` is a rank-2 RR-VGLM, then `lvplot(fit2)` will produce a scatter plot of the fitted latent variables $\widehat{\nu}_{i2}$ versus $\widehat{\nu}_{i1}$ (see (5.2)). A convex hull can be overlaid on each group. By default, all the observations belong to one group.

5.4.4 Miscellaneous Notes

Here are some miscellaneous notes.

1. The methods function for `summary()` computes the asymptotic standard errors of Sect. 5.3.2. Unfortunately, sometimes the estimated variance-covariance matrix (5.9) is not positive-definite, and if so, then this indicates an ill-posed model (e.g., one where the intercepts are part of $\boldsymbol{x}_2$ rather than $\boldsymbol{x}_1$) and a warning is issued. In such cases, some tactics to overcome this setback include the following.

   (i) Set `numerical = TRUE` to use numerical derivatives.
   (ii) Fit a slightly perturbed version of the model such as omitting or adding variables.
   (iii) Increase or decrease the rank.
   (iv) Choose another vector for `Index.corner` because some elements of $\widehat{\mathbf{A}}$ may be very large or very small.

   Alternatively, there are the arguments `omit13` and `kill.all` that allow an 'inferior' variance-covariance matrix to be returned, for example, one where $\mathbf{A}$ is fixed or $\mathbf{C}$ is fixed. Then the standard errors will be biased downwards.

2. The methods function for `summary()` only works with corner constraints (5.5). In the output, the elements of $\widetilde{\mathbf{A}}$ in (5.5) come first. They are labelled with the prefix `"I(latvar.mat)"`. The elements are enumerated going down each column starting from the first column, as in $\mathrm{vec}(\widetilde{\mathbf{A}})$. Then comes the fitted coefficients $\widehat{\boldsymbol{\beta}}^*_{(k)}$ corresponding to $\mathbf{B}_1$, followed by $\mathrm{vec}(\mathbf{C}^T)$. Actually, $\mathbf{B}_1$ and $\mathrm{vec}(\mathbf{C}^T)$ are intermingled, depending on the order of the variables in the original formula.

5.5 RR-VGLMs with Two Linear Predictors

A special subfamily of RR-VGLMs are models with $M = 2$ and $R = 1$. For these, it is easy to obtain a simple formula for the coupling between the two parameters θ_1 and θ_2 using the rank-1 constraint. The result is surprisingly useful because the parameters are usually treated separately, and there is additional flexibility due to the choice of link functions. Indeed, quite a number of special cases proposed in isolated contexts by various authors belong to this subfamily. This section draws heavily upon Yee (2014) and omitted details can be found there. We will show that, while dimension reduction is customarily applied to high-dimensional problems, even a drop from two dimensions to one dimension can be very useful.

5.5.1 Two-Parameter Rank-1 RR-VGLMs

From (5.2), we apply RRR to all explanatory variables, except possibly for the intercept. As we restrict ourselves to models with $M = 2$ and $R = 1$, corner constraints give us $\mathbf{A} = (1, a_{21})^T$ so that there is only one parameter to estimate. Then

Table 5.3 Expressions for θ_2 as a function of θ_1 for several parameter link functions (Eq. (5.15)). The rows are for θ_1, columns for θ_2. The parameters t_1, K_j and a_{21} are unknown and are to be estimated, and the K_j are positive. Notes: (i) For Variant II, $t_1 = 0$ (so $K_1 = 1$), and for Variant I, $t_1 \neq 0$ and is to be estimated. (ii) $K_1 = \exp(t_1)$, $K_2 = \exp(a_{21})$ and $K_3 = \exp(K_1)$. (iii) The power link is $g_j(\theta_j) = \theta_j^{s_j}$ for a prespecified s_j. (iv) $\mathcal{A} = \theta_1/(1 - \theta_1)$ for the $\theta_1 = $ logit row. Source: Yee (2014).

	Identity	Log	Logit	Power
Identity	$t_1 + a_{21}\,\theta_1$	$K_1 \cdot K_2^{\theta_1}$	$\dfrac{K_1 \cdot K_2^{\theta_1}}{1 + K_1 \cdot K_2^{\theta_1}}$	$(t_1 + a_{21}\,\theta_1)^{1/s_2}$
Log	$t_1 + a_{21}\log\theta_1$	$K_1 \cdot \theta_1^{a_{21}}$	$\dfrac{K_1 \cdot \theta_1^{a_{21}}}{1 + K_1 \cdot \theta_1^{a_{21}}}$	$(t_1 + a_{21}\log\theta_1)^{1/s_2}$
Logit	$t_1 + a_{21}\operatorname{logit}\theta_1$	$K_1\,\mathcal{A}^{a_{21}}$	$\dfrac{K_1\,\mathcal{A}^{a_{21}}}{1 + K_1\,\mathcal{A}^{a_{21}}}$	$(t_1 + a_{21}\operatorname{logit}\theta_1)^{1/s_2}$
Power	$t_1 + a_{21}\,\theta_1^{s_1}$	$K_1 \cdot K_2^{\theta_1^{s_1}}$	$\dfrac{K_1 \cdot K_2^{\theta_1^{s_1}}}{1 + K_1 \cdot K_2^{\theta_1^{s_1}}}$	$\left(t_1 + a_{21}\,\theta_1^{s_1}\right)^{1/s_2}$

$$
\begin{aligned}
g_1(\theta_1) \;=\; \eta_1 \;&=\; \beta_{(1)1} + \boldsymbol{c}^T \boldsymbol{x}_2, & (5.12) \\
g_2(\theta_2) \;=\; \eta_2 \;&=\; \beta_{(2)1} + a_{21}\left(\boldsymbol{c}^T \boldsymbol{x}_2\right) \\
&=\; \left(\beta_{(2)1} - a_{21} \cdot \beta_{(1)1}\right) + a_{21}\,\eta_1 \\
&=\; t_1 + a_{21} \cdot \eta_1, \quad \text{say.} & (5.13)
\end{aligned}
$$

There are two variants: the RRR may or may not involve the intercept. The latter (called Variant I) has $\mathbf{H}_1 = \mathbf{I}_2$ and $\mathbf{H}_k = (1, a_{21})^T$ for $k = 2, \dots, p$. The former (called Variant II) has all $\mathbf{H}_k = (1, a_{21})^T$ so that $t_1 = 0$ in (5.13). Thus for Variant I,

$$
\begin{pmatrix} g_1(\theta_1) \\ g_2(\theta_2) \end{pmatrix} = \boldsymbol{\eta} = \begin{pmatrix} \eta_1 \\ \eta_2 \end{pmatrix} = \mathbf{B}_1^T \boldsymbol{x}_1 + \mathbf{A}\,\mathbf{C}^T \boldsymbol{x}_2 = \begin{pmatrix} \beta_{(1)1} \\ \beta_{(2)1} \end{pmatrix} + \begin{pmatrix} 1 \\ a_{21} \end{pmatrix} \nu \qquad (5.14)
$$

where $\nu = \boldsymbol{c}^T \boldsymbol{x}_2$.

The link g_j is invertible, therefore an expression for θ_2 as a function of θ_1 is

$$
\theta_2 \;=\; g_2^{-1}\left(t_1 + a_{21} \cdot g_1(\theta_1)\right). \qquad (5.15)
$$

This is the central equation. It provides the coupling of η_1 and η_2, inducing a relationship between two parameters that can be useful, for example, for modelling a mean-variance relationship. The choice of link functions determines the form of the coupling between the two parameters. Some pairs of commonly used link functions and the coupling they produce are given in Table 5.3. For example, the RR-NB (which coincides with what some people call the NB-P) model of Sect. 5.5.2.3 is based on the log–log coupling. There is additional flexibility too, because the user can often choose between alternative link functions, e.g., for parameters in $(0, 1)$ one can select a link such as the logit, probit and complementary log–log (Table 1.2 lists the link functions currently available in **VGAM**). Furthermore, (5.15) provides a unified approach to a large number of nonlinear regression models found in the

Table 5.4 Summary of some reduced-rank two-parameter families for the normal, negative binomial, inverse Gaussian and gamma distributions. Parameters μ, t_1, a_{21}, K_1, δ_1 and δ_2 are unknown and to be estimated. The final column is the `family` argument of the call `rrvglm(y ~ x2 + x3 + ···, family = ...)` in **VGAM**. Variant I is used here throughout. Source: Yee (2014).

μ	$V(\mu) = \mathrm{Var}(Y)$	family =
$\boldsymbol{\beta}^T\boldsymbol{x}$	$K_1^2 \cdot K_2^{2\mu}$	`uninormal(zero = NULL)`
$e^{\boldsymbol{\beta}^T\boldsymbol{x}}$	$\delta_1\,\mu^{\delta_2} = K_1^2 \cdot \mu^{2a_{21}}$	`uninormal(lmean = loge, zero = NULL)`
$\boldsymbol{\beta}^T\boldsymbol{x}$	$(t_1 + a_{21}\cdot\mu)^2$	`uninormal(lsd = identitylink, zero = NULL)`
$\boldsymbol{\beta}^T\boldsymbol{x}$	$t_1 + a_{21}\cdot\mu$	`uninormal(lvar = identitylink, zero = NULL, var = TRUE)`
$e^{\boldsymbol{\beta}^T\boldsymbol{x}}$	$\mu + \delta_1\,\mu^{\delta_2} = \mu + K_1^{-1}\,\mu^{2-a_{21}}$	`negbinomial(zero = NULL)`
$e^{\boldsymbol{\beta}^T\boldsymbol{x}}$	$\delta_1\,\mu^{\delta_2} = K_1^{-1}\,\mu^{3-a_{21}}$	`inv.gaussianff(zero = NULL)`
$e^{\boldsymbol{\beta}^T\boldsymbol{x}}$	$\delta_1\,\mu^{\delta_2} = K_1^{-1}\,\mu^{2-a_{21}}$	`gamma2(zero = NULL)`

statistical literature for which classification is not straightforward, e.g., the models described in Sect. 5.5.2 are special cases of (5.15).

Before giving some examples, it is noted that it is certain that there are other useful but as yet unidentified RR-variants amongst the myriads of documented statistical distributions.

5.5.2 Some Examples

5.5.2.1 RR-Normal

Under the usual LM we have link functions $\boldsymbol{\eta} = (\mu, \log\sigma)^T$. While it has homoscedasticity as an assumption, one can test for heteroscedasticity immediately through $H_0 : a_{21} = 0$ versus $H_1 : a_{21} \neq 0$ because $K_2 = e^{a_{21}}$ and the variance function $\mathrm{Var}(Y) \equiv V(\mu) = K_1^2 \cdot K_2^{2\mu}$. Other combinations of the link functions can be used (Table 5.4).

There is a very large literature on heteroscedastic linear regression, e.g., Smyth et al. (2001), Carroll and Ruppert (1988). The latter lists three "off-the-shelf" models for a hormone assay data set (`hormone` in **VGAM**):

$$\log\sigma = \theta_0^* + \theta_1^*\,x + \theta_2^*\,x^{-1}, \tag{5.16}$$

$$\log\sigma = \theta_0^* + \theta_1^*\,\mu, \tag{5.17}$$

$$\log\sigma = \theta_0^* + \theta_1^*\,\log\mu, \tag{5.18}$$

the latter being known as the *power-of-the-mean* model. The first two may be fitted as VGLMs or RR-VGLMs.

Incidentally, `uninormal()` has a logical argument `var` that allows σ^2 to replace σ as the parameter of interest. Then, e.g., using an identity link for η_2, modelling $\sigma^2 = \theta_0^* + \theta_1^*\,\mu$ is possible, cf. (5.17).

5.5.2.2 RR-Zero-Inflated Poisson

The zero-inflated Poisson (ZIP) described in Chap. 17 is a mixture distribution of a Poisson and a structural or degenerate zero. There are two processes generating this type of data: one that generates the zeros and one that generates the Poisson counts, therefore there are two linear/additive predictors η_1 and η_2. Usually these processes are considered separately, however in ecology, the two may be coupled and bear some systematic relationship. Liu and Chan (2010) give an example involving trawl survey studies, where the spatio-temporal aggregation of fish due to schooling results in the probability of a positive catch being a monotonic function of the mean. Another example by the same authors is grasshopper species' abundances affected by swarming, due to suitable environmental conditions becoming available. In their development they have named these types of models 'constrained zero-inflated generalized additive models', or COZIGAMs, and there is an R package by the same name. Section 17.4 shows that COZIGAMs are RR-ZIPs and can be easily fitted using regression splines.

For the above situations, a *reduced-rank zero-inflated Poisson* (RR-ZIP) model may be suitable. It is given by

$$\log \mu \;=\; \eta_1 \;=\; \boldsymbol{\beta}_1^T \boldsymbol{x} \tag{5.19}$$

$$\text{logit } \phi^* \;=\; \eta_2 \;=\; \beta_{(2)1} + a_{21} \cdot \eta_1 \tag{5.20}$$

where $\boldsymbol{\beta}_1$, $\beta_{(2)1}$ and a_{21} are to be estimated. The model has $P(Y = y) = I(y = 0) \cdot (1 - \phi^*) + \phi^* \, e^{-\mu} \mu^y / y!$ with $\boldsymbol{\eta} = (\log \mu, \text{logit } \phi^*)^T$ as default link functions, and $\eta_2 = t_1 + a_{21} \eta_1$ because in some applications the probability of a non-zero value has a monotonic relationship with the Poisson mean.

Fitting an RR-ZIP is very easy in **VGAM**. As an example, suppose there are two covariates plus an intercept. Then such might be fitted by code of the form

```
rr.zip <- rrvglm(y ~ x2 + x3, zipoissonff(zero = NULL), data = zdata)
```

Setting `zero = NULL` annuls the intercept-only default for ϕ, to allow for (5.20). It works because the corner constraint $\mathbf{A} = (1, a_{21})^T$ is default (Sect. 5.2.2). Here, one might wish to replace the linear predictors by additive predictors using regression splines as illustrated in Sect. 17.4.

As a closing comment, where any theoretical justification is lacking, the use of this model should be checked empirically.

5.5.2.3 RR-Negative Binomial (NB-P)

In the nomenclature of Hilbe (2011), the negative binomial regression (1.15)–(1.16) is called the NB-H ("H" for heterogeneous) because it models the k parameter with all the covariates $\boldsymbol{x}$. If we apply the RRR idea to the NB-H model with the default links $\boldsymbol{\eta} = (\log \mu, \log k)^T$, then some simple algebra with (5.15) reveals the RR-NB (strictly, the RR-NB-H) has variance function

$$V(\mu(\boldsymbol{x})) \;=\; \mu(\boldsymbol{x}) + \delta_1 \, \mu(\boldsymbol{x})^{\delta_2} \tag{5.21}$$

where the δ_j are to be estimated. This coincides with the so-called NB-P ("P" for parameterized) subfamily. The NB-P is important because (5.21) is quite an adaptable function—it includes the NB-1 and NB-2 as special cases, and it can be thought of as somewhat data-driven rather than model-driven. An NB-2 model is the NB regression (1.15)–(1.16), but with the k parameter being intercept-only.

The typical coding to fit the RR-NB is of the convenient form

```
rrnb <- rrvglm(y ~ x2 + x3 + x4, family = negbinomial(zero = NULL), data = ndata)
```

Then quantities such as

```
a21 <- Coef(rrnb)@A["loge(size)", 1]
beta11 <- Coef(rrnb)@B1["(Intercept)", "loge(lambda)"]
beta21 <- Coef(rrnb)@B1["(Intercept)", "loge(size)"]
delta1 <- exp(-beta21 + a21 * beta11)
delta2 <- 2 - a21
```

can be extracted and computed. Currently the accessor function `Confint.rrnb()` facilitates these calculations.

Here are some further comments on (5.21).

(i) If $1 < \delta_2 < 2$, then the RR-NB can be loosely thought of as an intermediary model between NB-1 and NB-2. The RR-NB describes a continuum or family of NB-type models.

(ii) If $\delta_2 > 2$, then this suggests overdispersion with respect to the NB-2 distribution when μ is large ($\mu > 1$).

(iii) Taylor's power law in ecology states that $\text{Var}(Y) = a\,\mu^b$ for constants a and b (Taylor, 1961). Here, a is a 'sampling' parameter and b is an index of aggregation characteristics of the species. So if μ is large then $V(\mu(\boldsymbol{x})) \approx \delta_1\,\mu(\boldsymbol{x})^{\delta_2}$ so that the RR-NB can be used to approximate estimating b by estimating δ_2. Note that the rightmost term of (5.21) is the same as for a Tweedie distribution.

(iv) As stated above, confidence intervals for δ_2 are available, therefore it allows one to choose between NB-1 and NB-2 if this is really desired. If a 95% confidence interval for δ_2 covers unity, then an NB-1 (see also (11.9)) would be implied. If the interval for δ_2 covers 2, then an ordinary NB-2 would be suggested. If $\delta_2 = 1.5$, say, then it would be quite appropriate to retain it as an RR-NB and say it appears neither as an NB-1 nor a NB-2 model.

NB regression is covered in greater detail in Sect. 11.3 including an RR-NB example in Sect. 11.3.4.3. In particular, many important variants can be fitted within the VGLM/RR-VGLM framework.

5.5.2.4 RR 1-Parameter Families

Although (5.15) directly applies to a single distribution with two parameters, it is possible to apply the RRR to two 1-parameter distributions provided the **VGAM** family function can handle multiple responses. For example,

```
rrvglm(cbind(y1, y2) ~ x2 + x3, poissonff, data = pdata)
```

results in the coupling $\mu_2(\boldsymbol{x}) = K_1 \cdot [\mu_1(\boldsymbol{x})]^{a_{21}}$. Similarly,

```
rrvglm(cbind(y1, y2) ~ x2 + x3, binomialff(multiple.responses = TRUE), data = bdata)
```

results in

$$\frac{p_2(\boldsymbol{x})}{1 - p_2(\boldsymbol{x})} \;=\; K_1 \cdot \left(\frac{p_1(\boldsymbol{x})}{1 - p_1(\boldsymbol{x})}\right)^{a_{21}} \qquad\qquad (5.22)$$

where y1 and y2 are binary with values 0 and 1, and $p_j = P(Y_j = 1)$. The multiple.responses argument indicates the response matrix is made up of multiple responses; if multiple.responses = FALSE, then the first/second columns would be interpreted as the number of successes/failures of a single response variable.

5.6 RR-VGLM Examples

5.6.1 RR-Multiple Binomial Model

Sometimes it is informative to perform logistic regressions on several binary responses simultaneously, using a latent variable as explanatory. For example, in xs.nz, is there is any association between diseases such as asthma and cancer, and psychological variables such as worry, depression and nervousness? Here, we perform an analysis with the idea behind Sect. 5.5.2.4 in mind. Note, however, that this regression can be argued as being flawed, because the psychological variables might be considered responses rather than explanatory.

In the following, we have responses asthma, cancer, diabetes, heartattack, and stroke, and adjusting for ethnicity, sex, age50, smokenow, fh.cancer and fh.heartdisease, we treat ν as the real variable of interest. The latent variable ν is a linear combination of the 11 psychological variables depressed, ..., worrier.

```
> xs.nz <- transform(xs.nz, age50 = age - 50)  # Intercepts will be important
> rr.binom <-
     rrvglm(cbind(asthma, cancer, diabetes, heartattack, stroke) ~
            ethnicity + sex + age50 + smokenow + fh.cancer + fh.heartdisease +
            depressed + embarrassed + fedup + hurt + miserable +  # 11 psychological
            nofriend + moody + nervous + tense + worry + worrier, # variables
            noRRR = ~ ethnicity + sex + age50 + smokenow +
                   fh.cancer + fh.heartdisease,
            binomialff(multiple.responses = TRUE), data = xs.nz)
```

The centred variable age50 was created so that the intercepts are more meaningful.

Let's look at some selected output. Firstly,

```
> round(sort(concoef(rr.binom)[, 1]), digits = 2)
```

worry	hurt	moody	nervous	embarrassed	nofriend
-0.07	-0.06	-0.04	-0.01	0.01	0.03
miserable	fedup	tense	worrier	depressed	
0.09	0.12	0.15	0.15	0.24	

Those coefficients which are largest in magnitude are positive, therefore we interpret $\widehat{\nu}$ as a measure of psychological ill-being as opposed to well-being. Variables worry and hurt exhibit quite large negative signs but are dismissed as random error. To gauge the scale of $\widehat{\nu}$, a person exhibiting TRUE to all the psychological variables compared to somebody with all FALSEs would have a difference in $\widehat{\nu}$ of

```
> sum(concoef(rr.binom))

  [1] 0.6138
```

The matrix $\widehat{\mathbf{A}}$ is

```
> Coef(rr.binom)@A

                         latvar
    logit(E[asthma])     1.0000
    logit(E[cancer])     0.7140
    logit(E[diabetes])   0.2778
    logit(E[heartattack]) 1.6232
    logit(E[stroke])     1.1768
```

and it measures the slopes of each response on the logit scale with respect to $\widehat{\nu}$. It suggests that any effect on the log-odds of a disease, for a fixed change in $\widehat{\nu}$, is greatest for heart disease and least for diabetes. Not surprisingly, all elements are positive, so that there appears to be a positive association between the diseases and psychological ill-being.

We can apply the antilogit function to the fitted intercepts, to give the estimated prevalences:

```
> intercepts <- Coef(rr.binom)@B1["(Intercept)", ]
> sort(intercepts)   # Sorted intercepts

  logit(E[heartattack])      logit(E[stroke])    logit(E[diabetes])
               -5.625                -5.352                -3.902
       logit(E[cancer])       logit(E[asthma])
               -2.820                -2.337

> prevalences <- logit(intercepts, inverse = TRUE)
> names(prevalences) <- colnames(depvar(rr.binom))
> sort(prevalences)   # Sorted prevalences

  heartattack      stroke     diabetes       cancer       asthma
     0.003594    0.004715     0.019808     0.056243     0.088076
```

For a person with $\mathbf{x}_1 = \mathbf{0}$ and $\nu = 0$, this suggests that asthma has the highest prevalence among all the diseases, and surviving a heart attack the lowest prevalence. Such a person corresponds to a 50-year-old European-type female who does not currently smoke, has no family history of cancer or heart disease, is psychologically healthy in that she does not suffer from depression, is not miserable, etc. For this type of person, one may plot the estimated probabilities $\widehat{P}(Y_j = 1|\widehat{\nu})$. This is given in Fig. 5.1a and was produced by

```
> grid.len <- 200
> Latvar.fit <- latvar(rr.binom)
> Latvar.grid <- seq(min(Latvar.fit), max(Latvar.fit), length = grid.len)
> eta.vals <- outer(rep(1, grid.len), Coef(rr.binom)@B1["(Intercept)", ]) +
              cbind(Latvar.grid) %*% t(Coef(rr.binom)@A)
> probmat <- logit(eta.vals, inverse = TRUE)
> matplot(Latvar.grid, probmat, type = "l", lwd = 2, lty = 1:3, las = 1,
          xlab = "Latent psychological variable", main = "(a)",
          ylab = "Probability")
> legend("topleft", lwd = 2, lty = 1:3, col = 1:npred(rr.binom), bty = "n",
         legend = colnames(depvar(rr.binom)))
```

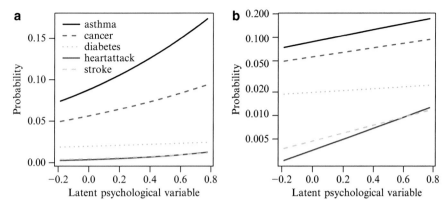

Fig. 5.1 (a) RR-VGLM-binomial model applied to several disease responses. The x-axis is $\widehat{\nu}$, a linear combination of 11 binary psychological variables. The y-axis is disease prevalence. (b) The same on a log scale.

Plotting the fitted values on a log-scale gives Fig. 5.1b. Both the relative prevalences and slopes of the diseases with respect to $\widehat{\nu}$ can be seen quite clearly.

5.6.2 Reduced Rank Regression for Time Series

Consider the multivariate autoregressive AR(L) model

$$\boldsymbol{Y}_t \;=\; \sum_{j=1}^{L} \boldsymbol{\Phi}_j \boldsymbol{Y}_{t-j} + \boldsymbol{\varepsilon}_t, \qquad \boldsymbol{\varepsilon}_t \sim (\boldsymbol{0}, \boldsymbol{\Omega}) \text{ independently}, \quad t = 1, \ldots, n, \quad (5.23)$$

where $\boldsymbol{Y}_t$ is $M \times 1$, and $\boldsymbol{\Phi}_j$ is $M \times M$ and to be estimated. As the number of lags L increases, the number of parameters involved grows rapidly. Ahn and Reinsel (1988) proposed the *nested reduced-rank autoregressive model* where $\boldsymbol{\Phi}_j$ is replaced by a matrix of rank R_j. One has the R_js being a nonincreasing sequence and $\boldsymbol{\Phi}_j = \mathbf{A}_j \mathbf{C}_j^T$ and range($\mathbf{A}_j$) $\supset$ range($\mathbf{A}_{j+1}$). It is nested because, with $R_j \equiv R < M$ and $\mathbf{A}_j = \mathbf{A}$, the special model $\boldsymbol{Y}_t = \mathbf{A} \sum_{j=1}^{L} \mathbf{C}_j^T \boldsymbol{Y}_{t-j} + \boldsymbol{\varepsilon}_t$ is obtained. Ahn and Reinsel (1988) gave a canonical form and computational details, and showed that a Newton-Raphson-like algorithm could be implemented, using standard software for GLS regression.

Strictly speaking, model (5.23) lies outside the RR-VGLM framework because *each* $\boldsymbol{\Phi}_j$ is of reduced rank, not the combined matrix $(\boldsymbol{\Phi}_1^T, \boldsymbol{\Phi}_2^T, \ldots, \boldsymbol{\Phi}_L^T)^T$ $(= \boldsymbol{\Phi}_*$, say). Nevertheless, the VGAM family function `rrar()` has been written to implement this model. It takes in an $n \times M$ matrix response, and the explanatory variables should just be an intercept term. The argument `Ranks` in `rrar()` specifies the ranks and must be of length L. Convergence is slow (much slower than a second-order rate) because $\boldsymbol{\Omega}$ has to be estimated. We use $\widehat{\boldsymbol{\Omega}} = n^{-1} \sum_{t=1}^{n} \widehat{\boldsymbol{\varepsilon}}_t \widehat{\boldsymbol{\varepsilon}}_t^T$.

As a numerical example, we mimic the analysis presented in Ahn and Reinsel (1988) who considered data consisting of monthly averages of grain prices in the United States for wheat flour, corn, wheat and rye for the period January 1961–October 1972. The units are dollars per 100 pound sack for wheat flour, and per bushel for corn, wheat and rye. The model

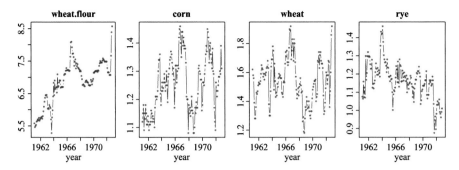

Fig. 5.2 Monthly average prices of Grain series, January 1961–October 1972, in data frame **grain.us**.

$$Y_t \;=\; \boldsymbol{\Phi}_1\,Y_{t-1} + \boldsymbol{\Phi}_2\,Y_{t-2} + \varepsilon_t$$

was fitted to **grain.us**, where $\mathrm{rank}(\boldsymbol{\Phi}_1) = 4$ and $\mathrm{rank}(\boldsymbol{\Phi}_2) = 1$ was chosen. This lag $L = 2$ model can be parameterized

$$Y_t \;=\; \mathbf{A}_{k_1}\left(\mathbf{C}_1^T\,Y_{t-1} + \mathbf{D}_2\,\mathbf{C}_2^T\,Y_{t-2}\right) + \varepsilon_t.$$

We have

```
> year <- seq(1961 + 1/12, 1972 + 10/12, by = 1/12)
> for (j in 1:4)
    plot(grain.us[, j] ~ year, main = names(grain.us)[j],
         type = "b", pch = "*", ylab = "", col = "blue")
```

This produces Fig. 5.2.

Now, some of the results of Ahn and Reinsel (1988) can be obtained as follows.

```
> colMeans(grain.us)     # a row vector

wheat.flour         corn        wheat         rye
      6.850        1.251        1.543        1.164

> cgrain <- scale(grain.us, scale = FALSE)   # Centre the time series only
> grain.rrar <- vglm(cgrain ~ 1, rrar(Ranks = c(4, 1)))
> print(grain.rrar@misc$Ak1, digits = 2)

        [,1] [,2] [,3] [,4]
[1,]   1.000    0    0    0
[2,]  -0.017    1    0    0
[3,]   0.327    0    1    0
[4,]   0.191    0    0    1

> print(grain.rrar@misc$C[[1]], digits = 3)

        [,1]     [,2]     [,3]     [,4]
[1,]   0.986   0.0271  -0.2890  -0.2000
[2,]  -0.411   0.7951  -0.1184  -0.0996
[3,]   0.576   0.0559   0.8236   0.0448
[4,]  -0.452  -0.0167   0.0466   0.8088
```

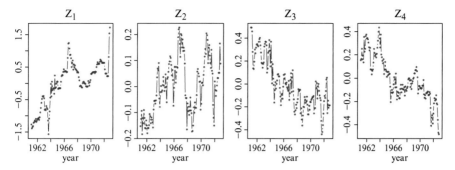

Fig. 5.3 Canonical variables of the Grain Price series, January 1961–October 1972.

and the reader is left to see what output arises from

```
summary(grain.rrar)
print(grain.rrar@misc$C, digits = 3)
print(grain.rrar@misc$D, digits = 3)
print(grain.rrar@misc$omega, digits = 3)
print(grain.rrar@misc$Phi, digits = 2)
```

The results differ slightly from Ahn and Reinsel (1988), possibly because we have used $t = 1 + L, \ldots, n$ instead of $t = 1, \ldots, n$, i.e., we have ignored the first L observations for simplicity.

Finally, the canonical variables are also returned. This transformed series $\boldsymbol{Z}_t$ whose components Z_{it} are arranged such that, with increasing values of the index i, less past information on $\boldsymbol{Z}_t$ is necessary to explain the present value of subvector Z_{it}. They may be plotted using

```
> for (j in 1:4)
      plot(grain.rrar@misc$Z[, j] ~ year, type = "b", pch = "*", col = "blue",
           main = eval(substitute(expression(Z[ .j ]), list(.j = j))), ylab = "")
```

to give Fig. 5.3. The y-limits do not correspond to the original series because of the centring. These y-limits corresponding to the original series can be obtained by using

```
(as.matrix(grain.us) %*% t(solve(grain.rrar@misc$Ak1)))[, i]
```

instead. More information can be found in Ahn and Reinsel (1988).

5.7 Row–Column Interaction Models

Row–column interaction models (RCIMs) are a subfamily of models applied to a matrix response $\mathbf{Y}$. They are essentially RR-VGLMs with certain row and column dummy variables set-up. As with VGLMs, the response $\mathbf{Y}$ may be continuous, counts, proportions, etc., however there is no explicit $\mathbf{X}$ except for what is determined from the row and column locations of the Y_{ij} values.

Define RCIMs as an RR-VGLM applied to $\mathbf{Y}$ with

$$g_1(\theta_1) \equiv \eta_{1ij} = \beta_0 + \alpha_i + \gamma_j + \sum_{r=1}^{R} c_{ir}\, a_{jr}. \qquad (5.24)$$

This means that the first parameter of a statistical model relating to a response matrix is, after a suitable transformation, equal to the sum of an intercept, a row and column main effect plus optional interaction term of the familiar form $\mathbf{CA}^T$. In (5.24), the parameters α_i and γ_j may be called the *row* and *column scores*, respectively. Identifiability constraints are needed for these, such as corner constraints, e.g., $\alpha_1 = \gamma_1 = 0$. The parameters a_{jr} and c_{ir} also need constraints, e.g., $a_{1r} = c_{1r} = 0$ for $r = 1, \ldots, R$. We can write (5.24) as

$$\eta_{1ij} \;\;=\;\; \beta_0 + \alpha_i + \gamma_j + \delta_{ij}$$

where the $n \times M$ matrix $\boldsymbol{\Delta} = [(\delta_{ij})]$ of interaction terms is approximated by a reduced rank regression.

Equation (5.24) makes sense if the first parameter is some measure of location such as a mean or median. For example, if $\mathbf{Y} = [(y_{ij})]$ is a $n \times M$ matrix of counts and $Y_{ij} \sim \mathrm{Poisson}(\mu_{ij})$ independently then

$$\log \mu_{ij} \;\;=\;\; \beta_0 + \alpha_i + \gamma_j + \sum_{r=1}^{R} c_{ir}\, a_{jr}, \quad i = 1, \ldots, n; \;\; j = 1, \ldots, M, \qquad (5.25)$$

is known as Goodman's RC(R) association model (GRC; Goodman, 1981). The model is saturated when $R = \min(n, M)$.

For most RCIMs, each column of $\mathbf{Y}$ is treated as one response so that, for the ith row, all the responses are modelled simultaneously by an $\boldsymbol{\eta}_i$. Exceptions to this can be handled by a careful selection of arguments of `rcim()` , e.g., Sect. 5.7.2.3 fits the multinomial logit model as a RCIM.

Note that (5.24) applies to the *first* linear/additive predictor; for models with $M > 1$, one can leave $\eta_2, \ldots, \eta_M$ unchanged or intercept-only. Of course, choosing η_1 for (5.24) is done only for convenience, and the software can allow the RRR to be applied to any other η_j instead. Then $g_j^{-1}(\widehat{\eta}_j)$ may be returned as the fitted values of the model via `fitted(rcim.object)`, and the result should be the same dimension as the two-way table.

To summarize, RCIMs in general are RR-VGLMs where one of the linear predictors is modelled as the sum of a row effect, a column effect, and an optional interaction effect expressed as an RRR. Table 5.5 summarizes a few possible RCIMs. Some more details may be found in Yee and Hadi (2014).

5.7.1 *rcim()*

The modelling function `rcim()` calls `vglm()` if the rank is zero, otherwise `rrvglm()`. In both cases the dummy variables and constraint matrices are set-up beforehand corresponding to (5.27) below. The `family` argument of `rcim()` is passed as an argument of the same name into `vglm()`/`rrvglm()`.

Table 5.5 Some **VGAM** family functions potentially useful in conjunction with `rcim()`. "GRC" stands for Goodman's RC model.

Family name	Comments
`alaplace2(0.5)`	Median polish (Mosteller and Tukey, 1977) when rank-0, (5.26)
`binomialff(multiple.responses = TRUE)`	Rasch conditional (fixed effects) model (Sect. 5.7.2.2)
`negbinomial()`	GRC-type model with overdispersion with respect to the Poisson
`poissonff()`	GRC model
`uninormal()`	Two-way ANOVA (one observation per cell)
`uninormal("explink")`	Quasi-variances when rank-0 (Sect. 5.7.3)
`zipoissonff()`	GRC-type model with lots of 0s and/or structural 0s

Currently, it is important that the first linear/additive predictor η_1 corresponds to the mean or some parameter measuring central location. Consequently, `zipoissonff()` should be used rather than `zipoisson()` because the latter models the complement of the probability of a structural zero in η_1, whereas the former models the mean of the Poisson distribution. All other parameters are generally fitted with intercept-only, for example, the k parameter for the negative binomial $\mathrm{NB}(\mu, k)$.

Ideally every possible **VGAM** family function will work with `rcim()`, however, currently not all family functions handle multiple responses. For those that do, the resulting RCIM may not make sense or be sensible, for example, the Kumaraswamy distribution has two positive shape parameters a and b so that $\eta_1 = \log a$ equals the RHS of (5.24) and b is intercept-only for

```
silly1 <- rcim(Y, kumar, Rank = 1)
```

The mean of this distribution is $b \cdot Be(1 + 1/a, b)$, hence such a fit is unlikely to be meaningful. In contrast, a useful model might be a median polish-type fit of the form

```
medpol0 <- rcim(Y, alaplace1(tau = 0.5), Rank = R)
```

This is because

$$\widetilde{\mu}_{ij} = \beta_0 + \alpha_i + \gamma_j + \sum_{r=1}^{R} c_{ir}\, a_{jr} \tag{5.26}$$

for the median of the (i, j) cell of **Y**. Scoring is not a reliable method for estimating the location parameter of an asymmetric Laplace distribution (Sect. 15.3.2), therefore if it does not fail then

```
fitted(medpol0)
```

should be the fitted values $\widehat{\widetilde{\mu}}_{ij}$.

Sometimes a `summary()` of a `"grc"` object fails because the estimated covariance matrix is not positive-definite. This is often due to numerical ill-conditioning: the position of the corner constraint for $\mathbf{I}_R$ in **A** causes some of the elements of $\widetilde{\mathbf{A}}$

to be very large or small in magnitude. The easiest fix is to try choose different values for the `Index.corner` argument in `grc()`, e.g., if

```
summary(grc(y, Rank = 2, Index.corner = 2:3))
```

fails, something like

```
summary(grc(y, Rank = 2, Index.corner = c(3, 5)))
```

might succeed.

Internally, the manner in which the row and column dummy variables are set up as follows. Starting from (5.2), $\mathbf{B}_1^T \boldsymbol{x}_{1i} =$

$$\begin{pmatrix} \beta_0 \mathbf{1}_M & \alpha_2 \mathbf{1}_M & \cdots & \alpha_n \mathbf{1}_M & \left(\mathrm{Diag}(\gamma_1, \ldots, \gamma_M)_{[-1,]}\right)^T \end{pmatrix} \begin{pmatrix} 1 \\ \boldsymbol{e}_{[-1]i} \\ \mathbf{1}_{M-1} \end{pmatrix}. \quad (5.27)$$

This shows, for example, that the intercept and row score variables have $\mathbf{1}_M$ as their constraint matrices. Similarly, because $\mathbf{B}_2$ is approximated by $\mathbf{CA}^T$, the ith row of $\boldsymbol{\Delta}$ will be approximated by $\boldsymbol{x}_{2i}^T \mathbf{CA}^T$, or equivalently,

$$\boldsymbol{\Delta} \approx \begin{pmatrix} \boldsymbol{x}_{21} \\ \vdots \\ \boldsymbol{x}_{2n} \end{pmatrix} \mathbf{C}\,\mathbf{A}^T.$$

The desired reduced-rank approximation of $\boldsymbol{\Delta}$ can be obtained if $\boldsymbol{x}_{2i} = \boldsymbol{e}_i$ so that $\mathbf{I}_{p_2}\mathbf{C}\,\mathbf{A}^T = \mathbf{C}\,\mathbf{A}^T$. Note that

$$\boldsymbol{\Delta} = \begin{pmatrix} 0 & \mathbf{0}^T \\ \mathbf{0} & \tilde{\boldsymbol{\Delta}} \end{pmatrix} \approx \mathbf{C}\,\mathbf{A}^T = \begin{pmatrix} \mathbf{0}^T \\ \mathbf{C}_{[-1]} \end{pmatrix} \begin{pmatrix} \mathbf{0} & \left(\mathbf{A}_{[-1]}\right)^T \end{pmatrix},$$

that is, the first row of $\mathbf{A}$ consists of structural zeros which are 'omitted' from the RRR of $\boldsymbol{\Delta}$. To effect this, if $R > 0$, then the argument `str0` defaults to 1. The argument `str0` may be assigned any vector of values from $\{1, 2, \ldots, M\}$ excluding those rows constituting the corner constraints—and such rows of $\mathbf{A}$ are set to be equal to $\mathbf{0}_M^T$. For example, if we set `str0 = 4` when $M = 5$ and $R = 2$, then

$$\mathbf{A} = \begin{pmatrix} 1 & 0 \\ 0 & 1 \\ a_{31} & a_{32} \\ 0 & 0 \\ a_{51} & a_{52} \end{pmatrix} \quad (5.28)$$

because `Index.corner = 1:Rank` by default. Elements $a_{31}, a_{51}, a_{32}, a_{52}$ are estimated.

Thus VGAM can fit (5.24) by setting up indicator variables, etc. before calling `rrvglm()`. Additionally, the function `grc()` has been written to fit Goodman's RC model easily. It accepts a matrix as its first argument, and most of the other arguments are fed into `rrvglm.control()`. Function `rcim()` has $R = 0$ as a default whereas `grc()` defaults to $R = 1$.

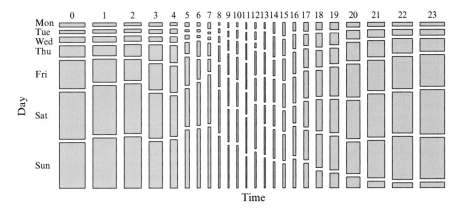

Fig. 5.4 Mosaic plot of `alcoff`; the area sizes are proportional to the counts.

5.7.2 Examples

For simplicity, in the following we mainly fit no-interaction models because the row and column effects cannot be interpreted when there are such.

5.7.2.1 Alcohol Offences

We use the matrix `alcoff` from **VGAM** which is 24×7. It records the number of alcohol offenders caught by the police from breath screening drivers, during 2009 in New Zealand A third of the data set can be seen by

```
> rbind(head(alcoff, 4), tail(alcoff, 4))

    Mon Tue Wed Thu  Fri  Sat  Sun
0   121  98 165 324  827 1379 1332
1    97  92 157 278  619 1327 1356
2    60  69 107 229  410  979 1011
3    55  60  75 238  401  693  718
20   74 135 283 508  591  490  166
21   84 154 326 610  866  754  131
22   90 143 345 765  976 1026  114
23  110 169 363 899 1265 1179  159
```

Here, the first row is from midnight to 1am, and the last row is for 11pm to midnight. Figure 5.4 is a mosaic plot of these data. The plot and perusal of the counts confirms what is expected: the greatest number of alcohol-related offences occur late on Friday and Saturday nights.

We fit a rank-0 Goodman's RC model to `alcoff` but first preprocess the data by offsetting the time of the day: we say the *effective* day starts at 6am, say, since partying late at night often spills over to the early morning. Hence effective **Monday** starts at 6am and finishes on Tuesday at the same hour. The function `Rcim()` and/or `moffset()` enables us to create the effective day variable. The GRC model is fitted by

```
> grc0.alcoff <- rcim(moffset(alcoff, "6", postfix = "*"),
                      rprefix = "Hour.24.", cprefix = "Day.")
```

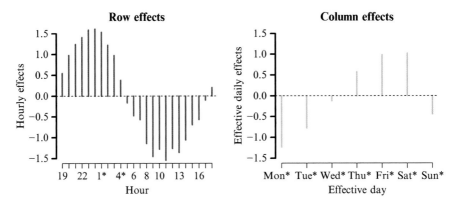

Fig. 5.5 Hourly and effective daily effects of a rank-0 Goodman's RC model fitted to `alcoff`. This is output from `plot(grc0.alcoff)`.

Alternatively, we could have used `grc()` with `Rank = 0`.

A plot of the fitted main effects was obtained from

```
> plot(grc0.alcoff, rcol = "blue", ccol = "orange", rfirst = 14, cfirst = 1,
        rtype = "h", ctype = "h", lwd = 3, ylim = c(-1.5, 1.5), las = 1,
        cylab = "Effective daily effects", rylab = "Hourly effects",
        rxlab = "Hour", cxlab = "Effective day") -> plot.grc0.alcoff
```

to give Fig. 5.5. The results agree with what is expected: as well as Friday and Saturday nights showing the greatest numbers of alcohol-related offences (and their following morning), there is a gradual increase from Sunday/Monday to these peak days. Also, they are at their lowest in the late morning to lunchtime period.

The estimated row and column effects, some raw, may be extracted as follows.

```
> round(plot.grc0.alcoff@post$raw.col.effects, digits = 2)

  Col.1 Day.2 Day.3 Day.4 Day.5 Day.6 Day.7
   0.00  0.46  1.11  1.81  2.22  2.26  0.79

> t(round(plot.grc0.alcoff@post$col.effects, digits = 2))

       Col.1 Day.2 Day.3 Day.4 Day.5 Day.6 Day.7
[1,]   -1.23 -0.78 -0.12  0.58  0.98  1.02 -0.45
attr(,"scaled:center")
[1] 1.234
```

5.7.2.2 Rasch Model

Latent trait or *item response theory* models have traditionally been applied in the fields of psychological and attainment testing. The Rasch model considered here is the simplest of a group of similar models. We wish to fit

$$\text{logit}\, P(Y_{ij} = 1 | \alpha_i) \;=\; \alpha_i - d_j \qquad (5.29)$$

to an $n \times M$ matrix $\mathbf{Y}$ consisting of 0s and 1s, where the probability is for the ith individual's correct response on item (question) j, and α_i denotes an individual's ability parameter, and d_j is the difficulty of the jth item. Ability is assumed to be an unobserved or latent trait. We will treat all parameters in (5.29) as fixed effects whereas a random effects approach would be more practical for large data sets (they typically assume $\alpha_i \sim N(0, \tau^2)$ independently). It is customary to have a minus sign for d_j because it represents item difficulty rather than ease, and it is common for a probit link to be used.

The original Rasch (1961) model treated the α_i as fixed effects, and one can use conditional maximum likelihood estimation by conditioning on their sufficient statistics, e.g., as dealing with matched pairs (Agresti, 2013, Sect.11.2.3).

The matrix **exam1** available in **VGAMdata** is a fictional data set involving 35 students given an 18-item ability test. We wish to determine which items are easiest and hardest, as well as who has the least and most ability. The first few students have results

```
> head(exam1, 3)

          q01 q02 q03 q04 q05 q06 q07 q08 q09 q10 q11 q12 q13 q14 q15 q16 q17 q18
Richard    1   1   1   1   1   1   1   0   0   0   0   0   0   0   0   0   0   0
Tracie     1   1   1   1   1   1   1   1   1   1   0   0   0   0   0   0   0   0
Walter     1   1   1   1   1   1   1   1   1   0   0   1   0   0   0   0   0   0
```

For this analysis it is necessary to remove items that were answered totally correct or wrong by all students, and students who scored 0 or 100% in the items, as these cases cause the $\widehat{d}_j$ and $\widehat{\alpha}_i$ to diverge to $\pm\infty$.

```
> Exam1 <- exam1[, colMeans(exam1) > 0]   # Delete questions that are too hard
> Exam1 <- Exam1[, colMeans(Exam1) < 1]   # Delete questions that are too easy
> Exam1 <- Exam1[rowMeans(Exam1) > 0, ]   # Delete people that are too weak
> Exam1 <- Exam1[rowMeans(Exam1) < 1, ]   # Delete people that are too smart
> rfit <- rcim(Exam1, family = binomialff(multiple.responses = TRUE))
```

Then

```
> plot.rfit <- plot(rfit, rcol = "blue", ccol = "orange", cylab = "Item effects",
                rylab = "Person effects", rxlab = "", cxlab = "", lwd = 2)
```

produces Fig. 5.6. The plot shows that the questions as a whole become increasingly more easy and the existing order of the students shows no particular pattern. Finally,

```
> names(plot.rfit@post)

  [1] "row.effects"     "col.effects"      "raw.row.effects" "raw.col.effects"

> order(plot.rfit@post$col.effects)   # Same as order(plot.rfit@post$raw.col.effects)

  [1] 14 13 12 11  9 10  8  7  5  3  6  4  2  1
```

ranks the difficulty of the items in increasing order.

One must be careful with fixed-effects Rasch models, because plain maximum likelihood estimation leads to inconsistent estimates because the number of parameters increases with increasing number of elements of **Y**.

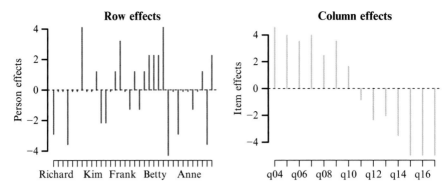

Fig. 5.6 Rasch fixed effects model to `exam1`, Eq. (5.29). Only a few people and items are labelled.

5.7.2.3 Multinomial Logit Model

Here, we fit a multinomial logit model (MLM) to a table of counts collected from Harvard University. Ethnicity is explanatory and the education program is response. A few programs have been omitted for simplicity.

```
> (Hued <- hued[, -c(1, 4, 5, 12)])
```

	GSAS	Business	Divinity	Education	Government	Law	Medical	PHealth
AsianPI	80	120	4	63	36	49	48	42
Black	18	41	10	47	24	67	17	22
Hispanic	28	32	7	30	33	32	12	12
IntStudent	281	285	10	103	238	199	20	200
NatAmer	3	5	2	3	7	3	1	3
White	350	311	99	364	166	310	77	130
NAOther	95	122	23	75	72	101	5	70

We can fit a rank-1 MLM-RCIM as follows.

```
> rcim1 <- rcim(Hued, family = multinomial, Rank = 1,
                cindex = 2:(ncol(Hued)-1), M = ncol(Hued)-1)
```

As there are 8 columns, $M = 8-1$ and the model is $\log(p_{ij}/p_{i8}) = \beta_0 + \alpha_i + \beta_j + \delta_{ij}$ for $i = 1, \ldots, 7$ and $j = 1, \ldots, 7$. Here, $\beta_1 \equiv 0$ because the argument `cindex` stands for column index, and these point to the columns of the response which are part of the vector of linear/additive predictor main effects.

5.7.3 Quasi-Variances

One useful application of RCIMs is the *quasi-variances* methodology as developed by Firth and de Menezes (2004) and implemented in `qvcalc`. Recall that, in R, a regression model with an explanatory factor has its first level's coefficients set to 0 by default ("treatment contrasts" in Table 1.5)—the baseline or reference category. Suppose there are L levels and the coefficient for the ith level is β_i. Upon fitting, we have $\beta_1 \equiv 0$, and $\widehat{\beta}_i$ and $\mathrm{SE}(\widehat{\beta}_i)$ for $i = 2, \ldots, L$, as the 'conventional' output, e.g., for publications. However, with these, it is only possible to compare $\widehat{\beta}_i$ with β_1

since $\widehat{\beta}_i = \widehat{\beta}_i - \beta_1$. Plotting the conventional output gives the familiar form of Fig. 5.7a. To compare $\widehat{\beta}_i - \widehat{\beta}_j$ for *all* i and j it is necessary to have the complete variance-covariance matrix, however most publications omit all but its diagonal because of space constraints.

Quasi-variances ameliorate these shortcomings. Their purpose is to be able to quickly make inferences about contrasts of the parameters without having to re-evaluate standard errors after reparameterization. The methodology attempts to assign standard errors to *all* levels such that *all* pairwise levels may be compared. They allow for a plot such as Fig. 5.7b, where each level has a *quasi-standard error* (QSE) associated with it. In a nutshell, quasi-variances are based on an approximation which summarizes all of the covariances among $\widehat{\beta}_1, \ldots, \widehat{\beta}_L$ at the cost of an additional number; there are L quasi-standard errors compared to $L-1$ 'conventional' standard errors. They are also known as 'floating absolute risks' in epidemiology.

The quasi-variance methodology finds constants $q_1, \ldots, q_L$ such that

$$\mathrm{Var}\left(\boldsymbol{c}^T \widehat{\boldsymbol{\beta}}\right) \approx \sum_{i=1}^{L} c_i^2 \, q_i \tag{5.30}$$

for all contrasts $\boldsymbol{c} = (c_1, \ldots, c_L)^T$. Recall that a contrast is a linear combination $\sum c_j \beta_j$ with $\sum c_j = 0$. In practice, we restrict ourselves to the simple but very commonly used contrasts of the form $\widehat{\beta}_i - \widehat{\beta}_j$, so that

$$v_{ij} \equiv \mathrm{Var}\left(\widehat{\beta}_i - \widehat{\beta}_j\right) \approx q_i + q_j. \tag{5.31}$$

Consequently one treats $\beta_1, \widehat{\beta}_2, \ldots, \widehat{\beta}_L$ as uncorrelated and with quasi-standard errors $\sqrt{q_1}, \ldots, \sqrt{q_L}$, respectively. Then (5.31) has a simple Pythagorean interpretation. Applied to the `ships` data in MASS, one can see that Fig. 5.7b is superior to Fig. 5.7a because of its more general interpretation, its shorter intervals and its overall greater information content.

Quasi-variances are not variances in general, and unfortunately negative values are possible. The RHS of (5.31) will almost always be non-negative, therefore only one or zero of the L quasi-variances can be negative.

5.7.3.1 Estimation

How can quasi-variances be estimated? Starting from (5.31) firstly observe that minimizing the relative errors is preferred to absolute errors, e.g., an error of 0.1 is small if the standard errors are 10, say, but large for 0.01. To this end, the quantity

$$\sum_{i < j} \left\{\log v_{ij} - \log\left(q_i + q_j\right)\right\}^2 \tag{5.32}$$

is minimized, which is akin to dealing with the quantity $(q_i + q_j)/v_{ij} - 1$ rather than $v_{ij} - q_i - q_j$.

It is easy to see that minimization of (5.32) can be performed by fitting an RCIM to a Gaussian-GLM. The only novelty is the use of an exponential link function, `explink()`: $g_1(\mu_{ij}) = \eta_{ij} = \exp(\mu_{ij})$ where $\eta_{ij} = q_i + q_j$, and the 'response' is $y_{ij} = \log v_{ij}$.

5.7.3.2 Software Usage

If `fit` is a `vglm()` object with a factor as explanatory then minimizing (5.32) can be achieved by calls such as

```
fit.qvar1 <- rcim(Qvar(fit, "factorname"), family = uninormal("explink"))
fit.qvar2 <- rcim(Qvar(fit, coef.indices = c(0, 7:9)), fam = uninormal("explink"))
```

Here, the processing function `Qvar()` takes as input a `vglm()` object and information directing it to the factor of interest. Then the relevant subset of the estimated variance-covariance matrix is extracted (and if necessary, a column of 0s and a row of 0s is appended to the LHS and top), and then an estimate of the LHS of (5.31) is computed for each pairwise combination, and then the logarithm is taken.

Behind the scenes, what is also fed into `rcim()` is an $L \times L$ matrix of prior weights upon which the objective function (5.32) is minimized using least squares. This square matrix of prior weights is of order-L. It comprises 1s, except that its diagonal elements are some small positive number, i.e., $\mathbf{1}_L \mathbf{1}_L^T - (1 - \varepsilon)\mathbf{I}_L$ for $\varepsilon \approx 0$ with $\varepsilon > 0$. The function `Qvar()` constructs this matrix, attaches it to the response, and then it is picked up specifically by `uninormal()` later.

All the input is symmetric, therefore upon convergence the ith row effect and ith column effect should be equal. Although the RCIM is estimated using corner constraints as in (5.25), the matrix of predicted values has ith diagonal elements $2q_i$ so that quasi-variances may be computed with:

```
QuasiVar <- exp(diag(fitted(fit.qvar1))) / 2        # Version 1
QuasiVar <- diag(predict(fit.qvar2)[, c(TRUE, FALSE)]) / 2   # Version 2
QuasiSE  <- sqrt(quasiVar)
```

Both expressions for the quasi-variance yield identical results. For convenience, one may use `qvar()`:

```
QuasiVar <- qvar(fit.qvar2)   # Equivalent to Versions 1 and 2
```

A function is available to produce plots similar to Fig. 5.7b,c. These have been called *comparison intervals* to emphasize that they are constructed for inference about differences.

Note that quasi-variances for explanatory factors should be possible for any VGLM that has *one* linear predictor. Things become complicated when the factor appears in more than one η_j.

Another note is that the result of the quasi-variance approximation can be treated as uncorrelated, so that the Pythagorean rule holds ((5.31), for simplicity), but the approximation itself makes use of the full variance-covariance matrix in its calculation.

5.7.3.3 The Accuracy of the Approximation

Quasi-variances effectively approximate a matrix of simple contrasts that has been formed from a subset of $\widehat{\mathrm{Var}(\boldsymbol{\beta})}$ by a rank-0 RCIM with normal errors on a log scale. The approximation tends to work well in practice. It may fail, however, with $q_i < 0$ (but this is not very common) and only one q_i may be negative. Also, when the $\widehat{\beta}_i$ are highly correlated, the approximation becomes less accurate.

A measure of the approximation's accuracy is

$$\mathrm{RE}(\boldsymbol{c}) \equiv \sqrt{\frac{c_1^2 \, q_1 + \cdots + c_L^2 \, q_L}{\boldsymbol{c}^T \, \mathrm{Var}(\widehat{\beta}_1, \ldots, \widehat{\beta}_L) \, \boldsymbol{c}}} - 1 = \left(\frac{\mathrm{Var}\left(\boldsymbol{c}^T \mathbf{Q}\right)}{\mathrm{Var}(\boldsymbol{c}^T \widehat{\boldsymbol{\beta}})} \right)^{\frac{1}{2}} - 1, \qquad (5.33)$$

where $\mathbf{Q} = \mathrm{diag}(q_1, \ldots, q_L)$, known as the *relative error* for any contrast $\boldsymbol{c}$. Two statistics are commonly reported to gauge the error in the approximation. The first is the worst possible errors—i.e., the minimum and maximum values of $\mathrm{RE}(\boldsymbol{c})$ over all $\boldsymbol{c}$. The second is $\mathrm{RE}(\boldsymbol{c})$ over all simple contrasts $\widehat{\beta}_i - \widehat{\beta}_j$. Firth (2003) suggests that relative errors of up to ± 10 percent may be considered not very serious.

5.7.3.4 Least Significant Difference Plots

An enhancement to comparison intervals based on quasi-variances is to add *least significant difference* (LSD) intervals. This is motivated by the difficulty of interpreting confidence intervals in one-way ANOVA. It is well-known that if two 95% confidence intervals (one for β_i and another for β_j) do not overlap then the difference is significantly different at the 5% level of significance. Compared to the standard method (computing a confidence interval for the difference), the overlap method (rejecting H_0 if and only if there is no overlap) is more conservative and less powerful. That is, $H_0 : \beta_i = \beta_j$ is rejected less often when H_0 is true, and H_0 is rejected less frequently when the null hypothesis is false.

The above difficulty also applies to when plotting $\widehat{\beta}_i \pm 2\,\mathrm{QSE}(\widehat{\beta}_i)$ where $\mathrm{QSE}(\widehat{\theta}_i) = \sqrt{q_i}$ as in Fig. 5.7b. Additionally, it has the drawback that it is ineffective for making immediate decisions: ideally one would want intervals that do not overlap if and only if they are statistically significantly different. Motivated by ANOVA, we apply the LSD intervals technique here.

The usual ANOVA formula for LSD intervals is

$$\widehat{\theta}_i \; \pm \; \frac{t_{df}(\alpha/2)}{\sqrt{2}} \, \mathrm{SE}(\widehat{\theta}_i)$$

where $df = \min(df_1, df_2, \ldots)$. However, LSD intervals for factors are appropriate with quasi-variances, since each level now has one. So replacing the conventional standard error by the quasi-standard error, and simplifying the t-multiplier to $z(\alpha/2)$, yields the limits

$$\widehat{\theta}_i \; \pm \; \frac{z(\alpha/2)}{\sqrt{2}} \, \mathrm{QSE}(\widehat{\theta}_i). \qquad (5.34)$$

Plotting these with $\alpha = 0.05$ displays the so-called 5% LSD intervals, which provides an approximate but quick visual check on whether the difference between two estimates is significant at the 5% level of a 2-sided test. Its derivation assumes $q_i \approx q_j$.

The plotting methods function `qvplot()` represents LSD intervals by arrows. The function has arguments which can be set for adjustments, e.g., default

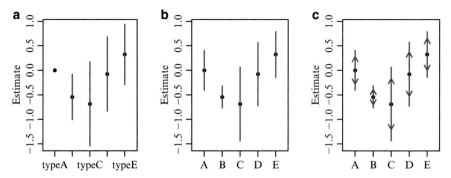

Fig. 5.7 Quasi-variances computed for the **ships** data in **MASS**. The fitted model is a quasi-Poisson-GLM fitted to the ship data (McCullagh and Nelder, 1989) with respect to ship types (A–E) on the damage rate on a log scale. (**a**) Confidence intervals for contrasts with type A ships based on conventional standard errors; (**b**) Comparison intervals based on quasi-variances; (**c**) 5% LSD intervals (*arrows*) based on quasi-variances overlaid on (**b**). For (**a**)–(**c**), the formulas are $\widehat{\beta}_i \pm 2 \operatorname{SE}(\widehat{\beta}_i)$, $\widehat{\beta}_i \pm 2 \sqrt{q_i}$ and $\widehat{\beta}_i \pm z(0.025) \sqrt{q_i/2}$, respectively.

`conf.level = 0.95` produces $\alpha = 5\%$ LSD intervals but `conf.level = NA` suppresses them. Since argument `interval.width = 2` by default, the arrows reside within the ± 2 QSE bands. A warning is issued if the QSEs are deemed very different, e.g., $\max(q_i)/\min(q_i)$ is large.

As an example, the 5% LSD intervals in Fig. 5.7c for the first two levels for the ship data almost cross, therefore we expect their *p*-value to be close to 0.05. Indeed, testing the second coefficient from the original fit, one has a Wald statistic of -1.96, which borders on exactly the 5% level of significance. Visually, two other pairs appear significantly different, viz. B versus E, and C versus E.

5.7.3.5 Example

We illustrate how quasi-variances may be fitted by use of the **ships** data in **MASS**.

```
> data("ships", package = "MASS")
> ships <- within(ships, { year   <- as.factor(year)
                           period <- as.factor(period) })
> Shipmodel <- vglm(incidents ~ type + year + period, quasipoissonff, data = ships,
                    subset = (service > 0), offset = log(service))
```

The basic RCIM and plots in Fig. 5.7b,c are from

```
> rcim.ship <- rcim(Qvar(Shipmodel, "type"), fam = uninormal("explink"), maxit = 99)
> quasiVar <- qvar(rcim.ship)
> quasiSE <- sqrt(quasiVar)
> qvplot(rcim.ship, scol = "blue", ylim = c(-1.5, 1), main = "(b)",
         pch = 16, slwd = 1.5, conf.level = NA)    # Suppress arrows (LSD intervals)
> qvplot(rcim.ship, scol = "blue", ylim = c(-1.5, 1), main = "(c)",
         pch = 16, slwd = 1.5, length.arrows = 0.07)    # Has LSD arrows
```

Plots such as these are sometimes called *centipede plots*—especially when the number of levels of the factor L is large and the levels of the factors are arranged according to the sorted values of the estimates.

Bibliographic Notes

Reinsel and Velu (1998) is a treatment of classical RRR applied to the normal-errors case. RR-VGLMs were proposed in Yee and Hastie (2003), and the special cases of models with $M = 2$ and $R = 1$ were elucidated in Yee (2014). RCIMs were first described in Yee and Hadi (2014). A modern treatment of many classical multivariate techniques, including RRR, is Izenman (2008).

There are many books on item response theory such as Baker and Kim (2004); a small description is given in Agresti (2013) and de Gruijter and Van der Kamp (2008, Chaps.9–10) is a good introduction.

LSD intervals are described in Andrews et al. (1980) and Schenker and Gentleman (2001). The R packages **gnm** and **qvcalc** fit some models described in this chapter.

Exercises

Ex. 5.1. Similar to Sect. 5.6.1, fit a RR-bivariate odds ratio model to the **xs.nz** data with **asthma** and **cancer** as responses. Use ν as a linear combination of the 11 psychological variables. Can you interpret $\widehat{\nu}$? How does the association between the responses change as a function of $\widehat{\nu}$?

Ex. 5.2.   Other Couplings
Using Table 1.2 as a reference, add the following links to Table 5.3, i.e., specify the additional rows and columns.

(a) log–log link,
(b) complementary log–log link,
(c) reciprocal link.

Ex. 5.3.   RR-MLM Fitted to Vowel Data   [(Yee and Hastie, 2003)]
Consider the **vowel.train** and **vowel.test** data frames in **ElemStatLearn**.

(a) Fit RR-multinomial logit models to **vowel.train** of ranks 1,...,10. Use all the covariates **x.1**–**x.10** as x_2. Based on the AIC, which dimension seems to be preferred?
(b) For each model in (a), use **vowel.test** to estimate the probability of misclassification. Based on this criterion, what is the optimal rank?
(c) Given a set of I models with estimates of their performance in terms of an 'error' (e.g., prediction error, or probability of misclassification) and associated SE ($\widehat{e}_i$ and SE($\widehat{e}_i$), say), the *1-SE rule* chooses the simplest model that is within one SE of the best model. That is, the least complex model whose error estimate is no more than $\widehat{e}_j + \text{SE}(\widehat{e}_j)$ where $j = \arg\min_{i=1,...,I} e_i$. The 1-SE rule attempts to choose a simpler model which is essentially indistinguishable from the 'best' model, by taking into account the statistical variability in the performance estimates. The rule can help reduce any overfitting. Apply the 1-SE rule to (b); is there any change to your answer?

Ex. 5.4. Continuing on somewhat from Ex. 5.3(a), consider the rank-2 RR-MLM fitted to the training data. Show that a multinomial logit model treating $\widehat{\mathbf{A}}$

as known and fixed produces standard errors (for all the intercepts and elements
of $\mathbf{C}$) that are smaller than those of the rank-2 RR-MLM. Of course, your fitted
coefficients should be identical to the RR-MLM.

Ex. 5.5. Couplings
Verify the following entries in Table 5.4.

(a) `negbinomial(zero = NULL)`.
(b) `inv.gaussianff(zero = NULL)`.
(c) `gamma2(zero = NULL)`.
(d) Repeat (a)–(c) with Variant II.

Ex. 5.6. RR-Normal
Consider models (5.16)–(5.18).

(a) Explain how (5.16) may be fitted as a VGLM or RR-VGLM.
(b) Explain how (5.17) may be fitted as a VGLM or RR-VGLM.
(c) Suppose $g_1(\mu) = \log \mu$. Explain how (5.18) may be fitted as a VGLM or RR-
   VGLM, and within **VGAM**.
(d) Fit (5.16) to the `hormone` data.
(e) Fit (5.17) to the `hormone` data.

Ex. 5.7. Quasi-Variances and LSD Intervals

(a) If $q_i \approx q_j$ $(= q)$, say, verify that an approximate $100(1 - \alpha)\%$ CI for $\beta_i - \beta_j$
   is $\widehat{\beta}_i - \widehat{\beta}_j \pm z(\alpha/2)\sqrt{2q}$.
(b) Suppose $\widehat{\beta}_i > \widehat{\beta}_j$. We want a multiplier λ so that the LSD intervals only just
   overlap, i.e., $\widehat{\beta}_i - \lambda\sqrt{q} = \widehat{\beta}_j + \lambda\sqrt{q}$. Deduce from (a) that $\lambda = z(\alpha/2)/\sqrt{2}$.

Ex. 5.8. Quasi-Variances: `miserable` and `marital` Example
A sociologist is interested in the association of the `miserable` variable as a response
and `marital` status, after adjusting for gender and age, in the `xs.nz` data frame.

(a) Run the following logistic regression

```
> mdata <- na.omit(xs.nz[, c("miserable", "marital", "age", "sex")])
> mfit <- vglm(miserable ~ marital + age + sex, binomialff, data = mdata)
```

   and, for the `marital` variable, compute all simple contrasts and their standard
   errors from the complete variance-covariance matrix.
(b) Compute the quasi-variances within **VGAM**, and obtain a plot of them. Can
   anything be learnt from this?

Ex. 5.9. RR 1-Parameter Families
Consider fitting RR 1-parameter families as in Sect. 5.5.2.4.

(a) What mathematical formula results from

```
rrvglm(cbind(y1, y2, y3, y4) ~ x2 + x3, poissonff, data = pdata)
```

   Express your answer in a similar manner to (5.22).

(b) Do the same as (a) for

```
rrvglm(cbind(y1, y2, y3, y4) ~ x2 + x3, binomialff(multiple.responses = TRUE),
       data = bdata)
```

Ex. 5.10. A RR-VGLM Variant: 'transposed' RRR

Reinsel and Velu (2006) consider RRR where only a subset of the components of $\boldsymbol{\eta}$ have the $\mathbf{CA}^T$ formulation. In particular, suppose that one wants to perform RRR on a response $\boldsymbol{y} = (\boldsymbol{y}_1^T, \boldsymbol{y}_2^T)^T$, but with

$$\boldsymbol{\eta} = \begin{pmatrix} \boldsymbol{\eta}_1 \\ \boldsymbol{\eta}_2 \end{pmatrix} = \begin{pmatrix} \mathbf{B}_1^T \boldsymbol{x}_1 \\ \mathbf{AC}^T \boldsymbol{x}_2 \end{pmatrix}$$

where the dimension of $\boldsymbol{y}_j$ and $\boldsymbol{\eta}_j$ is M_j, with $M_1 + M_2 = M$. Here, $\mathbf{B}_1$ is of full rank, and $\mathbf{A}$ is $M_2 \times R$, and $\mathbf{C}$ is $p_2 \times R$.

(a) Work out the constraint matrices associated with every variable in $\boldsymbol{x}_1$ and $\boldsymbol{x}_2$.
(b) As a specific example, if $M = 6$, $M_1 = 3 = M_2$, $\boldsymbol{x}_1 = (x_1, x_2, x_3)^T$, $\boldsymbol{x}_2 = (x_4, x_5, x_6)^T$ and $R = 1$, then complete the following call:

```
rrvglm(ymatrix ~ -1 + x1 + x2 + x3 + x4 + x5 + x6,
       VGAMfamilyFunction(zero = ???), noRRR = ~ ???,
       Index.corner = ???, Rank = ???, constraints = ???, str0 = ???)
```

Ex. 5.11. Analysis of Crash Data

(a) Fit a rank-0 GRC model (e.g., using rcim()) to the crashp data frame (crash data involving pedestrians.) Plot the row and column main effects and then try interpreting them.
(b) Do the same as (a) using a zero-inflated Poisson distribution. Is there much quantitative and qualitative difference?
(c) Repeat (a) applied to crashtr (crashes involving trucks).
(d) Repeat (a) applied to crashi (crashes involving injuries).
(e) All of the above fitted no-interaction models. Comment.

Ex. 5.12. More general than (5.2) is

$$\boldsymbol{\eta} = \mathbf{B}_1^T \boldsymbol{x}_1 + \sum_{k=2}^{K} \mathbf{A}_k \mathbf{C}_k^T \boldsymbol{x}_k,$$

where $K \geq 2$ is any specified positive integer, $\mathbf{A}_k$ is $M \times R_k$ and unknown, and $\boldsymbol{x}_k$ is a p_k-vector with $p_1 + \cdots + p_K = p$. Although this is currently not yet implemented in VGAM, discuss the computational details behind estimating such a model.

Ex. 5.13. Fatal Crash Data (Two Countries)

Consider the data frames crashf and crashf.au.

(a) Create a data frame called crashf.nz from crashf having the same dimension and format as crashf.au.
(b) Fit rank-0 Goodman's RC models to crashf.au and crashf.nz separately.

(c) † Fit a rank-0 Goodman's RC model to both data sets simultaneously subject to the constraint $\mu_{ij}^{au} = \phi \, \mu_{ij}^{nz}$, where ϕ is some additional parameter to be estimated. Show that $\widehat{\phi} = \sum_{ij} y_{ij}^{au} / \sum_{ij} y_{ij}^{nz} = 10\frac{2}{3}$.
Hint: extract out the VLM matrix of both fits and feed it into `vglm.fit()`. Some hacking may be required to get it working.

(d) † For your answer in (c) calculate the standard errors for the row effects, column effects and $\widehat{\phi}$.

Ex. 5.14. † Search the literature for an unimplemented model that has many parameters for which a reduced-rank fit would be a good idea. Write a **VGAM** family function for the model, and run it under `rrvglm()` on a suitable data set. Does a biplot of your fit add to your understanding of the data?

Ex. 5.15. **Olympic Games Medal Counts**

(a) Fit a rank-0 GRC model to the first 10 countries in the data frame `olym08`. Can you interpret the coefficients? By looking at a barplot of these select countries choose a subset of 5 of them and repeat the fitting—and justify your choice of countries. Are the new coefficients more accurate or meaningful?

(b) Repeat (a) with `olym12`.

(c) Do the results of (a) and (b) differ appreciably? Try giving a reason for this.

Ex. 5.16. Suppose one runs

```
rrvglm(y ~ bs(x2, 4), family = multinomial,
       Rank = 1, Corner = TRUE,
       data = mdata)
```

Show that this is a multinomial logit model with

$$\eta_j = \beta_{(j)1} + a_j f(x_2), \qquad a_1 = 1, \quad j = 1, \dots, M,$$

where f is a smooth function of x_2 estimated by a B-spline.

Ex. 5.17. **Variance Functions**

(a) For

```
rrvglm(y ~ x2 + x3, gamma2(zero = NULL), data = gdata)
```

show that $\mathrm{Var}(Y) = K_1 \mu^\delta$ for some parameters K_1 and δ.

(b) Show that the same form of the variance function results from

```
rrvglm(y ~ x2 + x3, inv.gaussianff(zero = NULL), data = gdata)
```

How can this code be adapted to have $\mathrm{Var}(Y) = \mu$?

(c) What is the variance function for

```
vglm(y ~ x2 + x3, gamma2(parallel = TRUE, zero = NULL), data = gdata)
```

Ex. 5.18. Consider the NB-1, whose variance function is $V(\mu) = \phi \, \mu$.

(a) Use (11.2) to write down the density function $f(y; \mu, \phi)$ of the NB-1.

(b) Take the limit as $\phi \to 1$ of $f(y; \mu, \phi)$ and show that it gives the Poisson(μ) probability function (1.7). Hint: use (A.46) and (A.47).

Ex. 5.19. Consider the `machinists` data set.

(a) Fit an NB-1 and NB-2 model to the data set. Explain why the two are equivalent. Do you think an ordinary Poisson model would suffice?
(b) Now fit an NB-2 subject to $\mu = k$. Test to see whether such a model is plausible.
(c) Write down the probability function of the model corresponding to (b).

Ex. 5.20.   Median Polish
Consider the `USPersonalExpenditure` data set in `datasets`.

(a) Apply `medpolish()` to `log10(USPersonalExpenditure)`. Obtain a matrix of fitted values.
(b) Try to obtain a similar result using RCIMs. Compare the fitted values of the two models.

Ex. 5.21.   Rasch Model to LSAT Data
Bock and Leiberman (1970, Table 1) present data on the Law School Admission Test (LSAT) for $n = 1000$ students. These students were drawn from a larger sample of students applying for admission to law schools at various universities in USA. We shall concern ourselves with "Section 6" of the test. It is a 5-item multiple-choice test, where a 1 for each item means the correct answer, else 0.

(a) Fit a simple binomial Rasch model to this data set (`lsat` data frame in SMIR; use `wt6` for the frequencies).
(b) Rank the items according to difficulty.

Ex. 5.22.   RR-Normal and Quasi-Poisson Models

(a) Consider the following models.

```
fit1 <- vglm(y ~ x2 + x3 + x4, quasipoissonff, data = qdata)
fit2 <- vglm(y ~ x2 + x3 + x4, data = qdata,
             uninormal(var = TRUE, parallel = TRUE ~ x2 + x3 + x4 - 1,
                       zero = NULL, lmean = "loge"))
```

Explain why both models would be expected to give similar results when $\mu > 10$, say.
(b) What advantage does `fit2` have over `fit1`?
(c) Explain why

```
fit3 <- rrvglm(y ~ x2 + x3 + x4,
               uninormal(lmean = "loge", imethod = 2, var = TRUE, zero = NULL),
               data = qdata)
```

might be a good idea first for these data.

Ex. 5.23.   What relationship between μ_1 and μ_2 does (5.15) imply for the 2-stage (sequential) binomial model (10.35)? Tailor your answer to hold specifically for the call

```
rrvglm(cbind(y1, y2) ~ x2 + x3, seq2binomial, weights = size1, data = sdata)
```

Ex. 5.24.   **RR-Adjacent Categories Model**

Let y be a categorical response taking on levels $\{1, 2, 3\}$, and consider the call

```
rrvglm(y ~ x2 + x3 + x4, acat, data = adata)
```

Show that this model has $p_3/p_1 \propto \exp\{(1 + a_{21})\nu\}$ where $\nu = (x_2, x_3, x_4) \cdot \mathbf{c}$. Then show that $P(Y = 1) = p_1 = \left[1 + \exp\{\beta_{(1)1} + \nu\} + \exp\{\beta_{(1)1} + \beta_{(2)1} + (1 + a_{21})\nu\}\right]^{-1}$.

The vulgar are found in all ranks, and are not to be distinguished by the dress they wear.
—Lucius Annaeus Seneca

Chapter 6
Constrained Quadratic Ordination

For this cause left I thee in Crete, that thou shouldest set in order the things that are wanting, and ordain elders in every city, as I had appointed thee:...
Titus 1:5

6.1 Introduction

6.1.1 Ordination

In this chapter, RR-VGLMs are extended to give the Quadratic RR-VGLM (QRR-VGLM) class. This enables one to perform *constrained quadratic ordination* (CQO). The word "ordination" has several meanings, but its use here is based on older English for ordering, and in statistics it is traditionally the name given to a collection of multivariate techniques that order or arrange certain aspects of data. We will mainly describe ordination applied to the field of plant ecology, whereby species and environmental data are ordered and displayed on an ordination diagram to reveal their interrelationships. For example, which species are similar? And how do specific environmental variables affect all the species? Indeed, ordination remains one of the most common tasks of community ecologists who perform statistical analyses on field data.

Notationally, the species and environmental data are held in matrices $\mathbf{Y}$ (known as a *site-by-species matrix*) and $\mathbf{X} = (\mathbf{X}_1, \mathbf{X}_2)$, which are $n \times S$ and $n \times p$, respectively. There are n sites, p environmental variables of which the first is an intercept term, and S species. For example, the data might be collected inside 100 m^2 quadrants randomly distributed over a very large region encompassing a wide range of environmental types such as forest, desert, alpine and hinterland. At each site, the abundance of S species might be measured as counts (or simply as presence/absence), and variables in $\mathbf{X}$ recorded at each site too. In practice, any climate variables might need to be estimated, therefore suffer from the complication of measurement error—however, we shall assume that this has not occurred. The special case of the elements of $\mathbf{Y}$ equalling 1/0 for presence/absence, respectively, is known as an *incidence matrix*.

© Thomas Yee 2015
T.W. Yee, *Vector Generalized Linear and Additive Models*,
Springer Series in Statistics, DOI 10.1007/978-1-4939-2818-7_6

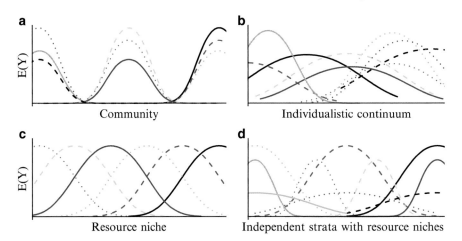

Fig. 6.1 Four hypothetical models for community ecology, along a gradient or latent variable. Plot (c) corresponds to a species packing model.

One of the key concepts in community ecology is that of an ecological *gradient*. Clearly, species are affected by environmental variables such as temperature, soil acidity, amount of rainfall and solar radiation, etc., and these may be combined to produce an ecological gradient that 'explains' the species' distributions. Representing a proxy or conglomerate variable, an ecological gradient is usually modelled as a linear combination of the environmental variables, i.e., (5.1). They are what some statisticians call latent variables. One can theorize that there is a dominant gradient which explains species' abundance the most. Then there is a secondary gradient that explains what was left unexplained by the first, etc. Subservient gradients are assumed to be independent of the ones before it, so that the overall estimation and interpretation is easier.

It may be argued that response curves should be unimodal in nature because very low and very high values of many types of environmental variables, such as temperature and certain nutrients, lead to species being absent, as opposed to an optimal environment for the species being somewhere in between. Expanding on the nutrient example, in plant ecology, it is well-known that while the deficiency of a particular nutrient stunts growth, the other extreme leads to it becoming toxic. That is, too much of a good thing can be detrimental to a species. This has been called the *Goldilocks principle* in many disciplines such as developmental psychology, economics and engineering. A first-stab approximation is to assume a symmetric bell-shape, like the density of a normal distribution. Indeed, such curves/surfaces have become a tenet in community ecology.

In keeping with RR-VGLMs, the first environmental variable is an intercept term of ones because we partition the environmental variables at a site by $x = (x_1^T, x_2^T)^T$, and usually $x_1 = 1$ and x_2 contains the abiotic environmental variables. That is, $\mathbf{X}_2$ are the 'real' environmental variables one wants to use in the ordination, but we want to do so after adjusting for the explanatory variables $\mathbf{X}_1$—which is often just an intercept term.

As in the previous chapter, let R be the *rank* (or *dimension*) of the ordination, $\boldsymbol{\nu} = (\nu_1, \dots, \nu_R)^T = \mathbf{C}^T x_2$ is the vector of R latent variables or underlying environmental gradients or trends. We write $\nu = c^T x_2$ when $R = 1$. The matrix $\mathbf{C}$ holds the *constrained* (or *canonical*) *coefficients*, and $\boldsymbol{\nu}_i$ are the

Table 6.1 Does a QRR-VGLM make the four CCA assumptions? The first three constitute a species packing model. Note: an RR-VGAM (Chap. 7) makes none of the four CCA assumptions, nor the bottom one.

CCA assumption	QRR-VGLM for CQO
1. Equal tolerances	No and yes. No: the ordination diagram needs contours (if $R \geq 2$). Yes: the ordination diagram is very easy to interpret
2. Equal maximums	No
3. Species' optimums uniformly distributed over the latent variable space	No
4. Species' site scores uniformly distributed over the latent variable space	No
Symmetric bell-shaped responses	Yes

site scores. The elements of $\mathbf{C}$ are interpreted as weights or coefficients, e.g., if $\boldsymbol{c}^T = (1.03, 0.05, -0.95, -0.08)$ and $\boldsymbol{x}_2 = $ (maximum annual temperature, soil pH, minimum annual temperature, rainfall)^T (scaled), then ν is essentially annual temperature seasonality. As another example, if $\boldsymbol{c}^T = (1.03, 0.05, -0.05, -0.08)$, then we would treat ν as maximum annual temperature, and one might view this as variable selection. It is the role of the data analyst to ascribe the human interpretation of ν based on the value of $\boldsymbol{c}$, if possible.

It should be noted that if the site scores are connected to the $\boldsymbol{x}_2$ via $\boldsymbol{\nu}_i = \mathbf{C}^T \boldsymbol{x}_{2i}$ then this is known as *direct gradient analysis*. Most of this chapter is devoted to this case. Where the site scores are totally free parameters, because no environmental variables $\boldsymbol{x}_2$ were measured, this is known as *indirect gradient analysis*, and it is the subject of Sect. 6.7.

There are several popular ecological conjectures on how the species' response curves are distributed along the (primary) ecological gradient, such as those of Fig. 6.1, e.g., it has been postulated that species tend to develop maximally separated niches over limiting environmental resources. If so, then the optimal constrained coefficients should be in a direction in which the species' response functions are the most different. The suppositions of response functions' properties belong to the so-called niche theory, but we will not involve ourselves with that here.

In practice we estimate one or two, or maybe even three, gradients. The first will have the most explanatory power, followed by the second, etc. so that the first is known as the dominant gradient. If there are two, then an ordination diagram, where the axes are the fitted gradients, may be plotted so that two species close to each other imply their distributions are similar, etc.

Since the 1930s, a number of ordination techniques have been proposed, mostly based on heuristic arguments. Consequently many of them were defective and were limited in scope, e.g., the 'horseshoe' effect, and not suitable for presence/absence or count species data. Probably the most popular technique is *canonical correspondence analysis* (CCA; ter Braak, 1986); however, it was developed as a heuristic approximation to CQO described in this chapter. The reasons put forth at the time for using CCA instead of CQO were mainly because of CQO's higher computational complexity and expense. Ironically, the two raw ingredients of CQO—GLM software and a general optimizer—were readily available in the 1980s. CCA assumes a *species packing model* (Fig. 6.1c and Table 6.1) whereas CQO is more flexible.

One can think of CQO as a more rigorous way to perform ordination. Based on GLM ideas, QRR-VGLMs have several advantages compared to CCA such as its simpler formulation, greater flexibility due to fewer assumptions, the ability to extract out more ecologically meaningful information, and it possesses greater soundness of statistical theory. It operates on common types of species data such as presence/absence data and Poisson counts, whereas CCA treats all data-types the same—this is like using a baseball bat to play golf, badminton and tennis. One advantage that CCA does have over QRR-VGLMs is its robustness; a reason for this is that QRR-VGLMs are estimated by MLE, and therefore are sensitive to departures from the model formulation. Consequently, they are *much more* challenging to fit, and this chapter attempts to offer some suggestions in this regard.

This chapter concentrates on 1-parameter distributions such as the Poisson and binomial, therefore $M = S$. Then both j and s are interchangeably used to index the species from $\mathcal{S}$, where $\mathcal{S} = \{1, \dots, S\}$ is the set of all species or responses.

To summarize thus far, we wish to fit symmetric bell-shaped response curves or surfaces, as a function of ν_1 (and possibly ν_2) to species' data, especially counts and presences/absences. The resulting ordination is something formerly called *canonical Gaussian ordination*, or CGO, by Yee (2004a). This acronym was abandoned because the word "Gaussian" has two meanings that may cause confusion. The first meaning is that the response is normally distributed or continuous. The other is that of a bell-shaped curve. Of course, it is the latter that was intended, hence "quadratic" is more informative, provided it is understood that it is on the η-scale. Also, "constrained" is more meaningful than "canonical" to most non-mathematicians because the site scores are constrained to be linear combinations of the explanatory variables.

6.1.2 Prediction and Calibration

Table 6.2 gives the nomenclature for the methods advocated in this chapter, and it is largely built upon the ideas of ter Braak and Prentice (1988). It arises by cross-classifying two factors:

(i) constrained (C) or unconstrained (U);
(ii) the shape of the response to the latent variable on the (transformed) η-scale (linear (L), quadratic (Q), or just smooth (A)).

The rationale behind this classification scheme is that ecologists have long debated what shape species' responses have along a dominant gradient—the second factor gives the three most important cases. The first factor acknowledges that $\mathbf{X}$ may or may not have been collected from a field study—an important demarcation—i.e., direct or indirect gradient analysis.

Table 6.2 shows therefore that all three "constrained" methods have an "unconstrained" counterpart whereby the site scores $\boldsymbol{\nu}_i$ are not constrained to be linear combinations of $\boldsymbol{x}_2$. For example, UQO is CQO but where the site scores are largely free parameters. Although this chapter is primarily about constrained ordination, Sect. 6.7 looks at the unconstrained ordination problem.

Table 6.2 Nomenclature for 6 ordination methods based on GLM-type regression techniques. The shapes are on the η-scale. In each cell, the first bold line is the biological application or ordination method, the second line is the statistical class of models, and the remaining lines are some special cases. Abbreviations: A = additive, C = constrained (preferred) or canonical, L = linear, O = ordination, Q = quadratic, R = regression, RDA = redundancy analysis, RR = reduced-rank, U = unconstrained.

	Constrained	Unconstrained
Linear	CLO	ULO
	RR-VGLM (Yee and Hastie, 2003)	U-VGLM
	e.g., (Gaussian) RRR (= RDA) (Anderson, 1951), RR-MLM (Yee and Hastie, 2003)	e.g., Goodman's RC model (Goodman, 1981)
Quadratic	CQO	UQO
	QRR-VGLM (Yee, 2004a)	QU-VGLM (Yee, 2006)
	e.g., Gaussian logit ordination (ter Braak and Prentice, 1988)	e.g., Gaussian ordination (Gauch et al. (1974), Kooijman (1977))
Smooth	CAO	UAO
	RR-VGAM (Yee, 2006)	U-VGAM (Yee, 2006)
	e.g., projection pursuit regression (Friedman and Stuetzle, 1981), constrained principal curves (De'ath, 1999)	e.g., principal curves (Hastie and Stuetzle, 1989)

The regression methods of Table 6.2 have the 3 major application areas listed in Table 6.3, of which heuristic ordination methods such as CCA are unsuitable for the latter two. In contrast, *all* the methods in Table 6.2 can perform *all* 3 tasks, and Sect. 6.4 looks at a few specific types of analyses to perform these tasks. Note that the three tasks refer to ν instead of x_2 because x_2 might not exist, or else if it does, since $\nu = c^T x_2$ implies it is impossible to estimate the value of a variable in x_2 unless one assigns values to the other variables in x_2. Because of the 3-fold potential use, "O" in the table could be replaced by "P" for prediction and "C" for calibration, e.g., "constrained quadratic prediction", or CQP, might be coined for a QRR-VGLM whose primary purpose is to predict species compositions at sites.

6.2 Quadratic RR-VGLMs for CQO

QRR-VGLMs are a class of models that can fit bell-shaped curves or surfaces to a set of latent variables. They extend RR-VGLMs (5.2) by adding on a quadratic form in ν. For one-parameter distributions such as the Poisson and binomial,

$$\boldsymbol{\eta} \equiv \begin{pmatrix} \eta_1 \\ \vdots \\ \eta_M \end{pmatrix} = \mathbf{B}_1^T \boldsymbol{x}_1 + \mathbf{A}\boldsymbol{\nu} + \begin{pmatrix} \boldsymbol{\nu}^T \mathbf{D}_1 \boldsymbol{\nu} \\ \vdots \\ \boldsymbol{\nu}^T \mathbf{D}_M \boldsymbol{\nu} \end{pmatrix} \tag{6.1}$$

$$= \boldsymbol{\alpha} - \frac{1}{2} \begin{pmatrix} (\boldsymbol{\nu} - \boldsymbol{u}_1)^T \mathbf{T}_1^{-1} (\boldsymbol{\nu} - \boldsymbol{u}_1) \\ \vdots \\ (\boldsymbol{\nu} - \boldsymbol{u}_M)^T \mathbf{T}_M^{-1} (\boldsymbol{\nu} - \boldsymbol{u}_M) \end{pmatrix}, \tag{6.2}$$

where $\mathbf{D}_j$ are $R \times R$ symmetric matrices, and $\boldsymbol{\alpha}$ is some vector depending on $\boldsymbol{x}_1$, the *optimums* $\boldsymbol{u}_j$ and the *tolerance matrices* $\mathbf{T}_j$. It is easy to show (Ex. 6.1) that

$$\eta_j(\boldsymbol{\nu}) = \left\{ \frac{1}{2} \boldsymbol{u}_j^T \mathbf{T}_j^{-1} \boldsymbol{u}_j + \boldsymbol{\beta}_j^T \boldsymbol{x}_1 \right\} - \frac{1}{2} (\boldsymbol{\nu} - \boldsymbol{u}_j)^T \mathbf{T}_j^{-1} (\boldsymbol{\nu} - \boldsymbol{u}_j), \quad (6.3)$$

for $j = 1, \ldots, M$, which gives the expression for α_j.

The jth response surface in (6.2) is bell-shaped in the latent variables $\boldsymbol{\nu}$ if and only if $\mathbf{D}_j$ is negative-definite, i.e., if and only if $\mathbf{T}_j = -\frac{1}{2} \mathbf{D}_j^{-1}$ is positive-definite. The matrices $\mathbf{D}_j$ and $\mathbf{T}_j$ control the (ellipsoidal) contours of the bell-shaped surface in the R-dimensional latent variable space, e.g., if they are diagonal, then the axes of the ellipsoids are parallel to the ν_r $(r = 1, \ldots, R)$ (ordination) axes. The contours represents points that have the same fitted value (e.g., abundance or probability). The fitted $\widehat{\mathbf{C}}$ are called the (estimated) *constrained coefficients*, the jth column of $\widehat{\mathbf{C}}$ are called the jth constrained coefficients (which are interpreted as weights), and the optimum of Species j is

$$\boldsymbol{u}_j = \mathbf{T}_j \, \boldsymbol{a}_j. \quad (6.4)$$

The representation (6.2) is preferred to (6.1), because of the ecological interpretations that can be ascribed to the parameters. In particular, in one dimension,

- the tolerance measures how wide the response curve is, i.e., how much deviation the species can tolerate from its optimal environment, so that if t_j is large then the response curve decreases slowly as ν moves away from its optimum, implying that the species can still flourish under a variety of different environments. Technically, the large t_j case refers to a *stenoecous* species, whereas the small t_j case is called a *eurycous* species. Thus the tolerance is a measure of niche width, and the parameter acts like the standard deviation of a normal distribution;

- the parameter u_j is the value of the gradient (in the ecological rather than the mathematical sense) in which $\mu_j(\nu)$ is a maximum; therefore, it is the optimal environment for Species j. For example, with Poisson counts, $\mu_j(u_j) = e^{\alpha_j}$ so that α_j is directly related to the prevalence/abundance of the species at its optimum (Eq. (6.3)). We call $\mu_j(u_j)$ the *maximum* of Species j.

As a specific rank-1 example of a QRR-VGLM, consider Poisson data with S species. The model for Species s is the Poisson regression

$$\log \mu_s(\nu) = \eta_s(\nu) = \beta_{(s)1} + \beta_{(s)2} \, \nu + \beta_{(s)3} \, \nu^2 \quad (6.5)$$

$$= \alpha_s - \frac{1}{2} \left(\frac{\nu - u_s}{t_s} \right)^2, \quad s = 1, \ldots, S. \quad (6.6)$$

Much of classical ecological theory can be expressed in terms of the three parameters α_s, u_s and t_s, e.g., Shelford's law of tolerance, Liebig's law of the minimum, and Gause's law. Equation (6.5) is a quadratic in ν so that the response curve $E(Y_s|\nu)$ is symmetric bell-shaped provided that $\beta_{(s)3} < 0$. More generally, $\eta(\boldsymbol{\nu})$ ought to be a concave function in $\boldsymbol{\nu}$. The coefficient $\beta_{(s)3}$ in (6.5) is important, because a negative value implies that the response curve is unimodal about the optimum u_s. If $\beta_{(s)3} = 0$, then the curve is sigmoid, and if $\beta_{(s)3} > 0$, then the curve is "u"-shaped—something ecologically unrealistic.

Table 6.3 Three major application areas of the regression methods advocated in this chapter.

Use	Comment
Ordination	Order the response curves/surfaces as functions of $\boldsymbol{\nu}$. This is an arrangement of the species with each other, and with the environmental variables, if any
Prediction	Predict $\boldsymbol{y}$, given $\boldsymbol{\nu}$. This is possible because standard regression models are used
Calibration	Estimate $\boldsymbol{\nu}$, given $\boldsymbol{y}$. Given a site with species composition $\boldsymbol{y}_0 = (y_{01}, \ldots, y_{0S})^T$, say, the generic calibration problem is to estimate $\boldsymbol{\nu}_0$, the value of the site score there, e.g., bio-monitoring pollution levels based on species' responses, and reconstructions of climate history based on species compositions. Calibration is the subject of Sect. 6.4.5

For the practitioner, the *equal-tolerances assumption* is of particular importance:

$$\mathbf{T}_1 = \mathbf{T}_2 = \cdots = \mathbf{T}_S. \tag{6.7}$$

Ideally, the analyses should be made with and without that assumption, and the results compared. Some self-proclaimed experts and proponents of niche theory have indicated strongly that most species do not have equal niche width. However, practitioners who assume equal tolerances have several good reasons for doing so. Firstly, because ordination diagrams (Sect. 6.4.2) for $R = 2$ are much easier to interpret—contours are unneeded. Secondly, often $\widehat{\mathbf{T}}_s$ can be non-positive-definite for some species (i.e., not bell-shaped), but estimating one common tolerance matrix over all species might at least give sensible results overall. This is likened to borrowing strength over the species, and it often results in greater numerical stability, provided that most of the data exhibits unimodality.

QRR-VGLMs are fitted using the modelling function `cqo()`, which may one day be renamed to `qrrvglm()`. By default, it assumes equal tolerances, however, the estimated tolerance matrix may not be positive-definite; see Sect. 6.4.1. The reasons given in Sect. 5.2.1 for RRR also hold for CQO.

6.2.1 An Example: Hunting Spiders Data

To illustrate the basic ideas early on, consider the hunting spiders data set, called **hspider**. It is well known in the ordination literature, and is detailed in ter Braak (1986). Briefly, these data were collected in a Dutch dune area over a 60-week period, and consists of abundances (numbers trapped) of 12 species of hunting spiders. There were 6 environmental variables (water, bare sand, twigs, cover moss, cover herbs, and light reflection) measured at the $n = 28$ sites. We standardize the environmental variables in our analyses here.

In our first model, we try a rank-1 Poisson CQO with unequal tolerances.

```
> hspider[, 1:6] <- scale(hspider[, 1:6])   # Standardized environmental variables
> set.seed(1234)   # For reproducibility of the results
> p1ut.hs <- cqo(cbind(Alopacce, Alopcune, Alopfabr, Arctlute, Arctperi,
                Auloalbi, Pardlugu, Pardmont, Pardnigr, Pardpull,
                Trocterr, Zoraspin) ~
            WaterCon + BareSand + FallTwig + CoveMoss + CoveHerb + ReflLux,
            poissonff, data = hspider, eq.toler = FALSE, trace = FALSE)
```

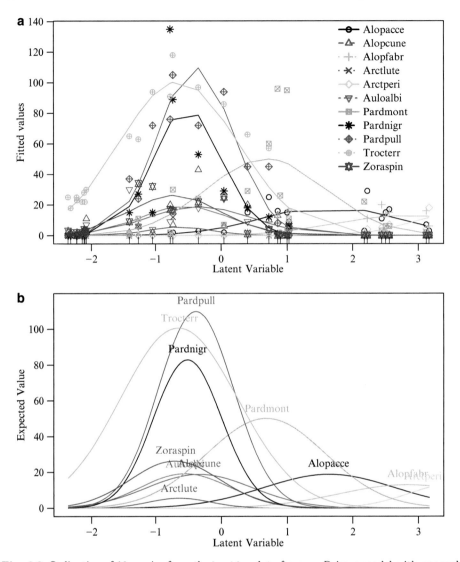

Fig. 6.2 Ordination of 11 species from the `hspider` data frame; a Poisson model with unequal tolerances, called `pIut.hs.2`. (**a**) `lvplot()` output. (**b**) `persp()` output, which is a perspective plot and a 'continuous' version of (**a**).

The estimated tolerances, $\widehat{t}_s$, are

```
> Tol(pIut.hs)[1, 1, ]

Alopacce Alopcune Alopfabr Arctlute Arctperi Auloalbi Pardlugu Pardmont
 1.00000  0.72101  1.09723  0.22434  0.71902  0.51714 -2.39571  0.89509
Pardnigr Pardpull Trocterr Zoraspin
 0.27949  0.35603  0.86867  0.50134
```

The scaling is such that the first species' tolerance is unity. This suggests that attempting to fit a bell-shaped curve to the species `Pardlugu` has failed because its $\widehat{\nu}^2$ coefficient is positive. Let's omit that species and refit the model.

```
> set.seed(1234)  # For reproducibility of the results
> p1ut.hs.2 <- cqo(cbind(Alopacce, Alopcune, Alopfabr, Arctlute, Arctperi, Auloalbi,
                    Pardmont, Pardnigr, Pardpull, Trocterr, Zoraspin) ~
                WaterCon + BareSand + FallTwig + CoveMoss + CoveHerb + ReflLux,
                poissonff, data = hspider, Crow1positive = FALSE, eq.toler = FALSE,
                trace = FALSE)
> S <- ncol(depvar(p1ut.hs.2))  # Number of species
> clr <- (1:(S+1))[-7]  # Omits yellow
> lvplot(p1ut.hs.2, main = "(a)", y = TRUE, lcol = clr, pch = 1:S, pcol = clr)
> legend("topright", leg = colnames(depvar(p1ut.hs.2)), col = clr,
            pch = 1:S, merge = TRUE, bty = "n", lty = 1:S, lwd = 2)
```

This gives Fig. 6.2a. It can be seen that there is a mixture of dominant and low-abundance species. This plot has the observed values added, which may clutter the plot when n and/or S are large, as well as lines joining the fitted values that give the appearance of discreteness. An alternative plot which does not have these features is `persp()`. The output of

```
> persp(p1ut.hs.2, main = "(b)", col = clr, label = TRUE)  # Perspective plot
```

appears in Fig. 6.2b. All the fitted bell-shaped curves can be seen—but there is always the danger with this type of plot of not knowing how well the fitted curves fit the observed data. The constrained coefficients are

```
> round(concoef(p1ut.hs.2), digits = 2)
```

	latvar
WaterCon	-0.33
BareSand	0.46
FallTwig	-0.57
CoveMoss	0.23
CoveHerb	-0.27
ReflLux	0.68

These signs agree with the CCA analysis of ter Braak (1986), who interpreted $\widehat{\nu}$ as a moisture gradient.

Now we can try fitting an equal-tolerances model to all the species, just to see if it makes any qualitative difference.

```
> set.seed(1234)  # For reproducibility of the results
> p1et.hs <- cqo(cbind(Alopacce, Alopcune, Alopfabr, Arctlute, Arctperi,
                Auloalbi, Pardlugu, Pardmont, Pardnigr, Pardpull,
                Trocterr, Zoraspin) ~
            WaterCon + BareSand + FallTwig + CoveMoss + CoveHerb + ReflLux,
            poissonff, data = hspider, Crow1positive = FALSE, eq.toler = TRUE,
            trace = FALSE)
> lvplot(p1et.hs, main = "(a)", y = TRUE, lcol = clr, pch = 1:S, pcol = clr, las = 1)
> legend("topright", leg = colnames(depvar(p1et.hs)), col = clr,
        pch = 1:S, merge = TRUE, bty = "n", lty = 1:S, lwd = 2)
> persp(p1et.hs, main = "(b)", col = clr, label = TRUE, las = 1)  # Perspective plot
```

This gives Fig. 6.3. Not surprisingly, the two models are very similar. The species `Pardlugu` is shown to be a minor species appearing at the boundary of the environmental space. This explains why it did not exhibit unimodality—its signal has been overwhelmed by random error and weakened by an edge effect. This ill-conditioning is commonly the case for species distributed away from the convex hulls of the site scores. The same is seen in the rank-2 ordination (Fig. 6.6).

We can look at the constrained coefficients, optimums, tolerances, etc. by using `Coef()`:

```
> Coef(p1et.hs)

  C matrix (constrained/canonical coefficients)
                latvar
  WaterCon  -0.35750
  BareSand   0.55314
  FallTwig  -0.91928
  CoveMoss   0.31630
  CoveHerb  -0.30516
  ReflLux    0.70390

  B1 and A matrices
                     (Intercept)        A
  loge(E[Alopacce])      1.09373  1.98832
  loge(E[Alopcune])      2.84707 -0.44120
  loge(E[Alopfabr])     -2.24519  3.10397
  loge(E[Arctlute])      0.92198 -0.72167
  loge(E[Arctperi])     -8.69297  4.83720
  loge(E[Auloalbi])      2.57346 -0.66305
  loge(E[Pardlugu])     -0.67625 -2.55136
  loge(E[Pardmont])      3.58278  0.89167
  loge(E[Pardnigr])      3.76112 -0.58679
  loge(E[Pardpull])      4.21020 -0.40758
  loge(E[Trocterr])      4.44445 -0.84106
  loge(E[Zoraspin])      2.76993 -0.85949

  Optimums and maximums
            Optimum  Maximum
  Alopacce  1.98832  21.5513
  Alopcune -0.44120  18.9992
  Alopfabr  3.10397  13.0937
  Arctlute -0.72167   3.2622
  Arctperi  4.83720  20.2118
  Auloalbi -0.66305  16.3345
  Pardlugu -2.55136  13.1768
  Pardmont  0.89167  53.5343
  Pardnigr -0.58679  51.0744
  Pardpull -0.40758  73.2049
  Trocterr -0.84106 121.2850
  Zoraspin -0.85949  23.0875

  Tolerance
           latvar
  Alopacce      1
  Alopcune      1
  Alopfabr      1
  Arctlute      1
  Arctperi      1
  Auloalbi      1
  Pardlugu      1
  Pardmont      1
  Pardnigr      1
  Pardpull      1
  Trocterr      1
  Zoraspin      1

  Standard deviation of the latent variables (site scores)
  latvar
   2.371
```

The constrained coefficients are qualitatively the same as the unequal-tolerances model. From the bottom of the output, we have $\texttt{sd(latvar(p1et.hs))} = \widehat{\mathrm{Var}}(\widehat{\nu}_i) \approx 2.371$. It is left to the reader to reconcile the remaining output with Fig. 6.3.

It is a good idea to see how often the best model was chosen out of all those fitted:

```
> sort(deviance(p1et.hs, history = TRUE))   # A history of all the iterations

 [1] 1585.1 1585.1 1585.1 1585.1 1585.1 1585.1 2472.1 2472.1 2472.1 2472.1
```

Thus the best model was achieved 6 out of 10 times. This suggests that there does seem to be a global solution that is relatively easy to converge to, compared with several local solutions. One would be worried if the best model was only obtained once, because this suggests there may be some other better solution not found yet.

Lastly for now, we try fitting a rank-2 equal-tolerances Poisson model. We cheat a little here by using arguments that will be explained a little later, as well as knowing what the true global optimal model is.

```
> set.seed(555)   # For reproducibility of the results
> p2et.hs <- cqo(cbind(Alopacce, Alopcune, Alopfabr, Arctlute, Arctperi,
                  Auloalbi, Pardlugu, Pardmont, Pardnigr, Pardpull,
                  Trocterr, Zoraspin) ~
              WaterCon + BareSand + FallTwig + CoveMoss + CoveHerb + RefLux,
              poissonff, data = hspider, Crow1positive = FALSE, Rank = 2,
              I.toler = TRUE, Bestof = 3, isd.latvar = c(2.1, 0.9))
> if (deviance(p2et.hs) > 1127) warning("suboptimal fit obtained")
> persp(p2et.hs, xlim = c(-6, 5), ylim = c(-6, 3), theta = 120, phi = 20)
```

This gives Fig. 6.4, which blends in all the silhouettes of the bell-shaped surfaces of each species combined. The dominant species mask the rarer species, and some of the species distributions are outside the range of the environmental data.

Table 6.4 lists some particularly useful arguments which users of `cqo()` should heed.

6.2.2 Normalizations for QRR-VGLMs

For QRR-VGLMs, it is more convenient to abandon the corner constraints of RR-VGLMs (5.5). Instead, upon examination of (6.1)–(6.2), there is quite some scope to choose normalizations that make interpretation easier. In particular, $\mathbf{T}_s^{-1}(\boldsymbol{\nu} - \boldsymbol{u}_s) = \mathbf{T}_s^{-1}\mathbf{M}_1\mathbf{M}_1^{-1}(\boldsymbol{\nu} - \boldsymbol{u}_s)$ and $\mathbf{A}\boldsymbol{\nu} = \mathbf{A}\mathbf{M}_2\mathbf{M}_2^{-1}\boldsymbol{\nu}$ for any nonsingular matrices $\mathbf{M}_1$ and $\mathbf{M}_2$. Two nice properties to have are as follows.

- **Property (A)**

$$\widehat{\mathbf{T}}_s = \mathbf{I}_R. \tag{6.8}$$

That is, the niche width in every dimension is unity. Additionally, since the matrix is diagonal, the Mahalanobis distance and Euclidean distance coincide in every direction. This normalization is similar to a multivariate normal distribution with variance-covariance matrix $\mathbf{I}_R$.

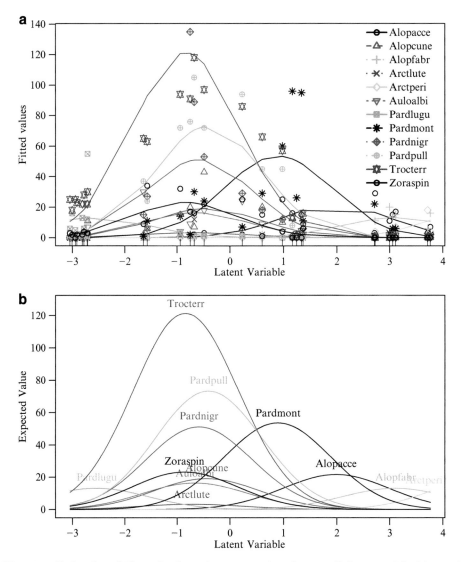

Fig. 6.3 Ordination of 12 species from the `hspider` data frame; a Poisson model with equal tolerances, called `p1et.hs`. (a) `lvplot()` output. (b) `persp()` output, which is a perspective plot and a 'continuous' version of (a).

- **Property (B)**

$$\widehat{\mathrm{Var}}(\widehat{\boldsymbol{\nu}}_i) = \mathbf{I}_R. \tag{6.9}$$

That is, the latent variables have unit standard deviation, and are uncorrelated. Uncorrelated latent variables are a good idea because they can be loosely be thought of as unrelated to each other. With two ordination axes of uncorrelated latent variables, one can think of the second axis as being unrelated to, and less important than, the first axis, which represents the dominant gradient.

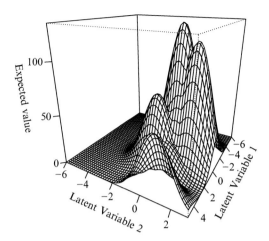

Fig. 6.4 Perspective plot of a rank-2 CQO fitted to the `hspider` data frame, called `p2et.hs`. All 12 species are fitted, using an equal-tolerances Poisson model, but only the response surfaces of the dominant species may be seen. See also Fig. 6.6 for an ordination diagram of the above.

We can choose $\mathbf{M}_1$ and $\mathbf{M}_2$ to obtain Property (A) or (B), but not both. Whatever we choose, a compromise on the other property can be obtained (in terms of the matrix just being diagonal). Because the following results are so important, we state them as a theorem.

Theorem　For QRR-VGLMs (6.1)–(6.2), provided $\exists\, s \in \mathcal{S}$ such that $\widehat{\mathbf{T}}_s$ is positive-definite, it is possible to scale the $\widehat{\boldsymbol{\nu}}_i$ so that

1. (6.8) is satisfied for at least one $s \in \mathcal{S}$, and
2. $\widehat{\mathrm{Var}}(\widehat{\boldsymbol{\nu}}_i)$ is diagonal.

Furthermore, (6.8) holds for *all* $s \in \mathcal{S}$ if `I.tolerances = TRUE` or if `eq.tolerances = TRUE`.　　　　　　□

The proof is left as an exercise to the reader (Ex. 6.2).

　　Currently, the defaults are `eq.tolerances = TRUE` and `I.tolerances = FALSE`, which means that if the pooled tolerance estimate is positive-definite, then (6.8) holds for all $s \in \mathcal{S}$, and $\widehat{\mathrm{Var}}(\widehat{\boldsymbol{\nu}}_i)$ is *diagonal*.

　　Conversely, it is easy to show that if (6.9) is desired, then one needs to relax (6.8) by making $\mathbf{T}_s$ diagonal.

　　In the case of unequal tolerances, the argument `refResponse` can be used to select the s value such that (6.8) holds. This argument applies to `Coef()`, `lvplot()` and other generic functions. See Sect. 6.4.1 for details.

6.3 Fitting QRR-VGLMs

Table 6.7 describes the 3 cases of how QRR-VGLMs may be fitted. They depend critically on the 2 arguments `I.tolerances` and `eq.tolerances`. These arguments also relate directly to the theorem of Sect. 6.2.2. Having either of them **TRUE** results in an equal-tolerances model (6.7), however, they are subtlety different in 2 ways.

Table 6.4 Upper table: certain `qrrvglm.control()` arguments of particular relevance to the user. Some defaults are given. Lower table: other arguments used by some methods functions. Recall $s \in \mathcal{S}$ denotes one selected response.

Argument	Comment
`Bestof = 10`	Number of models fitted. The one with the minimum deviance is selected. Helps safeguard against local solutions. Increase this value for greater certainty, in conjunction with `set.seed()` for reproducibility
`Cinit = NULL`	An initial $\mathbf{C}$ matrix $(p_2 \times R)$. If inputted, only one model is fitted. Fitting a stereotype model is one possibility for Poisson counts (Ex. 6.7); see also Sect. 6.3.2
`Crow1positive = TRUE`	An R-vector of logicals describing whether the signs of the first row of $\mathbf{C}$ are positive, i.e., $c_{11}, \ldots, c_{1R}$. If the rth value is `FALSE`, then the solution is reflected about the rth ordination axis
`eq.tolerances = TRUE`	By default, an equal-tolerances model (6.7) is fitted, however, the $\widehat{\mathbf{T}}_s$ may not be positive-definite. If `FALSE`, then each species is modelled separately, and one of the species has $\widehat{\mathbf{T}}_s \equiv \mathbf{I}_R$ (provided at least one species has a positive-definite tolerance matrix). This option is memory-hungry. See Sect. 6.2.2 and Table 6.7
`isd.latvar`	See Sects. 6.3.1 and 6.3.2
`I.tolerances = FALSE`	If `TRUE`, then another way to fit an equal-tolerances model (6.7) is used; then $\widehat{\mathbf{T}}_s \equiv \mathbf{I}_R \ \forall s \in \mathcal{S}$, therefore all the variables in $\boldsymbol{x}_2$ should be `scale()`d. This is computationally the fastest method on simulated data. See Sects. 6.3.1, 6.3.2 and Table 6.7
`MUXfactor = rep(7, length = Rank)`	See Sect. 6.3.2
`noRRR = ~ 1`	A formula indicating which variables comprise $\boldsymbol{x}_1$
`Rank = 1`	R, the dimension of the ordination. A rank-1 model should always be attempted first. Rank-2 models are much harder to fit, and certainly not all data sets are expected to support this
`trace = TRUE`	Prints out some output as a record of the computational details, in real time
`Use.Init.Poisson.QO = TRUE`	Logical. Use an algorithm based on an equal-tolerances Poisson QO to find initial values? If `FALSE`, then `Cinit` should be inputted
`refResponse`	A value $s \in \mathcal{S}$ specifying the response to be considered as the reference species. It will have $\widehat{\mathbf{T}}_s \equiv \mathbf{I}_R$. Mainly used if `eq.tolerances = FALSE`
`varI.latvar`	Logical. If `TRUE`, then $\widehat{\mathrm{Var}}(\widehat{\boldsymbol{\nu}}_i) = \mathbf{I}_R$ (Eq. (6.9)), else $\widehat{\mathrm{Var}}(\widehat{\boldsymbol{\nu}}_i)$ is diagonal. See Sect. 6.2.2

Firstly, in the algorithm used—one is fast but requires scaling of the environmental variables—while the other is memory-hungry and slower (at least with simulated data). Secondly, though (6.7) holds in both cases, the tolerance matrices are only guaranteed to be positive-definite if `I.tolerances = TRUE`; if `eq.tolerances = TRUE`, then the $\widehat{\mathbf{T}}_s$ may or may not be positive-definite. Some more computational details about these arguments are given in Sect. 6.3.1.

Currently, only the **VGAM** family functions listed in Table 6.5 are supported by `cqo()`. The `multiple.responses` argument for the binomial-variants indicates that the response matrix is composed of multiple binary responses, i.e., there are S species. This is necessary to retain upward compatability, e.g., `binomialff()` with a 2-column matrix response is interpreted as a matrix of successes and fail-

Table 6.5 Summary of some properties of statistical distributions in regression-based ordina-tion. The ϕ_j are overdispersion parameters. The argument `mul = TRUE` is an abbreviation for `multiple.responses = TRUE`.

Family function	$\text{Var}(Y_j)$	Support	η_j	Notes
`binomialff(mul = TRUE)`	$\mu_j(1-\mu_j)$	$\{0,1\}$	$\text{logit}\,\mu_j$	`link = "cloglog"` may be preferable
`gaussianff()`	σ_j^2	$(-\infty,\infty)$	μ_j	Constant variance in the LM
`poissonff()`	μ_j	$0(1)\infty$	$\log \mu_j$	
`quasibinomialff(mul = TRUE)`	$\phi_j\,\mu_j(1-\mu_j)$	$\{0,1\}$	$\text{logit}\,\mu_j$	`link = "cloglog"` may be preferable
`quasipoissonff(mul = TRUE)`	$\phi_j\,\mu_j$	$0(1)\infty$	$\log \mu_j$	

Table 6.6 Methods functions currently for CQO and CAO objects in **VGAM**. The bottom section are plotting functions. See also Table 8.7.

R function	Purpose
`calibrate()`	Calibration: estimate $\boldsymbol{\nu}$ from $\boldsymbol{y}$. See Sect. 6.4.5
`Coef()`	$\widehat{\mathbf{A}}$, $\widehat{\mathbf{B}}_1$, $\widehat{\mathbf{C}}$, $\widehat{\mathbf{D}}$, $\widehat{\boldsymbol{u}}_s$, $\widehat{\mathbf{T}}_s$, $\widehat{\boldsymbol{\nu}}_i$, etc. (Eq. (6.1))
`concoef()`	Constrained (canonical) coefficients $\widehat{\mathbf{C}}$ (Eq. (6.1))
`is.bell()`	Are the species' response curves/surfaces bell-shaped? (Eq. (6.1))
`latvar()`	Matrix of latent variables $\widehat{\boldsymbol{\nu}}_i = \widehat{\mathbf{C}}^T \boldsymbol{x}_{2i}$ (site scores; Eq. (5.1)). Is $n \times R$
`Max()`	Maximums $E[Y_s|\widehat{\boldsymbol{u}}_s] = g^{-1}(\widehat{\alpha}_s)$ (Eq. (6.2)). If $\boldsymbol{x}_1 \neq 1$, then the maximum of a species is undefined as it depends on values of variables in $\boldsymbol{x}_1$. Consequently, an `NA` will be returned
`Opt()`	Optimums $\widehat{\boldsymbol{u}}_s$ (species scores; Eq. (6.2))
`predict()`	Prediction: estimate $\boldsymbol{y}$ from $\boldsymbol{x}$
`Rank()`	Rank R
`resid()`	Residuals (e.g., working, response, . . .)
`summary()`	Summary of the object
`Tol()`	Tolerances $\widehat{\mathbf{T}}_s$ (Eq. (6.2))
`biplot()`	Same as `lvplot()`
`lvplot()`	Latent variable plot (ordination diagram; for $R = 1$ or 2). See Sect. 6.4.2
`persp()`	Perspective plot (for $R = 1$ or 2). See Sect. 6.4.3
`trplot()`	Trajectory plot (for $R = 1$ only). See Sect. 6.4.4

ures rather than two species. If `multiple.responses = TRUE` then the response (matrix) must contain 0s and 1s only. The use of a complementary log–log link for presence/absence data may be preferred, because of its connection with the Poisson distribution (Ex. 1.2).

6.3.1 Arguments `I.tolerances` and `eq.tolerances`

These arguments reside in `qrrvglm.control()`. However, their differences should be understood. Basically, the reason is computational.

Choosing between an equal-tolerances and unequal-tolerances model is a trade-off between interpretability and quality of fit. In real life, it is unrealistic assumption. But then it can be argued that bell-shaped curves/surfaces are an unrealistic assumption too. Certainly, for $R = 2$, equal tolerances make interpretation much easier because elliptical contours need not be added to the ordination diagram. With $R = 1$, one doesn't need an equal-tolerances assumption so much because one can gauge how large the tolerances are by applying a function such as `persp()`.

So how are the arguments `I.tolerances` and `eq.tolerances` related? And how are algorithms for fitting the models affected by these? The answers to these questions are given in Table 6.7. The argument `eq.tolerances` refers to whether $\mathbf{T}_s = \mathbf{T}$ for all $s \in \mathcal{S}$, for some order-R matrix $\mathbf{T}$. Note that $\mathbf{T}$ may or may not be positive-definite; ideally, it is. In contrast, the argument `I.tolerances` *is* positive-definite; it is more directed at specifying the algorithm used, and if `TRUE`, offsets (Sect. 3.3) of the form $-\frac{1}{2}\nu_{ir}^2$ are used in the algorithm because $\mathbf{T}_s = \mathbf{I}_R$ by definition. Note that setting `I.tolerances = TRUE` *forces* bell-shaped curves/surfaces on the data regardless of whether this is appropriate or not. Having `I.tolerances = TRUE` implies `eq.tolerances = TRUE`, but not vice versa.

Computationally, any offset values which are large will cause numerical problems. Therefore it is highly recommended that all numerical variables in $\boldsymbol{x}_2$ be standardized to mean 0 and unit variance. This will result in the site scores being centred at 0 because

$$E(\boldsymbol{\nu}) \;=\; E(\mathbf{C}^T\boldsymbol{x}_2) \;=\; \mathbf{C}^T E(\boldsymbol{x}_2) \;=\; \mathbf{C}^T \mathbf{0} \;=\; \mathbf{0}.$$

Standardizing variables can be achieved with `scale()`, hence something like

```
cqo(cbind(spp1, spp2, spp3) ~
    scale(temperature) + scale(rainfall) + scale(log1p.nitrogen),
    poissonff, data = pdata, I.tolerances = TRUE)
```

is probably a good idea for species counts.

In practice, scaling $\boldsymbol{x}_2$ and setting `I.tolerances = TRUE` is the recommended way of fitting an equal-tolerance model because they are computed more efficiently. Each species can be fitted separately, and the number of parameters is low, e.g., with $R = 1$, there are 2 parameters per species, and for $R = 2$ there are 3 parameters per species (In general, there are $R + 1$ parameters for a rank-R problem). This contrasts with `eq.tolerances = FALSE`, where there are 3 and 6 parameters, respectively, for $R = 1$ and 2.

On simulated data, it is often the case that Case 1 is the fastest, followed by Case 3 and then Case 2. However, with real data, this ordering can easily be scrambled because of the lack of convergence problems, etc. for a particular species.

However, if `I.tolerances = TRUE` fails because of numerical problems, then the next best thing to do is to set `I.tolerances = FALSE` and `eq.tolerances = TRUE`. This will result in a different algorithm being used, which will usually be much slower but there is less risk of numerical problems. The memory requirements, however, will be much larger for this choice.

Table 6.7 The relationship between the arguments `eq.tolerances` and `I.tolerances`. It is assumed that the constraint matrices of all variables in $\boldsymbol{x}_1$ is $\mathbf{I}_S$. "Separate estimation" means each species can be fitted separately, otherwise "Joint estimation" means fitting one big model involving all the species. The index $s \in \mathcal{S}$ indexes species. The significance of the 3 cases is discussed in Sect. 6.4.1.

	`eq.tolerances = TRUE`	`eq.tolerances = FALSE`
`I.tolerances = TRUE`	Case 1 (Sect. 6.4.1)	Error message
	$\mathbf{T}_s \equiv \mathbf{I}_R \ \forall s \in \mathcal{S}$	
	Separate estimation	
	Computationally the fastest	
	The x_k should be scaled	
`I.tolerances = FALSE`	Case 2 (the default)	Case 3
	$\widehat{\mathbf{T}}_s = \widehat{\mathbf{T}} \ \forall s \in \mathcal{S}$, but may not be positive-definite	$\widehat{\mathbf{T}}_s$ are unequal, but may not be positive-definite
	Joint estimation	Separate estimation
	Computationally memory-hungry and the slowest	

6.3.2 Initial Values and the `isd.latvar` Argument

Initial values require some comment. It is possible to efficiently obtain an initial $\mathbf{C}$ based on an equal-tolerances Poisson model. This method is the default because of `Use.Init.Poisson.QO = TRUE`. The user can bypass this by assigning to the `Cinit` argument a $p_2 \times R$ matrix. If `Use.Init.Poisson.QO = FALSE` and `Cinit` is not assigned a value, then VGAM will choose some random normal variates. Users should therefore use `set.seed()` with different seeds before running the same code, and thus try to ensure the global solution is obtained.

The solution of a lower-rank QRR-VGLM might be used for initial values for a higher-rank model. For example, for a rank-R model, try $\mathbf{C}_R^0 = (\widehat{\mathbf{C}}_{R-1}, \boldsymbol{\varepsilon})$ where $\boldsymbol{\varepsilon} \sim N_{p_2}(\mathbf{0}, \sigma^2 \mathbf{I}_{p_2})$.

The `isd.latvar` argument specifies the initial standard deviation of the latent variable values $\boldsymbol{\nu}_i$. It is used to scale the columns of the initial $\mathbf{C}$, and is used only if `I.tolerances = TRUE`. Because all species' tolerances are unity, it is easy to picture how spread out the site scores are relative to the response curves. For example, the effect of several values of `isd.latvar` is illustrated in Fig. 6.5. It can be seen that as `isd.latvar` increases, the range of the sites scores increases relative to species' tolerances. That is, there is more environmental range in the data as `isd.latvar` increases. An another example, Fig. 1.1a has a value of about 5 whereas the value for Fig. 1.1b is about 0.3.

In practice, reasonable values of `isd.latvar` might be between 2 to 10, say. Values less than 2 correspond to little range of the environmental space relative to the species' distributions, and in such cases, CQO is very difficult and probably should not be attempted. The argument should actually be of length R, and it is recycled to this length if necessary. Each successive value should be less than the previous one, e.g., `c(4, 2)` might be appropriate for a rank-2 problem. This is because the first ordination axis should have the greatest spread of site scores. Each successive ordination axis will have less explanatory power compared to previous axis, hence a decreasing `isd.latvar` sequence. If convergence failure occurs, then try varying this argument somewhat, e.g., `isd.latvar = c(6, 2)` or

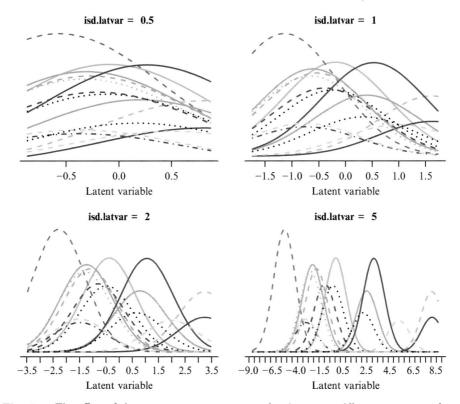

Fig. 6.5 The effect of the argument `isd.latvar` on the site scores. All response curves have unit tolerances (because $\mathbf{T}_s = \mathbf{I}_R \ \forall s$), and the optimums are located the same relative distance from each other. The site scores are uniformly distributed over the latent variable space, and have been scaled to have a standard deviation `isd.latvar`. The tick marks are at the same values.

`isd.latvar = c(2, 1)`, because good initial values are usually needed for QRR-VGLMs. Big data sets with a lot of species collected over a wide range of environments should warrant larger values of `isd.latvar`.

A related argument is `MUXfactor`, which stands for 'multiplicative factor'. If any offset value are greater than `MUXfactor * isd.latvar[r]` in absolute value, then the rth ordination axis is scaled so that the standard deviation of the site scores is `MUXfactor * isd.latvar[r]`. This is why it is a good idea for the site scores to be centred at 0—and this can be achieved if all the variables in $\boldsymbol{x}_2$ are centred at 0. The reason for `MUXfactor` is that `optim()` may perform a line search at a value of $\mathbf{C}$ that gives a very large spread of site scores. If values are too large, then numerical difficulties will occur. Usually a value of `MUXfactor` between 3 or 4 should be alright. If not, then the value should be decreased slightly.

6.3.3 Estimation

QRR-VGLMs are estimated using an algorithm different from RR-VGLMs, and it is more difficult for a number of reasons. Firstly, the log-likelihood may contain local maximums, therefore a local solution may be obtained instead of the global solution. Thus the model should be fitted several times with different starting values, to increase the chances of obtaining *the* solution. Consequently, if the solution does not look right, then try fitting the model several more times and/or adjusting some arguments and/or transforming the environmental variables. The argument `Bestof` specifies how many different initial values are to be used, and this should be assigned a reasonable integer value (e.g., at least 10). Secondly, the estimation is more prone to numerical difficulties, especially as S increases. Thirdly, it can be many times more numerically intensive than RR-VGLMs, therefore `trace = TRUE` is currently the default (so that the function does not appear to freeze while running).

Internally, VGAM currently employs `optim()`, which is a general-purpose optimization function. It uses the BFGS quasi-Newton method to minimize the total deviance over all the species, as a function of the elements of **C**. Given **C**, a GLM is fitted to each species with x_1 and ν as explanatory variables. The code calls C functions to evaluate the logistic and Poisson regressions more quickly. The software design is not ideal, and it may be improved in the future. Care in programming must be taken to ensure that the GLMs converge even when the optimizer attempts to evaluate the objective function at an extreme value of **C** because numerical problems can easily occur.

It is mentioned in passing here that an equal-tolerances Poisson QRR-VGLM can be approximated by fitting a reduced-rank multinomial logit model (RR-MLM or stereotype model; Sect. 5.2.3).

Standard errors for $\widehat{\mathbf{A}}$, $\widehat{\mathbf{C}}$, $\widehat{\mathbf{B}}_1$ and $\widehat{\mathbf{C}}$ are presently too difficult to compute; bootstrapping may be a solution here.

6.4 Post-Fitting Analyses

Fitted `cqo()` objects have class `"qrrvglm"`. Once a QRR-VGLM has been fitted, the generic functions listed in Table 6.6 may be applied to the fit. Their use should be seen in the light of the three uses of Table 6.3.

6.4.1 Arguments `varI.latvar` *and* `refResponse`

Section 6.2.2 described two favourable normalization properties that a CQO might have, called Properties (A) and (B). Sometimes it is desired to renormalize a fitted model, e.g., for an unequal-tolerances model, one of the species has a tolerance matrix of $\mathbf{I}_R$, but it is desired to change the species that has this. Rather than refitting a new model, there are two arguments `varI.latvar` and `refResponse` (Table 6.4) which allow a renormalization of the original fit. These arguments are available in some of the generic/methods functions such as `Coef()` and `persp()`.

The `varI.latvar` argument specifies whether (6.9) holds or not. If it has the value `TRUE`, then the site scores are uncorrelated, and have a standard deviation along each ordination axis equal to unity. With this option, species' tolerances can be compared with the amount of variability of the data set. If `FALSE`, then $\widehat{\text{Var}}(\widehat{\nu}_i)$ is simply diagonal, i.e., the site scores are uncorrelated, but each ordination axis has a different standard deviation of sites scores. The ordination axes are sorted so that the standard deviation of sites scores never increases. In other words, $\widehat{\text{Var}}(\widehat{\nu}_i)$

is always a diagonal matrix, and the elements along the diagonal are either all 1s or a decreasing sequence.

The argument `refResponse` specifies which species or response is to be chosen as the *reference species*. This designated species then has a tolerance matrix equal to $\mathbf{I}_R$. Setting a value different from what the original fit chose will only have an effect for Case 3, which is when each species has its own tolerance matrix. In practice, the reference species could be chosen as the dominant species, or some species that all other species can be compared with. The default value of `NULL` means that the software searches from the first to the last species, and chooses the first one with a positive-definite tolerance matrix.

For all three cases, the general algorithm is, in sequential order, as follows.

 (i) If necessary, find a reference species.
 (ii) If possible, transform it so that its tolerance matrix is $\mathbf{I}_R$.
(iii) Transform, by rotation, the site scores to be uncorrelated (i.e., $\widehat{\text{Var}}(\widehat{\boldsymbol{\nu}}_i)$ is diagonal).
(iv) If `varI.latvar = TRUE`, then scale the ordination axes so that the standard deviation of the site scores are unity (i.e., (6.9)), and the reference species' tolerance matrix is diagonal.

Let's look at what this algorithm does in the three individual cases.

$\boxed{\text{Case 1}}$ This is the nicest of all cases. All tolerance matrices are equal and positive-definite because they are all $\mathbf{I}_R$. The general algorithm gives a unique solution, and the first ordination axis has the greatest spread of the sites scores (as measured by the standard deviation), followed by the second ordination axis, etc. That is, (6.9) does not hold—the matrix is diagonal only. The general algorithm results in an ordination diagram where distances have their intuitive meanings. If the distribution of the sites scores and optimums are spread out much more on the first ordination axis compared to the second axis, then this suggests that a rank-1 ordination should suffice.

$\boxed{\text{Case 2}}$ This case is equivalent to Case 1 if the (common) estimated tolerance matrix is positive-definite. Ideally, this is so. If not, then the general algorithm will only return uncorrelated ordination axes. If `varI.latvar = TRUE`, then (6.9) will hold.

$\boxed{\text{Case 3}}$ This case is the most arbitrary. Each species has its own tolerance matrix, which may or may not be positive-definite. The reference species ends up with an $\mathbf{I}_R$ tolerance matrix, and the site scores are uncorrelated but have a different standard deviation along each ordination axis. Choosing the reference species does make a difference for $R = 2$: the ordination diagram is rotated so that the elliptical contours of the reference species has semi-major and semi-minor axes parallel to the ordination axes. Other species will generally have semi-major and semi-minor axes that are not parallel to the ordination axes.

A species whose tolerance matrix is not positive-definite will not have an optimum or maximum—they are assigned an `NA` value.

6.4.2 Ordination Diagrams

A major goal of CQO is to produce an ordination diagram (or *latent variable plot*), which is practical for ranks $R = 1$ and 2. They enable one to explore relationships among the site scores $\widehat{\boldsymbol{\nu}}_i$, the environmental variables $\boldsymbol{x}_2$, and the optimums $\widehat{\boldsymbol{u}}_j$. Relative abundances can also be read off CQO diagrams. In VGAM, latent variable plots are implemented by the generic function `lvplot()`.

A CQO latent variable plot for rank $R = 1$ such as Fig. 6.2a is straightforward: the x-axis is $\widehat{\nu}$ and the y-axis is $\widehat{\mu}$ or $\widehat{\eta}$. For a CQO diagram for a rank $R = 2$ model, the x-axis is $\widehat{\nu}_1$ and the y-axis is $\widehat{\nu}_2$. However, ideally, one would want directions and Euclidean distances to have a natural meaning, as well as latent variables that are uncorrelated. All these ideals are met when an equal-tolerances assumption is made. To see this, note that distances between points on an ordination diagram must be viewed in terms of the quantity $(\boldsymbol{\nu} - \boldsymbol{u}_j)^T \mathbf{T}_j^{-1} (\boldsymbol{\nu} - \boldsymbol{u}_j)$ in (6.2). This term is like a squared Mahalanobis distance, therefore proximities must be viewed with respect to the contours associated with the bell-shaped response surfaces defined by (6.2). This is because, for example, the abundance or probability of occurrence of a species decreases with distance from its optimum. We wish the Mahalanobis distance and Euclidean distance to coincide by having $\mathbf{T}_j = \mathbf{I}_R$.

If `eq.tolerances = FALSE` is used for a rank-2 model, then it is necessary to interpret the latent variable plot with reference to the elliptical contours. Then the user would have to compute Mahalanobis distances to correctly interpret distances!

As an example of a $\mathbf{T}_j = \mathbf{I}_R$ ordination diagram, consider one for `p2et.hs` that is given in Fig. 6.6. It is resultant from

```
> lvplot(p2et.hs, ellipse = 0.95,  label = TRUE, xlim = c(-3, 5.7),
         C = FALSE, Ccol = "brown", sites = TRUE, scol = "gray50",
         pcol = "blue", pch = "+", chull = TRUE, ccol = "gray50", main = "(a)")
> lvplot(p2et.hs, ellipse = FALSE, label = TRUE, xlim = c(-3, 5.7),
         C = TRUE, Ccol = "brown", sites = TRUE, scol = "gray50",
         pcol = "blue", pch = "+", chull = TRUE, ccol = "gray50", main = "(b)")
```

If $\widehat{\boldsymbol{u}}_s$ is outside the convex hull of the site scores, then the ordination becomes more difficult, and there is a good reason for dropping that species from the regression. For example, the optimum of `Arctperi` is clearly seen to lie outside the convex hull, therefore there would be a lot of uncertainty (statistical error) associated with this species. It should probably be dropped. The species `Pardnigr` also lies outside the convex hull, albeit only a little and only with respect to one ordination axis. This species is still difficult to model accurately in 2-dimensions, and its inclusion in an analysis would be more subjective; possibly it could be considered in a rank-1 ordination only.

Now

```
> (allvariance <- diag(latvar(p2et.hs)))

  [1] 0.72092 0.35305

> sqrt(allvariance)

  [1] 0.84907 0.59418
```

therefore one might say that 67.1 percent of the variance in the site scores is explained by the first ordination axis.

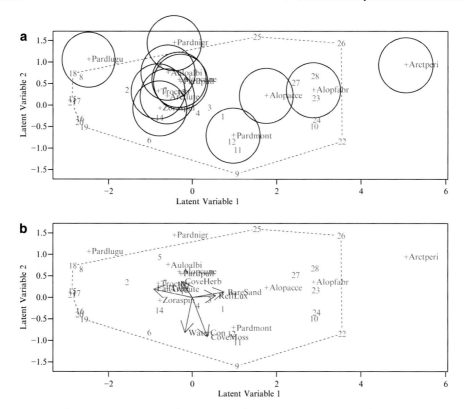

Fig. 6.6 Ordination diagrams of 12 species from the `hspider` data frame; the Poisson model `p2et.hs` with equal tolerances. A convex hull surrounds the site scores. In (**a**), the *circles* indicate the abundance of each species at 95% of its maximum abundance. In (**b**), the *arrows* display the contribution of each environmental variable towards each of the ordination axes.

For a rank-2 equal-tolerances model, the default output of `lvplot()` will give a 'natural' latent variable plot, in the sense that distances between points will be subject to their intuitive interpretation (the closer they are, the more similar). This is because the contours of the ellipses are scaled so that they are circular. Consequently, $\widehat{\mathrm{Var}}(\widehat{\boldsymbol{\nu}}_i)$ will be diagonal. In order for the latent variable plot to not look misleading, the aspect ratio of the graph should be unity, i.e., the sides of the graph must be scaled so that the circular contours do actually appear circular. On a computer screen, this is easy since it simply entails resizing the graphics window using the mouse. If `eq.tolerances = TRUE`, then these latent variable plots are computed by rotating the species so that their tolerance matrices are diagonal, and then the canonical axes are stretched/shrunken so that the estimated $\mathbf{T}_j$ are now $\mathbf{I}_R$.

One of the beauties about CQO is that there is a lot of information that can be gleaned from the fit. Here are some examples applying to Fig. 6.6.

1. The latent variable plot suggests that Sites 1, 11 and 12 have similar and high abundances of `Pardmont`; indeed, the counts are 60, 95 and 96, respectively. The model gives this species a maximum of 70.6, therefore the (response) residuals are a mixture of small and large values.

2. From (6.6), if $\widehat{\mathbf{T}} = \mathbf{I}_2$ in a Poisson CQO, then the estimated mean abundance one tolerance unit away from the optimum is about 40 percent less than the species' maximum. This comes about by, e.g.,

```
> exp(-0.5 * (1:3)^2)

  [1] 0.606531 0.135335 0.011109
```

More generally, the *absolute* abundance of a species at a location that is a distance k away (in latent variable units) from the species' optimum is $\exp(-\frac{1}{2}k^2)$ multiplied by the species' maximum. For example, $\|\widehat{\boldsymbol{\nu}}_6 - \widehat{\boldsymbol{u}}_{\texttt{Pardmont}}\| \approx 2$ implies the absolute abundance of `Pardmont` at Site 6 has been reduced by a factor of about 86%. Since its maximum is 70.6, then it is expected to have about 10 counts. In fact, the observed value is 11.

3. In general, if Site i_2 is c times more far from a species optimum $\boldsymbol{u}$ than Site i_1 ($c > 1$), then

$$\frac{A_{i_1}}{A_{i_2}} = \exp\left\{\frac{c^2 - 1}{2} \|\boldsymbol{\nu}_{i_1} - \boldsymbol{u}\|^2\right\}, \tag{6.10}$$

where A_i is the absolute abundance of the species at site i.

4. *Relative* abundances of *one species at two sites* can be readily read off from the CQO biplots. For example, let A_3 and A_4 be the absolute abundance of `Trocterr` at Sites 3 and 4, respectively. Then, because $\widehat{\mathbf{T}} = \mathbf{I}_2$, we have $A_3 = \exp(\alpha_{\texttt{Trocterr}} - \frac{1}{2}\|\boldsymbol{\nu}_3 - \boldsymbol{u}_{\texttt{Trocterr}}\|^2)$ and $A_4 = \exp(\alpha_{\texttt{Trocterr}} - \frac{1}{2}\|\boldsymbol{\nu}_4 - \boldsymbol{u}_{\texttt{Trocterr}}\|^2)$, where the Euclidean distances are measured in latent variable units. In Fig. 6.6, it can be seen that $\|\widehat{\boldsymbol{\nu}}_4 - \widehat{\boldsymbol{u}}_{\texttt{Trocterr}}\| \approx 1.27$ and $\|\widehat{\boldsymbol{\nu}}_3 - \widehat{\boldsymbol{u}}_{\texttt{Trocterr}}\| \approx 1.07$, therefore $\widehat{A}_4 / \widehat{A}_3 \approx \exp\left\{\frac{1}{2}(1.19^2 - 1^2)\right\} \approx 1.25$. That is, Site 4 is expected to have about 1.25 times the counts of `Trocterr` than Site 3. In fact, the actual ratio is $86/66 \approx 1.3$.

5. In the same vein, relative abundances of *two species at one site* can be readily read off from the CQO diagrams. Let A'_s be the absolute abundance of Species s at Site i. Then, because $A'_s = \exp(\alpha_s - \frac{1}{2}\|\boldsymbol{\nu}_i - \boldsymbol{u}_s\|^2)$, we have

$$\frac{A'_{s_1}}{A'_{s_2}} = \exp(\alpha_{s_1} - \alpha_{s_2}) \cdot \exp\left\{\frac{1}{2}\left(\|\boldsymbol{\nu}_i - \boldsymbol{u}_{s_2}\|^2 - \|\boldsymbol{\nu}_i - \boldsymbol{u}_{s_1}\|^2\right)\right\}. \tag{6.11}$$

When the maximums of the two species s_1 and s_2 are equal (i.e., $\alpha_{s_1} = \alpha_{s_2}$), then the first exponential can be ignored. To give a simple example, it is noted that, from the output from the rank-1 model, the maximums of `Pardmont` and `Pardnigr` are very similar. Now $\|\widehat{\boldsymbol{\nu}}_{14} - \widehat{\boldsymbol{u}}_{\texttt{Pardmont}}\| \approx \|\widehat{\boldsymbol{\nu}}_{14} - \widehat{\boldsymbol{u}}_{\texttt{Pardnigr}}\| \approx 1.78$ so that Site 14 is expected to have approximately the same `Pardmont` and `Pardnigr` counts. In fact, the counts are 14 and 15.

6.4.3 Perspective Plots

For rank-2 models with $\boldsymbol{x}_1 = 1$, the response surface of any subset of the species can be plotted as a perspective plot using the generic function `persp()`.

For rank-1 models with $\boldsymbol{x}_1 = 1$, `persp()` will produce a plot similar to `lvplot()`, but with the fitted curves smoothed out. We saw previously an

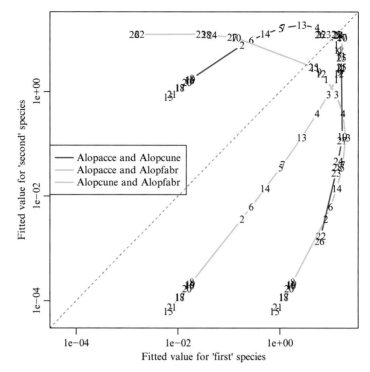

Fig. 6.7 Trajectory plot of three hunting spiders species. A rank-1 Poisson CQO is fitted to these. Site numbers have been placed on each curve.

example in Fig. 6.2b. The choice between `lvplot()` and `persp()` then depends on the purpose of plotting them; `lvplot()` is 'closer' to the data set, while `persp()` can easily be wrongly interpreted as the 'truth'.

6.4.4 Trajectory Plots

For rank-1 models, some authors make use of *trajectory plots* or *isocline plots*, whereby the estimated abundances of two species are plotted as a curve in two-dimensional space (sometimes referred to as "species space"). Trajectory plots are often employed in ecological theory. Here, they are suitable for rank-1 models, and they plot the fitted values of pairwise combinations of species. If S is large, then it is wise to select only a few species to plot. A log scale on both axes is often more effective. Here is an example.

```
> myxlim <- c(0.5e-4, 20)
> tr.hs <- trplot(plut.hs.2, which.species = 1:3, log = "xy", type = "b", lty = 1,
              col = mycols, lwd = 2, label = TRUE, xlim = myxlim, ylim = myxlim)
> legend("left", lwd = 2, lty = 1, col = mycols,
        with(tr.hs, paste(species.names[, 1], species.names[, 2], sep = " and ")))
> abline(a = 0, b = 1, lty = "dashed", col = "gray50")  # A useful reference line
```

The plot (Fig. 6.7) shows the trajectories of the first three species of an equal-tolerances Poisson CQO model fitted to the hunting spiders. The site labels have been added to the trajectories, and this enables the ordering of the sites along the gradient to be read off. It is important in trajectory plots that both x- and y-axes use identical scales for easier interpretation. To aid this, using a logarithmic scale is recommended.

In analogy to the curve $(x(t), y(t))$ parameterized by t, the mathematical equation for the path of the trajectory plot (parameterized by ν and assuming that both axes are on a log scale) is

$$(x(\nu), y(\nu)) = \left(\alpha_1 - \frac{1}{2} \left(\frac{\nu - u_1}{t_1} \right)^2, \quad \alpha_2 - \frac{1}{2} \left(\frac{\nu - u_2}{t_2} \right)^2 \right) \qquad (6.12)$$

for $A < \nu < B$, say. From (6.12), the furthest extent the curve reaches along the x-axis is at α_1, and it corresponds to $\nu = u_1$. Similarly, the highest point of the curve along the y-axis is at α_2, and it corresponds to $\nu = u_2$. It is thus easy to see which species has a higher maximum. (For example, the curve for "Alopcune and Alopfabr" gets closer to the RHS edge than to the top edge, therefore $\alpha_1 > \alpha_2$, i.e., Alopcune has a higher maximum than Alopfabr). The trajectory is anticlockwise and clockwise for $u_1 < u_2$ and $u_2 < u_1$, respectively. For example, it is clockwise for "Alopacce and Alopcune" and anticlockwise for the other two.

Note that if both species' response curves are identical ($\mu_1 = \mu_2$) then the curve will lie on the line $x = y$, i.e., have unit slope. This is the dashed line in Fig. 6.7.

Trajectory plots can, theoretically, be applied to 3 species at a time if a z-axis is added, but currently the software limit is 2 axes.

6.4.5 Calibration

Calibration is of vital importance in many applications in the biological sciences. Here, it refers to estimating values of the latent variables ν_i at a site to be estimated, given the species data y_i there. For example, when the environmental variables are of greater interest than the species data, e.g., bio-monitoring such as monitoring pollution levels based on species' responses. Here, the tenet is that "species automatically integrate environmental conditions over time". In other situations, it is impossible to measure the environmental variables, e.g., reconstruction of climate history from fossil records. Calibration can be particularly important because measuring species abundances is often much easier than measuring certain environmental variables. In this section, we show how the regression methods of Table 6.2 can be used to perform calibration. In particular, maximum likelihood calibration is implemented using the generic function calibrate(), where methods functions for CQO and CAO objects have been written for these.

With CQO we cannot estimate all the environmental variables there, x_0, in general unless $R = p_2$, and this is rarely the case. However, it is feasible to estimate ν_0, the site score, or value of the latent variables ν, at the site. An ecological application of calibration is to plot calibrated values onto an ordination diagram, to see how close they are. Values far apart imply that the joint species distributions are very different. Thus calibration can be used as a measure of the distance between two sites' species compositions. This is illustrated in the example below (Fig. 6.8).

We will use maximum likelihood calibration using fitted CQO response functions. It makes the fundamental assumption that the species are statistically independent—this will be quite unreasonable in some applications because of symbiotic relationships such as competition and mutualism. Regardless of its realism, there seems little one can do otherwise! In practice, it is usually impossible to know the dependency amongst the species, especially as S becomes large. That is, the interspecific associations cannot be modelled well, in general.

The basic idea is as follows. For discrete independent responses, the calibrated site score is defined as the solution to the problem

$$\widetilde{\nu}_0 \; = \; \max_{\boldsymbol{\nu}} \; P[\boldsymbol{Y} = \boldsymbol{y}_0 | \boldsymbol{\nu}] \; = \; \max_{\boldsymbol{\nu}} \sum_{s=1}^{S} \log P[Y_s = y_{0s} | \boldsymbol{\nu}]. \tag{6.13}$$

In VGAM, the generic function calibrate() can be thought of as the opposite of predict() because predict() starts from $\boldsymbol{x}_0$ and outputs $\widehat{\boldsymbol{\eta}}_0$ and $\widehat{\boldsymbol{\mu}}_0$, whereas calibrate() starts from $\boldsymbol{y}_0$ and outputs $\widetilde{\boldsymbol{\nu}}_0$ (and possibly $\widetilde{\boldsymbol{\eta}}_0$ and $\widetilde{\boldsymbol{\mu}}_0$, evaluated at $\widetilde{\boldsymbol{\nu}}_0$). An example of its use is

```
fit <- cqo(cbind(spp1, spp2, spp3) ~ rainfall + temperature + soilpH + humidity,
           poissonff, data = myframe, Rank = 2)
y0 <- data.frame(spp1 = c(5, 12), spp2 = c(45, 65), spp3 = c(0, 2))
calibrate(fit, y0)
```

which would return a 2-row matrix of $\widetilde{\boldsymbol{\nu}}_0$ estimates.

Maximum likelihood calibration has the nice theoretical property that, under regularity conditions, the calibrated value converges to the truth as the number of species increases indefinitely: $\widetilde{\boldsymbol{\nu}}_i \to \boldsymbol{\nu}_i$ as $S \to \infty$.

As an example, we fit a rank-1 equal-tolerances Poisson CQO to the hunting spiders data set, minus 2 randomly chosen sites i. Then we estimate the site scores there in two ways: (i) forming $\widehat{\boldsymbol{\nu}}_i = \widehat{\mathbf{C}}^T \boldsymbol{x}_{2i}$ from the CQO; (ii) calibration to give $\widetilde{\boldsymbol{\nu}}_i$. In the machine learning literature, one can consider the 2 sites as a test sample, and the remainder as the training data.

```
> set.seed(1234)   # For reproducibility
> hspider[, 1:6] <- scale(hspider[, 1:6])   # Standardize the environmental variables
> N <- 2   # Number of sites to calibrate for
> test.index <- sample(nrow(hspider), size = N)   # N randomly chosen sites
> p1et.hs.m2 <- cqo(cbind(Alopacce, Alopcune, Alopfabr, Arctlute,
                Arctperi, Auloalbi, Pardlugu, Pardmont,
                Pardnigr, Pardpull, Trocterr, Zoraspin) ~
            WaterCon + BareSand + FallTwig + CoveMoss + CoveHerb + ReflLux,
            quasipoissonff, data = hspider[-test.index, ], trace = FALSE)
> cal.p1et.hs <- calibrate(p1et.hs.m2, hspider[test.index, ])
>
> # Add the calibrated sites scores to a perspective plot
> S <- ncol(depvar(p1et.hs))   # Number of species
> clr <- rep(c(1:6, 8), len = S)   # Omits yellow
> persp(p1et.hs, las = 1, col = clr, label = TRUE)
> abline(v = cal.p1et.hs, col = 1:N, lty = 1:N, lwd = 1)   # Calibrated values
> C.matrix <- concoef(p1et.hs)   # Constrained coefficients C
> abline(v = as.matrix(hspider[test.index, rownames(C.matrix)]) %*% C.matrix,
        col = 1:N, lty = 1:N, lwd = 2)   # From CQO fit
```

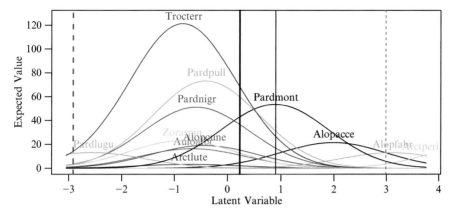

Fig. 6.8 Two calibrated sites from a rank-1 equal-tolerances Poisson QRR-VGLM fitted to the hunting spiders data without those 2 sites. The *thick vertical lines* are the CQO sites scores $\widehat{\nu}_i$, and the *thinner vertical lines* are the calibrated sites scores $\widetilde{\nu}_i$. Each i has the same colour and line type.

This gives Fig. 6.8. Now the species compositions at the calibrated sites are

```
> hspider[test.index, colnames(depvar(p1et.hs))]
```

	Alopacce	Alopcune	Alopfabr	Arctlute	Arctperi	Auloalbi	Pardlugu	Pardmont
4	2	6	0	1	0	24	1	7
17	0	0	0	0	0	0	2	0
	Pardnigr	Pardpull	Trocterr	Zoraspin				
4	29	94	86	25				
17	0	0	23	2				

and it is not difficult to explain the results. The calibrated value for Site 4, $\widetilde{\nu}_4$, is not too far from its CQO site score $\widehat{\nu}_4$ because of its high abundance of species such as `Trocterr` and `Pardpull`. In contrast, Site 17 has almost no hunting spiders apart from `Trocterr`, hence it appears at the very LHS of the plot, far away from $\widehat{\nu}_{17}$. This example illustrates the obvious: calibration can perform poorly when there is insufficient data.

6.5 Some Practical Considerations

CQO is not a robust methodology, and its sensitivity to departures from the underlying statistical assumptions is a major shortcoming. The models are also computationally expensive and prone to numerical difficulties. Consequently, fitting QRR-VGLMs requires finesse, and a lot of prior experience modelling with GLMs is beneficial. To help ameliorate these problems, some practical suggestions are given, which are roughly ordered sequentially. In general, using the technique requires care, preparation and thought—*much more* than other ordination methods.

1. **Data suitability** The data needs to cover a wide enough environmental range so as to exhibit unimodal behaviour, e.g., Fig. 1.1a, and not Fig. 1.1b. For example, are there sites where the temperature is too cold, and sites that are too hot for species to flourish? Otherwise a CLO might be possible, although this is not biologically sensible for inferences beyond the range of the data.

2. **Subsetting**   A small and careful selection of a subset of species and environmental variables is more likely to lead to success than feeding in all the data without careful pre-screening. Choose a small subset of dominant species and environmental variables which are likely to conform to the model reasonably well. For example, omit species that are rare or have very narrow tolerances. Later models can build upon the simple ones by the possible addition of species and variables.

   Especially for larger data sets, initially work with a simple random sample of the sites in order to reduce the computation. This is particularly true for CAO, but CQO is expensive too. For example,

   (i) The number of species should be kept reasonably low, e.g., 12 maximum. Feeding in 100+ species wholesale is a recipe for failure. Choose a few species carefully, e.g., 5 well-chosen species is better than 20 species thrown in willy-nilly.

   (ii) If the number of sites is large, then choose a smaller random sample to do the model building, e.g., $n = 500$ maximum. This will reduce the memory requirements and time expense of the computations, while constructing a suitable model.

   (iii) The number of explanatory variables should be kept low, e.g., 7 maximum. If the fitted constrained coefficient for a particular covariate is ≈ 0 in all CQO models, then consider dropping it permanently from future models.

3. **Pre-processing x_2**   Outliers and heavy skew, etc. will easily annul the method, hence we need to pre-process the environmental variables well. This involves:

   (a) Removal of outliers and influential sites (high-leverage points). This entails visually examining plots of each variable. Highly correlated variables should be avoided, e.g., with the `trap0` data analysis of Sect. 6.6, variables `Rain` and `LevelTW` are both measures of the water level, therefore only one of them is included in the model `p1ut.to`.

   (b) Scaling of each environmental variable is highly recommended, e.g., `scale()`d to mean 0 and unit variance. This, of course, applies to all the x_2 variables apart from the factors (grouping variables). If `I.tolerances = TRUE` is chosen, then scaling is almost necessary.

   (c) Transform variables in x_2 to reduce heavy skew and increase the symmetry. For example, suppose that $x_2 = (x_{21}, x_{22}, x_{23})^T = $ (temperature, soil pH, Nitrogen concentration)^T where each variable is centred and scaled. A CAO diagram shows that all species' response curves are right-skewed. Then replacing x_{23} by $\log(x_{23})$ is tried and the subsequent CAO diagram shows a marked improvement in symmetry of all response curves. Then a CQO model is fitted to the transformed data set to obtain better estimates of the optimums, maximums and tolerances, etc. This example is based on Palmer (1993), who advocated using the logarithm transformation for soil chemical data. In general, CAO is best used as an exploratory tool and to help CQO work better, and inferences based on CQO instead are recommended. The function `log1p()` for $\log(1 + x_k)$ may be more suitable than `log()` to handle 0s.

4. **Processing y**   Each species should be screened individually first, e.g., for presence/absence data, is the species totally absent or totally present at all

sites?—`sort(colMeans(data))` can help screen for such species. Species with only a handful of presences or a handful of absences should probably be omitted from a binomial CQO analysis. For counts, be on the lookout for outliers.

5. **Fit RR-VGAMs first** One should obtain a CAO (Chap. 7) first to see whether the species display unimodal responses. Some responses may be identified as unsuitable for CQO, and omitted. Unless the responses are unimodal and bell-shaped, it is senseless fitting QRR-VGLMs to these.

6. **On fitting QRR-VGLMs** The following is advised.

   (a) **Rank-1 first, possibly rank-2** A rank-1 model should always be attempted first. Often, a rank-1 model is all that is warranted, even after much work. Only if the data is very amenable should a rank-2 model be attempted. Obtaining a good rank-2 model is no easy task, in general, since there are a lot of things that can go wrong; suitable techniques are needed to alleviate them.

   If the 'truth' is of rank-R, then fitting a higher-rank model will give numerical problems, because the site scores will lie in an R-dimensional subspace. For example, if $R = 1$ but a rank-2 model is fitted, then the site scores and optimums will lie effectively on a line. (That is, provided one manages to fit the model in the first place). The tolerance and/or canonical coefficients for the second ordination axis would then be unstable and difficult to estimate.

   (b) **Fit an equal-tolerances model first** This can be done as Case 1 and Case 2 (Table 6.7). Ideally both should be attempted and shown to be the same.

   (c) **Fit equal-tolerances and unequal-tolerances models** It is a good idea to try fit equal-tolerances and unequal-tolerances models, and compare them. When $R = 2$, the ordination diagram of the equal-tolerances model is more easily interpreted because elliptical contours are required for the unequal-tolerances model—otherwise it would be susceptible to misinterpretation. The plot's aspect ratio should be adjusted correctly. More details are given in Sect. 6.4.2.

   (d) For `I.tolerances = TRUE`, as well as scaling the x_k, a careful choice of values for the argument `isd.latvar` is important, as well as `MUXfactor` (Sect. 6.3.2).

   (e) The liberal use of `set.seed()` be made to ensure reproducibility of results. As many starting values as possible should be used, to safeguard against obtaining local solutions.

7. **Practice on simulated data** It is advisable to gain experience with simulated data first, because the 'truth' is known. The experimental function `rcqo()` may be used to generate data conforming to the CQO model, with and without certain assumptions of the species packing model. Also, the hunting spiders data is a nice, realistic and small data set that is worth the time and effort to learn to model well using CQO and CAO, before attempting the analysis of other data sets.

8. **Read the online help and software changes** Unfortunately, the present implementation is not static, and hopefully it will be improved over time. Some of the software defaults may change.

9. **Closing comments** Fitting CQO models well require a substantial amount of GLM experience, as well as an understanding of the basic mathematics and

assumptions behind the models. Numerical problems often are indicative that the model and data do not agree. In the context of CQO, attempting to fit bell-shaped curves/surfaces to data that isn't is likely to be a frustrating experience.

6.6 A Further Example: Trout Data

The data frame `trap0` contains the counts of 2 fish species, rainbow (*Oncorhynchus mykiss*) and brown (*Salmo trutta*) trout, trapped at Lake Otamangakau in the central North Island of New Zealand. Brown trout were introduced from Europe via Tasmania, Australia, in the 19th century, whereas rainbows came from California, a state of USA adjacent to Mexico. Brown trout are actually quite wily, whereas rainbows are a lot dumber and easier to catch. In this example, we segregate the males and females in order to create 4 'species' or responses. The data were collected daily from about April to September each year during 2005–2012 inclusive.

We use the following as explanatory variables: minimum and maximum ambient temperatures (°C), water level (0 = none, 100 = flooding situation), day of the year (1 = January 1st, ..., 365 = December 31st), and the year represented as a factor. The non-factors are scaled prior to analysis to make them more comparable, at least in magnitude. It is shown in Sect. 7.3.2 that the 4 curves arising from a CAO are indeed approximately symmetric bell-shaped, hence a CQO is justified (fitting an RR-VGAM first is always recommended). Here, our intent is just fitting a CQO.

```
> trap0 <- transform(trap0, sc.doy     = scale(doy),
                            sc.LevelTW = scale(LevelTW),
                            sc.MinAT   = scale(MinAT),
                            sc.MaxAT   = scale(MaxAT),
                            f.year     = factor(Year))
> set.seed(123)
> p1ut.to <- cqo(cbind(BFTW, BMTW, RFTW, RMTW) ~
               sc.doy + f.Year + sc.MinAT + sc.MaxAT + sc.LevelTW,
               eq.tolerances = FALSE, family = poissonff,
               trace = FALSE, Bestof = 10, Crow1positive = TRUE,
               data = trap0)
```

Now the constrained coefficients of this rank-1 unequal-tolerances Poisson QRR-VGLM are

```
> round(concoef(p1ut.to), digits = 2)   # sc.doy is by far the largest coefficient

              latvar
sc.doy          2.21
f.Year2006      0.13
f.Year2007     -0.30
f.Year2008     -0.21
f.Year2009     -0.38
f.Year2010     -0.17
f.Year2011      0.29
f.Year2012      0.11
sc.MinAT        0.07
sc.MaxAT       -0.04
sc.LevelTW      0.11
```

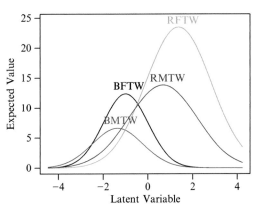

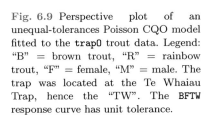

Fig. 6.9 Perspective plot of an unequal-tolerances Poisson CQO model fitted to the `trap0` trout data. Legend: "B" = brown trout, "R" = rainbow trout, "F" = female, "M" = male. The trap was located at the Te Whaiau Trap, hence the "TW". The `BFTW` response curve has unit tolerance.

The variable `sc.doy` has by far the largest coefficient, hence the fitted latent variable is interpreted largely as the time of the year. That its sign is positive means that the latent variable axis runs from left to right with the passage of time. As 2005 is the reference year, each dummy variable shows the annual shift compared to this particular year. For example, the fish tended to go upstream earlier in 2007, whereas 2011 appeared to have relatively late-season spawning runs as a whole. It would be interesting to see if these observations are correlated with the El Niña and El Niño weather cycle patterns affecting that part of the world.

A perspective plot from

```
> persp(p1ut.to, col = 1:4, label = TRUE)
```

produces Fig. 6.9. Several features that can be seen are well-known to fishermen who frequent the waters. These include the following.

(i) The brown trout peak earlier than rainbow trout. This is not surprising since the two species have quite different life cycles and ecology. As every Taupo region fisherman will tell the reader, brown trout are targeted a few months before rainbow trout.

(ii) The counts are much higher in the middle of the Julian year, viz. mid-winter. This is not surprising; starting in the autumn months (around April), the fish are known to congregate at the river mouths before heading upstream to spawn. The peak rainbow population is achieved around early Spring (October) and then it declines towards the summer months of December–February.

(iii) There are fewer jacks compared to hens. The reason is that the post-spawning survival is higher in females despite the largest biomass investment in females (c.25% of body weight in egg mass). Males invest less biomass in milt production ($< 10\%$), but they defend potential spawning grounds until complete exhaustion.

(iv) The coefficient of `sc.LevelTW` is positive. Not surprisingly, the large spawning runs are triggered by rain that raises the water level. Trout possess a magnetic strip along their sides that can measure the humidity in the air above, and this is used to help predict the likelihood of rain. Swimming through very shallow water is more hazardous!

(v) The jacks peak earlier than hens, for both species. Evidently, males defend their spawning grounds before and while the hens occupy them.

With $\widehat{\nu}$ being so dominantly comprised of one variable, it is left as an exercise to the reader to fit Poisson regressions to these data with a quadratic effect in doy only.

We now examine the tolerances:

```
> Tol(p1ut.to)[1, 1, ]

    BFTW   BMTW   RFTW   RMTW
  1.0000 1.2776 2.1519 2.3020

> sqrt(range(Tol(p1ut.to)[1, 1, ]))

  [1] 1.0000 1.5172
```

The widths of the response curves do seem to differ by a factor of about 1.5, i.e., an equal-tolerances assumption is unrealistic. Lastly, we look at the maximums.

```
> sort(round(Max(p1ut.to), digits = 1))

 BMTW BFTW RMTW RFTW
  6.6 12.4 13.8 23.5
```

The brown trout appear to have approximately half the numbers compared to rainbows at their respective optimums.

6.7 Unconstrained Quadratic Ordination

We saw from Sect. 6.1.2 that sometimes no environmental variables x_2 have been measured, hence setting $\nu_i = \mathbf{C}^T x_{2i}$ is impossible. Instead, the site scores are treated largely as free parameters which are able to take on any possible value. In this section, we show how they may be estimated by MLE. Even more so than CQO, the MLEs are very sensitive to departures from the model assumptions.

6.7.1 RCIMs and UQO

Solving for the optimal $\widehat{\nu}_i$ by maximum likelihood estimation sounds like a very difficult optimization problem, because the site scores are totally unconstrained and there are n of them. Indeed it *is* difficult, especially for binary responses and/or unequal species' tolerances. However, we now show that *unconstrained quadratic ordination* (UQO) can be performed generally by fitting an RCIM (Sect. 5.7) in order to obtain initial values, and then iterate between fitting a CQO and calibration. We might write this method by the following (which misuses notation):

$$\text{UQO} \;\approx\; \text{RCIM} + (\text{CQO} + \text{Calibration})^{\infty} . \tag{6.14}$$

The CQO may or may not assume equal tolerances, and the quality of the initial values from the RCIM is determined by how similar all the species' tolerances are (unequal tolerances means curvature when the estimated site scores are plotted against the true site scores, e.g., Fig. 6.10). In fact, simulations show that this method does not converge, and that often only one iteration is recommended.

For simplicity, consider Poisson count responses and a rank-1 model with equal tolerances ($t_j = 1$, say). Goodman's RC(1) model is

$$\eta_{ij} \;=\; \log \mu_j(\nu_i) \;=\; \beta_0 + \alpha_i + \gamma_j + c_i \, a_j, \tag{6.15}$$

(cf. (5.24)) for $i = 1, \dots, n$, and $j = 1, \dots, S$. For identifiability, let $\alpha_1 = \gamma_1 = 0$ and $a_2 = 1$ (corner constraints for $\mathbf{A}$) and $a_1 = c_1 = 0$ (structural zeros).

Now, the Poisson UQO model, with a slight change in notation, is

$$\log \mu_{ij} \;=\; \mathcal{A}_j - \frac{1}{2}\left(\frac{\nu_i - u_j}{t_j}\right)^2 \;=\; -\frac{1}{2}\nu_i^2 + \left(\mathcal{A}_j - \frac{1}{2}u_j^2\right) + \nu_i u_j. \tag{6.16}$$

Matching up terms of (6.15) with (6.16) suggests that

$$a_j \;=\; u_j \ \text{ and} \tag{6.17}$$
$$c_i \;=\; \nu_i \tag{6.18}$$

(actually, the RCIM is overparameterized but we focus on the cross-product term). With this parameterization, the parameters are scaled so that the second species' optimum is at unity, and the first site score is at the origin.

Here is the methodology illustrated using simulated data.

```
> set.seed(111)
> n <- 100; p <- 5; S <- 5
> pdata <- rcqo(n, p, S, es.opt = FALSE, eq.max = FALSE, eq.toler = TRUE,
                sd.latvar = 3/4)
> true.nu <- attr(pdata, "latvar")   # The 'truth'
> attr(pdata, "tolerances")[, 1]   # The tolerances

 y1 y2 y3 y4 y5
  1  1  1  1  1

> attr(pdata, "optimums")[, 1]   # The optimums

       y1         y2         y3         y4         y5
 -0.581651  -0.017276   1.335232  -0.160282  -0.337917

> Y <- Select(pdata, "y")   # Y matrix (n x S)
> uqo.rcim1 <- rcim(Y, Rank = 1)   # Traditional parameterization
> uqo.grc1 <- grc(Y)   # An equivalent simpler call
```

This uses the traditional parameterization of Goodman's RC model described in Chap. 5.7, which has $c_1 \equiv 0$.

The estimated species' optimums can be plotted against the true optimums as follows (ditto for the site scores).

```
> max(abs( fitted(uqo.grc1) - fitted(uqo.rcim1)))    # Should be 0

  [1] 1.3593e-05

> max(abs(predict(uqo.grc1) - predict(uqo.rcim1)))    # Should be 0

  [1] 4.1183e-06

>
> # Plot 1
> plot(attr(pdata, "optimums"), Coef(uqo.grc1)@A, col = 1:S, type = "p",
        main = "(a)")
> mylm <- lm(Coef(uqo.grc1)@A ~ attr(pdata, "optimums"))
> abline(coef = coef(mylm), col = "orange", lty = "dashed")
>
> # Plot 2
> fill.val <- 0   # Choose this for the traditional parameterization
> plot(attr(pdata, "latvar"), c(fill.val, Coef(uqo.grc1)@C),
        las = 1, col = "blue", type = "p", main = "(b)")
> mylm <- lm(c(fill.val, Coef(uqo.grc1)@C) ~ attr(pdata, "latvar"))
> abline(coef = coef(mylm), col = "orange", lty = "dashed")
```

This gives Fig. 6.10a,b. It may be seen that there is close correspondence between the fitted RCIM and the truth. The estimates are linearly related to the exact values, because of some scaling that has not yet been performed. In Fig. 6.10b, the correlation between the true site scores and the estimated UQO site scores is 0.93. Consequently, these results look quite acceptable.

Given a UQO fit, one can then fit a CQO using the $\hat{\nu}_i$ as explanatory variables, without making an equal-tolerances assumption. This can be done here with

```
> myform <- attr(pdata, "formula")
> p1ut <- cqo(myform, family = poissonff,
              eq.toler = FALSE, trace = FALSE, data = pdata)
> c1ut <- cqo(cbind(y1, y2, y3, y4, y5) ~ scale(latvar(uqo.rcim1)),
              family = poissonff, eq.toler = FALSE, trace = FALSE, data = pdata)
> lvplot(p1ut, lcol = 1:S, y = TRUE, pcol = 1:S, pch = 1:S, pcex = 0.5, main = "(c)")
> lvplot(c1ut, lcol = 1:S, y = TRUE, pcol = 1:S, pch = 1:S, pcex = 0.5, main = "(d)")
```

Here, p1ut is a CQO fit of the original data, and c1ut is a CQO fit of the UQO fit. The two plots are very similar, which is to be expected with such 'nice' data.

It is left as an exercise (Ex. 6.11) to show that the above UQO-RCIM argument holds for general rank-R, e.g., Goodman's RC(R) fits rank-R Poisson UQO. Once again, equal tolerances are assumed, and the algebra is simplest when $\mathbf{T}_j = \mathbf{I}_R$.

It was mentioned above that UQO is too difficult, in general, for presence/absence data. This is partly because binary data contains relatively little information. It is also partly because, e.g., a site with $\mathbf{y}_i = \mathbf{0}$ results in $\hat{\nu}_i \to \pm\infty$ because it wants to go to regions of the parameter space where the joint probability $P(Y_{i1} = 1, \ldots, Y_{iS} = 1)$ is zero.

Bibliographic Notes

Most books on statistical ecology that cover ordination only describe the older classical multivariate techniques, which are largely ignored here. However, Jongman et al. (1995) remains a useful overview of ordination; it describes techniques that

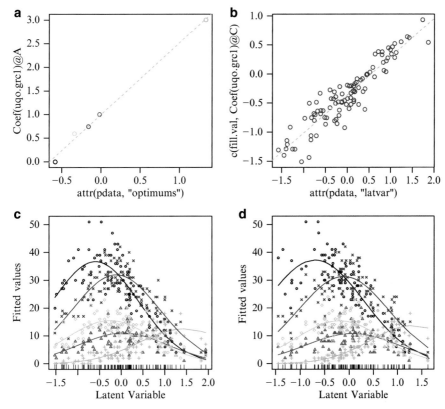

Fig. 6.10 (a)–(b) compares the UQO solution with the truth for a rank-1 Poisson simulated data set. (a) Estimated optimums $\widehat{u}_j$ versus u_j. (b) Estimated site scores $\widehat{\nu}_i$ versus ν_i. (c) CQO fitted to the original data. (d) CQO fitted to the scaled UQO site scores. In (a)–(b) the *dashed orange line* is a simple linear regression through the points.

overlap substantially with this chapter, as well as the older multivariate methods. An introduction to the subject of calibration is ter Braak (1995). This chapter is motivated by the pioneering work of ter Braak and Prentice (1988), and reflections on some topics in constrained and unconstrained ordination are given in ter Braak and Šmilauer (2015).

Zhu et al. (2005) propose an alternative algorithm for fitting CQO models, based on the equivalence of CCA and LDA (this equivalence is shown in, e.g., Takane et al. (1991), ter Braak and Verdonschot (1995)), and their algorithm can be generalized to allow for CAO. Zhang and Thas (2012) consider CQO with zero-inflated data.

The connection between Poisson UQO and Goodman's RC model is made in Yee and Hadi (2014). More generally, for a range of data types, an RCIM might be used to estimate the site scores and optimums of an equal-tolerances UQO, because both models share a common η_{ij}. A recent article on unconstrained ordination is Hui et al. (2015).

Exercises

Ex. 6.1.

(a) Show that (6.1) leads to (6.4).
(b) Show that (6.1) and (6.2) lead to (6.3).

Ex. 6.2.   Prove the theorem in Sect. 6.2.2.

Ex. 6.3.   Prove (6.10).

Ex. 6.4.   Consider the rank-1 Poisson CQO model (6.5)–(6.6).

(a) Express coefficients u_s, t_s, α_s as functions of $\beta_{(s)1}$, $\beta_{(s)2}$ and $\beta_{(s)3}$.
(b) Express the coefficients $\beta_{(s)1}$, $\beta_{(s)2}$ and $\beta_{(s)3}$ as functions of u_s, t_s, α_s.

Ex. 6.5.   Consider a rank-1 Poisson QRR-VGLM with $\boldsymbol{x}_1 = 1$. Let t_j be the tolerance of species j. Show that

$$\frac{\mu_j(u_j \pm k\,t_j)}{\mu_j(u_j)} \;=\; \exp\left\{-\frac{1}{2}k^2\right\}, \qquad (6.19)$$

where $k > 0$. This says that if you are k tolerance units away from the species' optimum then the ratio of the mean abundance there, relative to the species' maximum, declines exponentially in $-\frac{1}{2}k^2$. Show that the RHS of (6.19) is also the corresponding expression for the rank-2 equal-tolerances Poisson CQO model with $\mathbf{T}_j = \mathbf{I}_2$ when k is the Euclidean distance away from the species' optimum. Is this true for general rank-R?

Ex. 6.6.   Hunting Spiders Data—Quasi-Poisson CQO
Fit a rank-1 equal-tolerances quasi-Poisson CQO to the hunting spiders data frame `hspider`, but omit the species `Pardlugu`. Confirm that 10 of the 11 species indicate overdispersion relative to the Poisson model.

Ex. 6.7.   RR-Multinomial Logit Model and Poisson QRR-VGLMs

(a) Show that fitting an equal-tolerances Poisson QRR-VGLM to data $\mathbf{Y}$ and $\mathbf{X}$ is approximately the same as fitting a reduced-rank multinomial logit model (Sect. 5.2.3).
(b) To confirm (a) on a real data set, fit a rank-1 reduced-rank multinomial logit model to all the species and variables in the `hspider` data frame. Confirm that most of the `scale()`d constrained coefficients of the 2 models are largely in agreement.   [Yee (2006, App.C)]

Ex. 6.8.   Hunting Spiders Data—CQO

(a) Fit a rank-1 equal-tolerances Poisson CQO to the hunting spiders data, as in Sect. 6.2.1 for `p1et.hs`. Obtain a perspective plot.
(b) Convert the data into presence/absence, and fit a binomial QRR-VGLM that employs a complementary log–log link. Obtain a perspective plot.
(c) Compare the results of (a) and (b) with respect to: (i) the ordering of the optimums, (ii) the ordering of the maximums
(d) Attempt to fit a rank-2 equal-tolerances binomial CQO model. If successful, obtain an ordination diagram (and use colour to improve the presentation). Compare your plot to Fig. 6.6.

Ex. 6.9. Hunting Spiders Data—UQO

(a) Fit a rank-1 equal-tolerances Poisson CQO to the hunting spiders data, as in Sect. 6.2.1 for `p1et.hs`. Obtain a perspective plot.
(b) Fit a Goodman's RC(1) model to the same.
(c) Plot the site scores of both models. Is there a linear association? If not, can you explain why?

Ex. 6.10. `trap0` Analysis

(a) Following on from Sect. 6.6, fit a Poisson regression to each of the 4 responses with respect to `doy` only; allow for a quadratic effect so that effectively, `p1ut.to` is fitted but the gradient is the day of the year directly.
(b) For each species-sex combination, find the optimum, i.e., the day of the year that each maximum occurs. Ignore February 29.
(c) Suppose on a certain day that the following numbers of fish were trapped.

```
data.frame(BFTW = 10, BMTW = 5, RFTW = 20, RMTW = 10)
```

Suggest the most likely Julian date that this occurs on.

Ex. 6.11. Rank-R UQO by RCIMs

(a) Modify the argument in Sect. 6.7.1 to the rank-2 case. Assume $\mathbf{T}_j = \mathbf{I}_2$ for Species j.
(b) Extend the argument in (a) to the general rank-R case.

Ex. 6.12. Calibration and `hspider`

For each site separately in the `hspider` data frame, omit that site and fit a CQO to the remaining sites; then calibrate at the omitted site. Compare all the calibrated site scores with the site scores obtained by fitting one CQO to the entire data set. Comment. Note: for all of the above, fit equal-tolerances Poisson QRR-VGLMs.

Ex. 6.13. Calibration and `hspider` and Presence/Absence Data

Repeat Ex. 6.12 but convert the species data into presence/absence and fit equal-tolerances binomial QRR-VGLMs. Note: some species may have to be omitted because of numerical problems.

The reader should realise by now that, like other statistical methods, ordination techniques cannot be blindly used but require thought and experience to get the best results.
—Gower (1987)

Chapter 7
Constrained Additive Ordination

Nothing is well ordered which is hasty and precipitate.
—Lucius Annaeus Seneca

7.1 Introduction

In the previous chapter, response curves/surfaces were assumed to be bell-shaped and symmetric functions of underlying environmental gradients. The true shape of these has been alluded to by hundreds, if not thousands, of papers in the biological literature. Its determination has important implications for both continuum theory and community analysis, because many theories and models in community ecology assume that responses are symmetric unimodal (e.g., Fig. 6.1).

This chapter describes the class of *reduced-rank vector generalized additive models* (RR-VGAMs), which peforms *constrained additive ordination* (CAO). There is no need for the species packing model assumptions of Table 6.1. CAO is data-driven as opposed to CQO, which is model-driven. Allowing the data to 'speak for themselves', community ecologists can more readily explore how response curves behave as a function of the dominant gradient.

7.2 Constrained Additive Ordination

RR-VGAMs are a nonparametric extension of QRR-VGLMs, and can loosely be thought of as a GAM fitted to each species against a very small number of latent variables. The modelling function `cao()` currently implements CAO, but it is quite limited in its capabilities: to rank $R = 1$, Poisson and binary responses, and with known dispersion parameters (i.e., `poissonff()` and `binomialff()` families). Also, $\boldsymbol{x}_1$ must be an intercept term only, i.e., $\boldsymbol{x}_1 = 1$.

In more detail, suppose $R = 1$ and we have presence/absence responses. Then

```
cao(cbind(spp1, ..., sppS) ~ x2 + ... + xp,
    family = binomialff(multiple.responses = TRUE),
    data = bdata)
```

© Thomas Yee 2015
T.W. Yee, *Vector Generalized Linear and Additive Models*,
Springer Series in Statistics, DOI 10.1007/978-1-4939-2818-7_7

will fit the nonparametric logistic regressions

$$\text{logit}\,\mu_{is} \;=\; \eta_s \;=\; \beta_{(s)1} + f_s(\nu_i), \qquad s = 1, \ldots, S, \tag{7.1}$$

to optimally chosen site scores, where f_s are arbitrary smooth centred functions estimated by a smoothing spline. Actually, the current implementation absorbs the intercept $\beta_{(s)1}$ into the $f_s(\nu_i)$, as shown in Figs. 7.1 and 7.3. The variables `spp1`, ..., `sppS` should have values 1/0 for presence/absence, respectively. In (7.1), the estimated site scores are $\widehat{\nu}_i = \widehat{c}^T x_{2i}$ where $x_{2i} = (x_{i2}, \ldots, x_{ip})$. With Poisson abundance data, (7.1) has its logit link replaced by log, and the call would have `family = poissonff`.

In order to make the f_s and ν_i unique in (7.1), we can stipulate $\widehat{\text{Var}}(\nu_i) = 1$ (cf. (6.9); this is actually the **VGAM** default), and specify whether the first coefficient of $\widehat{c}$ is positive (the default) or negative. Choosing the negative sign reflects the ν axis at the origin.

With RR-VGAMs, there is no clear definition of the tolerance; compared to CQO, this is one disadvantage of CAO. Maximums and optimums can be defined, however,

$$u_s \;=\; \underset{\nu \in \mathcal{H}}{\arg\max}\ \eta_s(\nu) \tag{7.2}$$

where $\mathcal{H}$ is some region encompassing the data, such as a convex hull surrounding the site scores. At least in one dimension, the optimum should not be defined too close to the boundary, e.g., if $\widehat{u}_s = \min(\widehat{\nu}_i)$ or $\max(\widehat{\nu}_i)$ then an `NA` is returned. The maximum for Species s still retains its definition as before as $\mu_s(u_s)$.

Another example of a rank-2 Poisson CAO model is

$$\log \mu_{is} \;=\; \beta_{(s)1} + f_{1(s)}(\nu_{i1}) + f_{2(s)}(\nu_{i2}), \qquad s = 1, \ldots, S, \tag{7.3}$$

where the smooth functions $f_{r(s)}(\nu_r)$ are centred for identifiability. The two latent variables

$$\nu_1 \;=\; c_{(1)}^T x_2 \quad \text{and} \quad \nu_2 \;=\; c_{(2)}^T x_2$$

are ideally rotated to make them uncorrelated. Unfortunately, this model cannot be fitted easily yet. Similarly, (7.3) can be naturally extended to R gradients.

7.2.1 Controlling Function Flexibility

Importantly, the `df1.nl` argument controls how smooth the functions f_s are. An *effective nonlinear degrees of freedom* (ENDF; Sect. 2.4.7.4) value of 0 means that a function is linear, resulting in a RR-VGLM for that species and the performing of a constrained linear ordination (CLO). As the nonlinear degrees of freedom increases, the smooth can become more wiggly. A value between 0 and 3 (say) is suggested. However, a common mistake is to allow too much flexibility to the curves, especially when the data set is not too large. As the nonlinear degrees of freedom increases, the optimization problem becomes more difficult because of the increased number of local solutions. Crudely, a value of about 1.5 might give the approximate flexibility of a quadratic, hence `df1.nl = 1.5` might give approximately the same qualitative results as a CQO, but with the advantage that the model is data-driven. The `df1.nl` argument allows different species to have different nonlinear degrees of freedom, e.g., `df1.nl = c(1.5, spp8 = 2, spp4 = 1.75)` means that Species 8 and Species 4 have 2 and 1.75 nonlinear degrees of freedom, respectively, and all other species have 1.5 nonlinear degrees of freedom. Assigning a `df1.nl` value that is too low for a species may result in a lack-of-convergence problem in the IRLS algorithm. For this, a remedy is to assign a slightly larger value.

Overfitted models adapt too closely to the particular data set on hand and to its specific nuances, thus they do not generalize well to other future data sets. For many data sets, it should be possible to find a nonlinear degrees of freedom value somewhere between 0 and 3 that gives a suitable fit for each species. Data sets with more sites might be given slightly higher values.

7.2.2 Estimation

As with the internals of `cqo()`, VGAM currently employs `optim()`. Given c, the ν_i are computed, and a GAM fitted to each response separately. Like VGAMs, the optimization may be justified by maximizing a penalized log-likelihood similar to (4.47). The time cost of `cao()` is usually substantially more than `cqo()` for the same number of responses and explanatory variables.

Approximate standard error bands about the $\widehat{\eta}_s$ in (7.1) can be obtained by computing the estimated site scores and fitting an ordinary GAM to the species. However, the resulting standard error bands will be too narrow, due to not taking account of the variability in the constrained coefficients. An example is given in Sect. 7.3.3. Improved standard errors are an area for future work.

7.2.3 Practical Advice

Fitting CAO models requires as much finesse as fitting CQOs, and prior GAM experience is needed to avoid making poor choices. The following advice may be considered supplementary to Sect. 6.5.

(1) Keep the nonlinear degrees of freedom low, e.g., 0–2, and possibly up to 2.5 or 3. Excessive values make the optimization problem unstable and fraught with a proliferation of local solutions. Unfortunately, the constrained coefficients $\widehat{c}$ can be very sensitive to the amount of wiggliness afforded to the smooths, hence several attempts are generally required to choose reasonable values. Typically, the user cycles in a loop of plotting the component functions and refitting with revised smoothing parameters.

(2) Keep the number of species, sites and explanatory variables low, e.g., $S \leq 10$, $n \leq 500$ and $p \leq 5$. Because the computations are even more expensive than CQO, sites may be sampled and a few species chosen judiciously during the model-building process. The full data set may be used towards the end of the analysis.

(3) Try a large range of initial values. This is because the log-likelihood has increasingly more local solutions as the smooths are allowed to be more flexible.

(4) As with cqo(), set.seed() should be used beforehand if reproducible answers
are to be obtained—the initial values depend on random numbers. To get a
guage on how stable the solution is, try sort(deviance(caoModel, history
= TRUE)).

7.3 Examples

7.3.1 Hunting Spiders

We perform a rank-1 Poisson CAO to the hunting spiders data set, to confirm that
the unequal-tolerances CQO of Sect. 6.2.1 is reasonably justified. This might be
done in an indirect manner by examining the residuals of p1ut.hs.2, however we
do it more directly by

```
set.seed(1)
hspider[, 1:6] <- scale(hspider[, 1:6])   # Standardize the environmental variables
p1cao.hs <- cao(cbind(Alopacce, Alopcune, Alopfabr, Arctlute, Arctperi, Auloalbi,
              Pardlugu, Pardmont, Pardnigr, Pardpull, Trocterr, Zoraspin) ~
           WaterCon + BareSand + FallTwig + CoveMoss + CoveHerb + ReflLux,
           poissonff, data = hspider, Rank = 1,
           df1.nl = 2, Bestof = 10, Crow1positive = FALSE)
```

Then the estimated component functions can be seen (Fig. 7.1) by

```
> plot(p1cao.hs, lcol = "blue", lwd = 2, ylim = c(-5, 5), xlab = "", ylab = "")
```

Most species' component functions appear quadratic and concave, therefore they
would be suitable for inclusion in a CQO. The species **Pardlugu** might be omitted, as well as **Arctperi**, since they appear to deviate from the parabolic shape
expected in a QRR-VGLM. The estimated constrained coefficients are

```
> round(concoef(p1cao.hs), digits = 2)

              latvar
    WaterCon  -0.15
    BareSand   0.23
    FallTwig  -0.41
    CoveMoss   0.16
    CoveHerb  -0.14
    ReflLux    0.26
```

and not surprisingly, these largely agree with the 'equivalent' CQO with respect
to their signs (p1ut.hs.2 in Sect. 6.2.1).

7.3.2 Trout Data

In Sect. 6.6 a rank-1 unequal-tolerances Poisson QRR-VGLM was fitted to
the trap0 data with a few environmental variables and the day of the year.
There were 4 responses: rainbow and brown trout, crossed with males and females.

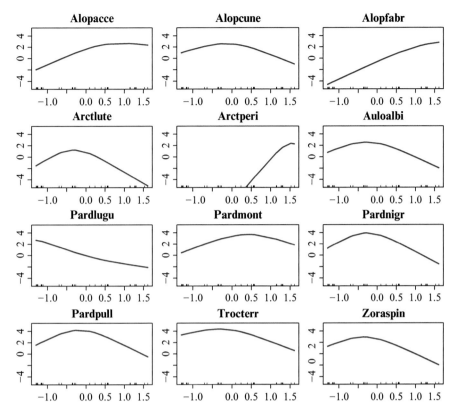

Fig. 7.1 Estimated (uncentred) component functions of `p1cao.hs`, a rank-1 Poisson RR-VGAM fitted to the `hspider` data. The x-axis is $\widehat{\nu}$.

They are the counts of such fish, trapped daily during the non-summer months. In this section, we confirm that the response curves are indeed symmetric bell-shaped with respect to the dominant gradient. As before, we scale all covariates to zero mean and unit variance, so that the relative importance of the x_k can be compared, with at least a little justification, when looking at $\widehat{c}$. The following is quite numerically intensive.

```
> trap0 <- transform(trap0, sc.doy    = scale(doy),
                            sc.LevelTW = scale(LevelTW),
                            sc.MinAT   = scale(MinAT),
                            sc.MaxAT   = scale(MaxAT),
                            f.year     = factor(Year))
> set.seed(123)  # For reproducibility
> p1cao.to <- cao(cbind(BFTW, BMTW, RFTW, RMTW) ~
                  sc.doy + f.Year + sc.MinAT + sc.MaxAT + sc.LevelTW,
                  family = poissonff, df1.nl = 2.5, trace = FALSE, data = trap0)
```

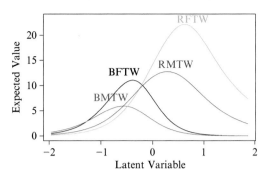

Fig. 7.2 Perspective plot of a rank-1
Poisson RR-VGAM fitted to the trap0
data. The latent variable of this CAO,
which is predominantly the day of the
year, has unit variance. The responses are
combinations of male and female rainbow and brown trout, all captured at the
Te Whaiau Trap of Lake Otamangakau.
See also Fig. 6.9.

Then

```
> Coef(p1cao.to)

  C matrix (constrained/canonical coefficients)
              latvar
  sc.doy        0.9711
  f.Year2006    0.0448
  f.Year2007   -0.1362
  f.Year2008   -0.0884
  f.Year2009   -0.1719
  f.Year2010   -0.0682
  f.Year2011    0.1335
  f.Year2012    0.0537
  sc.MinAT      0.0282
  sc.MaxAT     -0.0127
  sc.LevelTW    0.0532


  Optimums and maximums
        Optimum Maximum
  BFTW    -0.39     11.1
  BMTW    -0.56      5.9
  RFTW     0.62     22.2
  RMTW     0.28     12.7


  Nonlinear degrees of freedom
        df1.nl
  BFTW     1.5
  BMTW     1.5
  RFTW     1.5
  RMTW     1.5
```

The overall interpretation is in conformity with the biology of the data. The bulk of
the constrained coefficients is weighted on the day of the year, hence the seasonality
effect is by far the largest component. One can see this in the perspective plot
(Fig. 7.2)

```
> index <- 1:ncol(depvar(p1cao.to))
> persp(p1cao.to, col = index, label = TRUE, las = 1)
```

The shapes are clearly unimodal with a great deal of symmetry, therefore a CQO would be quite acceptable for these data.

7.3.3 Diseases in a Cross-Sectional Study

In Sect. 5.6.1 an RR-multiple binomial model was fitted to some common diseases in a large cross-sectional study of working New Zealanders. In this example, we allow for smoothing to explore the effect of the binary psychological variables. Because of the inability of `cao()` to handle $x_1 \neq 1$, we restrict this analysis to a more homogenous subset of middle-aged European-type males who were not current smokers, and had no family history of cancer or heart disease.

```
> xs.nz.em <- subset(xs.nz, 45 < age  & age < 55 & sex == "M" & smokenow == 0 &
            ethnicity == "European" & fh.cancer == 0 & fh.heartdisease == 0)
> sort(colMeans(xs.nz.em[, c("asthma","cancer","diabetes","heartattack","stroke")],
            na.rm = TRUE))  # Disease prevalences

      stroke heartattack    diabetes        cancer       asthma
   0.0034483   0.0068493   0.0137457     0.0547945    0.0756014
```

Then

```
> set.seed(123)
> b1cao.xs <-
    cao(cbind(asthma, cancer, diabetes, heartattack, stroke) ~
        depressed + embarrassed + fedup + hurt + miserable +    # 11 psychological
        nofriend + moody + nervous + tense + worry + worrier,  # variables
        df1.nl = 1.25, Crow1positive = FALSE, trace = FALSE,
        binomialff(multiple.responses = TRUE), data = xs.nz.em)
> nrow(depvar(b1cao.xs))  # n

  [1] 276

> sort(deviance(b1cao.xs, history = TRUE))

  [1] 328.14 328.14 328.14 328.14 328.14 328.14 328.14 328.15 328.19 328.65
```

The answer seems quite stable here. The estimated constrained coefficients are

```
> round(sort(concoef(b1cao.xs)[, 1]), digits = 2)

   depressed      nervous       fedup embarrassed        worry     nofriend
       -1.42        -0.99       -0.69       -0.34        -0.13        -0.08
      worrier         hurt   miserable        tense        moody
        0.03         0.15        0.90         1.19         1.56
```

This is more difficult to interpret than in Sect. 5.6.1, since there is a mixture of large positive and negative loadings. A plot of the fitted component functions from

```
> plot(b1cao.xs, ylim = c(-10, 0), lcol = "blue", las = 1)
```

appears as Fig. 7.3. The interpretion of $\hat{c}$ is made easier by this plot, because the slopes for the diseases are mainly positive and negative. It appears that `miserable`, `tense` and `moody` are positively associated with `cancer`, `diabetes` and `stroke`. Likewise, `depressed`, `nervous` and `fedup` appear positively associated with `heartattack` and `asthma`. Approximate pointwise ± 2 SE bands about the functions are given in Fig. 7.4.

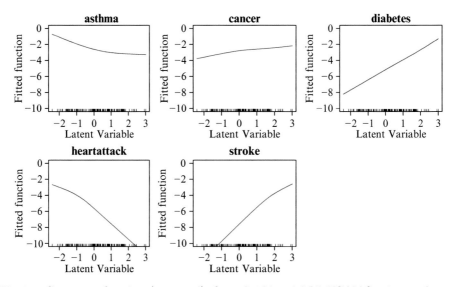

Fig. 7.3 Component functions (uncentred) of a rank-1 binomial RR-VGAM fitted to a subset of the **xs.nz** data. The latent variable of this CAO is a linear combination of 11 binary psychological variables.

We have already seen that plotting the component functions of a CAO to see if they are quadratic in shape is an informal way to select species that might be suitable for CQO analysis. It is possible to do better. If the functions were indeed quadratic, then we would expect the first derivative of the fitted functions to be linear with respect to $\widehat{\nu}$. This can be done as follows.

```
> lvdata <- data.frame(latvar1 = c(latvar(b1cao.xs)))
> check.b1cao.xs <-
      vgam(depvar(b1cao.xs) ~ s(latvar1, df = 1.25),  # Must be same as the original
           binomialff(multiple.responses = TRUE), data = lvdata, trace = FALSE)
> plot(check.b1cao.xs, lcol = "blue", scol = "orange", se = TRUE, ylim = c(-4, 4),
       las = 1)
> plot(check.b1cao.xs, lcol = "blue", deriv = 1)
```

The idea is to fit a VGAM to the species against the site scores, thus mimicking the CAO. But for VGAMs, one can plot the first derivatives of the component functions. The bottom row of Fig. 7.4 shows $\widehat{f}_j'(\widehat{\nu})$, although they are likely to be affected by considerable uncertainty.

7.4 Some Afterthoughts

CAO requires more powerful computing and fairly specialized training to master, and can exhibit fragility with dirty data. However, it provides the ability to explore the data. For inference, CQO models are preferred over CAO in an analogous manner that GLMs are preferred over GAMs for the same reason. CAO models can help obtain better CQO models because it allows the data to "speak for itself". From the examples, it is readily apparent that CAO diagrams allow a data-driven,

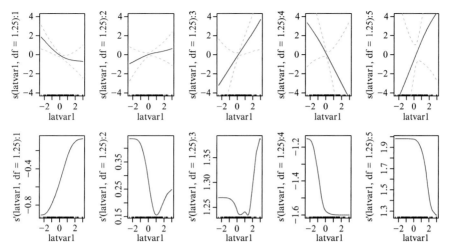

Fig. 7.4 *Top row*: component functions (centred) of a rank-1 binomial RR-VGAM fitted to a subset of the `xs.nz` data, cf. Fig. 7.3. The latent variable of this CAO is a linear combination of 11 binary psychological variables. *Bottom row*: first derivatives of the respective fitted functions.

investigative and interactive examination of the data. Altogether, RR-VGAMs are a natural class of models that allow constrained ordination to be performed in an exploratory manner.

RR-VGAMs are related to other regression models proposed in the literature. In particular, single-index models (Ichimura, 1993) have

$$Y = f(\beta_1 x_1 + \cdots + \beta_p x_p) + \varepsilon,$$

for some smooth function f. Here, $x_1 \neq 1$ because f includes any shift in location and level, and the coefficients β_p are subject to some normalization, for uniqueness. This might be called a rank-1 RR-VAM. The generalization of this to count and presence/absence data and beyond would be a rank-1 RR-VGAM.

Bibliographic Notes

RR-VGAM-like models have been used a lot in econometrics, e.g., Yatchew (2003). Schimek (2000) also surveys a number of such models; see also Harezlak et al. (2015).

Constrained additive ordination models were proposed by Yee (2006), with specific applications to plant ecology. The alternative algorithm of Zhu et al. (2005) potentially allows CAO to be performed.

Exercises

Ex. 7.1. Consider (7.1) where $\nu = (x_2, x_3, x_4) \cdot c$, say. Obtain expressions for $\partial \mu_{is}/\partial x_k$ and $\partial^2 \mu_{is}/\partial x_k^2$, treating f_s and c as fixed.

Ex. 7.2.   Create a data frame from **hspider** comprising presence/absence data, and fit a rank-1 Bernoulli RR-VGAM with a complementary log-log link. How do your constrained coefficients compare to the results of Sect. 7.3.1?

Ex. 7.3.   Plot the fitted component functions for the **trap0** example with approximate pointwise ± 2 SE bands about the functions. Also, plot the first derivatives as a function of the site scores. Your plots should be something like Fig. 7.4. Comment.

This then, is a proof of a well-trained mind, to delight in what is good, and to be annoyed at the opposite.
—Marcus Tullius Cicero

Chapter 8
Using the VGAM Package

Results from **VGAM** *...should be treated with caution.*
—Freedman and Sekhon (2010)

8.1 Introduction

This chapter attempts to describe the **VGAM** package for **R** as a whole, and tie it in with the preceeding theory. We mainly start from scratch so that there is some duplication of previous content. Section 1.5.2 reviewed a few important facets of the S language, but more details can be found in the references at the end of this chapter. Other **VGAM** details, aimed more at programmers, appear in Chap. 18. Note that the software details presented here are subject to change.

8.1.1 Naming Conventions of VGAM Family Functions

On the face of it, **VGAM** offers a bewildering set of family functions. However, it can be seen from Fig. 1.2 that there is a lot of structure! Many of them cluster naturally in groups, e.g., as described in Chaps. 11–12. Because of such numbers, it is useful to have some naming conventions to help navigate among them. Table 8.1 gives a brief summary. Some character strings appear at the beginning of the name, others end with certain characters, including digits. While not all family functions adopt these conventions, most do.

8.1.2 Naming Conventions of Arguments

To further make **VGAM** easier for practitioners, the arguments of almost all **VGAM** family functions adhere to additional conventions. They are listed below, in no particular order. Distributions involving the form

© Thomas Yee 2015   249
T.W. Yee, *Vector Generalized Linear and Additive Models*,
Springer Series in Statistics, DOI 10.1007/978-1-4939-2818-7_8

Table 8.1 Some naming conventions for **VGAM** family functions. Prefixes appear in the upper table; suffixes in the lower table. Note: not all **VGAM** family functions follow these conventions.

Prefix	Comments
a	Asymmetric, e.g., `alaplace1()`, Sect. 15.3.2
aml	Asymmetric maximum likelihood, e.g., `amlnormal()`, Sect. 15.4
bi	Bivariate, e.g., `binormal()` for $N_2(\mu_1, \mu_2, \sigma_{11}, \sigma_{22}, \sigma_{12})$
cens.	Censored, e.g., `cens.gumbel()`, `cens.normal()`, `cens.poisson()`, `cens.rayleigh()`
d, e, p, q, r	Density, expectiles, cumulative distribution, quantile, random variates, Sect. 11.1.1
double.	Double, e.g., `double.cens.normal()`
fold	Folded, e.g., `foldnormal()`
gen	Generalized, e.g., `gengamma.stacy()`, `genpoisson()`
inv.	Inverse, e.g., `inv.binomial()`, `inv.lomax()`
loglin	Loglinear, e.g., `loglinb2()`
mix	Mixture, e.g., `mix2poisson()`, `mix2normal()`
neg	Negative, e.g., `negbinomial()`
pos	Positive (Sect. 17.1), e.g., `posbinomial()`, `pospoisson()`
rec.	Records, e.g., `rec.normal()`
sc.	Scaled, e.g., `sc.studentt2()`
seq	Sequential, e.g., `seq2binomial()`
skew	Skewed, e.g., `skewnormal()`
sm.	Smart (Table 8.3), e.g., `sm.bs()`, `sm.ns()`, `sm.poly()` `sm.scale()`
trunc	Truncated, e.g., `truncgeometric()`
uni	Univariate, e.g., `uninormal()` for $N_1(\mu, \sigma)$
za	Zero-altered (hurdle) (Sect. 17.1)
zi	Zero-inflated (Sect. 17.1)

Suffix	Comments
1, 2, 3,…	Number of parameters, e.g., `studentt()`, `studentt2()`, `studentt3()`
61, …	Year of an article, e.g., `freund61()` for Freund (1961)
cop	Copula (Sect. 13.3), e.g., `binormalcop()`
ff	"family function", e.g., `poissonff()` to avoid interference with `poisson()`, `zetaff()` rather than `zeta()` for $\zeta(x)$ (Sect. A.4.3)
I, II, III, …	Type I, Type II, …e.g., `paretoII()`
R	Two purposes: (1). 'Raw' parameter(s) such as the scale and shape (as opposed to the mean) that match R's `dpqr`-type function, e.g., `betaR()`, `gammaR()`. (2). Also, the order of the arguments may not match R's unless setting `lss = FALSE`, e.g., `weibullR()`.
.or	Odds ratio, e.g., `binom2.or()`
.rho	ρ, e.g., `binom2.rho()`
.vcm	Varying-coefficients model, e.g., `normal.vcm()`

$$\left(\frac{y-a}{b}\right)^s \tag{8.1}$$

have a as the *location* parameter, b as the positive *scale* parameter, and s as the *shape* parameter. Almost always, b has the log link, `loge()`, as the default. Note that some authors define a scale parameter as b^{-1} in (8.1), i.e.,

$$(B[y-a])^s \tag{8.2}$$

where $B = b^{-1}$ might be called a rate parameter. Dealing directly with B is possible using the `negloge()` link function, because $\log(1/\theta) = -\log\theta$ is the negative log link so that the regression coefficients would be merely negated compared to the default.

Here are the main conventions. Not all family functions fully comply, however, the vast majority do.

1. The argument names `zero`, `parallel` and `exchangeable` all have the same effect over all VGAM family functions (Table 18.6). While these are used to construct constraint matrices conveniently, there are limitations and room for conflict.
2. Scale and shape parameters are called `scale` and `shape` if the model/distribution has only one of them. If there are more, then they are usually called `scale1`, `scale2`, ..., and `shape1`, `shape2`, ...
3. Argument names for parameter link functions begin with the letter "l". For example, `lscale` or `lscale1`, `lscale2`, ... for the scale parameter(s), and `lshape` or `lshape1`, `lshape2`, ... for the shape parameter(s). The default has a character value, e.g., `lscale = "loge"` means a $\log_e$ link is default.
4. Argument names for initial values begin with the letter "i". For example, `ishape`, `ishape1`, `ishape2`, ... A default value of `NULL` means that an initial value for that parameter is computed internally (self-starting). Often the location and scale parameters are estimated from the initial shape parameter.
5. The location parameter is slightly different. When there is one, it is usually called `location` with link function `llocation`. When there is more than one location parameter, they are called `loc1`, `loc2`, ..., and serviced by `lloc1`, `lloc2`, ..., and `iloc1`, `iloc2`, ..., for initial values. If a location parameter is known or must be specified, then the default value is given in the arguments of the function.
6. With respect to the definition of $\boldsymbol{\eta}$, the default order is always location, scale, and then shape parameters. This also holds in terms of the order of the arguments. For link function and initial values they are grouped in blocks of "l" and "i".
7. In terms of the documentation, the Greek letters for denoting the location, shape, scale parameters are flexible and can be chosen according to the convention of the area. For example, in extreme value modelling, ξ is often used for the shape parameter for the GEV distribution (16.2). Another example is quantile regression where τ is used, while ω is used for expectile regression.
8. For a model/distribution with only one parameter, the argument `link` or `link.parameterName` is quite acceptable. Similarly, an argument `init` or `init.parameterName` is common too.

8.2 Basic Usage of VGAM

VGAM centres on the functions `vglm()` and `vgam()`. The former is similar to `glm()`, i.e., which has as its central ideas data frames, families, IRLS, and the formula language described in Chambers and Hastie (1991). Here are some examples:

1. `vglm(y ~ 1, family = chisq, data = cdata)`

   solves for the MLE of the parameter ν of a random sample $y_1, y_2, \ldots$ assumed from a χ^2_ν distribution. As $\nu > 0$, $\eta = \log \nu$ is the default.

2. `vglm(y ~ x2 + x3, family = cumulative(parallel = TRUE), data = cdata)`

   fits a proportional odds model (1.22) to a factor response `y` and explanatory variables `x2` and `x3`, found in a data frame called `cdata`. Whereas in `glm()` where `y` is usually a vector, `vglm()` often allows `y` to be a matrix. There is no `x1` in the formula because it is implicitly there as the intercept ($x_1 = 1$). In general, all variables ought to be stored in a data frame.

3. `vgam(y ~ s(x2) + x3 + s(x4, df = c(5, 1)), multinomial, data = mdata)`

   fits a nonparametric multinomial logit model; the "`s()`" denotes a vector smooth term which is a generalization of a cubic smoothing spline (`smooth.spline()`). The function `s()` should only be used in conjunction with `vgam()`, and then its use is only symbolic because it does not directly result in any computation. If `s(x2)` is used in the formula of a `vglm()` call, say, then it will simply return its argument, here, `x2`.

   Many options available with `glm()`/`gam()` are also available with `vglm()`/`vgam()`, e.g., `subset`, `na.action`, `trace`. Importantly, a **VGAM** object should be manipulated as much as possible via generic functions such as `coef()`, `predict()` and `summary()`. In general, the raw components of a **VGAM** object may be misleading, and they should be extracted by extractor or accessor functions where possible (Tables 8.5, 8.6, 8.7).

8.2.1 Some Miscellaneous Arguments

Some miscellaneous arguments are as follows.

extra
:   This argument enables one to pass any additional information into the innards of the IRLS algorithm. It is suitable for big structures such as vectors and matrices. In particular, `extra` appears as an argument in `@deviance()` and `@loglikelihood` as well as `@linkfun()`, and `@linkinv()`. Some family functions require select input to be inputted through `extra`.

form2
:   Unlike `lm()` and `glm()`, `vglm()` and `vgam()` have *two* formula arguments, called `formula` and `form2`. Some **VGAM** family functions require the model to be specified using two formulas, such as `normal.vcm()`. Table 8.5 describes a few details about the model matrix and response for `form2`. Sometimes the response for `form2` is optional.

Table 8.2 Certain `vglm.control()` arguments.

Argument	Comments
`criterion =` `"deviance"`	The criterion used for testing convergence. Usually is `"loglikelihood"` or `"coefficients"` in practice. Assigning `"coefficient"` will often result in a more stringent convergence criterion, giving a more accurate answer, however, half-stepping will be unavailable
`epsilon = ` 10^{-7}	Positive convergence tolerance ε in (8.3). Roughly speaking, the Fisher-scoring iterations are assumed to have converged when two successive (and scaled) `criterion` values are within `epsilon` of each other
`half.stepsizing =` `TRUE`	Is half-stepsizing is allowed? If **TRUE** and $\ell^{(a)} < \ell^{(a-1)}$ then a half-step (or $\frac{1}{4}$ or $\frac{1}{8}$ ...) is taken until an improvement is made. The argument is ignored if `criterion = "coefficients"`
`maxit = 30`	Maximum number of IRLS iterations allowed. Usually the default is adequate, however, if `stepsize` and/or `half.stepsizing` are set then this argument should be assigned a higher value
`stepsize = 1`	Usual stepsize to be taken between each Fisher-scoring iteration (α in (3.55)). It should be a value in $(0, 1]$, where a value of unity corresponds to an ordinary step. A value of 0.5 means half-steps are taken. Setting a value near zero will cause convergence to be generally slow, but may help increase the chances of successful convergence for some family functions such as the negative binomial with canonical link (Sect. 11.3.3) and the asymmetric Laplace distribution (Sect. 15.3.2)
`trace = FALSE`	Assigning **TRUE** is recommended in general. At each IRLS iteration some output is printed out that is helpful for monitoring convergence. This is particularly important for some families whose MLEs are intrinsically more difficult to estimate
`xij`	See Sect. 3.4

Argument `form2` is mandatory if the `xij` argument is used. Then `form2` should contain the union of all terms in `xij` and `formula`. It transpires that, from the large model matrix $\mathbf{X}_{\text{form2}}$, select columns are extracted in order to build $\mathbf{X}_{\text{VLM}}$.

8.2.2 Constraints

As described in Sect. 3.3, the `constraints` argument may be assigned a list with a constraint matrix $\mathbf{H}_k$ for each term. Here are some examples.

1.
```
fit.npom <- vglm(y ~ x2, cumulative(parallel = FALSE), data = cdata)
```

   stops the parallelism constraint from being applied to any of the explanatory variables. The result is a fully non-proportional odds model, which is susceptible to numerical problems if the fitted η_j do intersect inside the data cloud. Equivalently,

```
fit.npom <- vglm(y ~ x2, cumulative(parallel = TRUE ~ -1), data = cdata)
fit.npom <- vglm(y ~ x2, cumulative(parallel = TRUE ~  0), data = cdata)
```

—these are actually more recommended as they are more informative: one cannot determine whether a constraint applies to an intercept or not from a single logical.

2. Both

```
myform <- y ~ x2 + s(x3, 3) + s(x4) + x5
vgam(myform, binom2.or(exchangeable = TRUE  ~ s(x3, 3) + x5), data = bdata)
vgam(myform, binom2.or(exchangeable = FALSE ~ x2 + s(x4) - 1), data = bdata)
```

make the effect of X_3 and X_5 to be the same for both marginal probabilities. Explicitly, the model they fit is

$$
\begin{aligned}
\text{logit } p_1 &= \beta_{(1)1}^* + \beta_{(1)2}^* x_2 + f_{(1)3}^*(x_3) + f_{(1)4}^*(x_4) + \beta_{(1)5}^* x_5, \\
\text{logit } p_2 &= \beta_{(1)1}^* + \beta_{(2)2}^* x_2 + f_{(1)3}^*(x_3) + f_{(2)4}^*(x_4) + \beta_{(1)5}^* x_5, \\
\log \psi &= \beta_{(2)1}^* + \beta_{(3)2}^* x_2 + f_{(2)3}^*(x_3) + f_{(3)4}^*(x_4) + \beta_{(2)5}^* x_5.
\end{aligned}
$$

3. ```vgam(y ~ x2 + s(x3), binom2.or(zero = 3), data = bdata)```

constrains the third linear/additive predictor to equal an intercept term only, i.e., $\eta_3 = \beta_{(3)1}$. In this case the odds ratio is simply a point estimate and not a function of the covariates. In general, **zero** may be assigned a vector of integers in the range 1 to M. If multiple responses are handled, then **zero** may be assigned a vector of negative integers in the range -1 to $-M_1$ which is applied to *each* response. Note that the exchangeability constraint applies to the intercepts, whereas the parallelism constraint usually doesn't.

4. What happens when there is a contradiction of arguments such as

```
cm <- diag(M)
fit <- vglm(ymatrix ~ x2 + x3, myVGAMfamfun(parallel = TRUE),
            constraints = list("(Intercept)" = cm, x2 = cm, x3 = cm))
```

i.e., which constraint is applied to x_2, say? The answer depends on the model and what the programmer feels has higher precedence. However, since it is more difficult to specify **constraints**, this should override values produced by the constraints vector **parallel**. Note that, at present, VGAM requires the constraints list to be fully specified, i.e., *all* terms must be included in the list. Unfortunately, there is little to no internal consistency checking at this stage. This deficiency might be addressed in the future.

It should be noted that the arguments **parallel**, **exchangeable** and **zero** are merely quick methods of constructing the $\mathbf{H}_k$ and assigning it to the **constraints** argument internally—this is used in a similar manner **contrasts** is used for factors in `lm()`/`glm()`/`gam()`—but for VGAMs it is used to constrain the functions. See Sect. 3.3 for more details.

8.2.3 Control Functions

The functions `vglm()` and `vgam()` come with `vglm.control()` and `vgam.control()`, respectively. These provide default values for algorithmic variables to test for convergence, e.g., **maxit** is the maximum number of IRLS iterations, **epsilon** is the tolerance in the convergence criterion between two successive iterations. Table 8.2 gives some details about the most useful `vglm.control()` arguments.

Users can supply values when invoking `vglm()`/`vgam()`, e.g.,

```
vgam(y ~ s(x2), aVGAMfamily, data = adata, maxit = 4, epsilon = 1e-9)
vgam(y ~ s(x2), aVGAMfamily, data = adata,
     control = vgam.control(maxit = 4, epsilon = 1e-9))
```

8.2.4 Convergence Criteria

In contrast to `glm()`, which iterates until the change in *deviance* is sufficiently small, VGAM allows a number of alternative quantities for testing convergence. The most common three are

- deviance, D, chosen by `criterion = "deviance"`,
- log-likelihood, ℓ, chosen by `criterion = "loglikelihood"`,
- coefficients, $\boldsymbol{\beta}$, chosen by `criterion = "coefficients"`.

The reason for allowing a variety of criteria is because many models do not have an expression for the deviance. For a few others, a log-likelihood does not exist. Testing convergence by examining the change in $\boldsymbol{\beta}^{(a)}$ is the one method that is applicable to all VGAM models, and therefore is the last choice if the family function can compute the others. Should they exist, the deviance is computed in `@deviance` of the family function, and log-likelihood in `@loglikelihood`. VGAM will look in the above order for a function with the specified name in the family function. As a last resort, it will use the regression coefficients if there is no objective function.

Some of these arguments can be seen in Table 8.2:

```
> args(vglm.control)

  function (checkwz = TRUE, Check.rank = TRUE, Check.cm.rank = TRUE,
      criterion = names(.min.criterion.VGAM), epsilon = 1e-07,
      half.stepsizing = TRUE, maxit = 30, noWarning = FALSE, stepsize = 1,
      save.weights = FALSE, trace = FALSE, wzepsilon = .Machine$double.eps^0.75,
      xij = NULL, ...)
  NULL
```

Regardless of the convergence criterion used, the deviance and log-likelihood are computed at the final iteration, if possible, and their values stored on the object. They should be retrieved using the appropriate accessor function (Tables 8.5, 8.6, 8.7).

Currently, the exact criterion for testing the convergence of the regression coefficients is when the maximum over $k = 1, \ldots, p$ of the following is satisfied:

$$\frac{\left|\beta_{(j)k}^{*(a)} - \beta_{(j)k}^{*(a-1)}\right|}{\texttt{epsilon} + \left|\beta_{(j)k}^{*(a)}\right|} < \texttt{epsilon} \tag{8.3}$$

for all $j = 1, \ldots, \texttt{ncol}(\mathbf{H}_k)$. Here, a is the iteration number. This will ensure that each $\widehat{\beta}_{(j)k}^{*}$ will be correct to approximately d significant figures, where d is the number of zero decimal places in `epsilon`, e.g., $d \approx 3$ if `epsilon = 0.0001`. The reason for the `epsilon` in the denominator is to avoid division by zero.

In general, for a scalar quantity A such as the deviance and log-likelihood, currently VGAM will terminate when the criterion

$$\frac{\left|A^{(a)} - A^{(a-1)}\right|}{\texttt{epsilon} + \left|A^{(a-1)}\right|} < \frac{\texttt{epsilon}}{\sqrt{n}} \tag{8.4}$$

is satisfied. The justification for the $\sqrt{n}$ is that for the most common case of A, which is the log-likelihood, the normalizing constant means that the value of ℓ grows with increasing sample size. Having the $\sqrt{n}$ partially compensates for this and helps avoid premature convergence for very large sample sizes.

To round off this section, it is mentioned that the exact S expressions for (8.3) and (8.4) are the `convergence` components of `vglm.control()` and `vgam.control()`. Of course, it is possible for programmers to replace these by their own code.

8.2.5 Smart Prediction

Models involving terms with data-dependent functions such as `scale()`, `bs()` and `poly()` can sometimes give wrong predictions. To see why, consider the snippet

```
> fit <- lm(y ~ scale(x2) + bs(x3), data = ldata)
> pfit <- predict(fit, newdata)
```

The data-dependent parameters of `fit` are the mean and standard deviation of `x2`, and the knot locations of `x3`. In the prediction, these data-dependent parameters must be reused instead of naïvely computing the terms based on the new data, e.g., `scale(x2)` gets evaluated on `newdata$x2` so that its mean and standard deviation might be used instead.

Admittedly, the technology is sufficient here so that `pfit` is correct, however, counter-examples that users need to be wary of are models such as

```
fit1 <- lm(y ~ I((x - mean(x)) / sqrt(var(x))), data = ldata)
fit2 <- glm(y ~ I((x - min(x))^2), poisson, data = pdata)
fit3 <- glm(y ~ I(scale(x2)) + bs(scale(x3)) + poly(scale(x4), 2, raw = TRUE),
            binomial, data = bdata)
wrong.prediction1 <- predict(fit1, newdata1)
wrong.prediction2 <- predict(fit2, newdata2)
wrong.prediction3 <- predict(fit3, newdata3)
```

Indeed, any function such as `min()`, `max()`, `var()`, `range()` or functions computing quantiles, etc. can potentially lead to incorrect predictions. This may occur because the data-dependent parameters, when applied to the new data, will be in general different from their values with the original data. In statistical jargon, the problem is that the basis functions change but the old coefficients are used. Specifically, during the `predict()` call, the terms of `fit` are reused to construct a model frame and then a model matrix $\widetilde{\mathbf{X}}$, say, so that the new $\widehat{\boldsymbol{\eta}} = \widetilde{\mathbf{X}}\,\widehat{\boldsymbol{\beta}}$ may be computed. Unless the original data-dependent parameters are used in the computation of `pfit`, the wrong basis functions will be used by the prediction, i.e., $\widetilde{\mathbf{X}}$ is wrong.

Most calls are correctly handled using R's *safe prediction* (based on Chambers and Hastie, 1991, pp.108,288–9). Its inner workings are a little complicated. Technically, one has an S3 `makepredictcall()` methods function that will modify a term's call so that the data-dependent parameters can be passed in via arguments. These data-dependent parameters are extracted from the attributes attached to

Table 8.3 Smart functions supplied in **VGAM** for users. They currently work with `vglm()` and `vgam()`, etc., but not with `lm()` and `glm()`, etc. The bottom-most subtable is a utility function.

Smart function	Comments
`sm.bs()`	B-splines. Smart version of `bs()` in splines
`sm.ns()`	Natural cubic splines. Smart version of `ns()` in splines
`sm.poly()`	Orthogonal polynomials. Smart version of `poly()`
`sm.scale()`	Standardizing: centring and scaling. Smart version of `scale()`
`sm.cut()`, `sm.max()`, `sm.mean()`, `sm.min()`, `sm.sd()`, `sm.var()`	Not written yet—see Ex. 18.8
`is.smart()`	Logical. Is a function smart? e.g., `is.smart(sm.bs)` returns `TRUE`

the value of the function. However, terms such as those of **fit3** exposes safe prediction as inadequate because the attributes associated with the inner function call are lost in nested calls.

Fortunately, **VGAM** provides *smart* functions (Table 8.3) which potentially handles the most complex of terms. The resulting scheme is called *smart prediction*. Although most users need not be concerned with the technical details behind the scheme, they will have to be aware of terms that lead to erroneous predictions, and then call (and if necessary, write—see Sect. 18.6) the appropriate smart function. By convention, the names of these smart functions begin with an `sm.`, and presently work only in **VGAM** with modelling functions such as `vglm()` and `vgam()`. These modelling functions have a `smart = TRUE` argument because there may be times when it is certain that smart prediction is neither needed or desired, e.g., a **VGAM** family function that calls `vlm()`. The data-dependent parameters are stored on the object's `smart.prediction` slot.

As an example, the equivalent of **fit3** is

```
Fit3 <- vglm(y ~ I(sm.scale(x2)) + sm.bs(sm.scale(x3)) +
             sm.poly(scale(x4), 2, raw = TRUE), binomialff, data = bdata)
right.prediction3 <- predict(Fit3, newdata3)
```

Historically, when `lm()` was first written for **S-PLUS**, it was very difficult to decide which was better for prediction—a general solution that worked reasonably well all the time, or a specific solution (like presented here) which works perfectly well most of the time. The former option was chosen. By "most" of the time, smart prediction has a lurking danger—users can get lulled into complacency and then let an unsmart function slip in unnoticed and make errors without being aware. A careful examination of the `smart.prediction` slot is one way of checking, but this is prone to error if there are many data-dependent terms in the formula. Another check is to predict on several subsets of the original data and then compare these to the original predictions, e.g.,

```
fit <- vglm(y ~ sm.ns(x2), acat, data = adata)   # Suppose adata has more than 6 rows
max(abs(head(predict(fit)) - predict(fit, head(adata))))   # Should be 0
max(abs(tail(predict(fit)) - predict(fit, tail(adata))))   # Should be 0
```

Admittedly, this can be a little laborious and it is not entirely fool-proof.

8.3 More Advanced Usage of VGAM

8.3.1 Initial Values

VGAM family functions having the `imethod` argument usually correspond to a model or distribution that is relatively hard to fit successfully, therefore care is needed to ensure that good initial values are obtained. To exhaust all algorithms currently implemented, this argument accepts an integer with value 1 or 2 or 3, ... For example, `imethod = 1` might be the method of moments, and `imethod = 2` might be based on a grid search. For many VGAM family functions, it is advisable to try `imethod` with all possible values, and in conjunction with `trace = TRUE`, in order to safeguard against problems such as converging to a local solution.

8.3.1.1 More General Arguments

In contrast to arguments for initial values specific to a VGAM family function, there are three arguments that are common to `vglm()` and `vgam()`. They are summarized in Table 8.4. All three are related directly to VGAM output, for example, if

```
fit <- vglm(...)
```

has converged, then for most family functions,

```
vglm(..., etastart = predict(fit), trace = TRUE)
vglm(..., coefstart = coef(fit), trace = TRUE)
vglm(..., mustart = fitted(fit), trace = TRUE)
```

should result in the same model within one or two iterations. The one with `etastart` assigned is most likely to require the least cost to re-converge. Section 8.5 describes a trick whereby `etastart` may be assigned the predicted values from a simpler model in order for a more complex model to be fitted.

For many models such as univariate discrete distributions, let $\widetilde{\mu}$ be some positive measure of central tendency, e.g., the weighted mean or the median of $y_i + \frac{1}{8}$. One method is to use (Fig. 8.1)

$$\mu_i^{(0)} = s\,\widetilde{\mu} + (1 - s)\,y_i, \qquad 0 \le s \le 1. \tag{8.5}$$

Practical experience has shown that this often works quite well, for $s \approx 0.95$, say. That is, initial values for the fitted mean are a slight perturbation of the central measure towards the actual values. For some distributions, setting $s = 1$ actually gives poor performance. The argument `ishrinkage` in some family functions is the adjustment factor s.

Table 8.4 Three general arguments peculiar to `vglm()` and `vgam()`, which allow the input of initial values.

Argument	Comments
`etastart`	n-vector or $n \times M$ matrix of η_i^Ts. This is the preferred of the three
`coefstart`	$\beta^{(0)}$ in (3.9) and (3.23). Must be the correct length and order
`mustart`	n-vector or n-row matrix of μ_i^T's. This is the least preferred because it only works for a few families with $M_1 = 1$

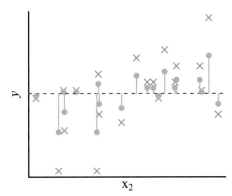

Fig. 8.1 Response y_i ($\times$) in a simple linear regression, with shrinkage initial values ($\bullet$) based on $s = \frac{1}{2}$ in (8.5). The dashed horizontal line is at $\bar{y}$, and s is argument `ishrinkage`.

8.3.2 Speeding Up the Computations

It is well-known that `glm()` calls the method `glm.fit()` to perform the actual IRLS computations. The front-end function `glm()` expends a considerable amount of resources to process the formula and data frame into the model frame and then the model matrix $\mathbf{X}_{LM}$. When doing simulations and other expensive tasks requiring a very large number of GLMs, it is possible to eliminate a lot of overhead by calling `glm.fit()` directly with model matrix and response, etc. fed in as preprocessed input.

In the same spirit, VGLMs can be fitted more quickly by calling `vglm.fit()`, since it has argument `method = "vglm.fit"`. However, this demands greater sophistication and a high level of hacking—certainly not recommended for even semi-skilled users. Now

```
> args(vglm.fit)

  function (x, y, w = rep(1, length(x[, 1])), X.vlm.arg = NULL,
      Xm2 = NULL, Ym2 = NULL, etastart = NULL, mustart = NULL,
      coefstart = NULL, offset = 0, family, control = vglm.control(),
      criterion = "coefficients", qr.arg = FALSE, constraints = NULL,
      extra = NULL, Terms = Terms, function.name = "vglm", ...)
  NULL
```

It is recommended that a VGLM be fitted the usual (slower) way once in order to set things up before attempting to call it the faster way. Here's a simple example. First generate some data and fit a model the ordinary manner.

```
> mdata <- data.frame(x2 = sort(runif(n <- 100)))   # Generate some data
> mdata <- transform(mdata, y = rnorm(n, mean = 1 + 2 * x2))
> fit0 <- vglm(y ~ x2, uninormal, data = mdata)   # Fit a model the slow usual way
```

Now extract the LM matrix with some necessary attributes and call `vglm.fit()`.

```
> quickfit <- vglm.fit(y = depvar(fit0),
#                      etastart = predict(fit0),   # Faster convergence if used
                       x = model.matrix(fit0, type = "lm"),
                       family = fit0@family,
                       Terms = terms(fit0),   # This may need hacking
                       constraints = constraints(fit0, type = "term"),
                       control = fit0@control)
```

Often convergence is faster by uncommenting the `etastart` argument.

One can measure the execution time using `system.time()`. Applying it in this case with `etastart` uncommented shows that the call to `vglm.fit()` on the author's machine was about 2.2 times faster.

One disadvantage is that the output of `vglm.fit()` is only semi-packaged compared to `vglm()`. Here, the object `quickfit` is a list with components

```
> names(quickfit)

 [1] "assign"        "coefficients"  "constraints"   "df.residual"
 [5] "df.total"      "effects"       "fitted.values" "offset"
 [9] "rank"          "residuals"     "R"             "terms"
[13] "loglikelihood" "predictors"    "contrasts"     "control"
[17] "crit.list"     "extra"         "family"        "iter"
[21] "misc"          "post"          "ResSS"         "x"
[25] "y"
```

Consequently, all the generic functions cannot be used, and further hacking may be needed to extract out the relevant components, e.g., $\widehat{\boldsymbol{\beta}}$ can be accessed as `quickfit$coefficients`. A second disadvantage is the inherent danger that calling `vglm.fit()` directly can bring—users are advised to test their code and examine their answers carefully.

The reader might well wonder why `constraints(fit0, type = "term")` is used. The answer is that terms such as `bs(x2)` will generate multiple columns in the LM matrix, whereas `vglm.fit()` expects one constraint matrix per term.

8.4 Some Details on Selected Methods Functions

Upon estimation, the fitted object should be manipulated using generic functions where possible. This section provides some details on a few such methods functions. A summary is given in Tables 8.5, 8.6, 8.7, which lists the results of many generic functions applied to fitted **VGAM** objects.

8.4.1 The fitted() Generic

Some **VGAM** family functions possess the argument `type.fitted`, which accepts a vector of character strings and offers a variety of fitted values. The first choice is always the default. For example,

```
> args(zipoissonff)

function (llambda = "loge", lonempstr0 = "logit", type.fitted = c("mean",
    "pobs0", "pstr0", "onempstr0"), ilambda = NULL, ionempstr0 = NULL,
    imethod = 1, ishrinkage = 0.8, zero = -2)
NULL
```

means that this family function returns the mean $\widehat{\phi}_i \, \widehat{\lambda}_i$ as the fitted values, by default (cf. (17.7)).

Table 8.5 **VGAM** generic functions applied to a `vglm()`/`vgam()` object called `fit`. Most are extractor functions. Those marked "‡" are the default. Many quantities are linked to equations described elsewhere. Here, $i = 1, \ldots, n$, $j = 1, \ldots, M$, $k = 1, \ldots, p$, $t = 1, \ldots, \tau$, where τ is the number of terms. Also, $S = Q/Q_1$ is the number of responses. All estimated quantities are computed at the final IRLS iteration. See also Tables 8.6, 8.7, and Table A.2- for a glossary. Let `Y` be the response(s).

Function	Value
`AIC(fit)`	Akaike information criterion (Table 9.1)
`AICc(fit)`	AIC, corrected for finite samples (Table 9.1)
`BIC(fit)`	Bayesian (Schwarz) information criterion (Table 9.1)
`coef(fit)`	$\widehat{\boldsymbol{\beta}}^*$ (Eq. (3.30)); $\widehat{\beta}^*_{(j)k}$ in (3.28) is labelled like `xk:j`
`coef(fit, matrix = TRUE)`	$\widehat{\mathbf{B}}$ (Eq. (3.29))
`Coef(fit)`	$\widehat{\boldsymbol{\theta}} = \left(\ldots, g_j^{-1}(\widehat{\beta}_{(j)1}), \ldots \right)^T$, i.e., for `Y` $\sim$ `1` only
`constraints(fit, type = "lm")`‡	$\mathbf{H}_k$, $k = 1, \ldots, p_{\mathrm{LM}}$, constraint matrices (Eqs. (3.27), (3.29), (3.38)), a list
`constraints(fit, type = "term")`	$\mathbf{H}_t$, $t = 1, \ldots, \tau$, constraint matrices per term (Eq. (3.25))
`constraints(fit, matrix = TRUE)`	$(\mathbf{H}_1 \vert \mathbf{H}_2 \vert \cdots \vert \mathbf{H}_{p_{\mathrm{LM}}})$, constraint matrices side-by-side, is $M \times p_{\mathrm{VLM}}$
`depvar(fit)`	Dependent/response variables $\mathbf{Y} = (\boldsymbol{y}_1, \ldots, \boldsymbol{y}_n)^T$ (cf. Eq. (3.2)), is $n \times Q$
`depvar(fit, type = "lm2")`	Dependent/response variables for argument `form2`
`deviance(fit)`	Deviance $D = \sum\limits_{s=1}^{S} \sum\limits_{i=1}^{n} d_{is}$, Sect. 3.5.2. Is `NULL` if unimplemented or undefined
`dfbeta(fit)`	$\widehat{\boldsymbol{\beta}}^* - \widehat{\boldsymbol{\beta}}^*_{[-i]}$, $n \times p$, Sect. 3.7.5
`df.residual(fit, type = "vlm")`‡	Residual (VLM) degrees of freedom $= n_{\mathrm{VLM}} - p_{\mathrm{VLM}}$ (Eq. (3.56)). Printed in `summary(vglmObject)`
`df.residual(fit, type = "lm")`	Residual (LM) degrees of freedom $= n_{\mathrm{LM}} - p_{(j)\mathrm{LM}}$ (Eq. (3.57)), i.e., for each η_j
`fitted(fit)`	$[(\widehat{\mu}_{ij})]$, Sect. 8.4.1
`hatvalues(fit)`	$\mathrm{diag}(\boldsymbol{\mathcal{H}})$, $n \times M$ (Eq. (3.63))
`is.parallel(fit)`	Are $\mathbf{H}_k = \mathbf{1}_M$? Section 3.3.1.2
`is.zero(fit)`	Are $\eta_j = \beta_{(j)1}$ (intercept-only)? Section 3.3.1.1
`linkfun(fit)`	g_j, the link functions for each η_j
`logLik(fit)`	Log-likelihood $\ell = \sum\limits_{i=1}^{n} w_i\, \ell_i$ (Eq. (3.7)), $\ell = \sum\limits_{s=1}^{S} \sum\limits_{i=1}^{n} w_{is}\, \ell_{is}$ for $S > 1$ responses (Eq. (3.51))
`logLik(fit, summation = FALSE)`	Log-likelihood elements $[(w_{is}\, \ell_{is})]$

Table 8.6 Continuation of Table 8.5. See also Table 8.7.

Function	Value
`model.matrix(fit, type = "vlm")`‡	VLM matrix ($nM \times p_{\mathrm{VLM}}$) $\mathbf{X}_{\mathrm{VLM}}$ ($= \mathbf{X}^*$) (Eq. (3.20))
`model.matrix(fit, type = "lm")`	LM matrix ($n \times p$) $\mathbf{X}$ ($= (\boldsymbol{x}_1, \ldots, \boldsymbol{x}_n)^T = \mathbf{X}_{\mathrm{LM}}$) (Eq. (3.18))
`model.matrix(fit, "lm", linp = j)`	Subset of $\mathbf{X}_{\mathrm{VLM}}$ corresponding to η_j
`model.matrix(fit, type = "lm2")`	VLM matrix for `form2`, $\mathbf{X}_{\mathtt{form2}}$, has elements $[(x_{2ij})]$
`nobs(fit, type = "lm")`‡	n ($= n_{\mathrm{LM}}$)
`nobs(fit, type = "vlm")`	nM ($= n^* = n_{\mathrm{VLM}}$) $= \mathtt{nrow}(\mathbf{X}_{\mathrm{VLM}})$
`nparam(fit)`	Number of parameters (Table 9.2)
`npred(fit)`	M (Eqs. (3.2), (3.5)) $= \dim((\eta_1, \ldots, \eta_M)^T) =$ total number of linear/additive predictors
`npred(fit, type = "one")`	M_1: M for one response
`nvar(fit, type = "vlm")`‡	p_{VLM} ($= p^* = \sum_{k=1}^{p} \mathtt{ncol}(\mathbf{H}_k)$)
`nvar(fit, type = "lm")`	p ($= p_{\mathrm{LM}}$)
`predict(fit, type = "link")`‡	$[(\widehat{\eta}_{ij})]$ is $n \times M$ (Eq. (3.40))
`predict(fit, type = "response")`	$[(\widehat{\mu}_{ij})]$, same as `fitted(fit)`
`QR.Q(fit)`	The $\mathbf{Q}$ matrix of a QR decomposition of $\mathbf{X}_{\mathrm{VLM}}^{**(a-1)}$ in (3.23) at the final IRLS iteration
`QR.R(fit)`	The $\mathbf{R}$ matrix of a QR decomposition of $\mathbf{X}_{\mathrm{VLM}}^{**(a-1)}$ in (3.23) at the final IRLS iteration
`resid(fit, type = "working")`‡	$\boldsymbol{z}_i - \boldsymbol{\eta}_i = \mathbf{W}_i^{-1}\boldsymbol{u}_i$, $n \times M$ (Eq. (3.58))
`resid(fit, type = "deviance")`	$\sqrt{w_i}\,\mathrm{sign}(y_i - \widehat{\mu}_i)\,\sqrt{d_i}$ or $\sqrt{w_{is}}\,\mathrm{sign}(y_{is} - \widehat{\mu}_{is})\,\sqrt{d_{is}}$. Is `NULL` if deviance is non-existent (Eq. (3.62))
`resid(fit, type = "pearson")`	$\sqrt{w_i}\,\mathbf{W}_i^{-1/2}\boldsymbol{u}_i$, $n \times M$ (Eq. (3.59))
`resid(fit, type = "response")`	$[(y_{ij} - \widehat{\mu}_{ij})]$ (Eq. (3.61))
`simulate(fit, nsim = 1)`	Simulate $\widehat{\boldsymbol{y}}_i$ (Sect. 8.4.3). Available for selected families only
`vcov(fit)`	$\widehat{\mathrm{Var}}(\widehat{\boldsymbol{\beta}}^*)$ (Eq. (3.21))
`vcov(fit, untransform = TRUE)`	$\widehat{\mathrm{Var}}(\widehat{\boldsymbol{\theta}})$ where $g_j(\theta_j) = \beta_{(j)1}$, i.e., for `Y ~ 1` only
`weights(fit, type = "prior")`‡	Prior weights w_i (usually the `weights` argument of `vglm()`) (Eq. (3.7)), or $[(w_{is})]$
`weights(fit, type = "working")`	$w_i\mathbf{W}_i$ (in matrix-band format is the ith row) (Eq. (3.11)), see Sect. 18.3.5 for $S > 1$ case

Table 8.7 Continuation of Tables 8.5, 8.6. These functions apply more to RR-VGLMs, QRR-VGLMs and/or CAOs. Here, S is the number of species/responses, and $s = 1, \ldots, S$. See also Table 6.6.

Function	Value
calibrate(fit)	Calibrate, estimate $\widehat{\boldsymbol{\nu}}_i$ from $\boldsymbol{y}_i$, Sect. 6.4.5
Coef(fit)	$\widehat{\mathbf{B}}_1$, $\widehat{\mathbf{A}}$, $\widehat{\mathbf{C}}$ for RR-VGLMs (Eq. (5.2)), plus $\widehat{\mathbf{T}}_s$, $\widehat{\boldsymbol{u}}_s$ for QRR-VGLMs (Eq. (6.2))
concoef(fit)	Constrained (canonical) coefficients $\widehat{\mathbf{C}}$ (Eq. (6.1))
hatplot()	Plot of the hat matrix diagonals (Sect. 3.7.5)
is.bell(fit)	Are the species' response curves/surfaces bell-shaped? (Eq. (6.1))
latvar(fit)	Latent variables (site scores), $(\widehat{\boldsymbol{\nu}}_1, \ldots, \widehat{\boldsymbol{\nu}}_n)^T$ where $\widehat{\boldsymbol{\nu}}_i = \widehat{\mathbf{C}}^T \boldsymbol{x}_{2i}$, $n \times R$ (Eq. (5.2))
lvplot()	Latent variable plot (ordination diagram; for $R = 1$ or 2) (Sect. 6.4.2)
Max(fit)	Maximums $\widehat{\mu}_s(\widehat{\boldsymbol{u}}_s)$ for QRR-VGLMs, S-vector, (Eq. (6.2))
Opt(fit)	Optimums $\widehat{\boldsymbol{u}}_s$ for QRR-VGLMs, $R \times S$ (Eqs. (6.1), (6.2))
persp()	Perspective plot (for $R = 1$ or 2) (Sect. 6.4.3)
Rank(fit)	R, rank (Eq. (5.2))
Tol(fit)	Tolerance matrices, for QRR-VGLMs, $R \times R \times S$ array, $\widehat{\mathbf{T}}_s$ (Eq. (6.1))
trplot(fit)	Trajectory plot (Sect. 6.4.4)

After fitting a `vglm()`/`vgam()` object, it is sometimes possible to obtain the other types of fitted values by using a call of the form

```
fitted(vglm.object, type.fitted = "another.choice")
```

In this case, the appropriate `fitted()` methods function will compute (not extract in this case) the new type of fitted value using the model's $\widehat{\boldsymbol{\eta}}_i$ values. The value of the argument here should match one in the **VGAM** family function. So continuing with this example,

```
fit <- vglm(y ~ x2 + x3, zipoissonff, data = zdata)
fitted(fit, type.fitted = "pobs0")
```

returns a 1-column matrix with elements equal to the probability of an observed 0, i.e., $\widehat{P}(Y_i = 0) = 1 - \widehat{\phi}_i + \widehat{\phi}_i \exp(-\widehat{\lambda}_i)$.

8.4.2 The summary() Generic

All good `summary()` methods functions return an object of a certain class that is printed out by some other methods function. In the case of `summary()` being applied to a `vglm()` object, the result is an object of class `"summary.vglm"`, and this is displayed by the `show()` methods function.

When displayed, the output appears similar in form to the output of a `summary()` of a `glm()` object. At the heart of the output is a 4-column matrix whose columns are: the estimates $\widehat{\boldsymbol{\beta}}^*_{(j)k}$, their standard errors $\mathrm{SE}(\widehat{\boldsymbol{\beta}}^*_{(j)k})$, the Wald statistics $(\widehat{\boldsymbol{\beta}}^*_{(j)k} - 0)/\mathrm{SE}(\widehat{\boldsymbol{\beta}}^*_{(j)k})$, and (usually) the 2-sided p-values. Currently, all the Wald statistics are assumed to follow a standard normal distribution, so that these correspond to approximate asymptotic tests. Specifically, they test $H_0 : \beta^*_{(j)k} = 0$ versus $H_1 : \beta^*_{(j)k} \neq 0$. Occasionally, for some models, the p-values are suppressed because the model's regularity conditions are not met. The matrix should be extracted by applying **coef()** to the **summary()**, e.g., **coef(summary(vglmObject))**.

Usually, the very rightmost part of the matrix is embellished with the *significance stars* characters: '.', '*', '**' and '***', depending on how diminutive the p-values are. The use of significance stars is controversial, and it has led towards statistical malpractice. They can be suppressed by, e.g.,

```
> options(show.signif.stars = FALSE)   # Global effect
> summary(vglmObject)
> summary(vglmObject, signif.stars = FALSE)   # Local effect
```

The overall residual degrees of freedom, given by **df.residual(vglmObject, type = "vlm")**, is printed towards the bottom of the output. As can be seen from Table 8.5, it is possible to obtain the residual degrees of freedom corresponding to each η_j separately by **df.residual(vglmObject, type = "lm")**.

8.4.3 The *simulate()* Generic

Many **VGAM** family functions with a corresponding **r**-type function have a **@simslot** slot so that **simulate()** can be run on the model. This allows random variates to be generated for each observation at the MLE, i.e., $\widehat{y}_i$ based on $(x_i, \widehat{\theta})$. Now

```
> args(simulate)

function (object, nsim = 1, seed = NULL, ...)
NULL
```

The most basic call to **simulate()** returns an n-row data frame with 1-column if the usual fitted value is a 1-column matrix and the model has a single response. More generally, usually there are **nsim** columns and multiples of n rows. Let $\mathcal{N}$ be the number of simulations (**nsim**).

Care must be given for models that return a multi-column matrix as fitted values. Let $\mathcal{F} = $ **ncol(fitted(fit))** be the number of columns of the fitted values of the object **fit**. Models with $\mathcal{F} > 1$ can arise from, e.g.,

- a multivariate model or distribution, e.g., **binormal()**, **dirichlet()**, and/or
- those with multiple responses.

Depending on the family function, the elements may be ordered either as in an array of dimension $n \times \mathcal{N} \times \mathcal{F}$ or $n \times \mathcal{F} \times \mathcal{N}$.

An $n \times \mathcal{N} \times \mathcal{F}$ example is the bivariate normal distribution, where $\mathcal{F} = 2$. Most **VGAM** family functions return the mean as the fitted value, therefore one would expect the means over all simulations to be approximately equal to their respective fitted values, provided that **nsim** is large.

```
> set.seed(123); n <- 100
> bdata <- data.frame(x2 = runif(n), x3 = runif(n))
> bdata <- transform(bdata, y1 = rnorm(n, 1 + 2 * x2),
                            y2 = rnorm(n, 3 + 4 * x2))
> fit1 <- vglm(cbind(y1, y2) ~ x2, binormal(eq.sd = TRUE), data = bdata)
> nsim <- 1000   # Number of simulations for each observation
> my.sims <- simulate(fit1, nsim = nsim)
> dim(my.sims)   # A data frame

  [1]   200 1000

> aaa <- array(unlist(my.sims), c(n, nsim, ncol(fitted(fit1))))   # n by N by F
> summary(rowMeans(aaa[, , 1]) - fitted(fit1)[, 1])   # Should be all 0s

      Min.    1st Qu.    Median      Mean    3rd Qu.      Max.
  -0.065900 -0.024300 -0.000190  0.000947  0.023900  0.080200

> summary(rowMeans(aaa[, , 2]) - fitted(fit1)[, 2])   # Should be all 0s

      Min.    1st Qu.    Median      Mean    3rd Qu.      Max.
  -0.05440  -0.01580   0.00209   0.00494   0.02470   0.09010
```

Note that the order of elements is such that the left-most index varies fastest for arrays (Sect. 1.5.2.7).

Here is an $n \times \mathcal{F} \times \mathcal{N}$ example, based on a zero-inflated Poisson distribution.

```
> n <- 100; set.seed(111); nsim <- 1000
> zdata <- data.frame(x2 = runif(n))
> zdata <- transform(zdata, lambda1 = loge(-0.5 + 2 * x2, inverse = TRUE),
                            lambda2 = loge( 0.5 + 2 * x2, inverse = TRUE),
                            pstr01  = logit( 0,           inverse = TRUE),
                            pstr02  = logit(-1.0,         inverse = TRUE))
> zdata <- transform(zdata, y1 = rzipois(n, lambda = lambda1, pstr0 = pstr01),
                            y2 = rzipois(n, lambda = lambda2, pstr0 = pstr02))
> fit.zip  <- vglm(cbind(y1, y2) ~ x2, zipoissonff, data = zdata, crit = "coef")
> my.sims <- simulate(fit.zip, nsim = nsim)
> dim(my.sims)   # A data frame

  [1]   200 1000

> aaa <- array(unlist(my.sims), c(n, ncol(fitted(fit.zip)), nsim))   # n by F by N
> summary(rowMeans(aaa[, 1, ]) - fitted(fit.zip)[, 1])   # Should be all 0s

      Min.    1st Qu.    Median      Mean    3rd Qu.      Max.
  -0.151000 -0.025300  0.001040  0.000171  0.017200  0.115000

> summary(rowMeans(aaa[, 2, ]) - fitted(fit.zip)[, 2])   # Should be all 0s

      Min.    1st Qu.    Median      Mean    3rd Qu.      Max.
  -0.32300  -0.04350   0.00602  -0.00352   0.04060   0.22700
```

8.4.4 The plot() Generic

The component functions of a **vgam()** object may be plotted using the generic function **plot()**. Table 8.8 is a brief summary of some of the more useful arguments. Here are some additional notes about this methods function.

Table 8.8 Some useful arguments when plotting a `vgam()` object. Some default values are given.

Argument	Comment
`deriv = 0`	Derivatives can be plotted, e.g., for 1 and 2 too (but currently only for linear and s() terms), e.g., Fig. 2.11
`lcol, scol, llwd, slwd, llty, slty`	Line colours, line width and line types for the component functions and their pointwise standard errors bands
`noxmean = FALSE`	Standard errors of the linear component of each component function at the mean value of x_k is 0 by definition. By default, this property is accentuated by explicitly adding this point (called A in Fig. 8.2a)
`overlay = FALSE`	Component functions can be plotted on the same graph or separately. If TRUE then the use of `which.cf` and/or `which.term` may be useful to select the component functions to be overlaid
`raw = TRUE`	By default, the $\widehat{f}^*_{(j)k}$ are plotted, else the $\widehat{f}_{(j)k}$ are plotted. See Sect. 8.4.5
`residuals = FALSE`	The type of residuals specified by argument `type.residuals` are plotted too if set to TRUE. This may be more informative than having `rug = TRUE`
`rugplot = TRUE`	Add a rugplot? If TRUE then these are jittered 'rug' values at the bottom of each x_k plot that denote the approximate x_{ik} values. In Fig. 8.2 there is so much data that the rug appears solid
`scale`	Makes the y-axes comparable to all component functions. Setting a (positive) value means that `ylim` will be at least `scale` in range, in all plots. An alternative is to use `ylim` directly
`se = FALSE`	Plot the pointwise ± 2 standard errors? See Sect. 4.2.3.2
`which.cf = NULL`	Specifies which component functions are to be plotted, i.e., the j for $\widehat{f}^*_{(j)k}$, e.g., `which.cf = 2` for the second one. The default value means all of them
`which.term = NULL`	Specifies which terms are to be plotted, e.g., `which.term = c("s(age)", "s(height)")`. The default value means all of them. May be numeric or a character string. If character then the white spaces must be exactly correct, e.g., `"s(age, df = 3)"` might be alright but `"s(age, df=3)"` wouldn't (it follows similarly to Sect. 3.3.1.3)

1. One can plot the component functions of a `vglm()` object as if it were a VGAM. To do this, coerce the object into an object of class `"vgam"`, e.g.,

```
plot(as(vglmObject, "vgam"), se = TRUE)
```

2. The pointwise SEs of smooth terms are currently stored in the **preplot** slot, after a call to `plot()`. A simple example involving $\eta = f_{(1)2}(x_2)$ is given in Sect. 17.2.3; its derivation is based on the decomposition (4.12).

3. The argument `noxmean = FALSE` means, by default, the SE for the linear component of a smooth function at $\overline{x}_k$ is 0. In the case that x_k is *binary* and numeric, such as for variable `sex01` in the following example, it is a good idea to set `noxmean = TRUE`. (This might be implemented in the future automatically.) The two plots, for `sex01` only, are in Fig. 8.2.

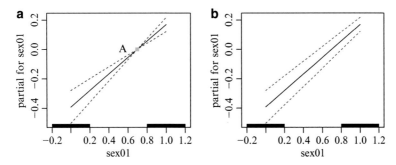

Fig. 8.2 Fitted logistic regression VGAM (the response is `nofriend`) to European-type people in `xs.nz` with (**a**) `noxmean = FALSE`, (**b**) `noxmean = TRUE`. In (**a**), the point A, which has 0 as its SE, has been explictly added to the plot, whereas it has been omitted in (**b**). The x-coordinate of A is at the mean of the variable `sex01`. The y-coordinate is 0 because component functions are centred.

```
xs.nz.e <- subset(xs.nz, ethnicity == "European")
xs.nz.e <- na.omit(xs.nz.e[, c("nofriend", "age", "sex", "ethnicity")])
xs.nz.e <- transform(xs.nz.e, sex01 = as.numeric(sex) - 1)   # 0 == female
fit1.nof <- vgam(nofriend ~ s(age, df = 3) + sex01, binomialff, data = xs.nz.e)
plot(fit1.nof, se = TRUE, scol = "blue", main = "(a)", which.term = 2) -> pfit1

# Add point A
term2 <- pfit1@preplot["sex01"]
term2 <- term2[[1]]
points(mean(term2$x), 0, col = "orange", pch = 15)
text(mean(term2$x), 0, "A", adj = 3)

plot(fit1.nof, se = TRUE, scol = "blue", main = "(b)", which.term = 2,
     noxmean = TRUE)
```

Currently, the `plot()` methods function for `"vgam"` objects calls the `predict()` methods function with `type = "terms"`.

8.4.5 The predict() Generic

The methods function for `predict()` for `"vgam"` objects allows `deriv` = a nonnegative integer value. It works only with `type = "terms"`, and currently, it only works for linear and `s()` terms.

Note that for VGAMs (cf. (3.25)),

$$\boldsymbol{\eta}_i \;=\; \boldsymbol{o}_i + \sum_{k=1}^{p} \boldsymbol{f}_k(x_{ik}) \;=\; \boldsymbol{o}_i + \sum_{k=1}^{p} \mathbf{H}_k\, \boldsymbol{f}_k^*(x_{ik}), \tag{8.6}$$

and if `raw = TRUE` then

$$\boldsymbol{\eta}_i \;=\; \boldsymbol{o}_i + \sum_{k=1}^{p} \mathbf{H}_k^{**}\, \boldsymbol{f}_k^*(x_{ik}) \tag{8.7}$$

where

$$\mathbf{H}_k^{**} \;=\; \begin{pmatrix} \mathbf{I}_{\texttt{ncol}(\mathbf{H}_k^{**})} \\ \mathbf{O} \end{pmatrix}.\qquad\qquad(8.8)$$

That is, when `raw = TRUE`, artificial constraint matrices $\mathbf{H}_k^{**}$ are constructed to pick out the *raw* component functions comprising $\boldsymbol{f}_k^*$.

If `raw = FALSE`, then the smooth functions have been premultiplied by the constraint matrices. Thus the $\widehat{f}_{(j)k}$ are plotted instead of the $\widehat{f}_{(j)k}^*$.

When using `predict(as(vglmObject, "vgam"), ..., type = "terms")`, an attribute called `attr(, "vterm.assign")` is returned. This is needed for termwise information and prediction. As well, `attr(, "constant")` is returned if `deriv = 0`: it is based on

$$\eta_j \;=\; \sum_{k=1}^p \beta_{(j)k}\, x_k \;=\; \left(\beta_{(j)1} + \sum_{k=2}^p \beta_{(j)k}\, \overline{x}_k \right) + \sum_{k=2}^p \beta_{(j)k}\,(x_k - \overline{x}_k).\qquad(8.9)$$

Because of how `"vgam"` objects `plot()`s standard errors, the model matrix is centred when `type = "terms"`; hence the modified constants. Actually, (8.9) only holds when there are trivial constraints.

8.5 Some Suggestions for Fitting VGLMs and VGAMs

Fitting VGLMs/VGAMs well requires experience and skill. For example, Zuur et al. (2012, p.86) construct constraint matrices with columns equal to $\mathbf{0}_M$—a fundamental mistake—and not surprisingly, they report "We encounter numerical optimisation problems..." (Zuur et al., 2012, p.84).

Here are some suggestions and tricks. Some of them apply to general regression modelling as a whole.

1. Examine each x_k separately to check for features such as heavy skew and outliers. Often a transformation such as a `sqrt()`, `log()` or `log1p()` is begging. The function `pairs()` is useful here.

2. Monitor convergence: set `trace = TRUE`. Many models/distributions are intrinsically difficult to fit, therefore monitoring convergence is a good idea. Increasing the maximum number of iterations allowed is usually not needed, unless the stepsize is chosen to be less than unity. Difficulty obtaining convergence is symptomatic of the data and model not agreeing. It may be due to gross overfitting or underfitting, multicollinearity, outliers or influential points, etc.

```
slowfit <- vglm(y ~ x2 + x3, aVGAMfamilyFunction, data = adata, trace = TRUE,
                maxit = 100, stepsize = 0.5)  # Sometimes more reliable
```

3. Use a stepping-stone model: fit a simpler model to provide initial values for a complex model.

```
fit.simple <- vglm(y ~ 1, aVGAMfamilyFunction, data = adata, trace = TRUE)
vglm(y ~ x2 + x3, aVGAMfamilyFunction, data = adata,
     etastart = predict(fit.simple), trace = TRUE)
```

A specific example is the partial proportional odds model (Sect. 14.4.1)

```
fit.pom  <- vglm(y ~ x2 + x3 + x4 + x5, cumulative(parallel = TRUE), data = cdata)
fit.ppom <- vglm(y ~ x2 + x3 + x4 + x5, cumulative(parallel = FALSE ~ x2),
                 etastart = predict(fit.pom), data = cdata, trace = TRUE)
```

Here, the parallelism assumption is not made for x2. However, if the lines intersect within the data range of x2, then numerical problems will occur.

Another example is setting, e.g., zero = 2:3 first and then relaxing zero = NULL later with the use of etastart from the first model. Specifically, for zinegbinomial(),

```
fit0 <- vglm(y ~ x2 + x3, zinegbinomial(zero = c(1, 3)), data = ndata)
fit1 <- vglm(y ~ x2 + x3, zinegbinomial(zero = NULL), data = ndata,
             etastart = predict(fit0), trace = TRUE)
```

Here, fit0 is a simpler model because two of its parameters are intercept-only. Then fit1, which has all covariates affecting each of the three parameters, is less likely to suffer from converge problems. Incidentally, fit1 would not be justified unless the data size were large and all the linear effects had been checked by fitting smoothers.

4. Multicollinear x_k: it is not a good idea to enter pairs of highly correlated variables. Selecting one sometimes suffices. The functions pairs() and cor() are useful here, as well as examining the hat values and using other diagnostics.
5. After fitting, check the coefficients in matrix form and/or the constraint matrices.

```
coef(fit1, matrix = TRUE)   # See all the linear predictors (a good idea)
constraints(fit1)
head(depvar(fit1))   # Garbage in ==> garbage out
```

6. Perform diagnostic plots using the residuals (Sect. 3.7).
7. Allow for smoothing if the data set is not tiny. That is, vgam() may be the better function to see what's going on first before switching to vglm(). The component functions often suggest a parametric shape such as a quadratic, or a transformation on the x_k to make it parametric.
8. Read the help files! Especially, see the default for zero.
9. Fit several models and compare them.

8.5.1 Doing Things in Style

To promote good style and more portable code, the following are recommended.

1. Use extractor/accessor functions where possible such as coef(), fitted(), depvar(), predict(), etc., and not fit@coefficients, fit@fitted.values, fit@y, fit@predictors, etc. Tables 8.5, 8.6, 8.7 lists many such functions.
2. Have all the variables in a data frame, and use the data argument of vglm()/vgam(), etc. As well as providing safety, prediction is more likely to be correct.

3. Monitor convergence, e.g., set `trace = TRUE`. If convergence fails, set `imethod` = 2, then `imethod = 3`, then `imethod = 4,...`, where possible. For example, `vglm(y ~ ..., uninormal(lmean = "loge", imethod = 2))` currently is more likely to work if `y` contains negative values.

   Use arguments such as `ilocation`, `iscale`, `ishape`, etc. when available to safeguard against local solutions.

8.5.2 Some Useful Miscellanea

1. Fitting a VGAM as a VGLM with regression splines: `bs()` and `ns()`.

```
Fit.vglm <- vglm(y ~ ns(x2) + ns(x3, df = 5), ...)
Fit.vgam <- vgam(y ~  s(x2) +  s(x3, df = 5), ...)
```

   should be similar. Both `s()` and `ns()` use natural cubic splines, therefore generally handle the boundaries better than `bs()`. The advantages of `Fit.vglm` are the half-stepping feature and more standard inference. Its disadvantages include being more prone to initial-value problems.

2. Here is a fictional example of constraint matrices when the formula contains an interaction term.

```
H1 <- diag(3)[, -3]
H2 <- diag(3)[,  1, drop = FALSE]
Hlist <- list("(Intercept)" = diag(3), "Area" = H1,
              "Year" = H1, "Area:Year" = H2, "Length" = H1)
fit.zinb <- vglm(Intensity ~ Area * Year + Length, stepsize = 0.5,
                 zinegbinomial(zero = NULL, nsimEIM = 500),
                 constraints = Hlist, data = zdata, trace = TRUE)
```

   The interaction term `Area * Year` has been expanded out manually into its main effects and pairwise-interaction term—this is reflected in `Hlist`.

3. Here is an example involving regression splines where η_j is a function of $f_{(j)2}(x_2)$ for $j = 1, 2$, but the component functions have differing flexibilities.

```
> xs.nz.f <- subset(xs.nz, sex == "F")
> xs.nz.f <- na.omit(xs.nz.f[, c("babies", "age", "ethnicity")])
> Hlist1 <- list("(Intercept)"   = diag(2),
                 "bs(age, df = 3)" = rbind(1, 0),
                 "bs(age, df = 4)" = rbind(0, 1),
                 "ethnicity"      = diag(2))
> fit1.baby <- vglm(babies ~ bs(age, df = 3) + bs(age, df = 4) + ethnicity,
                    zipoisson(zero = NULL), data = xs.nz.f, constraints = Hlist1)
```

   The reason why a different `bs()` term is needed to fit each $f_{(j)2}(x_2)$ is because its `df` argument cannot accept a vector. In contrast, one can fit a similar model to `fit1` easier with `vgam()`, using

```
> fit2.baby <- vgam(babies ~ s(age, df = c(3, 4)) + ethnicity,
                    zipoisson(zero = NULL), data = xs.nz.f)
```

   because `s()` *can* be assigned a vector for `df`.

4.
```
fit.1 <- vglm(...)
Hlist <- constraints(fit.1, type = "term")
fit.2 <- vglm(..., constraints = Hlist)
```

should result in `fit.1` and `fit.2` being identical. If the family function of `fit.2` has a `zero` argument, then set `zero = NULL` in order to effectively disable it.

5. Adjusted dependent vectors ($n \times M$) and working weight matrices ($M \times M \times n$) (Sect. 3.2) may be obtained as follows.

```
zmat <- predict(fit) + resid(fit, type = "working") # Matrix of working responses
wz.array <- m2a(weights(fit, type = "working", M = npred(fit))) # M x M x n
```

8.6 Slots in `vgam()` Objects

R implements two object systems, known as S3 and S4. S4 methods, which are more formal and rigorous, are adopted by **VGAM**. However, probably more readers will be more familiar with S3, therefore some brief notes are now given.

While accessor functions such as `coef()` and `fitted()` are recommended for enquiring or interrogating information about a statistical model fit, occasionally users will require to access information stored in less obvious parts of an object. For this, the function `slotNames()` is helpful to list the slot names of objects. Slots are similar to components of a list, and they are accessed using the operator "`@`" (not "`$`"). For example,

```
> pneumo <- transform(pneumo, let = log(exposure.time))
> fit <- vgam(cbind(normal, mild, severe) ~ s(let), propodds, data = pneumo)
> slotNames(fit)
 [1] "Bspline"          "nl.chisq"         "nl.df"              "spar"
 [5] "s.xargument"      "var"              "extra"              "family"
 [9] "iter"            "predictors"       "assign"             "callXm2"
[13] "contrasts"        "df.residual"      "df.total"           "dispersion"
[17] "effects"          "offset"           "qr"                 "R"
[21] "rank"             "ResSS"            "smart.prediction"   "terms"
[25] "Xm2"              "Ym2"              "xlevels"            "call"
[29] "coefficients"     "constraints"      "control"            "criterion"
[33] "fitted.values"    "misc"             "model"              "na.action"
[37] "post"             "preplot"          "prior.weights"      "residuals"
[41] "weights"          "x"                "y"
```

Hence, e.g., one uses `slot(fit, "post")` or `fit@post`, instead of `fit$post`.

Table 8.9 gives a few details about some of these. Here are some further notes:

1. Many of these slots should not be accessed directly. Rather, Tables 8.5, 8.6, 8.7 should be used where possible.
2. VGAM objects have been designed with economy in size in mind. This is because many models have multiple η_j, multiple responses and/or a multivariate responses, therefore the data structures can be naturally large. One of the largest data structures is `wz`, the working weights. By default, these are

Table 8.9 Some slots in a `vgam()` object and a few details. Almost all also hold for a `vglm()` object except the daggered (†) ones.

Slot name	Comment
`@Bspline` †	The B-spline coefficients, knot locations, and the minimum and maximum of the design points, are stored here. This list comprises sub-components identical to `vsmooth.spline()@fit` (for prediction purposes)
`@family`	The complete evaluated family function is saved here. This means, e.g., that `@weight` can usually be empty. If present, the `infos()` function can also be evaluated too, e.g., `fit@family@infos()`, which is data-independent
`@misc`	A list where miscellaneous pieces of information may be placed. Writers of family functions should assign values to `misc` in `@last`
`@qr`	The output of LINPACK implementing the QR decomposition at the final IRLS iteration. Possibly this may be dropped in the future, and/or only one of it and `object@x` kept. Currently, `QR.Q(vglmObject)` accesses this slot
`@R`	Part of the QR decomposition too. Currently `QR.R(vglmObject)` returns this slot
`@weight`	Working weights held here, in matrix-band format (Sect. 18.3.5). Usually this slot is empty and is recomputed later when needed since, by default, `save.weights = FALSE` in `vglm.control()`. This is because `@weights` tends to be one of the largest slots on the object. In general, it is $n \cdot O(M^2)$ in size, which is very large for $M \gg 1$. An exception is when simulated Fisher scoring is used; then they are usually saved on the object

usually not assigned to `@weights` if they can be recomputed easily. These $\mathbf{W}_i$ come from the final IRLS iteration, and can be saved on the object by choosing `save.weights = TRUE`. By default, `save.weights = FALSE` in the `vglm()` and `vgam()` control functions. To obtain `@weights` from an object, type `weights(object, type = "working")`. This will be in matrix-band format.

8.7 Solutions to Some Specific Problems

8.7.1 Obtaining the LM-Type Model Matrix for η_j

How can one obtain the LM-type model matrix for η_j? That is, corresponding to the jth row of (3.27). It is very easy for trivial constraints because it is simply $\mathbf{X}_{\text{LM}}$. When there are constraints, one can invoke

```
model.matrix(fit, linpred.index = j, type = "lm")
```

It constructs $\mathbf{X}_{\text{VLM}}$ and then extracts the relevant subset. The LM-type model matrix for η_j is useful for, e.g., seemingly unrelated regressions (Sect. 10.2.3).

Note that, in Table 8.5, the `formula` and `form2` arguments of `vglm()` give rise to $\mathbf{X}_{\text{LM}}$ and $\mathbf{X}_{\text{form2}}$, respectively. The latter should be extracted by

```
model.matrix(vglmObject, type = "lm2")
```

8.7.2 Predicting at $\bar{x}$

How can predictions from an 'average' value $\bar{x}$ be made as if it were an observation? The solution is to add it to the data frame and give it some very small positive prior weight. Then its fitted and predicted values can be used. For example, since **pneumo** is a data frame with wholly numerical variables, the following fits a non-proportional odds model.

```
> mydata <- transform(pneumo, prior.wts = 1, let = log(exposure.time))
> mydata <- data.frame(rbind(mydata, colMeans(mydata)))  # A new (last) row
> N <- nrow(mydata)  # The last row
> mydata$prior.wts[N] <- 1e-7  # A tiny positive value
> mydata$normal[N] <- mydata$normal[1]  # A junk response
> mydata$mild[N]   <- mydata$mild[1]     # Ditto
> mydata$severe[N] <- mydata$severe[1]  # Ditto
>
> fit1.ave <- vglm(cbind(normal, mild, severe) ~ let,
                   cumulative, data = mydata, weight = prior.wts)
> fit2.ave <- vglm(cbind(normal, mild, severe) ~ let,
                   cumulative, data = mydata[-N, ], weight = prior.wts)
> predict(fit2.ave, mydata[N, ], type = "response")  # Average response

    normal    mild   severe
 9 0.78564 0.11941 0.094944

> fitted(fit1.ave)[N, ] -
    predict(fit2.ave, mydata[N, ], type = "response") # Should be 0

      normal        mild        severe
 9 9.7225e-09 -5.3954e-09 -4.3271e-09
```

Thus the last row of **mydata** corresponds to an 'average' value of the data set. Its inclusion in **fit1.ave** has very little effect on the fit itself.

Bibliographic Notes

There are many resources on the internet for learning the deeper aspects of R (http://www.R-project.org). Currently there is an explosion of applied statistics books based on R, and a few general titles include Dalgaard (2008), Maindonald and Braun (2010), de Vries and Meys (2012). An even more elementary text is Zuur et al. (2009), and Spector (2008) focuses on data manipulation.

For linear and generalized linear modelling, try Faraway (2006), Fox and Weisberg (2011), Faraway (2015). Other general books on statistical modelling in R include Venables and Ripley (2002), Davison (2003), Crawley (2005), Aitkin et al. (2009), Jones et al. (2014).

Altman et al. (2004) cover many topics that the serious regression modeller should know. Although focused on the social sciences, the book covers core topics such as convergence problems, numerical problems, sensitivity analysis, etc.

Exercises

Ex. 8.1. Run the code underlying Fig. 8.2, but plot the component function in **age**. Interpret all the component functions, e.g., what does this say about older versus younger people, and males versus females?

Ex. 8.2. **Simulation**

(a) Run the following code

```
> set.seed(123); n <- 1000; nsim <- 20
> ddata <- data.frame(rdiric(n = n, shape = exp(c(y1 = -1, y2 = 1, y3 = 0))),
                       x2 = runif(n))
> dfit <- vglm(cbind(y1, y2, y3) ~ x2, dirichlet, data = ddata, crit = "coef")
> head(fitted(dfit), 2)  # Each row sums to unity

        y1       y2      y3
1 0.092607 0.66398 0.24341
2 0.087349 0.66967 0.24298

> twenty.sims <- simulate(dfit, nsim)
```

and show that all the simulated fitted values for the first observation ($i = 1$) sum to unity. (This is because the Dirichlet distribution has response vectors and fitted values which sum to unity.)

(b) Compute the MLEs for each of the 20 simulation data sets, and work out the standard deviation of each regression coefficient over the simulation data. Compare them to the original SEs, i.e., of **dfit**.

Ex. 8.3. For each of **fit1.baby** and **fit2.baby** in Sect. 8.5.2, replace any one of the component functions in **age** by a linear function.

Ex. 8.4. **Smart Prediction**

(a) Run the following code and verify that prediction for the model is wrong.

```
set.seed(123); n <- 20; pdata <- data.frame(x2 = sort(runif(n)))
pdata <- transform(pdata, y1 = rpois(n, exp(1 + 2 * x2)))
bad.pred <- glm(y1 ~ bs(scale(x2), df = 3), poisson, data = pdata)
```

(b) Using smart prediction, fit a model named **good.pred** that is the equivalent of **bad.pred**. Show that **good.pred** results in correct prediction.

(c) Identify 5 more (ordinary) R functions, additional to those given in Table 8.3, which will give prediction problems unless smart or safe.

Ex. 8.5. † **Calling vglm.fit()**
Suppose $x_2 \sim \text{Unif}(0,1)$ independently, $\mu = \exp(1 + 2x_2)$, and $Y \sim \text{Poisson}(\mu)$. We want to estimate $\text{SE}(\widehat{\beta}_{(1)1})$ and $\text{SE}(\widehat{\beta}_{(1)2})$ from $\eta = \log \mu$ using simulation.

(a) Suppose that each data set has $n = 50$. Let $N = 1000$. Call **vglm.fit()** N times, each with a new data set generated as above. Then work out the standard deviation of the estimates of each regression coefficient.

(b) Repeat (a) but call **vglm()** instead. How much faster is calling **vglm.fit()**?

Ex. 8.6. For a proportional odds model fitted to the **pneumo** data mimicking McCullagh and Nelder (1989, p.179),

```
pneumo <- transform(pneumo, let = log(exposure.time))
pom.pneumo <- vglm(cbind(normal, mild, severe) ~ let, propodds, data = pneumo)
```

obtain the $n \times M$ matrix of first derivatives of the log-likelihood function, i.e., the ith row is $\boldsymbol{u}_i^T$ as in (3.14). Check that the column sums are 0 (cf. (A.17)).

Casually perusing scholarly journals, and briefly scanning those articles that conduct nonlinear estimation, will convince the reader of two things. First, many researchers run their solvers with the default settings. This, of course, is a recipe for disaster...
—Altman et al. (2004)

Chapter 9
Other Topics

Surely you do not think that we can keep ourselves supplied with something to say every day on such a variety of topics, unless we thoroughly cultivate our minds by study, or that our minds could endure such strain, unless we should relax them by the same study?
—Marcus Tullius Cicero

9.1 Introduction

This chapter looks at a few topics not conveniently placed previously. Section 9.2 is more computational in nature, and describes some details about how working weights can be computed and some alternative algorithms. The remaining two sections are more practical: Sect. 9.3 briefly describes information criteria such as AIC and BIC which users often resort to for model selection, and Sect. 9.4 sketches the details for bias-reduction estimation, currently for the GLM class, an approach which can be useful when n is small.

9.2 Computing the Working Weights †

Usually the greatest impediment in implementing a VGLM is the computing of adequate working weight matrices $\mathbf{W}_i$ (3.11). The EIM, leading to Fisher scoring, is the first choice because it is usually positive-definite over a large region of the parameter space, and consequently it is preferred over the OIM (Newton-Raphson algorithm). Given reasonable initial values, Fisher scoring tends to work well when the model is appropriate for the data—slow convergence is usually indicative of something not quite right, hence the general recommendation to set `trace = TRUE` to monitor convergence. The **VGAM** package mainly uses the EIM, else it resorts to a technique called *simulated Fisher scoring* (SFS). This section describes SFS and a few other techniques.

It is emphasized that each *individual* working weight matrix must be computed, and they all need to be positive-definite—it is not sufficient that their

© Thomas Yee 2015

T.W. Yee, *Vector Generalized Linear and Additive Models*,
Springer Series in Statistics, DOI 10.1007/978-1-4939-2818-7_9

sum is so. It is also remarked that software with symbolic algebra facilities, such as mathStatica (Rose and Smith, 2002, 2013) and Maple, are powerful tools for obtaining expressions for EIMs. Even R's deriv() may also be useful for simple log-likelihoods involving cumbersome differentiation. However, even if a closed expression for an EIM element is available, being able to evaluate it may be too difficult, e.g., it may require numerical integration.

9.2.1 Weight Matrices Not Positive-Definite

Sometimes a subset of the $\mathbf{W}_i$ are not positive-definite, such as those whose fitted values are far away from the optimal solution during the first few IRLS iterations. There are many possible remedies, and a simple one currently adopted by VGAM is to apply a Greenstadt modification to each $\mathbf{W}_i$. This involves computing its spectral decomposition $\mathbf{W}_i = \mathbf{P}_i \, \boldsymbol{\Lambda}_i \, \mathbf{P}_i^T$ and replacing negative eigenvalues λ_i by $|\lambda_i|$, and then recomputing a new $\mathbf{W}_i^*$. If $\lambda_i \approx 0$, then it is replaced by some small positive value. The intent is to obtain a quick fix to get away from the fringes of the parameter space and into its interior. If the last IRLS iteration employs a Greenstadt modification, then the fitted model must be treated with caution, e.g., the estimated variance-covariance of the regression coefficients may be misleading or inaccurate (this is another good reason for monitoring convergence, e.g., setting trace = TRUE).

Another technique for ensuring the $\mathbf{W}_i$ are positive-definite is to make them diagonally dominant, e.g., by adding a matrix diag($\boldsymbol{\varepsilon}$) where each element ε_j is some small positive quantity. This can be done absolutely or relatively: replace w_{jj} by $w_{jj} + \varepsilon_j$; or w_{jj} by $w_{jj}(1 + \varepsilon_j)$. Probably the relative method is to be preferred in general, and the family function posbernoulli.tb() is an example of this idea. This function has arguments ridge.constant and ridge.power to describe a common value of ε_j, which decays quickly to 0 as the IRLS iterations progress. Specifically, at iteration a, a positive value $\omega K a^\pi$ is added to the diagonals, where K and π correspond to the two arguments, and ω is the mean of elements of such working weight matrices. At present, ridge.power has value $\pi = -4$. Upon convergence, the ridge factor should play a negligible role.

9.2.2 Simulated Fisher Scoring

SFS approximates

$$\mathrm{Var}\left(\frac{\partial \ell_i}{\partial \boldsymbol{\theta}}\right) \tag{9.1}$$

by simulation at the current iteration $\boldsymbol{\theta}^{(a)}$, cf. (18.15) and (A.18). SFS requires three ingredients:

  (i) a model satisfying the usual MLE regularity conditions,
 (ii) a tractable and computable score vector, and
(iii) the ability to generate random variates from the model, i.e., an r-type function.

In (9.1), repeated realizations of the score vector are generated and the sample variance is computed. This is done for all i by a vectorized computation. If the model is intercept-only, then the variance can be averaged over all i as well, in order to obtain a more accurate estimate. The method will fail if all the random variates have the same value so that the sample variance and therefore working weight is zero.

Families which use simulated Fisher scoring have a `nsimEIM` argument. This specifies the number of simulations to be performed, so that increasing its value should lead to a more accurate approximation.

As a specific example, currently `negbinomial()` employs two algorithms, one of which is SFS since `rnbinom()` comes in base R and

$$\frac{\partial \ell_i}{\partial k_i} = \psi(y_i + k_i) - \psi(k_i) - \frac{y_i + k_i}{\mu_i + k_i} + 1 + \log \frac{k_i}{\mu_i + k_i} \tag{9.2}$$

is amenable. Clearly, $-E[\partial^2 \ell_i / \partial k_i^2]$ involves an infinite series of trigamma functions, hence it is not trivial to compute (Sect. 11.3.1). But something like the code snippet

```
ysim <- rnbinom(n = n * S, mu = mu, size = kmat)   # mu & kmat are current estimates
dl.dk <- digamma(ysim + kmat) - digamma(kmat) - (ysim + kmat) / (mu + kmat) +
         1 + log(kmat / (kmat + mu))   # Random score vectors
run.varcov <- run.varcov + dl.dk^2   # Unscaled variance (element-wise)
```

can be performed `nsimEIM` times, and then the element-wise mean of the variable `run.varcov` can be taken to estimate (9.1). Here, `S` is S, the number of responses, and the matrices `mu` and `kmat` are both $n \times S$. In this example though, SFS may fail if $\mu_i^{(a)}$ is close to 0, because all random variates can be zero with high probability.

9.2.3 The BHHH Method

IRLS is flexible enough to incorporate various different sub-algorithms. For example, one popular technique for maximum likelihood estimation, called the Berndt-Hall-Hall-Hausman (BHHH; Berndt et al., 1974) method, involves using the mean of the outer products of the score vectors (gradients) to approximate the negative Hessian. Applied to the log-likelihood (1.31), the approximation is

$$-\overline{\mathbf{W}} = -\left(\sum_{i=1}^{n} w_i\right)^{-1} \sum_{i=1}^{n} w_i \frac{\partial \ell_i}{\partial \boldsymbol{\eta}_i} \frac{\partial \ell_i}{\partial \boldsymbol{\eta}_i^T} \tag{9.3}$$

evaluated at the current iteration. Clearly, one advantage is that its implementation does not require any second derivatives. Another is that the approximation is positive-semidefinite—a better property than possibly being negative-definite as in the case of the Newton-Raphson algorithm because the problem is not as severe. It is common to allow for stepsizing to be applied in the BHHH algorithm, and this entails multiplying (9.3) by the step, which can then be determined by a line search. Applying the BHHH algorithm to the VGLM/VGAM framework, the working weight for observation i can be $w_i \overline{\mathbf{W}}$.

9.2.4 Quasi-Newton Updates

Another method for obtaining the working weight matrices is to apply a quasi-Newton update at each IRLS iteration where possible. Quasi-Newton methods, also known as variable metric methods, usually apply a rank-1 or rank-2 update to the Hessian matrix as the optimization proceeds. In this way, it is hoped that the curvature in ℓ becomes better and better approximated as the algorithm iterates. There are two common quasi-Newton variants, called the DFP and BFGS (named after their proposers, Davidon-Fletcher-Powell and Broyden-Fletcher-Goldfarb-Shanno), with general consensus that the latter is superior. Importantly, only first-order derivative information is used. One can start out with $\mathbf{W}_i^{(0)} = \mathbf{W}_i^{(1)} = w_i \mathbf{I}_M$, which are steepest-descents steps for the first two steps.

One can apply the BFGS update to *each* working weight matrix $\mathbf{W}_i^{(a-1)}$ at iteration a of the IRLS algorithm. Let

$$q_i^{(a-1)} = -\left(u_i^{(a-1)} - u_i^{(a-2)} \right), \quad s_i^{(a-1)} = \eta_i^{(a-1)} - \eta_i^{(a-2)}, \qquad (9.4)$$

where u_i is defined in (3.14). The reason for the negative sign for $q_i^{(a-1)}$ is that the formula (9.5) is for minimizing an objective function, whereas we are maximizing a log-likelihood. Symmetry and positive-definiteness are assured, provided that $s_i^{(a-1)T} q_i^{(a-1)} > 0$. We have, for each i and at iteration a,

$$\mathbf{W}_i^{(a-1)} = \mathbf{W}_i^{(a-2)} + \frac{q_i^{(a-1)} q_i^{(a-1)T}}{s_i^{(a-1)T} q_i^{(a-1)}} - \frac{\mathbf{W}_i^{(a-2)} s_i^{(a-1)} s_i^{(a-1)T} \mathbf{W}_i^{(a-2)}}{s_i^{(a-1)T} \mathbf{W}_i^{(a-2)} s_i^{(a-1)}}. \quad (9.5)$$

Practical experience with this method shows that it does not perform as well as simulated Fisher scoring. The latter gives the user direct control over how accurately each EIM is to be computed, but the unmodified quasi-Newton method lacks this ability. That the working weights (9.5) are not very accurate at the MLE is reflected in the property that quasi-Newton does not produce very accurate standard errors, e.g., as acknowledged by other authors such as Greene (2012, p.1139). Hence, summary() cannot be expected to yield accurate SEs and Wald statistics, etc. if the family function uses (9.5).

9.2.5 Numerical Derivatives

Occasionally, numerical derivatives of a function are computed for want of an analytical expression, e.g., some blocks of (5.9). In one variable, the naïve forward finite-difference approximation $f'(x) \approx [f(x + h) - f(x)]/h$ for $h \approx 0$, $h \neq 0$, has error $O(h)$, which is better replaced by the central difference formula $f'(x) \approx [f(x + h/2) - f(x - h/2)]/h$ whose error is $O(h^2)$. Similarly,

$$f''(x) \approx \frac{f'(x + h/2) - f'(x - h/2)}{h} \approx \frac{f(x + h) - 2f(x) + f(x - h)}{h^2}$$

has error $O(h^2)$.

Table 9.1 Some information criterion (IC). See text for notation. Here, n = number of observations, p = number of parameters, ℓ = log-likelihood. See Table 9.2 for the value of p for each class in the VGLM/VGAM framework.

Acronym	Function	Formula	Name
AIC	`AIC()`	$-2\ell + 2p$	Akaike's information criterion. Akaike (1973)
AIC_c	`AICc()`	$-2\ell + 2p + \dfrac{2p(p+1)}{n-(p+1)}$	AIC with a finite sample size correction
BIC	`BIC()`	$-2\ell + p\log n$	Bayesian information criterion. Also known as the Schwarz (1978) IC or SIC

The ideal choice of h is affected by the higher-order derivatives of f, and by floating-point arithmetic, e.g., when h is too small then subtracting two nearly equal quantities results in catastrophic cancellation and a large loss of significance. A generally recommended choice is $h = \sqrt{\epsilon}x$ where ϵ is known as the machine epsilon: the smallest value of the form 2^k for which $1 + \epsilon > 1$. It can be accessed by `.Machine$double.eps` in R, and is typically around 10^{-16}.

Finite-difference formulas for partial derivatives are also readily available, e.g., Cheney and Kincaid (2012). Package numDeriv provides methods for calculating (usually) accurate numerical first and second order derivatives.

9.3 Model Selection by AIC and BIC

The VGLM/VGAM framework fits many models, therefore some form of model selection is very useful. Given a set of candidate models, a common model-selection technique is to choose the one with the minimum value of some information criterion (IC). Among such, the Akaike IC (AIC) and Bayesian IC (BIC) are the most common. Their formula follows the simple penalty function idea outlined in Sect. 1.5.1, by which the log-likelihood is contrasted with the number of parameters in the model. More generically, this is the goodness-of-fit of the model penalized by its complexity.

While the likelihood ratio test is applicable to nested models, ICs are commonly applied to non-nested models. They measure the relative qualities of statistical models, not their absolute qualities, so it is their ranking that matters.

The AIC has been shown to be inconsistent: as $n \to \infty$, it tends to overfit. In contrast, BIC is a consistent estimator of the true model, which roughly means that it will pick the truly best model, from a set of candidate models, with probability one, asymptotically.

The AIC has been corrected for bias in finite samples by Hurvich and Tsai (1989), commonly called AIC_c. Its derivation is based on an LM. Some authors such as Burnham and Anderson (2002) recommend using AIC_c over AIC when the sample size is small, or when the number of parameters is large. When n is large, their difference tends to be negligible.

The Bayesian or Schwarz IC is closely related to the AIC, but it has a greater penalty term for model complexity, making it a popular alternative to the AIC, especially when n is large. It too has been applied widely to maximum likelihood-based models, but it has a tendency to underfit when n is small because of its heavy penalty term.

Methods functions for computing AIC and BIC for VGLMs, VGAMs and other variants are available in **VGAM** (Table 9.1). To compute these, the number of parameters p for each class of models is needed. Table 9.2 gives formulas for p, and for simplicity, some of the formulas are given for special cases of $M_1 = 1$ models only. For VGAMs with vector smoothing splines, the effective degrees of freedom is a heuristic quantity, therefore model selection for these models is approximate.

When comparing certain models with others, it is necessary to set the argument `omit.constant = TRUE`, e.g., `posbinomial(omit.constant = TRUE)` versus `posbernoulli.tb()`. This is because the log-likelihood functions based on (17.12) and $f_{\mathrm{B}}(y)$ in Table 17.6 differ by constants of the form $\log \binom{N}{N y_i}$. Setting the argument as such omits the constants so that the ℓ, and consequently AIC and BIC, become comparable.

In closing, we note that in **stats**, the methods function `step()` allows an automated stepwise search of variables for `lm()` and `glm()` objects, based on the AIC. Both forward and backward directions are allowed for at any particular step. Users of stepwise regression for variable selection should be mindful of its potential perils, e.g., Miller (2002), who writes

> I hope that it will disturb many readers and awaken them to the dangers of using automatic packages that pick a model and then use least squares to estimate regression coefficients using the same data.

A more fully featured function called `stepAIC()` is available in **MASS**. At present, there is no `step()` equivalent for VGLMs/VGAMs.

9.4 Bias-Reduction

While maximum likelihood estimation results in estimates that are asymptotically unbiased, for small samples there may be a sizeable bias that warrants some form of correction. This section describes a bias-reduction technique that can be easily embedded within the IRLS algorithm of VGLMs. We will find some of its technicalities are very much related to the hat matrix $\mathcal{H}$ described in Sect. 3.7.5. Recall from Sect. 2.3.6.3 the simple example of complete separation and quasi-complete separation.

To reemphasize the need and effect of bias-reduction, consider the simple example shown in Fig. 9.1 of $n = 20$ equally spaced points from -1 to 1 including the endpoints, with $y_i = 0$ for negative x_i, else $y_i = 1$. This completely separable data set can easily arise from an underlying logistic regression model when the sample size is small. Figure 9.1 plots the naïve logistic regression fit—its estimated slope diverges to infinity but the algorithm is deemed to have converged before it explodes. Overlaid are the bias-reduced binary regression models with logit and probit links—the slopes are finite and monitoring the convergence of the models displays no unusual behaviour. To obtain a finite estimate, bias-reduction has had the effect of shrinking the regression coefficients towards $\mathbf{0}$.

Figure 9.1 points towards an increasingly common application: in an era of 'big data' that includes large genomic studies in bioinformatics, fitting multinomial

Table 9.2 Number of parameters for models fitting within the VGLM/VGAM/etc. framework. Several equivalent formulas are given for some models. Notes: (i) R is the rank for several of the models; and S is the number of responses/species too. (ii) The `edf.nl` refers to the effective nonlinear degrees of freedom (ENDF) of a smooth (0 = linear function). (iii) Only special case models of QRR-VGAMs and RR-VGAMs are given here.

Model	Number of parameters
VGLMs	$p_{\text{VLM}} = p^* = \sum_{k=1}^{p} \text{ncol}(\mathbf{H}_k) = \dim(\boldsymbol{\beta}^*) = \text{ncol}(\mathbf{X}_{\text{VLM}})$
VGAMs	$\sum_{k=1}^{p} \sum_{j=1}^{\text{ncol}(\mathbf{H}_k)} \left\{ 1 + \text{edf.nl}(f^*_{(j)k}(x_k)) \right\}$
RR-VGLMs	$\sum_{k=1}^{p_1} \text{ncol}(\mathbf{H}_k) + (M - R + p_2)R$
RCIMs ($M_1 = 1$)	$1 + (R+1)(n + M - 2) - R^2$
QRR-VGLMs (CQO) ($M_1 = 1$, equal tolerances)	$\sum_{k=1}^{p_1} \text{ncol}(\mathbf{H}_k) + \{S - R + p_2 + (R+1)/2\}R$
QRR-VGLMs (CQO) ($M_1 = 1$, unequal tolerances)	$\sum_{k=1}^{p_1} \text{ncol}(\mathbf{H}_k) + \{S - R + p_2 + S(R+1)/2\}R$
RR-VGAMs (CAO) ($M_1 = 1$ and $R = 1$)	$\sum_{k=1}^{p_1} \text{ncol}(\mathbf{H}_k) + \sum_{s=1}^{S} \left\{ 1 + \text{edf.nl}(f_{(s)1}(\nu_1)) \right\}$

logit models to sparse data in a high-dimensional $\boldsymbol{x}$-space often results in infinite estimates in some $\hat{\beta}_k$. Bias-reduction is a technique that will always give a finite answer, hence a 'sensible' solution to some people.

For a regular parametric model, the maximum likelihood estimator $\hat{\boldsymbol{\beta}}$ has an asymptotic bias of the form

$$\boldsymbol{b}(\boldsymbol{\beta}) = \frac{\boldsymbol{b}_1(\boldsymbol{\beta})}{n} + \frac{\boldsymbol{b}_2(\boldsymbol{\beta})}{n^2} + \frac{\boldsymbol{b}_3(\boldsymbol{\beta})}{n^3} + \cdots. \tag{9.6}$$

To reduce this, Firth (1993) gave a general method removing the $O(n^{-1})$ term. It involves a modified score vector with components

$$U_k^*(\boldsymbol{\beta}) = U_k(\boldsymbol{\beta}) + A_k(\boldsymbol{\beta}), \qquad k = 1, \ldots, p, \tag{9.7}$$

in which A_k is allowed to depend on the data and is $O_p(1)$ as $n \to \infty$. Here,

$$U_k(\boldsymbol{\beta}) = \frac{\partial \ell}{\partial \beta_k} = \frac{\partial \ell}{\partial \eta} \frac{\partial \eta}{\partial \beta_k} = \sum_{i=1}^{n} (\boldsymbol{u}_i)_k \, \boldsymbol{x}_i, \quad \text{cf. (3.14)}. \tag{9.8}$$

Let $\tilde{\boldsymbol{\beta}}$ be the solution of the adjusted score function

$$\boldsymbol{U}^*(\boldsymbol{\beta}) = \boldsymbol{U}(\boldsymbol{\beta}) + \boldsymbol{A}(\boldsymbol{\beta}) = \boldsymbol{0}. \tag{9.9}$$

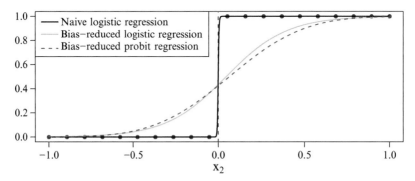

Fig. 9.1 Fitted curves for three binary regression models fitted to completely separable data, with $n = 20$. A grey vertical line at $x_2 = 0$ is plotted. If there were an additional two points at $(0, 0)$ and $(0, 1)$, then the data would have quasi-complete separation.

Firth showed that $\tilde{\boldsymbol{\beta}}$ has $O(n^{-2})$ bias, if $\boldsymbol{A}$ is chosen to be

$$\boldsymbol{A}^{(E)} \;=\; \frac{1}{2}\,\mathrm{trace}\big(\boldsymbol{\mathcal{I}}_E^{-1}(\mathbf{P}_k + \mathbf{Q}_k)\big)\,, \tag{9.10}$$

where $\mathbf{P}_k = E(\boldsymbol{U}\boldsymbol{U}^T U_k)$ and $\mathbf{Q}_k = E(-\boldsymbol{\mathcal{I}}_O\, U_k)$ are higher-order joint moments of the derivatives of ℓ. Another expression for $\boldsymbol{A}^{(E)}$ is $-\boldsymbol{\mathcal{I}}_E(\boldsymbol{\beta})\,\boldsymbol{b}_1(\boldsymbol{\beta})/n$. The reason for the superscript "(E)" is that this is one of two popular choices for $\boldsymbol{A}$—based on the EIM—with the other being based on the OIM and usually denoted by superscript "(O)". Solving (9.9) may be achieved by modifying the usual VGLM Fisher scoring algorithm to

$$\tilde{\boldsymbol{\beta}}^{(a)} \;=\; \tilde{\boldsymbol{\beta}}^{(a-1)} + \left(\boldsymbol{\mathcal{I}}_E^{(a-1)}\right)^{-1}\boldsymbol{U}^{(a-1)*},$$

and standard errors for $\tilde{\boldsymbol{\beta}}$ can be obtained in the usual way by taking the square roots of $\mathrm{diag}(\boldsymbol{\mathcal{I}}_E^{-1})$ evaluated at $\tilde{\boldsymbol{\beta}}$.

Unfortunately, obtaining closed-form expressions for $\partial\boldsymbol{A}/\partial\boldsymbol{\beta}$ involves tedious algebra even for the simplest models. To address this, Kosmidis and Firth (2010) developed a new generic algorithm, which unifies bias-reducing methods previously proposed for specific models. This new algorithm uses a series of iterative bias corrections, e.g., using the $\boldsymbol{A}^{(E)}$ adjustment, $\boldsymbol{\beta}^{(a)} = \hat{\boldsymbol{\beta}}^{(a)} - \boldsymbol{b}(\boldsymbol{\beta}^{(a-1)})$ where $\hat{\boldsymbol{\beta}}^{(a)}$ is the candidate value for the MLE based on taking a single Fisher scoring step from the current value of the bias-reduced estimate, and $\boldsymbol{b}(\boldsymbol{\beta}^{(a-1)})$ is the $O(n^{-1})$ bias evaluated at $\boldsymbol{\beta}^{(a-1)}$.

For GLMs, (9.9)–(9.10) are tractable and are as follows. The adjusted score function reduces to

$$\boldsymbol{U}^*(\boldsymbol{\beta}) \;=\; \sum_{i=1}^{n} \frac{w_i}{d_i}\left(y_i + \frac{h_{ii}\,d_i'}{2\,w_i} - \mu_i\right)\boldsymbol{x}_i \tag{9.11}$$

where $d_i = d\mu_i/d\eta_i$, $d_i' = d^2\mu_i/d\eta_i^2$, and h_{ii} is the diagonal of the hat matrix (3.63). Equation (9.11) suggests simply using the pseudo-response

$$y_i^* = y_i + \frac{h_{ii}\, d_i'}{2\, w_i} \tag{9.12}$$

instead of y_i in the IRLS algorithm. Two cases of this, with logit and probit links for a binary response, are seen in Fig. 9.1.

9.4.1 Binary Case

The special case of logistic regression warrants further comment. For this, there are three configurations of the sample points $(\boldsymbol{x}_i, y_i)$: *completely separation*, *quasi-complete separation* and *overlap* (Albert and Anderson, 1984). MLEs only exist for the latter case. The definitions of the separation cases can be formulated by considering a multinomial logit model with $J = M + 1$ groups. Let the last group be baseline, say, and let $\boldsymbol{\beta}^\dagger$ be the vector of all the regression coefficients as in (3.8). Then groups $\mathcal{H}_1, \ldots, \mathcal{H}_J$ are completely (quasi-completely) separated for the sample $\boldsymbol{x}_1, \ldots, \boldsymbol{x}_n$, if there exists an $\boldsymbol{\beta}^\dagger \neq \boldsymbol{0}$ such that for all $\boldsymbol{x}_i \in \mathcal{H}_j$ and $j, k = 1, \ldots, J$ $(j \neq k)$

$$\left(\boldsymbol{\beta}_j - \boldsymbol{\beta}_k\right)^T \boldsymbol{x}_i > (\geq)\, 0$$

then $\hat{\boldsymbol{\beta}}^\dagger$ does not exist. Its effects are that some regression coefficients diverge to $\pm\infty$. With such data, many software implementations give warnings that are vague, if any at all. In R, the safeBinaryRegression package can be used to detect (but not remedy) this problem.

9.4.2 Software Implementation

At present, VGAM implements bias-reduction for only a few families, e.g., `poissonff()` and `binomialff()`. They have a default argument `bred = FALSE` which can be assigned `TRUE` to obtain $\tilde{\boldsymbol{\beta}}$. Another package with bias-reducing capabilities is brglm by I. Kosmidis, and it has a function `separation.detection()` that can be used also to detect separation. A third package, specifically for logistic regression and based on Firth (1993), is logistf.

Some users might be tempted to use always the bias-reduced method, as it seems to offer a free lunch compared to the ordinary method. This is not advised because bias-reduction may inflate the asymptotic variance substantially, e.g., in the gamma model. In particular, Wald-type confidence intervals for $\tilde{\boldsymbol{\beta}}$ perform poorly in small samples.

Bibliographic Notes

Osborne (2006) looks at several relevant topics such as the connection between least squares and Fisher scoring, especially for ML problems belonging to the exponential family (3.1) and monitoring convergence; see also Osborne (1992). Lange (2010, Chap.14) and Lange (2013, Chap.10) cover some of the basic aspects of scoring.

Simulation-based estimation methods are described more generally in Greene (2012, Chap.15). The Greenstadt modification is described in Kennedy and Gentle (1980), as well as other numerical topics. For more about non-invertible Hessians, see Gill and King (2004). Yee and Stephenson (2007) suggested the use of the quasi-Newton update method of Sect. 9.2.4 within the IRLS algorithm.

Detailed treatments of IC-based model selection can be found, e.g., in Sakamoto et al. (1986), Burnham and Anderson (2002), Miller (2002), Claeskens and Hjort (2008), Konishi and Kitagawa (2008). A few aspects of model selection are given in Ripley (2004). For practical examples of variable selection and stepwise regression for LMs/GLMs, try Venables and Ripley (2002), Fox and Weisberg (2011), Faraway (2015).

Bias-correction as a whole now has a sizeable literature, and Kosmidis (2014a) is a review paper that gives context to the bias-reduction technique described here. A reference for separable data is Lesaffre and Albert (1989).

Exercises

Ex. 9.1. Run the `mix2poisson()` example on the London Times data. Fit several other similar models, but with different settings for the `nsimEIM` argument. Does this make much difference to the solution? Answer by comparing the MLEs and comparing their standard errors.

Ex. 9.2. For a standard Poisson regression model, how large does `nsimEIM` have to be in order that the sample variance of randomly generated $\partial \ell_i / \partial \eta$ is within ± 0.1 of the true value $\mu_i^{(a)} = 10$, with 95 % probability?

Ex. 9.3. **Low Birthweights: Model Selection**
Consider the data frame `birthwt` in MASS, which concerns a medical study about low birth weight (less than 2.5 kg) babies, and several risk factors for such. About a third are classified as being low weight. We shall fit logistic regressions to the response `low`, with candidate explanatory variables `lwt`, `age` and (transformed) `race`. [Claeskens and Hjort (2008)]

(a) Create a data frame from `birthwt` which has a variable `f.race`, say, that is a factor version of the `race` variable (currently it is numeric).
(b) It is thought that `lwt` is an important (protected) variable, therefore must always be in a logistic regression. The other two, `age` and `f.race`, are optional. Fit several combinations of models and use AIC to choose the best one.
(c) Repeat (b) using the BIC. Are the 'best' models different?

Ex. 9.4. **Complete Separation for Binary Responses**
Consider Fig. 9.1.

(a) Suppose $P(Y_i = 1 | x_i) = \frac{1}{3}$ independently, and the x_i are equally spaced in $[-1, 1]$ including the endpoints. Compute the probability of complete separation, given that the data are not all 0s or all 1s. The separation may occur anywhere in $(-1, 1)$.
(b) Repeat (a) for general n; obtain a formula for the probability of complete separation. Plot the probabilities for $n = 4, 6, 8, \ldots, 20$. Comment.

Ex. 9.5.   Bias-Reduction: Adjusted Responses

Show that bias-reduction, when applied to the following situations, results in the following pseudo-responses. Hint: use (9.12) and some links are in Table 1.2.

(a) Poisson regression: $y_i^* = y_i + h_{ii}/2,$ [(Kosmidis and Firth, 2009, Table 1)]
(b) Poisson regression with $\eta = \mu$: $y_i^* = y_i,$
(c) logistic regression: $y_i^* = y_i + h_{ii}\left(\frac{1}{2} - p_i\right),$
(d) probit analysis: $y_i^* = y_i - h_{ii}\, p_i(1 - p_i)\eta_i/[2\phi(\eta_i)].$
(e) For binary regression with a complementary log-log link: derive y_i^*.

Ex. 9.6.   Search for a binary response in **xs.nz** which is completely separable for some subset of the data, but having at least $n \geq 100$. Then fit a bias-reduced logistic regression.

Now this is not the end, nor is it even the beginning of the end, but it is, perhaps, the end of the beginning.
—Winston Churchill, November 1942

Part II
Some Applications

Chapter 10
Some LM and GLM Variants

Experience has shown, and a true philosophy will always show, that a vast,
perhaps the larger portion of the truth arises from the seemingly irrelevant.
—Edgar Allan Poe, *The Mystery of Marie Rogêt*

10.1 Introduction

In this chapter we survey a few miscellaneous models that may be considered extensions of LMs and binomial GLMs, and which are accommodated within the VGLM/VGAM framework.

10.2 LM Variants

The standard linear model has been embellished in more ways than any other regression model. While the variants are almost endless, we obtain a flavour by considering a few here which naturally succumb to the framework. In particular, they are varying-coefficient models where the regression coefficients are modelled as functions of some explanatory variables, the Tobit model which is a censored LM, seemingly unrelated regressions where we have a set of LMs tied together at the random-error level, and the first-order autoregressive time series model which allows for correlation between successive y_i values. These models are summarized in Table 10.1.

10.2.1 Varying-Coefficient Models

Rather than having fixed regression coefficients as an LM, *varying-coefficient models* (VCMs) allow the regression coefficients to be modelled as functions of other explanatory variables (known as *effect modifiers*), here with the help of parameter

© Thomas Yee 2015
T.W. Yee, *Vector Generalized Linear and Additive Models*,
Springer Series in Statistics, DOI 10.1007/978-1-4939-2818-7_10

Table 10.1 LM variants described in Sect. 10.2.

Model	Main specification	VGAM family function
Varying coefficient model	Eq. (10.1)	`normal.vcm()`
Tobit model	Eqs. (10.6)–(10.7)	`tobit()`
Seemingly unrelated regressions	Eq. (10.13)	`SURff()`
AR(1) time series model	Eqs. (10.27)–(10.29)	`AR1()`

link functions. For example, we may want a subset of regression coefficients to be positive, and another subset to be positive and add to unity. Although VCMs can be generalized to, e.g., the GLM class, we restrict our attention to the LM here.

Notationally, we augment "$(\boldsymbol{x}_i, y_i)$" in this section by $(\boldsymbol{x}_i, \boldsymbol{x}_{2i}, y_i)$ to represent our data, for $i = 1, \ldots, n$. Here, the usual LM applies to Y_i as a linear function of $\boldsymbol{x}_{2i} = (x_{2i1}, \ldots, x_{2ip_2})^T$, and the regression coefficients are allowed to vary with $\boldsymbol{x}_i$. We write the VCM as $Y_i \sim N(\mu_i, \ \sigma^2)$ independently, where

$$\mu_i \;=\; \sum_{k=1}^{p_1} \gamma_k(\boldsymbol{x}_i) \cdot x_{2ik} + \sum_{k=1+p_1}^{p_2} \gamma_k(\boldsymbol{x}_i) \cdot x_{2ik}, \tag{10.1}$$

coupled with $\eta_M = g_M(\sigma)$ or $\eta_M = g_M(\sigma^2)$. The RHS of (10.1) has two parts and this is intentional, because the right-most part applies specifically to an optional `multilogit()` link that enables the regression coefficients $\gamma_{1+p_1}, \ldots, \gamma_{p_2}$ to be positive and sum to unity. These γ_k might be interpreted as proportions, e.g., asset allocations in a financial portfolio, after adjusting for covariates $x_{21}, \ldots, x_{2p_1}$.

The effect modifiers $\boldsymbol{x}_i$ sometimes contain a time variable so that the model captures the temporal changes of the response with respect to the $\boldsymbol{x}_{2i}$. Linearity between $E(Y)$ and the $\boldsymbol{x}_{2i}$ is still retained, but the additional flexibility in the coefficients allows for the modelling of a special type of interaction. Consequently, VCMs have found applications in such areas as nonlinear time series, functional data analysis and longitudinal studies, and financial modelling.

As a VGLM/VGAM, the vector of linear/additive predictors for (10.1) is $\boldsymbol{\eta}^T \equiv (\eta_1, \ldots, \eta_M)^T =$

$$\left(g_1^{-1}(\gamma_1), \ \ldots, \ g_{p_1}^{-1}(\gamma_{p_1}), \ \log\!\left(\frac{\gamma_{1+p_1}}{\gamma_{p_2}}\right), \ \ldots, \ \log\!\left(\frac{\gamma_{p_2-1}}{\gamma_{p_2}}\right), \ \log \sigma \right), \tag{10.2}$$

for parameter link functions g_j. Thus one can model each of the regression coefficients $\gamma_k(\boldsymbol{x}_i)$ nonparametrically as a GAM. With no multi-logit links, and identity links with $\eta_j(\boldsymbol{x}_i) = \boldsymbol{\beta}_j^T \boldsymbol{x}_i$ throughout, the VCM (10.2) can easily be fitted using any LM software. A multi-logit link applied to several proportions has one less independent parameters, therefore if there is such a link then $M = p_2$, otherwise $M = p_1 + 1 = p + 1$.

The VCM (10.1)–(10.2) is implemented by the family function `normal. vcm()`. Covariates $\boldsymbol{x}_{2i}$ are serviced by the argument `form2`, which is assigned the usual LM formula for the mean. How each regression coefficient γ_k is modelled as a function of the $\boldsymbol{x}_i$ is entered through the argument `formula`. Here is an example.

```
vglm(y1 ~ 1 + x2, form2 = ~ 1 + x2.2 + x2.3 + x2.4 + x2.5,
     normal.vcm(link.list = list("(Intercept)" = "multilogit",
                                 "x2.2"        = "multilogit",
                                 "x2.3"        = "loge",
                                 "x2.4"        = "logoff",
                                 "(Default)"   = "identitylink",
                                 "x2.5"        = "multilogit"),
               earg.list = list("(Intercept)" = list(),
                                 "x2.2"        = list(),
                                 "x2.3"        = list(),
                                 "x2.4"        = list(offset = 1),
                                 "(Default)"   = list(),
                                 "x2.5"        = list()),
               zero = c(1:2, 5)), data = ndata, trace = TRUE)
```

which has $\boldsymbol{x}_i = (1, x_{i2})^T$ as effect modifiers, and it fits

$$\mu_i = \sum_{k \,\in\, \mathcal{C}_1} \gamma_k(\boldsymbol{x}_i) \cdot x_{2ik} + \sum_{k \,\in\, \mathcal{C}_2} \gamma_k(\boldsymbol{x}_i) \cdot x_{2ik}$$

where $\gamma_1 + \gamma_2 + \gamma_5 = 1$ with each $\gamma_j > 0$. Here, the sets $\mathcal{C}_1 = \{3, 4\}$, $\mathcal{C}_2 = \{1, 2, 5\}$, $\gamma_3 = \exp(\eta_3)$ and $\gamma_4 = \exp(\eta_4) - 1$. The last sorted level of $\mathcal{C}_2$ is the baseline/reference group and this is dropped from $\boldsymbol{\eta}$, hence $M = 5$. The default for η_5 is $\log \sigma$. The order of the variables here has not been arranged as (10.1), and variables 1, x2.2 and x2.5 comprise $\boldsymbol{x}_{2i}$. The use of zero = c(1:2, 5) means σ and the multi-logit link parameters are modelled as intercept-only.

A default link function can be assigned using "(Default)" as a list component. This mirrors the naming style "(Intercept)" for 1 in a formula. As a default, normal.vcm() uses the identity link for all γ_k, i.e., $\gamma_k = \beta_{(k)1} x_1 + \beta_{(k)2} x_2 + \cdots$. Extra arguments such as offsets can be passed in using the earg.list argument, and it requires names just like list.link to match up the respective term. An ordinary LM is fitted to obtain initial values for the γ_k, and to ensure that they satisfy the range restrictions pertaining to each parameter link function, some of the most common link functions in Table 1.2 are checked explicitly. However, should that fail, there are arguments icoefficients and etastart.

Technically, the working weight matrices are strictly of rank-1. To see this, suppose that there are no multilogit terms in (10.1) for simplicity, then

$$-E\left(\frac{\partial^2 \ell_i}{\partial \eta_j \, \partial \eta_k}\right) = -E\left(\frac{\partial^2 \ell_i}{\partial \mu^2}\right) \frac{\partial \mu_i}{\partial \gamma_j} \frac{\partial \mu_i}{\partial \gamma_k} \frac{\partial \gamma_j}{\partial \eta_j} \frac{\partial \gamma_k}{\partial \eta_k}$$

$$= \frac{1}{\sigma_i^2} \cdot x_{2ij} \cdot x_{2ik} \cdot \frac{\partial \gamma_j}{\partial \eta_j} \frac{\partial \gamma_k}{\partial \eta_k} \tag{10.3}$$

so that

$$\mathbf{W}_i \equiv -E\left(\frac{\partial^2 \ell_i}{\partial \boldsymbol{\eta} \, \partial \boldsymbol{\eta}^T}\right) \propto \boldsymbol{x}_{2i} \, \boldsymbol{x}_{2i}^T. \tag{10.4}$$

A remedy for this problem is to apply the BHHH method described in Sect. 9.2.3. When an ordinary LM is fitted as the VCM, this remedy leads to the LM solution (Ex. 10.2). The BHHH method pools the $\boldsymbol{x}_{2i} \, \boldsymbol{x}_{2i}^T$ over the entire data set. Specifically, the outer product in the very most-RHS of (10.4) is approximated by $n^{-1} \mathbf{X}_2^T \mathbf{X}_2$. Partly as a result of this approximation, normal.vcm() can exhibit fragility if disparate link functions are used.

10.2.1.1 Engine Example

As an illustration, we mimic part of an analysis described in Hastie and Tibshirani (1993) concerning a data set of $n = 88$ observations from a single-cylinder engine. The response is $Y = \mathrm{NO}_x$, the nitric oxide and nitrogen dioxide concentration when fuelled by ethanol. The explanatory variables are E for the engine's equivalence ratio, and C for its compression ratio.

A plot of Y versus E, stratified by C into three groups (low, medium and high values), is given in Fig. 10.1a. Likewise, a plot of Y versus E for three strata of C is given in Fig. 10.1b, and this plot reveals interactions because systematic differences occur between them, as also can be seen in Fig. 10.2, which is Fig. 10.1b with least squares lines fitted to each stratum. These lines have different slopes. After fitting several models (10.55)–(10.58), they focus on

$$\mathrm{NO}_x \;=\; \beta_1(E) + \beta_2(E) \cdot C + \varepsilon, \tag{10.5}$$

which is a VCM applied to the simple linear regression of Y versus C, where smoothing splines are used to estimate the intercept and slope as a function of E. This is justified from Fig. 10.2, because a simple linear regression of Y versus C is not unreasonable for each stratum.

We can fit the VCM (10.5) with

```
> data("ethanol", package = "locfit")
> fit.vcm <- vgam(NOx ~ s(E, df = 5.8), normal.vcm,data = ethanol, form2 = ~ C)
```

Here, to provide about 8 degrees of freedom as Hastie and Tibshirani (1993), the empirical formula $df \approx \frac{5}{4}\mathrm{trace}(\mathbf{S}) - \frac{1}{2}$, (4.46), was used, hence $\mathrm{trace}(\mathbf{S}) \approx 6.8$; this contains an intercept and a slope. Then each (centred) component function can be plotted, with some labelling, by

```
> working.resids <- resid(fit.vcm, type = "working")[, 1:2]
> plot(fit.vcm, se = TRUE, scol = "blue", which.cf = 1,
       resid = working.resids, pcol = "orange", main = expression(hat(beta)[1](E)))
> plot(fit.vcm, se = TRUE, scol = "blue", which.cf = 2,
       resid = working.resids, pcol = "orange", main = expression(hat(beta)[2](E)))
```

to give Fig. 10.1c,d. Not surprisingly, they are very similar to their solution (Hastie and Tibshirani, 1993, Figs. 3(a)–(b)). It can be seen from the scale of the y-axis that the effect of $\hat{\beta}_1(E)$ is much larger than $\hat{\beta}_2(E)$, and that both functions are nonlinear.

10.2.2 The Tobit Model

Figure 10.3a displays what appears to be an LM fitted to some fictional scatter plot data. The response is the amount spent per month by households on a certain type of good such as "luxuries", regressed against X = household income. What makes this different from an ordinary LM is the presence of $y_i = 0$ values, which are plotted with a different colour and symbol. If the $y_i = 0$ observations were treated at face-value, then $\hat{\boldsymbol{\beta}}$ from an ordinary LM fit would yield a downwards-biased estimate of the slope coefficient and an upward-biased estimate of the intercept. It could be surmised that the mean response is

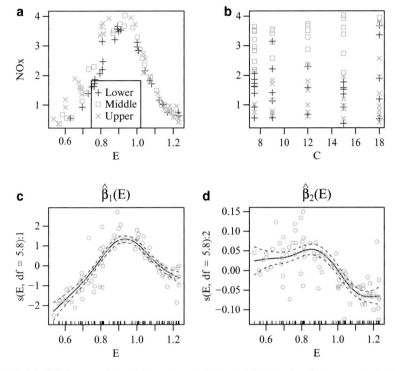

Fig. 10.1 (**a**)–(**b**) Scatter plots of the `ethanol` data stratified by `C` and `E`, respectively. In (**a**) `C` has been split into low, medium and high subgroups. In (**b**) `E` has been split into three similar subgroups—see Fig. 10.2 for an expansion. Plots (**c**)–(**d**) are the estimated (centred) component functions of the VCM (10.5) where both the intercept and slope are a smoothing spline of the variable `E`. The dashed lines are pointwise ± 2 SEs about the $\hat{\beta}_j(E)$. The points about the curves are the working residuals.

zero until the household income exceeds a certain level. One could envisage an underlying LM whose negative points have been replaced by zero values—so that it satisfies the constraint that $y_i \geq 0$ (expenditures are non-negative because this was prior to the widespread use of credit cards!). Tobin (1958) first considered such a model, and this has since been referred to as the (standard) Tobit model.

The Tobit model is essentially a censored normal distribution. The *standard* Tobit is the multiple linear regression

$$y_i^* = \boldsymbol{\beta}^T \boldsymbol{x}_i + \varepsilon_i, \quad \varepsilon_i \sim N(0, \sigma^2) \text{ independently,} \quad (10.6)$$

coupled with

$$y_i = \begin{cases} y_i^*, & \text{if } y_i^* > 0; \\ 0, & \text{if } y_i^* \leq 0. \end{cases} \quad (10.7)$$

With more generality, the Tobit model has observations $y_i = \min\{\max(y_i^*, L), U\}$ where $L < U$ are two known censoring points ($L = 0$ and $U = \infty$ in the standard model.) That is, the response is as usual if it lies between L and U, else has value L or U if 'out of range'. The censoring points are determined by physical limitations/meanings.

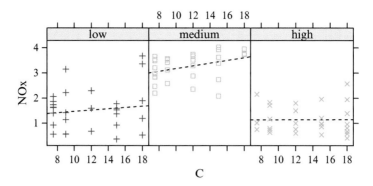

Fig. 10.2 Expansion of Fig. 10.1b with a least squares line added to each subset.

Historically, the word "Tobit" was coined by Goldberger (1964) as an amalga-
mation of "Tobin's probit." Probit models share similarities with the Tobit model.

Many refer to Y^* as a latent (i.e., unobserved) variable. Tobit modelling is
synonymous with Y being a *limited* dependent variable, i.e., a dependent variable
subject to a known upper and/or lower bound. As well as being widely used
in econometrics, the Tobit model has potential applications in every field where
the ordinary LM is used, e.g., an academic test that is too difficult/easy so that
many students score the lowest/highest possible mark, respectively, a water-testing
device for lead poisoning that is applied to households but is unable to detect lead
concentrations below a certain threshold.

Note that $P(Y = L) = P(Y^* \leq L)$ and $P(Y = U) = P(Y^* \geq U)$, so that the
likelihood function comprises a mixture of one continuous and two discrete parts:

$$\prod_{i:y_i=L} \Phi\left(\frac{L - \mu_i^*}{\sigma_i}\right) \prod_{i:L<y_i<U} \sigma_i^{-1}\, \phi\left(\frac{y_i - \mu_i^*}{\sigma_i}\right) \prod_{i:y_i=U} \Phi\left(\frac{-(U - \mu_i^*)}{\sigma_i}\right), \quad (10.8)$$

where $\mu_i^* = \boldsymbol{x}_i^T \boldsymbol{\beta}$. The middle term corresponds to a *doubly truncated* normal
(10.59), and the EIM for the standard Tobit is given in Amemiya (1985). The
MLE of a correctly specified Tobit model is consistent.

The **VGAM** family **tobit()** implements Tobit model estimation. Not surprisingly, it is similar to **uninormal()** in that it has

$$g_1(\mu^*) \;=\; \eta_1 \;=\; \mu \;=\; \boldsymbol{\beta}_1^T \boldsymbol{x} \qquad\qquad (10.9)$$
$$g_2(\sigma) \;=\; \eta_2 \;=\; \log \sigma \;=\; \beta_{(2)1} \qquad\qquad (10.10)$$

as the default. Here is a simple simulated example (which serves as the basis of
Fig. 10.3).

```
> tdata <- data.frame(Income = seq(0, 2, length = (n <- 100))); set.seed(1)
> Meanfun <- function(x) -2 + 3 * x
> tdata <- transform(tdata, Spending = rtobit(n, mean = Meanfun(Income)))
> tfit <- vglm(Spending ~ Income, tobit, data = tdata)
> coef(tfit, matrix = TRUE)

                   mu loge(sd)
(Intercept) -1.9940 -0.12474
Income       3.0581  0.00000
```

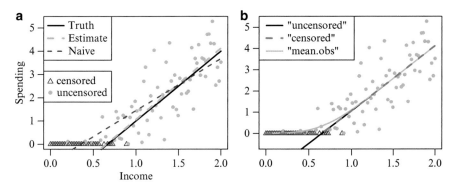

Fig. 10.3 Tobit model fitted to some simulated data. This mimics the type of problem motivating (Tobin, 1958), viz. spending is non-negative, and is linear beyond a certain income. Values of zero are plotted with a different colour and symbol for clarity. (**a**) The purple dashed line is a naïve fit that treats all values as if they were 'real'. The estimate and the truth are similar. (**b**) The 3 types of fitted values currently distinguished by the argument `type.fitted`.

There are currently three types of means that `tobit()` may return as fitted values, as determined by the argument `type.fitted` (they are illustrated in Fig. 10.3b):

`"uncensored"` $\hat{\mu}_i^*$, the mean corresponding to (10.6), which is linear with respect to $\boldsymbol{x}_i$, and unbounded. That is, the estimate of $E(Y_i^*|\boldsymbol{x}) = \boldsymbol{x}_i^T\hat{\boldsymbol{\beta}}$ (the default).

`"censored"` $\min\{\max(\hat{\mu}_i^*, L), U\}$, the censored mean. These fitted values are constrained to lie in $[L, U]$.

`"mean.obs"`   The expected value $E(Y|\boldsymbol{x}_i) =$

$$\mu_i^*\{\Phi_U - \Phi_L\} + \sigma_i\{\phi_L - \phi_U\} + L\,\Phi_L + U(1 - \Phi_U) \qquad (10.11)$$

where $\Phi_U = \Phi((U - \mu_i^*)/\sigma_i)$, $\phi_L = \phi((L - \mu_i^*)/\sigma_i)$, etc. This is the mean of the observed values y_i. These fitted values are also constrained to lie in $[L, U]$. This version asymptotes smoothly where the regression line meets the threshold, whereas the `"censored"`-typed fitted value has a discontinuous first derivative at the intersection.

To finish up here, it is mentioned that Amemiya (1985) distinguishes between 5 variants of the Tobit model, called Types I–V, depending on where and when censoring occurs. Family function `tobit()` implements the Type I. The other types are more complicated, involving more than one latent variable or response.

10.2.3 Seemingly Unrelated Regressions

Seemingly unrelated regressions (SUR) were proposed by Zellner (1962) and that has become one of the most enduring innovations in the field of econometrics. The basic model is a set of LMs tied together at the error-term level, and it can be written

$$\boldsymbol{y}^{(j)} = \mathbf{X}_j\,\boldsymbol{\beta}^{(j)} + \boldsymbol{\varepsilon}^{(j)}, \qquad j = 1, \ldots, M, \qquad (10.12)$$

where $\boldsymbol{y}^{(j)}$ is an $n \times 1$ vector of observations on the jth response variable, $\mathbf{X}_j$ is an $n \times p_j$ model matrix with full column-rank, $E(\boldsymbol{\varepsilon}^{(j)}) = \mathbf{0}$, and $\mathrm{Cov}(\boldsymbol{\varepsilon}^{(j)}, \boldsymbol{\varepsilon}^{(k)}) = \sigma_{jk}\,\mathbf{I}_n$, for $j, k = 1, \ldots, M$. If the observations correspond to different time points, then the model implies that the errors in different equations are correlated at each point in time but are uncorrelated over time. The entire system may be written

$$\begin{pmatrix} \boldsymbol{y}^{(1)} \\ \boldsymbol{y}^{(2)} \\ \vdots \\ \boldsymbol{y}^{(M)} \end{pmatrix} = \begin{pmatrix} \mathbf{X}_1 & \mathbf{0} & \cdots & \mathbf{0} \\ \mathbf{0} & \mathbf{X}_2 & \ddots & \vdots \\ \vdots & \ddots & \ddots & \mathbf{0} \\ \mathbf{0} & \cdots & \mathbf{0} & \mathbf{X}_M \end{pmatrix} \begin{pmatrix} \boldsymbol{\beta}^{(1)} \\ \boldsymbol{\beta}^{(2)} \\ \vdots \\ \boldsymbol{\beta}^{(M)} \end{pmatrix} + \begin{pmatrix} \boldsymbol{\varepsilon}^{(1)} \\ \boldsymbol{\varepsilon}^{(2)} \\ \vdots \\ \boldsymbol{\varepsilon}^{(M)} \end{pmatrix} \tag{10.13}$$

or $\boldsymbol{y} = \mathbf{X}\boldsymbol{\beta} + \boldsymbol{\varepsilon}$, say, where $E(\boldsymbol{\varepsilon}) = \mathbf{0}$ and $\mathrm{Cov}(\boldsymbol{\varepsilon}) = \boldsymbol{\Sigma}_0 \otimes \mathbf{I}_n$.

If $\boldsymbol{\Sigma}_0 \; (= [(\sigma_{jk})])$ is known, then the single equation ordinary and generalized least squares estimators are

$$\hat{\boldsymbol{\beta}}_{\mathrm{OLS}} = (\mathbf{X}^T\mathbf{X})^{-1}\mathbf{X}^T\boldsymbol{y} \quad \text{and} \tag{10.14}$$

$$\hat{\boldsymbol{\beta}}_{\mathrm{GLS}} = \left[\mathbf{X}^T(\boldsymbol{\Sigma}_0^{-1} \otimes \mathbf{I}_n)\mathbf{X}\right]^{-1} \mathbf{X}^T(\boldsymbol{\Sigma}_0^{-1} \otimes \mathbf{I}_n)\,\boldsymbol{y}. \tag{10.15}$$

The OLS estimator is consistent, but it does not take into account the correlation structure of the disturbances across equations. Consequently, it is generally less efficient than $\hat{\boldsymbol{\beta}}_{\mathrm{GLS}}$. The GLS estimator (also known as Aitken's GLS estimator) is the best linear unbiased estimator (BLUE) for $\boldsymbol{\beta}$. Zellner (1962) observed that $\hat{\boldsymbol{\beta}}_{\mathrm{GLS}}$ in (10.15) is the Gauss-Markov estimator, given $\boldsymbol{\Sigma}_0$, of $\boldsymbol{\beta}$. He showed that $\hat{\boldsymbol{\beta}}_{\mathrm{GLS}}$ reduces to $\hat{\boldsymbol{\beta}}_{\mathrm{OLS}}$ when $\mathbf{X}_1 = \cdots = \mathbf{X}_M$ and/or $\boldsymbol{\Sigma}_0$ is diagonal. In the latter case, the regressions are truly unrelated.

The equations in (10.12) can be stacked side-by-side to give

$$\left(\boldsymbol{y}^{(1)} \Big| \cdots \Big| \boldsymbol{y}^{(M)}\right) = \left(\mathbf{X}_1\,\boldsymbol{\beta}^{(1)} \Big| \cdots \Big| \mathbf{X}_M\,\boldsymbol{\beta}^{(M)}\right) + \left(\boldsymbol{\varepsilon}^{(1)} \Big| \cdots \Big| \boldsymbol{\varepsilon}^{(M)}\right) \tag{10.16}$$

$$= \left(\mathbf{X}_1 \Big| \cdots \Big| \mathbf{X}_M\right) \mathrm{Diag}(\boldsymbol{\beta}^{(1)}, \ldots, \boldsymbol{\beta}^{(M)}) + \left(\boldsymbol{\varepsilon}^{(1)} \Big| \cdots \Big| \boldsymbol{\varepsilon}^{(M)}\right),$$

where $\mathrm{Cov}(\boldsymbol{\varepsilon}^{(j)}, \boldsymbol{\varepsilon}^{(k)}) = \sigma_{jk}\,\mathbf{I}_n$. Continuing to assume that $\boldsymbol{\Sigma}_0$ is known, the ith row of (10.16) is

$$\boldsymbol{y}_i^T = \left(\boldsymbol{x}_{i1}^T\,\boldsymbol{\beta}^{(1)}, \ldots, \boldsymbol{x}_{iM}^T\,\boldsymbol{\beta}^{(M)}\right) + \boldsymbol{\varepsilon}_i^T \tag{10.17}$$

where $E(\boldsymbol{\varepsilon}_i) = \mathbf{0}$, $\mathrm{Cov}(\boldsymbol{\varepsilon}_i) = \boldsymbol{\Sigma}_0$, and the $\boldsymbol{\varepsilon}_i$ are independent. We can write the transpose of (10.17) as $\boldsymbol{y}_i = \boldsymbol{\eta}_i + \boldsymbol{\varepsilon}_i$ and interpret this as a VLM, so that constraint matrices $\mathbf{H}_k$ specify which response variables each x_k is a regressor for. If $\boldsymbol{\Sigma}_0$ is unknown, then we can still treat (10.17) as a VLM, provided that an estimate of it is available. Indeed, the model can be be fitted with `vglm()`, and the `SURff()` family. Its arguments are

```
> args(SURff)

function (mle.normal = FALSE, divisor = c("n", "n-max(pj,pk)",
    "sqrt((n-pj)*(n-pk))"), parallel = FALSE, Varcov = NULL,
    matrix.arg = FALSE)
NULL
```

As well as constraining the coefficients, the VGLM/VGAM framework allows the xij facility to operate on (10.17) too. It is noted, however, that $\mathbf{X}_{\mathrm{VLM}} \neq \mathbf{X}$ here, because the ordering of the response taken by VGAM differs from the order of the response vector in (10.13).

Suppose now that $\boldsymbol{\Sigma}_0$ in the seemingly unrelated regressions model is unknown (the usual case). Zellner's idea is to replace $\boldsymbol{\Sigma}_0$ by a consistent estimator, $\tilde{\boldsymbol{\Sigma}}_0$, say. Then one can utilize the estimated GLS estimator

$$\widehat{\widehat{\boldsymbol{\beta}}} = \left[\mathbf{X}^T \left(\tilde{\boldsymbol{\Sigma}}_0^{-1} \otimes \mathbf{I}_n \right) \mathbf{X} \right]^{-1} \mathbf{X}^T \left(\tilde{\boldsymbol{\Sigma}}_0^{-1} \otimes \mathbf{I}_n \right) \boldsymbol{y}, \qquad (10.18)$$

where $\tilde{\boldsymbol{\Sigma}}_0$ is based on OLS residuals $\tilde{\boldsymbol{\varepsilon}}_j = \boldsymbol{y}^{(j)} - \mathbf{X}_j \hat{\boldsymbol{\beta}}_{\mathrm{OLS}}^{(j)}$ and has elements

$$\tilde{\sigma}_{jk} = n^{-1} \cdot \tilde{\boldsymbol{\varepsilon}}_j^T \tilde{\boldsymbol{\varepsilon}}_k, \qquad j, k = 1, \ldots, M. \qquad (10.19)$$

Then $\widehat{\widehat{\boldsymbol{\beta}}}$, defined by (10.18) and (10.19), is frequently referred to as *Zellner's SUR estimator* and is very commonly used. Other names include *Zellner's (asymptotically) efficient estimator* and *Zellner's 2-stage (Aitken) estimator*. It is a two-stage estimator because it entails M OLS regressions followed by the GLS regression. Zellner (1962) showed that the bias of (10.18) is at most $O(n^{-1})$, and that it is asymptotically efficient. Under general conditions, its asymptotic distribution is normal.

A second estimator is to continue past the 2-stage estimator by iterating between estimating $\boldsymbol{\Sigma}_0$ from the latest residuals and re-estimating $\boldsymbol{\beta}$ by GLS, until final convergence is achieved. This is known as the *iterative GLS* (IGLS) method (another common name is *Zellner's iterative (Aitken) estimator*).

A third estimator, obtainable by setting the argument mle.normal = TRUE, is the MLE made under the assumption that the errors have a multivariate normal distribution. Then

$$\tilde{\boldsymbol{\beta}}_{\mathrm{ML}} = \left[\mathbf{X}^T \left(\tilde{\boldsymbol{\Sigma}}_{0,\mathrm{ML}}^{-1} \otimes \mathbf{I}_n \right) \mathbf{X} \right]^{-1} \mathbf{X}^T \left(\tilde{\boldsymbol{\Sigma}}_{0,\mathrm{ML}}^{-1} \otimes \mathbf{I}_n \right) \boldsymbol{y}, \qquad (10.20)$$

$$\tilde{\boldsymbol{\Sigma}}_{0,\mathrm{ML}} = n^{-1} \cdot (\tilde{\boldsymbol{u}}_1 \, \tilde{\boldsymbol{u}}_2 \, \cdots \, \tilde{\boldsymbol{u}}_M)^T (\tilde{\boldsymbol{u}}_1 \, \tilde{\boldsymbol{u}}_2 \, \cdots \, \tilde{\boldsymbol{u}}_M), \qquad (10.21)$$

where $\tilde{\boldsymbol{u}}_j = \boldsymbol{y}^{(j)} - \mathbf{X}_j \tilde{\boldsymbol{\beta}}_{\mathrm{ML}}^{(j)}$ are vectors of residuals. Under general conditions, $\tilde{\boldsymbol{\beta}}_{\mathrm{ML}}$ has the usual properties of consistency, asymptotic efficiency and asymptotic normality, and having asymptotic variance-covariance matrix

$$\left(\mathbf{X}^T \left[\boldsymbol{\Sigma}_0^{-1} \otimes \mathbf{I}_n \right] \mathbf{X} \right)^{-1}. \qquad (10.22)$$

In fact, all three estimators have the same asymptotic properties.

If the variables in $\mathbf{X}_j$ are all the same and $\boldsymbol{\beta}_1 = \cdots = \boldsymbol{\beta}_M$, then this is known as the *pooled model*. It corresponds to constraint matrices $\mathbf{H}_1 = \cdots = \mathbf{H}_p = \mathbf{1}_M$, which is the usual parallelism assumption. An example is given below. A slight relaxation of this model is to allow $\mathbf{H}_1 = \mathbf{I}_M$ instead.

SUR amounts to *estimated GLS* (EGLS) or *feasible GLS* (FGLS)—this is GLS with a variance-covariance matrix that has been estimated consistently. The 2-stage SUR estimator is unbiased in small samples, assuming that the error terms

Table 10.2 General Electric (GE) and Westinghouse (WE) data (**gew**). Regressors x_2 & x_4 are the capital stocks for GE, and x_3 & x_5 are the market values for WE. The response variables y_1 & y_2 may be regarded as investment figures for the companies. The year is shown for GE only.

General Electric							Westinghouse					
year y_1	x_2	x_3	year	y_1	x_2	x_3	y_2	x_4	x_5	y_2	x_4	x_5
1934 33.1	97.8	1170.6	1944	93.6	319.6	2007.7	12.93	1.8	191.5	39.27	92.4	737.2
1935 45.0	104.4	2015.8	1945	159.9	346.0	2208.3	25.90	0.8	516.0	53.46	86.0	760.5
⋮	⋮	⋮	⋮	⋮	⋮	⋮	⋮	⋮	⋮	⋮	⋮	⋮
1942 61.3	319.9	1749.4	1952	179.5	800.3	2371.6	37.02	84.4	617.2	90.08	174.8	1193.5
1943 56.8	321.3	1687.2	1953	189.6	888.9	2759.9	37.81	91.2	626.7	68.60	213.5	1188.9

have a symmetric distribution; in large samples, it is consistent and asymptotically normal with limiting distribution

$$\sqrt{n}\left(\hat{\boldsymbol{\beta}}-\boldsymbol{\beta}\right) \xrightarrow{\mathcal{D}} N_p\left(\mathbf{0},\left(n^{-1}\mathbf{X}^T\left[\boldsymbol{\Sigma}_0^{-1}\otimes \mathbf{I}_n\right]\mathbf{X}\right)^{-1}\right).$$

Rather than having n as the divisor as in (10.19), some alternatives have been suggested, such as

$$n-\max(p_i,p_j) \quad \text{and} \quad \sqrt{(n-p_i)(n-p_j)}. \tag{10.23}$$

The estimate of σ_{jk} is unbiased for $p_j = p_k$. The argument **divisor** offers these choices, but the default is n.

10.2.3.1 SUR Example

As a simple seemingly unrelated regressions example, consider the well-known General Electric and Westinghouse data, called **gew** (Table 10.2). Many texts and articles illustrate SUR by fitting the simple model

$$\begin{aligned}
Y_{i1} &= \beta_{(1)1}+\beta_{(1)2}\,x_{i2}+\beta_{(1)3}\,x_{i3}+\varepsilon_{i1}, \\
Y_{i2} &= \beta_{(2)1}+\beta_{(2)4}\,x_{i4}+\beta_{(2)5}\,x_{i5}+\varepsilon_{i2}, \quad i=1,\dots,20.
\end{aligned} \tag{10.24}$$

Variables in **gew** are postfixed with ".g" and ".w" to specify the two companies: $x_2 = $ **capital.g**, $x_3 = $ **value.g**, $x_4 = $ **capital.w**, and $x_5 = $ **value.w**. It is very common in SUR analyses to enter in all the variables linearly, but we shall see below that an additive model of this data actually shows a lot of curvature. The responses are $y_1 = $ **invest.g** and $y_2 = $ **invest.w**.

Note that $E(\boldsymbol{y}_i) = \boldsymbol{\eta}_i =$

$$\mathbf{I}_2\begin{pmatrix}\beta_{(1)1}\\\beta_{(2)1}\end{pmatrix}+\begin{pmatrix}1\\0\end{pmatrix}\beta_{(1)2}\,x_{i2}+\begin{pmatrix}1\\0\end{pmatrix}\beta_{(1)3}\,x_{i3}+\begin{pmatrix}0\\1\end{pmatrix}\beta_{(2)4}\,x_{i4}+\begin{pmatrix}0\\1\end{pmatrix}\beta_{(2)5}\,x_{i5}$$

$$=\mathbf{I}_2\begin{pmatrix}\beta_{(1)1}^*\\\beta_{(2)1}^*\end{pmatrix}+\begin{pmatrix}1\\0\end{pmatrix}\beta_{(1)2}^*\,x_{i2}+\begin{pmatrix}1\\0\end{pmatrix}\beta_{(1)3}^*\,x_{i3}+\begin{pmatrix}0\\1\end{pmatrix}\beta_{(1)4}^*\,x_{i4}+\begin{pmatrix}0\\1\end{pmatrix}\beta_{(1)5}^*\,x_{i5}$$

corresponding to the notation established in Sect. 3.3. We can fit the Zellner model (10.24) as follows.

```
> Hlist <- list("(Intercept)" = diag(2),
                "capital.g"    = rbind(1, 0),
                "value.g"      = rbind(1, 0),
                "capital.w"    = rbind(0, 1),
                "value.w"      = rbind(0, 1))
> zef <- vglm(cbind(invest.g, invest.w) ~ capital.g + value.g + capital.w + value.w,
              SURff(divisor = "sqrt"), data = gew,
              constraints = Hlist, maxit = 1, epsilon = 1e-11)
```

By setting `maxit = 1`, the Zellner's 2-stage estimator is computed. The estimates and SEs are

```
> coef(zef, matrix = TRUE)

               invest.g  invest.w
 (Intercept) -27.71932 -1.251988
 capital.g     0.13904  0.000000
 value.g       0.03831  0.000000
 capital.w     0.00000  0.063978
 value.w       0.00000  0.057630

> round(sqrt(diag(vcov(zef))), digits = 3)  # SEs

 (Intercept):1 (Intercept):2    capital.g      value.g    capital.w
        29.321         7.545        0.025        0.014        0.053
        value.w
          0.015
```

The estimate of Σ_0 at the final IRLS iteration can be obtained by inverting one of the working weight matrices (because they are identical). As they are in matrix-band format, some processing is needed.

```
> Sigma0.inv.mb <- head(weights(zef, type = "work"), 1)  # All identical
> Sigma0.inv <- m2a(Sigma0.inv.mb, M = npred(zef))[,, 1]  # All identical
> (Sigma0.hat <- chol2inv(chol(Sigma0.inv)))

          [,1]    [,2]
 [1,]  811.08  224.28
 [2,]  224.28  105.96
```

Its correlation coefficient is 0.77, which is non-negligible. The number of regression parameters for each of the M LMs can be obtained by, e.g.,

```
> nobs(zef, type = "lm") - df.residual(zef, type = "lm")

 invest.g invest.w
        3        3
```

which performs an $n - (n - p_j)$-type calculation.

If one leaves `maxit` at its default, then the usual IRLS iterations will mean the IGLS estimate is computed. Let's do this and use the square root estimator in (10.23) as the divisor.

```
> igls <- vglm(cbind(invest.g, invest.w) ~ capital.g + value.g + capital.w + value.w,
               SURff(divisor = "sqrt"), data = gew, constraints = Hlist)
```

For this, $\hat{\Sigma}_0$ is

```
> sigma0.hat

           [,1]    [,2]
    [1,] 826.16 229.83
    [2,] 229.83 107.00
```

which is essentially the same as `zef` (correlation coefficient is 0.77.)

Now to illustrate the mechanics of fitting a pooled model, this requires the use of the `xij` argument to stack values on top of each other, plus $\mathbf{H}_1 = \cdots = \mathbf{H}_p = \mathbf{1}_M$ obtained by setting `parallel = TRUE` or using the `constraints` argument. The following is not suitable for the `gew` data, but it is used for illustrative purposes only.

```
> Gew <- transform(gew, Capital = capital.g, Value = value.g)
> fitp1 <- vglm(cbind(invest.g, invest.w) ~ Capital + Value,
                SURff(parallel = TRUE), data = Gew, maxit = 1,
                xij = list(Capital ~ capital.g + capital.w - 1,
                           Value   ~ value.g   + value.w   - 1),
                form2 = ~ capital.g + value.g + capital.w + value.w +
                          Capital + Value)
> Hlist <- list("(Intercept)" = rbind(1, 1),
                "Capital"     = rbind(1, 1),
                "Value"       = rbind(1, 1))
> fitp2 <- vglm(cbind(invest.g, invest.w) ~ Capital + Value,
                SURff, data = Gew, maxit = 1, constraints = Hlist,
                xij = list(Capital ~ capital.g + capital.w - 1,
                           Value   ~ value.g   + value.w),
                form2 = ~ capital.g + value.g + capital.w + value.w +
                          Capital + Value)
```

Both `fitp1` and `fitp2` are identical. Of course, it is possible here to allow $\mathbf{H}_1 = \mathbf{I}_2$ instead—this is left as an exercise (Ex. 10.4).

While SUR traditionally enters variables linearly as in (10.24), there is no reason why an additive model cannot be fitted (e.g., Smith and Kohn (2000))—the VGLM/VGAM framework naturally accommodates this. For example, the additive model extension of (10.24) is

$$Y_{i1} = \beta_{(1)1} + f_{(1)2}(x_{i2}) + f_{(1)3}(x_{i3}) + \varepsilon_{i1} \tag{10.25}$$
$$Y_{i2} = \beta_{(2)1} + f_{(2)4}(x_{i4}) + f_{(2)5}(x_{i5}) + \varepsilon_{i2}, \quad i = 1, \dots, 20. \tag{10.26}$$

To fit such a model in **VGAM** using regression splines and smoothing splines, try

```
> Hlist3 <- Hlist4 <- list("(Intercept)"   = diag(2),
                           "bs(capital.g)" = rbind(1, 0),
                           "bs(value.g)"   = rbind(1, 0),
                           "bs(capital.w)" = rbind(0, 1),
                           "bs(value.w)"   = rbind(0, 1))
> names(Hlist4) <- c("(Intercept)", "s(capital.g)", "s(value.g)", "s(capital.w)",
                     "s(value.w)")
> fit.rs <- vglm(cbind(invest.g, invest.w) ~
                 bs(capital.g) + bs(value.g) + bs(capital.w) + bs(value.w),
                 SURff, data = gew, constraints = Hlist3, maxit = 1)
> fit.ss <- vgam(cbind(invest.g, invest.w) ~
                 s(capital.g) + s(value.g) + s(capital.w) + s(value.w),
                 SURff, data = gew,
                 constraints = Hlist4, maxit = 1, bf.maxit = 100) # Suppress warning
> plot(as(fit.rs, "vgam"), se = TRUE, scol = "blue")
```

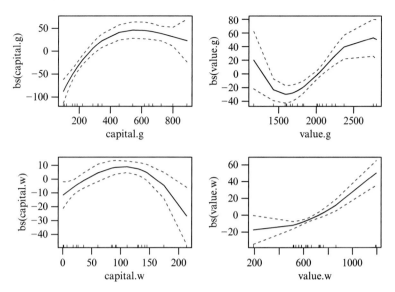

Fig. 10.4 Smooths of a SUR model applied to the `gew` data. The fitted model is called `fit.rs`.

which gives Fig. 10.4. The plots and the summary suggest nonlinearity in about 3 of the 4 plots! Hence this casts doubt on the validity of (10.24): most of the variables $x_2, \ldots, x_5$ should *not* be entered linearly. This example illustrates a point in data analysis that does not seem to be well-heeded: employing a sophisticated method to obtain a small improvement, such as avoiding a small loss of efficiency, is misguided if more fundamental errors are made, such as overlooking gross nonlinearity.

For a pooled nonparametric model, one is restricted to regression splines, however, the technique described in Sect. 3.4.4 can be used.

10.2.4 The AR(1) Time Series Model

The first-order autoregressive process time series model AR(1) provides a simple example where the EIMs are not all identical even when the model is intercept-only. For data y_i, $i = 1, \ldots, n$, it is common for the AR(1) model to be defined by

$$Y_1 \sim N(\mu, \sigma^2/(1 - \rho^2)), \tag{10.27}$$

$$Y_i = \mu^* + \rho Y_{i-1} + \varepsilon_i, \quad i = 2, \ldots, n, \tag{10.28}$$

$$\varepsilon_i \sim N(0, \sigma^2) \text{ independently}, \tag{10.29}$$

where $|\rho| < 1$. Some books omit the drift parameter μ^* from (10.28), which corresponds to $E(Y_i) = 0$. Having this parameter therefore allows a test of stationarity in the form of $H_0 : \mu^* = \mu_i^*$ versus $H_1 : \mu^* \neq \mu_i^*$ for all i, i.e., is the scaled mean intercept-only?

From (10.27)–(10.29) it is easy to see that the conditional distribution

$$Y_i|Y_{i-1} \sim N(\mu^* + \rho Y_{i-1}, \sigma^2) \tag{10.30}$$

can be used to calculate the joint distribution: $f(\boldsymbol{y}; \mu^*, \sigma, \rho) =$

$$f_{Y_1}(y_1) \cdot \prod_{i=2}^{n} f_{Y_i|Y_{i-1}}(y_i|y_{i-1}) \;=\; \sqrt{\frac{1-\rho^2}{2\pi\sigma^2}} \exp\left\{-\frac{1}{2}\frac{(y_1-\mu)^2}{\sigma^2}(1-\rho^2)\right\} \cdot$$

$$\prod_{i=2}^{n} \frac{1}{\sqrt{2\pi\sigma^2}} \exp\left\{-\frac{1}{2}\frac{(y_i-\mu^*-\rho\,y_{i-1})^2}{\sigma^2}\right\},$$

which is the product of the marginal likelihood with the conditional likelihood.

Some very basic properties arising from (10.27)–(10.29), for intercept-only models, are:

$$E(Y_i) \;\equiv\; \mu \;=\; \frac{\mu^*}{1-\rho}, \qquad i = 2, \ldots, n,$$

$$\mathrm{Var}(Y_i) \;=\; \frac{\sigma^2}{1-\rho^2} \qquad \text{for all } i = 1, \ldots, n,$$

$$\mathrm{Cov}(\varepsilon_i, Y_{i-j}) \;=\; 0 \qquad \text{for } j > 0,$$

$$\mathrm{Corr}(Y_i, Y_{i-j}) \;=\; \rho^j \qquad \text{for } j = 0, 1, 2, \ldots.$$

Let $\boldsymbol{\theta} = (\mu^*, \sigma, \rho)^T$ be the parameters to be estimated. Then

$$\ell(\boldsymbol{\theta}; \boldsymbol{y}) \;=\; \ell_1(\boldsymbol{\theta}; y_1) + \sum_{i=2}^{n} \ell_i(\boldsymbol{\theta}; y_i|y_{i-1}). \tag{10.31}$$

Sometimes, practitioners maximize the conditional log-likelihood function only, while others use the full (exact) log-likelihood function (10.31). These two options are reflected in the `type.likelihood` argument of the VGAM family function `AR1()`, which has default value `c("exact", "conditional")`, with the first choice being the default method. If `type. likelihood = "conditional"`, then the MLE of μ^* and ρ may be obtained by the simple linear regression (10.28), while if `type.likelihood = "exact"` then iteration *is* really needed to compute the MLE of $\boldsymbol{\theta}$. Computationally, if `type.likelihood = "conditional"`, then the first observation is effectively deleted by setting its prior weight equal to some small positive number, e.g., $w_1 \approx 0$ in (3.7). This crude way of ignoring the first observation may be replaced by actual deletion in future versions of the software.

The EIM for the first observation is

$$\boldsymbol{\mathcal{I}}_{E1}(\mu_1^*, \sigma_1, \rho_1) \;=\; \begin{pmatrix} \dfrac{1+\rho_1}{1-\rho_1}\dfrac{1}{\sigma_1^2} & 0 & 0 \\[2ex] 0 & \dfrac{2}{\sigma_1^2} & \dfrac{2\rho_1}{\sigma_1(1-\rho_1^2)} \\[2ex] 0 & \dfrac{2\rho_1}{\sigma_1(1-\rho_1^2)} & \dfrac{2\rho_1^2}{(1-\rho_1^2)^2} \end{pmatrix}. \tag{10.32}$$

It is easy to show that this is of rank-2, and that it might be 'fixed up' by multiplying the off-diagonal element by a number slightly less than 1, e.g., 0.99.

For $i > 1$, the EIM for the ith observation is

$$\boldsymbol{\mathcal{I}}_{Ei}(\mu_i^*, \sigma_i, \rho_i) \;=\; \frac{1}{\sigma_i^2} \begin{pmatrix} 1 & 0 & \mu_i \\[1ex] 0 & 2 & 0 \\[1ex] \mu_i & 0 & \mu_{i-1}^2 + \dfrac{\sigma_{i-1}^2}{1-\rho_{i-1}^2} \end{pmatrix}, \tag{10.33}$$

where a subscript i is needed. When the individual EIMs are added, and μ^* is disregarded, and an intercept-only model is fitted, then the EIM is given by the well-known matrix

$$
\boldsymbol{\mathcal{I}}_E(\sigma, \rho) = \begin{pmatrix} \dfrac{2n}{\sigma^2} & \dfrac{2\rho}{\sigma(1-\rho^2)} \\ \dfrac{2\rho}{\sigma(1-\rho^2)} & \dfrac{2\rho^2}{(1-\rho^2)^2} + \dfrac{n-1}{1-\rho^2} \end{pmatrix}.
\tag{10.34}
$$

10.3 Binomial Variants

Variant models of the binomial distribution that have been implemented in **VGAM** are summarized in Table 10.3, and most are cast into the VGLM/VGAM framework in this section. For example, Sect. 10.3.1 describes a model for two sequential binomial distributions, and Sects. 10.3.3–10.3.4 describe models for bivariate binary responses (one is based on odds ratios, one is based on an underlying N_2 distribution; and there is also a family of loglinear models.)

10.3.1 Two-Stage Sequential Binomial

Crowder and Sweeting (1989) considered a 'bivariate binomial' model where there are Y_1^* successes arising from a binomial distribution, and then Y_2^* successes out of the initial number of successes. Their example was that each of w spores has a probability μ_1 of germinating. Of the Y_1^* spores that germinate, each has a probability μ_2 of bending in a particular direction. Let Y_2^* be the number that bend in the specified direction.

This model can be described as a 2-stage (sequential) binomial distribution, and the **VGAM** family function `seq2binomial()` implements this. Let $Y_1 = Y_1^*/w$ and $Y_2 = Y_2^*/Y_1^*$ be the respective sample proportions. The joint probability function is $P(y_1, y_2; \mu_1, \mu_2) =$

$$
\binom{w}{wy_1} \mu_1^{wy_1}(1-\mu_1)^{w(1-y_1)} \cdot \binom{wy_1}{wy_1y_2} \mu_2^{wy_1y_2}(1-\mu_2)^{wy_1(1-y_2)},
\tag{10.35}
$$

where independence is assumed throughout. Here, the support is $y_1 = 0(w^{-1})1$ and $y_2 = 0((wy_1)^{-1})1$. The default links are $\boldsymbol{\eta} = (\text{logit } \mu_1, \text{logit } \mu_2)^T$, and the fitted values returned are $(\hat{\mu}_1, \hat{\mu}_2)$. The `parallel` argument allows the constraint $\mu_1 = \mu_2$. As with other binomial-based family functions, the prior weights w give the number of trials, here for the first stage only.

Table 10.3 Binomial variants currently implemented in **VGAM**. Most are described in Sect. 10.3. Index j takes on values 1 and 2.

Model	Main specification	VGAM family
Bivariate odds ratio model	Eqs. (1.18)–(1.19)	`binom2.or()`
Bivariate probit model	Eqs. (10.40)–(10.41)	`binom2.rho()`
Bradley-Terry model	Eq. (10.36)	`brat()`
Double-exponential binomial	Eq. (10.54)	`double.expbinomial()`
Loglinear bivariate binomial	Eq. (10.46)	`loglinb2()`
Loglinear trivariate binomial	Eq. (10.47)	`loglinb3()`
Two-stage sequential binomial	Eq. (10.35)	`seq2binomial()`

10.3.2 The Bradley-Terry Model

The standard Bradley-Terry model is applicable to pairwise-comparisons data where there are two outcomes, such as win/lose or better/worse. Typical examples are sports competitions between two individuals or teams, and comparing two food items (e.g., wines, olive oils, cheeses) at a time—in food-tasting experiments, judges can more easily decide between two items rather than comparing more than two items at a time. Also known as the Bradley-Terry-Luce model, they have also been used to rank different journals, e.g., Journal A citing Journal B is a 'loss' to A and a 'win' to B. The possibility of a third outcome—ties—is deferred till later.

Suppose that the comparison between items T_i and T_j is conducted n_{ij} times for $i = 1, \ldots, I = M + 1$, e.g., there could be n_{ij} judges or sports matches. Let $N = \sum\sum_{i<j} n_{ij}$ be the total number of pairwise comparisons, and assume independence for ratings of the same pair by different judges, and for ratings of different pairs by the same judge. If $\alpha_i > 0$ is the *ability* or *worth* of item T_i, then

$$P[T_i > T_j] = p_{i/ij} = \frac{\alpha_i}{\alpha_i + \alpha_j}, \quad i \neq j,$$

where "$T_i > T_j$" means i is preferred over j. Hence the probability that T_i is superior to T_j in any given trial is their relative values to each other. Let Y_{ij}^* be the number of times that T_i is preferred over T_j in the n_{ij} comparisons of the pairs. Then $Y_{ij}^* \sim \mathrm{Bin}(n_{ij}, p_{i/ij})$ but, as usual, we operate on the sample proportions $Y_{ij} = Y_{ij}^*/n_{ij}$ (also known as the *scaled binomial* distribution). Then

$$L(\alpha_1, \ldots, \alpha_M) \propto \prod_{i=1}^{M} \prod_{j=i+1}^{M+1} \binom{n_{ij}}{n_{ij}\, y_{ij}} \left(\frac{\alpha_i}{\alpha_i + \alpha_j} \right)^{n_{ij} y_{ij}} \left(\frac{\alpha_j}{\alpha_i + \alpha_j} \right)^{n_{ij}(1-y_{ij})}$$

and the **VGAM** family function `brat()` maximizes this. By default, $\alpha_{M+1} \equiv 1$ is used for identifiability, however, this can be changed very easily to some other baseline category and/or value. Note that

$$\mathrm{logit}\left(\frac{\alpha_i}{\alpha_i + \alpha_j} \right) = \log\left(\frac{\alpha_i}{\alpha_j} \right) = \lambda_i - \lambda_j, \tag{10.36}$$

say.

In its current implementation, `brat()` can only handle intercept-only models because the response is taken to be a $1 \times M(M+2)$ matrix of counts, hence effectively $n = 1$. By default, it has

$$\eta_j = \lambda_j = \log \alpha_j = \beta_{(j)1}, \quad j = 1, \ldots, M. \tag{10.37}$$

When fitting the Bradley-Terry model, one may use the preprocessing function `Brat()` to convert the square matrix of counts into the format used internally by `brat()`. The typical usage is

```
vglm(Brat(counts.matrix) ~ 1, family = brat)
```

Here, the rows of `counts.matrix` are 'winners', and the columns are 'losers'.

To close, we mention two extensions of the standard Bradley-Terry model, the first of which is implemented in **VGAM**. Firstly, ties (e.g., a draw or equal preference) are allowed for some types of comparison. Of several common proposals to handle ties, one is

$$
\begin{aligned}
P(T_i > T_j) &= \alpha_i/(\alpha_i + \alpha_j + \alpha_0), \\
P(T_i < T_j) &= \alpha_j/(\alpha_i + \alpha_j + \alpha_0), \\
P(T_i = T_j) &= \alpha_0/(\alpha_i + \alpha_j + \alpha_0),
\end{aligned}
\tag{10.38}
$$

where α_0 (> 0) is an extra parameter. This is implemented by the family function `bratt()`. It has $\boldsymbol{\eta} = (\log \alpha_1, \ldots, \log \alpha_{M-1}, \log \alpha_0)^T$ by default, where there are now M competitors, and $\alpha_M \equiv 1$. Like `brat()`, one can choose a different baseline group and value. A count matrix of ties can be fed in as the second argument of `Brat()`.

Secondly, in some applications, it is realistic to postulate a 'home team advantage', e.g., the contest occurs at one of the team's location. Another example is where the order of the comparison makes a difference, e.g., in wine tasting competitions, it is well-known that the first wine tasted is usually thought of as better than the other. Then (10.36) might be generalized to

$$\text{logit}\,(\alpha_i/(\alpha_i + \alpha_j)) = \log\,(\alpha_i/\alpha_j) = \alpha_h + \lambda_i - \lambda_j. \tag{10.39}$$

The home team advantage parameter α_h is positive if indeed there is a real home advantage.

10.3.2.1 Rugby Example

Let's fit a Bradley-Terry model with ties to the **rugby** and **rugby.ties** data, which give the wins, losses and ties of international rugby union matches up to late-2013 of about 10 countries. We'll only look at a few selected countries.

```
> countries <- c("Australia", "England", "France", "New Zealand", "South Africa")
> (Rugby.small <- rugby[countries, countries])

              loser
  winner        Australia England France New Zealand South Africa
    Australia          NA      24     23          41           33
    England            16      NA     53           7           12
    France             17      37     NA          12           11
    New Zealand       101      27     39          NA           50
    South Africa       44      22     21          34           NA

> (Rugby.ties.small <- rugby.ties[countries, countries])

              loser
  winner        Australia England France New Zealand South Africa
    Australia          NA       1      2           6            1
    England             1      NA      7           1            2
    France              2       7     NA           1            6
    New Zealand         6       1      1          NA            0
    South Africa        1       2      6           0           NA
```

Here are two models, the latter allowing for ties (draws).

```
> rugger.fit  <- vglm(Brat(Rugby.small) ~ 1, brat(refgp = 1))
> rugger.fitt <- vglm(Brat(Rugby.small, Rugby.ties.small) ~ 1, bratt(refgp = 1))
```

The 'abilities' of each team may be computed as follows.

```
> abilities <- c(exp(0), Coef(rugger.fitt))
> names(abilities) <- c(countries,
                        if (familyname(rugger.fitt) == "brat") NULL else "alpha0")
> round(abilities, 2)

     Australia      England       France  New Zealand South Africa       alpha0
          1.00         0.77         0.62         2.33         1.38         0.10

> round(sort(abilities), 2)

        alpha0       France      England    Australia South Africa  New Zealand
          0.10         0.62         0.77         1.00         1.38         2.33
```

These results rank the countries according to their rugby abilities. The value **alpha0** is a measure of the frequency of draws.

Here is some output for the model without ties.

```
> check <- InverseBrat(fitted(rugger.fit))   # Probabilities of winning
> round(check, 2)

                Australia England France New Zealand South Africa
    Australia          NA    0.56   0.61        0.30         0.41
    England          0.44      NA   0.55        0.25         0.36
    France           0.39    0.45     NA        0.22         0.31
    New Zealand      0.70    0.75   0.78          NA         0.62
    South Africa     0.59    0.64   0.69        0.38           NA
```

This shows each pair of countries and the probability that the **winner** country beats the **loser** country. It is left to the reader to check that **check + t(check)** is a matrix whose off-diagonal elements are all 1s.

So, based on these data, who might win the next Rugby World Cup, say? Based on historical data over all test matches, one might be led to believe that New Zealand has the best chance. However, as sports followers and investors know, spectacular past performance does not necessarily mean a similar future result.

10.3.3 Bivariate Responses: The Bivariate Probit Model

Pairs of binary responses are a frequent form of data. Such commonly arise in medical and biological studies, e.g., ophthalmic studies where each eye is a response, and measurements on pairs such as twins. Write $\boldsymbol{Y} = (Y_1, Y_2)^T$, where Y_1 and Y_2 takes only the values 0 and 1 (denoting "failure" and "success", respectively). Let $\mu_{rs} = P(Y_1 = r, Y_2 = s)$, $r, s = 0, 1$, be the joint probabilities, and $\mu_j = P(Y_j = 1)$, $j = 1, 2$, be the marginal probabilities.

We saw in Sect. 1.2.3 that a commonly used model for bivariate responses is the bivariate odds ratio model. Another popular model is the *bivariate probit model* (BPM; Ashford and Sowden, 1970), which can be written

$$P(Y_j = 1|\boldsymbol{x}) = \Phi(\eta_j(\boldsymbol{x})), \qquad j = 1, 2, \tag{10.40}$$

$$P(Y_1 = 1, Y_2 = 1|\boldsymbol{x}) = \Phi_2(\eta_1(\boldsymbol{x}), \eta_2(\boldsymbol{x}); \rho(\boldsymbol{x})). \tag{10.41}$$

Essentially, the two binary responses are mapped onto a standardized bivariate normal distribution with correlation parameter ρ, where the 4 quadrants intersect at (η_1, η_2). The BPM can be simply interpreted in terms of latent variables:

$$y_{1i}^* = \eta_1 + \varepsilon_{1i}, \tag{10.42}$$

$$y_{2i}^* = \eta_2 + \varepsilon_{2i}, \tag{10.43}$$

where the errors have a standardized bivariate normal distribution

$$\begin{pmatrix} \varepsilon_{1i} \\ \varepsilon_{2i} \end{pmatrix} \sim N_2 \left(\begin{pmatrix} 0 \\ 0 \end{pmatrix}, \begin{pmatrix} 1 & \rho_i \\ \rho_i & 1 \end{pmatrix} \right), \tag{10.44}$$

independently. Then the observed responses are generated by

$$y_{si} = \begin{cases} 1, & y_{si}^* > 0, \\ 0, & y_{si}^* \leq 0, \end{cases} \quad s = 1, 2. \tag{10.45}$$

The standardized version is used for identifiability and to enhance interpretability, e.g., the zero means that the intercepts for the marginals are adjusted accordingly.

Note that while the bivariate odds ratio model may be fitted with different links for the marginals, the BPM is theoretically tied to the bivariate normal distribution and so it does not offer such flexibility: each marginal is modelled as a "probit analysis". The multivariate probit model, of which the BPM is a special case, is generally applicable to $M \geq 3$ binary responses, however, it is computationally difficult to estimate because it requires integration of an N_M density.

Also note that, as Fisher scoring is implemented, no Y_j is allowed as an explanatory variable otherwise the EIM becomes invalid (Freedman and Sekhon, 2010).

As $-1 < \rho < 1$, the default link for $\eta_3 = \log((1+\rho)/(1-\rho))$ (called `"rhobit"`) and therefore $\rho = (\exp\{\eta_3(\boldsymbol{x})\}-1)/(\exp\{\eta_3(\boldsymbol{x})\}+1)$ satisfies the range restrictions. While ρ may be modelled as a function of covariates, it is recommended that it remain as intercept-only (the default) unless the data set or its effect is large. The same is true for the association parameter for the bivariate odds ratio model: it has an argument `oratio`, denoting the odds ratio, which is set to intercept-only because `zero = 3`, by default, too.

The bivariate odds ratio model has several other advantages over the BPM, e.g., it is computationally simpler, and odds ratios are preferred to correlation coefficients when describing the association between two binary variables, because they are more interpretable and have less severe range-restriction problems.

10.3.4 Binary Responses: Loglinear Models

Models for multivariate binary responses can be constructed from loglinear models. The bivariate case is

$$\log P(Y_1 = y_1, Y_2 = y_2 | \boldsymbol{x}) = u_0(\boldsymbol{x}) + u_1(\boldsymbol{x})\, y_1 + u_2(\boldsymbol{x})\, y_2 + u_{12}(\boldsymbol{x})\, y_1\, y_2, \quad (10.46)$$

where $y_j = 0$ or 1, and $\boldsymbol{\eta} = (u_1, u_2, u_{12})^T$. Here, u_0 is a normalizing parameter equal to $-\log(1 + e^{u_1} + e^{u_2} + e^{u_1+u_2+u_{12}})$. The trivariate case has

$$\log P(Y_1 = y_1, Y_2 = y_2, Y_3 = y_3) = u_0 + u_1\, y_1 + u_2\, y_2 + u_3\, y_3 +$$
$$u_{12}\, y_1\, y_2 + u_{13}\, y_1\, y_3 + u_{23}\, y_2\, y_3 \quad (10.47)$$

with $\boldsymbol{\eta} = (u_1, u_2, u_3, u_{12}, u_{13}, u_{23})^T$. Family functions `loglinb2()` and `loglinb3()` fit these models, e.g.,

```
vgam(cbind(y1, y2) ~ s(x2, df = c(4, 2)),
     loglinb2(exchangeable = TRUE, zero = NULL), data = ldata)
```

should fit (10.46) subject to $u_1 = u_2$. Here, $u_1(x_2)$ and $u_{12}(x_2)$ are assigned 4 and 2 degrees of freedom, respectively. An identity link for each of the u_s and u_{st} is chosen, because the parameter space is unconstrained. For the reason given below, parameters u_{st} are intercept-only by default. As another example of exchangeable errors, `loglinb3(exch = TRUE)` would fit (10.47) subject to $u_1 = u_2 = u_3$ and $u_{12} = u_{13} = u_{23}$. Exchangeable models with 3 binary responses are less commonly encountered than the 2-response case.

A **VGAM** convention is that the fitted values correspond to the matrix with columns $\hat{\mu}_{stv}$ where $(s, t, v, \ldots) = (0, 0, \ldots, 0), (0, \ldots, 0, 1), \ldots, (1, \ldots, 1)$. In particular, for $Q = 2$ responses, these are $(0, 0), (0, 1), (1, 0), (1, 1)$. The input of the response is easiest fed in as a Q-column matrix of 1s and 0s.

Now suppose more generally that there are Q binary responses. For loglinear models, it is often a good idea to force $u_{stv} \equiv 0$, and similarly for other higher-order associations (done for the $Q = 3$ case above). There are several reasons why this is so. Unless the data contain all fitted combinations, such assumptions are often necessary because the estimates become unbounded. It also reduces the complexity of the problem, furthermore, higher-order associations become increasingly more difficult to interpret. And if smoothed, higher-order associations should be assigned less flexibility than lower-order associations. The general loglinear model is

$$\log P(Y_1 = y_1, \ldots, Y_Q = y_Q | \boldsymbol{x}) = u_0(\boldsymbol{x}) + \sum_{s=1}^{Q} u_s(\boldsymbol{x}) \, y_s + \sum_{s<t} u_{st}(\boldsymbol{x}) \, y_s \, y_t +$$

$$\sum_{s<t<v} u_{stv}(\boldsymbol{x}) \, y_s \, y_t \, y_v + \ldots . \qquad (10.48)$$

The normalizing parameter u_0 satisfies $e^{-u_0} = 1 + \sum_{s=1}^{Q} e^{u_s} + \sum_{s<t} e^{u_s + u_t + u_{st}} +$

$$\sum_{s<t<v} e^{u_s + u_t + u_v + u_{st} + u_{sv} + u_{tv}} + \cdots + \exp\left(\sum_{s=1}^{Q} u_s + \sum_{s<t} u_{st} + \cdots \right).$$

One has $\boldsymbol{\eta} = (u_1, \ldots, u_S, u_{12}, u_{13}, \ldots, u_{123\ldots Q})^T$ which grows very quickly in dimension with respect to Q, hence the need to set high-order interaction terms to 0. With IRLS, it may be shown for this model that Newton-Raphson coincides with Fisher scoring. Although a simpler algorithm called the *iterative proportional fitting procedure* is still used for estimating loglinear models, our limited experience has indicated that IRLS works well.

As an example, we try to mimic the results of McCullagh and Nelder (1989, Sect.6.6). We'll coerce the initial table of counts into a 2-column matrix of 0s and 1s that will be weighted by a variable `Frequency`, which will be uses as input. Below, the matrix `Counts` has 4 columns ordered as $(y_1 = 0, y_2 = 0)$, $(0, 1)$, $(1, 0)$ and $(1, 1)$.

```
> coalminers <- transform(coalminers, Age = (age - 42) / 5)
> fit.temp <- vglm(cbind(nBnW, nBW, BnW, BW) ~ Age, binom2.or, data = coalminers)
> Counts <- round(c(weights(fit.temp, type = "prior")) * depvar(fit.temp))
> newminers <- data.frame(breathlessness = c(0, 0, 1, 1),   # Values recycle
                          wheeze         = c(0, 1, 0, 1),   # Values recycle
                          Frequency      = c(t(Counts)),
                          Age            = with(coalminers, rep(Age, each = 4)))
> newminers <- subset(newminers, Frequency > 0)   # Not needed here, actually
> fit.coal <- vglm(cbind(breathlessness, wheeze) ~ Age, loglinb2(zero = NULL),
                   weight = Frequency, data = newminers)
> coef(fit.coal, matrix = TRUE)

              u1      u2       u12
(Intercept) -3.4778 -2.0090  3.05948
Age          0.5154  0.2006 -0.16615
```

As an exercise (Ex. 10.14), it is left to the reader to reconcile this with the following conditional probabilities (McCullagh and Nelder, 1989):

$$\begin{aligned}
\text{logit } P(Y_1 = 1 | Y_2 = 1, x_2) &= -0.418 + 0.349 \, x_2, & (10.49) \\
\text{logit } P(Y_2 = 1 | Y_1 = 1, x_2) &= 1.051 + 0.034 \, x_2, & \\
\log \psi &= 3.059 - 0.166 \, x_2, & (10.50)
\end{aligned}$$

where ψ is the odds ratio. Conditional probabilities are a natural outcome when the joint probabilities are modelled via loglinear models.

10.3.5 Double Exponential Models

Overdispersion is a common characteristic of data and one potential method for handling it is to consider the class of *double exponential distributions* (Efron, 1986). The basic idea is that an ordinary one-parameter exponential family allows the addition of a second parameter θ which varies the dispersion of the family without changing the mean. The extended family behaves like the original family with sample size changed from N to $N\theta$. Then $0 < \theta < 1$ corresponds to overdispersion. The extended family is an exponential family in μ when N and θ are fixed, and an exponential family in θ when N and μ are fixed.

The formal definition

$$\tilde{f}_{\mu,\theta,N}(y) \;=\; c(\mu,\theta,N)\,\theta^{1/2}\,\{g_{\mu,N}(y)\}^{\theta}\,\{g_{y,N}(y)\}^{1-\theta}\,[dG_N(y)] \qquad (10.51)$$

is called a double exponential family with parameters μ, θ and N. Here,

$$g_{\mu,N}(y) \;=\; e^{N[\eta y - \psi(\mu)]} \cdot [dG_N(y)] \qquad (10.52)$$

is the density of a given exponential family. The expectation parameter $\mu = \int_{-\infty}^{\infty} y\, g_{\mu,N}(y)\, dG_N(y)$, y is the natural statistic, η is the canonical or natural parameter, $\psi(\mu)$ is a normalizing function, and $G_N(y)$ is the *carrier measure* for the exponential family.

The normalizing constant $c(\mu,\theta,n)$ in (10.52) nearly equals 1, so (10.52) is approximated by

$$f_{\mu,\theta,N}(y) \;\approx\; \theta^{1/2}\,\{g_{\mu,N}(y)\}^{\theta}\,\{g_{y,N}(y)\}^{1-\theta}\,[dG_N(y)]\,. \qquad (10.53)$$

Approximately, the mean of Y is μ. The *effective sample size* is the dispersion parameter multiplied by the original sample size, i.e., $N\theta$.

At present, only one double exponential family is implemented, and that is `double.expbinomial()`. The default model has $\boldsymbol{\eta} = (\text{logit}\,\mu, \text{logit}\,\theta)^T$, which restricts both parameters to lie between 0 and 1—although the dispersion parameter can be modelled over a larger parameter space by assigning the link function argument `ldispersion`. The approximate double-binomial family based on (10.53) gives

$$f_{\mu,\theta}(y) \;\propto\; \theta^{1/2}\,\mu^{Ny\theta}(1-\mu)^{N(1-y)\theta}y^{Ny(1-\theta)}(1-y)^{N(1-y)(1-\theta)} \qquad (10.54)$$

for $y = 0(N^{-1})1$. Maximum likelihood estimation is used for the two (mean and dispersion) parameters. In fact, Fisher scoring is used, and the two estimates are asymptotically independent because the EIM is diagonal. Indeed, it is approximated by $\text{diag}(N\theta/V(\mu), \frac{1}{2}\theta^{-2})$ where $\text{Var}(Y) = V(\mu)$.

For overdispersed binomial data, the most commonly used full-likelihood model is the beta-binomial (Sect. 11.4). Likewise, for counts that are overdispersed with respect to the Poisson distribution, negative binomial regression is most commonly employed (Sect. 11.3).

Bibliographic Notes

Varying-coefficient models are described in, e.g., Hastie and Tibshirani (1993) and Fan and Yao (2003). A recent VCM review is Park et al. (2015).

Some treatments on SUR include Srivastava and Dwivedi (1979), Srivastava and Giles (1987), Greene (2012). The package systemfit can fit SUR models as well as two-stage least squares (2SLS), weighted two-stage least squares (W2SLS), and three-stage least squares (3SLS) models.

Tobit models are described in e.g., Amemiya (1984) and Smithson and Merkle (2013), and most econometrics books such as Greene (2012) and Wooldridge (2006). Software for fitting censored regression models in R include censReg, and survreg() in survival.

The Bradley-Terry model is also described in, e.g., Agresti (2013), and can be fitted with packages BradleyTerry2 and prefmod.

A general reference for bivariate binomial data is McCullagh and Nelder (1989). The bivariate odds ratio model is also described in Palmgren (1989). Other software exists for fitting bivariate probit-like models. The MNP package fits the multinomial probit model via Markov chain Monte Carlo. Package SemiParBIVProbit fits semiparametric BPMs, while SemiParSampleSel allows for semiparametric sample selection modelling with continuous response (using copula sample selection models). General books with loglinear model content include McCullagh and Nelder (1989), Christensen (1997), Agresti (2013), and von Eye and Mun (2013).

With some similarity to double exponential models, package dglm fits a family of models called double-GLMs, as described in Smyth (1989).

Exercises

Ex. 10.1.   VCMs Fitted to the ethanol Data
Here, some alternative models to (10.5) are fitted.

(a) Produce scatter plots of Y versus E and C, as in Fig. 10.1a,b. Note that each stratum has approximately equal sample sizes.
(b) Fit the following VCMs and compute their residual sum of squares. From these values and the degrees of freedom, suggest which model is to be preferred. Here, estimate the $\beta_j(E)$ by smoothing splines using s(E, df = 6.8 - 1) within vgam().

$$\mathrm{NO}_x = \beta_1(E) + \beta_2 \cdot C + \varepsilon, \qquad (10.55)$$
$$\mathrm{NO}_x = \beta_1(E) + \beta_2 \cdot C \cdot E + \varepsilon, \qquad (10.56)$$
$$\mathrm{NO}_x = \beta_1(E) + \beta_2 \cdot C + \beta_3 \cdot C \cdot E + \varepsilon, \qquad (10.57)$$
$$\mathrm{NO}_x = \beta_1(E) + \beta_2(E) \cdot C + \varepsilon. \qquad (10.58)$$

Note that (10.5) and (10.58) are identical.
(c) Replace both functions in (10.58) by poly(E, 2) and obtain the value of E which gives the maximum value. Are these replacement functions acceptable?

Ex. 10.2.   VCM Fitting an Ordinary LM

Consider a VCM used to fit an ordinary LM, i.e., `formula = y ~ 1` and `family = normal.vcm`. Show that the BHHH method leads to the ordinary LM normal equations solution and therefore has the same standard errors. For simplicity, do this for $\eta_1, \ldots, \eta_{M-1}$ only and ignore $\eta_M = \log \sigma$.

Ex. 10.3.   Observed Mean of the Tobit Model

A random variable V has a *doubly truncated* normal distribution if its probability density function is   [Johnson et al. (1994)]

$$\sigma^{-1} \, \phi\left(\frac{v - \mu}{\sigma}\right) \left\{\Phi\left(\frac{B - \mu}{\sigma}\right) - \Phi\left(\frac{A - \mu}{\sigma}\right)\right\}^{-1}, \qquad A < v < B. \qquad (10.59)$$

Here, A and B are the *lower* and *upper truncation points*.

(a) Show that

$$E(V) \;=\; \mu + \sigma \, \frac{\phi[(A - \mu)/\sigma] - \phi[(B - \mu)/\sigma]}{\Phi[(B - \mu)/\sigma] - \Phi[(A - \mu)/\sigma]}.$$

(b) The Tobit model is where Y has a rescaled distribution of V, augmented by $P(Y = A) = P(Y^* < A)$ and $P(Y = B) = P(Y^* > B)$, where $Y^* \sim N(\mu, \sigma^2)$. Show that $E(Y)$ equals Eq. (10.11).

(c) Use (10.11) to show that, for the standard Tobit model,

$$E(Y|\boldsymbol{x}_i) \;=\; \Phi(\mu_i^*/\sigma_i) \, [\mu_i^* + \sigma_i \, \lambda(-\mu_i^*/\sigma_i)]$$

where

$$\lambda(u) \;=\; \frac{\phi(u)}{1 - \Phi(u)} \;=\; \frac{\phi(-u)}{\Phi(-u)}$$

is known as the *inverse Mills ratio*.

Ex. 10.4.   Relaxed Pooled SUR

Adapt `fitp1` and `fitp2` in Sect. 10.2.3.1 to allow $\mathbf{H}_1 = \mathbf{I}_2$, i.e., a relaxation of the pooled model. Confirm that your solutions are the same.

Ex. 10.5.   Pooled Nonparametric SUR

For illustration's sake only, suppose $x_2 = x_4$ and $x_3 = x_5$ in the `gew` data, i.e., the capital stock variables are the same, and ditto for the market values. Fit (10.25)–(10.26) using regression splines, subject to the functions being equal, i.e., a pooled model.

Ex. 10.6.   OLS and GLS in Seemingly Unrelated Regressions

Show that OLS and the GLS estimators (10.14)–(10.15) are equivalent when any one of the following conditions hold.

(a) $\boldsymbol{\Sigma}$ is diagonal.
(b) $\mathbf{X}_1 = \mathbf{X}_2 = \cdots = \mathbf{X}_M$.

Ex. 10.7.   Derive the mean and variance of Y_1^* and Y_2^* in the sequential binomial model (10.35). Derive the EIM with respect to μ_1 and μ_2.

Ex. 10.8.   `seq2binomial()`

Suppose that during WW2, of 376 heavy bombers that were dispatched to raid an enemy target, 60 were lost. Of those that returned and were sent on another mission, 54 were lost. Use `seq2binomial()` to test whether the probabilities of returning from each operation was equal. Discuss the realism of the assumptions behind this short analysis.

Ex. 10.9.   Bradley-Terry Model with Rugby Data
Repeat the analysis of Sect. 10.3.2.1 but use the entire data set. Do the results change much for those select countries originally considered? Comment on all the results as a whole.

Ex. 10.10.   Consider a 2×2 table for a bivariate binomial response vector. Suppose that three cells are mapped onto $\boldsymbol{\eta} = (\eta_1, \eta_2, \eta_3)^T$ in the order $(y_1 = 0, y_2 = 0)$, $(y_1 = 0, y_2 = 1)$, $(y_1 = 1, y_2 = 0)$, and a multinomial logit model is fitted with constraint matrices

$$\mathbf{H}_k \;=\; \begin{pmatrix} 1 & 0 \\ 0 & 1 \\ 0 & 1 \end{pmatrix}, \qquad k = 1, \ldots, p.$$

Hence the $(y_1 = 1, y_2 = 1)$ cell is baseline. Compare this model with that of `binom2.or(exchangeable = TRUE, ZERO = NULL)`: is there any difference?

Ex. 10.11.   Thurstone Model
In what is known as Thurstone's model for paired comparisons, suppose the worth of two options A_j are $N(\mu_j, \sigma_j^2)$ distributed, for $j = 1, 2$, and that a judge decides that A_1 is better than A_2 with probability $P(A_1 > A_2)$. Let's initially make allowance for a nonzero correlation between A_1 and A_2, so that $A_1 - A_2 \sim N(\mu_* \equiv \mu_1 - \mu_2, \; \sigma_*^2 \equiv \sigma_1^2 + \sigma_1^2 - 2\rho\,\sigma_1\sigma_2)$.

(a) Show that $P(A_1 > A_2) = \Phi(\mu_*/\sigma_*)$.
(b) Under the simplifying assumptions that $\rho = 0$ and $\sigma_1 = \sigma_2 = 1$, say, deduce that

$$\hat{\mu}_* \;=\; \sqrt{2}\; \Phi^{-1}(\hat{\pi}) \tag{10.60}$$

   where $\hat{\pi}$ is the sample proportion of people preferring A_1 over A_2.
(c) Deduce that Thurstone's model is like a Bradley-Terry model, but $A_1 - A_2$ is normally distributed rather than logistically distributed, i.e., it uses a probit link rather than a logit link.

Ex. 10.12.   Consider a Bradley-Terry-type model generalized for triple comparisons, i.e., let

$$\theta_{ijk} \;=\; P(T_i > T_j > T_k)$$

be the probability that T_i is preferred to T_j, and T_j is preferred to T_k. One way is to choose

$$\theta_{ijk} \;=\; \frac{\alpha_i}{\alpha_i + \alpha_j + \alpha_k} \cdot \frac{\alpha_j}{\alpha_j + \alpha_k}.$$

Write down an expression for the likelihood function. Compute its first derivatives with respect to the parameters.

Ex. 10.13.   † Loglinear Model for 4 Binary Responses
(a) Consider a loglinear model for four binary responses, as a special case of (10.48). Enumerate $\boldsymbol{\eta}$ as $(u_1, \ldots, u_4, u_{12}, \ldots, u_{14}, u_{23}, \ldots, u_{34})^T$, where higher-order interactions have been set to 0. Write down the log-likelihood, its first derivatives with respect to the parameters, and its EIM.

(b) Write a **VGAM** family function for the model, called `loglinb4()`. Allow for the `exchangeable` and `zero` arguments, and set the u_{st} to be intercept-only, by default.

Ex. 10.14.   Reconcile the fit `fit.coal` in Sect. 10.3.4 with the conditional probabilities (10.49)–(10.50).

Ex. 10.15.   Fit a double exponential binomial model to the `xs.nz` data, as follows. Consider the 11 psychological variables of Sect. 5.6.1. We are interested in the proportion of these psychological variables which are present for any particular person. Adjust for age, ethnicity, gender, current smoking, family history of cancer, and family history of heart disease. Does the regression model confer anything additional to an ordinary logistic regression?

Double, double toil and trouble; Fire burn, and caldron bubble.
—William Shakespeare, *Macbeth*

Chapter 11
Univariate Discrete Distributions

I count him braver who overcomes his desires than him who conquers his enemies; for the hardest victory is over self.
—Aristotle

11.1 Introduction

This chapter and the next summarize a collection of univariate discrete and continuous distributions that are presented as VGLMs/VGAMs, and have been implemented as **VGAM** family functions. There are currently over 70 such distributions in total. Most can be classified as belonging to at least one subclass, and some of these subclasses are described here. Due to space limitations, most appear only as entries in the appropriate table, and it is only possible to provide some sketchy notes on a handful of distributions, e.g., negative binomial (Sect. 11.3), and beta-binomial (Sect. 11.4). Some family functions appear in multiple tables. Zero-inflated, zero-altered and positive variants are deferred to Chap. 17. The notes primarily related to univariate continuous distributions in Sect. 12.2.3 also apply to the tables at the end of this chapter.

Many **VGAM** family functions are accompanied by **dpqr**-type functions for the density, the distribution function, the quantile function, and generation of random deviates, respectively, e.g., `dzipois()`, `pzipois()`, `qzipois()`, `rzipois()` for the zero-inflated Poisson family function `zipoisson()` (or `zipoissonff()` is preferable). For example, these appear in Table 17.7. Often the **dpqr**-type function names are abbreviated, while the **VGAM** family function name is fuller. Another example is `dnaka()`, `pnaka()`, `qnaka()`, `rnaka()` for the Nakagami family function `nakagami()`, collectively represented as "`nakagami(dpqr)`" in Table 12.8.

© Thomas Yee 2015

T.W. Yee, *Vector Generalized Linear and Additive Models*,
Springer Series in Statistics, DOI 10.1007/978-1-4939-2818-7_11

Table 11.1 S conventions for functions associated with distributions and random variates, i.e., dpqr-type functions. These are adopted by **VGAM**, although some arguments such as `lower.tail` and `log.p` may not be implemented yet. Arguments specific to the distribution are denoted by .... The bottom half of the table correspond to bivariate distributions (see Chap. 13).

Function, with defaults	Default values
d*<distribution>*(x, ..., log = FALSE)	$f(x)$
p*<distribution>*(q, ..., lower.tail = TRUE, log.p = FALSE)	$F(q) = P(X \leq q)$
q*<distribution>*(p, ..., lower.tail = TRUE, log.p = FALSE)	$\min_{q:p \leq F(q)} q = F^{-1}(p)$
r*<distribution>*(n, ...)	$y_i \sim F$ independently
d*<distribution>*(x1, x2,, ..., log = FALSE)	$f(x_1, x_2)$
p*<distribution>*(q1, q2, ..., log.p = FALSE)	$F = P(X_1 \leq q_1, X_2 \leq q_2)$
r*<distribution>*(n, ...)	$\boldsymbol{y}_i \sim F$ independently

The standard **dpqr**-type functions available in the **stats** package of R appear in Table 11.6 (discrete) and Table 12.3 (continuous). Entries there have, e.g., "`negbinomial([dpqr])`", where the "[" and "]" indicate that the **dpqr**-type functions are part of **stats**. To offer greater compatibility with these pre-existing R functions, the argument `lss` appears in some **VGAM** family functions because the order of the arguments is not location, scale and then shape—see Sect. 12.2.1 for details.

In keeping with R, the word "density" is used loosely here to denote f in general. More strictly, discrete distributions call $f(y) = P(Y = y)$ the *probability function* or the *probability mass function* (PMF), whereas the *probability density function* (PDF) is reserved for continuous Y. With further looseness, although this book uses Y as the response, we sometimes deviate here and use X to keep it compatible with **d**-type functions with first argument x.

It is added that the CRAN Task Views on 'Probability Distributions' lists many other R packages relevant to the subject of this chapter and the next.

11.1.1 dpqr-Type Functions

R comes with about 20 in-built distributions that are supplied in the **stats** package. Most whose regularity conditions are satisfied can have parameter estimation performed by full MLE by a **VGAM** family function. They are listed in Tables 11.6 (discrete) and 12.3 (continuous). As well as providing a **VGAM** family function, some distributions have a selection of associated **dpqr**-type functions.

VGAM follows standard S conventions for **dpqr**-type functions as given in Table 11.1. Here are some notes concerning univariate distributions (the upper half of the table).

1. The **d**-type functions return the PMF or PDF $f(x) = P(X = x)$, **p**-type functions return the value of the CDF $p = P(X \leq q) = F(q)$, **q**-type functions for continuous distributions return the quantile q such that $q = F^{-1}(p)$, **r**-type

functions return n random variates from that distribution. Clearly, the p-type and q-type functions are 'inverses' of each other, e.g., `pnaka(qnaka(p = probs, shape = 2), shape = 2)` should return `probs`.

For discrete distributions, q-type functions define the quantile as the smallest value x such that $p \leq F(x)$. That is, quantiles are right-continuous. This is in agreement with the general definition of a quantile (15.1).

2. The logical argument `log` in d-type functions indicate whether the natural logarithm of the density is returned. When the log density is needed, setting `d<distribution>(x, ..., log = TRUE)` is recommended over the naïve call `log(d<distribution>(x, ..., log = FALSE))` because a more accurate answer may be returned, as well as being less likely to suffer from numerical problems such as overflow, e.g.,

```
> dnorm(40, log = TRUE)    # More accurate

  [1] -800.92

> log(dnorm(40))   # May be inaccurate

  [1] -Inf
```

3. In a similar way, setting the logical argument `log.p = TRUE` in p-type functions returns $\log F(q)$ instead of $F(q)$ (the default). If needed, it is better practice to set `log.p = TRUE` rather than taking the logarithm of the ordinary answer, because the former may be more accurate. For example,

```
> pnorm(9, log.p = TRUE)    # More accurate

  [1] -1.1286e-19

> log(pnorm(9))   # May be less accurate

  [1] 0
```

for the upper tail of a standard normal distribution. Similarly, some q-type functions also have a `log.p` argument, and if it is set `TRUE`, then the p argument should have negative values because it is interpreted as being log p instead of p. For example,

```
> log.prob <- -1e-20    # This means prob is very close, but not equal to, 1
> qexp(exp(log.prob), log.p = FALSE)    # Infinite solution

  [1] Inf

> qexp(log.prob, log.p = TRUE)    # Finite solution

  [1] 46.052
```

at the upper tail of a standard exponential distribution.

4. Some p-type functions have a `lower.tail` argument; if so then it allows more precise results when the default, `lower.tail = TRUE`, would return 1. Hence setting `lower.tail = FALSE`, which returns $P(X > x)$, is a good idea when this upper tail probability is smaller than machine precision (`.Machine$double.eps`; about 2×10^{-16} on many machines). For example,

```
> pexp(40, lower.tail = FALSE)  # More accurate

  [1] 4.2484e-18

> 1 - pexp(40)  # May be inaccurate

  [1] 0
```

5. For **r**-type functions, argument **n** may be a single integer or else a vector whose length is used instead. Also, for distributions with an **r**-type function, the generic function `simulate()` can often be applied to fitted **VGAM** families corresponding to that distribution—see Sect. 8.4.3.

6. For **r**-type functions, usually 1 of 2 algorithms are commonly employed. They both can be applied to a continuous distribution, however, the acceptance-rejection method can be adapted to handle discrete distributions. They are:

   (i) *The inverse transform method.* If $U \sim \text{Unif}(0,1)$, then the random variable $F^{-1}(U)$ has PDF $f(y)$. Practically, this method requires Y to be continuous and F to be algebraically invertible. The inverse transform method also forms the basis for many **q**-type functions. A simple example is the Rayleigh distribution where $F(y;b) = 1 - \exp\{-\frac{1}{2}(y/b)^2\}$. It is easily inverted to $F^{-1}(p) = b\sqrt{-2\log(1-p)}$. Consequently, random variates generated by `rrayleigh()` use $b\sqrt{-2\log U}$.

   (ii) *The acceptance-rejection method.* A simplified version is as follows. Suppose we can find a density g that we can sample from, which has the property that we can find some constant $C \geq 1$ such that $f(y) \leq C\,g(y)$ for all y. Then:
   (a) generate $T_1 \sim g(y)$,
   (b) generate $U_2 \sim \text{Unif}(0,1)$,
   (c) if $u_2 \leq f(t_1)/(C\,g(t_1))$ then $Y = t_1$ is used (accept), else go to (a) (reject).

   As a specific example, consider Fig. 11.1a where random variates from a Beta(2, 4) distribution, say, are generated using $g = $ `dunif()`. For this, it is easy to show that the mode occurs at $y = 0.25$, and that the density value there is $C = $ `dbeta(0.25, 2, 4)` ≈ 2.109. Generating 300 random points in the overall rectangle results in 149 acceptances—a rejection rate of about 50 percent.

   As another example, suppose that we wish to generate random variates from Kumaraswamy distribution (Table 12.11) with shape parameters 3 and 4, say, based on already having `dbeta()` as g, and access to `rbeta()`, i.e., using the Beta distribution. From Fig. 11.1b, we can see that a Beta(3, 2) density overshadows it when multiplied by a constant $C = 1.5$. Then

```
> shape1 <- 3; shape2 <- 4; Constant <- 1.5
> set.seed(1); N <- 200
> t1 <- rbeta(N, shape1, shape2 / 2)
> u2 <- runif(N)
> u2.scaled <- u2 * Constant * dbeta(t1, shape1, shape2 / 2)
> Accept <- (u2.scaled < dkumar(t1, shape1, shape2))
> y.rv <- t1[Accept]  # Approximates rkumar(n, shape1, shape2)
> length(y.rv) / N  # Proportion of acceptances; length(y.rv) is effectively n

  [1] 0.695
```

The points plotted under and above the curve are from

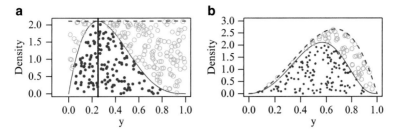

Fig. 11.1 Acceptance-rejection method for generating random variates. The *solid blue points* are accepted; the hollow orange points are rejected. (**a**) The density $g = $ `dunif()` is used to generate Beta(2, 4) random variates. The *vertical line* at $y = 0.25$ denotes the position of the mode, thus defining $C = f(0.25)$. (**b**) A Kumaraswamy(3, 4) density (f; *blue*) is overshadowed by a scaled Beta(3, 2) density ($C \cdot g(y)$; *purple dashed*); the scaling constant is $C = 1.5$.

```
> points(y.rv, u2.scaled[Accept], pch = 16, cex = 0.5, col = "blue")
> points(t1[!Accept], u2.scaled[!Accept], col = "orange")
```

The value of C used here was not determined optimally, but chosen simply because it worked. In practice, given g, the smallest possible value of C might be determined to improve the efficiency of the algorithm so that the rejection rate is minimized. The acceptance-rejection method can be adapted for discrete distributions by choosing a discrete distribution for g.

7. For zero-altered, zero-inflated and positive distributions, there are straightforward techniques to express the formulas for the **dpqr**-type functions, based on those of their parent distribution.

11.2 Dispersion Models

This subclass contains continuous and discrete distributions, but are described here, based on Jørgensen (1997). The central notion is that the location and scale are generalized to position and dispersion, which are parameterized by μ and σ^2, respectively. The density of a reproductive dispersion model has the form

$$f(y; \mu, \sigma^2) \;=\; a(y; \sigma^2) \cdot \exp\left\{ \frac{-d(y; \mu)}{2\,\sigma^2} \right\}, \tag{11.1}$$

where d is known as a unit deviance function, and $\sigma > 0$. If a log-likelihood $\ell(y; \mu)$ is such that the MLE of μ is y, then the unit deviance can be written $d(y; \mu) \propto \ell(y; y) - \ell(y; \mu)$, provided that $d(y; \mu) > 0$ for all $y \neq \mu$. Of course, the normal distribution is the obvious example of (11.1) having $d = (y - \mu)^2$, a squared distance. Only special functions are allowed for a and d because (11.1) must be a density for all values of μ and σ^2.

Jørgensen (1997) discusses many models lying within a dispersion model framework. Some of these are GLMs, and they are listed in Table 2.3. Others which are discrete and implemented by **VGAM** are listed in Table 11.7; others which are continuous are listed in Table 11.8.

A special subclass of dispersion models is called *natural exponential family* (NEF) or *exponential dispersion models*. Morris (1982) showed that there are 6 exponential dispersion models which have a variance function that is a polynomial function of μ of degree 2 or less. These are called NEF-QVF (quadratic variance function), and they are the binomial, Poisson, negative binomial, normal, gamma, and NEF-GHS (generalized hyperbolic secant distribution). The first three are discrete and the remainder are 2-parameter continuous distributions. The normal has constant variance function, the Poisson has a linear variance function, and the remaining four are quadratics in the mean. VGAM implements all except the NEF-GHS (Tables 11.7 and 11.8)—they are `binomialff()`, `poissonff()`, `negbinomial()`/`polyaR()`, `uninormal()`, `gamma2()`.

11.3 Negative Binomial Regression

The negative binomial distribution (NBD) is commonly used for count regression, because of its ability to accommodate overdispersion with respect to the Poisson distribution. Several different scenarios give rise to it, and consequently there are various common parameterizations. Since μ is of central interest for GLMs, especially when compared to a Poisson regression, we concern ourselves mainly with

$$
\begin{aligned}
f(y;\mu,k) &= \frac{\Gamma(y+k)}{\Gamma(y+1)\,\Gamma(k)} \left(\frac{\mu}{\mu+k}\right)^y \left(\frac{k}{k+\mu}\right)^k, \quad \mu > 0,\ k > 0, \quad (11.2) \\
&= \binom{y+k-1}{y} \left(\frac{\mu}{\mu+k}\right)^y \left(\frac{k}{k+\mu}\right)^k, \quad y = 0,1,2,\ldots, \quad (11.3)
\end{aligned}
$$

which is implemented by the family function `negbinomial()`. This parameterization is particularly popular in ecology, and it has mean μ and variance $\mu + \mu^2/k$. As $k \to \infty$ the probability function (11.2) converges to the Poisson PMF (1.7). It is natural to choose the link $\eta_1 = \log \mu$, as in Poisson regression (also the software default).

That the NBD can model overdispersion relative to the Poisson is a not surprising but important consequence of the fact that it arises from a Poisson distribution whose mean parameter is gamma distributed, i.e., a Poisson-gamma mixture. Specifically, suppose that $Y|\Lambda = \lambda \sim \text{Poisson}(\lambda)$ and that $\Lambda \sim \text{Gamma}(\mu, s)$, as in `gamma2()` (Table 12.13 with s being used as the shape parameter). Then

$$
P(Y = y; \mu, k) = \int_0^\infty P(Y = y | \Lambda = \lambda)\, f_\Lambda(\lambda)\, d\lambda \qquad (11.4)
$$

can readily be shown to be equal to (11.3) with $k = s$.

While the NBD can accommodate overdispersion relative to the Poisson distribution, it cannot for underdispersion because $\text{Var}(Y) \geq \mu$. The parameter k is known as an *index parameter*. Its reciprocal, $1/k$ ($= \alpha$, say), is also known as an index parameter, but it is more commonly known as the *dispersion* parameter. Other names for α include the *ancillary* and *heterogeneity* parameter. Others describe α as an aggregation or clumping parameter, with large values meaning more clumping. With k being positive too, the software default link is $\eta_2 = \log k$,

Table 11.2 A summary of some NBD variants implemented in **VGAM**. Their variance functions $\text{Var}(Y) \equiv V(\mu)$ are included, and omission of "$(\boldsymbol{x})$" after a parameter indicates that the parameter is intercept-only. See also Tables 5.4 and 11.3.

Variant	Variance function $V(\mu(\boldsymbol{x}))$	Comment
NB-1	$\phi\,\mu(\boldsymbol{x})$	NB-2 with no exponent
NB-2	$\mu(\boldsymbol{x}) + \dfrac{\mu(\boldsymbol{x})^2}{k}$	The `negbinomial()` default. Elsewhere, the most common software implementation
NB-C_1-H	NB-H with $\eta = \log\left(\dfrac{\mu(\boldsymbol{x})}{\mu(\boldsymbol{x}) + k(\boldsymbol{x})}\right)$	"C_1-H" for canonical link with $k(\boldsymbol{x})$ known
NB-C_2-2	NB-2 with $\eta_1 = \log\left(\dfrac{\mu(\boldsymbol{x})}{\mu(\boldsymbol{x}) + k}\right)$	"C_2-2" for canonical link with k unknown
NB-G	$\mu(\boldsymbol{x}) + \mu(\boldsymbol{x})^2$, i.e., $k \equiv 1$	"G" for geometric
NB-H	$\mu(\boldsymbol{x}) + \dfrac{\mu(\boldsymbol{x})^2}{k(\boldsymbol{x})}$	"H" for heterogeneous
NB-P	$\mu(\boldsymbol{x}) + \delta_1\,\mu(\boldsymbol{x})^{\delta_2}$	"P" for exponent parameterized

however, $\eta_2 = \beta_{(2)1}$ is intercept-only for `negbinomial()`. The parameters μ and k are asymptotically independent because the EIM is diagonal (Sect. A.1.2.3), and the distribution of $\widehat{k}$ is very skewed so that the standard error of $\log \widehat{k}$ is more useful.

The NBD provides an example of a variance function containing an unknown parameter that is not a dispersion parameter, in the usual GLM sense. That the Poisson distribution is the limiting case as $k \to \infty$ is one extreme; another extreme is the logarithmic distribution, which is the limiting distribution of a positive NBD as $k \to 0^+$ (`logff()` in Table 11.10; Ex. 17.1).

NB regression as a whole has now become a sizeable subject, due to the number of variants proposed, e.g., Hilbe (2011, Table 8.1) lists 22 of them. Table 11.2 enumerates those NB variants implemented in **VGAM** with $\eta_1 = g_1(\mu)$. As Table 11.3 shows, they are all easily fitted by using a combination of appropriate constraint matrices, link functions, and choice between a VGLM versus a RR-VGLM.

11.3.1 Computational Details

For a given data set, currently `negbinomial()` uses an exact method and/or simulated Fisher scoring (Sect. 9.2.2) to compute the (k, k)-EIM element. SFS may fail for $\widehat{\mu} \approx 0$ and $\widehat{k} \gg 1$, however, in this region of the parameter space the exact method works well (Fig. 11.2). The exact method naïvely computes

$$-E\left[\frac{\partial^2 \ell_i}{\partial k_i^2}\right] \approx \frac{1}{\mu_i + k_i} - \frac{1}{k_i} + \psi'(k_i) - \sum_{y=0}^{U_i} f(y; \mu_i, k_i) \cdot \psi'(y + k), \quad (11.5)$$

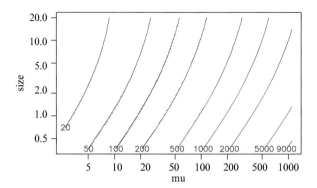

Fig. 11.2 Contour plot of `qnbinom(0.995, size = size, mu = mu)`, where μ and k are on a log-scale. Regions of the (μ, k)-space with a value less than 1000, say, might have the EIM computed by the exact method (11.5). Some of the contour levels appear jagged due to the discrete nature of `qnbinom()`.

for sufficiently large U_i. The upper bound U_i might be computed by `qnbinom(0.995, size = size, mu = mu)`, say, and if sufficiently small (< 1000, say), then this method might be adopted, otherwise SFS used. Some other computational details for (11.2) are given in Lawless (1987).

The shrinkage method for initial values (Sect. 8.3.1) is implemented in the code.

11.3.2 A Second Parameterization—polyaR()

Another popular NBD parameterization, which is implemented by `polyaR()`, is

$$f(y_2; k, p) \;=\; \binom{y_2 + k - 1}{y_2} p^k (1 - p)^{y_2}, \quad 0 < p < 1, \quad k > 0, \quad (11.6)$$

for $y_2 = 0, 1, 2, \ldots$. If k is a (positive) integer, then $Y_2 =$ the number of failures which occur in a sequence of independent Bernoulli trials until k successes are reached. For k being a positive real as in (11.2), the mean and variance of Y_2 are $\mu = k(1-p)/p$ and $k(1-p)/p^2 = \mu + \mu^2/k$. Strictly, (11.6) holds for $0 < p \le 1$ but we omit both endpoints as possibilities.

Note that some practitioners reserve "negative binomial" strictly for the integer k case (where it is also known as the Pascal distribution). These people then refer to the NBD with the real k case as the Pólya distribution. In VGAM, we always treat k as real, and use the two names to distinguish between the two parameterizations with tongue in cheek. Regardless, the geometric distribution corresponds to $k = 1$; and the sum of k independent geometric random variables has probability function $f(y_2; p) = p(1 - p)^{y_2}$ for $y_2 = 0, 1, 2, \ldots$.

Incidentally, a third parameterization, very similar to the second, is

$$f(y_3; k, p) = \binom{y_3 - 1}{k - 1} p^k (1-p)^{y_3 - k}, \quad y_3 = k, k+1, \dots \quad (11.7)$$

for $0 < p < 1$ and positive integer k. This models the number of Bernoulli trials until the kth success occurs, and has $Y_3 = Y_2 + k$.

11.3.3 Canonical Link

The use of the NBD with its canonical link $g_{c,\text{NB}} = \log(\mu/(\mu + k))$ is more for theoretical interest than for practice. With known k, fitting an NB regression with $g_{c,\text{NB}}$ amounts to fitting a GLM in the usual sense (Ex. 11.9). However, in the VGLM/VGAM framework, if k is known then it is really more accurate to write k as k_i (or k_{ij} if there are more than one response) to reflect the possibility of it being inputted as a general vector (or matrix if there are multiple responses). For this reason, "$k(\boldsymbol{x})$" is actually used, and this leads to the notation adopting an "H". A second comment about the framework is that there is no need to be restricted to its canonical link in the first place: the family function `negbinomial.size()` performs NB regression with the full range of links for $\eta = g(\mu)$ subject to known $k(\boldsymbol{x})$, and its first argument receives the $k(\boldsymbol{x})$, which may be a vector or a matrix of the same dimension as the response. Thus the special case NB-C_1-H may be fitted by, e.g., `negbinomial.size(size, lmu = "nbcanlink")` which has $\eta_{ij} = g_{c,\text{NB}}(\mu_{ij}, k_{ij})$.

NB-C_2 models are more difficult to fit than the NB-C_1. Both pose dangers due to $\eta_1 < 0$, whereas an ordinary linear/additive predictor η_j ought to be unbounded. Since the NB-C_2 has both μ and k to be estimated, `negbinomial(lmu = "nbcanlink")` is not recommended in general. Practical experience has shown that k is best left as an intercept-only (the default)—this model is therefore called the NB-C_2-2. Also, one should actively take smaller steps to try to improve its estimation reliability. Also, as Hilbe (2011, pp.210,309) notes, having k in the link and variance can result in estimation difficulties, with it being sensitive to initial values and having tedious convergence with NR-type algorithms. Currently, setting something like `imethod = 3`, `stepsize = 0.5`, `maxit = 100` can help, which is a combination of ambling to the solution and using different initial values.

11.3.4 Fitting Other NB Variants

11.3.4.1 Quasi-Likelihood Poisson

The two most common methods of handling counts are *quasi(-likelihood)-Poisson* (QLP) and NB regression. Both share the loglinear relationship

$$\log \mu(\boldsymbol{x}) = \eta_1(\boldsymbol{x}) = \boldsymbol{\beta}_1^T \boldsymbol{x}. \quad (11.8)$$

The QLP model has $\widehat{\boldsymbol{\beta}}_1$ being the Poisson MLE and the variance function

$$\text{Var}(Y|\boldsymbol{x}) = \phi\,\mu(\boldsymbol{x}), \quad (11.9)$$

Table 11.3 How **VGAM** can fit NB variants and other associated models. All parameters receive full maximum likelihood estimation, and k is an intercept-only (scalar) unless specified. Argument `size` is the NB parameter k. The NB-P is also known as an RR-NB (Sect. 5.5.2.3). See also Tables 11.2 and 17.6–17.7.

NB variant	Var(Y)	Modelling function	VGAM family function
NB-1	$(1 + \delta_0^{-1})\,\mu(\boldsymbol{x}) = \phi\,\mu(\boldsymbol{x})$	vglm()	negbinomial(parallel = TRUE, zero = NULL)
NB-2	$\mu(\boldsymbol{x}) + \mu(\boldsymbol{x})^2/k$	vglm()	negbinomial()
NB-C_1-H	$\mu(\boldsymbol{x}) + \mu(\boldsymbol{x})^2/k(\boldsymbol{x})$	vglm()	negbinomial.size(size, lmu = "nbcanlink")
NB-C_2-2	$\mu(\boldsymbol{x}) + \mu(\boldsymbol{x})^2/k$	vglm()	negbinomial("nbcanlink")
NB-G	$\mu(\boldsymbol{x}) + \mu(\boldsymbol{x})^2$	vglm()	negbinomial.size(size = 1)
NB-H	$\mu(\boldsymbol{x}) + \mu(\boldsymbol{x})^2/k(\boldsymbol{x})$	vglm()	negbinomial(zero = NULL)
NB-P	$\mu(\boldsymbol{x}) + \delta_1\,\mu(\boldsymbol{x})^{\delta_2}$	rrvglm()	negbinomial(zero = NULL)
Poisson	$\mu(\boldsymbol{x})$	vglm()	negbinomial.size(size = Inf)
Poisson	$\mu(\boldsymbol{x})$	vglm()	poissonff()
Quasi-Poisson	$\phi\,\mu(\boldsymbol{x})$	vglm()	quasipoissonff()

where ϕ is a scale or dispersion parameter that is usually estimated by the method of moments (Eq. (2.30)). Overdispersion corresponds to $\phi > 1$, and the case $\phi = 1$ is known as *equidispersion*. Confidence intervals for ϕ have been a research problem but they are available for the NB-1 fitted as a VGLM (Sect. 11.3.4.2 below).

How might one choose between the QLP and NB-1? One solution is to fit the RR-NB (5.21), which has the flexible variance function $\mu + \delta_1\mu^{\delta_2}$, where δ_1 and δ_2 are all positive parameters that are estimated by full MLE. We can conduct tests of $H_0 : \delta_2 = 1$ and $H_0 : \delta_2 = 2$. For example, if we do not reject $\delta_2 = 1$, then full MLE of (11.9) is easy via a VGLM. Confidence intervals for ϕ are then available (whereas they are unavailable or nontrivial for QLP). If a 95 % confidence interval for ϕ covers unity, then an ordinary Poisson regression could be reasonable.

11.3.4.2 Fitting a NB-1

How can one fit a negative binomial subject to Var(Y) $\propto \mu$? The answer is to choose $k = \delta_0\,\mu$, say. Then Var(Y) = $(1 + \delta_0^{-1})\,\mu$ which is of the form (11.9). Also, $\log k = \log \delta_0 + \log \mu$ so that $\eta_2 = \log \delta_0 + \eta_1$, i.e., η_2 and η_1 differ only by $\log \delta_0$. To force the coefficients of $\log \mu$ to be equal to those of $\log k$ for every explanatory variable in $\boldsymbol{x}$ except for the intercept, a parallelism assumption is used so that $\mathbf{H}_2 = \cdots = \mathbf{H}_p = (1, 1)^T$ in (3.27). The intercepts are estimated without constraints: $\mathbf{H}_1 = \mathbf{I}_2$.

The call in **VGAM** for the NB-1 is of the form

```
nb1 <- vglm(y ~ x2 + x3, negbinomial(parallel = TRUE, zero = NULL), data = ndata)
```

where argument `parallel` is a shortcut for constructing the $\mathbf{H}_k$. Setting `zero = NULL` annuls that argument. Then

```
1 + exp(-diff(coef(nb1, matrix = TRUE)["(Intercept)", ]))
```

is $\widehat{\phi}$ because the `parallel` argument here does not apply to the intercept term by default. Thus the first row of `coef(nb1, matrix = TRUE)` is $\left(\widehat{\beta}_{(1)1}, \widehat{\beta}_{(2)1}\right)$. The interpretation of an NB-1 fitted in this manner is as an NB regression which is subject to a quasi-Poisson-type variance function constraint (11.9).

11.3.4.3 Fitting a NB-P

As a numerical illustration, we fit an NB-P to the `azpro` data frame in COUNTS (Yee, 2014). These data comprise 3589 patients entering a hospital in 1991 in Arizona, USA, to receive one of two standard cardiovascular treatments (CABG $= 0$, PTCA $= 1$), called variable `procedure`. The other variables being adjusted for are `sex` (M $= 1$, F $= 0$), `admit` ($0 =$ elective, $1 =$ urgent/emergency), and `age75` (0 if age < 75 years, otherwise 1). The response is the length of hospital stay (`los`, in days).

```
> data(azpro, package = "COUNT")
> rrnb.azpro <- rrvglm(los ~ procedure + sex + admit + age75,
                     negbinomial(zero = NULL), data = azpro)
> unlist(Confint.rrnb(rrnb.azpro))  # Neither a NB-1 nor NB-2

   a21.hat beta11.hat beta21.hat    CI.a211    CI.a212 CI.delta21 CI.delta22
  0.588840   1.442765   1.390036   0.403762   0.773918   1.226082   1.596238
    delta1     delta2 SE.a21.hat
  0.582470   1.411160   0.094429
```

The fitted NB-2 (Hilbe, 2011) purports a variance function of

$$\widehat{\mu} + \frac{\widehat{\mu}^2}{6.25} \approx \widehat{\mu} + 0.16 \cdot \widehat{\mu}^2.$$

In fact, the fitted RR-NB has $\widehat{\mu} + 0.58 \cdot \widehat{\mu}^{1.41}$, and an approximate 95 % confidence interval for δ_2 is $[1.23, 1.6]$, therefore neither an NB-1 or NB-2 is strictly appropriate. This is supported by a likelihood ratio test of the RR-NB versus NB-2 with p-value 1.5×10^{-12}. The profile log-likelihood as a function of $\widehat{a}_{21}$ (Fig. 11.3) is well approximated by a quadratic, and it reveals that both NB-1 and NB-2 models can be seen as inappropriate compared to the likelihood ratio confidence limits (A.24).

11.3.5 Some Practical Suggestions

Here are some practical suggestions when using VGAM to perform NB regression.

- The parameters of (11.2) match with the arguments of `[dpqr]nbinom(size, prob, mu)` by `size` = k and `mu` = μ, while `prob` = p in the Pólya parameterization (11.6) described in Sect. 11.3.2.
- Both `negbinomial()` and `polyaR()` handle multiple responses. By default, the k are modelled as intercept-only (as `zero = -2`). As k is positive, a number of

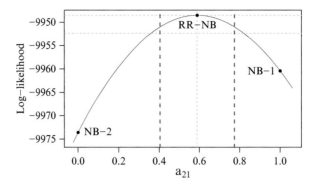

Fig. 11.3 Profile log-likelihood $\ell(a_{21})$ for `rrnb.azpro`; $\ell(\widehat{a}_{21})$ is the highest point. The MLE and likelihood-ratio confidence limits are the *orange horizontal lines*. The Wald confidence limits are the *grey vertical lines*. The point at $a_{21} = 0$ is ℓ for NB-2 (intercept-only for k). The point at $a_{21} = 1$ is ℓ for NB-1.

links are reasonable choices including `reciprocal()`, which might be more numerically stable than just k or even $\log k$. Alternatively, the `loglog()` is feasible for $\widehat{k} \gg 1$. If `lsize = negloge`, then $\eta_2 = -\log k = \log \alpha$.

- Although the fitting of an NB-2 (default) is often a reasonable first model, difficulties will occur with Poisson data or underdispersion relative to the Poisson distribution. Both cases have $\widehat{k} \to \infty$ so that numerical problems will be encountered. In these situations, possibly using a `reciprocal` or `loglog` link might help; else simply fit a quasi-Poisson model having $\widetilde{\phi} < 1$. It is always a good idea to monitor convergence, e.g., set `trace = TRUE`, and try tricks such as those described in Sect. 8.5.

- Another reasonable first model is the RR-NB. One can think of the NB-P as a data-driven NB model. The δ_2 parameter and its confidence interval can be examined to see if it covers the values 1 or 2 for the NB-1 or NB-2, respectively. If so, then these simpler models should be fitted.

- Currently, the deviance of a NB-2 regression,

$$2 \sum_{i=1}^{n} w_i \left\{ y_i \log\left(\frac{y_i}{\mu_i}\right) - (y_i + k) \log\left(\frac{y_i + k}{\mu_i + k}\right) \right\} \tag{11.10}$$

is not returned unless specifically requested, and half-stepping is effectively turned off. This is because the deviance does not necessarily decrease at each IRLS step, since both μ and k are being estimated. Half-stepping can be suppressed and the deviance computed by using a call of the form

```
vglm(y ~ x2 + x3, negbinomial(deviance = TRUE), criterion = "coef", data = ndata)
```

or

```
vglm(y ~ x2 + x3, negbinomial(deviance = TRUE), half.step = FALSE, data = ndata)
```

The deviance is calculated only *after* convergence, and then it is attached to the object. Of course, the `deviance()` generic function should be applied to the fitted object to obtain its value.

11.4 The Beta-Binomial Model

The beta-binomial model is a random-effects binomial model that is commonly used for the analysis of teratological data. Proposed by Williams (1975), the classical example of this type of experiment involves a group of n pregnant rats which are randomized and exposed to a chemical. The ith rat gives birth to a litter of size N_i, of which y_i^* are malformed or die within a specified time period. Table 11.4 gives a small data set of this description. The scientific objective is to determine whether the risk of malformation differs between groups. Here, there were 32 pregnant rats which were randomized to receive a placebo or a chemical, and sacrificed prior to the end of gestation or pregnancy. Each fetus was examined and a binary response indicating the presence or absence of a particular malformation was recorded. It is well-known that such data is very likely to exhibit the so-called *litter effect*, whereby offspring from the same litter tend to respond more alike than offspring from different litters. We will find below that such an effect is typically reflected by a positive correlation parameter ρ. If $\rho = 0$, then an ordinary logistic regression should suffice. The beta-binomial model has been shown to provide a much better fit to many such data sets than the simple binomial model. Another example of the litter effect, from dental science, is the number of teeth with caries, Y_i^*, out of N_i teeth for individual i.

The beta-binomial distribution assumes that a random malformation probability Π_i in cluster i comes from a beta distribution with mean μ_i. Such an assumption is based more on mathematical convenience than any biological justification. Given $\Pi_i = \pi_i$, the number of malformations Y_i^* within the ith cluster follows a Binomial(N_i, π_i) distribution. Note that $\Pi \sim \text{Beta}(\alpha, \beta)$ implies

$$f(\pi; \alpha, \beta) = \frac{\pi^{\alpha-1}(1-\pi)^{\beta-1}}{Be(\alpha, \beta)}, \quad 0 < \pi < 1, \quad 0 < \alpha, \quad 0 < \beta, \qquad (11.11)$$

with $E(\Pi) = \alpha/(\alpha+\beta)$ and $\text{Var}(\Pi) = \alpha\beta/[(\alpha+\beta)^2(\alpha+\beta+1)]$. Then the marginal distribution of Y_i^* is

$$P(Y_i^* = y_i^*) = \binom{N_i}{y_i^*} \frac{Be(y_i^* + \alpha_i, N_i - y_i^* + \beta_i)}{Be(\alpha_i, \beta_i)}, \quad y_i^* = 0(1)N_i. \qquad (11.12)$$

Write $Y_i^* = Z_{i1} + \cdots + Z_{i,N_i}$ where $Z_{ij} = 0$ or 1, and let $\rho_i = \text{Corr}(Z_{ij}, Z_{ik})$. It then follows from (A.34) that $E(Y_i^*) = N_i\alpha_i/(\alpha_i + \beta_i) = N_i\mu_i$, say, and $\rho_i = (\alpha_i + \beta_i + 1)^{-1}$, as well as (A.36) leading to

$$\text{Var}(Y_i^*) = N_i\mu_i(1 - \mu_i)[1 + (N_i - 1)\rho_i]. \qquad (11.13)$$

Provided that $N_i > 1$, it can be seen that the beta-binomial allows overdispersion with respect to the binomial distribution when $\rho > 0$. As an illustration, taking the square root of (11.13) with $N_i = 10$, say, the standard deviation of Y_i^* is plotted in Fig. 11.4. It can be seen how, as ρ increases from 0, $\text{Var}(Y_i^*)$ increases relative to the binomial. Thus the beta-binomial distribution is a way of modelling overdispersion relative to the binomial distribution.

The beta-binomial model can be fitted by the family function `betabinomial()`. However, like `binomialff()`, the response Y is a *proportion* rather than the number of successes. Thus $Y_i = Y_i^*/N_i$ so that $E(Y_i) = \mu_i$, and the N_i are assimilated as prior weights. The family function has default links $\boldsymbol{\eta} = (\text{logit } \mu, \text{logit } \rho)^T$,

Table 11.4 Toxological experiment data (Williams (1975); in data frame `prats`). The subjects are fetuses from 2 randomized groups of 16 pregnant rats each, and they were given a placebo or chemical treatment. For each litter, the number of pups alive at 4 days (the litter size) and at the 21 day lactation period were recorded. Bold values in the boxes represent 2 such litters.

Control	13/13, 12/12, **9/9** , **8/8** , 12/13, 11/12, **9/10** , 8/9, 11/13, 4/5, 5/7, **7/10**
Treated	12/12, 11/11, 10/10, 9/9, 10/11, **9/10** , **8/9** , 4/5, 7/9, 4/7, 5/10, 3/6, 3/10, 0/7

because both parameters lie in the unit interval. Also, an intercept-only model for ρ is the default, in order to provide greater numerical stability: `betabinomial (zero = 2)`.

Actually, Prentice (1986) pointed out that the correlation coefficient ρ need not be positive as had been previously thought. He showed that the lower bound for ρ_i is

$$\max\left\{\frac{-\mu_i}{N_i - \mu_i - 1}, \frac{-(1 - \mu_i)}{N_i + \mu_i - 2}\right\}.$$

VGAM cannot handle this type of constraint directly, however, users might try `lrho = "rhobit"` (Table 1.2; allows $-1 < \rho < 1$) and hope that there are no numerical difficulties resulting from estimates falling below the lower bounds during estimation.

Here are some further notes.

1. An alternative VGAM family for the beta-binomial, `betabinomialff()`, deals directly with the shape parameters, and has default links $\boldsymbol{\eta} = (\log \alpha, \log \beta)^T$.
2. The basic model assumes all covariates are cluster-specific, e.g., treatment and mothers' characteristics. At present, individual-specific covariates cannot be handled because multiple responses are not supported. If it were, then its usage would be something like `betabinomial(multiple.responses = TRUE)`, and each column of the response matrix would be a different rat (and padded by `NA`s), and the `xij` argument detailed in Sect. 3.4 would be operable.
3. Writing $\theta_i = \rho_i/(1 - \rho_i) = 1/(\alpha_i + \beta_i)$, it follows from (11.13) that

$$\mathrm{Var}(Y_i^*) \;\; = \;\; N_i\,\mu_i(1 - \mu_i)(N_i\,\theta_i + 1)/(1 + \theta_i).$$

Thus, the binomial is a special case of the beta-binomial with $\theta_i = 0$. Note that choosing `lrho = "logit"` means that $\theta = \exp(\eta_2)$. It is usual to test for homogeneity of proportions (in the presence of common dispersion): $H_0 :$ $\mu_1 = \cdots = \mu_n$ versus $H_1 :$ not all the μ_i are equal, with the assumption that $\theta_1 = \cdots = \theta_n = \theta$ which is unknown and unspecified.

11.5 Lagrangian Probability Distributions

This class of probability distributions is detailed in Consul and Famoye (2006), which this description is based on. Lagrangian distributions are divided into three subclasses, and are based on the Lagrange transformation $u = z/g(z)$ where $g(z)$ is an analytic function on $[-1, 1]$ with $g(0) \neq 0$:

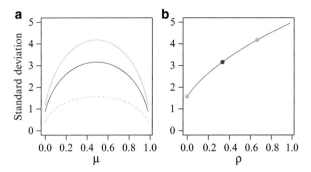

Fig. 11.4   Standard deviation, $SD(Y_i^*)$, in the beta-binomial distribution, from (11.13) with $N_i = 10$. (a) As a function of μ, for $\rho = 0.33$ (*blue*), $\rho = 0.67$ (*green*) and $\rho = 0$ (*orange dashed*). (b) As a function of ρ, where $\mu_i = 0.5$.

$$z \;=\; \sum_{k=1}^{\infty} \frac{u^k}{k!} \left\{ D^{k-1}(g(z))^k \right\}_{z=0}. \tag{11.14}$$

A function is analytic iff its Taylor series about x_0 converges to the function in some neighbourhood for every x_0 in its (open) domain. Consequently, they are infinitely differentiable there. Equation (11.14) expresses $g(z)$ as a power series in u. Here, D is the differentiation operator, e.g., $Df(x) = df(x)/dx$, $D^2f(x) = d^2f(x)/dx^2$. We say that the *probability generating function* (PGF) of a random variable Y is $g(t)$, say, where $g(t) = E(t^Y) = \sum_{y=0}^{\infty} t^y P_y$ and $P_y = P(Y = y)$ are the probabilities.

The significance of

$$u\,g(z) \;=\; z \tag{11.15}$$

is that, in a multiplicative process, $g(z)$ is the PGF of the number of segments from any vertex in a rooted tree with numbered vertices, and u is the PGF for the number of vertices in the rooted tree. Indeed, the number of vertices after n segments in a tree can be interpreted as the number of members in the nth generation of a branching process. Application areas include population growth, the spread of epidemics and rumors, and nuclear chain reactions.

The three subclasses are called basic, delta and general Lagrangian distributions. For brevity, we will only mention the first type. Let $g(z)$ be as above with the additional constraint that $g(1) = 1$. Then the smallest root $z = \ell(u)$ of the transformation $z = u\,g(z)$ defines a PGF $z = \psi(u)$, and the Lagrangian expansion (11.14) in powers of u as

$$z \;=\; \psi(u) \;=\; \sum_{y=1}^{\infty} \frac{u^y}{y!} \left\{ D^{y-1}(g(z))^y \right\}_{z=0}, \tag{11.16}$$

if $D^{y-1}(g(z))^y|_{z=0} \geq 0$ for all values of y. Then the corresponding PMF of the basic Lagrangian distribution is

$$P(Y = y) \;=\; \frac{1}{y!} \left\{ D^{y-1}(g(z))^y \right\}_{z=0}, \qquad y \in \mathbb{N}^+. \tag{11.17}$$

Table 11.5 Lagrangian probability distributions currently supported by **VGAM**. Notes: (i) the geometric distribution is a special case of the Borel distribution, which is itself a special case of the Borel-Tanner distribution with $Q = 1$. (ii) `geometric()` actually has support $0(1)\infty$.

Distribution	Density $f(y; \boldsymbol{\theta})$	Support	Range of $\boldsymbol{\theta}$	Mean	VGAM family, and some $g(z)$
Borel	$\dfrac{y^{y-2} a^{y-1}}{(y-1)!} e^{-ay}$	$1(1)\infty$	$(0,1)$	$\dfrac{1}{1-a}$	`borel.tanner(dr, Q = 1)`, $\exp\{a(z-1)\}$
Felix	$\dfrac{y^{(y-3)/2} a^{(y-1)/2}}{((y-1)/2)! \, e^{ay}}$	$1(2)\infty$	$0 < a < \frac{1}{2}$	$\dfrac{1}{1-2a}$	`felix(d)`, $\exp\{a(z^2-1)\}$
Generalized Poisson	$\dfrac{\theta(\theta + y\lambda)^{y-1}}{y! \, e^{y\lambda+\theta}}$	$0(1)\infty$	$0 < \lambda < 1,$ $0 < \theta$	$\dfrac{\theta}{1-\lambda}$	`genpoisson()`
Geometric	$(1-p)^{y-1} p$	$1(1)\infty$	$0 < p < 1$	$\dfrac{1-p}{p}$	`geometric([dpqr])`, $1 - p + pz$

VGAM family functions belonging to the Lagrangian class are listed in Table 11.5, along with $g(z)$ for the basic sublass entries. Other basic Lagrangian distributions result from other choices of $g(z)$.

Bibliographic Notes

Johnson et al. (2005) is a general encyclopaedic book for univariate discrete distributions. There are some general books that summarize many common distributions, e.g., Balakrishnan and Nevzorov (2003) and Forbes et al. (2011). Leemis and McQueston (2008) describes the relationships between over 70 univariate distributions. More details about random variate generation can be found in, e.g., Devroye (1986), Fishman (1996), Hörmann et al. (2004), Jones et al. (2014).

Fuller accounts of the negative binomial distribution can be found in Johnson et al. (2005, Chap.5), Hilbe (2011) and Cameron and Trivedi (2013). The notation of Hilbe (2011) is largely adopted in this chapter.

Jørgensen (1997) is an authoritative treatment of dispersion models. Lagrangian distributions are also described in Johnson et al. (2005, Sect.7.2).

Exercises

Ex. 11.1.   Use `runif(n)` and the acceptance-rejection method to generate random variates from the beta distribution with 3 and 5 as the shape parameters. That is, to generate the equivalent of `rbeta(n, 3, 5)`.

Ex. 11.2.   Find the optimal value of C for an acceptance-rejection method applied to a standard normal density, based on having **rlaplace()** and **dlaplace()**. That is, we wish to write **rnorm()** so that $f = \phi$ and $g(y) = 0.5 \exp(-|y|)$. For simplicity, fix the location/mean and scale/standard deviation parameters at 0 and 1 respectively.   [Devroye (1986)]

Ex. 11.3.   Show that $\widehat{\mu} = y$ for the Poisson PMF, and then obtain an expression for the unit deviance function.

Ex. 11.4.   Suppose that $Y \sim$ truncated geometric(L, U) where $y = L, L + 1, \ldots, U$, and L and U are known. Explain why this model can be fitted by something like

```
vglm(y - L ~ x2 + x3, truncgeometric(upper.limit = U - L), data = tdata)
```

Ex. 11.5.   Suppose that we want to simultaneously regress two species' counts called **y1** and **y2**, each with an NB-2, to covariates **x2** and **x3**. Write down $\boldsymbol{\eta}$, the $\mathbf{H}_k$, and the syntax for fitting the model in **VGAM** under the following constraints (treated separately).

(a) Subject to their k parameters being equal and intercept-only.
(b) Subject to their μ parameters being equal, and each k parameter being intercept-only.

Ex. 11.6.   Consider the variable **pubtotal** (total number of publications, from the MathSciNet database) of the **profs.nz** data frame.

```
> ooo <- order(with(profs.nz, pubtotal), decreasing = TRUE)
> head(profs.nz[ooo, -4], 3)

   pubtotal cites initials firstyear      ID pub1stAuthor ARPtotal institution
1        84    80      CDL     1977 109295           38       86          MU
21       58    48      JAJ     1965 234899           31       NA          UW
7        56   827      DVJ     1957 203111           33       59          VU

> tail(profs.nz[ooo, -4], 3)

   pubtotal cites initials firstyear      ID pub1stAuthor ARPtotal institution
3         7     1      MJA     1999 650954            5       NA          MU
13        6    25      CMT     1977 273433            0       NA          UA
15        1     0      JMC     2006 791650            0       NA          UA
```

(a) Obtain the five number summary, mean and variance. Comment.
(b) Fit NB-1, NB-2, NB-H and NB-P models, with $\texttt{I(2014 - firstyear)}$ as explanatory. Why use this variable and what are its limitations?
(c) Decide which model in (b) seems best by applying some goodness-of-fit test, i.e., justify your reasoning.
(d) Repeat (b)–(c) but using the variable **pub1stAuthor** (the number of first-author publications) as the response.

Ex. 11.7.   Doodlebugs
Fit a Poisson regression to the **V1** data set. Do the same with a quasi-Poisson model. Which is preferable? Try fitting an NBD—hint: the **reciprocal()** link function may be better for the k parameter since its value will be large if the data is Poisson.

Ex. 11.8.   Geometric-Binomial Relationship

(a) Generate a random sample of 100 random variates from a geometric distribution with probability $\frac{1}{4}$ of success. Show that fitting a logistic regression with response of the form `cbind(1, y1)` gives the same regression coefficient (an intercept-model will do) as using the family function `geometric()` on `y1` directly.

(b) Use log-likelihood contributions ℓ_i to explain the result.

(c) Apply `logLik()` to your fits—are they different? Compute this difference explicitly.

Ex. 11.9.   Show that, if k is known, the NB regression $g_{c,\mathrm{NB}}(\mu) = \eta$ results in a GLM with its canonical link.

Ex. 11.10.   Derive the deviance formula (11.10) for the NB-2.

Ex. 11.11.   Negative Binomial Computations

(a) Obtain an expression for the log-likelihood function of the NBD probability function (11.2). Show that the off-diagonal element of the EIM is 0.

(b) Show that $\partial \ell_i / \partial k = \log p + \psi(y_i + k) - \psi(k) - (y_i - \mu)/(\mu + k)$.

(c) How might this be computed, even without access to `digamma()`?

Ex. 11.12.   NBD Arising from a Poisson-Gamma Mixture
Starting at (11.4), complete the steps to show that the PMF of the NBD, (11.3), results with $\mu = \mu$ and $k = s$.

Ex. 11.13.   Beta-Binomial Distribution

(a) Given the conditional distribution of Y_i^* and the distribution of Π_i outlined in Sect. 11.4, show that the marginal distribution of Y_i^* is (11.12).

(b) Show that, for $y^* = 0, 1, \ldots, N$,

$$P(Y^* = y^*) = \frac{\binom{N}{y^*} \prod_{r=0}^{y^*-1} (\mu + r\theta) \prod_{r=0}^{N-y^*-1} (1 - \mu + r\theta)}{\prod_{r=0}^{N-1} (1 + r\theta)}. \qquad (11.18)$$

(c) Fit a simple beta-binomial model to the `prats` data of Table 11.4. Obtain an approximate 95 % confidence interval for ρ. Would an ordinary binomial model suffice? Quantify any difference between the control and treatment groups by using the odds ratio.

Ex. 11.14.   Borel-Tanner Distribution

(a) From its PMF in Table 11.9, show that the mean of the Borel-Tanner distribution is as specified. Show that its variance is $Qa/(1-a)^3$.

(b) Derive its EIM.

(c) Let $Q = 1$, i.e., the Borel distribution. Show that underdispersion and overdispersion (relative to the Poisson) occurs for $0 < a < (3-\sqrt{5})/2$ and $(3-\sqrt{5})/2 < a < 1$, respectively.   [Consul and Famoye (2006)]

Ex. 11.15. † Write a **VGAM** family function for the positive beta-binomial distribution, defined as a beta-binomial distribution with zeros excluded.

Ex. 11.16. The following data concerns accidents to 647 women working on high explosive shells during a 5-week period during WW1, and was reported by Greenwood and Yule in 1920. The number of accidents was $y = 0, 1, \ldots, 5$, with respective frequencies $447, 132, 42, 21, 3, 2$ (actually, the 5 here was a "≥ 5"). Fit the standard Poisson and NB-2 models, plus any other two discrete distributions. Use some measure of goodness of fit to decide which of your models is best.

Ex. 11.17. **2008 World Fly Fishing Championships**
Consider the `wffc` data set.

(a) Fit a negative binomial regression to the number of trout caught per session, using the sector, time of day (morning and afternoon), and the day as covariates.
(b) Does the data appear to be overdispersed relative to a Poisson distribution? Interpret the fitted coefficients.
(c) Estimate the catch per unit effort (CPUE; number of fish caught per hour of fishing) for an angler on the Waihou River during the afternoon of Day 1.   [Yee (2010b)]

Chance is always powerful: let your hook always be cast. In a pool where you least expect it, there will be a fish.
—Publius Ovidius Naso

Table 11.6 Discrete distributions in the stats package of R, and their corresponding VGAM family functions. Notes: (i) All have dpqr-type functions except for multinomial(), which has dr-type only. (ii) See also Table 12.3 for continuous distributions.

Distribution	PMF $f(y;\boldsymbol{\theta})$	Support	Range of $\boldsymbol{\theta}$	Mean	VGAM family
Binomial	$\binom{N}{Ny}\mu^{Ny}(1-\mu)^{N(1-y)}$	$0(1/N)N$	$0<\mu<1,\ N\in\mathbb{Z}^+$	μ	binomialff()
Geometric	$(1-p)^y p$	$0(1)\infty$	$0<p<1$	$\dfrac{1-p}{p}$	geometric()
Hypergeometric	$\dfrac{\binom{N-M}{n-y}\binom{M}{y}}{\binom{N}{n}}$	$\min(0,n+M-N)(1)\max(M,n)$	$N\in\{0,1,\ldots\},$ $M\in\{0,1,\ldots,N\},$ $n\in\{0,1,\ldots,N\}$	$\dfrac{nM}{N}$	
Multinomial	$\dfrac{N!}{y_1!\,y_2!\cdots y_{M+1}!}\prod_{j=1}^{M+1} p_j^{y_j}$	$y_j\in 0(1)N,\ \sum_{j=1}^{M+1} y_j = N$	$0<p_j<1,\ p_{M+1}=$ $1-p_1-\cdots-p_M,\ N=$ $y_1+\cdots+y_{M+1}>0$	$\boldsymbol{p}$	multinomial([dr])
Pólya (an NBD variant)	$\binom{y+k-1}{y}p^k(1-p)^y$	$0(1)\infty$	$0<k,\ 0<p<1$	$\dfrac{k(1-p)}{p}$	polyaR()
Poisson	$\dfrac{e^{-\lambda}\lambda^y}{y!}$	$0(1)\infty$	$0<\lambda$	λ	poissonff()
Signed rank	—	$0(\Delta)\dfrac{n(n+1)}{2}$	$n\in\mathbb{Z}^+$	$\dfrac{n(n+1)}{4}$	
Wilcoxon rank sum statistic	—	$0(1)mn$	$m\in\mathbb{Z}^+,\ n\in\mathbb{Z}^+$	$\dfrac{mn}{2}$	

Table 11.7 Discrete dispersion models (Jørgensen, 1997) implemented in VGAM. The means are returned as the fitted values, by default.

Distribution	Probability mass function $f(y;\boldsymbol{\theta})$	Support	Range of $\boldsymbol{\theta}$	Mean	VGAM family
Binomial	$\binom{N}{Ny}\mu^{Ny}(1-\mu)^{N(1-y)}$	$0(1/N)N$	$0<\mu<1,\ N\in\mathbb{Z}^+$	μ	binomialff([dqpr])
Generalized Poisson	$\dfrac{\theta(\theta+y\lambda)^{y-1}}{y!}\,e^{-y\lambda-\theta}$	$0(1)\infty$	$0\leq\lambda<1,\ 0<\theta$	$\dfrac{\theta}{1-\lambda}$	genpoisson()
Inverse binomial	$\dfrac{\lambda\,\Gamma(2y+\lambda)\,\{\rho(1-\rho)\}^y\,\rho^\lambda}{\Gamma(y+1)\,\Gamma(y+\lambda+1)}$	$0(1)\infty$	$\frac{1}{2}<\rho<1,\ 0<\lambda$	$\dfrac{\lambda(1-\rho)}{2\rho-1}$	inv.binomial()
Negative binomial	$\binom{y+k-1}{y}\left(\dfrac{\mu}{\mu+k}\right)^y\left(\dfrac{k}{k+\mu}\right)^k$	$0(1)\infty$	$0<\mu,\ 0<k$	μ	negbinomial([dqpr])
Pólya (a NBD variant)	$\binom{y+k-1}{y}p^k(1-p)^y$	$0(1)\infty$	$0<k,\ 0<p<1$	$\dfrac{k(1-p)}{p}$	polyaR([dpqr])
Poisson	$\dfrac{e^{-\lambda}\lambda^y}{y!}$	$0(1)\infty$	$0<\lambda$	λ	poissonff([dpqr])

Table 11.8 Continuous dispersion models (Jørgensen, 1997) implemented in VGAM. The means are returned as the fitted values, by default. See Table A.1 for Bessel function definitions.

Distribution	Density $f(y;\boldsymbol{\theta})$	Support	Range of $\boldsymbol{\theta}$	Mean	VGAM family
Gamma	$\dfrac{\lambda(\lambda y/\mu)^{\lambda-1}\exp(-\lambda y/\mu)}{\mu\,\Gamma(\lambda)}$	$(0,\infty)$	$0<\mu,\,0<\lambda$	μ	gamma2([dpqr])
Hyperbolic secant	$\dfrac{\exp\{\theta y+\log(\cos(\theta))\}}{2\cosh(\pi y/2)}$	$(-\infty,\infty)$	$\theta\in\left(-\dfrac{\pi}{2},\dfrac{\pi}{2}\right)$	$\tan\theta$	hypersecant()
Hyperbolic secant	$\dfrac{\cos(\theta)}{\pi}\,y^{-1/2+\theta/\pi}\,(1-y)^{-1/2-\theta/\pi}$	$(0,1)$	$\theta\in\left(-\dfrac{\pi}{2},\dfrac{\pi}{2}\right)$	$\dfrac{1}{2}+\dfrac{\theta}{\pi}$	hypersecant.1()
Inverse Gaussian	$\left(\dfrac{\lambda}{2\pi y^3}\right)^{\frac{1}{2}}\exp\left\{-\dfrac{\lambda}{2\mu^2}\dfrac{(y-\mu)^2}{y}\right\}$	$(0,\infty)$	$0<\mu,\,0<\lambda$	μ	inv.gaussianff(dpr)
Leipnik (transformed)	$\dfrac{\{y(1-y)\}^{-\frac{1}{2}}}{Be\left(\dfrac{\lambda+1}{2},\dfrac{1}{2}\right)}\left[1+\dfrac{(y-\mu)^2}{y(1-y)}\right]^{-\lambda/2}$	$(0,1)$	$-1<\lambda$	μ	leipnik()
Normal	$\dfrac{1}{\sqrt{2\pi\sigma^2}}\exp\left\{-\dfrac{1}{2}\left(\dfrac{y-\mu}{\sigma}\right)^2\right\}$	$(-\infty,\infty)$	$0<\sigma$	μ	uninormal([dpqr])
Reciprocal inverse Gaussian	$\sqrt{\dfrac{\lambda}{2\pi y}}\exp\left\{-\dfrac{\lambda(y-\mu)^2}{2y}\right\}$	$(0,\infty)$	$0<\lambda$	μ	rigff()
Simplex	$\dfrac{\exp\left\{-\dfrac{(y-\mu)^2}{2\sigma^2\,y(1-y)\,\mu^2\,(1-\mu)^2}\right\}}{\sqrt{2\pi\sigma^2\{y(1-y)\}^3}}$	$(0,1)$	$0<\mu<1,\,0<\sigma$	μ	simplex(dr)
von Mises	$\dfrac{\exp\{\kappa\cos(y-\alpha)\}}{2\pi I_0(\kappa)}$	$[0,2\pi]$	$0<\alpha<2\pi,\,0<\kappa$	α	vonmises()

Table 11.9 More models implemented in **VGAM**. Some special functions are described in Sect. A.4.

Distribution	Density $f(y; \boldsymbol{\theta})$	Support	Range of $\boldsymbol{\theta}$	Mean	VGAM family
Borel-Tanner	$\dfrac{Q\, y^{y-Q-1}\, a^{y-Q}}{(y-Q)!}\, e^{-ay}$	$Q(1)\infty$	$(0,1)$	$\dfrac{Q}{1-a}$	`borel.tanner(dr)`
Haight's zeta	$(2y-1)^{-\alpha} - (2y+1)^{-\alpha}$	$1(1)\infty$	$(0,\infty)$	$\zeta(\alpha) \cdot (1 - 2^{-\alpha})$	`hzeta(dpqr)`
Negative binomial (k known)	$\dbinom{y+k-1}{y} \left(\dfrac{\mu}{\mu+k}\right)^{y} \left(\dfrac{k}{k+\mu}\right)^{k}$	$0(1)\infty$	$\mu > 0$	μ	`negbinomial.size([dpqr])`
Yule-Simon	$\rho\, Be(y, \rho+1)$	$1(1)\infty$	$\rho > 0$	$\dfrac{\rho}{\rho-1}$ if $\rho > 1$	`yulesimon(dpr)`

Table 11.10 More discrete distributions currently supported by VGAM. All but `dirmultinomial()` are univariate. Notes: (i) For `betabinomial()`, $\mathcal{A} = \mu(\rho^{-1} - 1)$ and $\mathcal{B} = (1 - \mu)(\rho^{-1} - 1)$. (ii) For `dirmultinomial()`, $N_* = y_1 + \cdots + y_M$. (ii) For `zipf()`, $H_{n,m} = \sum_{i=1}^{n} i^{-m}$ is known as the nth generalized harmonic number.

Distribution	Probability mass function $f(y; \boldsymbol{\theta})$	Support	Range of $\boldsymbol{\theta}$	Mean	VGAM family
Beta-binomial	$\binom{N}{Ny} \dfrac{Be(\mathcal{A} + Ny, \mathcal{B} + N(1 - y))}{Be(\mathcal{A}, \mathcal{B})}$	$0 \left(\frac{1}{N}\right) 1$	$\begin{array}{l} 0 < \mu < 1, \\ 0 < \rho < 1 \end{array}$	μ	`betabinomial(dpr)`
Beta-binomial	$\binom{N}{Ny} \dfrac{Be(\alpha + Ny, \beta + N(1 - y))}{Be(\alpha, \beta)}$	$0 \left(\frac{1}{N}\right) 1$	$0 < \alpha, 0 < \beta$	$\dfrac{\alpha}{\alpha + \beta}$	`betabinomialff(dpr)`
Beta-geometric	$\dfrac{Be(\alpha + 1, y + \beta)}{Be(\alpha, \beta)}$	$0(1)\infty$	$0 < \alpha, 0 < \beta$	$\dfrac{\beta}{\alpha - 1}$ if $\alpha > 1$	`betageometric(dpr)`

Logarithmic	$\dfrac{a \cdot c^y}{y}, \quad a = \dfrac{-1}{\log(1-c)}$	$1(1)\infty$	$0 < c < 1$	$\dfrac{ac}{1-c}$	logff(dpr)
Dirichlet-multinomial	$\dbinom{N_*}{y_1,\ldots,y_M}\dfrac{\prod_{j=1}^{M}\prod_{r=1}^{y_j}(\pi_j(1-\phi)+(r-1)\phi)}{\prod_{r=1}^{N_*}(1-\phi+(r-1)\phi)}$	$y_j \in 0(1)N_*$	$0 < \pi_j$	$E(Y_j) = N_*\pi_j$	dirmultinomial()
Mixture of 2 Poissons	$\phi\,\dfrac{e^{-\lambda_1}\lambda_1^y}{y!} + (1-\phi)\,\dfrac{e^{-\lambda_2}\lambda_2^y}{y!}$	$0(1)\infty$	$\begin{array}{l}0 < \phi < 1,\\ 0 < \lambda_j\end{array}$	$\phi\lambda_1 + (1-\phi)\lambda_2$	mix2poisson()
Truncated geometric	$\dfrac{(1-p)^y p}{1-(1-p)^{U+1}}$	$0(1)U$	$0 < p < 1$	$\dfrac{1-p}{p} - \dfrac{U(1-p)^U}{1-(1-p)^U}$	truncgeometric(upper = U)
Zeta	$\left[y^{p+1} \cdot \zeta(p+1)\right]^{-1}$	$1(1)\infty$	$0 < p$	$\dfrac{\zeta(p)}{\zeta(p+1)}$ if $p > 1$	zetaff(d)
Zipf	$y^{-s}\big/\sum_{i=1}^{N} i^{-s}$	$1(1)N$	$0 < s$	$\dfrac{H_{N,s-1}}{H_{N,s}}$	zipf(dp)

Chapter 12
Univariate Continuous Distributions

On the other hand, it is impossible for a cube to be written as the sum of two cubes or a fourth power to be written as a sum of two fourth powers or, in general, for any number which is a power greater than the second to be written as a sum of two like powers. I have a truly marvellous demonstration of this proposition, which this margin is too narrow to contain.
—Pierre de Fermat

12.1 Introduction

As in the previous chapter, many univariate distributions are amenable to the VGLM/VGAM framework due to (i) regularity conditions holding, and (ii) EIMs that can be computed or approximated easily. This chapter enumerates the current crop of univariate continuous distributions implemented in **VGAM** (Tables 12.3–12.14). As with the previous chapter, they are arranged in overlapping groups, and users are directed to the online help and references for details (Table 12.1). Unfortunately, space limitations do not permit any details about any particular distribution here, apart from a handful in Sect. 12.3.

For those univariate continuous distributions already 'supported' in R (in the `stats` package) via the **dpqr**-type functions, Table 12.3 lists the corresponding **VGAM** family functions. For these, the parameter names of the **VGAM** family function are the same as the argument names, e.g., '`rate`' in `rexp()` is the parameter name used by `exponential()`. Other examples are '`size`' and '`mu`' for `negbinomial()`, and '`mean`' and '`sd`' for `uninormal()`. Of course, this does not apply to those distributions which are fitted by `glm()`; for these, **VGAM** adds '`ff`' to the end of the family function name to avoid conflict, and these family functions attempt to replicate the functionality of their `glm()`-type counterparts.

T.W. Yee, *Vector Generalized Linear and Additive Models*,
Springer Series in Statistics, DOI 10.1007/978-1-4939-2818-7_12

Table 12.1 Some references for some specific distributions (upper table), and classes of distributions (lower table), relevant to **VGAM**—some classes are relevant to other chapters.

Distribution	References
Beta	Gupta and Nadarajah (2004)
Exponential	Balakrishnan and Basu (1995), Pal et al. (2006), Marshall and Olkin (2007), Ahsanullah and Hamedani (2010)
Gamma	Bowman and Shenton (1988), Marshall and Olkin (2007)
Laplace	Kotz et al. (2001), Kozubowski and Nadarajah (2010)
Lagrangian	Consul and Famoye (2006)
Pareto	Arnold (2015)
Weibull	Murthy et al. (2004), Rinne (2009)

Types of distributions	References
General distributions	Balakrishnan and Nevzorov (2003), Leemis and McQueston (2008), Forbes et al. (2011)
Univariate continuous	Johnson et al. (1994), Johnson et al. (1995), Kotz and van Dorp (2004)
Incomes/actuarial/loss models	Kleiber and Kotz (2003), Chotikapanich (2008), Kaas et al. (2008), Klugman et al. (2012), Richards (2012), Klugman et al. (2013), Nadarajah and Bakar (2013)
Dispersion models	Jørgensen (1997)
Mixture	Everitt and Hand (1981), Titterington et al. (1985), Lindsay (1995), McLachlan and Peel (2000), Frühwirth-Schnatter (2006)
Skew normal	Azzalini (2014)
Survival	Lawless (2003), Marshall and Olkin (2007)
Bivariate distributions and copulas (ch.13)	Joe (2014), Trivedi and Zimmer (2005), Nelsen (2006), Balakrishnan and Lai (2009), Mai and Scherer (2012), Schepsmeier and Stöber (2014)

The other tables in this chapter correspond to

- their support, e.g.,

  – $(-\infty, \infty)$: Tables 12.4, 12.5,
  – (A, ∞): Tables 12.6, 12.7, 12.8, 12.9,
  – (A, B) for finite A and B: Table 12.10,

- the 'parent' distribution they are based on, e.g.,

  – beta-type: Table 12.11,
  – exponential-type: Table 12.12,
  – gamma-type: Table 12.13,

- their focus, e.g.,

  – size distributions: Table 12.14.

Table 12.2 Some special cases of the Pearson system of distributions, and **VGAM** family functions for fitting them. The normal distribution arises as limiting cases of all types.

Pearson Type	Distribution	VGAM family
I	Beta	`betaff()`, `betaR()`
VI	Beta-prime	`betaprime()`
IV	Cauchy	`cauchy1()`, `cauchy()`
III	Chi-squared	`chisq()`
III	Exponential	`exponential()`. See also Table 12.12
III	Gamma	`gammaff()`, `gamma2()`, `gammaR()`. See also Table 12.13
VI	F	`fff()`
All types	Normal	`gaussianff()`, `uninormal()`
XI	Pareto	`paretoff()`
XII	Pareto	`truncpareto()`
VII	Student t	`studentt()`, `studentt2()`, `studentt3()`

12.2 Some Basics

12.2.1 Location, Scale and Shape Parameters

Scale parameters play a larger role with continuous distributions than with discrete ones. A (positive) scale parameter b (sometimes called θ_2 here) can be defined as satisfying

$$F(y; b, \boldsymbol{\theta}^*) \;=\; F(y/b; 1, \boldsymbol{\theta}^*) \tag{12.1}$$

where $\boldsymbol{\theta}^*$ denotes the remaining parameters. Consequently,

$$f(y; b, \boldsymbol{\theta}^*) \;=\; \frac{1}{b}\, f(y/b; 1, \boldsymbol{\theta}^*). \tag{12.2}$$

Additionally, a location parameter a (sometimes called θ_1 here) is often used in the form $y - a$ to shift the distribution to the right by the amount a. Combining both location and shift, along with a shape parameter s (sometimes called θ_3, or s_j if there are more than one) more generally, if one has a random variable Y, then one may also consider the random variable $T = [(Y - \theta_1)/\theta_2]^s$ for suitable θ_1 and $\theta_2 > 0$. Then

$$f(t; \theta_1, \theta_2, s) \;=\; \frac{s\,|y - \theta_1|^{s-1}}{\theta_2^s}\, f_0\!\left(\left[\frac{y - \theta_1}{\theta_2}\right]^s\right) \tag{12.3}$$

relates both densities (A.30). To provide a more unified front, most **VGAM** family functions represent θ_1 and θ_2 by the arguments `location` and `scale`, and with `shape`, or `shape1`, `shape2`, ..., for any shape parameters. As seen many times before, the link function for `scale` is always `"loge"`, by default, due to its positivity.

Unfortunately, in some software including R itself, the terms 'intercept', 'scale' and 'shape' are used very loosely, e.g., `survreg()`'s `scale` is the reciprocal of `rweibull()`'s `shape`, and `survreg()`'s intercept is the logarithm of `rweibull()`'s `scale`. Note that `[dpqr]weibull()` parameterizes the density as $f(x) = (s/b)(x/b)^{s-1} \exp(-(x/b)^s)$ for `shape` parameter s and `scale` parameter b.

Many 3-parameter distributions are not implemented in **VGAM** because their support is defined by the location parameter, therefore the usual regularity conditions are not met. For example, the 3-parameter Weibull having support (a, ∞) has been a topic considered by more than a few workers, e.g., Harper et al. (2011). Consequently, **VGAM** family functions for only some distributions have a 1-parameter variant (location only), a 2-parameter variant (location and scale only), and a 3-parameter variant (location, scale and shape). These often attach a 1, 2 and 3 to the end of the function name, and sometimes the default for a distribution is so obvious that the number is omitted, e.g., `studentt()` estimates the degrees of freedom ν, which is its shape parameter.

Since the EIM is so important for the VGLM/VGAM framework, the following result (see, e.g., Lehmann and Casella, 1998) is useful. For the location-scale families with PDF

$$f\left(\frac{y - \theta_1}{\theta_2}\right),\tag{12.4}$$

if $f(y) > 0$ and $f'(y)$ exists for all y, $\theta_1 \in \mathbb{R}$ and $\theta_2 > 0$, then the EIM elements are

$$\mathcal{I}_{E,11} = \frac{1}{\theta_2^2} \int_{-\infty}^{\infty} \left[\frac{f'(y)}{f(y)}\right]^2 f(y)\, dy,\tag{12.5}$$

$$\mathcal{I}_{E,22} = \frac{1}{\theta_2^2} \int_{-\infty}^{\infty} \left[\frac{yf'(y)}{f(y)} + 1\right]^2 f(y)\, dy,\tag{12.6}$$

$$\mathcal{I}_{E,12} = \frac{1}{\theta_2^2} \int_{-\infty}^{\infty} y \left[\frac{f'(y)}{f(y)}\right]^2 f(y)\, dy,\tag{12.7}$$

assuming that all the integrals are finite. Other authors such as Shao (2005) express the off-diagonal element as

$$\mathcal{I}_{E,12} = \frac{1}{\theta_2^2} \int_{-\infty}^{\infty} \left[\frac{yf'(y)}{f(y)} + 1\right] f'(y)\, dy.\tag{12.8}$$

While many distributions have the form (12.4), others such as the exponential and gamma customarily use the reciprocal of the scale parameter b instead, e.g.,

$$\lambda(y - a) = \frac{y - a}{b},$$

where λ is sometimes referred to as the *rate* parameter. If so, then the parameter name is often called `rate` rather than `scale`, and then the `"negloge"` link function is a way of interchanging between the two because it operates on the very simple relationship $\log \lambda = -\log b$. Here's simple example involving the Erlang distribution, which is a special case of the gamma distribution with an integer-valued shape parameter. To add a twist, there are multiple responses.

```
> n <- 1000
> set.seed(123)
> rate1 <- exp(2); rate2 <- exp(4)
> edata <- data.frame(y1 = rgamma(n, shape = 3, rate = rate1),   # Generate Erlang
                      y2 = rgamma(n, shape = 5, rate = rate2))   # random variates
> fit1 <- vglm(cbind(y1, y2) ~ 1, erlang(shape = c(3, 5)), data = edata)
> fit2 <- vglm(cbind(y1, y2) ~ 1, erlang(shape = c(3, 5), link = "negloge"),
               data = edata)
```

```
> coef(fit1, matrix = TRUE)

              loge(scale1)  loge(scale2)
  (Intercept)      -2.0405       -4.0121

> coef(fit2, matrix = TRUE)

              negloge(scale1)  negloge(scale2)
  (Intercept)          2.0405           4.0121
```

Note that, for multiple responses, the parameter names are postfixed by the characters 1, 2, ..., according to their order, e.g., in this example, `scale1` and `scale2` are used.

Incidentally, because the shape parameter is integer-valued, how might one estimate it?—because using a family function based on the gamma distribution returns a real-valued estimate. In this example, we can compute ℓ based on the two integers either side of $\hat{s}$, as follows. For simplicity, we do this for one of the responses.

```
> fit3 <- vglm(y1 ~ 1, gammaR(lss = FALSE), data = edata, trace = FALSE)
> coef(fit3, matrix = TRUE)   # Order of parameters is the same as rgamma()

              loge(shape)  loge(rate)
  (Intercept)      1.1458      2.0877

> shape.hat <- Coef(fit3)["shape"]   # Can do this for intercept-only models
> c(shape.hat, floor = floor(shape.hat), ceiling = ceiling(shape.hat))

        shape   floor.shape  ceiling.shape
        3.145         3.000          4.000

> fit4 <- vglm(y1 ~ 1, erlang(shape =   floor(shape.hat)), data = edata)
> fit5 <- vglm(y1 ~ 1, erlang(shape = ceiling(shape.hat)), data = edata)
> logLik(fit4) > logLik(fit5)

  [1] TRUE
```

Hence `fit4` is the better fitting model, corresponding to the correct shape parameter of 3.

12.2.1.1 Order of Arguments

Almost all **VGAM** family functions order their arguments as: `location`, `scale`, and `shape` (or `rate`), i.e., $\boldsymbol{\theta} = (a, b, s_1, s_2, \ldots)^T$. For some families it is necessary to have the argument `lss` for the purpose of matching the order of some pre-existing R functions residing in **stats** (`dpqr`-type functions). Here are two examples.

```
> args(rweibull)

  function (n, shape, scale = 1)
  NULL

> args(weibullR)

  function (lscale = "loge", lshape = "loge", iscale = NULL, ishape = NULL,
      lss = TRUE, nrfs = 1, probs.y = c(0.2, 0.5, 0.8), imethod = 1,
      zero = ifelse(lss, -2, -1))
  NULL
```

```
> args(dgamma)

  function (x, shape, rate = 1, scale = 1/rate, log = FALSE)
  NULL

> args(gammaR)

  function (lrate = "loge", lshape = "loge", irate = NULL, ishape = NULL,
      lss = TRUE, zero = ifelse(lss, -2, -1))
  NULL
```

Hence `weibullR(lss = FALSE)` and `gammaR(lss = FALSE)` correspond to $\boldsymbol{\theta} = (s, b)^T$ and $\boldsymbol{\theta} = (s, \lambda)^T$, respectively, i.e., matching their respective `dpqr`-type R functions. One may see from the output that `zero` always points to the shape parameter as being intercept-only, by default.

Whenever it appears, the default is `lss = TRUE`, and setting `lss = FALSE` is necessary to match the order of pre-existing R functions.

12.2.2 Initial Values

Because of the large number of distributions, initial values require much more attention in general compared to GLMs because their convergence properties are far less well-behaved.

For illustration, consider a simple example involving a 1-parameter Cauchy distribution with a known scale parameter of unity.

```
> n <- 10; set.seed(91)
> cdata <- data.frame(y = rcauchy(n))
> cfit <- vglm(y ~ 1, cauchy1, data = cdata, trace = FALSE)   # TRUE is a good idea
> coef(cfit, matrix = TRUE)

                location
  (Intercept)   0.16096
```

The log-likelihood ℓ is plotted in Fig. 12.1 as a function of a. It is evident that ℓ is multimodal, and that poor initial values may lead to a local solution. For certain families such as this, VGAM chooses initial values based on a grid search. But the user should make liberal use of arguments such as `imethod`, `ilocation`, `iscale`, etc. Incidentally, for the Cauchy distribution, if both location and scale parameters are to be estimated (`cauchy()`), then there is a unique maximum likelihood solution provided that $n > 2$ and less than half the data are located at any one point. By the way, occurrences of `trace = FALSE` in this chapter are to conserve space; it is recommended readers use `trace = TRUE` to monitor convergence.

As the number of parameters increases, it often becomes more difficult to safeguard against the multimodal ℓ problem. For example, the 4-parameter generalized beta II distribution of Table 12.14 is *much* harder to estimate than special cases of itself. Large sample sizes are needed, in general, to justify the estimation of its 3 parameters. And if so, then probably these (shape) parameters should be intercept-only. For some distributions such as the GEV (Chap. 16), it often takes n in the order of thousands in order to reasonably model the shape parameter as a function of even one covariate. Having intercept-only shape parameters often leads to greater stability in the estimation because of fewer numerical problems that

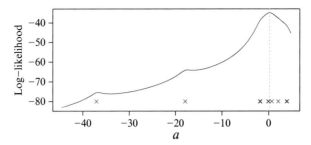

Fig. 12.1 Log-likelihood $\ell(a)$ as a function of the location parameter, for a random sample of size 10 from a standard Cauchy distribution. The *dashed vertical line* denotes $\hat{a}$, and the purple $\times$ denotes the data.

arise when they are allowed to become too flexible. The **zero** argument should be used to limit the number of parameters which are allowed to vary as a function of the covariates. Setting **zero = NULL** is quite unjustified unless there is much data and/or there is a strong signal and/or only a few variables.

From both a user's and developer's point of view, it is worth knowing that many distributions have several common parameterizations. Some complex densities such as the generalized hyperbolic distribution are critically dependent on the best choice of these—for these the EIM is more well-conditioned. Ideally, EIMs are best diagonal where possible because the parameters are asymptotically uncorrelated, therefore the estimation can more easily be broken down to separate subproblems.

As with all **VGAM** family functions, monitoring convergence is a very good idea. If the data and model agree, then usually 4–8 iterations are all that are necessary for convergence. More than 10 iterations, say, suggests something is awry and that corrective action might be taken, else that distribution abandoned.

As an example of using a simpler model to provide initial values for a more complex one, consider the Fisk distribution which is a special case of the Singh-Maddala distribution with shape parameter $q = 1$ (Table 12.14). In the following, **fit0** is nested in **fit1** and is used to provide initial values for the more flexible model.

```
> set.seed(123); sdata <- data.frame(x2 = runif(n <- 1000))
> sdata <- transform(sdata, eta.shape1.a  =  2,
                             eta.scale.b   = -1 + x2,
                             eta.shape3.q  =  1)
> sdata <- transform(sdata, shape1.a = exp(eta.shape1.a),
                             scale.b  = exp(eta.scale.b),
                             shape3.q = exp(eta.shape3.q))
> sdata <- transform(sdata, y1 = rsinmad(n, shape1.a = shape1.a,
                            scale = scale.b, shape3.q = shape3.q))
> fit0 <- vglm(y1 ~ x2, fisk(lss = FALSE), data = sdata)  # A 2-parameter problem
> init.shape3.q <- 1  # Try this initial value for shape3.q
> fit1 <- vglm(y1 ~ x2, sinmad(zero = c(1, 3), lss = FALSE), data = sdata,
            etastart = cbind(predict(fit0), log(init.shape3.q)))
> coef(fit1, matrix = TRUE)  # A 3-parameter problem

               loge(shape1.a) loge(scale) loge(shape3.q)
 (Intercept)          2.0313     -1.0121        0.87166
 x2                   0.0000      0.9706        0.00000
```

Here, it is important to match up each column of `etastart` so that they refer to the same parameter, and having the same link function too. Incidentally, `fit1` took 6 iterations to converge; this value is presently stored as `fit1@iter`.

12.2.3 About the Tables

As most of this chapter consists of the tables at the end, here are some general notes regarding them—they hold for *most* VGAM family functions but not necessarily all of them. Some of the notes also hold for the discrete distributions of the previous chapter.

1. For location-scale families, there are two forms:

   (i) (12.4) is the most common, where a or θ_1 is used to refer to the location parameter, and b or θ_2 the positive scale parameter. Often the argument and parameter names are called `location` and `scale`. If the distribution's support depends on a, then its value must be specified by the user.
   (ii) For completeness, the second form is stated:

$$f(\lambda (y - \theta_1)), \tag{12.9}$$

   i.e., $\lambda = b^{-1}$. Then λ is referred to a rate parameter, although r may be used instead. Then the argument and parameter names are often called `location` and `rate`.

   The order of the parameters is, if relevant, location, followed by scale, followed by shape parameters. Some families allow for known a to be inputted—and if so, then it is common for the lower support limit to be a so that $a < y_i$ for all i.

2. Usually the mean is returned as the fitted values, by default. If not, then the median $\widetilde{\mu}$ is returned instead.
3. The parameters $\theta_j \in \mathbb{R}$ unless specified otherwise.
4. Section A.4 may need to be consulted for details about special functions such as those of Bessel, digamma and trigamma.
5. Some family functions have for accompaniment the full range of `dpqr`-type functions; others a smattering, and others none at all (currently). Functions written in the future will be added to VGAM incrementally.
6. The VGAM family functions of continuous distributions do not allow data lying on the boundaries of their support, e.g., `chisq()` will crash for $y_i = 0$. Of course, `Inf`-valued responses are not allowed in general either.
7. Of a, b and s, the highest probability of an intercept-only parameter, by default, is for the shape parameter s, followed by the scale parameter b. The location parameter is never intercept-only, by default. Section 12.2.5 attempts to justify the default values of `zero` based on this ordering. Recall that setting `zero = NULL` makes none of the parameters intercept-only, however, this should only be used if there is a lot of data at hand and x is very low-dimensional.
8. Distributions based on mathematical functions (e.g., `beta()`, `gamma()`, `zeta()`, `lgamma()`) often have a postfix "ff" or "R" appended to the name, e.g., `gammaR()`, `betaff()`. The "R" means it tries to match up with some pre-existing R function, e.g., `gammaR()` is based on `dgamma()`'s shape and scale

parameters. However, most applied statisticians are more interested in directly the mean, therefore will find `gamma2()` more useful. In this case, `gammaff()` is the `Gamma()`-imitation that runs with `glm()`. When written, `weibull()` will use the parameterization of (12.15)–(12.16).

9. For those distributions with **dpqr**-type functions, sometimes $\widehat{\mu}$, $\widetilde{\mu}$ or general quantiles may be returned as fitted values. Then the arguments `type.fitted` and `percentiles` are relevant, e.g., `family = VGAMFamilyFn(type.fitted = "quantile", perc = 50)` will return the median. Then a vector for `perc` is allowable, e.g., `perc = c(25, 50, 75)` should return a 3-column matrix of fitted values. In general, `percentiles` can be assigned a value in $(0, 100)$. If the mean is supported, then `type.fitted = "mean"` is the way to obtain this.

10. Because some distributions are intrinsically more difficult to estimate, especially those with more parameters, the estimation reliability of the VGAM family functions vary. Over time, it is hoped that all of them will be improved in terms of their reliability and choice of initial values.

12.2.4 Q-Q Plots

With the multitude of distributions that may be fitted, the ability to compare the fits of several models against the data is important. A QQ-plot may be used to compare two distributions by plotting the quantiles (the "Q") against each other. If the distributions are the same, then we should expect the points to lie approximately on the $x = y$ line. This method is applicable if a VGAM family function is accompanied by a **p**-type or **q**-type function. Also, the method is restricted to intercept-only models.

In the following toy example, the response `y1` follows a Birnbaum-Saunders distribution, and `y2` is Fréchet distributed. Fitting a Birnbaum-Saunders distribution to both, the resulting QQ-plots confirm that the fit to `y1` is reasonable but the fit to `y2` isn't (Fig. 12.2).

```
> n <- 100; set.seed(1)
> bdata <- data.frame(x1 = rep(1, n))    # bdata has n rows
> bdata <- transform(bdata, Scale = exp(1), Shape = exp(0.5))   # Recycling
> bdata <- transform(bdata, y1 =    rbisa(n, scale = Scale, shape = Shape),
                            y2 = rfrechet(n, scale = Scale, shape = Shape))
> fit1 <- vglm(y1 ~ 1, bisa, data = bdata, trace = FALSE)
> fit2 <- vglm(y2 ~ 1, bisa, data = bdata, trace = FALSE)
> pfit1 <- predict(fit1, untransform = TRUE)
> pfit2 <- predict(fit2, untransform = TRUE)
> bdata <- transform(bdata, scale1.hat = pfit1[, "scale"],
                            shape1.hat = pfit1[, "shape"],
                            scale2.hat = pfit2[, "scale"],
                            shape2.hat = pfit2[, "shape"])
> pvector <- ppoints(n)   # n equally-spaced points on (0, 1)
>
> plot(qbisa(pvector, scale = scale1.hat, shape = shape1.hat) ~ sort(y1),
       data = bdata, col = "blue", log = "xy", ylab = "Estimated quantiles",
       main = "(a); q-type function")
> abline(a = 0, b = 1, col = "orange", lty = "dashed")
>
```

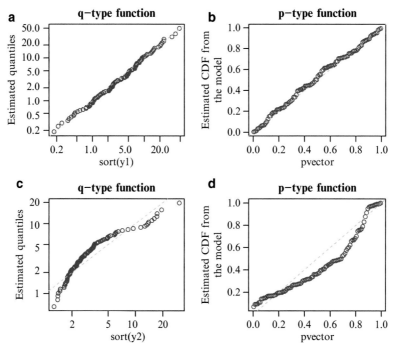

Fig. 12.2 QQ-plots of 2 simulated data sets. A Birnbaum-Saunders distribution is fitted to both. (**a**)–(**b**) Data from a Birnbaum-Saunders distribution; (**c**)–(**d**) data from a Fréchet distribution. Plots (**a**) and (**c**) are based on the q-type function, and both axes are on a log-scale. Plots (**b**) and (**d**) use the p-type function. An $x = y$ line appears in all plots.

```
> plot(sort(pbisa(y1, scale = scale1.hat, shape = shape1.hat)) ~ pvector,
       data = bdata, col = "blue", ylab = "Estimated CDF from the model",
       main = "(b); p-type function")
> abline(a = 0, b = 1, col = "orange", lty = "dashed")
>
> plot(qbisa(pvector, scale = scale2.hat, shape = shape2.hat) ~ sort(y2),
       data = bdata, col = "blue", log = "xy", ylab = "Estimated quantiles",
       main = "(c); q-type function")
> abline(a = 0, b = 1, col = "orange", lty = "dashed")
>
> plot(sort(pbisa(y2, scale = scale2.hat, shape = shape2.hat)) ~ pvector,
       data = bdata, col = "blue", ylab = "Estimated CDF from the model",
       main = "(d); p-type function")
> abline(a = 0, b = 1, col = "orange", lty = "dashed")
```

To explain a few details behind this, the R function `ppoints(n)` is used to form an equally spaced vector of length **n** on the interval $(0, 1)$, e.g., the plotting positions are something similar to $(i + \frac{1}{2})/(1 + n)$ $(= p_i, \text{say})$. If $Y_1, \ldots, Y_n$ are a random sample from some continuous distribution with CDF $F_Y(y; \boldsymbol{\theta})$, then $U_i = F_Y(y_i; \boldsymbol{\theta})$ has a standard uniform distribution. Let $Y_{(i)}$ be the ith order statistic, so that $Y_{(1)} \leq Y_{(2)} \leq \cdots \leq Y_{(n)}$. Then $F_Y(Y_{(i)}; \boldsymbol{\theta}) = U_{(i)} \approx p_i$. Plots (a) and (c) check to see whether $F^{-1}(p_i; \widehat{\boldsymbol{\theta}}) \approx y_{(i)}$, while plots (b) and (d) check to see whether $F(y_{(i)}; \widehat{\boldsymbol{\theta}}) \approx p_i$. For the interested reader, the ideas here are borrowed from, e.g., `qqnorm()` and `qqplot()`.

A common mistake amongst amateur data analysts is the over-interpretation of QQ-plots. When n is small, there is a substantial amount of statistical variation amongst the order statistics, hence it is unreasonable to expect the plotted points to lie very close to the $x = y$ line. It is common to quantify the agreement of a fitted distribution with the observed data by citing a correlation coefficient.

12.2.5 Scale and Shape Parameters

It was recommended above that usually shape parameters should be modelled more simply compared to scale parameters, as often reflected in the default of the `zero` argument of VGAM family functions. In this section, we offer a little justification for this recommendation; and if accepted, it must be taken with a grain of salt (or perhaps several pinches).

Consider a 2-parameter gamma distribution parameterized by positive scale and shape parameters b and s (equivalent to `gammaR(lrate = "negloge")`):

$$f(y; b, s) = \frac{e^{-y/b}\, y^{s-1}}{b^s\, \Gamma(s)}, \quad 0 < y. \tag{12.10}$$

We use this as illustrative of the general principle that often there is less information content on the shape parameter compared to the scale parameter. For one observation, it is easy to show that the inverse of its EIM is

$$\frac{b^2}{s\,\psi'(s) - 1} \begin{pmatrix} \psi'(s) & -1/b \\ -1/b & s/b^2 \end{pmatrix},$$

upon which the SEs can be obtained. Figure 12.3a plots these SEs as a function of s, for 3 values of b: e^{-1}, $e^0 = 1$, e^1. A log-scale is used in plot (b). With these b values, the densities for 2 values of s appear in Fig. 12.3c,d, under the original scale and a log-scale. It may be seen that, for fixed b, $\text{SE}(\widehat{s}) > \text{SE}(\widehat{b})$ for s sufficiently large. With such greater variability in $\widehat{s}$, it pays to model such parameters more simply, e.g., as intercept-only. This recommendation, however, is not without its dangers: $\text{SE}(\widehat{s}) \ll \text{SE}(\widehat{b})$ when $s \approx 0$. Ideally, the user will know that, for low s and large b, there is the ability to model this parameter more accurately than b.

12.3 A Few Groups of Distributions

12.3.1 Size Distributions

Table 12.14 summarizes a subset of distributions on $(0, \infty)$ described in Kleiber and Kotz (2003), called statistical size distributions. These are concerned with economic and actuarial size phenomena of various types such as income and wealth. This topic started with the work of Pareto around the turn of the 20th century, and other applications include the size of cities and companies, the number of publications of authors, and word frequencies in text. They classify size distribution into 3 types according to their tail behaviour as $y \to \infty$: Pareto-type ($f(y) \sim y^{-\alpha}$),

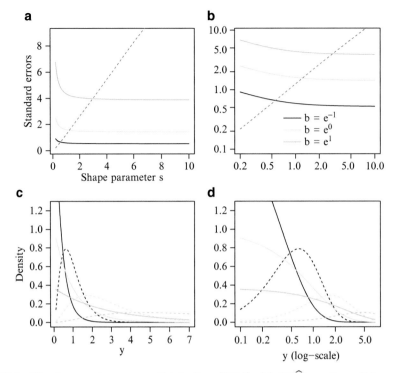

Fig. 12.3 The 2-parameter gamma distribution (12.10). (a) SE$(\widehat{b})$ are the *solid curves*, for various values of the scale parameter b given in the legend of (b). The *purple dashed line* is SE$(\widehat{s})$. (b) The same with both axes on a log-scale. (c) The densities with shape parameters $s = e^0$ (*solid lines*) and e^1 (*dashed*). (d) The same as (c) with one axis on a log-scale.

lognormal-type ($f(y) \sim \exp\{-(\log y)\,\alpha\}$), gamma-type ($f(y) \sim e^{-\alpha y}$). Indeed, all the entries are special cases of a generalized beta II distribution, and parameterized by a scale parameter (b, say) and one or more shape parameters called a, p, q. All 4 parameters are positive, and are called `scale`, `shape1.a`, `shape2.p`, `shape3.q`. The 4- and 3-parameter distributions, being so flexible, can be especially difficult to estimate reliably due to multiple local solutions, particularly for small samples. The tricks given in Sect. 8.5 are very relevant and recommended for use.

12.3.2 Pearson System

Karl Pearson's system of continuous distributions is based on the differential equation

$$\frac{df}{dy} = \frac{(y - a)\,f(y)}{b_0 + b_1\,y + b_2\,y^2}. \tag{12.11}$$

There are 13 members, of which several currently can be fitted with **VGAM** family functions. Note that for some members this may be no easy task, e.g., Johnson et al. (1994, p.19) state that fitting Type IV by MLE is very difficult and is rarely attempted. The normal density corresponds to $b_1 = b_2 = 0$, and some people refer to it as Type V even though it is the limit of all types. Table 12.2 lists some examples falling in the Pearson system and **VGAM** family functions for fitting them.

12.3.3 Poisson Points in the Plane and Volume

A problem from geometric probability involves measuring the distance between a fixed point P inside an areal region or volume and its nearest neighbours. Suppose the number of points in any region of area A of the plane has a Poisson($\mu = \lambda A$) distribution so that λ is the *density* of the points. Let $Y_{(j)}$ be the distance from P to its jth nearest neighbour, so that it is the jth order statistic in terms of distances. Then its PDF is

$$f_{Y_{(j)}}(y) = \frac{2(\lambda\pi)^j}{(j-1)!} \, y^{2j-1} \, \exp\left\{-\lambda\pi y^2\right\}, \quad 0 < y. \tag{12.12}$$

A similar formula holds when considering a volume V with number of points generated from a Poisson($\mu = \lambda V$) distribution. Then the PDF for $Y_{(j)}$ is

$$f_{Y_{(j)}}(y) = \frac{3\left(\frac{4}{3}\lambda\pi\right)^j}{(j-1)!} \, y^{3j-1} \, \exp\left\{-\frac{4}{3}\lambda\pi y^3\right\}, \quad 0 < y. \tag{12.13}$$

The **VGAM** family function `poisson.points()` estimates the density λ in both cases. It has argument `ostatistic` which is assigned the single value j, and an argument `dimension` which has value 2 or 3. While `ostatistic` stands for "order statistic" the argument need not be assigned an integer, e.g., the value 1.5 coincides with the Maxwell distribution when the dimension is 2. Note that the value 1 coincides with the Rayleigh distribution when the dimension is 2.

Exercises

Ex. 12.1.   EIMs for Location-Scale Distributions

(a) Given (12.7), show that (12.8) is an identical expression.
(b) Use (12.5)–(12.7) to obtain the EIM for the $N(\mu, \sigma)$ distribution.
(c) Use (12.5)–(12.7) to obtain the EIM for the 2-parameter logistic distribution (Table 12.3).

Ex. 12.2.   Normal Distribution EIM
For parameters (μ, σ) and for one observation, show that the EIM is equal to $\sigma^{-2} \cdot \mathrm{diag}(1, 2)$. What is the EIM for parameterization (μ, σ^2)? For symmetric distributions about a location parameter θ_1, of the form (12.4), deduce that the off-diagonal EIM element is 0.

Ex. 12.3. ALD EIM
For κ known and fixed, why cannot (12.5)–(12.8) be used to obtain the top-left 2×2 submatrix of the EIM of the asymmetric Laplace distribution (15.12) given by (15.14)?

Ex. 12.4. Random Variates from a Triangular Distribution
Write a simple R function `rTriangle(n, theta, lower = 0, upper = 1)` which expects single-valued arguments satisfying $L < \theta < U$, where $L = $ `lower` and $U = $ `upper`. Implement the acceptance-rejection method described in Sect. 11.1.1 based on a uniform distribution covering it. You may make use of `dtriangle()` in VGAM. Test your function out on several θ values including $\frac{1}{2}$, 1 and $1\frac{1}{2}$, fixing $L = -1$ and $U = 2$.

Ex. 12.5. Generalized Gamma Distribution

(a) Generate 1000 observations from a generalized gamma (Stacy, 1962) distribution with parameter values $b = e^2$, $d = e^0$, $k = e^{2.5}$.
(b) Fit some 2-parameter gamma distribution, and use it for initial values to fit the 3-parameter distribution.
(c) Does (b) improve the convergence rate? What about the reliability for successful convergence?

Ex. 12.6. Mixture of 2 Exponential Distributions
Consider the mixture of two exponential distributions specified in Table 12.12.

(a) Derive the CDF $F(y)$.
(b) Does this mixture distribution possess the memoryless property of the exponential distribution? i.e., does $P(Y > s + t | Y > t) = P(Y > s)$?
(c) Generate 1000 observations from the distribution with $\phi = \text{expit}(-\frac{1}{2})$, $\lambda_1 = e^2$ and $\lambda_2 = e^{-1}$. (The expit function is the inverse of the logit). Then estimate the parameters thereof.

Ex. 12.7. Singh-Maddala Distribution
The 2-parameter paralogistic distribution is a special case of the 3-parameter Singh-Maddala distribution with parameter constraint $a = q$.

(a) Generate 1000 observations from a paralogistic distribution with $a = e^1$ and unit scale parameter, and then estimate its parameters with `paralogistic()`.
(b) Use the `constraints` argument of `vglm()` and the family function `sinmad()` to estimate its parameters. Confirm that the log-likelihood, fitted values and standard errors of the estimates are the same as (a).

Ex. 12.8. Dagum Distribution
The 2-parameter inverse paralogistic distribution is a special case of the 3-parameter Dagum distribution with parameter constraint $a = p$.

(a) Generate 1000 observations from an inverse paralogistic distribution with $a = e^1$ and unit scale parameter, and then estimate its parameters with the VGAM family function `inv.paralogistic()`.
(b) Use the `constraints` argument of `vglm()` and the family function `dagum()` to estimate its parameters. Confirm that the log-likelihood, fitted values and standard errors of the estimates are the same as (a).

Ex. 12.9. Weibull EIM

Consider the Weibull distribution in Table 12.3 whose EIM is given by (for $\boldsymbol{\theta} = (b, s)^T$; see also (A.49))

$$\begin{pmatrix} (s/b)^2 & -(1-\gamma)/b \\ -(1-\gamma)/b & [\pi^2 + 6(\gamma-1)^2]/(6s^2) \end{pmatrix}. \tag{12.14}$$

(a) Show that the EIM is positive-definite for all positive b and s.
(b) Consider the parameterization

$$\eta_1 = \log \mu, \tag{12.15}$$
$$\eta_2 = \log s. \tag{12.16}$$

Derive $-E[\partial^2 \ell_i/(\partial\boldsymbol{\eta}\,\partial\boldsymbol{\eta}^T)]$.

Ex. 12.10. Kumaraswamy Distribution

Jones (2009) gives EIM of the Kumaraswamy distribution for $\boldsymbol{\theta} = (\alpha, \beta)^T$ where $f(y; \boldsymbol{\theta}) = \alpha\beta y^{\alpha-1}(1-y^\alpha)^{\beta-1}$ is the density (α is shape1, etc., see Table 12.11). Then the elements of the EIM are

$$\mathcal{I}_{E,11} = \alpha^{-2}\left(1 + [\beta/(\beta-2)]\left\{[\psi(\beta)-\psi(2)]^2 - [\psi'(\beta)-\psi'(2)]\right\}\right),$$
$$\mathcal{I}_{E,22} = \beta^{-2},$$
$$\mathcal{I}_{E,12} = -\left\{\psi(1+\beta)-\psi(2)\right\}/[\alpha(\beta-1)].$$

Identify any singularities, and propose any corrections for computing these quantities.

Ex. 12.11. Maxwell and Rayleigh Distributions—and Poisson Points

(a) Show that Maxwell's distribution and the 2-dimensional Poisson points distribution coincide with $s = \frac{3}{2}$ and $a = 2\pi\lambda$.
(b) Show that Rayleigh distribution and the 2-dimensional Poisson points distribution coincide ... what is the value of b?
(c) Derive the score vectors and EIMs of the Maxwell and Rayleigh distributions.
(d) Repeatedly (500 times) generate 100 random points uniformly distributed on the unit square and record the distance from $(\frac{1}{2}, \frac{1}{2})$ to the closest point. Then feed your data into the appropriate VGAM family function to estimate the density. Obtain an approximate 95 % confidence interval for the density. Does it contain the truth?

Ex. 12.12. Pareto Distribution

Show that if $Y|\Lambda$ has an exponential distribution with rate parameter λ and Λ has a gamma distribution, then the marginal distribution of Y has a shifted Pareto-like distribution. Hint: use the gammaR() density, and show the unconditional density of Y is close to that of paretoff().

Ex. 12.13. Hypersecant Distribution

Look up hypersecant() and hypersecant01() in the tables, and use the change of variable technique (A.30) to show that $\pi y = \text{logit}\, u$ ties the two together.

Ex. 12.14. Pearson Type I and II Densities

Consider the Pearson system (12.11). [Balakrishnan and Lai (2009)]

(a) The Pearson I is a shifted beta density

$$f(y; s_1, s_2) \; \propto \; (1+y)^{s_1-1} \, (1-y)^{s_2-1}, \quad y \in (-1, 1). \tag{12.17}$$

Derive an expression for the constant of proportionality, and then evaluate its value for $s_1 = 1$ and $s_2 = e$.

(b) Generate 1000 observations from (12.17). Then estimate the s_j.

(c) The Pearson II has density

$$f(y; s) \; \propto \; \left(1 - y^2\right)^{s-1}, \quad y \in (-1, 1) \tag{12.18}$$

(called the *symmetric beta distribution*). Derive an expression for the constant of proportionality, and then evaluate its value for $s = \exp(\frac{1}{2})$.

(d) Generate 1000 observations from (12.18). Then estimate s.

Ex. 12.15.   Lindley Distribution
Derive the EIM for the 1-parameter Lindley distribution. Derive an expression for its CDF.

Ex. 12.16.   Triangle Distribution
Derive the EIM for the triangle distribution (Table 12.10). Derive an expression for its CDF.

A thought is an idea in transit.
—Pythagoras

Table 12.3 Continuous distributions in the **stats** package of R (in the form of **dpqr**-type functions), and their corresponding VGAM family functions. Notes: (i) Setting argument `lss = FALSE` is needed to match their order of arguments. (ii) Functions `[pq]tukey()` and `[dpqr]unif()` have been omitted. (iii) Some families such as `studentt()` have 1-, 2- and 3-parameter variants. (iv) See also Table 11.6 for discrete distributions.

Distribution	Density $f(y;\boldsymbol{\theta})$	Support	Range of $\boldsymbol{\theta}$	Mean (or median $\widetilde{\mu}$)	VGAM family
Beta	$\dfrac{y^{s_1-1}(1-y)^{s_2-1}}{Be(s_1,s_2)}$	$(0,1)$	$s_1>0,\,s_2>0$	$\dfrac{s_1}{s_1+s_2}$	`betaR()`
Cauchy	$\left\{\pi b\left(1+\left[\dfrac{y-a}{b}\right]^2\right)\right\}^{-1}$	$(-\infty,\infty)$	$b>0$	$\widetilde{\mu}=a$	`cauchy()`
Chi-square	$\dfrac{y^{\frac{\nu}{2}-1}\exp\{-y/2\}}{2^{\nu/2}\,\Gamma(\nu/2)}$	$(0,\infty)$	$\nu>0$	ν	`chisq()`
Exponential	$\lambda\,e^{-\lambda y}$	$(0,\infty)$	$\lambda>0$	λ^{-1}	`exponential()`
Fisher's F	See Table 12.6.	$(0,\infty)$	$\nu_1>1,\,\nu_2>1$	$\dfrac{\nu_2}{\nu_2-2}$ if $\nu_2>2$	`fff()`
Gamma	$\dfrac{\exp(-\lambda y)\,y^{s-1}\,\lambda^s}{\Gamma(s)}$	$(0,\infty)$	$\lambda>0,\,s>0$	$\dfrac{s}{\lambda}$	`gammaR(lss = TRUE)`
Logistic	See Table 12.4	$(-\infty,\infty)$	$b>0$	a	`logistic()`
Log normal	$\dfrac{1}{\sqrt{2\pi}\,\sigma y}\exp\left\{\dfrac{-(\log y-\mu)^2}{2\sigma^2}\right\}$	$(0,\infty)$	$\sigma>0$	$\exp\{\mu+\sigma^2/2\}$	`lognormal()`
Normal	$\dfrac{1}{\sqrt{2\pi\sigma^2}}\exp\left\{-\dfrac{1}{2}\left(\dfrac{y-\mu}{\sigma}\right)^2\right\}$	$(-\infty,\infty)$	$\sigma>0$	μ	`uninormal()`
Student t	$\dfrac{\nu^{-1/2}}{Be\left(\dfrac{\nu}{2},\dfrac{1}{2}\right)}\left[1+\dfrac{y^2}{\nu}\right]^{-(\nu+1)/2}$	$(-\infty,\infty)$	$\nu>0$	0	`studentt()`
Weibull	$\dfrac{s\,y^{s-1}}{b^s}\exp\left\{-\left(\dfrac{y}{b}\right)^s\right\}$	$(0,\infty)$	$b>0,\,s>0$	$b\,\Gamma(1+s^{-1})$	`weibullR(lss = TRUE)`

Table 12.4 Univariate continuous distributions implemented in VGAM with support on $(-\infty, \infty)$; Part 1. Note: ϕ_1 is used here as the standard normal PDF, to avoid conflict with the parameter ϕ.

Distribution	Density function $f(y; \theta)$	Support	Range of θ	Mean (or median $\widetilde{\mu}$)	VGAM family
Cauchy (1-parameter)	$\left\{\pi b\left(1 + \left[\dfrac{y-a}{b}\right]^2\right)\right\}^{-1}$, b known	$(-\infty, \infty)$		$\widetilde{\mu} = a$	cauchy1([dpqr])
Cauchy (2-parameter)	$\left\{\pi b\left(1 + \left[\dfrac{y-a}{b}\right]^2\right)\right\}^{-1}$	$(-\infty, \infty)$	$0 < b$	$\widetilde{\mu} = a$	cauchy([dpqr])
Hyperbolic secant	$\dfrac{\exp\{\theta y + \log(\cos(\theta))\}}{2\cosh(\pi y/2)}$	$(-\infty, \infty)$	$\theta \in \left(-\dfrac{\pi}{2}, \dfrac{\pi}{2}\right)$	$\tan\theta$	hypersecant()
Log F	$\dfrac{e^{s_1 y}}{Be(s_1, s_2) \cdot [1 + \exp(y)]^{s_1 + s_2}}$	$(-\infty, \infty)$	$0 < s_1,\ 0 < s_2$	$\psi(s_1) - \psi(s_2)$	logF(d)
Logistic (b known for logistic1())	$\dfrac{\exp\left\{\dfrac{-(y-a)}{b}\right\}}{b\left(1 + \exp\left\{\dfrac{-(y-a)}{b}\right\}\right)^2}$	$(-\infty, \infty)$	$0 < b$	a	logistic[1]([dpqr])
Mixture of 2 normals	$\dfrac{\phi}{\sigma_1}\phi_1\left(\dfrac{y-\mu_1}{\sigma_1}\right) + \dfrac{1-\phi}{\sigma_2}\phi_1\left(\dfrac{y-\mu_2}{\sigma_2}\right)$	$(-\infty, \infty)$	$0 < \phi < 1$	$\phi\mu_1 + (1-\phi)\mu_2$	mix2normal()

Table 12.5 Univariate continuous distributions implemented in VGAM with support on $(-\infty, \infty)$; Part 2. For $\texttt{tikuv}()$, $z = (y - \mu)/\sigma$.

Distribution	Density function $f(y; \boldsymbol{\theta})$	Support	Range of $\boldsymbol{\theta}$	Mean	VGAM family
Skew-normal	$2\,\phi(y)\,\Phi(\alpha y)$	$(-\infty, \infty)$		$\alpha\sqrt{\dfrac{2}{\pi(1+\alpha^2)}}$	$\texttt{skewnormal(dr)}$
Student t (1-parameter)	$\dfrac{\nu^{-1/2}}{Be\left(\frac{\nu}{2}, \frac{1}{2}\right)}\left[1 + \dfrac{y^2}{\nu}\right]^{-(\nu+1)/2}$	$(-\infty, \infty)$	$0 < \nu$	0 if $\nu > 1$	$\texttt{studentt([dpqr])}$
Student t (2- and 3-parameters)	$\dfrac{\nu^{-1/2}}{b\,Be\left(\frac{\nu}{2}, \frac{1}{2}\right)}\left[1 + \dfrac{\{(y-a)/b\}^2}{\nu}\right]^{-(\nu+1)/2}$	$(-\infty, \infty)$	$0 < \nu,\ 0 < b$	a if $\nu > 1$	$\texttt{studentt[23]([dpqr])}$
Akkaya and Tiku (2008)	$\dfrac{K}{\sqrt{2\pi}\,\sigma}\left[1 + \dfrac{1}{2h}\left(\dfrac{y-\mu}{\sigma}\right)^2\right]^2 \exp\left\{-\dfrac{1}{2}\left(\dfrac{y-\mu}{\sigma}\right)^2\right\}$	$(-\infty, \infty)$	$0 < \sigma$	μ	$\texttt{tikuv(dpqr)}$

Table 12.6 Univariate continuous distributions implemented in VGAM with support on (A, ∞) with A known; Part I. Notes: (i) The mean for foldnormal(dpqr) is $\sigma\sqrt{2/\pi}\exp\left\{-\mu^2/(2\sigma^2)\right\} + \mu\left[1 - 2\Phi(-\mu/\sigma)\right]$. (ii) See also Table 12.12 for the exponential distribution, Table 12.13 for the gamma distribution, and Table 12.14 for size distributions.

Distribution	PDF $f(y;\theta)$ or CDF $F(y;\theta)$	Support	Range of θ	Mean (or median $\widetilde\mu$)	VGAM family
Benini	$2s\exp\left\{-s\left[\left(\log\left(\frac{y}{b}\right)\right)^2\right]\right\}\dfrac{\log(y/b)}{y}$, b known	(b,∞)	$0 < s$	$\widetilde\mu = b\exp\sqrt{\log 2}\,^{1/s}$	benini1(dpqr)
Birnbaum-Saunders	$F = \Phi\left(\dfrac{\xi(y/b)}{s}\right)$, $\xi(t) = \sqrt{t} - \dfrac{1}{\sqrt{t}}$	$(0,\infty)$	$0 < b,\, 0 < s$	$b\left(1 + s^2/2\right)$	bisa(dpqr)
Chi-square	$\dfrac{y^{\nu/2-1}\exp\{-y/2\}}{2^{\nu/2}\,\Gamma(\nu/2)}$	$(0,\infty)$	$0 < \nu$	ν	chisq([dpqr])
Erlang	$\dfrac{\exp(-y/b)\,y^{s-1}}{b^s\,(s-1)!}$, $s\in\mathbb{N}^+$ is known	$(0,\infty)$	$0 < b$	sb	erlang(shape = s)
Fisher's F	$\dfrac{\left(\frac{\nu_1}{\nu_2}\right)^{\nu_1/2}y^{(\nu_1-2)/2}}{Be\left(\frac{\nu_1}{2},\frac{\nu_2}{2}\right)}\left(1+\dfrac{\nu_1 y}{\nu_2}\right)^{-(\nu_1+\nu_2)/2}$	$(0,\infty)$	$1 < \nu_1,\, 1 < \nu_2$	$\dfrac{\nu_2}{\nu_2 - 2}$ if $\nu_2 > 2$	fff([dpqr])
Folded normal	$\dfrac{1}{\sigma}\left\{\phi\left(\dfrac{\mu+y}{\sigma}\right) + \phi\left(\dfrac{\mu-y}{\sigma}\right)\right\}$	$(0,\infty)$	$0 < \mu,\, 0 < \sigma$	See below	foldnormal(dpqr)
Fréchet	$\dfrac{sb}{(y-a)^2}\left[\dfrac{b}{y-a}\right]^{s-1}\exp\left\{-\left(\dfrac{b}{y-a}\right)^s\right\}$, a known	(a,∞)	$0 < b,\, 0 < s$	$a + b\,\Gamma(1 - s^{-1})$ if $s > 1$	frechet(dpqr)

Table 12.7 Univariate continuous distributions implemented in VGAM with support on (A, ∞); Part II.

Distribution	Density function $f(y;\theta)$	Support	Range of θ	Mean (or median $\tilde{\mu}$)	VGAM family
Inverse Gaussian	$\sqrt{\dfrac{\lambda}{2\pi y^3}}\exp\left\{\dfrac{-\lambda(y-\mu)^2}{2\mu^2 y}\right\}$	$(0,\infty)$	$0<\mu,\,0<\lambda$	μ	`inv.gaussianff(dpr)`
Gumbel-II	$\dfrac{s\,y^{s-1}\exp\{-(y/b)^s\}}{b^s}$	$(0,\infty)$	$0<b,\,0<s$	$b\,\Gamma(1-1/s)$	`gumbelII(dpqr)`
Lévy	$\sqrt{\dfrac{b}{2\pi}}\dfrac{1}{(y-a)^{3/2}}\exp\left\{\dfrac{-b}{2(y-a)}\right\}$, a known	(a,∞)	$0<b$	$\tilde{\mu}=a+\dfrac{b/2}{\left(\text{erfc}^{-1}\frac{1}{2}\right)^2}$	`levy(dpqr)`
Log normal	$\dfrac{1}{\sqrt{2\pi}\,\sigma y}\exp\left\{\dfrac{-(\log y-\mu)^2}{2\sigma^2}\right\}$	$(0,\infty)$	$0<\sigma$	$\exp\left\{\mu+\dfrac{1}{2\sigma^2}\right\}$	`lognormal(dpqr)`
Lindley	$\dfrac{\theta^2}{1+\theta}(1+y)e^{-\theta y}$	$(0,\infty)$	$0<\theta$	$\dfrac{\theta+2}{\theta(\theta+1)}$	`lindley(dpr)`
Maxwell	$\sqrt{\dfrac{2}{\pi}}\,\lambda^{3/2}y^2\exp\{-\lambda y^2/2\}$	$(0,\infty)$	$0<\lambda$	$\sqrt{\dfrac{8}{\pi\lambda}}$	`maxwell(dpqr)`
Poisson points (plane)	$\dfrac{2(\lambda\pi)^j}{(j-1)!}\,y^{2j-1}\exp\{-\lambda\pi y^2\}$	$(0,\infty)$	$0<\lambda$	$\dfrac{\Gamma(j+\frac{1}{2})}{\sqrt{\lambda}\,\pi\,\Gamma(j)}$	`poisson.points(j, dim = 2)`
Poisson points (volume)	$\dfrac{3\left(\frac{4}{3}\lambda\pi\right)^j}{(j-1)!}\,y^{3j-1}\exp\left\{-\dfrac{4}{3}\lambda\pi y^3\right\}$	$(0,\infty)$	$0<\lambda$	$\dfrac{\Gamma(j+\frac{1}{3})}{(4\lambda/3)^{1/3}\,\Gamma(j)}$	`poisson.points(j, dim = 3)`

Table 12.8 Univariate continuous distributions implemented in VGAM with support on (A, ∞); Part III.

Distribution	PDF $f(y; \boldsymbol{\theta})$ or CDF $F(y; \boldsymbol{\theta})$	Support	Range of $\boldsymbol{\theta}$	Mean (or median $\widetilde{\mu}$)	VGAM family
Nakagami	$2\left(\dfrac{s}{b}\right)^s y^{2s-1}\dfrac{\exp\left\{-sy^2/b\right\}}{\Gamma(s)}$	$(0,\infty)$	$0 < b,\, 0 < s$	$\dfrac{\sqrt{b/s}\;\Gamma(s+\frac{1}{2})}{\Gamma(s)}$	nakagami(dpqr)
Pareto I	$\dfrac{s\,b^s}{y^{s+1}}$, b known	(b,∞)	$0 < s\ (0 < b)$	$\dfrac{b\,s}{s-1}$, if $s > 1$	paretoff(dpqr)
Pareto II	$F = 1 - \left[1 + \left(\dfrac{y-a}{b}\right)\right]^{-s}$, a given	(a,∞)	$0 < b,\, 0 < s$	$\widetilde{\mu} = a + b(2^{1/s} - 1)$	paretoII(dpqr)
Pareto III	$F = 1 - \left[1 + \left(\dfrac{y-a}{b}\right)^{1/g}\right]^{-1}$, a given	(a,∞)	$0 < b,\, 0 < g$	$\widetilde{\mu} = a + b$	paretoIII(dpqr)
Pareto IV	$F = 1 - \left[1 + \left(\dfrac{y-a}{b}\right)^{1/g}\right]^{-s}$, a given	(a,∞)	$0 < b,\, 0 < g,\, 0 < s$	$\widetilde{\mu} = a + b(2^{1/s} - 1)^{g}$	paretoIV(dpqr)
Rayleigh	$\dfrac{y}{b^2}\exp\left\{-\dfrac{1}{2}\left(\dfrac{y}{b}\right)^2\right\}$	$(0,\infty)$	$0 < b$	$b\sqrt{\dfrac{\pi}{2}}$	rayleigh(dpqr)
Reciprocal inverse Gaussian	$\sqrt{\dfrac{\lambda}{2\pi y}}\exp\left\{-\dfrac{\lambda(y-\mu)^2}{2y}\right\}$	$(0,\infty)$	$0 < \lambda$	μ	rigff()

Table 12.9 Univariate continuous distributions implemented in VGAM with support on (A, ∞); Part IV.

Distribution	PDF $f(y; \boldsymbol{\theta})$	Support	Range of $\boldsymbol{\theta}$	Mean	VGAM family
Rice	$\dfrac{y}{\sigma^2} \exp\left\{ \dfrac{-(y^2 + v^2)}{2\sigma^2} \right\} I_0\left(\dfrac{yv}{\sigma^2} \right)$	$(0, \infty)$	$0 < \sigma,\, 0 < v$	$\sigma\sqrt{\dfrac{\pi}{2}} e^{z/2}\left[(1 - z) I_0\left(\dfrac{-z}{2} \right) - z I_1\left(\dfrac{-z}{2} \right) \right],\ z \equiv \dfrac{-v^2}{2\sigma^2}$	`riceff(dpqr)`
Wald	$\left(\dfrac{\lambda}{2\pi y^3} \right)^{\frac{1}{2}} \exp\left\{ -\dfrac{\lambda}{2} \dfrac{(y - 1)^2}{y} \right\}$	$(0, \infty)$	$0 < \lambda$	1	`waldff(dpr)`
Weibull	$\dfrac{s\, y^{s-1}}{b^s} \exp\left\{ -\left(\dfrac{y}{b} \right)^s \right\}$	$(0, \infty)$	$0 < b,\, 0 < s$	$b\, \Gamma(1 + s^{-1})$	`weibullR([dpqr]lss = TRUE)`

Table 12.10 Univariate continuous distributions implemented in VGAM with support on (A,B), for finite A and B. See also Table 12.11 for distributions related to the beta distribution.

Distribution	PDF $f(y;\boldsymbol{\theta})$	Support	Range of $\boldsymbol{\theta}$	Mean (or median $\widetilde{\mu}$)	VGAM family
Cardioid	$\dfrac{1}{2\pi}\{1+2\rho\cos(y-\mu)\}$	$(0,2\pi)$	$0<\mu<2\pi,\ -\tfrac{1}{2}<\rho<\tfrac{1}{2}$	$\pi+\dfrac{\rho}{\pi}\Big[(2\pi-\mu)\sin(2\pi-\mu)+\cos(2\pi-\mu)-\mu\sin(\mu)-\cos(\mu)\Big]$	cardioid(dpqr)
Hyperbolic secant	$\dfrac{\cos(\theta)}{\pi}\,u^{-\frac{1}{2}+\frac{\theta}{\pi}}\,(1-u)^{-\frac{1}{2}-\frac{\theta}{\pi}}$	$(0,1)$	$-\dfrac{\pi}{2}<\theta<\dfrac{\pi}{2}$	$\dfrac{1}{2}+\dfrac{\theta}{\pi}$	hypersecant01()
McCullagh (1989)	$\dfrac{\{1-y^2\}^{\nu-1/2}}{Be(\nu+\frac{1}{2},\frac{1}{2})\,(1-2\theta y+\theta^2)^\nu}$	$(-1,1)$	$-1<\theta<1,\ -\tfrac{1}{2}<\nu$	$\dfrac{\theta\,\nu}{1+\nu}$	mccullagh89()
Simplex	$\dfrac{\exp\left\{\dfrac{-(y-\mu)^2}{2\sigma^2\,y\,(1-y)\,\mu^2\,(1-\mu)^2}\right\}}{\sqrt{2\,\pi\,\sigma^2\,y^3\,(1-y)^3}}$	$(0,1)$	$0<\mu<1,\ 0<\sigma$	μ	simplex(dr)
Triangle	$\dfrac{I[A\le y\le\theta]}{B-A}\cdot\dfrac{2(y-A)}{\theta-A}+\dfrac{I[\theta\le y\le B]}{B-A}\cdot\dfrac{2(B-y)}{B-\theta}$	(A,B)	$A<\theta<B$	$\dfrac{A+B+\theta}{3}$	triangle(dpqr)
Truncated Pareto (I)	$\dfrac{s\,L^s}{[1-(L/U)^s]\,y^{s+1}}$, L and U given	(L,U), with $0<L$	$0<s$	$\dfrac{s\,L^s[U^{1-s}-L^{1-s}]}{[1-(L/U)^s]\,(1-s)}$, $\ s\neq 1$	[dpqr]truncpareto (lower = L, upper = U)

Table 12.11 More distributions currently supported by VGAM: those related to the beta distribution. Notes: (i) For betabinomial(), $\mathcal{A} = \mu(\rho^{-1} - 1)$ and $\mathcal{B} = (1 - \mu)(\rho^{-1} - 1)$. (ii) betaff() can handle a (A, B) support, for known A and B. (iii) See also Table 12.10 for distributions defined on (A, B), for finite A and B.

Distribution	PDF or PMF $f(y; \boldsymbol{\theta})$	Support	Range of $\boldsymbol{\theta}$	Mean	VGAM family function
Beta	$\dfrac{y^{s_1-1}(1-y)^{s_2-1}}{Be(s_1, s_2)}$	$(0, 1)$	$0 < s_1,\ 0 < s_2$	$\dfrac{s_1}{s_1 + s_2}$	betaR([dpqr])
Beta	$\dfrac{y^{\mu\phi-1}(1-y)^{(1-\mu)\phi-1}}{Be(\mu\phi, (1-\mu)\phi)}$	$(0, 1)$	$0 < \mu < 1,\ 0 < \phi$	μ	betaff([dpqr])
Beta-binomial	$\dbinom{N}{Ny} \dfrac{Be(\mathcal{A} + Ny, \mathcal{B} + N(1-y))}{Be(\mathcal{A}, \mathcal{B})}$	$0 \left(\frac{1}{N}\right) 1$	$0 < \mu < 1,\ 0 < \rho < 1$	μ	betabinomial(dpr)
Beta-binomial	$\dbinom{N}{Ny} \dfrac{Be(s_1 + Ny, s_2 + N(1-y))}{Be(s_1, s_2)}$	$0 \left(\frac{1}{N}\right) 1$	$0 < s_1,\ 0 < s_2$	$\dfrac{s_1}{s_1 + s_2}$	betabinomialff(dpr)
Beta-prime	$\dfrac{y^{s_1-1}(1+y)^{-s_1-s_2}}{Be(s_1, s_2)}$	$(0, 1)$	$0 < s_1,\ 0 < s_2,$	$\dfrac{s_1}{s_2 - 1}$ if $s_2 > 1$	betaprime()
Kumaraswamy	$s_1 s_2 y^{s_1-1}(1 - y^{s_1})^{s_2-1}$	$(0, 1)$	$0 < s_1,\ 0 < s_2$	$s_2\, Be(1 + s_1^{-1}, s_2)$	kumar(dpqr)
Generalized beta (Libby and Novick, 1982)	$\dfrac{\lambda^{s_1} y^{s_1-1}(1-y)^{s_2-1}}{Be(s_1, s_2)\{1 - (1-\lambda)y\}^{s_1+s_2}}$	$(0, 1)$	$0 < s_1,\ 0 < s_2,\ 0 < \lambda$	$\tilde{\mu}$	lino(dpqr)

Table 12.12 Univariate continuous distributions implemented in VGAM: those related to the exponential distribution. See also Table 12.13 for the gamma distribution, and Tables 12.6–12.9 for distributions defined on (A, ∞).

Distribution	Probability density function $f(y; \boldsymbol{\theta})$	Support	Range of $\boldsymbol{\theta}$	Mean (or median $\widetilde{\mu}$)	VGAM family
Exponential	$\lambda e^{-\lambda(y-a)}$, with a known	(a, ∞)	$0 < \lambda$	$a + \lambda^{-1}$	`exponential([dpqr])`
Exponentiated exponential	$s\lambda(1 - \exp(-\lambda y))^{s-1}\exp(-\lambda y)$	$(0, \infty)$	$0 < \lambda,\, 0 < s$	$\dfrac{\psi(\alpha+1) - \psi(1)}{\lambda}$	`expexpff()`
Exponential geometric	$\dfrac{1-s}{b} e^{-y/b}\left[1 - se^{-y/b}\right]^{-2}$	$(0, \infty)$	$0 < b,\, 0 < s < 1$	$\dfrac{b(s-1)}{s}\log(1-s)$	`expgeometric(dpqr)`
Exponential logarithmic	$\dfrac{(1-s)\,e^{-y/b}}{b\,(-\log p)\left[1 - (1-s)\,e^{-y/b}\right]}$	$(0, \infty)$	$0 < b,\, 0 < s < 1$	$\widetilde{\mu} = b\log(1 + \sqrt{s})$	`explogff(dpqr)`
Generalized Rayleigh	$\dfrac{2\,s\,y}{b^2} e^{-(y/b)^2}\left(1 - e^{-(y/b)^2}\right)^{s-1}$	$(0, \infty)$	$0 < b,\, 0 < s$	$\widetilde{\mu} = b\sqrt{-\log\left(1 - 2^{-1/s}\right)}$	`genrayleigh(dpqr)`
Mixture of 2 exponentials	$\phi\lambda_1 e^{-\lambda_1 y} + (1-\phi)\lambda_2 e^{-\lambda_2 y}$	$(0, \infty)$	$0 < \phi < 1,\, 0 < \lambda_j$	$\dfrac{\phi}{\lambda_1} + \dfrac{1-\phi}{\lambda_2}$	`mix2exp()`

Table 12.13 More distributions currently supported by VGAM: those related to the gamma distribution. See also Table 12.12 for the exponential distribution, and Tables 12.6–12.9 for distributions defined on (A,∞).

Distribution	Probability density function $f(y;\boldsymbol{\theta})$	Support	Range of $\boldsymbol{\theta}$	Mean	VGAM family
Gamma (1-parameter)	$\dfrac{y^{s-1}\exp(-y)}{\Gamma(s)}$	$(0,\infty)$	$0<s$	s	gamma1[dpqr]
Gamma (2-parameter)	$\dfrac{s\,(sy/\mu)^{s-1}\,\exp(-sy/\mu)}{\mu\,\Gamma(s)}$	$(0,\infty)$	$0<\mu,\,0<s$	μ	gamma2[dpqr]
Gamma (2-parameter)	$\dfrac{\exp(-\lambda y)\,y^{s-1}\,\lambda^s}{\Gamma(s)}$	$(0,\infty)$	$0<\lambda,\,0<s$	$\dfrac{s}{\lambda}$	gammaR[dpqr] lss = TRUE
Gamma (Stacy, 1962)	$\dfrac{d\,b^{-dk}\,y^{dk-1}\,\exp(-(y/b)^d)}{\Gamma(k)}$	$(0,\infty)$	$0<b,\,0<d,\,0<k$	$\dfrac{b\,\Gamma(k+1/d)}{\Gamma(k)}$	gengamma.stacy(dpqr)
Log gamma (1-parameter)	$\dfrac{\exp\{sy-e^y\}}{\Gamma(s)}$	$(-\infty,\infty)$	$0<s$	$\psi(s)$	lgamma1(dpqr)
Log gamma (3-parameter)	$\dfrac{\exp\left\{\dfrac{s(y-a)}{b}-\exp\left(\dfrac{y-a}{b}\right)\right\}}{b\,\Gamma(s)}$	$(-\infty,\infty)$	$0<b,\,0<s$	$a+b\,\psi(s)$	lgamma3(dpqr)
Log gamma (Prentice, 1974)	$\dfrac{|s|\exp(w/s^2-e^w)}{b\,\Gamma(s^{-2})}$, $\;w=\dfrac{(y-a)s}{b}+\psi(s^{-2})$	$(-\infty,\infty)$	$0<b$	a	prentice74()

Table 12.14 Statistical size distributions implemented in VGAM, based on Kleiber and Kotz (2003). Notes: (i) All are special cases of the 4-parameter generalized beta II, and their common support is $(0,\infty)$. (ii) VGAM uses $\theta^T = (b, s_1, s_2, s_3) = (b, a, p, q)$, where b is the scale parameter, and the s_j are shape parameters; they correspond to lss = TRUE. In contrast, θ_*^T is the ordering and notation used by Kleiber and Kotz (2003); they correspond to lss = FALSE. (iii) All functions except betaII() and genbetaII() have dpqr-type functions. (iv) See also Tables 12.6–12.9 for distributions defined on (A,∞).

Distribution	PDF $f(y;\theta)$	θ_*^T	Mean (or median $\widetilde{\mu}$)	VGAM family
Beta II	$\dfrac{y^{p-1}}{b^p\,Be(p,q)\,\{1+y/b\}^{p+q}}$	$(a=1,b,p,q)$	$\dfrac{b\,\Gamma(p+1)\,\Gamma(q-1)}{\Gamma(p)\,\Gamma(q)}$ if $1<q$	betaII()
Dagum	$\dfrac{ap\,y^{ap-1}}{b^{ap}\,\{1+(y/b)^a\}^{p+1}}$	$(a,b,p,q=1)$	$\dfrac{b\,\Gamma(p+a^{-1})\,\Gamma(1-a^{-1})}{\Gamma(p)}$ if $-ap<1<a$	dagum(lss = FALSE)
Fisk	$\dfrac{a\,y^{a-1}}{b^a\,\{1+(y/b)^a\}^2}$	$(a,b,p=1,q=1)$	$b\,\Gamma(1+a^{-1})\,\Gamma(1-a^{-1})$ if $1<a$	fisk(lss = FALSE)
Generalized beta II	$\dfrac{a\,y^{ap-1}}{b^{ap}\,Be(p,q)\,\{1+(y/b)^a\}^{p+q}}$	(a,b,p,q)	$\dfrac{b\,\Gamma(p+a^{-1})\,\Gamma(q-a^{-1})}{\Gamma(p)\,\Gamma(q)}$ if $-ap<1<aq$	genbetaII(lss = FALSE)
Inverse Lomax	$\dfrac{p\,y^{p-1}}{b^p\,\{1+y/b\}^{p+1}}$	$(a=1,b,p,q=1)$	$\widetilde{\mu}=b\left[2^{1/p}-1\right]^{-1}$	inv.lomax()
Inverse paralogistic	$\dfrac{a^2\,y^{a^2-1}}{b^{a^2}\,\{1+(y/b)^a\}^{a+1}}$	$(a,b,p=a,q=1)$	$\dfrac{b\,\Gamma(a+a^{-1})\,\Gamma(1-a^{-1})}{\Gamma(a)}$ if $1<a$	inv.paralogistic(lss = FALSE)
Lomax	$\dfrac{q}{b\,\{1+y/b\}^{1+q}}$	$(a=1,b,p=1,q)$	$\dfrac{b}{q-1}$ if $1<q$	lomax()
Paralogistic	$\dfrac{a^2\,y^{a-1}}{b^a\,\{1+(y/b)^a\}^{1+a}}$	$(a,b,p=1,q=a)$	$\dfrac{b\,\Gamma(1+a^{-1})\,\Gamma(a-a^{-1})}{\Gamma(a)}$ if $1<a$	paralogistic(lss = FALSE)
Singh-Maddala	$\dfrac{a\,q\,y^{a-1}}{b^a\,\{1+(y/b)^a\}^{1+q}}$	$(a,b,p=1,q)$	$\dfrac{b\,\Gamma(1+a^{-1})\,\Gamma(q-a^{-1})}{\Gamma(q)}$ if $-a<1<aq$	sinmad(lss = FALSE)

Chapter 13
Bivariate Continuous Distributions

Two are better than one; because they have a good reward for their labour... And if one prevail against him, two shall withstand him; and a threefold cord is not quickly broken.
Ecclesiastes 4:9,12

13.1 Introduction

A number of bivariate distributions for $(Y_1, Y_2)^T$ responses whose EIMs are easily computed have been implemented as VGLMs/VGAMs. This chapter summarizes the current offerings (Tables 13.1, 13.2). These include a special class of bivariate distributions called *copulas*. VGAM does not currently have any discrete bivariate distributions—these tend to involve infinite sums which are not computationally easy to handle. The typical usage is, e.g.,

```
fit <- vglm(cbind(y1, y2) ~ x2, family = binormal, data = bdata)
```

whereupon `fitted(fit)` is a 2-column matrix made up of the mean of each response, although not always.

Notationally, we let $F(y_1, y_2; \boldsymbol{\theta}) = P(Y_1 \leq y_1, Y_2 \leq y_2; \boldsymbol{\theta})$ be the CDF, which is dependent on parameters $\boldsymbol{\theta}$. Its PDF is $f(y_1, y_2; \boldsymbol{\theta})$.

13.1.1 Bivariate Distribution Theory—A Short Summary

Before encountering copulas, a few basic properties and results for bivariate distributions are briefly reviewed. Subscripts such as "Y_1, Y_2" are omitted unless necessary. Firstly, some fundamental properties: the PDF satisfies the two properties $f(y_1, y_2) \geq 0$ for all y_1 and y_2; and

$$\int_{-\infty}^{\infty} \int_{-\infty}^{\infty} f(y_1, y_2) \, dy_1 \, dy_2 = 1. \qquad (13.1)$$

© Thomas Yee 2015
T.W. Yee, *Vector Generalized Linear and Additive Models*,
Springer Series in Statistics, DOI 10.1007/978-1-4939-2818-7_13

Then
$$P(a_1 \leq Y_1 \leq b_1, a_2 \leq Y_2 \leq b_2) = \int_{a_1}^{b_1} \int_{a_2}^{b_2} f(y_1, y_2) \, dy_1 \, dy_2.$$

The PDF and CDF relate by
$$f(y_1, y_2) = \frac{\partial^2}{\partial y_1 \, \partial y_2} F(y_1, y_2), \tag{13.2}$$

and then
$$P(Y_1 \leq y_1) = F_{Y_1}(y_1) = F(y_1, \infty), \tag{13.3}$$
$$P(Y_2 \leq y_2) = F_{Y_2}(y_2) = F(\infty, y_2), \tag{13.4}$$

are the marginal CDFs. Also, $F(y_1, -\infty) = F(-\infty, y_2) = 0$. The marginal PDF for, e.g., Y_1 is
$$f_{Y_1}(y_1) = \int_{-\infty}^{\infty} f(y_1, y_2) \, dy_2,$$

and the conditional PDF of Y_2, given $Y_1 = y_1$, is
$$f_{Y_2|Y_1}(y_2|y_1) = \frac{f_{Y_1, Y_2}(y_1, y_2)}{f_{Y_1}(y_1)}$$

for $f_{Y_1}(y_1) > 0$.

Random variables Y_1 and Y_2 are independent iff $f_{Y_1, Y_2}(y_1, y_2) = f_{Y_1}(y_1) \cdot f_{Y_2}(y_2)$ for all y_1 and y_2. Under independence, $E(Y_1 \cdot Y_2) = E(Y_1) \cdot E(Y_2)$, but more generally, $E[H_1(Y_1) \cdot H_2(Y_2)] = E[H_1(Y_1)] \cdot E[H_2(Y_2)]$ for any functions H_1 and H_2. Regarding expectations, for any function H,
$$E[H(Y_1, Y_2)] = \int_{-\infty}^{\infty} \int_{-\infty}^{\infty} H(y_1, y_2) \, f(y_1, y_2) \, dy_1 dy_2. \tag{13.5}$$

Hence, for $j = 1, 2$,
$$\mu_j \equiv E(Y_j) = \int_{-\infty}^{\infty} \int_{-\infty}^{\infty} y_j \, f(y_1, y_2) \, dy_1 dy_2. \tag{13.6}$$

Related to (13.6) is the iterated expectation formula (A.35); see Sect. A.2.4 for similar formulas relating to the variance. The pq-th joint moment of (Y_1, Y_2) is $\mu_{pq} = E[Y_1^p Y_2^q]$ for $p = 0, 1, 2, \ldots$ and $q = 0, 1, 2, \ldots$ but excluding $p = q = 0$.

A common measure of the dependency between Y_1 and Y_2 is their covariance
$$\text{Cov}(Y_1, Y_2) = E[(Y_1 - \mu_1)(Y_2 - \mu_2)], \tag{13.7}$$

and the correlation between Y_1 and Y_2 is
$$\text{Corr}(Y_1, Y_2) = \frac{\text{Cov}(Y_1, Y_2)}{\sqrt{\text{Var}(Y_1) \, \text{Var}(Y_2)}}, \tag{13.8}$$

where $\text{Var}(Y_j) = \text{Cov}(Y_j, Y_j)$. Independence of Y_1 and Y_2 implies zero covariance, but not the converse. An exception is the multivariate normal distribution where $\text{Cov}(Y_s, Y_t) = 0$ iff Y_s and Y_t are independent.

13.1.2 dpr-Type Functions

Table 11.1 gives the conventions for **dpqr**-type functions, including those for bivariate distributions. Most functions for bivariate distributions are prefixed by **bi**. Those concerning copulas have suffix **cop** (Table 13.2), and it is common for these functions to have an argument called **apar**, generically standing for the association parameter.

Currently, **q**-type functions are not available for any bivariate distributions in VGAM.

13.2 Two Bivariate Distributions

Of the many bivariate distributions proposed, only two are briefly described here.

13.2.1 Bivariate Normal Distribution

Its density is $\phi_2(y_1, y_2; \mu_1, \mu_2, \sigma_1, \sigma_2, \rho) =$

$$\frac{1}{2\pi\sigma_1\sigma_2\sqrt{1-\rho^2}} \exp\left\{\frac{-1}{2(1-\rho^2)}\left[\left(\frac{y_1-\mu_1}{\sigma_1}\right)^2 - 2\rho\left(\frac{y_1-\mu_1}{\sigma_1}\right)\left(\frac{y_2-\mu_2}{\sigma_2}\right) + \left(\frac{y_2-\mu_2}{\sigma_2}\right)^2\right]\right\}, \quad (13.9)$$

for all real y_j and μ_j, and for $\sigma_j > 0$ and $-1 < \rho < 1$. It is more compactly written using vector notation as

$$\frac{1}{\sqrt{(2\pi)^d |\boldsymbol{\Sigma}|}} \exp\left\{-\frac{1}{2}(\boldsymbol{y}-\boldsymbol{\mu})^T \boldsymbol{\Sigma}^{-1}(\boldsymbol{y}-\boldsymbol{\mu})\right\}, \quad \boldsymbol{\Sigma} = \begin{pmatrix} \sigma_1^2 & \rho\,\sigma_1\sigma_2 \\ \rho\,\sigma_1\sigma_2 & \sigma_2^2 \end{pmatrix}, \quad (13.10)$$

for all $\boldsymbol{y}$, where $d = 2$. Here, $|\boldsymbol{\Sigma}| = \sigma_1^2\sigma_2^2(1-\rho^2)$ is the determinant of $\boldsymbol{\Sigma}$. It is common shorthand to write $\boldsymbol{Y} \sim N_2(\boldsymbol{\mu}, \boldsymbol{\Sigma})$. The *standard* N_2 distribution has means $\boldsymbol{\mu} = (0,0)^T$, and variances $\sigma_1^2 = \sigma_2^2 = 1$. The function `dbinorm()` returns $\phi_2(\cdot)$, effectively with $\rho = 0$ as the default. The N_2 CDF is commonly abbreviated $\Phi_2(\cdot)$, and may be computed using `pbinorm()`.

The properties of the bivariate normal distribution are well-known. The density is bell-shaped with elliptical contours. The marginals are $N(\mu_j, \sigma_j^2)$. The conditional density $Y_1|Y_2 = y_2$ is univariate normal. Figure 13.1 gives scatter plots of some randomly generated standard N_2 random vectors with various values of ρ. To generate these, let Z_1 and Z_1 be independent $N(0, 1)$ distributed, and let $Y_2 = \rho Z_1 + \sqrt{1-\rho^2} Z_2$. Then it easily shown (Ex. 13.1) that (Z_1, Y_2) has a standard N_2 distribution. From this, it is easily generalized to any N_2 distribution, as implemented in `rbinorm()`.

The family function `binormal()` estimates the parameters by Fisher scoring.

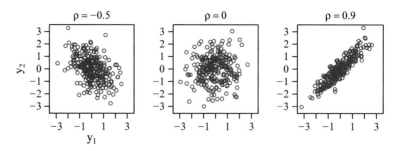

Fig. 13.1 250 random vectors generated from a standard bivariate normal distribution, with various values of ρ.

```
> args(binormal)

function (lmean1 = "identitylink", lmean2 = "identitylink", lsd1 = "loge",
    lsd2 = "loge", lrho = "rhobit", imean1 = NULL, imean2 = NULL,
    isd1 = NULL, isd2 = NULL, irho = NULL, imethod = 1, eq.mean = FALSE,
    eq.sd = FALSE, zero = 3:5)
NULL
```

It has $\boldsymbol{\eta} = (\mu_1, \mu_2, \log \sigma_1, \log \sigma_2, \text{rhobit} \, \rho)^T$ as the default. The `"rhobit"` link function (Table 1.2; (13.11) below) is a scaled version of Fisher's Z transformation. Given a formula containing covariates, the σ_j and ρ are intercept-only, by default. The arguments `eq.mean` and `eq.sd` constrain $\mu_1 = \mu_2$ and $\sigma_1 = \sigma_2$, provided that each parameter's link function is identical (strictly, $\eta_1 = \eta_2$ and $\eta_3 = \eta_4$ are enforced, respectively). Hence,

```
> fit.bvn <- vglm(cbind(y1, y2) ~ x2 + x3, family = binormal(eq.sd = TRUE), bdata)
```

corresponds to

$$\mu_j = \beta^*_{(j)1} + \beta^*_{(j)2} x_2 + \beta^*_{(j)3} x_3, \quad j = 1, 2,$$
$$\log \sigma_1 = \log \sigma_2 = \beta^*_{(3)1},$$
$$\log \frac{1+\rho}{1-\rho} = \beta^*_{(4)1}, \tag{13.11}$$

because

$$\mathbf{H}_1 = \begin{pmatrix} 1 & 0 & 0 & 0 \\ 0 & 1 & 0 & 0 \\ 0 & 0 & 1 & 0 \\ 0 & 0 & 1 & 0 \\ 0 & 0 & 0 & 1 \end{pmatrix}, \quad \mathbf{H}_2 = \mathbf{H}_3 = \begin{pmatrix} 1 & 0 \\ 0 & 1 \\ 0 & 0 \\ 0 & 0 \\ 0 & 0 \end{pmatrix}.$$

As a simple example, we consider something similar to the data of Fig. 4.1, but regressing $Y_1 = $ diastolic and $Y_2 = $ systolic blood pressures (DBP/SBP) versus $X = $ age with data for *all* (5649) male Europeans in that cross-sectional study. Firstly, we create a data frame of the subset:

```
> M.euro <- subset(xs.nz, sex == "M" & age < 70 & ethnicity == "European")
> M.euro <- na.omit(M.euro[, c("age", "dbp", "sbp")])
```

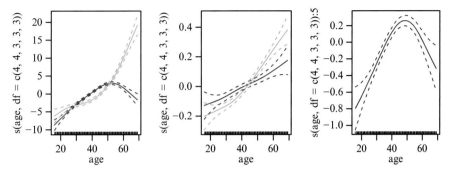

Fig. 13.2 Fitted (centred) component functions of a VGAM N_2 fitted to diastolic and systolic blood pressures data, versus age. The data set are 5649 male Europeans from `xs.nz`. From left to right, those for $\widehat{\mu}_1$ and $\widehat{\mu}_2$, those for $\log \widehat{\sigma}_1$ and $\log \widehat{\sigma}_2$, those for $\log((1 + \widehat{\rho})/ (1 - \widehat{\rho}))$.

Now we can fit an N_2 model to the means, allowing for smoothness for all functions including the correlation parameter—something that is feasible because the sample size is so large. It can be done as follows.

```
> fit.N2 <- vgam(cbind(dbp, sbp) ~ s(age, df = c(4, 4, 3, 3, 3)),
                 binormal(zero = NULL), data = M.euro)
> mycl <- c("blue", "limegreen")
> plot(fit.N2, which.cf = 1:2, overlay = TRUE, se = TRUE, lcol = mycl, scol = mycl)
> plot(fit.N2, which.cf = 3:4, overlay = TRUE, se = TRUE, lcol = mycl, scol = mycl)
> plot(fit.N2, which.cf = 5, se = TRUE, lcol = mycl, scol = mycl)
```

This gives Fig. 13.2. The functions for the means have been overlaid to help comparison, as also are those for the scale parameters. It is clear that the mean DBP and SBP functions are not parallel—this can be considered as an interaction between them with respect to age. For elderly people, the first plot suggests that DBP declines while SBP increases, with age. The second plot shows both component functions are monotonically increasing, which suggests that BP variability increases with age. The third plot suggests that the correlation between DBP and SBP increases from a young age, peaking for c.50-year-olds, and then declines.

These plots suggest that the standard deviation parameters, on a log scale, might be modelled linearly with respect to age; and that ρ should not be modelled as intercept-only. For models where ρ *is* intercept-only (the default), it is possible to test for the conditional independence between Y_1 and Y_2, given $\boldsymbol{x}$, by the p-value obtained from the Wald test $H_0 : \rho = 0$ versus $H_1 : \rho \neq 0$. This p-value is outputted by `summary()`.

13.2.2 Plackett's Bivariate Distribution

We saw in Sect. 1.2.3 that the odds ratio is a measure of association in a 2×2 contingency table. One scenario for this table is that it arises from some continuous bivariate distribution F having cutpoints on both axes, intersecting at point (z_1, z_2), say. Then $\mathbb{R}^2$ is partitioned into four quadrants, so that the two binary responses (Y_1, Y_2) could be thought of being generated with probabilities corresponding to those quadrants. For example, $(Y_1 = 0, Y_2 = 0)$ occurs with probability $F(z_1, z_2)$ (corresponding to the bottom left quadrant), while

$(Y_1 = 1, Y_2 = 1)$ occurs with probability $1 - F_1(z_1) - F_2(z_2) + F(z_1, z_2)$, the F_j being the marginal CDFs (upper right quadrant).

Given two marginal distributions, Plackett (1965) showed how to construct a 1-parameter class of bivariate distributions. The odds ratio ψ (also called the *cross product ratio*) was used to measure dependence:

$$\psi = \frac{F(z_1, z_2) \cdot (1 - F_1(z_1) - F_2(z_2) + F(z_1, z_2))}{(F_1(z_1) - F(z_1, z_2)) \cdot (F_2(z_2) - F(z_1, z_2))}. \tag{13.12}$$

By solving the quadratic $(\psi - 1)F^2 - TF + \psi F_1 F_2 = 0$ where $T = 1 + (\psi - 1)(F_1 + F_2)$, he obtained

$$F(y_1, y_2) = \frac{T - \sqrt{T^2 - 4\psi(\psi - 1)F_1 F_2}}{2(\psi - 1)}, \quad \psi > 0, \quad \psi \neq 1, \tag{13.13}$$

and $F(y_1, y_2) = y_1 y_2$ for $\psi = 1$. Upon differentiation of (13.13) with respect to both y_j (i.e., using (13.2)), the Plackett family PDF is

$$f(y_1, y_2) = \frac{\psi f_1 f_2 \left[1 + (\psi - 1)(F_1 + F_2 - 2F_1 F_2)\right]}{\{[1 + (\psi - 1)(F_1 + F_2)]^2 - 4\psi(\psi - 1)F_1 F_2\}^{3/2}}, \tag{13.14}$$

where the f_j are the marginal PDFs.

The Plackett family of copulas arises when the marginal distributions are the standard uniform, so its support is the unit square. The **VGAM** family function for this is `biplackettcop()` (Table 13.2).

13.3 Copulas

A special class of bivariate distributions are *copulas*. They have received much attention over the last two decades or so, and are now used widely in fields such as actuarial science, financial and economic modelling, survival analysis, extremes, environmental modelling, time series and biostatistics.

Informally, copulas are multivariate distributions whose univariate margins are Unif$[0, 1]$ distributed. It is a function that connects or "couples" multivariate distributions with their marginal distributions, consequently, they are important because they allow the dependency between random variables to be studied, separate from the effects of the marginal distributions. In areas such as economic modelling, there is usually much more information about the marginal distributions than about their joint behaviour, hence copulas are a way of modelling joint distributions from only marginal distributions.

A related reason for their importance is that multivariate distributions can be easily 'built' by way of copulas. One can choose the marginal distributions and the copula function, and the result is a multivariate distribution essentially of our choosing. A large number of copulas appear in the literature, and they differ by imposing different dependency structures. However, copulas are not a philosopher's stone; criticisms of them are highlighted in, e.g., Mikosch (2006) who wrote "It is not enough to introduce new copula families, give them a name and fit them to any kind of data."

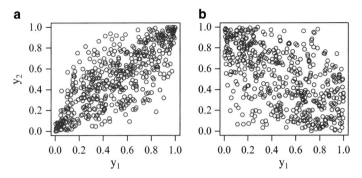

Fig. 13.3 500 random vectors generated from two copulas. (**a**) Bivariate Gaussian copula with $\alpha = \rho = 0.8$. (**b**) Bivariate Frank copula with $\alpha = 50$.

In this section we will restrict our treatment of copulas to the bivariate case. We also deviate in this section by using (U_1, U_2) in addition to (Y_1, Y_2), to reinforce the fact that the response vector lies in the open unit square $(0, 1)^2$.

Sklar (1959)'s theorem is the central result and starting point of copula theory: if F is the CDF of a random vector (Y_1, Y_2) with marginals F_1 and F_2, then there exists a corresponding copula C such that

$$F(y_1, y_2) = C\left(F_1(y_1), F_2(y_2); \boldsymbol{\alpha}\right) = C(u_1, u_2; \boldsymbol{\alpha}), \qquad (13.15)$$

say, for all $(y_1, y_2) \in \mathbb{R}^2$. We will assume that the F_j are continuous so that C is unique. One can view C as containing information on the dependence structure of F through the dependency parameters $\boldsymbol{\alpha}$.

Usually $\boldsymbol{\alpha}$ is only 1- or 2-dimensional, e.g., $\alpha = \rho$ for the bivariate Gaussian copula. For this particular copula,

$$C(u_1, u_2; \rho) = \Phi_2\left(\Phi^{-1}(u_1), \Phi^{-1}(u_2); \rho\right) \qquad (13.16)$$

for $-1 < \rho < 1$, and $0 \le u_j \le 1$ (Table 13.2), and a randomly generated sample is displayed in Fig. 13.3a. If $\rho = -0.8$ is plotted, then the points appear rotated by $90°$, i.e., they cluster about the line $y_2 = 1 - y_1$. The family function `binormalcop()` estimates ρ by Fisher scoring, and it has the associated **dpr**-type suite of functions.

Often the $\boldsymbol{\alpha}$ are of central interest, therefore the VGLM/VGAM framework should be suitable for this because each dependency parameter may be modelled like a GLM/GAM. Prior to this work, most analyses fitted the $\boldsymbol{\alpha}$ as intercept-only. However, it is crucial that the dependency parameters can be naturally modelled as functions of $\boldsymbol{x}$, either in a model-driven or data-driven manner, e.g., $\boldsymbol{x}$ might be the risk factors within a total-risk portfolio in financial mathematics.

In general, copulas have many interesting properties, and here are a few of them for the bivariate case:

1. $C(u_1, u_2) = 0$ if $u_1 = 0$ and/or $u_2 = 0$.
2. $C(u_1, 1) = u_1$ and $C(1, u_2) = u_2$.
3. For $j = 1, 2$, $C(u_1, u_2)$ is nondecreasing as u_j increases, keeping the other argument fixed.

4. To generate random variates from F, the following method may be used. It is based on the property just mentioned but, for simplicity, we will assume the more stronger assumption of $C(u_1, u_2)$ *increasing* as u_j increases, keeping the other argument fixed. The idea is to obtain a random variate from one of the marginal distributions, then the other random variate conditional on the first value, i.e., the inverse transform method described in Sect. 11.1.1.

   (i) Generate U_1 and U_2 independently from a standard uniform distribution.
   (ii) Let $Z_2 = C_{U_2|U_1}^{-1}(u_2)$ where $C_{U_2|U_1}(u_2) = \partial C(u_1, u_2)/\partial u_1 = P(U_2 \leq u_2 | U_1 = u_1)$ is the conditional CDF for U_2 given $U_1 = u_1$.
   (iii) Then $(U_1, Z_2) \sim F$.

   If the more general assumption of "nondecreasing" is used, then *quasi-inverses* are used—see Nelsen (2006, Sect.2.9) for details.
5. The unique copula corresponding to the continuous random vector $(Y_1, Y_2) \in \mathbb{R}^2$ remains unchanged when continuous monotonic transformations are applied to Y_1 and/or Y_2.

13.3.1 Dependency Measures

The *product copula* $C(u_1, u_2) = u_1 u_2$ corresponds to two independent random variables, and it serves as the benchmark of all other copulas. Hence, for all copulas bar this one, it is needful to be able to measure the dependence somehow. Many such measures have been proposed, and the most well-known is Pearson's linear correlation coefficient

$$\rho(X, Y) = \frac{\mathrm{Cov}(X, Y)}{\sqrt{\mathrm{Var}(X)\,\mathrm{Var}(Y)}} \tag{13.17}$$

for random variables[1] X and Y, equivalent to (13.8). Some advantages and properties of it are:

- $\rho \in [-1, 1]$;
- when $\rho = \pm 1$ then there exists an $a \in \mathbb{R}$ and $b > 0$ such that $Y = a \pm bX$;
- $\rho(X, Y) = \rho(f(X), g(Y))$ where f and g are both strictly increasing linear functions;
- when X and Y are independent then $\rho = 0$, but the converse is not necessarily true, e.g., one exception is $(X, Y) \sim N_2$.

However, its disadvantages include being only defined for finite variances, not being invariant to nonlinear strictly-increasing functions, and not necessarily being an optimal measure. It can be suitable for elliptical distributions (e.g., normal and t distributions), but not always.

   Instead, it is common to use dependency measures based on ranks. One such commonly used measure is Kendall's tau (ρ_τ). Another popular one is Spearman's rho (ρ_S), but this is not described here. Loosely, two random variables are *concordant* if 'large' values of one random variable are associated with 'large'

---

[1] In this subsection it is more convenient to use (X, Y) rather than (Y_1, Y_2) for the bivariate response.

values of the other random variable. Similarly, two random variables are *discordant* if large values of one random variable are associated with 'small' values of the other random variable. Kendall's tau is simply the difference in the probability of concordance and the probability of discordance:

$$\rho_\tau = P[(X_1 - X_2)(Y_1 - Y_2) > 0] - P[(X_1 - X_2)(Y_1 - Y_2) < 0], \qquad (13.18)$$

where (X_1, Y_1) and (X_2, Y_2) are two independent pairs of random variables from F.

With data (x_i, y_i), $i = 1, \ldots, n$, if $(x_i - x_j) \cdot (y_i - y_j) > 0$ then that comparison is concordant $(i \neq j)$. Similarly, $(x_i - x_j) \cdot (y_i - y_j) < 0$ for discordance. Out of $\binom{n}{2}$ comparisons, let c and d be the number of concordant and discordant pairs. Then Kendall's tau can be estimated by $(c - d)/(c + d)$. If there are t ties, then half the ties can be deemed concordant and the other half discordant, so that $\hat{\tau} = (c - d)/(c + d + t)$, say.

It can be shown that

$$\rho_\tau = 4 \int_0^1 \int_0^1 C(u_1, u_2) \, c(u_1, u_2) \, du_1 \, du_2 - 1, \qquad (13.19)$$

where c is the PDF, assuming it exists. Using this, it is sometimes possible to obtain an expression for ρ_τ as a function of α, e.g., Table 13.2, and the function inverted to provide initial values (Sect. 13.3.3).

The R function `cor()` computes Pearson, Kendall and Spearman correlation coefficients.

13.3.2 Archimedean Copulas

A very important class of copulas known as *Archimedean copulas* takes the form

$$C(u_1, \ldots, u_d) = \varphi^{-1}(\varphi(u_1) + \cdots + \varphi(u_d)), \qquad (13.20)$$

for all $(u_1, \ldots, u_d) \in [0, 1]^d$ (the bivariate case is $d = 2$). Archimedean copulas are popular largely because they can be easy to fit and there are lots of them. Here, φ is known as the *generator function*. Different generator functions yield several important families of copulas. However, (13.20) is a copula only if φ is a convex decreasing: $\varphi : (0, 1] \to [0, \infty)$ such that $\varphi(1) = 0$.

Some bivariate Archimedean copula families are implemented in VGAM (Table 13.2); these have prefix `bi` to reinforce that they are the special case of $d = 2$.

Another class of copulas are elliptical copulas. These correspond to certain well-known distributions whose densities have contours which are elliptical. Examples of this class are the Gaussian (normal) distribution and the Student t-distribution. Elliptical copulas are quite popular for fitting financial data, due to certain properties they possess.

13.3.3 Initial Values

In the past, estimation of $\boldsymbol{\alpha}$ was based on the method of moment estimators when it was possible. This was due to its simplicity and the lack of iteration needed. VGAM uses method of moment estimators for initial values for distributions whose $\boldsymbol{\alpha}$ involves a simple formula in ρ_τ or ρ_S, e.g., as in Table 13.2. For example, the bivariate Gaussian copula can have its correlation parameter ρ estimated by $\sin(\pi\widehat{\rho}_\tau/2)$ or $2\sin(\pi\widehat{\rho}_S/6)$. However, computing $\widehat{\rho}_\tau$ is an $O(n^2)$ operation, therefore prohibitive for large n. Under these situations, a small random sample of pairs can be taken.

Bibliographic Notes

Two texts on bivariate distributions are Marshall and Olkin (2007) and Balakrishnan and Lai (2009). Although not directly in the domain of this chapter, Kocherlakota and Kocherlakota (1992) covers discrete bivariate distributions, and Johnson et al. (1997) concerns discrete multivariate distributions.

Copulas are described in detail in Nelsen (2006) and Mai and Scherer (2012); see also Joe (2014), Trivedi and Zimmer (2005), Balakrishnan and Lai (2009). In R there are the copula, fCopulae and VineCopula packages, among others, for copula modelling.

Exercises

Ex. 13.1. N_2, **from Sect. 13.2.1**

(a) Fill in the details for generating random vectors first from a standard N_2 distribution, and then to a general N_2 distribution, all based on calls to `rnorm()`.
(b) If $\boldsymbol{Y} \sim N_2$ show that the conditional density of Y_1, given $Y_2 = y_2$, is $Y_1|Y_2 = y_2 \sim N(\mu_1 + \rho(\sigma_1/\sigma_2)(y_2 - \mu_2), \sigma_1^2(1 - \rho^2))$.
(c) If $Y_j \sim N_1(\mu_j, \sigma^2)$ independently, for $j = 1, 2$, then show that $Y_1 + Y_2$ and $Y_1 - Y_2$ are independent.
(d) Write down the $\mathbf{H}_k$ for `fit.bvn` if the following are used instead.

  (i) `family = binormal(eq.sd = FALSE)`
 (ii) `family = binormal(eq.mean = TRUE)`
(iii) `family = binormal(eq.mean = TRUE, eq.sd = TRUE)`
 (iv) `family = binormal(eq.sd = TRUE, zero = NULL)`

(e) One might define exchangeability for N_2 by $\mu_1 = \mu_2$ subject to $\sigma_1 = \sigma_2$. Under these assumptions, what $\mathbf{H}_k$ would arise for `cbind(y1, y2) ~ x2 + x3`? For simplicity, suppose that only the means are not intercept-only.

Ex. 13.2. **FGM Copulas**

(a) Show that when the standard exponential distribution is used as the marginals of the Farlie-Gumbel-Morgenstern (FGM) copula, then the result is the CDF of `bifgmexp()` (Table 13.1).
(b) Derive the CDF and PDF of the FGM copula with standard logistic marginal distributions.

Ex. 13.3. **Bivariate Normal Copulas—Simulated Data**

(a) Generate 500 random vectors from: $X_{i2} \sim \text{Unif}(0, 1)$ independently, $(Y_{i1}, Y_{i2}) \sim$ Bivariate normal copula, with $\eta = \text{rhobit } \rho_i = -1 + 3\,x_{i2}$.
(b) Apply smoothing to estimate η. Comment.
(c) Estimate the regression coefficients of this model, assuming linearity on the η-scale. Is the regression coefficient for x_2 significant?

Ex. 13.4. Starting at the CDF of the bivariate Gaussian copula in Table 13.2, derive the PDF $c(y_1, y_2; \rho)$. For one observation, compute the first derivative of ℓ and show it has mean zero, then derive the EIM and show that it equals $(1 + \rho^2)/(1 - \rho^2)^2$.

Couples are wholes and not wholes, what agrees disagrees, the concordant is discordant. From all things one and from one all things.
—Heraclitus

Table 13.1 Some bivariate distributions currently implemented in VGAM. Here, $\boldsymbol{y} = (y_1, y_2)^T$, and they have been separated according to their support.

Distribution	CDF $F(y_1, y_2; \boldsymbol{\theta})$ (or PDF $f(y_1, y_2; \boldsymbol{\theta})$))	Support	Range of $\boldsymbol{\theta}$	VGAM family
Freund (1961)'s exponential	$f(\boldsymbol{y}) = \alpha\beta' \exp\left\{-\beta'y_2 - (\alpha+\beta-\beta')y_1\right\}$	$0 < y_1 < y_2 < \infty$	$(\alpha, \alpha', \beta, \beta') \in (0,\infty)^4$	freund61()
	$f(\boldsymbol{y}) = \beta\alpha' \exp\left\{-\alpha'y_1 - (\alpha+\beta-\alpha')y_2\right\}$	$0 < y_2 < y_1 < \infty$	$(\alpha, \alpha', \beta, \beta') \in (0,\infty)^4$	freund61()
McKay's bivariate gamma	$f(\boldsymbol{y}) = \dfrac{y_1^{s_1-1}\,(y_2 - y_1)^{s_2-1}}{b^{s_1+s_2}\,\Gamma(s_1)\,\Gamma(s_2)}\exp(-y_2/b)$	$0 < y_1 < y_2 < \infty$	$(b, s_1, s_2) \in (0,\infty)^3$	bigamma.mckay()
Gamma hyperbola	$f(\boldsymbol{y}) = \exp\left\{-\dfrac{y_1}{\theta}\, e^{-\theta} - \theta y_2\right\}$	$(0,\infty) \times (1,\infty)$	$0 < \theta$	gammahyperbola()
Gumbel's Type I exponential	$\exp\left\{-y_1 - y_2 + \alpha y_1 y_2\right\} + 1 - e^{-y_1} - e^{-y_2}$	$(0,\infty)^2$	$\alpha \in \mathbb{R}$	bigumbelIexp()
FGM exponential	$e^{-y_1-y_2}\left[1 + \alpha\left(1 - e^{-y_1}\right)\left(1 - e^{-y_2}\right)\right] +$ $1 - e^{-y_1} - e^{-y_2}$	$(0,\infty)^2$	$-1 < \alpha < 1$	bifgmexp()
Bivariate logistic	$\left[1 + \exp\left\{-\left(\dfrac{y_1 - a_1}{b_1}\right)\right\} + \exp\left\{-\left(\dfrac{y_2 - a_2}{b_2}\right)\right\}\right]^{-1}$	$\mathbb{R}^2$	$0 < b_1, 0 < b_2$	bilogistic(dpr)
Bivariate normal, N_2	$\Phi_2(y_1, y_2; \mu_1, \mu_2, \sigma_1, \sigma_2, \rho)$; see Sect. 13.2.1	$\mathbb{R}^2$	$-1 < \rho < 1, 0 < \sigma_j$	binormal(dpr)
Bivariate Student-t	$f(\boldsymbol{y}) = \dfrac{1}{2\pi\sqrt{1-\rho^2}}\dfrac{(1 + y_1^2 + y_2^2 - 2\rho y_1 y_2)}{(\nu(1-\rho^2))^{(\nu+2)/2}}$	$\mathbb{R}^2$	$-1 < \rho < 1$	bistudentt(d)

Table 13.2 Bivariate copulas currently implemented by VGAM. Notes: (i) The support is $(u_1, u_2) \in [0,1]^2$, and α is the association parameter (**apar**). (ii) Non-Archimedean copulas have no generator functions φ (—). (iii) See (A.61) for the Debye function $D_n(x)$ definition. (iv) Much of this table was adapted from Trivedi and Zimmer (2005) and Nelsen (2006). (v) The EIMs of some of these families appear in Schepsmeier and Stöber (2014).

Copula	$C(u_1, u_2; \alpha)$	α-domain	ρ_S	ρ_τ	Generator $\varphi(t)$	VGAM family
AMH	$\dfrac{u_1\, u_2}{1 - \alpha(1-u_1)(1-u_2)}$	$(-1,1)$	Complicated	Complicated	$\log\dfrac{1-\alpha(1-t)}{t}$	biamhcop(dpr)
Clayton	$\left(u_1^{-\alpha} + u_2^{-\alpha} - 1\right)^{-1/\alpha}$	$(0,\infty)$	No closed form	$\dfrac{\alpha}{\alpha+2}$	$(1+t)^{-1/\alpha}$	biclaytoncop(dr)
Frank	$-\dfrac{1}{\alpha}\log\left(1 - \dfrac{(1-e^{-\alpha u_1})(1-e^{-\alpha u_2})}{1-e^{-\alpha}}\right)$	$\mathbb{R}\setminus\{0\}$	$1 - \dfrac{D_1(\alpha) - D_2(\alpha)}{\alpha/12}$	$1 - \dfrac{1-D_1(\alpha)}{\alpha/4}$	$-\log\left(\dfrac{e^{-\alpha t}-1}{e^{-\alpha}-1}\right)$	bifrankcop(dpr)
Gaussian	$\Phi_2\big(\Phi^{-1}(u_1), \Phi^{-1}(u_2); \rho=\alpha\big)$	$(-1,1)$	$\dfrac{6}{\pi}\sin^{-1}\left(\dfrac{\alpha}{2}\right)$	$\dfrac{2}{\pi}\sin^{-1}(\alpha)$	—	binormalcop(dpr)
FGM	$u_1 u_2\left[1 + \alpha(1-u_1)(1-u_2)\right]$	$(-1,1)$	$\dfrac{\alpha}{3}$	$\dfrac{2\alpha}{9}$	—	bifgmcop(dpr)
Plackett	Eq. (13.13) with $F_j = u_j$	$(0,\infty)$	$\dfrac{\alpha+1}{\alpha-1} - \dfrac{2\alpha\log\alpha}{(\alpha-1)^2}$	No closed form	—	biplackettcop(dpr)

Chapter 14
Categorical Data Analysis

Unless what we do be useful, vain is our glory.
—Phaedrus

14.1 Introduction

Iteratively reweighted least squares is a suitable algorithm for fitting regression models to a categorical response variable Y, and the VGLMs/VGAMs framework is convenient for handling nuances in this topic. We let Y take values labelled $1, 2, \ldots, M + 1$, i.e., there are $M + 1$ *levels*, *categories* or *classes*, and any observed Y_i takes on one of these values. The response is said to be *polytomous*. This chapter describes **VGAM** family functions for such a Y as this, and Table 14.1 summarizes those functions currently available. As in Agresti (2013, p.xiv), *categorical data analysis* is interpreted as methods for categorical response variables. It is convenient to consider separately the two main cases: when Y is *nominal* (no order) and *ordinal* (ordered). An example of the former is the choice of transport (air/bus/train/car) between two cities by 210 commuters, as described in Sect. 14.2.1. An example of the latter is Table 14.2, where the stages of a lung disease amongst coalminers are $Y = 1$ for none, $Y = 2$ for mild, and $Y = 3$ for severe symptoms; we use this data frame a number of times in this chapter for illustrative purposes.

14.1.1 The Multinomial Distribution

The foundation of categorical data analysis in this chapter is the multinomial distribution, a generalization of the binomial distribution to more than two classes. For a single individual, we drop the subscript i and denote $\boldsymbol{y}^* = (y_1^*, \ldots, y_{M+1}^*)^T$ as a vector of counts. We write $\boldsymbol{Y}^* \sim \text{Multinomial}(N, p_1, \ldots, p_M)$, having joint distribution

© Thomas Yee 2015
T.W. Yee, *Vector Generalized Linear and Additive Models*,
Springer Series in Statistics, DOI 10.1007/978-1-4939-2818-7_14

$$P(Y_1^* = y_1^*, \ldots, Y_M^* = y_M^*) = \binom{N}{y_1^*\, y_2^*\, \cdots\, y_M^*} p_1^{y_1^*} \cdots p_{M+1}^{y_{M+1}^*} \quad (14.1)$$

$$= \frac{N!}{y_1^*!\, y_2^*!\, \cdots\, y_{M+1}^*!} \prod_{s=1}^{M+1} p_s^{y_s^*},$$

where $p_{M+1} = 1 - p_1 - \cdots - p_M$, $0 < p_j < 1$, and $N = y_1^* + \cdots + y_{M+1}^*$ is the number of trials. The reason for having $M+1$ levels is that there are M independent probabilities to estimate. For convenience, let $J = M + 1$.

Some basic properties of the multinomial distribution are as follows. We let s and t both $\in \{1, \ldots, M\}$.

1. $E(Y_s^*) = N p_s$, and $\text{Cov}(Y_s^*, Y_t^*) = N\, p_s(1 - p_s)$ if $s = t$, and $-N\, p_s\, p_t$ if $s \neq t$. This may be written compactly as $\text{Var}(\boldsymbol{Y}^*) = N[\,\text{diag}(\boldsymbol{p}) - \boldsymbol{p}\,\boldsymbol{p}^T\,]$, where $\boldsymbol{p} = (p_1, \ldots, p_M)^T$. Its inverse is

$$N^{-1}\left[\,\text{diag}(1/\boldsymbol{p}) + p_{M+1}^{-1}\,\boldsymbol{1}_M\,\boldsymbol{1}_M^T\,\right], \quad (14.2)$$

   provided $0 < p_s < 1$ for all s, although this is not explicitly exploited by VGAM. The diagonal matrix in (14.2) has elements $1/p_s$.
2. $Y_s^* \sim \text{Binomial}(N, p_s)$ are the marginal distributions, for $s = 1, \ldots, J$. The quantity N is known as the *binomial index*.
3. If the $Y_j^* \sim \text{Poisson}(\lambda_j)$ independently, then

$$(Y_1^*, Y_2^*, \ldots, Y_J^* | Y_+^* = N) \sim \text{Multinomial}(N, \lambda_1/\lambda_+, \ldots, \lambda_M/\lambda_+) \quad (14.3)$$

   where $Y_+^* = Y_1^* + \cdots + Y_J^*$ and $\lambda_+ = \lambda_1 + \cdots + \lambda_J$. This result has been exploited by the "Poisson trick" (Sect. 14.3.1).
4. As with binomial GLMs, it is more convenient to treat the response as sample proportions rather than counts, and assimilate the N as prior weights. That is, $\boldsymbol{Y} = N^{-1}\boldsymbol{Y}^*$ so that $\boldsymbol{Y}^T\boldsymbol{1}_J = N$. Now for data $(\boldsymbol{x}_i, \boldsymbol{y}_i)$, $i = 1, \ldots, n$, we have

$$\ell = \sum_{i=1}^{n} w_i\, \ell_i = \sum_{i=1}^{n} N_i \sum_{j=1}^{M+1} y_{ij} \log p_{ij} \quad (14.4)$$

where $p_{ij} = p_j(\boldsymbol{x}_i) = P(Y_i = j | \boldsymbol{x}_i)$. Then, for $j = 1, \ldots, M$,

$$\frac{\partial \ell_i}{\partial p_{ij}} = \frac{y_{ij}}{p_{ij}} - \frac{y_{iJ}}{p_{iJ}} \quad \text{and} \quad -E\left(\frac{\partial^2 \ell_i}{\partial p_{ij}}\right) = \frac{1}{p_{ij}} + \frac{1}{p_{iJ}}. \quad (14.5)$$

14.2 Nominal Responses—The Multinomial Logit Model

The *multinomial logit model* (MLM) is the most common model in this case. We describe this below, as well as its reduced-rank variant called the *stereotype model*. The model, also known as the *multiple logistic regression model* or *polytomous logistic regression model*, is given by (1.28) It is a neural network—see, e.g., Ripley (1996)—and sometimes it is fitted as such, e.g., nnet() in MASS.

It is a good idea to centre the $\boldsymbol{x}_{[-1]}$, if possible, so that $\boldsymbol{x} = (1, \boldsymbol{0}^T)^T$ implies

$$e^{\beta_{(j)1}} \; = \; \frac{p_j(1, \boldsymbol{0})}{p_J(1, \boldsymbol{0})},$$

i.e., $\beta_{(j)1}$ is more interpretable because it measures the chance of landing in category j, relative to category $M + 1$, for an individual with $\boldsymbol{x}_{[-1]} = \boldsymbol{0}$.

Continuing on from (14.4) and (14.5), for $j \neq s$,

$$\frac{\partial p_{ij}}{\partial \eta_s} \; = \; \frac{\partial}{\partial \eta_s} \frac{e^{\eta_{ij}}}{\sum\limits_{t=1}^{J} e^{\eta_{it}}} \; = \; \frac{-e^{\eta_{ij}} e^{\eta_{is}}}{\left(\sum\limits_{t=1}^{J} e^{\eta_{it}}\right)^2} \; = \; -p_{ij}\, p_{is},$$

$$\frac{\partial p_{is}}{\partial \eta_s} \; = \; \frac{\left(\sum\limits_{t=1}^{J} e^{\eta_{it}}\right) e^{\eta_{is}} - e^{\eta_{is}} e^{\eta_{is}}}{\left(\sum\limits_{t=1}^{J} e^{\eta_{it}}\right)^2} \; = \; p_{is}(1 - p_{is}).$$

Using the chain rule formulas (18.10) and (18.11), then

$$\frac{\partial \ell_i}{\partial \eta_s} \; = \; \sum_{j=1}^{M+1} \frac{\partial \ell_{ij}}{\partial p_{ij}} \frac{\partial p_{ij}}{\partial \eta_s} \; = \; \frac{y_{is}}{p_{is}} p_{is}(1 - p_{is}) + \sum_{j \neq s} \frac{y_{ij}}{p_{ij}} (-p_{ij}\, p_{is}) \; = \; y_{is} - p_{is},$$

and for $s \neq t$,

$$-E\left[\frac{\partial^2 \ell_i}{\partial \eta_s^2}\right] \; = \; \sum_{j=1}^{M+1} -E\left[\frac{\partial^2 \ell_i}{\partial p_{ij}^2}\right] \left(\frac{\partial p_{ij}}{\partial \eta_s}\right)^2$$

$$= \; \frac{1}{p_{is}} p_{is}^2(1 - p_{is})^2 + \sum_{j \neq s} \frac{1}{p_{ij}} (-p_{ij}\, p_{is})^2$$

$$= \; p_{is}(1 - p_{is})^2 + p_{is}^2(1 - p_{is}) \; = \; p_{is}(1 - p_{is}),$$

$$-E\left[\frac{\partial^2 \ell_i}{\partial \eta_s\, \partial \eta_t}\right] \; = \; \sum_{j=1}^{M+1} -E\left[\frac{\partial^2 \ell_i}{\partial p_{ij}^2}\right] \left(\frac{\partial p_{ij}}{\partial \eta_s}\right) \left(\frac{\partial p_{ij}}{\partial \eta_t}\right)$$

$$= \; \frac{1}{p_{is}} p_{is}(1 - p_{is})(-p_{is}\, p_{it}) + \frac{1}{p_{it}} p_{it}(1 - p_{it})(-p_{is}\, p_{it}) +$$

$$\sum_{j \neq s,\, j \neq t} \frac{1}{p_{ij}} (-p_{ij}\, p_{is}) (-p_{ij}\, p_{it})$$

$$= \; p_{is}\, p_{it} \{1 - p_{is} - p_{it} - (1 - p_{is}) - (1 - p_{it})\} \; = \; -p_{is}\, p_{it}.$$

Thus the working weight matrices are $N_i \left\{\text{diag}(\widehat{\boldsymbol{\mu}}_i) - \widehat{\boldsymbol{\mu}}_i \widehat{\boldsymbol{\mu}}_i^T\right\}$, where $\boldsymbol{\mu}_i = \boldsymbol{p}_i$ (a slight change in notation) is an M-vector corresponding to $\boldsymbol{\eta}_i$.

Table 14.1 Quantities defined in **VGAM** for a categorical response Y taking values $1, \ldots, M+1$. Covariates $\boldsymbol{x}$ have been omitted for clarity. The LHS quantities are usually for $g^{-1}(\eta_j)$ or $g^{-1}(\eta_{j-1})$ for $j = 1, \ldots, M$ and $j = 2, \ldots, M+1$, respectively. All except for `multinomial()` are suited to ordinal Y. Notes: (i) † means a fixed link function g; (ii) `propodds()` is a shortcut to `cumulative(parallel = TRUE, reverse = TRUE, link = "logit")`; (iii) `reverse = FALSE` for all families except for `propodds()`; (iv) The link for `multinomial()` is known as the multilogit link (`multilogit()`).

Quantity	Default g	Range of j	VGAM family
$P(Y = j+1)/P(Y = j)$	log	$\{1, \ldots, M\}$	`acat()`
$P(Y = j)/P(Y = j+1)$	log	$\{2, \ldots, M+1\}$	`acat(reverse = TRUE)`
$P(Y > j \mid Y \geq j)$	logit	$\{1, \ldots, M\}$	`cratio()`
$P(Y < j \mid Y \leq j)$	logit	$\{2, \ldots, M+1\}$	`cratio(reverse = TRUE)`
$P(Y \leq j)$	logit	$\{1, \ldots, M\}$	`cumulative()`
$P(Y \geq j)$	logit	$\{2, \ldots, M+1\}$	`cumulative(reverse = TRUE)`
$P(Y = j)/P(Y = M+1)$	log †	$\{1, \ldots, M\}$	`multinomial()`
$P(Y = j)/P(Y = r)$	log †	$\{1, \ldots, M+1\}\backslash\{r\}$	`multinomial(refLevel = r )`
$P(Y \geq j)$	logit †	$\{2, \ldots, M+1\}$	`propodds()`
$P(Y \leq j)$	logit †	$\{1, \ldots, M\}$	`propodds(reverse = TRUE)`
$P(Y = j \mid Y \geq j)$	logit	$\{1, \ldots, M\}$	`sratio()`
$P(Y = j \mid Y \leq j)$	logit	$\{2, \ldots, M+1\}$	`sratio(reverse = TRUE)`

14.2.1 The xij Argument

Recall from Sect. 3.4 that the `xij` facility allows η_j to have a different value of a variable x_k and observation i. One common application of this is the multinomial logit model in a branch of econometrics called discrete choice modelling. Here, the concept of *utility* is central. The word 'utility' means usefulness and the ability of a 'product' to satisfy the needs or wants of a 'consumer'. It is a latent variable that cannot be measured or observed directly. An example is that it might be measured indirectly by the price somebody is willing to pay for a product, or by his/her preference as shown by choosing to buy one particular product from amongst a pool of several products, e.g., when somebody chooses to buy product B from products A–E, s/he is maximizing some utility function so that his/her desire or want is satisfied maximally. If product j's satisfaction to the person has some statistical distribution F_j as a function of the latent variable, then the maximum also has some distribution $1 - \prod_j (1 - F_j)$. An analogy of this is if $U_j \sim N(\mu_j, \sigma^2)$ then what is the distribution of $\max(U_1, \ldots, U_J)$? [Here, U_j is the utility of product j for $j = 1, \ldots, J$]. In the pioneering work of McFadden (1974), if the each product's utility is Gumbel (standard Type I extreme value; Gumbel($\mu = 0, \sigma = 1$) in Table 16.1) distributed, then the maximum has an extreme value distribution. That is, if

$$U_{ij} = \eta_{ij} + \varepsilon_{ij}, \quad \varepsilon_{ij} \sim \text{Gumbel independently},$$

where $\eta_{ij} = \boldsymbol{x}_{ij}^T \boldsymbol{\beta}$, and $Y_i = \arg\max_j \{U_{ij}\}$ then

$$P(Y_i = j \mid \boldsymbol{x}_{ij}) = P(\max(U_{i1}, \ldots, U_{iJ}) = U_{ij}) = \frac{\exp\{\eta_{ij}\}}{\sum_{s=1}^{J} \exp\{\eta_{is}\}}, \quad \text{cf. (1.28)}.$$

A more accessible proof of this is given in, e.g., Maddala (1983, Sect.3.1).

As a specific example of a random utility model, we mimic the 'conditional logit model' reported in Greene (2012, Table 18.3) by considering the `TravelMode` data frame in `AER`. This concerns the mode choice for travel between the Australian cities Sydney and Melbourne. The data set consists of 210 people's choice of transportation for travel between the two cities. Four choices of travel mode are `air`, `trn` (train), `bus` and `car`. The data set arises from case-control data: almost an equal number of each choice is represented. The explanatory variables are `gcost` (a measure of the generalized cost of the travel), `wait` (the terminal waiting time, 0 for car), and household `income`. The variables `gcost` and `wait` clearly differ for each travel mode. In contrast, variable `income` is individual-specific so that every person has the same fixed household income regardless what choice he/she made. The data frame `TravelMode` has a 'long' format or shape meaning the first 4 rows belong to individual 1, followed by the next 4 rows for individual 2, etc. Hence we need to manipulate the original data first in order to build a suitable data frame for using `vglm()`:

```
> data("TravelMode", package = "AER")
> air.df <- subset(TravelMode, mode == "air")   # Form 4 smaller data frames
> bus.df <- subset(TravelMode, mode == "bus")
> trn.df <- subset(TravelMode, mode == "train")
> car.df <- subset(TravelMode, mode == "car")
> TravelMode2 <- data.frame(income   = air.df$income,
                         wait.air  = air.df$wait  - car.df$wait,
                         wait.trn  = trn.df$wait  - car.df$wait,
                         wait.bus  = bus.df$wait  - car.df$wait,
                         gcost.air = air.df$gcost - car.df$gcost,
                         gcost.trn = trn.df$gcost - car.df$gcost,
                         gcost.bus = bus.df$gcost - car.df$gcost,
                         gcost     = air.df$gcost, # Value is unimportant
                         wait      = air.df$wait)  # Value is unimportant
> TravelMode2$mode <- subset(TravelMode, choice == "yes")$mode  # The response
> TravelMode2 <- transform(TravelMode2, inc.air = income, inc.trn = 0, inc.bus = 0)
> fit.travel <-
    vglm(mode ~ gcost + wait + income,
         multinomial(parallel = FALSE ~ 1), data = TravelMode2, trace = FALSE,
         xij    = list(gcost  ~ gcost.air + gcost.trn + gcost.bus,
                       wait   ~ wait.air + wait.trn + wait.bus,
                       income ~ inc.air + inc.trn + inc.bus),
         form2 = ~ gcost + wait + income +
                   gcost.air + gcost.trn + gcost.bus +
                   wait.air + wait.trn + wait.bus +
                   inc.air + inc.trn + inc.bus)
> coef(summary(fit.travel))

              Estimate Std. Error z value  Pr(>|z|)
(Intercept):1  5.20744   0.779054   6.684 2.320e-11
(Intercept):2  3.86904   0.443126   8.731 2.519e-18
(Intercept):3  3.16319   0.450265   7.025 2.138e-12
gcost         -0.01550   0.004408  -3.517 4.370e-04
wait          -0.09612   0.010440  -9.208 3.338e-20
income         0.01329   0.010262   1.295 1.954e-01
```

These results agree with Greene (2012, Table 18.3). The reason for subtracting `wait` and `gcost` of the `car`s option from the others is because cars are the baseline group, cf. (3.37). Section 3.4.4.1 'continues' this example by smoothing with respect to the cost variable.

Table 14.2 Period of exposure (years) and severity of pneumoconiosis amongst a group of coalminers. Source: Ashford, J. R. (1959), *Biometrics 15*(4), 573–581. The data are in pneumo.

Exposure time (ET)	Normal	Mild	Severe	ET	Normal	Mild	Severe
5.8	98	0	0	33.5	32	10	9
15.0	51	2	1	39.5	23	7	8
21.5	34	6	3	46.0	12	6	10
27.5	35	5	8	51.5	4	2	5

14.2.2 Marginal Effects and Deviance

The *marginal effects* for a model with a categorical response are the derivatives of the probabilities with respect to the explanatory variables. For an MLM (1.26) without constraints, the marginal effects are easily shown (Ex. 14.1) to be

$$\frac{\partial\, p_j(\boldsymbol{x}_i)}{\partial\, \boldsymbol{x}_i} \;\; = \;\; p_j(\boldsymbol{x}_i)\left\{\boldsymbol{\beta}_j - \sum_{s=1}^{M+1} p_s(\boldsymbol{x}_i)\,\boldsymbol{\beta}_s\right\}. \tag{14.6}$$

Care must be taken because the marginal effect for x_k need not have the same sign as $\beta_{(j)k}$ (they are for the $J = 2$ case, i.e., logistic regression), and they are suited for continuous x_k rather than discrete ones (especially binary x_k). Marginal effects are implemented in margeff(), which will accept a multinomial() VGLM as its first argument.

From (14.5), the MLE of p_{ij} is y_{ij} $(=\widetilde{p}_{ij}$, say). Hence the deviance is

$$D \;\; = \;\; 2\,\ell(\widetilde{\boldsymbol{p}};\boldsymbol{y}) - 2\,\ell(\widehat{\boldsymbol{p}};\boldsymbol{y}) \tag{14.7}$$

$$= \;\; 2\sum_{i=1}^{n} N_i \sum_{j=1}^{J} y_{ij}\log\widetilde{p}_{ij} - 2\sum_{i=1}^{n} N_i \sum_{j=1}^{J} y_{ij}\log\widehat{p}_{ij} \tag{14.8}$$

$$= \;\; 2\sum_{i=1}^{n} N_i \sum_{j=1}^{J} y_{ij}\log\left(\frac{y_{ij}}{\widehat{p}_{ij}}\right). \tag{14.9}$$

When the N_i (or more particularly the p_{ij}s) are sufficiently large, the deviance is approximately χ^2 with $nM - p$ degrees of freedom, which can be used as a measure of goodness of fit. The methods function deviance() returns D, where it is necessary for $0\log 0$ to be defined as 0 to handle $y_{ij} = 0$. This is justified by

$$\lim_{u\to 0^+} u \cdot \log u \;\; = \;\; 0. \tag{14.10}$$

14.2.3 Stereotype Model

As described in Sect. 5.2.3, this model is better called a *reduced-rank multinomial logit model* (RR-MLM). It was proposed by Anderson (1984), who described it as being suitable for *all* (ordered or unordered) categorical response variables. This was due to his rank-1 stereotype model restricting the elements of the **A** matrix to be ordered, however RR-VGLMs defined in Chap. 5 have no such constraints.

The basic idea of the RR-MLM is that if M and p are even moderately large, then the total number of regression coefficients in the MLM will be large. In such circumstances, the reduced-rank regression can decrease the number of parameters substantially if the rank is low.

As an imperfect example of a rank-2 RR-MLM, the marital status of the 9,000+ people in the `xs.nz` cross-sectional study is regressed against 11 psychological variables, adjusting for basic variables such as gender, age and ethnicity. One cause of the imperfection is due to the psychological variables being as much responses as explanatory. Therefore this example is to be viewed for its mechanical virtues than its interpretative vices.

```
> fit2.ms <-
    rrvglm(marital ~
             ethnicity + sex + age + smokenow + fh.cancer + fh.heartdisease +
             depressed + embarrassed + fedup + hurt + miserable +   # 11 psychological
             nofriend + moody + nervous + tense + worry + worrier,  # variables
             noRRR = ~ ethnicity + sex + age + smokenow + fh.cancer + fh.heartdisease,
             multinomial(refLevel = 2), data = xs.nz, Rank = 2)
> sort(round(concoef(fit2.ms)[, 1], 2))  # First latent variable

      worrier       moody    miserable          fedup        worry  embarrassed
        -0.17       -0.15        -0.06          -0.05        -0.04        -0.01
         hurt       tense     depressed       nervous     nofriend
         0.00        0.20          0.24          0.31         0.64

> sort(round(concoef(fit2.ms)[, 2], 2))  # Second latent variable

        worry  embarrassed       moody      worrier      nervous         hurt
        -0.26       -0.22        -0.12         -0.09         0.01         0.06
    miserable       tense        fedup     nofriend    depressed
         0.06        0.20         0.28          0.42         0.96
```

The first latent variable is interpreted as the absence of friends; the second mainly as being depressed, with a lack of friends manifesting itself again in more subdued role. Now the married group was chosen to be the baseline group because it is the largest group, and because all the other classes are linked to this group directly. For example, if the single group were baseline, then the widowed group could only connect to the single group via the married group, which is indirect and hence hinders the interpretation.

```
> head(depvar(fit2.ms), 2)   # Note the order of the response

    single married divorced widowed
  1      0       1        0       0
  2      0       1        0       0

> Coef(fit2.ms)@A

                      latvar1 latvar2
  log(mu[,1]/mu[,2]) 1.000000  0.0000
  log(mu[,3]/mu[,2]) 0.000000  1.0000
  log(mu[,4]/mu[,2]) 0.002523  0.4662
```

The corner constraints means that the latent variable $\hat{\nu}_1$ relates to the probability of being single, relative to being married/partnered; and $\hat{\nu}_2$ relates to the probability of being divorced/separated, relative to being married/partnered.

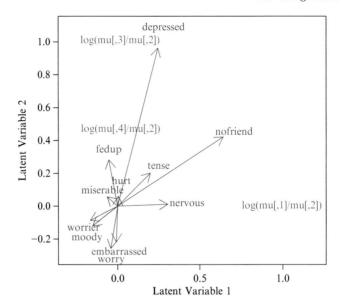

Fig. 14.1 Biplot of an RR-MLM fitted to `marital` status in `xs.nz`, with the latent variables being linear combinations of 11 binary psychological variables. The x- and y-axes are $\widehat{\nu}_1$ and $\widehat{\nu}_2$, respectively. The groups are: 1 single, 2 married/partnered (baseline), 3 divorced/separated, 4 widowed.

Being a widow appears to be a function of $\widehat{\nu}_2$ only. One might write the fitted model as

$$
\widehat{\boldsymbol{\eta}} \;=\;
\begin{pmatrix}
\log\left(\widehat{p}_{\text{single}}/\widehat{p}_{\text{married}}\right) \\
\log\left(\widehat{p}_{\text{divorced}}/\widehat{p}_{\text{married}}\right) \\
\log\left(\widehat{p}_{\text{widowed}}/\widehat{p}_{\text{married}}\right)
\end{pmatrix}
\;\approx\;
\begin{pmatrix}
\widehat{\boldsymbol{\beta}}_1^T \boldsymbol{x}_1 + \widehat{\nu}_1 \\
\widehat{\boldsymbol{\beta}}_3^T \boldsymbol{x}_1 + \widehat{\nu}_2 \\
\widehat{\boldsymbol{\beta}}_4^T \boldsymbol{x}_1 + \tfrac{1}{2}\widehat{\nu}_2
\end{pmatrix},
$$

for some nuisance $\widehat{\boldsymbol{\beta}}_j$, where

$$
\widehat{\nu}_1 \;\approx\; \frac{2}{3}\,\texttt{nofriend} + \frac{1}{3}\,\texttt{nervous}, \quad \text{and} \quad \widehat{\nu}_2 \;\approx\; \texttt{depressed} + \frac{2}{5}\,\texttt{nofriend}.
$$

Now the effect of

```
> biplot(fit2.ms, xlim = c(-0.3, 1.2), ylim = c(-0.3, 1.1),
         Ccol = "blue", Acol = "purple")
```

is seen in Fig. 14.1. The widower group appears midway between the origin and the divorced/separated group, and therefore it suggests that the ill-effects of bereavement are less severe than divorce/separation. The origin corresponds to the absence of all the (unhealthy) 11 emotional variables, hence denotes a psychological healthy person. The singles group, to the far right, is mainly associated with the lack of companionship. However, as mentioned above, all these 'results' must be taken with pinches of salt because, e.g., the psychological variables are likely to interact as both response and explanatory.

14.2.4 Summation Constraints

The identifiability constraints mentioned in Sect. 1.2.4 are, for *any* constant c, due to $p_j = e^{\eta_j + c}/\{\sum_{t=1}^{M+1} e^{\eta_t + c}\}$. The choice $c = -\eta_{M+1}$ leads to the default corner constraint $\eta_{M+1} \equiv 0$. A popular alternative to having the last level as baseline is to have summation constraints

$$\sum_{j=1}^{M+1} \beta_j^{\dagger} = \mathbf{0}. \tag{14.11}$$

This can be achieved by using offsets. Let $\boldsymbol{o} = (-\sum_{j=1}^{M+1} \boldsymbol{\beta}_j)/(M+1)$ from some MLM fit, and let $c_{ij} = \boldsymbol{o}^T \boldsymbol{x}_{ij}$ for the ith observation and η_j. Call this o_{ij}, say. The following simple example illustrates the method using the pneumoconiosis data set of Table 14.2.

```
> pneumo <- transform(pneumo, let = log(exposure.time))
> fit1 <- vglm(cbind(normal, mild, severe) ~ let, multinomial, data = pneumo)
>
> # Augment to get the complete B matrix; B becomes p x (M+1)
> B1 <- cbind(coef(fit1, matrix = TRUE), 0)
> mean.B1 <- rowMeans(B1)
> oij <-  matrix(model.matrix(fit1, type = "lm") %*% cbind(mean.B1),
                 nrow = nobs(fit1), ncol = npred(fit1))
>
> fit2 <- vglm(cbind(normal, mild, severe) ~ offset(oij) + let, multinomial, pneumo)
> B2 <- cbind(coef(fit2, matrix = TRUE), -mean.B1)
```

Objects `fit1` and `fit2` are two parameterizations of the same model. To check,

```
> c(rowSums(B2), diff = max(abs(fitted(fit1) - fitted(fit2))))  # Should be all 0s

  (Intercept)           let        diff
    3.486e-14    -1.038e-14   6.661e-16
```

are effectively all zeros. The two matrices, separated by `NA`s for readability, are

```
> both <- cbind(B1, NA, B2)
> colnames(both) <- NULL  # Column names are no longer correct
> both

                [,1]    [,2] [,3] [,4]    [,5]     [,6]    [,7]
(Intercept) 11.975  3.0391    0   NA   6.970 -1.9657 -5.005
let         -3.067 -0.9021    0   NA  -1.744  0.4211  1.323
```

Incidentally, some authors call (14.11) the *symmetric side constraint*, while $\boldsymbol{\beta}_{M+1} \equiv \mathbf{0}$ is called the *reference category side constraint*.

14.3 Using the Software

Here are some potentially useful notes.

- Some software chooses the first level of Y to be the reference level. To replicate this, use `multinomial(refLevel = 1)`.

- For the classification problem, multinomial-based models can classify observations to the level with the largest predicted or fitted probability. This is known as *Bayes' decision rule* (for minimum error). Code such as

```
> apply(fitted(fit1), 1, which.max)   # Highest probability

 1 2 3 4 5 6 7 8
 1 1 1 1 1 1 1 3

> colnames(fitted(fit1))[apply(fitted(fit1), 1, which.max)]   # Classification class

 [1] "normal" "normal" "normal" "normal" "normal" "normal" "normal" "severe"

> newdata <- pneumo   # For example
> apply(predict(fit1, newdata, type = "response"),
        1, which.max)   # Test sample classification

 1 2 3 4 5 6 7 8
 1 1 1 1 1 1 1 3
```

  can be useful here, provided that there are no tied probabilities. This data suggests that, after about 50 years coalmining, severe pneumoconiosis becomes the most likely class. This method requires care also if there are missing values. The code operates because the fitted values are a $(M + 1)$-column matrix.
- Input for categorical families may be

 (i) an $n \times (M + 1)$ matrix of counts. The columns are best labelled, and the jth column denotes $Y = j$.
 (ii) a factor vector. The unique values, when sorted (or levels, if a factor), denote the $M + 1$ levels from $1, \ldots, M + 1$. If a factor, it may be ordered or unordered. The functions `factor()`, `is.ordered()`, `levels()`, `ordered()`, are useful (Sect. 1.5.2.4). The ordinal regression models of Sect. 14.4 may issue a warning if the response is an unordered factor, and the same may be true if `multinomial()` operates on an ordered factor.

- Output for categorical families: if `fit` is an categorical **VGAM** object, then `fitted(fit)` is a $n \times (M + 1)$ matrix of fitted category probabilities (rows sum to unity), and `weights(fit, type = "prior")` contain the prior weights $N_i = \sum_{j=1}^{M+1} y_{ij}^*$. The call `depvar(fit)` returns the matrix of sample proportions y_{ij}.

14.3.1 The Poisson Trick

The conditional distribution (14.3) forms the basis of what is known informally as the "Poisson trick". It allows an MLM to be estimated by fitting a simpler Poisson regression with a suitable factor set up. Such was useful in the past when statistical software was less sophisticated. However, this trick is only really practical for grouped data because a regression coefficient is estimated for every i. They are treated as nuisance parameters. For completeness, we give a toy example here.

The sketch details are as follows. Suppose Y_{ij} are independently Poisson with means

$$\mu_{ij} = \exp\left\{\alpha_i + \boldsymbol{x}_i^T \boldsymbol{\beta}_j\right\} \tag{14.12}$$

for $j = 1, \ldots, M + 1$, and $\boldsymbol{\beta}_{M+1} = \mathbf{0}$. Then it can be shown that the Poisson likelihood factorizes the conditional distribution of $\boldsymbol{Y}_i$ given N_i (the multinomial model of interest) with N_i defined in (14.3). Hence the multinomial MLEs $\widehat{\boldsymbol{\beta}}_j$ can be obtained from the full unconditional Poisson likelihood. This result is due to the profile log-likelihood (using $\widehat{\alpha}_i$) being equal, up to an additive constant, to the MLM log-likelihood. The only twist to fitting (14.12) here is that while every other piece of software uses an nJ-vector response, we can fit this more naturally as J Poisson responses because `poissonff()` handles multiple responses.

As a numerical example, we fit an MLM to the `pneumo` data with log-exposure time as the covariate. This is particularly easy with **VGAM** because the interaction between the jth level of the response and $\boldsymbol{x}$ is easily handled, by forming the appropriate constraint matrices. In the following, the variable `ref.level` is the baseline reference level for the MLM, so it can take any value from $\{1, \ldots, M + 1\}$. The only slight complication is that the factor representing the dummy variables α_i sets $\alpha_1 = 0$, so that α_1 becomes the intercept of the reference group. Then the intercepts of the non-reference groups should be adjusted by setting the `ref.level`th column of $\mathbf{H}_1$ to $\mathbf{1}_M$.

```
pneumo <- transform(pneumo, let = log(exposure.time))   # Covariate
pneumo <- transform(pneumo, i.factor = factor(1:nrow(pneumo))) # Nuisance variables
Mplus1 <- 3   # The number of levels of the response
ref.level <- 2   # Choose any single value in the set {1:Mplus1}
H1 <- diag(Mplus1)
H1[, ref.level] <- 1   # Slight adjustment
clist <- list("(Intercept)" = H1,
              i.factor     = matrix(1, Mplus1, 1),   # Parallelism
              let          = diag(Mplus1)[, -ref.level, drop = FALSE])
pois.pneumo <- vglm(cbind(normal, mild, severe) ~ i.factor + let,
                    poissonff, data = pneumo, constraints = clist)   # Poisson trick
mlm.pneumo <- vglm(cbind(normal, mild, severe) ~ let,
               multinomial(refLevel = ref.level), data = pneumo)   # The 'answer'
```

Then the output can be compared as follows.

```
> coef(mlm.pneumo, matrix = TRUE)   # The 'answer'

                log(mu[,1]/mu[,2])  log(mu[,3]/mu[,2])
  (Intercept)               8.936              -3.0391
  let                      -2.165               0.9021

> coef(pois.pneumo)   # Hopefully the 'answer'

  (Intercept):1 (Intercept):2 (Intercept):3      i.factor2      i.factor3
         8.9360       -0.5519       -3.0391         1.3994         1.8567
      i.factor4      i.factor5      i.factor6      i.factor7      i.factor8
         2.3739        2.7043        2.5857         2.3977         1.5220
          let:1         let:2
        -2.1654        0.9021
```

To select out the MLM coefficients, we bypass the reference group's intercept and all the nuisance parameters:

```
> index <- (1:Mplus1)[-ref.level]
> B.hat <- coef(pois.pneumo, matrix = TRUE)
> as.vector(c(coef(pois.pneumo)[index], B.hat["let", -ref.level]))

  [1]  8.9360 -3.0391 -2.1654  0.9021
```

The purpose of `as.vector()` is to remove the misleading names. It is left as an exercise for the reader (Ex. 14.9) to show that the standard errors of the coefficients match and also to fit the same model using `glm()` with `family = poisson`.

14.4 Ordinal Responses

Here, the response $Y = 1, 2, \ldots, M + 1$, is ordered, e.g., low to high in some sense. Most of the regression models in this chapter are tailored for this common situation. One might be tempted to use a multinomial logit model, however this is inefficient because it ignores the information contained in the ordering. As reflected in Table 14.1, ordinal responses allow a greater variety of quantities to be modelled. Whatever quantity that is chosen should be based on the type of question being asked, e.g., if the J classes relate to time-order then a sequential model using stopping or continuation-ratios may be the more natural choice.

14.4.1 Models Involving Cumulative Probabilities

The most common models involve the *cumulative probabilities* $\gamma_j(\boldsymbol{x}) = P(Y \leq j|\boldsymbol{x})$, as popularized by McCullagh (1980). This gives rise to the class of *cumulative link models* of the form

$$g(\gamma_j(\boldsymbol{x})) \;=\; \eta_j(\boldsymbol{x}), \quad j = 1, \ldots, M, \tag{14.13}$$

where g is a suitable link function for a probability. They can be motivated by the underlying latent idea described in Sect. 14.4.1.1.

The most well-known example of (14.13) is not surprisingly with a logit link:

$$\text{logit } P(Y \leq j|\boldsymbol{x}) \;=\; \eta_j(\boldsymbol{x}), \quad j = 1, \ldots, M, \tag{14.14}$$

and is known as the *cumulative logit model*. Another example, with a complementary log-log link, is known as the *proportional hazards model* in discrete time and used in survival analysis. A third example, bearing a probit link, is often referred to as the *cumulative probit model* or *ordinal/ordered probit model*.

Some care is needed for (14.13)–(14.14). For fixed $\boldsymbol{x}$, the condition

$$\eta_1(\boldsymbol{x}) < \eta_2(\boldsymbol{x}) < \cdots < \eta_M(\boldsymbol{x}) \tag{14.15}$$

must be satisfied in order for the class probabilities $p_j(\boldsymbol{x})$ to be in range, because $p_j(\boldsymbol{x}) = \gamma_j(\boldsymbol{x}) - \gamma_{j-1}(\boldsymbol{x})$. Since the $\eta_j(\boldsymbol{x})$ are planes in p-dimensional space they will intersect unless parallel. In the VGLM formulation, the parallelism assumption amounts to having constraint matrices $\mathbf{H}_1 = \mathbf{I}_M$ and $\mathbf{H}_2 = \cdots = \mathbf{H}_p = \mathbf{1}_M$. Model (14.14) with a parallelism assumption is known as the *proportional-odds model*; it may be written

$$\text{logit } P(Y \leq j|\boldsymbol{x}) \;=\; \beta_{(j)1}^* + \boldsymbol{\beta}_{[-(1)]}^{*T} \boldsymbol{x}_{[-1]}, \quad j = 1, \ldots, M, \tag{14.16}$$

because $x_1 = 1$. It has the property that the intercepts are an increasing sequence in j, due to $P(Y \leq j|\boldsymbol{x})$ being an increasing function of j.

Another nice property of the proportional-odds model, called *strict stochastic ordering* (McCullagh, 1980), is that the effect of the covariates on the odds ratio is the same regardless of the division point j, hence its name. By separating out the intercept from two vectors $\boldsymbol{x}_1$ and $\boldsymbol{x}_2$, the result is that

$$\frac{\text{odds}(Y \leq j|(1, \boldsymbol{x}_{[-1]1}))}{\text{odds}(Y \leq j|(1, \boldsymbol{x}_{[-1]2}))} = \exp\left\{\boldsymbol{\beta}^{*T}_{[-(1)]}\left(\boldsymbol{x}_{[-1]1} - \boldsymbol{x}_{[-1]2}\right)\right\} \quad (14.17)$$

does not depend on j. Consequently, the interpretation of the regression coefficients is very similar to logistic regression: $\exp\{\beta^*_{(1)k}\}$ is the odds ratio for $Y \leq j$, from a change in x_k to $x_k + 1$, keeping all other variables fixed.

In practice, the parallelism assumption should be checked. One might allow non-parallelism to (i) some, or (ii) all of the x_k. With a logit link, the former case has been called a *partial proportional-odds* model, whereas the latter case a *non-proportional-odds* model. Regardless of the link, one would hope that the intersections of the η_j occur outside the $\boldsymbol{x}_i$ data cloud. If not then there will be problems due to (14.15) being violated for some observations i. Practically, for a final model that does satisfy (14.15) for all i in the first instance, there are some tricks available that can help avoid numerical problems *during estimation*. They are described in Sect. 8.5. Another alternative is to perform score tests—these do not require the fitting of any non-parallel model.

Formally, there are two common ways to test the parallelism assumption. The first is by a likelihood ratio test (LRT; Sect. A.1.4.2):

```
> fit.pom  <- vglm(cbind(normal, mild, severe) ~ let, propodds,  data = pneumo)
> fit.npom <- vglm(cbind(normal, mild, severe) ~ let, cumulative, data = pneumo)
> lrtest(fit.npom, fit.pom)

  Likelihood ratio test

  Model 1: cbind(normal, mild, severe) ~ let
  Model 2: cbind(normal, mild, severe) ~ let
    #Df LogLik Df Chisq Pr(>Chisq)
  1  12  -25.0
  2  13  -25.1  1  0.14       0.71
```

This tends to be the most accurate method for one model nested within another. The large p-value here indicates that a parallelism assumption is reasonable. The second method is using the Wald test:

```
> (cfit.npom <- coef(fit.npom))

  (Intercept):1 (Intercept):2         let:1         let:2
          9.593        11.105        -2.571        -2.744

> index <- 3:4  # These coefficients need testing for equality
> L.mat <- cbind(diag(npred(fit.npom) - 1), -1)  # Matrix of contrasts
> T.mat <- solve(L.mat %*% vcov(fit.npom)[index, index] %*% t(L.mat))
> W.stat <- t(L.mat %*% cfit.npom[index]) %*% T.mat %*% (L.mat %*% cfit.npom[index])
> W.stat

         [,1]
  [1,] 0.1576

> pchisq(W.stat, df = nrow(L.mat), lower.tail = FALSE)  # p-value

         [,1]
  [1,] 0.6914
```

Not surprisingly, the Wald statistic and its associated p-value closely resembles those of the LRT here, because of the large sample sizes. The theory behind this Wald test for parallelism for x_k, by testing $\beta_{(1)k}^* = \cdots = \beta_{(M)k}^*$, is to compute

$$W = \left(\mathbf{L}\,\widehat{\boldsymbol{\beta}}_{(k)}^*\right)^T \mathbf{T}(\widehat{\boldsymbol{\beta}}_{(k)}^*)\,\mathbf{L}\,\widehat{\boldsymbol{\beta}}_{(k)}^*$$

where $\mathbf{L} = (\mathbf{I}_{M-1} \mid -\mathbf{1}_{M-1})$ is a matrix of contrasts of $\boldsymbol{\beta}_{(k)}^*$. The matrix $\mathbf{T}(\boldsymbol{\beta}_{(k)}^*) = \left(\mathbf{L}\,\mathbf{F}^{\boldsymbol{\beta}_{(k)}^* \boldsymbol{\beta}_{(k)}^*}(\boldsymbol{\beta}^*)\,\mathbf{L}^T\right)^{-1}$ is the inverse of the variance-covariance matrix of the asymptotic distribution of $\mathbf{L}\,\widehat{\boldsymbol{\beta}}_{(k)}^*$, where $\mathbf{F}^{\boldsymbol{\beta}_{(k)}^* \boldsymbol{\beta}_{(k)}^*}(\boldsymbol{\beta}^*)$ is the $\boldsymbol{\beta}_{(k)}^*$-block of the inverse of the EIM. By the asymptotic normality of $\widehat{\boldsymbol{\beta}}^*$, the Wald test statistic W is asymptotically χ_{M-1}^2.

Informally, Harrell (2001) describes graphical methods for assessing the parallelism assumption, as well as an (anticonservative) score test.

14.4.1.1 Latent Variable Motivation

One reason for the proportional-odds model's popularity is its connection to the idea of a continuous latent response (cf. Chap. 5). That the model can be motivated by the categorical outcome Y being a categorized version of an unobservable (latent) continuous variable, Y', say, goes along the following lines (Fig. 14.2).

Suppose that we have a continuous variable Y' and some unknown *cutpoints* ordered as $\beta_{(1)1} < \beta_{(2)1} < \cdots < \beta_{(M)1}$. Additionally, with $-\infty \equiv \beta_{(0)1}$ and $\beta_{(M+1)1} \equiv \infty$ defined, then

$$Y = j \quad \text{if} \quad \beta_{(j-1)1} < Y' \le \beta_{(j)1}.$$

Suppose also that an LM explains Y', given $\boldsymbol{x}$:

$$Y' = -\boldsymbol{\beta}_{[-1]}^T \boldsymbol{x}_{[-1]} + \varepsilon, \tag{14.18}$$

where ε has distribution function $F(\cdot)$, and $\boldsymbol{x}$ is assumed to have an intercept. Then

$$P(Y \le j) = P\left(-\boldsymbol{\beta}_{[-1]}^T \boldsymbol{x}_{[-1]} + \varepsilon \le \beta_{(j)1}\right) = F\left(\beta_{(j)1} + \boldsymbol{\beta}_{[-1]}^T \boldsymbol{x}_{[-1]}\right) = F(\eta_j).$$

If ε has a logistic distribution, then

$$F(\varepsilon) = \frac{e^\varepsilon}{1 + e^\varepsilon}, \quad \text{and so} \quad P(Y \le j) = \frac{e^{\eta_j}}{1 + e^{\eta_j}}.$$

This is the proportional-odds model (14.16).

The above derivation can be seen to hold for other link-distribution pairs, e.g.,

- a probit link and normal distribution,
- a complementary log-log link and extreme value distribution,
- a cauchit link and Cauchy distribution.

To partly summarize, here are some advantages of the proportional-odds model.

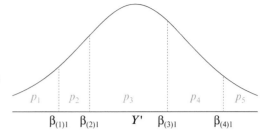

Fig. 14.2 Latent variable Y' interpretation for the cumulative logit model. Here, $\beta_{(j)1}$ are *thresholds* or *cutoff points* so that $P(Y \le j|\boldsymbol{x}) = P(Y' \le \beta_{(j)1})$. Then $p_j = P(Y = j|\boldsymbol{x})$. The distribution is a logistic (see (14.18)).

1. If the ordering of Y is reversed, then the model remains the same, because

$$P(Y > j) = 1 - \frac{e^{\eta_j}}{1 + e^{\eta_j}} = \frac{1}{1 + e^{\eta_j}} = \frac{e^{-\eta_j}}{1 + e^{-\eta_j}}.$$

Hence, effectively, only the sign of $\boldsymbol{\beta}$ changes. This is an advantage in data where the order is, in some sense, reversible. This property applies to other symmetric link functions such as the probit and cauchit, but not to the complementary log-log.
2. If adjacent categories are amalgamated, then the model has the same form and $\boldsymbol{\beta}$ is unchanged. This is conceptually appealing for data where the categories are arbitrary, and one could have used fewer or more.
3. The parallelism assumption results in fewer parameters to estimate and fewer numerical problems.

14.4.1.2 Software Use

VGAM fits cumulative link models (14.13) by the family function `cumulative()` (Table 14.1). It has an argument `link` whose default is `"logit"`. However, its `parallel` argument needs to be set to `TRUE` in order for the parallelism assumption to be applied to all the x_k except for the intercept. For this reason, there is less flexible shortcut called `propodds()` for fitting the proportional-odds model. As simple examples of its use,

```
vglm(y ~ x2, cumulative(link = probit, reverse = TRUE, parallel = TRUE), cdata)
```

fits the reversed ordered probit model

$$\Phi^{-1}\{P(Y \ge j|\boldsymbol{x})\} = \beta^*_{(j-1)1} + \beta^*_{(1)2}\, x_2, \qquad j = 2, \dots, M+1, \quad \text{and}$$

```
vgam(y ~ s(x2), cumulative(link = cloglog, reverse = FALSE, parallel = TRUE), cdata)
```

fits the proportional hazards model

$$\log(-\log P(Y > j|\boldsymbol{x})) = \beta^*_{(j)1} + f^*_{(1)2}(x_2), \qquad j = 1, \dots, M,$$

for some smooth function $f^*_{(1)2}$.

All categorical data family functions have the `parallel` argument, and most have `zero`. By default, `parallel = FALSE` (except for `propodds()`) and `zero = NULL` for all models. Unfortunately, the default value of `parallel` leads to a greater failure rate because of the intersecting η_j problem, and/or leads to a large number of parameters if it succeeds. Making a parallelism assumption requires explicit invocation, hence the existence of `propodds()` is a compromise solution. Table 14.3 summarizes whether `parallel = TRUE` results in the parallelism constraint being applied to the intercepts.

All the **VGAM** family functions in Table 14.1 for ordinal Y have a `reverse` argument which controls the direction of Y as a function of j, with respect to the η_j. Some authors use "backward" and "forward" for "reversed" and "non-reversed", respectively.

Marginal effects are also available for `cumulative()` fits. If `reverse = FALSE` then $p_j = \gamma_j - \gamma_{j-1} = h(\eta_j) - h(\eta_{j-1})$ where $h = g^{-1}$ is the inverse of the link function. Then `margeff()` will return the

$$\frac{\partial p_j(\boldsymbol{x})}{\partial \boldsymbol{x}} = h'(\eta_j) \boldsymbol{\beta}_j - h'(\eta_{j-1}) \boldsymbol{\beta}_{j-1} \tag{14.19}$$

as a $p \times (M + 1) \times n$ array.

Practically, if the proportional-odds assumption is inadequate then one might try using a different link function, or add interaction terms to the formula. Another strategy is the so-called partial proportional-odds model where the parallelism assumption is applied to a subset of the explanatory variables. Hence the proportional-odds assumption is relaxed for some variables while being retained for others. It is easy to fit partial proportional-odds models in **VGAM** using the syntax described in Sect. 3.3.1.2, e.g., the following are equivalent:

```
vgam(y ~ x2 + s(x3) + x4 + x5, cumulative(parallel = TRUE  ~ s(x3) + x5 - 1), cdata)
vgam(y ~ x2 + s(x3) + x4 + x5, cumulative(parallel = FALSE ~ x2 + x4), data = cdata)
```

14.4.1.3 Estimation Difficulties

As mentioned above, one of the advantages of assuming parallelism with respect to all the x_k (bar x_1) is that there are fewer numerical problems to contend with. Hence fitting a partial proportional-odds model requires care because of intersecting η_j during the IRLS iterations. Often there will be convergence failure if the data and the model are not in agreement, or if there is ill-conditioning in a loose sense. Here is a checklist of things to consider.

1. Are the η_j really linear and additive with respect to each x_k? If not, then try a transformation (e.g., log exposure time is used in the proportional-odds model for the `pneumo` data set), as well as interaction terms.
2. Consider amalgamating levels so there are a sizeable number of counts per level. An insufficient number of counts in a level aggravates the problem of intersecting η_j due to its relatively greater statistical variability.
3. Try several links, as these may have a noticeable effect on the parallelism.
4. Try other alternative types of ordinal models, e.g., continuation- and stopping-ratios models (Sect. 14.4.4), adjacent category models (Sect. 14.4.5).

Table 14.3 Summary of the default values of the `parallel` argument, and whether `parallel = TRUE` applies to the intercept.

family =	Default: parallel =	Applied to intercepts?
acat()	FALSE	No
cratio()	FALSE	No
cumulative()	FALSE	No
multinomial()	FALSE	Yes
propodds()	TRUE (implicit)	No
sratio()	FALSE	No

5. One can perform M logistic regressions by dichotomizing the response according to $Y \leq j$, and then the $\widehat{\beta}_{(j)k}$ compared (also in light of their SEs). Those which are similar suggest that a parallelism assumption may be valid.
6. The number of possible partially parallel models grows very quickly with the number of variables ($O(2^v)$ with v explanatory variables), therefore one strategy is to fit a fully parallel model and then try relaxing one x_k at a time.
7. Can better initial values be used? The stepping-stone trick of Sect. 8.5 may increase the chances of successful convergence if in fact they do not cross. Arguments `etastart` and `coefstart` may be useful here.
8. Allow the η_j to be more flexible with respect to x_k, e.g., using splines rather than as a linear term. However, do not allow too much flexibility because this will result in a higher chance that some of the η_js will cross.

14.4.2 Coalminers Example I

Here is a simple example relating to the **pneumo** data, which mimics some of the results of McCullagh and Nelder (1989, p.179). But we first estimate a more flexible model, a nonparametric nonparallel cumulative logit model, to check out the linearity and parallelism assumptions of η_1 and η_2. There are only 8 distinct values of the log-exposure time variable, therefore we do not allow the smooths to be very flexible.

```
> pneumo <- transform(pneumo, let = log(exposure.time))
> np.npom.pneumo <- vgam(cbind(normal, mild, severe) ~ s(let, df = 3),
                cumulative(rev = TRUE), data = pneumo)
> plot(np.npom.pneumo, se = TRUE, overlay = TRUE, lcol=3:4, scol=3:4, main = "(a)")
>
> matplot(with(pneumo, let), fitted(np.npom.pneumo), type = "l", col = 1:3,
          ylab = "Fitted value", xlab = "Log exposure time", main = "(b)")
> mycex <- sqrt(c(weights(np.npom.pneumo, type = "prior")) *
                depvar(np.npom.pneumo)) * 0.5
> Q <- ncol(depvar(np.npom.pneumo))
> for (j in 1:Q)
    points(depvar(np.npom.pneumo)[, j] ~ let, data = pneumo,
          cex = mycex[, j], col = j, pch = j)
>
> matplot(with(pneumo, let), predict(np.npom.pneumo, untransform = TRUE),
          type = "b", col = 1:3, ylab = "P(Y>=j),    j=1,2",
          xlab = "Log exposure time", main = "(c)")
```

Figure 14.3a shows that the functions appear to be linear and parallel. This gives us confidence in fitting the standard proportional odds model. The fitted class probabilities $\widehat{p}_j(\boldsymbol{x}_i)$ in plot (b), along with the raw sample proportions plotted approximately proportional to their size, show a close correspondence between the data and fitted model. Plot (c) shows the fitted reversed cumulative probabilities $\widehat{P}(Y \geq j)$ for $j = 2, 3$.

```
> pom.pneumo <- vglm(cbind(normal, mild, severe) ~ let, propodds, data = pneumo)
> coef(pom.pneumo, matrix = TRUE)

                 logit(P[Y>=2]) logit(P[Y>=3])
  (Intercept)           -9.676        -10.582
  let                    2.597          2.597

> margeff(pom.pneumo)[,, 1]

                 normal      mild    severe
  (Intercept)   0.05763 -0.031970 -0.025664
  let          -0.01547  0.009169  0.006298

> matplot(with(pneumo, let), t(margeff(pom.pneumo)["let",, ]), type = "b", col = 1:3,
          ylab = "Marginal effects", xlab = "Log exposure time", main = "(d)")
> abline(h = 0, col = "gray50", lty = "dashed")
```

To interpret just the severe group, a 10-year exposure has $\eta_3 \approx -10.582 + 2.597 \log 10 \approx -4.602$, meaning a probability of about 0.00993, or one chance in about 100. Doubling the exposure results in the odds (risk) of severe pneumoconiosis being multiplied by $2^{\widehat{\beta}^*_{(1)2}} \approx 6.05$, hence a 20-year exposure results in approximately one chance in 17 ($\approx 100/6$).

The marginal effects for the first time point with respect to let, which is exposure.time $= 5.8$, has a (larger) negative value for normal, and positive values for mild and severe. This is not surprising as Fig. 14.3b shows the normal curve falling faster and the others rising more slowly.

14.4.3 Coalminers Example II

We continue the previous example and illustrate how linearHypothesis() in car may be used to test for parallelism in each variable separately in a nonproportional odds model. To add a little complexity, we create the unrelated variable x3 and add it to the regression.

```
> library("car")
> set.seed(1); n <- nrow(pneumo)
> pneumo <- transform(pneumo, x3 =
runif(n))    # Added variable
> npom <- vglm(cbind(normal, mild, severe) ~ let + x3, cumulative, data = pneumo)
> coef(npom)    # Labelling is very important here

  (Intercept):1 (Intercept):2       let:1       let:2        x3:1
         9.6329        10.9482     -2.6064     -2.6581      0.1254
          x3:2
        -0.2129
```

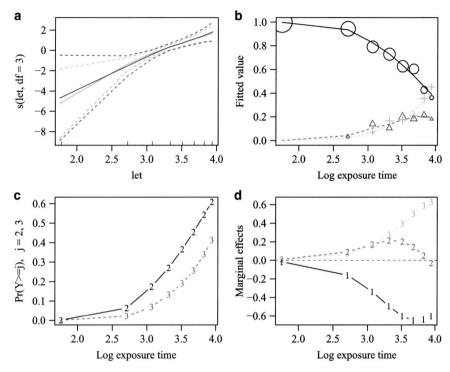

Fig. 14.3 Nonparametric nonparallel cumulative logit model fitted to the `pneumo` data: (**a**) the two centred functions are overlaid on to one plot; (**b**) the fitted values as a function of `let` with sample proportions plotted proportional to their counts; (**c**) reversed cumulative probabilities $\widehat{P}(Y \geq j)$ for $j = 2, 3$. (**d**) Proportional odds model: the marginal effects for variable `let` for each of the 3 levels.

```
> Terms <- attr(terms(npom), "term.labels")   # These match the coefficients
> for (Term in Terms) {
    lh <- paste(Term, ":", 1, " = ", Term, ":", 2, sep="")
    print(car::linearHypothesis(npom, lh)   )
}

Linear hypothesis test

Hypothesis:
let:1 - let:2 = 0

Model 1: restricted model
Model 2: cbind(normal, mild, severe) ~ let + x3

  Res.Df Df Chisq Pr(>Chisq)
1     11
2     10  1  0.01       0.91
Linear hypothesis test
```

```
  Hypothesis:
  x3:1 - x3:2 = 0

  Model 1: restricted model
  Model 2: cbind(normal, mild, severe) ~ let + x3

    Res.Df Df Chisq Pr(>Chisq)
  1     11
  2     10  1  0.45         0.5

> detach("package:car")
```

This example requires the fitting of a totally nonproportional odds model, which can be very difficult for some data sets. Then Wald tests are performed on each explanatory variable (term). In the above, the *p*-values indicate that there is no evidence against parallelism for any of the variables, keeping the coefficients for the other explanatory variable nonparallel.

14.4.4 Models Involving Stopping- and Continuation-Ratios

Quantities known as *continuation-ratios* are useful for the analysis of a sequential process where Y takes on successive values $1, 2, \ldots, M+1$ over time. Some examples of such are

(i) to ascertain the effect of a covariate on the number of children a couple have [$Y = 1$ (no children), 2 (1 child), 3 (2 children), 4 (3+ children)];
(ii) whether a risk factor is related to the progression of a certain type of cancer [$Y = 1$ (no disease), 2 (localized), 3 (widespread), 4 (terminal)].
(iii) to determine what variables (such as age, gender, ethnicity and marital status) influence the highest degree attained among university students [$Y = 1$ (Bachelors), 2 (Masters), 3 (Doctorate)].

For such data, two types of probabilities may be of interest: that of *stopping* at $Y = j$, or *continuing* past $Y = j$, given that Y has reached level j in the first place. Which of the two is more natural depends on the particular application. If there are time-varying covariates, as is the likely case in the examples, then the use of the `xij` argument is needed.

Unfortunately, continuation-ratios have been defined differently in the literature. Table 14.1 gives the **VGAM** definitions, which includes a new quantity termed the *stopping-ratio* to distinguish between the two types. For example, our (forward) stopping-ratio matches the definition of the continuation-ratio in Agresti (2013, pp.311–3), where continuation-ratio logits are defined as $\log(p_j/(p_{j+1} + \cdots + p_J))$ for $j = 1, \ldots, J - 1$.

Incidentally, continuation-ratios and stopping-ratios may be fitted by setting up the appropriate responses and indicator variables and using logistic regression. As an illustration, consider the model

$$\text{logit } P(Y = j | Y \geq j) \;=\; \text{logit } \frac{P(Y = j)}{P(Y \geq j)} \;=\; \log \frac{P(Y = j)}{P(Y > j)} \;=\; \eta_j, \quad j = 1, 2,$$

applied to `pneumo`. We fit 3 stopping-ratio models of varying degrees of parallelism.

In the following, the two η_js are effectively stacked together and estimated as one η; and the parallelism applies to the intercept too.

```
> fit.sr2 <- vglm(cbind(normal, mild, severe) ~ let,
                  sratio(parallel = FALSE ~ 0), data = pneumo)
> coef(fit.sr2, matrix = TRUE)

                 logit(P[Y=1|Y>=1]) logit(P[Y=2|Y>=2])
  (Intercept)                 9.012                9.012
  let                        -2.454               -2.454

> fit.lr2 <- vglm(cbind(c(normal, mild), c(mild + severe,  severe)) ~ c(let, let),
                  binomialff, data = pneumo)
> coef(fit.lr2, matrix = TRUE)

                 logit(prob)
  (Intercept)          9.012
  c(let, let)         -2.454
```

In the following, the parallelism does not apply to the intercept.

```
> fit.sr1 <- vglm(cbind(normal, mild, severe) ~ let,
                  sratio(parallel = FALSE ~ 1), data = pneumo)
> coef(fit.sr1, matrix = TRUE)

                 logit(P[Y=1|Y>=1]) logit(P[Y=2|Y>=2])
  (Intercept)                 8.734                8.051
  let                        -2.321               -2.321

> fit.lr1 <- vglm(cbind(c(normal, mild), c(mild + severe,  severe)) ~
                  -1 + gl(2, nrow(pneumo)) + c(let, let), binomialff,
                  data = pneumo)
> coef(fit.lr1, matrix = TRUE)

                         logit(prob)
  gl(2, nrow(pneumo))1         8.734
  gl(2, nrow(pneumo))2         8.051
  c(let, let)                 -2.321
```

(The function `gl()` generates levels of a factor. Here, it creates a factor variable of length 16 having 2 levels, the first 8 values having the first level, etc.) Finally, there is no parallelism at all in the following.

```
> fit.sr0 <- vglm(cbind(normal, mild, severe) ~ let,
                  sratio(parallel = TRUE ~ 0), data = pneumo)   # Same as sratio()
> coef(fit.sr0, matrix = TRUE)

                 logit(P[Y=1|Y>=1]) logit(P[Y=2|Y>=2])
  (Intercept)                 9.609                3.864
  let                        -2.576               -1.136

> fit.lr0 <- vglm(cbind(c(normal, mild), c(mild + severe,  severe)) ~
                  -1 + gl(2, nrow(pneumo)) + gl(2, nrow(pneumo)):c(let, let),
                  binomialff, data = pneumo)
> coef(fit.lr0, matrix = TRUE)

                                 logit(prob)
  gl(2, nrow(pneumo))1                 9.609
  gl(2, nrow(pneumo))2                 3.864
  gl(2, nrow(pneumo))1:c(let, let)    -2.576
  gl(2, nrow(pneumo))2:c(let, let)    -1.136
```

The interaction between a factor and a numerical vector means that there is a different slope for two `let` terms. It is left as an exercise to the reader (Ex. 14.6) to modify the above code in order to fit reversed continuation-ratio models, with varying degrees of parallelism, to these data.

14.4.5 Models Involving Adjacent Categories

An adjacent category model, which is not as popular as the proportional odds model, compares a level with the one immediately adjacent. This is very natural if the levels are enumerated over time, so that one can model successive changes. From Table 14.1, the reversed form is

$$g(p_j/p_{j+1}) = \eta_j, \qquad j = 1, \ldots, M, \tag{14.20}$$

where the logarithm is the most natural link function g because the ratio of interest is positive. For this model,

$$\log(p_j/p_{j+1}) = \log(p_j/p_{M+1}) - \log(p_{j+1}/p_{M+1}) = \eta_j^{MLM} - \eta_{j+1}^{MLM},$$

for $j = 1, \ldots, M$, therefore one can perform estimation by fitting an MLM and then subtracting the adjacent η_j. For example,

```
> pneumo <- transform(pneumo, let = log(exposure.time))   # Covariate
> fit.mlm <- vglm(cbind(normal, mild, severe) ~ let, multinomial, data = pneumo)
> cfit.mlm <- coef(fit.mlm, matrix = TRUE)
> cbind(cfit.mlm[, 1] - cfit.mlm[, 2], cfit.mlm[, 2])   # Same as cfit.acat

              [,1]    [,2]
 (Intercept)  8.936  3.0391
 let         -2.165 -0.9021

> fit.acat <- vglm(cbind(normal, mild, severe) ~ let, acat(rev = TRUE), pneumo)
> coef(fit.acat, matrix = TRUE)

              loge(P[Y=1]/P[Y=2])  loge(P[Y=2]/P[Y=3])
 (Intercept)                8.936                3.0391
 let                       -2.165               -0.9021
```

For the above, it may be shown that, under a parallelism assumption not applying to the intercepts, $\beta_{(j)k}$ is the log-odds ratio of falling into category j versus $j + 1$ when x_k increases by one unit, holding all other covariates in $\boldsymbol{x}$ fixed.

14.4.6 Convergence

The algorithm of McCullagh (1980) matches that of Chap. 3. He showed that a unique maximum of the likelihood is guaranteed for sufficiently large sample sizes, though infinite parameter values can arise with sparse data sets containing certain patterns of zeros.

However, one problem that can arise with categorical regression models is that the MLEs can be on the boundary of the parameter space with positive probability.

For example, the oenological data set `wine` of Randall (1989), examined by Christensen (2013) and Kosmidis (2014b), is

	temp	contact	bitter1	bitter2	bitter3	bitter4	bitter5
1	cold	no	4	9	5	0	0
2	cold	yes	1	7	8	2	0
3	warm	no	0	5	8	3	2
4	warm	yes	0	1	5	7	5

The `bitter` variables and `temp` are for the bitterness taste (1 = none, 5 = intense) and temperature. Variable `contact` is whether the juice came in contact with the skin for a specified period.

A partial proportional-odds model with separate slopes for the temperature variable exhibits such problems. When boundary problems occur, numerical problems may occur during the estimation and misleading standard errors result. In particular, Wald statistics $\to 0$ because the SEs diverge much faster than the estimates.

Some references for the convergence properties of several categorical models, and conditions that guarantee that the MLE exists, are given in Kosmidis (2014b).

14.5 Genetic Models

A number of common population genetics models based on the multinomial distribution are implemented in **VGAM**. Due to space limitations, this section does little more than just list them: Table 14.4 serves to direct the user to other tables of specific models. Two other distributions useful to genetics are also briefly touched upon.

As one example, for `ABO()`, the alleles A, B and O form six possible combinations (genotypes) consisting of AA, AO, BB, BO, AB, OO as in Table 14.5, where alleles A and B are dominant over allele O, while A and B are co-dominant, thus giving rise to four phenotypes (blood groups): A, B, AB, O. Let p, q and r be the probabilities for A, B and O, respectively, for a given population. We let $\boldsymbol{\eta} = (g_1(p), g_2(r))^T$, with the logit link for g_j being the default. The 4-column input has columns corresponding to A-B-AB-O blood types, respectively.

Some of the family functions allow the modelling of an *inbreeding coefficient* f when the argument `inbreeding = TRUE`. Weir (1996) describes three types of inbreeding coefficients, using other common notation found in the literature of that area:

- f or F_{IS};
- F or F_{IT}: the *total inbreeding coefficient*;
- θ or F_{ST}: the *coancestry coefficient between individuals*.

These measures of relationship between pairs of alleles are related by $f = (F - \theta)/(1 - \theta)$ where random mating implies $F = \theta$ and $f = 0$. The quantity f is one minus the observed frequency of heterozygotes over that expected from Hardy-Weinberg equilibrium (HWE), and if `inbreeding = FALSE` then $f \equiv 0$ so that the parameter is not estimated—this corresponds to HWE.

To conduct a hypothesis test for $H_0 : f = 0$ deserves some comment. The null hypothesis is nonstandard in that the hypothesized value is at the boundary

Table 14.4 Genetic models based on the multinomial distribution currently implemented by **VGAM** and their unique parameters. All (independent) parameters are probabilities, whose default link is the logit. Argument `inbreeding` refers to whether the inbreeding coefficient f is to be estimated ($f = 0$ if `inbreeding = FALSE`).

Family	Parameters	Response order	Table
`ABO()`	p, q	A, B, AB, O	Table 14.5
`AB.Ab.aB.ab()`	p	AB, Ab, aB, ab	Table 14.6
`AA.Aa.aa(inbreeding = FALSE)`	p_A	AA, Ab, aa	Table 14.7
`AA.Aa.aa(inbreeding = TRUE)`	p_A, f	AA, Ab, aa	Table 14.7
`A1A2A3(inbreeding = FALSE)`	p_1, p_2	A_1A_1, A_1A_2, A_2A_2, A_1A_3, A_2A_3, A_3A_3	Table 14.8
`A1A2A3(inbreeding = TRUE)`	p_1, p_2, f	G_1G_1, G_1G_2, G_1G_3, G_2G_2, G_2G_3, G_3G_3	Table 14.9
`MNSs()`	m_S, m_s, n_S	MS, Ms, MNS, MNs, NS, Ns	Table 14.10

of its parameter space. While χ^2 tests and exact tests are common, a standard likelihood ratio test (LRT) is *not* applicable because it is based on the assumption that $\widehat{\boldsymbol{\theta}}$ is an interior point of the parameter space (Sect. A.1.2.2). Consequently, the standard asymptotic results for MLEs and LRTs do not hold. Self and Liang (1987) examined this problem and gave examples where the limiting distributions of LRT statistics are mixtures of χ^2 distributions, e.g., 50 % χ_0^2 and 50 % χ_1^2. Here are two other examples of hypothesis testing of parameters at a boundary of its parameter space:

1. Testing a Poisson versus negative binomial distribution (NBD). From Sects. 1.2.2 and 11.3, an NBD with $k \to \infty$ approaches a Poisson distribution. Hilbe (2011, Sect.7.4.2) conducts boundary LRT on an NB-2 example.
2. Testing for no zero-inflation in a zero-inflated Poisson (ZIP) distribution. From Table 17.6, testing $H_0 : \phi = 0$ in a `zipoisson()` model is a test for no zero-inflation, i.e., an ordinary Poisson distribution. It may be seen that this testing of a boundary value is the same for the other zero-inflated distributions in that table. Likewise, for the `zipoissonff()` parameterization, testing no zero-inflation $H_0 : \phi^* = 1$ occurs at a boundary. For the ZIP, it is common to use a Vuong test (Vuong, 1989).

14.5.1 The Dirichlet Distribution

We now mention the Dirichlet distribution, a natural extension of the beta distribution, because it has been used in genetics. Suppose $\boldsymbol{Y} = (Y_1, \ldots, Y_M)^T$. We say the random vector $\boldsymbol{Y}$ has a Dirichlet distribution if $(Y_1, \ldots, Y_M)^T$ has density

$$\frac{\Gamma(\alpha_+)}{\prod\limits_{j=1}^{M} \Gamma(\alpha_j)} \prod_{j=1}^{M} y_j^{\alpha_j - 1} \tag{14.21}$$

Table 14.5 Probability table for the ABO blood group system. The genotype probabilities are derived under the assumption that Hardy-Weinberg equilibrium holds. Note: $r = 1 - p - q$. Source: Elandt-Johnson (1971, Table 14.1).

Phenotype (blood group)	A		B		AB	OO
Genotype	AA	AO	BB	BO	AB	OO
Probability	p^2	$2pr$	q^2	$2qr$	$2pq$	r^2

Table 14.6 Two-locus counts for F_2 population for linked loci. Allele 'A' is assumed dominant over 'a', and allele 'B' is assumed dominant over 'b'. Source: Weir (1996, Table 7.4).

Phenotypes	Probability	Genotypes	Probability
AB	$(2 + p^2)/4$	AABB	$p^2/4$
		AABb	$p(1 - p)/2$
		AaBB	$p(1 - p)/2$
		AaBb	$(1 - 2p(1 - p))/2$
Ab	$(1 - p^2)/4$	AAbb	$(1 - p)^2/4$
		Aabb	$p(1 - p)/2$
aB	$(1 - p^2)/4$	aaBB	$(1 - p)^2/4$
		aaBb	$p(1 - p)/2$
ab	$p^2/4$	aabb	$p^2/4$

Table 14.7 Probability table for the AA-Aa-aa genotypes, with ($f = 0$) and without the assumption of Hardy-Weinberg equilibrium. Source: Weir (1996, pp.56–58).

Genotype	AA	Aa	aa
Without inbreeding	p_A^2	$2p_A(1 - p_A)$	$(1 - p_A)^2$
With inbreeding	$p_A^2 + p_A(1 - p_A)f$	$2p_A(1 - p_A)(1 - f)$	$(1 - p_A)^2 + p_A(1 - p_A)f$

Table 14.8 Probability table and frequencies y_j for 3 alleles, which give rise to `A1A2A3(inbreeding = FALSE)`. This table corresponds to Table 14.9 with $f = 0$. See also Weir (1996, pp.61–3).

A_1A_1	A_1A_2	A_1A_3	A_2A_2	A_2A_3	A_3A_3
p_1^2	$2p_1p_2$	$2p_1(1 - p_1 - p_2)$	p_2^2	$2p_2(1 - p_1 - p_2)$	$(1 - p_1 - p_2)^2$

Table 14.9 Probability table for the `A1A2A3(inbreeding = TRUE)` family, and Brazilian genotypes data at the Haptoglobin locus (Lange, 2002, Table 3.2). The probability $p_3 = 1 - p_1 - p_2$.

Genotype	Genotype probability	Observed number
A_1/A_1	$f p_1 + (1 - f)p_1^2$	108
A_1/A_2	$2(1 - f)p_1 p_2$	196
A_1/A_3	$2(1 - f)p_1 p_3$	429
A_2/A_2	$f p_2 + (1 - f)p_2^2$	143
A_2/A_3	$2(1 - f)p_2 p_3$	513
A_3/A_3	$f p_3 + (1 - f)p_3^2$	559

where $\alpha_+ = \alpha_1 + \cdots + \alpha_M$, $\alpha_j > 0$, and the density is defined on the unit simplex

$$\Delta_M = \left\{ (y_1, \ldots, y_M)^T : y_1 > 0, \ldots, y_M > 0, \ \sum_{j=1}^{M} y_j = 1 \right\}.$$

Table 14.10 Probability table for the combinations of MNSs blood group system. Two probabilities are $p_{MS} = m_S^2 + 2m_S m_s$ and $p_{MNS} = 2(m_S m_s + m_s n_S + m_S n_s)$. Source: Elandt-Johnson (1971, Table 14.3).

Phenotype	MS	Ms	MNS	MNs	NS	Ns
Genotype	$L^{MS}L^{MS}$, $L^{MS}L^{Ms}$	$L^{Ms}L^{Ms}$	$L^{MS}L^{NS}$, $L^{Ms}L^{NS}$, $L^{MS}L^{Ns}$	$L^{Ms}L^{Ns}$	$L^{NS}L^{NS}$, $L^{NS}L^{Ns}$	$L^{Ns}L^{Ns}$
Probability	p_{MS}	m_s^2	p_{MNS}	$2m_s n_s$	$n_S^2 + 2n_S n_s$	n_s^2

The means are $E(Y_j) = \alpha_j / \alpha_+$. The family function `dirichlet()` estimates the α_j from an M-column response. The function `rdiric()` generates Dirichlet random vectors based on the property that $Y_i = G_i / \sum_{j=1}^{M} G_j$ where the G_j are independent gamma random variates of unit scale. This ensures that $\sum_{j=1}^{M} Y_j = 1$ and $Y_j \geq 0$.

14.5.2 The Dirichlet-Multinomial Distribution

Its density is given by $P(Y_1 = y_1, \ldots, Y_M = y_M) =$

$$\binom{2y_*}{y_1 \ y_2 \cdots y_M} \frac{\Gamma(\alpha_+)}{\Gamma(2y_* + \alpha_+)} \prod_{j=1}^{M} \frac{\Gamma(y_j + \alpha_j)}{\Gamma(\alpha_j)} \tag{14.22}$$

where $2y_* = \sum_{j=1}^{M} y_j$ and $\alpha_j > 0$. The motivation for this distribution is a Bayesian one—see Weir (1996) and Lange (2002). The posterior mean is $E(Y_j) = y_j + \alpha_j / (2y_* + \alpha_+)$. Note that y_j must be a non-negative integer.

The performance of the family function `dirmultinomial()` is found to deteriorate with large y_* (e.g., $y_* > 10^4$; Yu and Shaw, 2014). Such large values of y_* are motivated by problems found in contemporary high-throughput sequencing data sets in bioinformatics.

Bibliographic Notes

Some general references for categorical data analysis include Agresti (2010), Agresti (2013), Lloyd (1999), McCullagh and Nelder (1989), Powers and Xie (2008), Simonoff (2003), and Tutz (2012). An overview of models for ordinal responses is Liu and Agresti (2005), and Aitkin et al. (2009) gives some coverage on some of the topics of this chapter. Greene and Hensher (2010) has an emphasis towards choice modelling; other references are Greene (2012) and Hensher et al. (2014). Yee (2010a) complements this chapter. Additionally, Hilbe (2009), Smithson and Merkle (2013), Kateri (2014), Agresti (2015) and Bilder and Loughin (2015) are applied books useful for the practitioner.

Most of the genetic models of Sect. 14.5 are based on Elandt-Johnson (1971), Weir (1996), Lange (2002). The book van den Boogaart and Tolosana-Delgado (2013) gives an applied introduction to compositional data analysis based on R.

Exercises

Ex. 14.1. Marginal Effects

(a) Derive (14.6) for the multinomial logit model. Hint: consider $\partial \log p_j(\boldsymbol{x}_i)/\partial \boldsymbol{x}_i$.
(b) Derive (14.19) for the (nonparallel) cumulative logit model.
(c) Derive $\partial \log p_j(\boldsymbol{x}_i)/\partial \boldsymbol{x}_i$ for the `acat()` model.
(d) Derive $\partial \log p_j(\boldsymbol{x}_i)/\partial \boldsymbol{x}_i$ for the `cratio()` model.
(e) Derive $\partial \log p_j(\boldsymbol{x}_i)/\partial \boldsymbol{x}_i$ for the `sratio()` model.

Ex. 14.2. Prove (14.10).

Ex. 14.3. Wine Data
For the `wine` data, fit a partial proportional-odds model with separate slopes for the temperature variable. Show that the solution is found on the boundary of the parameter space.   [Kosmidis (2014b)]

Ex. 14.4. Pearson Chi-Squared Statistic
For the $M = 2$ case of the multinomial logit model, show that

$$(\boldsymbol{y}_i - \boldsymbol{p}_i)^T \boldsymbol{\Sigma}_i^{-1}(\boldsymbol{y}_i - \boldsymbol{p}_i) \; = \; \sum_{j=1}^{M+1} N_i \frac{(y_{ij} - p_{ij})^2}{p_{ij}},$$

where $\boldsymbol{\Sigma}_i = \mathrm{Var}(\boldsymbol{y}_i)$, and N_i is the number of counts, with $\boldsymbol{y}_i$ being a vector of sample proportions (this is Pearson's χ^2 statistic).   [Tutz (2012)]

Ex. 14.5. Show that the multinomial distribution belongs to the multivariate exponential family (3.1).

Ex. 14.6. Continuation-Ratio Models Fitted to `pneumo`
Fit the following backward continuation-ratio models to the `pneumo` data with $\boldsymbol{x} = (1, \mathtt{let})^T$:
$$\mathrm{logit}\, P(Y < j + 1 | Y \le j + 1) \; = \; \eta_j, \quad j = 1, 2,$$

where the parallelism (i) applies to all $\boldsymbol{x}$, (ii) applies to `let` only, (iii) applies to no covariates at all. Use both logistic regression and `cratio()` for (i)–(iii).

Ex. 14.7. Consider the following output (the models are for illustration only).

(a)
```
> pneumo <- transform(pneumo, let = log(exposure.time))
> coef(vglm(cbind(normal, mild, severe) ~ let, multinomial, data = pneumo),
      matrix = TRUE)

              log(mu[,1]/mu[,3]) log(mu[,2]/mu[,3])
(Intercept)               11.975             3.0391
let                       -3.067            -0.9021
```

Without running the following, what output is obtained?

```
coef(vglm(cbind(normal, mild, severe) ~ let, multinomial(refLevel = 2), pneumo),
     matrix = TRUE)
```

(b) Without running the following, what output is obtained?

```
coef(vglm(cbind(severe, mild, normal) ~ let, multinomial, data = pneumo),
     matrix = TRUE)
```

(c) Given

```
> coef(vglm(cbind(normal, mild, severe) ~ let, cumulative, data = pneumo),
       matrix = TRUE)

               logit(P[Y<=1]) logit(P[Y<=2])
  (Intercept)          9.593         11.105
  let                 -2.571         -2.744
```

without running the following, what output is obtained?

```
coef(vglm(cbind(normal, mild, severe) ~ let, cumulative(reverse = TRUE), pneumo),
     matrix = TRUE)
```

(d) Based on the model in (a), classify a coalminer who has an exposure time of 50 years with respect to pneumoconiosis.

Ex. 14.8. Poisson Trick

Consider fitting an MLM, based on estimating the Poisson regressions (14.12).

(a) Explain why the likelihood function L^*, say, is

$$L^* \propto \prod_{i=1}^{n} \prod_{j=1}^{J} \mu_{ij}^{y_{ij}} \exp\{-\mu_{ij}\}.$$

(b) Set the derivative of $\ell^* \equiv \log L^*$ (with respect to α_i) to zero, and then substitute $\widehat{\alpha}_i$ into the log-likelihood. Show that this is, up to an additive constant, the MLM log-likelihood.

Ex. 14.9. Poisson Trick for pneumo

(a) Show that the standard errors of the relevant coefficients of **pois.pneumo** and **mlm.pneumo** match (Sect. 14.3.1).
(b) Using **glm()** with **family = poisson**, fit a multinomial logit model to the **pneumo** data frame that is linear with respect to **log(exposure.time)** using the Poisson trick. Use the first level as the baseline group. Hint: create a 'long' version of the data frame **pneumo**, called **pneumo.long** say, using **reshape()**. Note: the first few rows of **pneumo.long** should look something like

```
> head(pneumo.long, 13)

           exposure.time  let i.factor     x3 y.factor obs.freq id
  1.normal           5.8 1.758        1 0.2655   normal       98  1
  2.normal          15.0 2.708        2 0.3721   normal       51  2
  3.normal          21.5 3.068        3 0.5729   normal       34  3
  4.normal          27.5 3.314        4 0.9082   normal       35  4
```

```
5.normal          33.5 3.512      5 0.2017    normal       32  5
6.normal          39.5 3.676      6 0.8984    normal       23  6
7.normal          46.0 3.829      7 0.9447    normal       12  7
8.normal          51.5 3.942      8 0.6608    normal        4  8
1.mild             5.8 1.758      1 0.2655      mild        0  1
2.mild            15.0 2.708      2 0.3721      mild        2  2
3.mild            21.5 3.068      3 0.5729      mild        6  3
4.mild            27.5 3.314      4 0.9082      mild        5  4
5.mild            33.5 3.512      5 0.2017      mild       10  5
```

Ex. 14.10. For `sratio(reverse = TRUE, parallel = FALSE)` applied to an ordinal response Y with levels $\{1, \ldots, M+1\}$, show that this fits $\eta_j = \log\{P(Y = j+1)/P(Y \leq j)\}$ for $j = 1, \ldots, M$. Write down η_j in terms of the coefficients of an intercept plus x_2, as explanatory.

Ex. 14.11. Show that Fisher scoring is equivalent to the Newton-Raphson algorithm for the multinomial logit model (1.28).

Ex. 14.12. † `rcumulative()`
Based on Sect. 14.4.1.1, write a simple R function `rcumulative()` that generates random variates from an intercept-only cumulative link model. Allow for at least the `"logit"`, `"probit"` and `"cauchit"` links. The associate distributions should be standardized, i.e., have location parameter 0 and unit scale parameter. It should have an argument `cutpoints` that receives the $\beta_{(j)1}$.

Ex. 14.13. **Derivatives of the Non-proportional-Odds Model**
For the non-proportional-odds model (14.14) applied to data $(\boldsymbol{x}_i, \boldsymbol{y}_i, N_i)$, $i = 1, \ldots, n$ with $N_i = N_i\, \boldsymbol{y}_i^T \mathbf{1}_{M+1} = 1$ being the prior weights, show that the first and expected second derivatives are

$$\left(\frac{\partial \ell_i}{\partial \eta_i}\right)_j = N_i\, \gamma_j(\boldsymbol{x}_i)[1 - \gamma_j(\boldsymbol{x}_i)]\left\{\frac{y_{ij}}{p_j(\boldsymbol{x}_i)} - \frac{y_{i,j+1}}{p_{j+1}(\boldsymbol{x}_i)}\right\},$$

$$\left(-E\left[\frac{\partial^2 \ell_i}{\partial \eta_i\, \partial \eta_i^T}\right]\right)_{jj} = N_i\, \gamma_j^2(\boldsymbol{x}_i)[1 - \gamma_j(\boldsymbol{x}_i)]^2\left\{\frac{1}{p_j(\boldsymbol{x}_i)} + \frac{1}{p_{j+1}(\boldsymbol{x}_i)}\right\},$$

$$\left(-E\left[\frac{\partial^2 \ell_i}{\partial \eta_i\, \partial \eta_i^T}\right]\right)_{t-1,t} = -N_i\, \frac{\gamma_{t-1}(\boldsymbol{x}_i)\,[1 - \gamma_{t-1}(\boldsymbol{x}_i)]\,\gamma_t(\boldsymbol{x}_i)\,[1 - \gamma_t(\boldsymbol{x}_i)]}{p_t(\boldsymbol{x}_i)},$$

and the other elements are zero, i.e., the EIM is tridiagonal. Here, $j = 1, \ldots, M$ and $t = 2, \ldots, M$.

Ex. 14.14. **Relationship with `binomialff()`**

(a) Suppose `y` is a vector of 0s and 1s, and one wants to perform a simple logistic regression $\eta = \text{logit}\, P(Y = 1|\boldsymbol{x})$. Determine which of the following will do this. If it doesn't, provide a simple adaptation to make it work. If it does, briefly explain why.

    (i)   `glm(y ~ ..., binomial, ...)`
    (ii)   `vglm(y ~ ..., binomialff, ...)`
    (iii) `vglm(cbind(1 - y, y) ~ ..., binomialff, ...)`
    (iv)   `vglm(1 - y ~ ..., multinomial, ...)`
    (v)   `vglm(y ~ ..., multinomial(refLevel = 1), ...)`

   (vi)   `vglm(cbind(1 - y, y) ~ ..., multinomial, ...)`
   (vii)   `vglm(cbind(y, 1 - y) ~ ..., multinomial, ...)`
   (viii)   `vglm(y ~ ..., cumulative(reverse = TRUE), ...)`

(b) Work out further calls of the above using `cratio()`, with and/or without `reverse = TRUE`.
(c) Work out further calls of the above using `sratio()`, with and/or without `reverse = TRUE`.
(d) Work out further calls of the above using `acat()`, with and/or without `reverse = TRUE`.

Ex. 14.15.   Derivatives for the `AA.Aa.aa()` Genotypes

(a) Assuming HWE, for this 1-parameter model given in Table 14.7, write down the log-likelihood and its first derivative with respect to p_A.
(b) Derive its second derivative and EIM.

Ex. 14.16.   Derivatives for the `AB.Ab.aB.ab()` Phenotypes

(a) For this 1-parameter model given in Table 14.6, write down the log-likelihood and its first derivative.
(b) Derive its second derivative and EIM.

Ex. 14.17.   Brazilian Genotypes at the Haptoglobin Locus
Table 14.9 gives data cited by Yasuda (1968) from 1948 people from north east Brazil.   [Lange (2002)]

(a) Apply the `A1A2A3()` family function to these data, with both `inbreeding = TRUE` and `inbreeding = FALSE`.
(b) Using a chi-squared test or otherwise, test for the goodness of fit.

Ex. 14.18.   Derivatives for the `ABO()` Blood Group

(a) For this 2-parameter model given in Table 14.5, write down the log-likelihood and its first derivatives.
(b) Derive its second derivatives and EIM.

Ex. 14.19.   Dirichlet Distribution
Use `dirichlet()` to fit a Dirichlet distribution to the data frame `ducklings`. Here, the response are the relative frequencies of serum proteins in white Pekin ducklings as determined by electrophoresis: p_1 = pre-albumin, p_2 = albumin, p_3 = globulins (proportions). Are the second and third shape parameters equal?

> *... Second, even researchers who do not rely on default options practically never attempt to verify the solution. One can only wonder how many incorrect nonlinear results have been published.*
> —Altman et al. (2004)

Chapter 15
Quantile and Expectile Regression

The percentile curves are usually computed one level at a time. Associated with great flexibility is the embarrassing phenomenon of quantile crossing.
—He (1997)

15.1 Introduction

A major deficiency of much of statistical modelling involving the regression function $E(Y|\boldsymbol{x})$ is the resultant information loss. In contrast, quantile regression (QR) allows for a complete picture by considering the (entire) conditional distribution of Y given $\boldsymbol{x}$. As such, there is no information loss because $F(y|\boldsymbol{x})$ contains all the information about the random variable Y at $\boldsymbol{x}$, whereas $E(Y|\boldsymbol{x})$ is but the first moment. This may be one reason why QR has become increasingly popular over the last decade or so.

This chapter mainly dwells on three topics. The first is a popular QR technique called the LMS method, which is amenable to IRLS (Sect. 15.2). Secondly, starting with what is called here the classical method, QR based on the *asymmetric Laplace distribution* (ALD) is described (Sect. 15.3). Note however that IRLS is not well-suited for solving for the location parameters of an ALD, nevertheless sometimes a reasonable solution may be obtained. Currently this work can be viewed as tentative and experimental. It will be seen that (i) a parallelism assumption ($\mathbf{H}_k = \mathbf{1}_M$) is a natural solution to the quantile-crossing problem; (ii) The log link applied to the ALD location parameter means that quantiles may be positive; (iii) A method called the 'onion' method may be used to perform QR, like estimating the layers of an onion—and it provides a second natural solution to the quantile-crossing problem. Thirdly, quantities somewhat similar to quantiles called expectiles are described (Sect. 15.4). As usual, some **VGAM** family functions are used for illustrative purposes—they are summarized in Table 15.2.

© Thomas Yee 2015
T.W. Yee, *Vector Generalized Linear and Additive Models*,
Springer Series in Statistics, DOI 10.1007/978-1-4939-2818-7_15

The subject of QR has received considerable research attention over the last decade, and there are now many application areas and quite a few proposed methods. Applications include medical studies (e.g., obesity and height versus age), ecology, economics (e.g., Fitzenberger et al., 2002), education and climatology (e.g., Fig. 15.13a of some Melbourne temperature data exhibits bimodal behaviour) to name just a few.

15.1.1 Some Notation and Background

In this chapter we use the words *quantile*, *centile*, and *percentile* interchangeably, and note that, for example, a 0.5-quantile is equivalent to the 50-percentile, which is the median. VGAM family functions for QR use an argument `percentiles` which should be assigned values between 0 and 100. Related terms are *quartiles*, *quintiles* and *deciles*, which divide the distribution into 4, 5 and 10 equal parts, respectively. Table 15.1 summarizes the notation used in this chapter.

Suppose a real-valued random variable Y has CDF $F(y) = P(Y \leq y)$. Then the τth-quantile of Y may be defined to be $Q_Y(\tau) = F_Y^{-1}(y) =$

$$Q(\tau) \quad = \quad \inf\{y : \tau \leq F(y)\}, \tag{15.1}$$

where $0 < \tau < 1$. Thus, for continuous F, a proportion τ lies below $Q(\tau)$ and $1 - \tau$ lies above $Q(\tau)$. In R, q-type functions satisfy (15.1) for discrete distributions as well as continuous ones. Like the CDF, the quantile function $Q(\tau)$ *completely* describes the distribution of Y; there is no information loss. This contrasts sharply with LMs and GLMs where only the mean of Y is modelled.

Quantiles are invariant to monotone transformations: if ψ is a nondecreasing function on $\mathbb{R}$, then

$$Q_{\psi(Y)}(\tau) \quad = \quad \psi(Q_Y(\tau)). \tag{15.2}$$

This means the quantiles of the transformed variable $\psi(Y)$ are the transformed quantiles of the original variable Y.

15.2 LMS-Type Methods

LMS-type quantile regression methods are based on three parameters called λ, μ and σ from which the method derives its name (i.e., the first letters of the Roman transcriptions of these Greek characters). For scatter plot data (x_i, y_i), $i = 1, \ldots, n$, with $y_i > 0$, the underlying idea is that a 3-parameter Box-Cox power transformation of the y_i, given x_i, has some parametric distribution, whose quantiles can be extracted. Then an inverse Box-Cox transformation of these quantiles will result in estimated quantiles on the original scale. In the VGLM/VGAM framework, potentially all 3 parameters are allowed to be smooth functions of x—and penalized MLE is used.

Table 15.1 Notation and nomenclature used in this chapter.

Notation and nomenclature	Comments
Y	Response. Has mean μ, CDF $F(y)$, PDF $f(y)$
$Q_Y(\tau) = \tau$-quantile of Y	$0 < \tau < 1$,
$\xi(\tau) = \xi_\tau = \tau$-quantile	Koenker and Bassett (1978), $\xi(\frac{1}{2}) = $ median
$\mu(\omega) = \mu_\omega = \omega$-expectile	$0 < \omega < 1$, $\mu(\frac{1}{2}) = \mu$, Newey and Powell (1987)
$\widehat{\xi}(\tau), \widehat{\mu}(\omega)$	Estimated quantiles and expectiles
Centile	Same as quantile and percentile here
Regression quantile	Koenker and Bassett (1978)
Regression expectile	Newey and Powell (1987)
Regression percentile	All forms of asymmetric fitting (Efron, 1992)
$\rho_\tau(u) = u \cdot (\tau - I(u < 0))$	Check function corresponding to $\xi(\tau)$, (15.6), Fig. 15.6a
$\rho_\omega^{[2]}(u) = u^2 \cdot \lvert \omega - I(u < 0) \rvert$	Check function corresponding to $\mu(\omega)$, (15.22), Fig. 15.6b
$\boldsymbol{\tau} = (\tau_1, \tau_2, \ldots, \tau_L)^T$	Vector of τ_j values, used for simultaneous estimation, Sect. 15.3.2.1

When LMS-Box-Cox QR is used, it is almost always meant that the transformation is made to a standard normal distribution. This is described in Sect. 15.2.1. Two lesser-known and experimental variants are sketched in Sect. 15.2.2.

15.2.1 The Box-Cox-Normal Distribution Version

The Box-Cox transformation of a positive Y is to $Z \sim N(0, 1)$ where

$$
Z = \begin{cases} \dfrac{\left(\dfrac{Y}{\mu(x)}\right)^{\lambda(x)} - 1}{\sigma(x)\,\lambda(x)}, & \lambda(x) \neq 0; \\[2em] \dfrac{1}{\sigma(x)}\,\log\left(\dfrac{Y}{\mu(x)}\right), & \lambda(x) = 0. \end{cases} \tag{15.3}
$$

The transformation is plotted in Fig. 15.1a for $\mu = \sigma = 1$. Approximately, one might interpret μ somewhat like a running median or mode of Y, since $Y/\mu \approx 1$ which gets mapped to the origin of the $N(0, 1)$ density. Of course, σ is the scale parameter and λ can be considered a shape parameter. For right-skewed data one would expect $\lambda < 1$ in order to bring its heavy tail in. The second equation of (15.3) handles the singularity at $\lambda = 0$. The parameter σ must be positive, therefore **VGAM** chooses $\boldsymbol{\eta}(x) = (\lambda(x), \mu(x), \log(\sigma(x)))^T$ as the default. Given $\widehat{\boldsymbol{\eta}}$, the α percentile (e.g., $\alpha = 50$ for median) can be estimated by inverting the Box-Cox power transformation to yield

$$
\widehat{\mu}(x) \left[1 + \widehat{\lambda}(x)\,\widehat{\sigma}(x)\,\Phi^{-1}(\alpha/100) \right]^{1/\widehat{\lambda}(x)}. \tag{15.4}
$$

One advantage of the LMS method is that it does not suffer from the quantile-crossing problem. This "disturbing" and "serious embarrassment" can occur for some QR methods (see, e.g., He (1997), Koenker (2005, Sect.2.5)), for example, where a point (x_0, y_0) may be classified as below the 20th but above the 30th

Table 15.2 Main functions for quantile and expectile regression in this chapter. The upper table are **VGAM** family functions; the lower table are generic functions. A "$\mathcal{L}^*$" denotes an asymmetric Laplace distribution (ALD).

lms.bcn()	Box-Cox transformation to normality (Sect. 15.2)
lms.bcg()	Box-Cox transformation to gamma distribution
lms.yjn()	Yeo-Johnson transformation to normality
laplace(dpqr)	Laplace distribution (Sect. 15.3)
alaplace1(dpqr)	$\mathcal{L}^*(\xi)$ with σ and κ (or τ) specified, Eq. (15.12)
alaplace2(dpqr)	$\mathcal{L}^*(\xi, \sigma)$ with κ (or τ) specified, Eq. (15.12)
alaplace3(dpqr)	$\mathcal{L}^*(\xi, \sigma, \kappa)$, Eq. (15.12)
sc.studentt2(dpqr)	Scaled Student t_2 distribution, $\sqrt{2}\,T_2$, Ex. 15.14
amlnormal([dpqr])	Asymmetric maximum likelihood—for normal (asymmetric least squares or ALS; Sect. 15.4.2)
amlbinomial([dpqr])	Asymmetric maximum likelihood—for binomial
amlpoisson([dpqr])	Asymmetric maximum likelihood—for Poisson
amlexponential([dpqr])	Asymmetric maximum likelihood—for exponential
deplot()	Density plot; of the fitted probability density functions (Sect. 15.2)
qtplot()	Quantile plots; of the fitted quantiles
cdf()	Cumulative distribution function $\widehat{F}(y_i)$
fitted()	Estimated quantiles for y_i

percentile! The methods of Sects. 15.3 and 15.4 suffer from this problem, however a parallelism constraint is a natural solution of the VGLM/VGAM framework. One can also use an onion method in conjunction with a log link.

It is worth noting the residuals. In the LMS method, the conditional distribution of the Z-scores defined in (15.3) was assumed to be standard normal. One can naturally define the residuals as the Z-scores themselves. Thus, fitting the λ, μ and σ curves with appropriate degrees of freedom, we can obtain the raw residuals as

$$z_i = (\sigma(x_i)\,\lambda(x_i))^{-1} \cdot \left\{ \left(\frac{y_i}{\mu(x_i)} \right)^{\lambda(x_i)} - 1 \right\}. \tag{15.5}$$

The availability of this definition of residual can be thought of as an advantage of the LMS method. In parametric or nonparametric QR methods it is not always possible to define residuals, particularly when quantiles are estimated separately. From a practical point of view, the residuals should be checked for standard normality, e.g., by a QQ-plot or overlaying the probability density function of a standard normal onto a histogram. Implementing these types of residuals in R is left as an exercise (Ex. 15.2).

One disadvantage of the LMS-BCN method is that it only handles unimodal Y at a given x, since the Box-Cox transformation to normality is monotonic. For example, the Melbourne temperatures data (Fig. 15.13a) is therefore not modelled well by this method. However, probably its greatest disadvantage is that the Box-Cox transformation to normality cannot be justified convincingly in many applications.

Here are some practical suggestions if this method is adopted.

1. Of the three functions, it is a good idea to allocate relatively more degrees of freedom to $\mu(x_2)$, because the two functions $\lambda(x_2)$ and $\sigma(x_2)$ usually vary as a function of x_2 more smoothly. These preferences can be easily be chosen in **VGAM**, e.g., by using

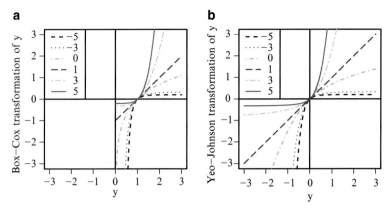

Fig. 15.1 (a) The Box-Cox transformation $(y^\lambda - 1)/\lambda$ for various values of λ. (b) The Yeo-Johnson transformation $\psi(\lambda, y)$.

```
vglm(y ~ bs(x2, df = c(2, 4, 2)), lms.bcn(zero = NULL), data = ldata,
     trace = TRUE)
vgam(y ~ s(x2, df = c(2, 4, 2)), lms.bcn(zero = NULL), data = ldata,
     trace = TRUE)
```

Assigning too much flexibility to the parameters often leads to numerical problems. Consequently the defaults for λ and σ are chosen to be intercept-only:

```
> args(lms.bcn)

function (percentiles = c(25, 50, 75), zero = c(1, 3),
    llambda = "identitylink", lmu = "identitylink", lsigma = "loge",
    idf.mu = 4, idf.sigma = 2, ilambda = 1, isigma = NULL, tol0 = 0.001)
NULL
```

and this more simple model is less likely to give convergence problems, provided that the amount of flexibility of $\mu(x)$ is chosen suitably.

2. The **VGAM** solution involves maximizing an *approximate* likelihood and using *approximate* derivatives. In fact, if too many iterations are performed, the solution may diverge and fail! For this reason it is a good idea to set **trace = TRUE** to monitor convergence. Using **vgam()**, sometimes setting, e.g., **maxit = 5**, is reasonable if the fit starts diverging after the 5th iteration. Also, successful convergence is sensitive to the choice of initial values.

The LMS QR method need not be restricted to a single covariate x_2. One can adjust for other explanatory variables such as gender and ethnicity. Section 15.2.4.1 gives such an example.

15.2.2 Other Versions

Two other LMS-type quantile regression methods are also available—one from a Box-Cox transformation to a gamma distribution, and one from a Yeo and Johnson (2000) transformation to a standard normal distribution. The latter transformation can be considered as a generalization of the Box-Cox transformation to the whole real line (Fig. 15.1). These methods are less likely to be used than the LMS-BCN, therefore the reader is referred to Lopatatzidis and Green (1998) and Yee (2004b).

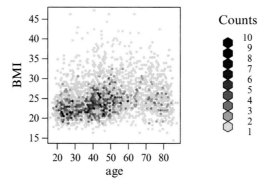

Fig. 15.2   Hexagonal binning plot of BMI versus age for European-type women in xs.nz.

15.2.3 Example

We fit an LMS-BCN quantile regression to a subset of 2614 European-type women from the xs.nz data frame. The response is body mass index (BMI; weight/height^2 in kg/m^2), a measure of obesity, regressed against age. As the data set is not small we use hexbin to obtain a hexagonal binning plot, as an alternative of the usual scatter plot. It appears in Fig. 15.2. The main features are that (i) BMI is increasing until about 60 years old followed by a decline, (ii) BMI is positively skewed at most ages, (iii) most of the data are of women aged about 55 or younger and is most dense for people in their 20s and 40s. Additionally, there appears an almost-linear trend from 20 to 50 years old. A fair proportion of women aged 40–50 are overweight because their BMI > 25. There appears to be two clusters linked together around age 30 plus another small cluster of women in their 70s.

Now to fit an LMS-Box-Cox-normal model let's try the following.

```
> women.eth0 <- subset(xs.nz, sex == "F" & ethnicity == "European")
> women.eth0 <- transform(women.eth0, BMI = weight/height^2)
> women.eth0 <- subset(women.eth0, !is.na(age) & !is.na(BMI))
> w0.LMS0 <- vgam(BMI ~ s(age, df = c(2, 4, 2)), trace = FALSE,
                  lms.bcn(zero = NULL), data = women.eth0)
> plot(w0.LMS0, se = TRUE, scol = "blue", rug = TRUE, cex.lab = 1.7)
```

The rationale for this first choice of df is to allow a little flexibility to λ and $\log \sigma$ so as to allow its true functional form to be seen. The component functions are plotted in Fig. 15.3. From this and summary(w0.LMS0), we model λ as a linear function of x:

```
> w0.LMS <- vgam(BMI ~ s(age, df = c(1, 4, 1.5)),
                 lms.bcn(zero = NULL, percentile = c(5, 25, 50, 75, 95)),
                 data = women.eth0, trace = FALSE)  # trace = TRUE is a good idea
```

is a suggested improvement. The corresponding quantile plot from

```
> qtplot(w0.LMS, pcol = "darkorange", lcol = "blue", tcol = "blue",
         ylab = "BMI", main = "", xlim = c(15, 100))
```

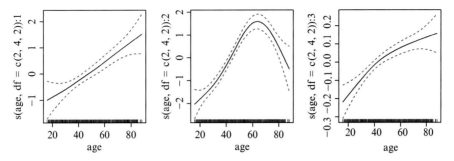

Fig. 15.3 Plot of the VGAM component functions of `w0.LMS0`, a `lms.bcn()` fit of `BMI` versus `age` for European-type women in the `xs.nz` data frame. These are (centred) $\widehat{\lambda}(x)$, $\widehat{\mu}(x)$, $\log \widehat{\sigma}(x)$.

is given in Fig. 15.4a. The three quartiles seem reasonable and appear to peak around the early 60s and exhibit right-skew. One explanation for the decline is due to overweight women dying prematurely. Or possibly they tended to leave the workforce while those who were not overweight had a greater chance of continuing on in the company.

To obtain a density plot with `deplot()` for $x = 20$-and 30-year-old women, say, try

```
> Ages <- c(20, 30); mycol <- c("blue", "limegreen")
> deplot(w0.LMS, x0 = Ages[1], y = seq(15, 35, by = 0.1), col = mycol[1],
         xlab = "BMI") -> dw0.LMS
> deplot(w0.LMS, x0 = Ages[2], y = seq(15, 35, by = 0.1), col = mycol[2],
         add = TRUE)
> abline(v = c(18.5, 24.9), col = "orange", lty = 2)  # Approximate healthy range
> legend("right", lty = 1, col = mycol, legend = as.character(Ages))
```

This gives Fig. 15.4b. Clearly, the distribution of BMI is positively skewed for both age groups and there is a tendency for older women to put on more weight. While most women have a healthy BMI there are a lot more overweight people than underweight ones.

One can check empirically the sample proportions below each quantile curve by

```
> 100 * colMeans(depvar(w0.LMS, drop = TRUE) < fitted(w0.LMS))  # Sample proportions

      5%      25%      50%      75%      95%
 4.7819  24.5983  50.5356  75.3634  94.4912
```

There is good agreement here. To finish up for now, some selected quantiles of BMI for 40-year-old European-type women, say, can be obtained from

```
> predict(w0.LMS, data.frame(age = 40), type = "response")

      5%      25%      50%      75%      95%
 1  19.584  21.861  23.922  26.588  32.262
```

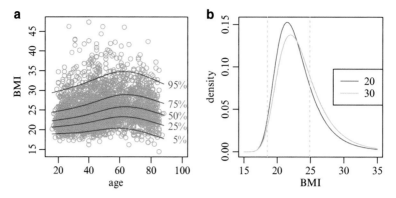

Fig. 15.4 **(a)** Quantile and **(b)** density plots of `w0.LMS`. In **(b)** the estimated densities are for 20 and 30 year olds, and the *vertical lines* denote an approximate healthy range.

15.2.4 More Advanced Usage of Some Methods Functions

With VGAM family functions for quantile regression the resulting fitted models can be plotted in more ways than just the generic function `plot()`. Here are some more details.

15.2.4.1 `qtplot()`

In theory the LMS QR method can operate with more than one 'primary' covariate such as age, e.g., adjust for other variables such as ethnicity and gender. To plot the results using `qtplot()`, however, is not easy, but possible by using its `newdata` argument. This allows prediction to occur for that data frame. Note that plotting the quantiles against the primary variable only makes sense if the non-primary variables have a common value.

Below is some code to illustrate this. We will use `women.eth02`, which consists of the women from `xs.nz` whose ethnicity traces to either Europe or the Pacific islands. That is, there is another variable which one adjusts for (the indicator variable is called `european`) and the 'primary' variable is `age`.

Figure 15.5 was obtained from running the simpler additive model

```
> fit3 <- vgam(BMI ~ s(age, df = 4) + european, data = women.eth02,
              lms.bcn(percentile = 50, zero = c(1, 3)), trace = FALSE)
> Age <- seq(18, 85, length = 100)   # Predict at these values
> half <- with(women.eth02, split(women.eth02, european))
> plot(BMI ~ age, women.eth02, type = "n")
>
> # First plot - for Pacific Islanders
> with(half[["0"]], points(jitter(age), BMI, pch = 15, cex = 0.7,
                    col = "green4"))
> Newdata <- data.frame(age = Age, european = 0)
> PI <- qtplot(fit3, newdata = Newdata, add = TRUE,
              lcol = "green4", tcol = "green4", lwd = 2, tadj = 0)
>
```

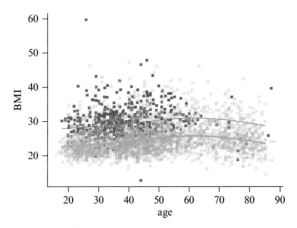

Fig. 15.5   Output of `qtplot()` when there are non-primary variables. The two groups are European and Pacific island women. The fitted 50-percentiles differ by a constant.

```
> # Second plot - for Europeans
> with(half[["1"]], points(jitter(age), BMI, pch = "o", cex = 0.7,
                            col = "orange"))
> Newdata <- data.frame(age = Age, european = 1)
> EU <- qtplot(fit3, newdata = Newdata, add = TRUE,
               lcol = "orange", tcol  = "blue", llwd = 3, tadj = 0)
```

Here, the two ethnicities' BMIs are assumed to only differ by some constant at any given age. Not surprising, Pacific Islander women have a more solid built than European-type women of the same age. In fact,

```
> range(PI@post$qtplot$fitted - EU@post$qtplot$fitted)

 [1] 5.2088 5.2088
```

shows that the median BMI of Pacific Island women is about 5.21 kg/m^2 higher than European women.

Here are some further notes about `qtplot()`.

1. When `newdata` is specified in `qtplot()` it is important to note that the function chooses the *first* term in the formula as the primary variable. If one had used

   ```
   fit <- vgam(BMI ~ european + s(age, df = 4), lms.bcn(zero = c(1, 3)), women.eth02)
   ```

   then `qtplot(fit, Newdata)` would attempt using `Newdata$european` as the primary variable and fail.
2. `qtplot()` uses the percentile values from the original model if the argument `percentiles` isn't specified.

15.2.4.2 `deplot()`

If `deplot()` is applied to an LMS-type QR model, then the `post` slot is assigned a list called `"deplot"` containing three components including the estimated densities. For example,

```
> names(dw0.LMS@post)

  [1] "cdf"     "deplot"

> names(dw0.LMS@post$deplot)

  [1] "newdata" "y"       "density"

> dw0.LMS@post$deplot$newdata

     age
  1   20
```

The components y and `density` are the x- and y-axes, respectively, of one of the curves in Fig. 15.4b.

15.2.4.3 cdf()

The `cdf()` methods function for an LMS-type QR model returns a vector with $\widehat{P}(Y \leq y_i | \boldsymbol{x}_i)$ for the ith observation, e.g.,

```
> head(cdf(w0.LMS))

         5        10        13        20        24        26
  0.011586  0.973394  0.015822  0.741580  0.054708  0.513065
```

This matches with

```
> head(fitted(w0.LMS), 3)

        5%      25%      50%      75%      95%
  5   18.960   20.738   22.379   24.570   29.627
  10  19.584   21.861   23.922   26.588   32.262
  13  19.852   22.276   24.453   27.239   33.029

> head(women.eth0[, c("age", "BMI")], 2)

      age    BMI
  5    18   18.053
  10   40   34.657
```

For example, a 40-year-old female European with a BMI of 34.66 corresponds to a cumulative probability of 0.973, i.e., as such she is at the top end of the BMI distribution of women her age.

15.3 Classical Quantile Regression by Scoring

In this section we explore how classical quantile regression might possibly be estimated by scoring an ALD. There are many types of ALDs, e.g., Kozubowski and Nadarajah (2010) review all known variations of the univariate Laplace distribution and identify over sixteen of them! Basically, the classical quantile regression approach of Koenker and Bassett (1978) estimates quantiles by linear programming techniques because both the objective function and constraints are linear functions of $\boldsymbol{x}$. However, it is known that the quantiles also coincide with the maximum likelihood solution of the location parameter in an ALD; see, e.g., Poiraud-Casanova and Thomas-Agnan (2000), Geraci and Bottai (2007)). This section is an attempt

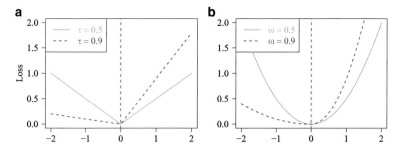

Fig. 15.6 Loss functions for: (a) quantile regression with $\tau = 0.5$ (L_1 regression) and $\tau = 0.9$; (b) expectile regression with $\omega = 0.5$ (least squares) and $\omega = 0.9$. Note: (a) is also known as the *asymmetric absolute loss function* or *pinball loss function*.

to direct the usual VGLM/VGAM Fisher scoring algorithm towards these models. The work described here is experimental and tosses about a few ideas, therefore the software is especially subject to change.

15.3.1 Classical Quantile Regression

The formulation of quantiles in terms of a regression problem is due to Koenker and Bassett (1978) and is referred here as the "classical" method. It is now described. As well as being a popular QR method it serves as the basis for expectile regression (Sect. 15.4). The classical QR method minimizes with respect to ξ the expectation of $\rho_\tau(Y - \xi)$, where

$$\rho_\tau(u) = u \cdot (\tau - I(u < 0)), \tag{15.6}$$

is known as a *check* function. This is plotted in Fig. 15.6a for two values of τ. The first value of $\tau = 0.5$ produces a symmetric loss function and corresponds to median estimation. The second case of $\tau = 0.9$ has slope τ on the RHS and slope $-(1 - \tau) = -0.1$ on the LHS; compared to the first case, this effectively shifts the solution to the RHS because negative deviations $Y - \xi$ are penalized less relative to positive ones. To minimize

$$E\left[\rho_\tau(Y - \xi)\right] = (\tau - 1) \int_{-\infty}^{\xi} (y - \xi)\, dF(y) + \tau \int_{\xi}^{\infty} (y - \xi)\, dF(y), \tag{15.7}$$

under regularity conditions, we set the derivative of (15.7) with respect to ξ equal to 0 to give

$$
\begin{aligned}
0 &= (\tau - 1)\left\{\left[(y - \xi) f(y)\right]_{y=-\infty}^{y=\xi} + \int_{-\infty}^{\xi} \frac{\partial}{\partial \xi}(y - \xi)\, dF(y)\right\} + \\
&\quad \tau \left\{\left[(y - \xi) f(y)\right]_{y=\xi}^{y=\infty} + \int_{\xi}^{\infty} \frac{\partial}{\partial \xi}(y - \xi)\, dF(y)\right\} \\
&= (\tau - 1)\left\{0 - F(\xi)\right\} + \tau \left\{0 - (1 - F(\xi))\right\} \\
&= F(\xi) - \tau,
\end{aligned}
\tag{15.8}
$$

so that $F(\widehat{\xi}) = \tau$ (cf. (15.1)). The solution $\widehat{\xi}(\tau)$ may not necessarily be unique.

The above dealt with population quantities. With random sample data $y_1, \ldots, y_n$ one can replace F by its empirical CDF

$$F_n(y) \;=\; \frac{1}{n} \sum_{i=1}^{n} I(y_i \leq y),$$

so that the τth sample quantile can be found by solving

$$\min_{\xi \in \mathbb{R}} \sum_{i=1}^{n} \rho_\tau(y_i - \xi) \;=\; \min_{\xi \in \mathbb{R}} \sum_{i=1}^{n} [(1 - \tau) \cdot (y_i - \xi)_- + \tau \cdot (y_i - \xi)_+]. \quad (15.9)$$

Equation (15.9) is more convenient than defining the τth sample quantiles in terms of the order statistics $y_{(1)}, \ldots, y_{(n)}$ because the optimization problem can be generalized to the situation where there are covariates $\boldsymbol{x}$. Specifically, if the regression model $y_i = \boldsymbol{x}_i^T \boldsymbol{\beta} + \varepsilon_i$, $\varepsilon_i \sim F$, is assumed, then the τth quantile is defined as any solution to the quantile regression minimization problem

$$\widehat{\boldsymbol{\beta}}(\tau) \;=\; \operatorname*{argmin}_{\boldsymbol{\beta} \in \mathbb{R}^p} \sum_{i=1}^{n} \rho_\tau\!\left(y_i - \boldsymbol{x}_i^T \boldsymbol{\beta}\right) \quad (15.10)$$

[cf. (15.9)]. This gives rise to the *linear conditional quantile function*

$$Q_Y(\tau | \boldsymbol{X} = \boldsymbol{x}) \;=\; \boldsymbol{x}^T \widehat{\boldsymbol{\beta}}(\tau). \quad (15.11)$$

Computationally, (15.9) is solved by linear programming because it involves minimizing the sum of *asymmetric least absolute deviations* (ALADs) which can be expressed as a linear combination of the $\boldsymbol{x}_i$ subject to linear constraints in $\boldsymbol{x}_i$. Details can be found in Koenker (2005, Chap.6).

As a whole, there are some compelling advantages which this classical methodology has over the LMS-type methods of Sect. 15.2. They include:

1. Rather than estimate these models by linear programming, one can (at least attempt to) use scoring which is the natural domain of the VGAM framework. Details are given in Sect. 15.3.
2. Inference is based on a well-established bed of theory, e.g., summarized in Koenker (2005, Chap.3). In contrast, inferences for LMS methods are ad hoc.
3. The idea can be generalized to other responses, e.g., counts and binary responses.
4. The idea can be generalized to expectile regression (Sect. 15.4) which lies more naturally in the domain of a scoring environment.

It is also argued that solving problem (15.7) is inadequate for four reasons. Firstly, it suffers from range restriction problems when applied to certain types of data, e.g., count data where the quantiles ought to be non-negative. Secondly, the rigid formulation (15.7) is solved by linear programming which requires special software and only solves that particular problem (Barrodale and Roberts, 1974). Thirdly, there is the ever-present quantile crossing problem. Fourthly, the VGLM/VGAM framework is simpler.

15.3.2 An Asymmetric Laplace Distribution

This section briefly outlines an ALD that can be used for quantile regression. A definitive treatment of ALDs is Kotz et al. (2001, Chap.3) where several ALD variants are described. They nominate a particular one as *the* ALD and label it as $\mathcal{L}^*(\xi, \sigma, \kappa)$. Its density is specified to be

$$
g(y; \xi, \sigma, \kappa) = \frac{\sqrt{2}}{\sigma} \frac{\kappa}{1 + \kappa^2}
\begin{cases}
\exp\left(-\frac{\sqrt{2}}{\sigma \kappa} |y - \xi|\right), & y \leq \xi, \\[2ex]
\exp\left(-\frac{\sqrt{2}\,\kappa}{\sigma} |y - \xi|\right), & y > \xi,
\end{cases}
\tag{15.12}
$$

for $\kappa > 0$. The special case $\kappa = 1$ corresponds to their (symmetric) Laplace distribution $\mathcal{L}(\xi, \sigma)$ which is also known as the *double exponential* distribution. An important property of $\mathcal{L}^*(\xi, \sigma, \kappa)$ is that

$$
P(Y \leq \xi) = \frac{\kappa^2}{1 + \kappa^2}.
\tag{15.13}
$$

VGAM's suite of `alaplace[123]()` functions are based on the variant $\mathcal{L}^*(\xi, \sigma, \kappa)$. The first derivatives of the log-likelihood function (3.7) with respect to its location and scale parameters are easily derived, in particular, setting the derivative of (15.12) with respect to ξ to zero implies

$$
\sum_{i:\, y_i \leq \xi} w_i \frac{\sqrt{2}}{\sigma_i \kappa_i} = \sum_{i:\, y_i > \xi} w_i \frac{\sqrt{2}\,\kappa_i}{\sigma_i}.
$$

Only if the σ_i are equal (e.g., σ is modelled as an intercept-only) would we expect the solution of `alaplace1()` and `alaplace2()` to be similar because the σ_i cancel out of both sides of the equation. However, it is a good idea to estimate ξ and σ because the estimated quantiles will be invariant to location and scale, i.e., a linear transformation of the response produces the same linear transformation of the fitted quantiles. This comes about because $(y - \xi)/\sigma = (cy - c\xi)/(c\sigma)$ for any constant $c \neq 0$. Thus it is safest to use `alaplace2()` to attempt quantile regression.

The EIM of $\mathcal{L}^*(\xi, \sigma, \kappa)$ given by Kotz et al. (2001, Eq.(3.5.1)) is

$$
\begin{pmatrix}
\dfrac{2}{\sigma^2} & 0 & \dfrac{-\sqrt{8}}{\sigma(1 + \kappa^2)} \\[2ex]
0 & \dfrac{1}{\sigma^2} & \dfrac{-(1 - \kappa^2)}{\sigma\kappa(1 + \kappa^2)} \\[2ex]
\dfrac{-\sqrt{8}}{\sigma(1 + \kappa^2)} & \dfrac{-(1 - \kappa^2)}{\sigma\kappa(1 + \kappa^2)} & \dfrac{1}{\kappa^2} + \dfrac{4}{(1 + \kappa^2)^2}
\end{pmatrix}.
\tag{15.14}
$$

From this, it might be decided that quantile regression can be performed by Fisher scoring. The computations are simplified because κ is specified (through τ)

so `alaplace2()` only deals with the 2×2 diagonal submatrix. Fortunately this submatrix is positive-definite over the entire parameter space meaning each iteration step is in an ascending direction.

If σ and κ are known, then the MLE of ξ based on a random sample of n observations is consistent, asymptotically normal and efficient, with asymptotic covariance matrix $(\sigma^2/2)$ (Kotz et al., 2001).

15.3.2.1 Parallelism of the Quantiles

The VGLM/VGAM framework naturally offers two solutions to the crossing quantile problem. The first is to choose the appropriate constraint matrices. The basic idea is to fit parallel curves on the transformed scale, an idea shared with the proportional odds model (1.23). The second solution is described in Sect. 15.3.4 and can be likened to estimating successive layers of an onion. However, while it does not make the strong parallelism assumption, it is numerically less stable.

For the first solution, suppose $\boldsymbol{\tau} = (\tau_1, \tau_2, \ldots, \tau_L)^T$ are the L values of τ of interest to the practitioner. Let ξ_s and σ_s be the corresponding τ_sth quantile and scale parameter in the ALD, $s = 1, \ldots, L$. For `alaplace2()` the elements of $\boldsymbol{\eta}$ are enumerated as

$$g_1(\xi_s(\boldsymbol{x})) \;=\; \eta_{2s-1} \;=\; \boldsymbol{\beta}_{2s-1}^T \boldsymbol{x} \tag{15.15}$$

$$g_2(\sigma_s(\boldsymbol{x})) \;=\; \eta_{2s} \;=\; \boldsymbol{\beta}_{2s}^T \boldsymbol{x}, \tag{15.16}$$

for links g_1 and g_2. By default, $\boldsymbol{\eta} = (\xi_1, \log \sigma_1, \ldots, \xi_L, \log \sigma_L)^T$ because

```
> args(alaplace2)

  function (tau = NULL, llocation = "identitylink", lscale = "loge",
      ilocation = NULL, iscale = NULL, kappa = sqrt(tau/(1 - tau)),
      ishrinkage = 0.95, parallel.locat = TRUE ~ 0, parallel.scale = FALSE ~
      0, digt = 4, idf.mu = 3, imethod = 1, zero = -2)
  NULL
```

It can be seen that the default sets the scale parameters to be the same and intercept-only: $\sigma_1 = \sigma_2 = \cdots = \sigma_L = \beta^*_{(L+1)1}$. The classical approach fits into the above framework by g_1 being the identity link: $\xi_s = \boldsymbol{\beta}_{2s-1}^T \boldsymbol{x}$.

The arguments `parallel.locat` and `parallel.scale` may be used to control the parallelism assumption with respect to the location and scale parameters. A parallelism assumption should always be made for the scale parameters. The default constraint matrices for `alaplace2()` are

$$\mathbf{H}_1 \;=\; \left(\mathbf{I}_L \otimes \begin{pmatrix} 1 \\ 0 \end{pmatrix} \;\middle|\; \mathbf{1}_L \otimes \begin{pmatrix} 0 \\ 1 \end{pmatrix} \right) \quad \text{and} \quad \mathbf{H}_k \;=\; \mathbf{I}_L \otimes \begin{pmatrix} 1 \\ 0 \end{pmatrix}, \quad k = 2, \ldots, p. \tag{15.17}$$

Setting `parallel.locat = TRUE` leaves $\mathbf{H}_1$ unchanged and sets

$$\mathbf{H}_k \;=\; \mathbf{1}_L \otimes \begin{pmatrix} 1 \\ 0 \end{pmatrix}, \quad k = 2, \ldots, p. \tag{15.18}$$

This has the effect that the location parameters are parallel, therefore the estimated quantiles never cross.

15.3.3 Example

To test out the methodology we use a simulated data set so that the 'truth' is known. Let's consider simple Poisson counts. A random sample of $n = 500$ observations were generated from $X_{i2} \sim \text{Unif}(0, 1)$ and

$$Y_i \sim \text{Poisson}\left(\mu(x_{i2}) = \exp\left\{-2 + \frac{6\sin(2x_{i2} - \frac{1}{5})}{\left(x_{i2} + \frac{1}{2}\right)^2}\right\}\right). \qquad (15.19)$$

Figure 15.7a is a jittered scatter plot of these data:

```
> set.seed(123); n <- 500
> adata <- data.frame(x2 = sort(runif(n)))
> mymu <- function(x) exp(-2 + 6 * sin(2 * x - 0.2) / (x + 0.5)^2)
> adata <- transform(adata, y = rpois(n, mymu(x2)))
```

Notice that there is a cluster of $y_i = 0$ at the LHS.

Let $\boldsymbol{\tau} = (\frac{1}{4}, \frac{3}{4})^T$ be the initial quantiles of interest. When faced with a typical data set, it is often a good idea to perform smoothing in order to suggest the functional form of the quantiles, especially when there is nonlinearity. So a VGAM with a nominal 4 degrees of freedom ($1 = \text{linear fit}$) to the vector smoothing spline was fitted here. The the model is

$$\log \xi_s(x_{i2}) = \beta^*_{(s)1} + f^*_{(s)2}(x_{i2}) \qquad (15.20)$$

where the $f^*_{(s)2}$ are effectively cubic smoothing splines. We can attempt to fit this model by

```
> Tau <- c(0.25, 0.75)
> afit <- vgam(y ~ s(x2, df = 4), data = adata,  # trace = TRUE,
               alaplace2(tau = Tau, llocation = "loge", parallel.locat = FALSE))
```

Setting `trace = TRUE` indicates that half-stepping was used at almost every iteration and that this clumsy behaviour is due to scoring being an unsuitable general algorithm for maximizing the likelihood. The behaviour is likened to hiccuping and crawling back and forth, taking two steps forward and one backward. Although the score vectors point in an upward direction, the combination of first and expected second derivatives usually do not lead to an improved step. Currently `vgam()` does not allow half-stepping, however `vglm()` does and that is a viable alternative.

Figure 15.7b plots the data and the fitted quantile curves (solid lines). Overlaid are the 'true' quantiles obtained by `qpois()` which appear as step functions. It may be seen that there is agreement over some parts of the fitted curves. The largest discrepancies are with $\tau = \frac{3}{4}$ at the $x \approx 0$ region. This lack of agreement may be attributed to a number of reasons—such as relative paucity of data there, insufficient flexibility given to the smoother (the remedy is to increase the degrees of freedom of the smoother) or intrinsic boundary-effects problems associated with the smoother. However, the two main points to take from the figure are that the quantile curves are always positive and that the smoothness of the fitted curves is more pleasing aesthetically than the discontinuity of step functions (which could not be obtained without knowing $\mu(x)$ in the first case).

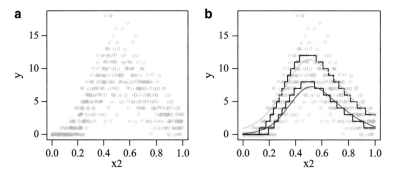

Fig. 15.7 (a) Scatter plot of simulated Poisson counts (15.19). The points have been jittered slightly. (b) Fitted $\mathcal{L}^*(\xi = \exp\{\beta^*_{(s)1} + f^*_{(s)2}(x_{i2})\}, \sigma = 1, \boldsymbol{\tau} = (\frac{1}{4}, \frac{3}{4})^T)$ model to the data. The smooth curves are the fitted $\tau = 0.25$ and 0.75 quantiles from a vector smoothing spline fit. The step functions are the output from `qpois()` at the corresponding τ values. The actual sample proportions lying below the curves are 35.2 and 79.6 %. See also Fig. 15.8b.

This example also serves to highlight a potential problem with quantile regression applied to count data, viz., for $\tau < P(Y = 0)$. In terms of sample data, if the sample proportion of zeros exceeds τ_s, then the optimization problem becomes ill-conditioned and the elements of the fitted regression coefficients $\widehat{\boldsymbol{\beta}_s}$ may diverge to $\pm\infty$.

Some selected output is

```
> afit@extra$percentile

  (tau = 0.25) (tau = 0.75)
        34.8         79.2
```

We wanted $\boldsymbol{\tau} = (\frac{1}{4}, \frac{3}{4})^T$, but the actual percentage of observations falling below the curves is 34.8 and 79.2. The desired and actual match directly only with an identity link.

What effect has a parallelism assumption on these results? This model is

$$\log \xi_s(x_{i2}) = \beta^*_{(s)1} + f^*_{(1)2}(x_{i2}), \tag{15.21}$$

where $f^*_{(1)2}$ is centred. We attempt to fit this with

```
> fitp <- vgam(y ~ s(x2, df = 4), data = adata,   # trace = TRUE,
               alaplace2(tau = Tau, llocation = "loge", parallel.locat = TRUE))
```

and Fig. 15.8b shows the fitted values. There is evidence of bias at the midpoints and edges of x; this lack of fit maybe the cost one must pay in order to ensure noncrossing quantiles. More formally, an approximate likelihood ratio test for H_0 : parallelism versus H_1 : nonparallelism has test statistic $2(\ell_1 - \ell_0) = X = 29$ which has an approximate χ^2_{df} distribution under the null hypothesis. Here, $df = 1993.76 - 1990.61 = 3.15$. An approximate p-value is $P(\chi^2_{df} > X) \approx 3 \times 10^{-6}$. There is very strong evidence against H_0. That is, it is concluded that the lower and upper quartiles are not parallel on the log scale. Viewed graphically, Fig. 15.8a overlays the estimates of the $f^*_{(s)2}$ from (15.20). The code is

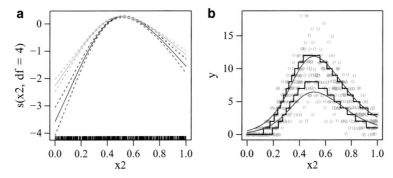

Fig. 15.8 (a) Fitted functions $\widehat{f}_{(s)2}(x_{i2})$ in (15.20) overlaid, with pointwise ± 2 SE bands. (b) Same as Fig. 15.7b except the quantiles (*purple curves*) are constrained to be parallel on the η-scale (log scale here). The actual sample proportion lying below the curves are 36.4 and 78.4 %.

```
my.col <- c("blue", "limegreen")
plot(as(afit, "vgam"), se = TRUE, scol = my.col, lcol = my.col, overlay = TRUE)
```

Clearly the two functions are roughly similar in shape, but they do not differ by a constant. From this too it may be concluded that a parallelism assumption is unreasonable for $\tau = 0.25$ and 0.75.

15.3.4 The Onion Method

Section 15.3.2.1 described one solution to the noncrossing problem: use parallelism constraint matrices so that the quantiles never intersect on the transformed scale. This assumption might be argued as being too restrictive. In this section we describe a second solution that is less restrictive. The method is called the *accumulative quantile method* (AQM). Informally, it can be called the 'onion' method since it can be likened to estimating successive layers of an onion.

Suppose we want the quantile solutions for a fixed τ where the elements are sorted into ascending order. The basic idea is to obtain the quantiles corresponding to τ_1 and then add a positive amount to get the quantile for τ_2. Then add another positive amount to get the quantile for τ_3, etc. The positive amounts are called corrections. The correction for τ_2 can be obtained by quantile regression using a log link applied to $r_{2i}^* = y_i - \widehat{\xi}_{1i}$ (which can be considered a "residual"). This gives an estimate $\widehat{\xi}_{2i}^*$, say, which is then added to $\widehat{\xi}_{1i}$ to give solution $\widehat{\xi}_{2i}$. We repeat the same process to $r_{3i}^* = y_i - \widehat{\xi}_{2i}$, etc. for the remaining τ_s.

A log link ensures that each successive quantile is greater than the previous quantile over all values of $\boldsymbol{x}$. The method gets its name because the solutions are accumulated sequentially by a series of $L-1$ corrections starting from the initial τ_1 quantile regression. Here is a basic numerical example.

```
> my.tau <- seq(0.2, 0.9, by = 0.1)   # Fixed, the desired values
> quant.mat <- matrix(0, nrow(adata), length(my.tau))   # Stores the quantiles
> adata <- transform(adata, offset.y = y * 0)
> use.tau <- my.tau
```

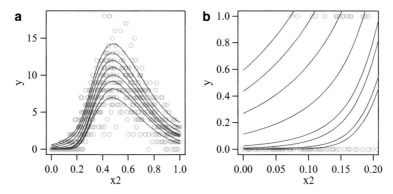

Fig. 15.9 (a) Quantile plot from the onion method applied to some simulated Poisson data. The quantiles are positive, nonparallel and noncrossing. Here, τ has values 0.2(0.1)0.9. (b) A zoom-in view of the LHS of (a).

```
> for (i in 1:length(my.tau)) {
    adata <- transform(adata, use.y = y - offset.y)
    iloc <- ifelse(i == 1, with(adata, median(y)), 1.0)  # well-chosen!
    mydf <- ifelse(i == 1, 5, 3)  # Maybe less smoothing will help
    fit3 <- vglm(use.y ~ ns(x2, df = mydf),
                 alaplace2(tau = use.tau[i], llocation = "loge", iloc = iloc),
                 data = adata, trace = FALSE)
    quant.mat[, i] <- (if (i == 1) 0 else quant.mat[, i-1]) + fitted(fit3)
    adata <- transform(adata, offset.y = quant.mat[, i])
}
```

Then Fig. 15.9a was produced by

```
> plot(y ~ x2, adata, col = "orange")
> with(adata, matlines(x2, quant.mat, col = "blue", lty = 1))
```

This particular example seems to work because the successive τ_s values are far enough apart. In practice numerical problems may occur when the values are too close, in which case good initial values become very important. Also, one can propose several enhancements to the above description. These include the following.

1. Start with the τ_s value closest to 0.5 in order to get a more stable solution and then apply the above description to all values of τ_s greater than 0.5, in sorted order. For $\tau_s < 0.5$ values apply a similar process to $-y_i$ using $1-\tau_s$ values. This enhancement starts in the centre and works its way out to both ends separately.
2. Allow for less flexibility to be given to the functions associated with the corrections. The first function may be more complicated, but subsequent ones should be simpler because they operate on residuals which are probably less complex in form.

Better initial values might be inputted using certain arguments in `alaplace2()` as well as `coefstart`, `etastart` and/or `mustart` of `vglm()`/`vgam()` (Sect. 8.3.1.1).

A variant of the above is an onion method, based on fitting expectiles and calibrating them so that a certain percentage falls below the curve. Expectiles are naturally estimated by scoring and therefore should exhibit less numerical problems compared to ALDs. A numerical example is given in Sect. 15.4.5.

15.4 Expectile Regression

15.4.1 Introduction

Rather than minimizing the expectation of $\rho_\tau(Y - \xi)$ in (15.6), one can instead consider minimizing the expectation of $\rho_\omega^{[2]}(Y - \mu)$ with respect to μ where

$$\rho_\omega^{[2]}(u) \;=\; u^2 \cdot |\omega - I(u < 0)|, \qquad 0 < \omega < 1. \tag{15.22}$$

This is plotted in Fig. 15.6b for two values of ω. As with τ in the ALAD case, as ω changes from 0.5 to 0.9, positive deviations $Y - \mu(w)$ are penalized more heavily compared to negative ones, therefore the expectile increases in value. The formula is a very natural alternative to (15.6), and the results will be shown below to be quite interpretable. The essential difference is that the u has been replaced by the second-order moment u^2.

Applying the same argument as (15.7)–(15.8), we minimize the *asymmetric least squares* (ALS) criterion

$$E\left[\rho_\omega^{[2]}(Y - \mu)\right] \;=\; (1 - \omega) \int_{-\infty}^{\mu} (y - \mu)^2 \, dF(y) + \omega \int_{\mu}^{\infty} (y - \mu)^2 \, dF(y), \tag{15.23}$$

and setting the derivative with respect to μ to zero gives

$$(1 - \omega) \int_{-\infty}^{\mu(\omega)} (y - \mu(\omega)) \, dF(y) \;+\; \omega \int_{\mu(\omega)}^{\infty} (y - \mu(\omega)) \, dF(y) \;=\; 0. \tag{15.24}$$

Newey and Powell (1987, Thm 1) showed that a unique solution exists if $E(Y) = \mu(0.5) = \mu$ exists, and they called the quantities *expectiles*. They used this name for regression surfaces obtained by ALS. This was deliberate, for the purpose of distinguishing them from the original *regression quantiles* of Koenker and Bassett (1978). Efron (1991, 1992) used the general name *regression percentile* to apply to all forms of asymmetric fitting.

In terms of interpretation, it can be seen that, given $X = x$, the quantile $\xi_\tau(x)$ specifies the position below which $100\tau\%$ of the (probability) mass of Y lies while the expectile $\mu_\omega(x)$ determines (again at $X = x$) the point such that $100\omega\%$ of the mean distance between it and Y comes from the mass below it. Thus expectiles are quite interpretable. Note that the 0.5-expectile $\mu(\tfrac{1}{2})$ is the mean μ while the 0.5-quantile $\xi(\tfrac{1}{2})$ is the median.

The above corresponds to the population. In terms of the sample, expectiles involve minimizing the asymmetrically weighted least-squares criterion

$$\sum_{i=1}^{n} (1 - \omega) \left[(y_i - \mu)_-\right]^2 + \omega \left[(y_i - \mu)_+\right]^2 \tag{15.25}$$

with respect to μ.

Although quantiles and expectiles are interrelated (see Sect. 15.4.1.1), each holds certain advantages and disadvantages so that neither is uniformly superior (it is like choosing between the conditional mean and median in conventional regression). The choice will usually depend on the particular application at hand. Here are some further sundry notes.

1. The ALS method is not as robust as the ALAD method against outliers. Quantiles depend only on local features of the distribution whereas expectiles have a more global nature. For example, increasing values in the upper tail of a distribution do not affect the quantiles of the lower tail but it affects the values of *all* the expectiles.
2. In view of the one-to-one mapping between expectiles and quantiles, Efron (1991) proposes that the τ-quantile be estimated by the expectile for which the proportion of in-sample observations lying below the expectile is τ. This provides justification for practitioners who use expectile regression to perform quantile regression. This approach is used in this section.
3. An ALS estimator is easier to compute, since least-squares calculations are more familiar to statisticians than linear programming. Furthermore, an ALS estimator is reasonably efficient under normality conditions (cf. Efron (1991)).
4. For quite a large class of nonlinear regression models, the conditional expectiles as functions of x are in a one-one correspondence with the conditional percentiles. Therefore, the ALS approach can be adapted to estimate conditional percentiles directly.
5. ALS regression is the least squares analogue of quantile regression.
6. Both quantile and expectile regression coincide with the MLE solutions where the data are assumed to be drawn from some specific distributions.

15.4.1.1 Interrelationship Between Expectiles and Quantiles

Quite generally, Newey and Powell (1987) stated that "expectiles have properties that are similar to quantiles". Jones (1994) explored this statement in greater detail and showed that the reason for this is that the expectiles of a distribution F are quantiles a distribution G which is related to F. The main details are as follows.

Let $P(s) = \int_{-\infty}^{s} y f(y) \, dy$ be the partial first moment,

$$\rho_{\tau}^{\{1\}}(u) = \tau - I(u < 0) \quad \text{and} \tag{15.26}$$

$$\rho_{\omega}^{\{2\}}(u) = (\omega - I(u < 0)) \cdot |u|. \tag{15.27}$$

From (15.7)–(15.8) we saw that one way of defining the ordinary τ-quantile of a continuous distribution with density f is as the value of ξ that equates $\int \rho_{\tau}^{\{1\}}(y - \xi) f(y) \, dy$ to 0. Similarly, for expectiles $\mu(\omega)$, (15.24) corresponds to the equation

$$\int_{-\infty}^{\infty} \rho_{\omega}^{\{2\}}(y - \mu(\omega)) f(y) \, dy = 0. \tag{15.28}$$

Then solving this equation shows immediately that $\omega = G(\mu(\omega))$ where

$$G(t) = \frac{P(t) - t F(t)}{2[P(t) - t F(t)] + t - \mu}, \tag{15.29}$$

a rearrangement of Newey and Powell (1987, Eq.(2.7)). Thus G is the inverse of the expectile function and its derivative is

$$g(t) = \frac{\mu F(t) - P(t)}{\{2[P(t) - t F(t)] + t - \mu\}^2}. \tag{15.30}$$

It can be shown that G is actually a distribution function so that g is its density function, i.e., the expectiles of F are precisely the quantiles of G defined here.

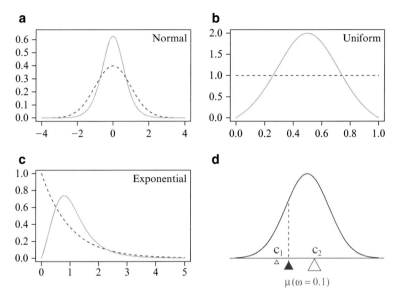

Fig. 15.10 (a)–(c) Density plots of expectile derived g (*orange solid lines*; (15.30)) for the original f of standard normal, standard uniform and standard exponential distributions (*blue dashed lines*). (d) Illustration of the interpretation of expectiles in terms of centres of balance, the *hollow triangles* at positions c_1 and c_2. Here, the *vertical dashed line* is at the 0.1-expectile, the *solid triangle* at $\mu(\omega = 0.1)$, which means that (15.33) is satisfied with $\omega = 0.1$.

Jones (1994) illustrated these results with the standard normal, standard uniform and standard exponential distributions (reproduced in Fig. 15.10a–c). For example, for the standard uniform,

$$g(t) = \frac{2t(1-t)}{\{2t(1-t)-1\}^2}, \qquad 0 \le t \le 1, \tag{15.31}$$

which is symmetric.

A few g and G have been written for **VGAM** in terms of the usual **dpqr**-type functions associated with a distribution (Table 15.3). An "e" has been added to signify its root as an expectile-defined distribution. For example, **deunif()** corresponds to the density (15.31). For some of the **q**-type functions a Newton-Raphson algorithm is used to solve for q satisfying $p = G(q)$. For such, numerical problems may occur when values of p are very close to 0 or 1.

Expectiles are interpretable, albeit not as simply as quantiles. Upon a rearrangement of (15.24), it can be seen that

$$\omega = \frac{\displaystyle\int_{-\infty}^{\mu(\omega)} [(\mu(\omega) - y)\, f(y)\, dy}{\displaystyle\int_{-\infty}^{\infty} \left|\mu(\omega) - y\right| f(y)\, dy}. \tag{15.32}$$

Loosely, the ω-expectile is the value $\mu(\omega)$ such that ω is the ratio of the total distances of randomly generated observations to the left of it compared to the total sum of distances of all randomly generated observations to $\mu(\omega)$. Of course these totals are influenced by the number of observations falling on the LHS and overall. The numerator of (15.32) is $E[\mu(\omega) - Y | Y \le \mu(\omega)] \times P(Y \le \mu(\omega))$. The denominator is $E[|Y - \mu(\omega)|]$, which is the sum of the numerator and the term $E[Y - \mu(\omega) | Y > \mu(\omega)] \times P(Y > \mu(\omega))$.

An even simpler method of interpretation is via centre of balances. Observe in Fig. 15.10d that the positions of c_1 and c_2 denote the centres of balance for the distributions to the left and right of the ω-expectile $\mu(\omega)$. Then

$$\omega = \frac{P[Y < \mu(\omega)] \cdot (\mu(\omega) - c_1)}{P[Y < \mu(\omega)] \cdot (\mu(\omega) - c_1) + P[Y > \mu(\omega)] \cdot (c_2 - \mu(\omega))}. \tag{15.33}$$

Here,

$$c_1 = E[Y | Y < \mu(\omega)] = \int_{-\infty}^{\mu(\omega)} y \, \frac{f(y)}{F(\mu(\omega))} \, dy, \tag{15.34}$$

and c_2 is similarly defined. Another important equation is

$$\mu = P[Y < \mu(\omega)] \cdot c_1 + P[Y > \mu(\omega)] \cdot c_2. \tag{15.35}$$

We'll see in Sect. 15.4.1.2 that c_1 is related to a quantity called the expected shortfall. In Fig. 15.10d the parent distribution happens to be a normal distribution with $\omega = 0.1$. Here, 10 percent of the centre of gravity at the expectile is found to its left due to c_1. Here's a simple numerical example to illustrate this.

```
> my.omega <- 0.25
> set.seed(1)
> y <- rnorm(n = 1000)
> (my.exp <- qenorm(my.omega))

 [1] -0.43633

> sum(my.exp - y[y <= my.exp]) / sum(abs(my.exp - y))    # Should be my.omega

 [1] 0.26158
```

This should be equal to `my.omega`. That is, ω should be the sum of all the distances (from observations to the expectile) left of the expectile, divided by the total sum of distances of all observations to the expectile. Equivalently:

```
> I1 <- mean(y < my.exp) * mean( my.exp - y[y < my.exp])
> I2 <- mean(y > my.exp) * mean(y[y > my.exp] - my.exp)
> I1 / (I1 + I2)    # Should be my.omega

 [1] 0.26158
```

or

```
> I1 <- sum( my.exp - y[y < my.exp])
> I2 <- sum(y[y > my.exp] - my.exp)
> I1 / (I1 + I2)    # Should be my.omega

 [1] 0.26158
```

It may be shown that

$$c_1 = \mu(\omega) + \{\omega/(2\omega - 1)\}(\mu - \mu(\omega))/F(\mu(\omega)), \tag{15.36}$$

$$c_2 = \mu(\omega) - \{(1 - \omega)/(2\omega - 1)\}(\mu - \mu(\omega))/[1 - F(\mu(\omega))]. \tag{15.37}$$

15.4.1.2 Expected Shortfall †

The *expected shortfall* (ES) is a concept used in financial mathematics to measure portfolio risk. It is also called the *Conditional Value at Risk* (CVaR), *expected tail loss* (ETL) and *worst conditional expectation* (WCE). The ES at the $100\tau\%$ level is the expected return on the portfolio in the worst $\tau\%$ of the cases. It is often defined as

$$\mathsf{ES}(\tau) = E(Y|Y < a), \tag{15.38}$$

where a is determined by $P(Y < a) = \tau$ and τ is the given threshold.

The ES is very much related to expectiles via (15.24) (Taylor, 2008). It can be shown that $\mu(\omega)$ satisfies

$$\left(\frac{1 - 2\omega}{\omega}\right) E\left[(Y - \mu(\omega)) \cdot I(Y < \mu(\omega))\right] = \mu(\omega) - E(Y), \tag{15.39}$$

which is another rearrangement of Newey and Powell (1987, Eq.(2.7)). Equation (15.39) can be rewritten $E\left[Y|Y < \mu(\omega)\right] =$

$$\left(1 + \frac{\omega}{(1 - 2\omega)\,F(\mu(\omega))}\right)\mu(\omega) - \frac{\omega}{(1 - 2\omega)\,F(\mu(\omega))}\,E(Y), \tag{15.40}$$

which is the same as (15.36). This provides a formula for the ES of the quantile that coincides with the ω-expectile. Referring to this as the τ-quantile, we can write $F(\mu(\omega)) = \tau$ and rewrite the expression as

$$\mathsf{ES}(\tau) = \left(1 + \frac{\omega}{(1 - 2\omega)\,\tau}\right)\mu(\omega) - \frac{\omega}{(1 - 2\omega)\,\tau}\,E(Y). \tag{15.41}$$

This equation relates the ES associated with the τ-quantile of the distribution of Y and the ω-expectile that coincides with that quantile. The equation is for the ES in the lower tail of the distribution. The equation for the upper tail of the distribution is produced by replacing ω and τ with $1 - \omega$ and $1 - \tau$, respectively.

Incidentally, another popular measure of financial risk is the *Value at Risk* (VaR). The VaR (ν_p, say) specifies a level of excessive *losses* such that the probability of a loss larger than ν_p is less than p (often $p = 0.01$ or 0.05 is chosen). Then the ES can be defined as the conditional expectation of the loss, given that it exceeds the VaR.

15.4.2 Expectiles for the Linear Model

Since quantiles are estimated by regression ("quantile regression") it comes as no surprise that expectiles are also estimated by regression and subsequently the method is called "expectile regression". The methodology is generally attributed to Aigner et al. (1976) and Newey and Powell (1987) and further developed in Efron (1991). For normally distributed responses, it is based on *asymmetric least squares* (ALS) estimation, a variant of OLS estimation. The method proposed by Koenker and Bassett (1978) is similar but is based instead on minimizing

Table 15.3 Density function, distribution function, expectile function and random generation for the distribution associated with the expectiles of several standardized distributions. These functions are available in **VGAM**.

Functions	Distribution
`[dpqr]eexp()`	Exponential
`[dpqr]sc.t2()`	(Scaled) $\sqrt{2}\,T_2$
`[dpqr]enorm()`	Normal
`[dpqr]eunif()`	Uniform

the ALAD (15.9). Then, Efron (1992) generalized ALS estimation to families in the exponential family, and in particular, the Poisson distribution. He called this *asymmetric maximum likelihood* (AML) estimation.

Consider the linear model $y_i = \boldsymbol{x}_i\boldsymbol{\beta} + \varepsilon_i$ and let $r_i(\boldsymbol{\beta}) = y_i - \boldsymbol{x}_i^T\boldsymbol{\beta}$ be a residual. The asymmetric squared error loss function for a residual r is r^2 if $r \leq 0$ and wr^2 if $r > 0$. Here w is a positive constant and is related to ω by $w = \omega/(1-\omega)$, a renormalization of (15.23). The solution is the set of regression coefficients that minimize the sum of these over the data set, weighted by the `weights` argument (so that it can contain frequencies). Written mathematically, the asymmetric squared error loss $S_w(\boldsymbol{\beta})$ is

$$S_w(\boldsymbol{\beta}) \;=\; \sum_{i=1}^{n} w_i\, Q_w^*(r_i(\boldsymbol{\beta})) \tag{15.42}$$

and Q_w^* is the asymmetric squared error loss function

$$Q_w^*(r) \;=\; \begin{cases} r^2, & r \leq 0, \\ w\,r^2, & r > 0. \end{cases} \tag{15.43}$$

The w_i are known prior weights, inputted using the `weights` argument of `vglm()`, etc. and retrievable afterwards as `weights(fit, type = "prior")`.

Here are some notes about ALS regression as implemented by the **VGAM** family function `amlnormal()`.

1. For quantile regression, usually the user will specify some desired value of the percentile and then the necessary w value is solved for numerically to obtain this. One useful property here is that the percentile is a monotonic function of w, meaning one can more easily solve for the root of a nonlinear equation. A numerical example is given in Sect. 15.4.3.
2. A rough relationship between w and the percentile 100α is given on in Efron (1991, p.102). Let $w^{(\alpha)}$ denote the value of w such that $\boldsymbol{\beta}_w$ equals $z^{(\alpha)} = \Phi^{-1}(\alpha)$, the 100α standard normal percentile point. If there are no covariates (intercept-only model) and y_i are standard normal, then

$$w^{(\alpha)} \;=\; 1 + \frac{z^{(\alpha)}}{\phi\!\left(z^{(\alpha)}\right) - (1-\alpha)z^{(\alpha)}}. \tag{15.44}$$

Some values are

```
> alpha <- c(1/2, 2/3, 3/4, 0.84, 9/10, 19/20)
> z.alpha <- qnorm(p = alpha)
> w.alpha <- 1 + z.alpha / (dnorm(z.alpha) - (1 - alpha) * z.alpha)
> round(cbind(alpha, w.alpha), digits = 2)
```

```
        alpha w.alpha
  [1,]   0.50    1.00
  [2,]   0.67    2.96
  [3,]   0.75    5.52
  [4,]   0.84   12.81
  [5,]   0.90   28.07
  [6,]   0.95   79.73
```

3. Some workers in the field are not very sympathetic to the parametric forms of quantile regression, and even less so to expectiles.
4. An iterative solution is required, and the Newton-Raphson algorithm is used. In particular, for Poisson regression with the canonical (log) link, following in from Efron (1992, Eq.(2.16)),

$$
\begin{aligned}
\boldsymbol{\beta}^{(a)} &= \boldsymbol{b}^{(a-1)} + d\boldsymbol{b}^{(a-1)} \\
&= \boldsymbol{b} - \ddot{\mathbf{S}}_w^{-1}\, \ddot{\boldsymbol{S}}_w \\
&= (\mathbf{X}^T\,(\mathbf{WV})\,\mathbf{X})^{-1}\mathbf{X}^T\,(\mathbf{WV})\left[\boldsymbol{\eta} + (\mathbf{WV})^{-1}\,\mathbf{W}\boldsymbol{r}\right]
\end{aligned}
$$

are the Newton-Raphson iterations (iteration number a suppressed for clarity), cf. (3.9). Here, $\boldsymbol{r} = \boldsymbol{y} - \boldsymbol{\mu}(\boldsymbol{b})$, $\mathbf{V} = \mathrm{diag}(v_1(\boldsymbol{b}),\ldots,v_n(\boldsymbol{b})) = \mathrm{diag}(\mu_1,\ldots,\mu_n)$ contains the variances of y_i and $\mathbf{W} = \mathrm{diag}(w_1(\boldsymbol{b}),\ldots,w_n(\boldsymbol{b}))$ with $w_i(\boldsymbol{b}) = 1$ if $r_i(\boldsymbol{b}) \le 0$ else w.
5. ALS quantile regression is consistent for the true regression percentiles $y^{(\alpha)}|\boldsymbol{x}$ in the cases where $y^{(\alpha)}|\boldsymbol{x}$ is linear in $\boldsymbol{x}$. Newey and Powell (1987) give a more general proof of this.
6. The ALS loss function (15.43) leads to an important invariance property: if the y_i are multiplied by some constant c, then the solution vector $\widehat{\boldsymbol{\beta}}_w$ is also multiplied by c. Also, a shift in location to $y_i + d$ means the estimated intercept increases by d too.

15.4.3 ALS Example

ALS quantile regression is implemented by the function **amlnormal()**, which has a **deviance** slot to compute the weighted asymmetric squared error loss (15.42) summed over all **w.aml** values. Here is a simple example that fits something similar to **w0.LMS** in Sect. 15.2.3. Recall the response is BMI regressed upon age for 2614 European women from the **xs.nz** data frame. Suppose we want to estimate using expectiles the same quantiles as **w0.LMS** in Sect. 15.2.3, viz. the 5, 25, 50, 75, 95 percentiles. Then we need to numerically search for the appropriate w for these and there are several ways of attempting this. One of them is to first define a new function

```
> find.w <- function(w, percentile = 50) {
    fit <- vglm(BMI ~ ns(age, df = 3), amlnormal(w = w), data = women.eth0)
    fit@extra$percentile - percentile
  }
```

Solving for the root as a function of w should hopefully give the desired percentile. The following code gives Fig. 15.11. It uses **uniroot()** since the percentile is a

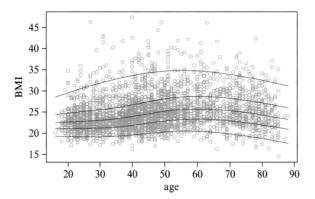

Fig. 15.11 'Quantile' plot from `amlnormal()` applied to `women.eth0`: 5, 25, 50, 75, 95 'percentile' curves. Each regression curve is a regression spline with 3 degrees of freedom (1 = linear fit).

monotonic function of w. A small complication is that `uniroot()` can only find one root at a time. We assume that each appropriate w lies between 0.0001 and 10000. After all the appropriate w values are found a model is fitted with all of these simultaneously and the curves are added to the plot.

```
> Tau <- c(5, 25, 50, 75, 95)
> w.aml <- numeric(length(Tau))   # Stores the appropriate w.aml values
> plot(BMI ~ age, women.eth0, col = "darkorange", las = 1)
> ooo <- with(women.eth0, order(age))
> women.eth0 <- women.eth0[ooo, ]   # Sort by age
> for (i in 1:length(Tau))
    w.aml[i] <- uniroot(f = find.w, interval = c(1/10^4, 10^4),
                        percentile = Tau[i])$root
> fit3 <- vglm(BMI ~ ns(age, df = 3), amlnormal(w = w.aml), data = women.eth0)
> matlines(with(women.eth0, age), fitted(fit3), col = "blue", lty = 1)
```

The result is Fig. 15.11. The values of w corresponding to each τ value and the empirical sample proportions below each curve are

```
> fit3@extra$w.aml

 [1]  0.0083459  0.1181113  0.6143516  3.1700633 50.8633500

> fit3@extra$percentile

  w.aml = 0.0083  w.aml = 0.1181  w.aml = 0.6144  w.aml = 3.1701 w.aml = 50.8633
          5.0115          24.9809          50.0000          75.0191          94.9885
```

Altogether, this appears to give a reasonable fit to the data, apart from boundary effects at low ages. Unfortunately the generic functions `qtplot()`, `deplot()`, etc. do not presently work on an `amlnormal()` fit.

15.4.4 Poisson Regression

ALS is a special case of AML estimation. The latter was used by Efron (1992) to obtain expectiles from the Poisson distribution. More generally,

$$S_w(\boldsymbol{\beta}) \;=\; \sum_{i=1}^{n} w_i \, D_w(y_i, \mu_i(\boldsymbol{\beta})) \tag{15.45}$$

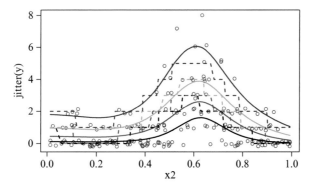

Fig. 15.12 Quantile plot from an `amlpoisson()` fit using simulated data. The dashed step functions are the `qpois()` quantiles calibrated with the true mean function and the proportion under the expectile curves. The response has been jitted to aid clarity.

is minimized (cf. (15.42)), where

$$D_w(\mu, \mu') = \begin{cases} D(\mu, \mu'), & \mu \le \mu', \\ w\,D(\mu, \mu'), & \mu > \mu'. \end{cases} \qquad (15.46)$$

Here, D is the deviance from a model in the exponential family $g_\eta(y) = e^{\eta y - \psi(\eta)}$.

The **VGAM** family function `amlpoisson()` is an implementation of AML estimation for the Poisson model. Here is a simple example.

```
> set.seed(1); pdata <- data.frame(x2 = sort(runif(n <- 200)))
> meanfun <- function(x) exp(-sin(8*x))
> pdata <- transform(pdata, y = rpois(n, meanfun(x2)))
> w.aml <- c(0.2, 1, 5, 50)  # An assortment of positive values
> fit.amlpois <- vgam(y ~ s(x2), amlpoisson(w.aml = w.aml), data = pdata)
```

The proportions under the expectile curves are

```
> p.hat <- colMeans(depvar(fit.amlpois, drop = TRUE) < fitted(fit.amlpois))
> 100 * p.hat
```

w.aml = 0.2	w.aml = 1	w.aml = 5	w.aml = 50
52.0	62.5	76.0	92.5

Using these values, one can compare the expectile curves to the Poisson quantiles as given by `qpois()`. Plotting these and the (discrete) quantiles together can be achieved by

```
> mycol <- 1:length(w.aml)
> plot(jitter(y) ~ x2, pdata, col = "purple", las = 1)
> with(pdata, matlines(x2, fitted(fit.amlpois), col = mycol, lwd = 2, lty = 1))
> for (i in 1:length(w.aml))
    lines(qpois(p = p.hat[i], lambda = meanfun(x2)) ~ x2,
          pdata, col = mycol[i], lty = 2, lwd = 2)
```

Closer agreement is observed when the mean function has larger values (Fig. 15.12).

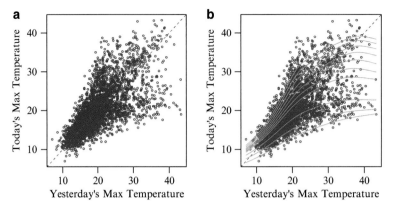

Fig. 15.13 (a) Melbourne maximum temperature data, called `melbmaxtemp`, in °C. (b) Onion method using `amlnormal()` expectiles applied to the data with an assortment of w_j values. The sample proportions below the curves are 1, 4.8, 20.2, 54.9, 65.2, 74.8, 82.7, 88.2, 91.7, 94.8 %.

15.4.5 Melbourne Temperatures and the Onion Method

Here is a second example of the onion method which is based on the computationally more suitable expectile regression. It is applied to the Melbourne maximum temperatures data and uses a vector of prespecified `w.aml` values.

```
> w.vector <- sort(c(1 / 10^(0:3), 2^(1:6)))   # An assortment of values
> quant.mat <- matrix(0, nrow(melb), length(w.vector))   # Stores the expectiles
> melb <- transform(melb, offset.y = today * 0)
> for (i in 1:length(w.vector)) {
    melb <- transform(melb, use.y = today - offset.y)
    init.val <- (if (i == 1) NULL else with(melb, max(use.y) / 10))
    onion.melb <- vgam(use.y ~ s(yesterday, df = 3),
                       amlnormal(w.aml = w.vector[i],
                                 imethod = 3,
                                 iexpectile = init.val,
                                 lexpectile = "loge"),
                       data = melb, trace = FALSE)
    quant.mat[, i] <- (if (i == 1) 0 else quant.mat[, i-1]) + fitted(onion.melb)
    melb <- transform(melb, offset.y = quant.mat[, i])
}
```

A plot of the results is given in Fig. 15.13. One weakness of this analysis is that the serial dependency is glossed over by an implicit independence assumption here. Despite this, the bimodal feature is somewhat detected and handled, albeit not very well. Evidently, the bimodality is due to a high-pressure system from the Australian mainland bringing high temperatures to Melbourne but is often followed by a cold front.

15.5 Discussion

Quantile regression is an important and useful type of regression that allows modelling the entire distribution of a response, rather than just through its mean function. It has become a large topic with a lot of research activity, directed especially at nonstandard types of data such as longitudinal and censored data, and now it has a wide range of applications.

Amazingly, as a consequence of an "Extremal Types Theorem", quantiles can also be modelled for data consisting of extremes, e.g., minimum temperatures and maximum flood levels. This is the subject of Chap. 16.

Bibliographic Notes

Koenker (2005) is a very accessible book on both the theory and practice of quantile regression and it has become a standard reference. His R package quantreg implements his generalized L_1 approach, and it has features enabling it to handle large data sets. Other QR books include Hao and Naiman (2007) and Davino et al. (2014). Fahrmeir et al. (2011) includes a chapter on quantile regression.

The LMS-Box-Cox-normal method is described in Green and Silverman (1994, Chap.6) in the same context of penalized likelihood and cubic spline regression that the VGLM/VGAM framework espouses.

As Ex. 15.14 shows, the Student t distribution with 2 degrees of freedom (Jones, 2002) is a particularly interesting distribution because a scaled version of it corresponds to a distribution whose expectiles and quantiles coincide. The package expectreg implements methods for expectile and quantile regression, and a data-analytic description of expectiles given in Schnabel and Eilers (2009).

Exercises

Ex. 15.1. † Show that the mean and median of a distribution cannot differ by more than one standard deviation.   [K. F. Yu]

Ex. 15.2. Consider the residuals defined by (15.5) for the LMS-Box-Cox-normal method. Write a small R function to calculate them for any lms.bcn() fit. Test out your code on w0.LMS from Sect. 15.2.3.

Ex. 15.3. Repeat the BMI analyses of Sect. 15.2.3 by using lms.bcg() instead. Are there any significant differences between the results?

Ex. 15.4.   The LMS-Box-Cox-Normal Method
Consider the model behind lms.bcn() as described in Sect. 15.2.1.

(a) Provide the details of how the bottom equation of (15.3) arises as $\lambda(x) \to 0$ in the top equation.
(b) Show that, for fixed x, the PDF of Y is given by

$$\frac{1}{\sqrt{2\pi}\,\mu\,\sigma}\left(\frac{y}{\mu}\right)^{\lambda-1}\exp\left\{-\frac{1}{2\lambda^2\sigma^2}\left[\left(\frac{y}{\mu}\right)^{\lambda}-1\right]^2\right\}. \qquad (15.47)$$

(c) Provide the details behind (15.4).

Ex. 15.5.   Show that $\rho_\tau(u)$ defined by (15.6) can be written as

$$\rho_\tau(u) \;=\; 0.5\,|u| + (\tau - 0.5)\,u.$$

Obtain a similar formulation for $\rho_\omega^{[2]}(u)$, as defined by (15.22) using $0.5u^2$ as the first term.

Ex. 15.6.   **Expectiles**

(a) Show that (15.35) holds.
(b) Show that c_1 and c_2 are given as (15.36) and (15.37).

Ex. 15.7.   **Quantiles and Expectiles**

(a) Verify that setting the derivative of $\int \rho_\tau(y - \xi)\,f(y)\,dy = 0$ with respect to ξ results in (15.8) for quantiles. Likewise that setting the derivative of $\int \rho_\omega^{[2]}(y - \mu(\omega))\,f(y)\,dy = 0$ with respect to μ implies (15.24) for expectiles.
(b) Now confirm (15.29). Then differentiate G to obtain g as in (15.30).
(c) Obtain expressions for $g(t)$ and $G(t)$ in (15.29)–(15.30) for the (i) standard normal, (ii) standard uniform, and (iii) standard exponential distributions. Show that two of these densities are symmetric. Plot them in R as in Fig. 15.10.

Ex. 15.8.   Show that if Y_1 and Y_2 are i.i.d. standard exponential distributions then $Y_1/p - Y_2/(1-p)$ has an ALD with unit scale parameter. [Kotz et al. (2001)]

Ex. 15.9.   Use `alaplace2()` to fit a QR that is similar to what `w0.LMS` is fitted to (described in Sect. 15.2.3). Compare the two models... do they differ appreciably?

Ex. 15.10. $\mathcal{L}^*(\xi, \sigma, \kappa)$

(a) Consider the $\mathcal{L}^*(\xi, \sigma, \kappa)$ density (15.12). Verify (15.13).
(b) Derive the 2×2 diagonal submatrix (15.14).   [Kotz et al. (2001)]

Ex. 15.11.   **Logit-Laplace and Log-Laplace Distributions**

(a) A random variable Y defined on $(0,1)$ is said to have a logit-Laplace distribution if $\operatorname{logit} Y \sim$ ALD. Suppose the ALD has the form   [Kotz et al. (2001)]

$$f(w; \xi, \alpha, \beta) \;=\; \frac{\alpha\beta}{\alpha + \beta} \begin{cases} \exp\left(\beta(w - \xi)\right), & w < \xi, \\ \exp\left(-\alpha(w - \xi)\right), & w \geq \xi, \end{cases} \tag{15.48}$$

for $\alpha > 0$ and $\beta > 0$. Derive the PDF of the logit-Laplace distribution.
(b) A random variable Y defined on $(0, \infty)$ is said to have a log-Laplace distribution if $\log Y \sim$ ALD. For an ALD having the form (15.48), derive the density of Y.

Ex. 15.12.   Yu and Zhang (2005) use an ALD whose density is

$$f(y; \xi, b, \tau) \;=\; \begin{cases} \dfrac{\tau(1 - \tau)}{b} \exp\left(-\dfrac{(1 - \tau)}{b}\left|y - \xi\right|\right), & y \leq \xi, \\[2ex] \dfrac{\tau(1 - \tau)}{b} \exp\left(-\dfrac{\tau}{b}\left|y - \xi\right|\right), & y > \xi, \end{cases} \tag{15.49}$$

with the usual support and parameter range restrictions. For want of a better name, we shall call this the "YZALD".

(a) Match the parameters of the $\mathcal{L}^*(\xi, \sigma, \kappa)$ distribution with the YZALD(ξ, b, τ) parameterization.

(b) Show that $E(Y) = \xi + b\,(1-2\tau)/(\tau(1-\tau))$, and obtain an expression for $\text{Var}(Y)$.

Ex. 15.13. Show that if $Y = a + b\,X$ for constants a and b then the ω-expectile of Y equals $a + b \times$(the ω-expectile of X). [Efron (1991)]

Ex. 15.14. (Scaled) $\sqrt{2}\,T_2$ Distribution

Here we derive a distribution whose expectiles and quantiles are the same.

(a) By equating $q(\alpha) = e(\alpha)$ in

$$\alpha \int_{e(\alpha)}^{\infty} (y - e(\alpha))\, dF(y) \;=\; (1 - \alpha) \int_{-\infty}^{e(\alpha)} (e(\alpha) - y)\, dF(y)$$

show that

$$F(y) \;=\; \begin{cases} \frac{1}{2} \cdot \left(1 + \sqrt{1 - 4/(4 + y^2)}\right), & y \geq 0, \\[2mm] \frac{1}{2} \cdot \left(1 - \sqrt{1 - 4/(4 + y^2)}\right), & y < 0. \end{cases} \tag{15.50}$$

(b) Confirm that the density function is

$$f(y) \;=\; \frac{2\,|y|}{(4 + y^2)^2\, \sqrt{1 - 4/(4 + y^2)}}. \tag{15.51}$$

What is $f(0)$?

(c) Verify that

$$\lim_{y \to \infty} \frac{-\log(1 - F(y))}{\log y} \;=\; 2.$$

(d) Show that this distribution is equivalent to $K = \sqrt{2}\,T_2$, where K has a standard `dsc.t2()` distribution and T_2 has a Student's t distribution with 2 degrees of freedom. [Koenker (1992)]

What I fail to see is any benefit derived from introducing the expectiles. Expectiles belong in the spittoon.
—Koenker (2013, p.332)

Chapter 16
Extremes

And to the C students, I say you, too, can be President of the United States.
—George W. Bush, Yale University commencement address, May 2001

16.1 Introduction

Most of applied statistics is concerned with what goes on in the centre of a distribution F, usually via the mean. In contrast, extreme value theory (EVT) is the branch of statistics concerned with inferences about the *tails* of F. EVT has important applications in many fields, such as environmental science (e.g., sea-levels, wind speeds, hydrology, peak flows of a river), reliability modelling (e.g., weakest-link-type models), finance and sport science (e.g., fastest running times). The subject began to mature during the 1950s with the pioneering work of E. J. Gumbel (1891–1966), and has emerged to become a sizeable field in its own right with very important applications because unusually large or small observations can be very influential, e.g., very large claims for an insurance company may result in bankruptcy, extremes in climate may result in environmental and human disasters including animal and plant species extinctions.

In this chapter, we depart from the notation of Chap. 12 and replace a, b and s by μ, σ and ξ for the location parameter, scale parameter and shape parameter, respectively—this is in keeping with most of the EVT literature. Also, the terms *quantiles* and *percentiles* are be used interchangeably, as in Chap. 15, so that, e.g., a 0.5-quantile is equivalent to percentile = 50, which is the median. VGAM family functions (Table 16.1) use the argument `percentile` to specify quantiles, so it should be assigned values between 0 and 100.

Historically, EVT was slow in adopting smoothing. This may have been due to most workers being theoreticians who were not confronted with routine data analyses. There were also computational hindrances because most of these people were not keen computer programmers. Estimation of the parameters were firstly intercept-only, then linear in one covariate, then linear with respect to multiple covariates, and then allowed to be additive models. Smoothing extremes data can still be a stumbling block to some, because many extreme value models are based

T.W. Yee, *Vector Generalized Linear and Additive Models*,
Springer Series in Statistics, DOI 10.1007/978-1-4939-2818-7_16

on asymptotics that require extrapolation, and thus are fundamentally model-driven. We believe, though, that a data-driven approach nested within this model-driven framework can be as useful as in any other application area.

The two most important distributions in EVT are the *generalized extreme value* (GEV) distribution and *generalized Pareto distribution* (GPD). The former applied to maximums is to EVT what the normal distribution, applied to sums, is to statistics, i.e., there is an analogy between EVT and the Central Limit Theorem based on the classical EV theory to follow. This chapter centres mainly on these two models.

16.1.1 Classical EV Theory

Let $M_N = \max(Y_1, \ldots, Y_N)$ where Y_i are i.i.d. random variables from a continuous cumulative distribution function F. Then

$$P(M_N \le y) \;=\; \prod_{j=1}^{N} P(Y_j \le y) \;=\; [F(y)]^N.$$

As $N \to \infty$, the RHS $\to 0$ so its distribution becomes degenerate and not very useful. But suppose that we can find normalizing constants a_N and $b_N > 0$ such that (this is convergence in distribution)

$$P\left(\frac{M_N - a_N}{b_N} \le y\right) \;\longrightarrow\; G(y) \tag{16.1}$$

as $N \to \infty$, where G is some proper distribution function. Then G becomes useful as it is a nondegenerate limiting distribution function. As a simple example, suppose that $Y_j \sim \mathrm{Exp}(\lambda = 1)$ independently, hence $F(y) = 1 - e^{-y}$. For $b_N = 1$ and $a_N = \log N$,

$$
\begin{aligned}
P(M_N - \log N \le y) &= [F(y + \log N)]^N = \left[1 - e^{-y-\log N}\right]^N \\
&= \left[1 - \frac{e^{-y}}{N}\right]^N \to \exp\left\{-e^{-y}\right\}.
\end{aligned}
$$

In a famous result called the "Extremal Types Theorem"[1] Fisher and Tippet in the 1920s showed that G is necessarily one of the following 3 possible types.

- **Weibull-type** ($\xi < 0$ in (16.2))

  $G(y) = \exp\{-(-y)^{-1/\xi}\}$ for $-\infty < y < 0$, and 1 otherwise, is the standard form. This has been used for wind speed, sea levels and temperature data.

- **Gumbel-type** ($\xi = 0$ in (16.2))

  $G(y) = \exp\{-e^{-y}\}$ for $-\infty < y < \infty$, is the standard form. This distribution was named after the pioneer Emil Gumbel, and is known simply as the *extreme*

---

[1] Also known as the *extreme value trinity theorem* or *three types theorem*.

Table 16.1 VGAM family functions for extreme value distributions. Notes: (i) All distributions have $\mu \in \mathbb{R}$, $\sigma > 0$, $\xi \in \mathbb{R}$. (ii) All family functions have associated dpqr-type functions. (iii) Plotting functions include guplot(), meplot(), qtplot(), rlplot(). (iv) The fitted values are percentiles by default, but the fitted mean is also available.

Distribution	CDF $F(y; \boldsymbol{\theta})$	Support	VGAM family
GEV(μ, σ, ξ)	$\exp\left\{ -\left[1 + \xi \left(\dfrac{y-\mu}{\sigma} \right) \right]_+^{-1/\xi} \right\}$	$(-\infty, \mu - \sigma/\xi)$ if $\xi < 0$, $(\mu - \sigma/\xi, \infty)$ if $\xi > 0$	gev()
Generalized Pareto(σ, ξ)	$1 - \left[1 + \xi \left(\dfrac{y-\mu}{\sigma} \right) \right]_+^{-1/\xi}$	$(\mu, \mu - \sigma/\xi)$ if $\xi < 0$, (μ, ∞) if $\xi \geq 0$	gpd()
Gumbel(μ, σ)	$\exp\left\{ -\exp\left[-\left(\dfrac{y-\mu}{\sigma} \right) \right] \right\}$	$(-\infty, \infty)$	gumbel()

value distribution. It is perhaps the most widely applied statistical distribution for climate modelling, and is also known as the Gumbel and log-Weibull distributions.

- **Fréchet-type** ($\xi > 0$ in (16.2))

  $G(y) = \exp\{-y^{-1/\xi}\}$ for $0 < y < \infty$, and 0 otherwise, is the standard form. This has been used for stream flow, rainfall, and economic analyses.

These are also known as Type III, I and II extreme value distributions, respectively, according to Gumbel (1958), and also as EV3, EV1 and EV2 by others.

Figure 16.1 displays some sample densities of the three extremal types. The Gumbel-type is characterized by a thin (some say medium) tail and positive-skew, whereas the Fréchet-type has a heavy tail and infinite higher-order moments. The Weibull-type has a short tail which is bounded at $\mu - \sigma/\xi$. For the latter model, it is therefore impossible to exceed some threshold—the fact that there is an upper bound is of special relevance for certain applications, e.g., building an engineering object to withstand the largest possible catastrophe. Some applications such as tensile strength have a natural lower bound of zero.

The above theory is very easily adapted to handle minimums by the relationship $\min(Y_1, \ldots, Y_N) = -\max(-Y_1, \ldots, -Y_N)$. Some applications of these are the fastest running times for the 100m sprint each year, and the study of monthly minimum temperatures.

In practice, it is common to block the data so that the three types theorem can be thought of as applying to each block. For example, if daily maximum temperatures are recorded, then each block is a calendar year, $N_i = 365$ or 366, $n =$ the number of years of data and $y_i =$ the annual maximum temperatures. This is in keeping with the notation $(\boldsymbol{x}_i, \boldsymbol{y}_i)$, $i = 1, \ldots, n$, used to represent data in the VGLM/VGAM framework. Later, when considering the r-largest order statistics, the response is $\boldsymbol{y}_i$ with elements satisfying $y_{i1} \geq y_{i2} \geq \cdots \geq y_{ir_i}$.

Care must be taken when blocking data to ensure independence. For example, if the response is wave heights or surge heights, then a big storm passing through in very late December is likely to induce dependencies between both calendar years.

The problem of trying to create a subset of hopefully independent data from dependent responses is often tackled by the *declustering* method. This is particularly true with exceedances data and GPD modelling (Sect. 16.3).

In the past, controversy would often result because practitioners chose one of these types—without much justification—to model their data.

16.2 GEV

It has been realized that it is more convenient to consider the generalized extreme value (GEV) distribution, a family which holds the three types as special cases. Its cumulative distribution function can be written

$$G(y; \mu, \sigma, \xi) = \exp\left\{-\left[1 + \xi\left(\frac{y - \mu}{\sigma}\right)\right]_+^{-1/\xi}\right\}, \qquad \sigma > 0, \qquad (16.2)$$

$-\infty < \mu < \infty$, and the subscript "+" means its support is y satisfying $1 + \xi(y - \mu)/\sigma > 0$. The μ, σ and ξ are the location, scale and shape parameters, respectively.

For the GEV distribution, the kth moment about the mean exists if $\xi < k^{-1}$. Provided that they exist, the mean and variance are given by $\mu + \sigma\{\Gamma(1-\xi)-1\}/\xi$ and $\sigma^2\{\Gamma(1 - 2\xi) - \Gamma^2(1 - \xi)\}/\xi^2$, respectively.

The GEV distribution can be fitted using the VGAM family function `gev()`, which handles multiple responses as well (see Sect. 16.2.1). It has

```
> args(gev)

function (llocation = "identitylink", lscale = "loge",
    lshape = logoff(offset = 0.5), percentiles = c(95, 99), iscale = NULL,
    ishape = NULL, imethod = 1, gshape = c(-0.45, 0.45), tolshape0 = 0.001,
    type.fitted = c("percentiles", "mean"), giveWarning = TRUE, zero = 2:3)
NULL
```

so that

$$\boldsymbol{\eta} = \left(\mu, \log \sigma, \log\left(\xi + \frac{1}{2}\right)\right)^T \qquad (16.3)$$

is its default. The first value of argument `type.fitted` indicates that percentiles are returned as the fitted values; for this, the values of argument `percentiles` are used. However, the mean can be returned instead if `type.fitted = "mean"`.

Why such a link for ξ in (16.3)? Smith (1985) established that if $\xi > -0.5$ then the MLEs are completely regular and the usual asymptotic properties apply. The $\frac{1}{2}$ in the above formula is represented by the default setting of the `lshape = logoff(offset = 0.5)` argument. For $-1 < \xi < -0.5$, the MLE generally exists and is superefficient. One might allow for $-1 < \xi$ by setting `lshape = logoff(offset = 1)`. If $\xi < -1$, then the MLE generally does not exist as it effectively becomes a 2-parameter problem; one might try `lshape = "identitylink"` but MLEs are unlikely to be obtainable for this case in general. In practice, it is quite uncommon to encounter the case $\xi < -\frac{1}{2}$, hence it should be a common requirement to have to change the default settings for `lshape`.

It is worth pointing out that it usually requires much data to estimate ξ accurately, typically, n of the order of hundreds or thousands of observations.

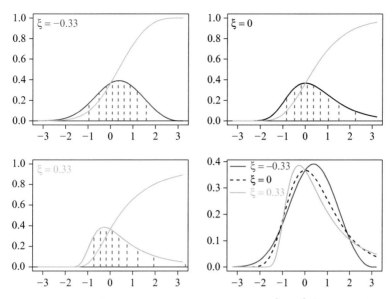

Fig. 16.1   GEV densities for values $\mu = 0$, $\sigma = 1$, and $\xi = -\frac{1}{3}$, 0, $\frac{1}{3}$ (Weibull-, Gumbel- and Fréchet-types, respectively). The *orange curve* is the CDF, the *dashed purple segments* divide the density into areas of $\frac{1}{10}$. The bottom RHS plot has the densities overlaid.

Consequently, the shape parameter should be modelled as simply as possible, and the `zero` argument of `gev()` reflects this by choosing intercept-only for this parameter. Simulation can be used to confirm this property.

Along the same lines, it can be seen from above that the scale parameter is modelled more simply than the location parameter, by default. This might not be a good idea because the variability trend, not the mean trend, drives extremes, at least for an upper tail of power-type or gamma-type (Withers and Nadarajah, 2009), i.e., the trend in scale dominates the trend in location. Consequently, modelling the scale parameter is not unimportant relative to the location parameter, and the software defaults might change in the future to reflect this.

In terms of quantiles, the inversion of (16.2) yields

$$y_p = \mu - \frac{\sigma}{\xi}\left\{1 - [-\log(1-p)]^{-\xi}\right\}, \qquad \xi \neq 0, \tag{16.4}$$

so that $G(y_p) = 1 - p$. In extreme value terminology, y_p is called the *return level* associated with the *return period* p^{-1}. For example, in terms of waiting times, if every year there is a probability p of an event occurring, and if the years are independent, then y_p is the level expected to be exceeded, on average, once every p^{-1} years. Alternatively, it may be interpreted as the mean number of events occurring within a unit time period, e.g., one year. The argument `percentiles` of `gev()` allows users to specify values of p by assigning the values $100(1-p)$, and the return levels can then be obtained by the `fitted()` methods function, as the example of Sect. 16.6.1 shows. Coles (2001, p.56) give details for $SE(\widehat{y}_p)$ based on the delta method; however, it is only for intercept-only models.

A common strategy is to test whether the data is Gumbel by fitting a GEV and conducting a likelihood ratio test of $H_0 : \xi = 0$ versus $H_1 : \xi \neq 0$. A Wald test is also common but is usually less accurate (Sect. 2.3.6.2). It is common to use

Table 16.2 Subset of the Venice sea levels data (data frame `venice`). For each year from 1931–1981, the 10 highest daily sea levels (cm) are recorded (except for 1935, with 6 values).

year	r1	r2	r3	r4	r5	r6	r7	r8	r9	r10	year	r1	r2	r3	r4	r5	r6	r7	r8	r9	r10
1931	103	99	98	96	94	89	86	85	84	79	1976	124	122	114	109	108	108	104	104	102	100
1932	78	78	74	73	73	72	71	70	70	69	1977	120	102	100	98	96	96	95	94	91	90
1933	121	113	106	105	102	89	89	88	86	85	1978	132	114	110	107	105	102	100	100	100	99
1934	116	113	91	91	91	89	88	88	86	81	1979	166	140	131	130	122	118	116	115	115	112
1935	115	107	105	101	93	91	NA	NA	NA	NA	1980	134	114	111	109	107	106	104	103	102	99
1936	147	106	93	90	87	87	87	84	82	81	1981	138	136	130	128	119	110	107	104	104	104

profile likelihoods in EVT to reduce the dimension of the optimization problem, however, this is not really needed in the VGLM/VGAM framework.

16.2.1 The r-Largest Order Statistics

Suppose now that instead of recording the maximum value, we record the most extreme r_i values (at a fixed value of $\boldsymbol{x}_i$). Thus this *block* data can be written $(\boldsymbol{x}_i, \boldsymbol{y}_i)^T$, where $\boldsymbol{y}_i = (y_{i1}, \ldots, y_{ir_i})^T$ with the property that $y_{i1} \geq y_{i2} \geq \cdots \geq y_{ir_i}$. Given $\boldsymbol{x}_i$, the data (not just the extremes) are assumed to be i.i.d. realizations from some distribution with continuous CDF F. Here are some examples of this type of data:

- Table 16.2 provides data where, for each year between $x = 1931$ and 1981, the 10 highest sea levels in Venice were measured (except for 1935 where the 6 highest were recorded). Thus $r_i = 10$ for all but one i.
- The top 10 runners in each age group in a school are used to estimate the 99 percentile of running speed as a function of age.

Let $Y_{(1)}, \ldots, Y_{(r)}$ be the r largest observations in any particular block, such that $Y_{(1)} \geq \cdots \geq Y_{(r)}$. Tawn (1988) showed that, for one block and for fixed r, as $N \to \infty$, the limiting joint distribution has density $f(y_{(1)}, \ldots, y_{(r)}; \mu, \sigma, \xi) =$

$$\sigma^{-r} \exp\left\{ -\left[1 + \xi\left(\frac{y_{(r)} - \mu}{\sigma}\right)\right]^{-1/\xi} - \left(1 + \xi^{-1}\right) \sum_{j=1}^{r} \log\left[1 + \xi\left(\frac{y_{(j)} - \mu}{\sigma}\right)\right] \right\},$$

for $y_{(1)} \geq \cdots \geq y_{(r)}$, and $1 + \xi(y_{(j)} - \mu)/\sigma > 0$ for $j = 1, \ldots, r$.

The `gev()` family function can handle this type of $r_i \geq 1$ data. If the r_i are not all equal, then the response should be a matrix padded with NAs. The data frame `venice` (Table 16.2) is an example of this. A special case of the block-GEV model is the block-Gumbel model described in the next section.

16.2.2 The Gumbel and Block-Gumbel Models

In terms of the three types theorem, the Gumbel distribution accommodates many commonly used distributions such as the normal, lognormal, logistic, gamma, exponential and Weibull. However, the rate of convergence of (16.1) varies enormously, e.g., the standard exponential converges quickly, but for the standard normal it is extremely slow. The Gumbel CDF is

$$G(y) = \exp\left\{-\exp\left[-\left(\frac{y-\mu}{\sigma}\right)\right]\right\}, \quad -\infty < y < \infty. \quad (16.5)$$

The mean and variance are given by $E(Y) = \mu + \sigma\gamma$ (where $\gamma \approx 0.5722$ is the Euler–Mascheroni constant (A.49)) and $\mathrm{Var}(Y) = \pi^2\sigma^2/6$. Inversion of (16.5) gives the quantiles

$$y_p = \mu + \sigma\left[-\log\left(-\log(1-p)\right)\right]. \quad (16.6)$$

Suppose that, for all blocks, the maximums are Gumbel distributed. For any particular block, let $Y_{(1)}, \ldots, Y_{(r)}$ be the r largest observations such that $Y_{(1)} \geq \cdots \geq Y_{(r)}$. Given that $\xi = 0$, the *joint* distribution of

$$\left(\frac{Y_{(1)} - a_N}{b_N}, \ldots, \frac{Y_{(r)} - a_N}{b_N}\right)^T$$

has, for large N, a limiting distribution having density

$$f(y_{(1)}, \ldots, y_{(r)}; \mu, \sigma) = \sigma^{-r} \exp\left\{-\exp\left(-\left[\frac{y_{(r)} - \mu}{\sigma}\right]\right) - \sum_{j=1}^{r}\left(\frac{y_{(j)} - \mu}{\sigma}\right)\right\},$$

for $y_{(1)} \geq \cdots \geq y_{(r)}$ (Smith, 1986). Upon taking logarithms, one can treat this as an approximate log-likelihood.

The **VGAM** family function `gumbel()` implements this model. The default is $\boldsymbol{\eta}(x) = (\mu(x), \log \sigma(x))^T$. Half-stepping may occur because of the approximate likelihood, therefore setting `criterion = "coeff"` may be a good idea to avoid any resultant peculiar behaviour.

Extreme quantiles for this *block-Gumbel* model can be calculated as follows. If the $y_{i1}, \ldots, y_{ir_i}$ are the r_i largest observations from a population of size R_i at $\boldsymbol{x}_i$, then a large $\alpha = 100(1 - c_i/R_i)\%$ percentile of F can be estimated by

$$\widehat{\mu}_i - \widehat{\sigma}_i \log c_i, \quad (16.7)$$

cf. (16.5). For example, for the Venice data, $R_i = 365$ (if *all* the data were collected, then there would be one observation per day each year resulting in 365 observations) and so a 99 percentile is obtained from $\widehat{\mu}_i - \widehat{\sigma}_i \log(3.65)$. When R_i is missing (which is the default, signified by the argument `R = NA`), then the $\alpha\%$ percentile of F can be estimated using $c_i = -\log(\alpha/100)$ in (16.7).

The *median predicted value* (MPV) for a particular year is the value for which the maximum of that year has an even chance of exceeding, i.e., the probability is $\frac{1}{2}$. It corresponds to $c_i = \log(\log(2)) \approx -0.367$ in (16.7). To obtain this, set `mpv = TRUE`.

From a practical point of view, one weakness of the block-Gumbel model is that often there is insufficient data to verify the assumption of $\xi = 0$. Also, it is recommended that $r_i \ll R_i$ because the convergence rate to the limiting joint distribution drops sharply as r_i increases.

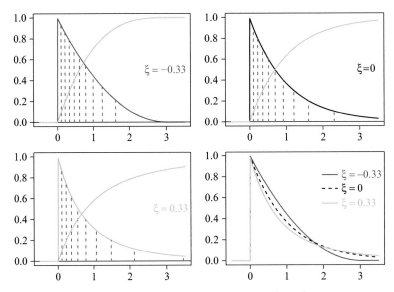

Fig. 16.2   GPD densities for values $\mu = 0$, $\sigma = 1$, and $\xi = -\frac{1}{3}$, 0, $\frac{1}{3}$ (beta-, exponential- and Pareto-types, respectively). The *orange curve* is the CDF, the *dashed purple segments* divide the density into areas of $\frac{1}{10}$. The bottom RHS plot has the densities overlaid.

16.3 GPD

The second important distribution in EVT is the generalized Pareto distribution (GPD). Giving rise to what is known as the threshold method, this common approach is based on *exceedances* over high thresholds. The idea is to pick a high threshold value u and study all the exceedances of u, i.e., values of Y greater than u. In extreme value terminology, $Y - u$ are the *excesses*. For deficits below a low threshold u, these may be converted to the upper tail by $-M_N = \max(-(Y_1 - u), \ldots, -(Y_N - u))$ where $-(Y_i - u)$ are known as deficits.

The GPD approach is considered superior to GEV modelling for several reasons. One is that it makes more efficient use of data: although the GEV can model the top r values (Sect. 16.2.1), the GPD models any number of observations above a certain threshold, therefore is more general. GPD modelling allows explanatory variables to be more efficiently used to explain the response, which is fully in line with the VGLM/VGAM framework. Compared to the GEV approach, this so-called *peaks over thresholds* (POT) approach also assumes $Y_1, Y_2, \ldots$ are an i.i.d. sequence from a marginal distribution F.

Suppose that Y has CDF F, and given $Y > u$, let $Y^* = Y - u$. Then

$$P(Y^* \leq y^*) \;=\; P(Y \leq u + y^* | Y > u) \;=\; \frac{F(u + y^*) - F(u)}{1 - F(u)}, \quad y^* > 0. \quad (16.8)$$

If $P(\max(Y_1, \ldots, Y_N) \leq y) \approx G(y)$ for G in (16.2), and for sufficiently large u, then the distribution of $Y - u | Y > u$ is approximately that of the GPD. That is, the GPD is the limiting distribution of (scaled) excesses above high thresholds.

The GPD was proposed by Pickands (1975), and has CDF

$$G(y; \mu, \sigma, \xi) \;=\; 1 - \left[1 + \xi\left(\frac{y-\mu}{\sigma}\right)\right]^{-1/\xi}, \qquad \sigma > 0, \tag{16.9}$$

for $1 + \xi(y-\mu)/\sigma > 0$. The μ, σ and ξ are the location, scale and shape parameters, respectively.

As with the GEV, there is a "three types theorem" to the effect that the following 3 cases can be considered, depending on ξ in (16.9).

- **Beta-type** $(\xi < 0)$

  G has support on $\mu < y < \mu - \sigma/\xi$. It has a short tail and a finite upper endpoint.

- **Exponential-type** $(\xi = 0)$

  $G(y) = 1 - \exp\{-(y - \mu)/\sigma\}$. The limit $\xi \to 0$ in the *survivor function* $1 - G$ gives the shifted exponential with mean $\mu + \sigma$ as a special case. This is a thin (some say medium) tailed distribution with the "memoryless" property $P(Y > a + b \,|\, Y > a) = P(Y > b)$ for all $a \geq 0$ and $b \geq 0$.

- **Pareto-type** $(\xi > 0)$

  $G(y) \sim 1 - cy^{-1/\xi}$ for $y > \mu$ and some $c > 0$. The tail is heavy, and follows Pareto's "power law" $g(y) \propto y^{-\alpha}$ for some α.

An example of each type can be seen in Fig. 16.2. The mean and variance are given by $\mu + \sigma/\{1 - \xi\}$ and $\sigma^2/\{(1 - 2\xi)(1 - \xi)^2\}$, provided that $\xi < 1$ and $\xi < \frac{1}{2}$, respectively.

In practice, choosing a threshold may be a delicate matter. The bias-variance tradeoff means that if u is too high, then the reduction in data means higher variance. Many applications of EVT do not have sufficient data anyway because extremes are often rare events, therefore information loss may be particularly costly.

Another practical consideration is the possible need to decluster the data. For example, very cold days are often followed by more very cold days, hence minimum daily temperatures are dependent. Then it would be dangerous to treat the $(\boldsymbol{x}_i, \boldsymbol{y}_i)$ as n independent observations. Declustering involves defining the clusters somehow so that they can be treated as being independent, selecting an observation from each cluster, and fitting a GPD to those observations.

It can be shown that the *mean excess function* of Y,

$$E(Y - u \,|\, u < Y) \;=\; \frac{\sigma + \xi u}{1 - \xi}, \tag{16.10}$$

which holds for any $u > \mu$, provided that $\xi < 1$. This gives a simple diagnostic for threshold selection: the *residual mean life* (16.10) should be linear with respect to u at levels for which the model is valid (Sect. 16.4.3). This suggests producing an empirical plot of the residual life plot and looking for linearity. If so, then the slope is $\xi/(1 - \xi)$. It involves plotting the sample mean of exceedances of u, versus u. This is known as a *mean life residual plot* or a *mean excess plot*. It is implemented in VGAM with `meplot()`.

VGAM fits the GPD by the family function `gpd()`, which accepts μ as known input, and internally operates on the excesses $y - \mu$. It has

```
> args(gpd)

function (threshold = 0, lscale = "loge", lshape = logoff(offset = 0.5),
    percentiles = c(90, 95), iscale = NULL, ishape = NULL, tolshape0 = 0.001,
    type.fitted = c("percentiles", "mean"), giveWarning = TRUE,
    imethod = 1, zero = -2)
NULL
```

as its defaults. Note that the working weight matrices $\mathbf{W}_i$ are positive-definite only if $\xi > -\frac{1}{2}$, and this is ensured with the default `lshape = logoff(offset = 0.5)` argument: $g_2(\xi) = \log(\xi + \frac{1}{2})$.

The fitted values of `gpd()` are percentiles obtained from inverting (16.9):

$$y_p = \mu + \frac{\sigma}{\xi}\left[p^{-\xi} - 1\right], \qquad 0 < p < 1. \tag{16.11}$$

If $\xi = 0$, then

$$y_p = \mu - \sigma \log(1 - p). \tag{16.12}$$

The mean is returned as the fitted value if `type.fitted = "mean"` is assigned.

In terms of the regularity conditions, the GPD is very similar to the GEV. Smith (1985) showed that for $\xi > -\frac{1}{2}$ the EIM is finite and the classical asymptotic theory of MLEs is applicable, while for $\xi \leq -\frac{1}{2}$, the problem is nonregular and special procedures are needed.

The POT approach is often linked with point processes (PP). Here, the probability of no event in $[0, T]$ is $e^{-\lambda t}$, and the mean number of events in this time period is λt, where λ is a rate parameter. Then a GEV(μ, σ, ξ) is related to a PP(λ, σ^*, ξ) via

$$-\log \lambda = -\frac{1}{2}\log\left(1 + \xi[(y - \mu)/\sigma]\right), \tag{16.13}$$

$$\sigma^* = \sigma + \xi(u - \mu), \tag{16.14}$$

where ξ coincides for both models.

16.4 Diagnostics

Based on the above theory, a number of diagnostic plots have been proposed in order to check some of the underlying assumptions. They apply to intercept-only models only. Some of the following are based on order statistics of the response $y_{(1)} \leq y_{(2)} \leq \cdots \leq y_{(n)}$.

16.4.1 Probability and Quantile Plots

Probability plots plot $\left(\widehat{G}(y_{(i)}), i/(n+1)\right)$ for $i = 1, \ldots, n$. Ideally, the points should hug the line $x = y$ if G is reasonable.

Quantile plots are similar and plot the points $\left(y_{(i)}, \widehat{G}^{-1}(i/(n+1))\right)$, for $i = 1, \ldots, n$. As with probability plots, the points should fall on the line $x = y$ if F has been specified correctly.

Some authors call probability and quantile plots by the names P-P and Q-Q plots.

16.4.2 Gumbel Plots

Gumbel plots are simply one special case of a quantile plot, using the Gumbel as the reference distribution. It is a good idea to perform one before more formal GEV regression fitting because curvature may indicate Weibull or Fréchet forms. As with the quantile plot in general, outliers may be detected. From its CDF (16.5), the Gumbel plot is a scatter plot of the points $\left(-\log(-\log[(i - \frac{1}{2})/n]), \ y_{(i)}\right)$, for $i = 1, \ldots, n$.

Although the `venice` data is looked at in some detail in Sect. 16.6.2, we take the liberty to Gumbel plot the highest and second highest annual sea levels.

```
> coords1 <- guplot(with(venice, r1), col = "blue", pch = 16)
> with(coords1, abline(lsfit(x, y), col = "gray50", lty = "dashed"))
> coords2 <- guplot(with(venice, r2), col = "blue", pch = 16)
> with(coords2, abline(lsfit(x, y), col = "gray50", lty = "dashed"))
```

This gives Fig. 16.3. A simple linear regression line has been added to each plot just to aid interpreting the trend. This is easily done, because `guplot()` returns a list with components `x` and `y`. The first plot is clearly linear, but the second plot shows a little nonlinearity, however, this does not raise much concern.

16.4.3 Mean Excess Plots

As seen from the mean excess function (16.10), the mean excess plot is a diagnostic plot for the GPD to help choose a suitable threshold, u_0, say. This involves plotting the mean excess over u versus values of u. If the data are GPD, then a straight line is expected beyond u_0. As seen from Fig. 16.8b, sometimes this is not an easy task because of the jaggedness, and the assessment of linearity is rather subjective.

16.5 Some Software Details

Here are some miscellaneous notes about general features and details regarding VGAM for EV data analysis.

1. For `gev()`, VGAM may fail because of the range restriction in (16.2). If this problem occurs at initialization, then the error may be irrecoverable. After the first iteration, with `vglm()` half-stepping avoids this problem, but `vgam()` does not presently implement half-stepping. If smoothing is required and a range restriction problem occurs, then try using `vglm()` with regression splines (Sect. 2.4.3).

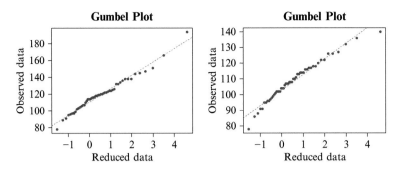

Fig. 16.3 Gumbel plot of the two highest annual sea levels of the Venice data (`venice`).

2. The three types theorem allows the shape parameter ξ to take on any sign or be equal to 0. In the situation that it is known that, e.g., $\xi < 0$ in the GEV, then `lshape = "logneg"` might be used to obtain $\eta_3 = \log(-\xi)$. Of course, if it is known that $\xi = 0$, then the `gumbel()` family function should be used.

3. Most **VGAM** family functions handle multiple (independent) responses such as `cbind(y1, y2, y3)`, and the vector of linear/additive predictors are, e.g., $(\eta_1^T, \eta_2^T, \eta_3^T)^T$. Since `gev()` and `gumbel()` handle matrix responses according to their *joint* distribution, we say that these family functions handle multivariate responses rather than multiple responses. Possibly in the future, `gevff()` and `gumbelff()` might be written to handle multiple responses. Then they might be applicable to, e.g., annual maximum temperatures data measured at 3 very far apart locations.

4. For the block-Gumbel model of Sect. 16.2.2, if $r_i > 1$ then the response for `vglm()`/`vgam()` is an $n \times \max(r_i)$ matrix, and padded with NAs if the r_i are unequal. If there are NAs, then one must use `na.action = na.pass`. However, NAs in the model matrix are disallowed. A few details about missing values are summarized in Sect. 1.5.2.6.

16.6 Examples

Some of the following examples use data sets from **ismev**.

16.6.1 Port Pirie Sea Levels: GEV Model

We will follow the analysis of the Port Pirie sea levels data described in Coles (2001, Sect.3.4.1). These data concern the annual maximum sea levels during 1923–1987 at Port Pirie, a sea port with an elevation of 4 m above sea level, and located about 200 km north of the city of Adelaide, South Australia. Of interest is to estimate the maximum sea levels over the next 100 years.

A scatter plot of the response versus year shows no particular trend, therefore we simply fit an intercept-only model as follows.

```
> data("portpirie", package = "ismev")
> fit1.pp <- vglm(SeaLevel ~ 1, gev, data = portpirie)
    # trace = TRUE is a good idea
> plot(depvar(fit1.pp) ~ Year, data = portpirie, col = "blue",
      ylab = "Sea level (m)", main = "(a)")
> matlines(with(portpirie, Year), fitted(fit1.pp), col = 1:2, lty = 2)
```

This produces Fig. 16.4a. The default quantiles are 95 % and 99 %, and these are
the dashed horizontal lines. Some diagnostic plots are given in Fig. 16.4b–d, from
the following code.

```
> n <- nobs(fit1.pp)
> params3 <- predict(fit1.pp, untransform = TRUE)
>
> plot(pgev(sort(depvar(fit1.pp)), loc = params3[, "location"],
            scale = params3[, "scale"], shape = params3[, "shape"]),
        ppoints(n), xlab = "Empirical", ylab = "Model", col = "blue",
        main = "(b)")
> abline(a = 0, b = 1, col = "gray50", lty = "dashed")
>
> plot(qgev(ppoints(n), loc = params3[, "location"],
            scale = params3[, "scale"], shape = params3[, "shape"]),
        sort(depvar(fit1.pp)), xlab = "Model", ylab = "Empirical", col = "blue",
        main = "(c)")
> abline(a = 0, b = 1, col = "gray50", lty = "dashed")
>
> hist(depvar(fit1.pp), prob = TRUE, col = "wheat", breaks = 9, main = "(d)")
> Range <- range(depvar(fit1.pp))
> Grid <- seq(Range[1], Range[2], length = 400)
> lines(Grid, dgev(Grid, loc = params3[1, "location"], scale = params3[1, "scale"],
                   shape = params3[1, "shape"]), col = "blue")
```

The model seems to fit satisfactorily. The estimated parameters and their standard
errors are

```
> Coef(fit1.pp)   # Only for intercept-only models

  location    scale    shape
   3.87475  0.19804 -0.05010

> (SEs <- sqrt(diag(vcov(fit1.pp, untransform = TRUE))))

  location    scale    shape
  0.027337 0.019468 0.082086
```

An approximate 95 % confidence interval for ξ is given by

```
> c(Coef(fit1.pp)["shape"] + 1.96 * c(-1, 1) * SEs["shape"])

  [1] -0.21099  0.11079
```

The value 0 is contained inside, hence it is not possible to distinguish which of
the 3 types of GEV it is. This is not surprising since $n = 65$ is rather small.

The above results agree with Coles (2001), whose SEs were 0.028, 0.020
and 0.098 for the 3 parameters, respectively. The difference is probably due to
the OIM being used instead of our EIM. His approximate 95 % confidence interval for ξ was $[-0.242, 0.142]$, and he obtained greater accuracy by repeating his
calculations using a profile likelihood and obtained $[-0.21, 0.17]$.

Now the estimates of the 10-year and 100-year return levels can be obtained by

```
> fit1.pp@extra$percentiles <- c(90, 99)
> head(fitted(fit1.pp, type.fitted = "percentiles"), 1)

        90%    99%
1  4.2962 4.6884
```

Details are not given here about how to construct an approximate 95 % confidence interval for these. But to predict 100 years in advance requires some serious thought, because such long-distance extrapolation into the future is based on the stationarity assumption made in the analysis, and that excludes the effect of future climate change.

To close up, Fig. 16.4d–e give return level plots produced by

```
> rlplot(fit1.pp, main = "(e)", pcol = "blue")
> rlplot(fit1.pp, main = "(f)", pcol = "blue", log = FALSE)
```

Linearity means a Gumbel model; convexity corresponds to $\xi < 0$ with an asymptote of $\mu - \sigma/\xi$ as $p \to 0$. If $\xi > 0$, then the plot is concave. Here, the plot looks linear, which is in agreement with the possibility that the GEV is of the Gumbel form.

16.6.2 Venice Sea Levels: The Block-Gumbel Model

We will fit a block-Gumbel model to the **venice** sea levels data of Table 16.2. Note that, in 1935, only the top 6 values are available. A preliminary VGAM fitted to all the data is

```
> fit1 <- vgam(cbind(r1, r2, r3, r4, r5, r6, r7, r8, r9, r10) ~ s(year,
               df = c(9, 3)),
               gumbel(R = 365, mpv = TRUE), data = venice, na.action = na.pass)
```

A large amount of flexibility is purposely allocated to the location parameter because a sneak preview of a scatter plot of these data confirms what is mentioned in Smith (1986) as two cycles—an astronomical one with a 19-year period, and an inexplicable one having an 11-year period. There is also a possible outlier with a value of almost 200. It is left to the reader (Ex. 16.5) to redo this analysis with **df = 3**, say, in the above, and deleting the outlier, to show how to gloss over the effect of these cycles and whether the results are sensitive to the outlier.

Figure 16.5 was produced by

```
> plot(fit1, se = TRUE, lcol = "blue", scol = "limegreen", slty = "dashed")
```

It appears that the first function, μ, is nonlinear, and that the second function, σ, may be a quadratic. The nonlinearity can more formally be tested by

```
summary(fit1)
```

which shows that both functions are significantly nonlinear. Let's fit such a model with

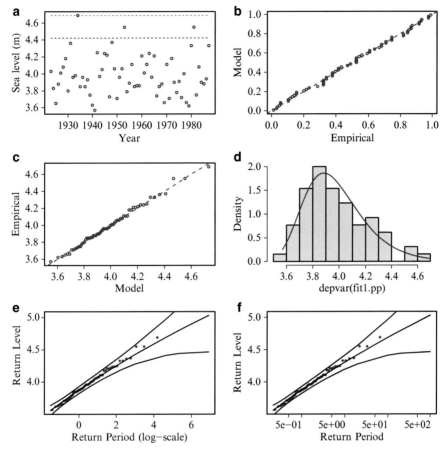

Fig. 16.4 Intercept-only GEV model fitted to the `portpirie` annual maximum sea levels data. (**a**) Scatter plot, and the *dashed horizontal lines* are the resulting 95 % and 99 % quantiles. (**b**) Probability plot. (**c**) Quantile plot. (**d**) Density plot. (**e**)–(**f**) Return level plots, with slight changes in the *x*-axis labelling.

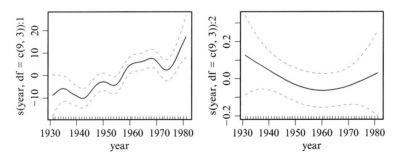

Fig. 16.5 VGAM fitted to the Venice sea level data (`fit1`).

```
> Hlist <- list("(Intercept)" = diag(2), "s(year, df = 9)" = rbind(1, 0),
```

```
                    "poly(year, 2)" = rbind(0, 1))
> fit2 <- vgam(cbind(r1, r2, r3, r4, r5, r6, r7, r8, r9, r10) ~
               s(year, df = 9) + poly(year, 2), gumbel(R = 365, mpv = TRUE),
               constraints = Hlist, data = venice, na.action = na.pass)
> head(fitted(fit2), 3)

       95%    99%    MPV
1  67.072 90.595 114.87
2  68.473 91.575 115.42
3  69.884 92.587 116.02
```

Then the quantile plot in Fig. 16.6a was produced by

```
> with(venice, matplot(year, cbind(r1, r2, r3, r4, r5, r6, r7, r8, r9, r10),
                       ylab = "Sea level (cm)", type = "p", main = "(a)",
                       ylim = c(70, 200), col = "blue", pch = 16))
> mycols <- c(1, 2, 3)
> with(venice, matlines(year, fitted(fit2), lty = 1, col = mycols))
```

Clearly, it is seen that there is a general increase in extreme sea levels over time (and/or that Venice is sinking), and that there is cyclical behaviour.

Now, for purely illustrative purposes, we obtain the quantile plot of an underfitted model where the location parameter is linear in **year** and the scale parameter is intercept-only. The purpose is to compare the quantile plots of the two models.

```
> fit3 <- vglm(cbind(r1, r2, r3, r4, r5, r6, r7, r8, r9, r10) ~ year,
               gumbel(R = 365, mpv = TRUE, zero = 2), data = venice,
               na.action = na.pass)
```

Following (16.7),

```
> qtplot(fit3, mpv = TRUE, lcol = mycols, tcol = mycols, pcol = "blue",
         pch = 16, tadj = -0.2, main = "(b)", ylim = c(70, 200), ylab = "")
```

This produces Fig. 16.6b. The quantiles of **fit3** are linear functions, and not surprisingly, they convey the main message of an increase during the whole time period.

Incidentally, **fit3** might be fitted equivalently by using

```
Select(venice, "r", sort = FALSE) ~ year
```

instead in the formula.

The plots include the 99 percentiles of the distribution. That is, for any particular year, we should expect $99\% \times 365 \approx 361$ observations below the line, or equivalently, 4 observations above the line. To check this, Fig. 16.7a was produced by

```
> plot(r4 ~ year, data = venice, ylab = "sea level",
       main = "(a)", ylim = range(r1), pch = "4", col = "blue")
> lines(fitted(fit2)[, "99%"] ~ year, data = venice, col = "orange")
```

The points of this plot are the fourth highest annual sea level recorded ($4/365 \approx 1\%$) and the fitted 99 percentile curve of the block-Gumbel model (same as Fig. 16.6a).

Finally, the MPV can be seen in Fig. 16.7b. It was produced by

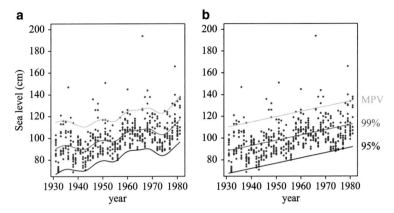

Fig. 16.6 (a) Quantile plot of the Venice sea level data (fit2). (b) Venice data overlaid with the fitted values of fit3. This model underfits.

```
> plot(r1 ~ year, data = venice, ylab = "sea level",
       main = "(b)", ylim = range(r1), pch = "1", col = "blue")
> lines(fitted(fit2)[ , "MPV"] ~ year, data = venice, col = "orange")
```

The MPV for a particular year is the value for which the maximum of that year has an even chance of exceeding. It is evident from this plot too that the sea level is increasing over time and/or Venice is sinking. As this data set is now rather truncated because its last year was over 30 years ago, it would be interesting to add data from 1981 to 2014, say, and repeat the analysis. This is especially relevant and interesting in an age of concern and uncertainty about global climate change and the rising of sea levels.

16.6.3 Daily Rainfall Data: The GPD Model

This is a quite large data set consisting of the daily rainfalls at a location in southwest England during 1914–1962 (Fig. 16.8a). There are 17531 observations in total, of which 47 percent have 0 values. We will fit a simple GPD intercept-only model to these data, mimicking the analysis of Coles (2001, Sect.4.4.1). Before fitting two equivalent models, we first place the data into a data frame—this is better practice than having one large vector, the original format of the data. As well, the mean excess plot was obtained (Fig. 16.8b):

```
> data("rain", package = "ismev")
> Rain <- data.frame(rain = rain, day = 1:length(rain))
> plot(rain ~ day, data = Rain, pch = 16, col = "blue", cex = 0.5, main = "(a)")
> meplot(with(Rain, rain), main = "(b)")
```

The plot appears to be piecewise-linear, with a knot around 60. Beyond this knot, there are very few observations and this is shown by the large 95 % confidence intervals. There might be a knot at around 30, because the segment running from 30 to 60 appears linear with a different slope to the segment running from 0 to 30.

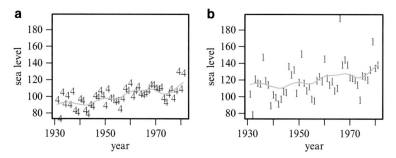

Fig. 16.7 (**a**) The fitted 99 percentile of `fit2` and the 99 percentile data values (4th highest sea level each year). (**b**) Fitted median predicted value (MPV) of the Venice sea level data from `fit2`. The points are the highest sea levels for each year.

Hence, choosing a threshold of 30 seems reasonable. We fit two equivalent GPD models with this choice:

```
> fit1.rain <- vglm(rain - 30 ~ 1, gpd, data = Rain, subset = rain > 30)
> fit2.rain <- vglm(rain ~ 1, gpd(threshold = 30), data = Rain,
    subset = rain > 30)
> length(depvar(fit2.rain))  # Effective sample size, n

[1] 152
```

In both cases, a subset of `Rain` is fed into `vglm()`. In `fit1.rain`, the threshold is explicitly subtracted from the response, whereas it is inputted as an argument in `fit2.rain`. The model `fit2.rain` is preferred, because it stores the value of the argument `threshold` on the object. Note that the saved responses of the 2 models differ by the threshold value:

```
> c(range(depvar(fit1.rain)), range(depvar(fit2.rain)),
    fit2.rain@extra$threshold[1])

[1]  0.2 56.6 30.2 86.6 30.0
```

The estimates and variance-covariance matrix are

```
> Coef(fit2.rain)

   scale   shape
 7.44047 0.18447

> round(VC2 <- vcov(fit2.rain, untransform = TRUE), digits = 5)

          scale     shape
scale   0.86263  -0.05799
shape  -0.05799   0.00923
```

These differ a little from Coles (2001, Sect.4.4.1) but are qualitatively the same. An approximate 95 % confidence interval for ξ is given by

```
> SEs <- sqrt(diag(VC2))
> c(Coef(fit2.rain)["shape"] + 1.96 * c(-1, 1) * SEs["shape"])

[1] -0.0038601  0.3728037
```

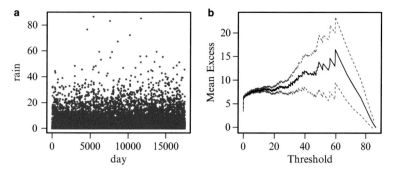

Fig. 16.8 The **rain** daily rainfall data. (**a**) Scatter plot. (**b**) Mean excess plot.

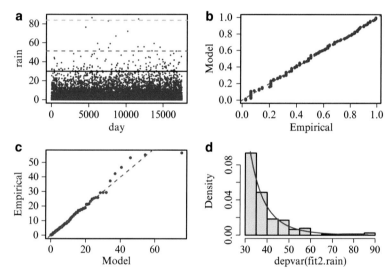

Fig. 16.9 Intercept-only GPD model fitted to the **rain** daily rainfall data. (**a**) Scatter plot, with the *solid black horizontal line* denoting the threshold at 30. The *dashed horizontal lines* are the resulting 90 % and 99 % quantiles. (**b**) Probability plot. (**c**) Quantile plot. (**d**) Density plot.

cf. his approximate 95 % confidence interval $[-0.014, 0.383]$. In both cases, the lower limit is only a little negative compared to a much larger positive upper limit. Since $\widehat{\xi} > 0$, it might be concluded that this implies that the support of the distribution is semi-infinite. The diagnostic plots given in Fig. 16.9 show that the fitted model is reasonable.

In Fig. 16.9a the fitted quantiles apply to the subset of data fed in, not the entire data set in this case. Hence, e.g., we should expect about 15.2 observations above the 90 % quantile, and there actually are 15 such observations.

Bibliographic Notes

There is now a substantial literature in EVT. An accessible introductory text is Coles (2001). Most EVT books tend to be much more theoretical in nature, and include technicalities and derivations not mentioned here; these books include Leadbetter et al. (1983), Embrechts et al. (1997), Kotz and Nadarajah (2000), Finkenstadt and Rootzén (2003), Smith (2003), Beirlant et al. (2004), Castillo et al. (2005), de Haan and Ferreira (2006), Reiss and Thomas (2007), Novak (2012). A recent review of univariate EVT is Gomes and Guillou (2015). Some further examples and details are given in Yee and Stephenson (2007). Specific important applied topics omitted in this chapter include multivariate extremes, spatial and temporal dependencies; these are covered in the above references.

Alternative software for fitting extremes are reviewed by Gilleland et al. (2013), and among the several R packages for such, there are extRemes, evd, evir, ismev, SpatialExtremes.

Exercises

Ex. 16.1.   Obtain expressions for (16.8) for the following distributions: (a) standard uniform, (b) standard exponential.

Ex. 16.2.   Consider (16.1).

(a) If the Y_j are standard uniformly distributed, $b_N = N^{-1}$ and $a_N = 1$, then find $G(y)$.
(b) Show that if the Y_j are standard Cauchy distributed, $b_N = N/\pi$ and $a_N = 0$, then $G(y) = e^{-1/y}$ (for positive y).

Ex. 16.3.   Max-Stability
A CDF F is said to be max-stable if, for all positive integers N, there exists constants a_N and $b_N > 0$ such that $[F(a_N + b_N\, y)]^N = F(y)$. Show that the standard version of the Gumbel distribution (16.5) is max-stable.

Ex. 16.4.

(a) From (16.2), obtain expressions for the density, median, and lower and upper quartiles of the 3-parameter GEV distribution.
(b) Repeat (a) for the 3-parameter GPD (16.9).

Ex. 16.5.   Venice Sea Levels Data—Without Underfitting and Overfitting
The analysis in Sect. 16.6.2 deliberately allowed some underfitting & overfitting.

(a) Make a copy of `venice` and set the outlier to `NA`.
(b) Fit a VGAM block-Gumbel model with `df = 3` for both parameters. Plot the fitted component functions with SEs. Comment.
(c) Obtain a quantile plot similar to Fig. 16.6. Comment.

Ex. 16.6.   Fitting GEV Models to Simulated Data

(a) Generate GEV data having the following parameters: $x_{i2} \sim \text{Unif}(0, 1)$, $\mu_i = x_{i2}$, $\sigma_i = 1$, $\log\left(\xi_i + \frac{1}{2}\right) = \frac{1}{4}\cos(2\pi x_{i2}) - 1$, for $n = 500$, $1000(1000)5000$. All observations are independent.

(b) Fit a VGAM GEV to each data set in (a). Comment on your results.

Ex. 16.7.   Maximum Daily Drinks Drunk in `xs.nz`

Consider the response variable `drinkmaxday` in the data frame `xs.nz`.

(a) Fit a VGAM GEV to these data, using age, sex and ethnicity as explanatory. Delete all the missing values from these variables first. For simplicity, model the scale and shape parameters as intercept-only.

(b) Plot the estimated component functions of your model and interpret them.

(c) Estimate the 25, 50, 75, 90 and 99 percentiles of the response for a 50-year-old male European. Compare your results with quantiles computed on the subset of male Europeans aged between 48 and 52 inclusive.

(d) Some of the values of `drinkmaxday` are unrealistically excessive. Delete those values which you think are not 'true', and repeat your analyses above. Does the deletion of the outliers make much of a difference to the results?

Ex. 16.8.   Longest Fish Caught per Competitor in `wffc.indiv`

Consider the response variable `length` in the data frame `wffc`, along with `wffc.indiv`.

(a) Create a data frame with each individual competitor as a row, and having at least the following 2 columns: the longest fish caught from the Waihou River, and the number of fish caught there.

(b) Obtain a scatter plot of the length of each competitor's biggest fish versus the number of fish caught. Comment.

(c) Fit a VGAM GEV to these data, with explanatory variable and response specified by the scatter plot. Overlay the 50, 75, 90, 95 percentiles on to the plot.

(d) Fit some quantile regression model to these data, and overlay the 50, 75, 90, 95 percentiles on to a scatter plot. Is there much difference between (c) and (d)?

Ex. 16.9.

(a) Show that the residual mean life (16.10) holds for the GPD.

(b) Derive an expression for (16.10) for the exponential-type GPD. Comment in terms of the memoryless property of the exponential distribution.

(c) Use simulation to demonstrate a simple example of (b).

Ex. 16.10.   Show that the limit as $\xi \to 0$ of G in (16.2) results in (16.5). Obtain the density from this.

Ex. 16.11.   Show that the difference between two independent standard Gumbel random variables has a standard logistic distribution, i.e., use $\mu = 0$ and $\sigma = 1$ in Table 16.1, and $a = 0$ and $b = 1$ in Table 12.4.

Ex. 16.12.   Show for the block-Gumbel model that

$$\frac{\partial \ell_i}{\partial \mu_i} = \sigma_i^{-1} \left\{ r_i - \exp\left(-\frac{y_{ir_i} - \mu_i}{\sigma_i} \right) \right\},$$

$$\frac{\partial \ell_i}{\partial \sigma_i} = \left\{ \sum_{k=1}^{r_i} \frac{y_{ik} - \mu_i}{\sigma_i^2} - \frac{r_i}{\sigma_i} - \frac{y_{ir_i} - \mu_i}{\sigma_i^2} e^{-(y_{ir_i} - \mu_i)/\sigma_i} \right\},$$

$$-E\left[\frac{\partial^2 \ell_i}{\partial \mu_i^2}\right] = \frac{r_i}{\sigma_i^2}, \qquad -E\left[\frac{\partial^2 \ell_i}{\partial \mu_i \partial \sigma_i}\right] = \frac{-(r_i \psi(r_i) + 1)}{\sigma_i^2},$$

$$-E\left[\frac{\partial^2 \ell_i}{\partial \sigma_i^2}\right] = \sigma_i^{-2} \left\{ 2\left[1 + r_i \psi(r_i)\right] - 2 \sum_{k=1}^{r_i - 1} \psi(k) + r_i \left[\psi'(r_i) + \psi^2(r_i) - 1\right] \right\}.$$

Ex. 16.13.   **Derivatives of the GEV Distribution**

(a) Starting at (16.2), derive the log-likelihood, and show that, as $\xi_i \to 0$,

$$\frac{\partial \ell_i}{\partial \mu} = \left(1 - e^{-z_i}\right)/\sigma_i, \qquad \frac{\partial \ell_i}{\partial \sigma} = \left[z_i \left(1 - e^{-z_i}\right) - 1\right]/\sigma_i, \qquad (16.15)$$

$$\frac{\partial \ell_i}{\partial \xi} = z_i \left[\frac{z_i}{2} \left(1 - e^{-z_i}\right) - 1\right], \qquad \text{where } z_i = (y_i - \mu_i)/\sigma_i.$$

(b) Now consider the EIM. Let $\varphi_1 = (1 + \xi)^2 \Gamma(1 + 2\xi)$ and $\varphi_2 = \Gamma(2 + \xi)$ $\{\psi(1 + \xi) + (1 + \xi)/\xi\}$. Then Prescott and Walden (1980) showed that (the subscript i will be dropped)

$$-E\left[\frac{\partial^2 \ell}{\partial \mu^2}\right] = \frac{\varphi_1}{\sigma^2}, \qquad -E\left[\frac{\partial^2 \ell}{\partial \sigma^2}\right] = \frac{1 - 2\Gamma(2 + \xi) + \varphi_1}{\sigma^2 \xi^2},$$

$$-E\left[\frac{\partial^2 \ell}{\partial \xi^2}\right] = \frac{1}{\xi^2} \left\{ \frac{\pi^2}{6} + \left(1 - \gamma + \frac{1}{\xi}\right)^2 - \frac{2\varphi_2}{\xi} + \frac{\varphi_1}{\xi^2} \right\}, \qquad (16.16)$$

$$-E\left[\frac{\partial^2 \ell}{\partial \mu \partial \sigma}\right] = \frac{-1}{\sigma^2 \xi} \left\{\varphi_1 - \Gamma(2 + \xi)\right\},$$

$$-E\left[\frac{\partial^2 \ell}{\partial \mu \partial \xi}\right] = \frac{-1}{\sigma \xi} \left(\varphi_2 - \frac{\varphi_1}{\xi}\right),$$

$$-E\left[\frac{\partial^2 \ell}{\partial \sigma \partial \xi}\right] = \frac{-1}{\sigma \xi^2} \left\{ 1 - \gamma + \frac{1 - \Gamma(2 + \xi)}{\xi} - \varphi_2 + \frac{\varphi_1}{\xi} \right\}.$$

Calculate the limit of these expressions as $\xi \to 0$. Note that (16.16) does not depend on the other 2 parameters, hence calculate its value to 2 decimal places numerically, as it is the most difficult case and appears intractable.

Ex. 16.14.   Consider the GPD applied to a response y_i, where $i = 1, \ldots, n$. Starting at (16.9), derive the log-likelihood, and show that, as $\xi_i \to 0$,

$$\frac{\partial \ell_i}{\partial \sigma_i} = \frac{1}{\sigma_i} \left[\frac{y_i}{\sigma_i} - 1\right], \qquad \frac{\partial \ell_i}{\partial \xi_i} = \frac{y_i}{\sigma_i} \left[\frac{y_i}{2\sigma_i} - 1\right]. \qquad (16.17)$$

Assume that $\mu_i = 0$.

Man's days are determined; you have decreed the number of his months and have set limits he cannot exceed.   —Job 14:5

Chapter 17
Zero-Inflated, Zero-Altered and Positive Discrete Distributions

Better to be a nobody and yet have a servant than pretend to be somebody and have no food.
—Proverbs 12:9

17.1 Introduction

Zero-truncated (*positive*), *zero-inflated* and *zero-altered* distributions are important extensions of discrete distributions, and the most common of such families occupy a happy niche within the VGLM/VGAM framework. This chapter describes those currently implemented in **VGAM**, which are summarized in Tables 17.6, 17.7. These three variants of an ordinary discrete distribution (called the *parent* here) defined on $0(1/n)1$ or $0(1)\infty$ involve very simple modifications to the parent. For example, if the parent distribution for a random variable Y^* has probability function $f(y) = P(Y^* = y)$ for $y = 0, 1, 2, 3, \ldots$ then a positive discrete distribution would arise by setting $P(Y = 0)$ to zero and the other probabilities scaled up, i.e., $Y \sim (Y^*|Y^* > 0)$ as portrayed in the right branch of Fig. 17.1a. The probability function is therefore

$$P(Y = y) \quad = \quad \frac{f(y)}{1 - f(0)} \quad \text{for } y = 1, 2, 3, \ldots. \tag{17.1}$$

Since $E[g(Y)] = (1 - f(0))^{-1} E[g(Y^*)]$ for a general function g, it follows that

$$E(Y) \quad = \quad \frac{\mu_*}{1 - f(0)} \quad \text{and} \tag{17.2}$$

$$\text{Var}(Y) \quad = \quad \frac{[1 - f(0)] \sigma_*^2 - f(0) \mu_*^2}{[1 - f(0)]^2}. \tag{17.3}$$

Some common examples include the positive Poisson (e.g., the size of crowds of people in a park, the positive number of traffic accidents recorded at an intersection, the number of items purchased per customer queuing in a supermarket checkout line) and positive binomial (e.g., the number of times individual animals are captured in a capture–recapture survey).

© Thomas Yee 2015
T.W. Yee, *Vector Generalized Linear and Additive Models*,
Springer Series in Statistics, DOI 10.1007/978-1-4939-2818-7_17

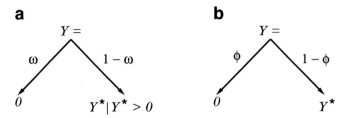

Fig. 17.1 Decision tree diagram for (**a**) zero-alteration versus (**b**) zero-inflation. They depict (17.4) and (17.7), respectively. Here, Y^* corresponds to a parent distribution such as the binomial or Poisson, and Y is the response of interest. The probabilities ω and ϕ dictate the decisions.

Zero-valued responses often require special treatment compared to the other positive values because a zero may arise from multiple sources, e.g., from a 'random' component and a 'fixed' component, for example, Y = the number of children born to an individual: males always have zero values but a female might possibly have a zero value. Another example is Y = the number of insects per leaf on a plant but a proportion ϕ of leaves are unsuitable for feeding and therefore do not have any insects. Empirically, count data is often characterized by an 'excess' of zeros relative to such distributions as the Poisson and negative binomial. There are two popular approaches for handling these (portrayed in Fig. 17.1).

1. *Zero-alteration.* Here $P(Y = 0)$ is modelled separately from $P(Y > 0)$. Then the probability function is

$$P(Y = y) = \begin{cases} \omega, & y = 0; \\ (1 - \omega) \dfrac{f(y)}{1 - f(0)}, & y = 1, 2, 3, \ldots, \end{cases} \quad (17.4)$$

with $0 < \omega < 1$ being the probability of an observed 0 (argument "pobs0" is used in VGAM). The bottom equation of (17.4) is the positive version of the parent distribution. Zero-altered distributions have been called *hurdle models* because the first of two processes, the one generating the zeros (the second is generating positive values), is a Bernoulli random variable that determines whether the response is zero or positive—in the latter case the "hurdle is crossed." Whereas zero-inflated distributions have been described as *overlapping* models because of the two sources of zeros, zero-altered distributions are known as *separated* models. It is straightforward to show (e.g., from (17.2)–(17.3)) that

$$E(Y) = \frac{(1 - \omega)\, \mu_*}{1 - f(0)} \quad \text{and} \quad (17.5)$$

$$\mathrm{Var}(Y) = \frac{1 - \omega}{1 - f(0)} \left\{ \sigma_*^2 + \frac{\mu_*^2\, [\omega - f(0)]}{1 - f(0)} \right\} \quad (17.6)$$

where μ_* and σ_*^2 pertain to the parent distribution of Y^*.

2. *Zero-inflation.* A (discrete) random variable Y has a zero-inflated distribution if it has value 0 with probability ϕ, otherwise it has some other distribution that also includes the value 0. The two processes generating this type of data

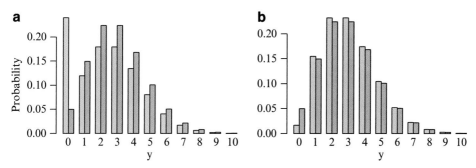

Fig. 17.2 Probability functions of a (a) zero-inflated Poisson with $\phi = 0.2$, (b) zero-deflated Poisson with $\omega = -0.035$. Both are compared to their parent distribution which is a Poisson($\mu = 3$) in *orange*.

implies that it is a mixture model of the parent distribution and a degenerate distribution with the probability concentrated at the origin. Thus

$$P(Y = y) = \begin{cases} \phi + (1 - \phi)\, f(0), & y = 0; \\ (1 - \phi)\, f(y), & y = 1, 2, 3, \ldots . \end{cases} \tag{17.7}$$

The ϕ parameter is sometimes called the probability of a *structural zero*, whereas a zero generated from the parent distribution is sometimes called a *sampling zero*. The former is represented by the argument "pstr0" in VGAM. Depending on the application, we might or might not be able to distinguish between the two types. As $\phi \to 0^+$, the zero-inflated distribution approaches its parent distribution. It is easy to show the zero-inflated model has

$$E(Y) = (1 - \phi)\, \mu_* \quad \text{and} \quad \text{Var}(Y) = (1 - \phi)\left[\sigma_*^2 + \phi\, \mu_*^2\right]. \tag{17.8}$$

A simple example of a zero-inflated Poisson distribution is Fig. 17.2a.

Which of the two models is preferable often depends on the application, e.g., zero-inflated models are used more often in ecology because their justification is stronger and there is greater interpretability.

Note that *zero-deflation* is also possible: $-f(0)/(1 - f(0)) \le \phi < 0$ in (17.7) still ensures $P(Y = 0) \ge 0$ but ϕ no longer retains its interpretation as a probability. For example,

$$P(Y = y) = I[y = 0]\, \phi + (1 - \phi)\, \frac{e^{-\lambda} \lambda^y}{y!}, \quad \phi \in (0, 1), \quad y = 0(1)\infty \tag{17.9}$$

is the usual zero-inflated Poisson (ZIP) density, but if $-(e^\lambda - 1)^{-1} \le \phi < 0$ then it is known as the zero-deflated Poisson distribution. The resulting probability of a zero count is *less* than its nominal parent's value. As $\phi \downarrow -f(0)/(1 - f(0))$, the distribution converges to its positive-distribution counterpart. A zero-deflated Poisson probability function is plotted in Fig. 17.2b.

17.1.1 Software Details

Tables 17.1, 17.6, 17.7 summarize current positive, ZA- and ZI-functions available in **VGAM**. The `za`-type and `zi`-type come in pairs: those ending in `ff` and those that don't. Each pair differs by the following:

1. Those ending in `ff` are parameterized by $\omega^* = 1 - \omega$ or $\phi^* = 1 - \phi$, which are intercept-only by default. Those that don't end in `ff` use ω or ϕ, which are not intercept-only.
2. The ω^* and ϕ^* correspond to η_M, the last linear/additive predictor. The ω and ϕ correspond to η_1.
3. Arguments `pobs0` and `pstr0` correspond to ω and ϕ, respectively. These are the probability of an observed 0 and the probability of a structural 0. Likewise, "onem" means "one minus" or $1-\cdot$, hence arguments `onempobs0` and `onempstr0` correspond to ω^* and ϕ^*, respectively.

As a specific example, `zipoisson()` and `zipoissonff()` differ in that $\boldsymbol{\eta} = (\text{logit}\,\phi, \log \lambda)^T$ in the former, and for the latter, the η_j are switched to $\boldsymbol{\eta} = (\log \lambda, \text{logit}\,\phi^*)^T$. Parameter ϕ^* is intercept-only whereas ϕ is not. Incidentally, reduced-rank regression may be applied to the ZIP; see Sect. 17.4.

Since those ending in `ff` have more parameters which are intercept-only, they are less likely to exhibit numerical problems and therefore are recommended more generally. Another reason the `ff`-type function is recommended is that they can be fed into `rcim()` more simply.

Internally, a `za`-type function is a merger of `binomialff()` and the `pos`-type function. The working weight matrices are block-diagonal, and Fisher scoring is implemented as usual.

The corresponding `[dpqr]`-type functions are available. By default, the `[dpqr]za`-type functions have `pobs0 = 0` so they behave like their respective positive distributions. Similarly, `pstr0 = 0` for the `[dpqr]zi`-type functions, hence they act like their ordinary parent distributions. Some of the `[dpqr]zi`-type functions handle zero-deflation; for these, the argument `pstr0` loses its interpretation as a probability.

17.1.2 A Zero-Inflated Poisson Example

The Poisson model is nested within the ZIP (17.9) by virtue of $\phi = 0$, therefore testing the common hypothesis of a ZIP versus Poisson can be done by a modified LRT. Note that $H_0 : Y \sim$ Poisson corresponds to a point on the boundary of the parameter space. A common alternative is a ZIP-adaption of one of Vuong (1989)'s general tests of nonnested models.

Sometimes it is required to estimate the probability that a zero response is due to a structural zero. That is, given $y_i = 0$, what is the probability its source is the structural zero? For `zipoisson()`, it involves dividing $\widehat{\phi}_i$ by $\widehat{P}(Y_i = 0) = \widehat{\phi}_i + (1 - \widehat{\phi}_i) \exp(-\widehat{\lambda}_i)$. The following code illustrates this.

Table 17.1 How **VGAM** can fit ZI-, ZA- and positive-type models. All parameters receive full maximum likelihood estimation. Most of the NB nomenclature follows Hilbe (2011). Legend: ZI = zero-inflated, ZA = zero-altered (hurdle), Pos = positive, and $\phi = P(Y = 0)$ for a structural zero in ZI-type models. Argument size is the NB parameter k. The quantities $\mathcal{A} = 1 - (1-p)^n$ and $\mathcal{B} = \{k/(k + \mu)\}^k$ apply to the ZAB and NB-variants, respectively. See also Tables 11.2, 17.6–17.7.

Variant (ZA, ZI, Pos)	$\mathrm{Var}(Y)$	Modelling function	VGAM family function
ZIB	$(1 - \phi)\, n\, p\, [1 + (\phi\, n - 1)p]$	vglm()	zibinomial()
ZIG	$(1 - \phi)\, \dfrac{1 - p}{p^2}\, [1 + \phi(1 - p)]$	vglm()	zigeometric()
ZINB	$(1 - \phi)\, \mu\, \left[1 + \mu(\phi + k^{-1})\right]$	vglm()	zinegbinomial()
ZIP	$(1 - \phi)\, \mu\, (1 + \phi\mu)$	vglm()	zipoisson()
ZAB	$\dfrac{1 - \phi}{\mathcal{A}} \left\{ \dfrac{p(1-p)}{n} + \dfrac{p^2\, (\phi - (1-p)^n)}{\mathcal{A}} \right\}$	vglm()	zabinomial()
ZAG	$(1 - \phi)(1 + \phi - p)/p^2$	vglm()	zageometric()
ZANB	$\dfrac{1 - \phi}{1 - \mathcal{B}} \left\{ \mu\left(1 + \dfrac{\mu}{k}\right) + \dfrac{\mu^2\, (\phi - \mathcal{B})}{1 - \mathcal{B}} \right\}$	vglm()	zanegbinomial()
ZAP	$\dfrac{(1 - \phi)\, \mu}{1 - e^{-\mu}} \left\{ 1 + \dfrac{\mu\, (\phi - e^{-\mu})}{1 - e^{-\mu}} \right\}$	vglm()	zapoisson()
Pos-B	$\dfrac{p(1-p)}{n\, \{1 - (1-p)^n\}} - \dfrac{p^2\, (1-p)^n}{\{1 - (1-p)^n\}^2}$	vglm()	posbinomial()
Pos-NB	$\dfrac{\mu + \mu^2/k}{1 - \mathcal{B}} - \dfrac{\mu^2\, \mathcal{B}}{(1 - \mathcal{B})^2}$	vglm()	posnegbinomial()
Pos-P	$\mu\, e^{\mu}(-1 + e^{\mu} - \mu)/(e^{\mu} - 1)^2$	vglm()	pospoisson()
COZIGAM	$\dfrac{K_1\, \mu^{1+a_{21}}\, (1 + \mu + K_1\, \mu^{a_{21}})}{[1 + K_1\, \mu^{a_{21}}]^2}$	rrvglm()	zipoissonff(zero = NULL)

```
> set.seed(123)
> zdata <- data.frame(x2 = runif(n <- 100))
> zdata <- transform(zdata, pstr0 = logit(0.5 + 1 * x2, inverse = TRUE),
                             lambda = loge(-0.5 + 2 * x2, inverse = TRUE))
> zdata <- transform(zdata, y1 = rzipois(n, lambda = lambda, pstr0 = pstr0))
> zipfit <- vglm(y1 ~ x2, zipoisson(zero = NULL), data = zdata)
> head(zdata, 1)

        x2  pstr0 lambda y1
1 0.28758 0.68731 1.0781  0
```

Table 17.2 Short summary of the notation used for the positive-Bernoulli distribution (Sect. 17.2) for capture–recapture experiments. Some additional details are in the text.

Symbol	Explanation
N	(Closed) population size to be estimated
$\mathbf{Y}$	Capture history matrix ($n \times \tau$) with values 1 (captured) and 0 (noncaptured). Each row has at least one 1
n	Total number of individual animals caught in the trapping experiment
τ	Number of sampling occasions, $\tau \geq 2$
"h"	Model $\mathcal{M}$ subscript, for *heterogeneity*
"b"	Model $\mathcal{M}$ subscript, for *behavioural* effects
"t"	Model $\mathcal{M}$ subscript, for *temporal* effects
z_{ij}	= 1 if animal i has been captured before occasion j, else = 0
p_{ij}	Probability that animal i is captured at sampling occasion j ($j = 1, \ldots, \tau$)
y_{0i}	Number of noncaptures before the first capture, for animal i
y_{r0i}	Number of noncaptures after the first capture, for animal i
y_{r1i}	Number of recaptures after the first capture, for animal i, $y_{0i} + 1 + y_{r0i} + y_{r1i} = \tau$
p_{cj}, p_{rj}	Probability that an animal is captured/recaptured at sampling occasion j
β^*	All the regression coefficients related to the capture
$Q_{s:t}$	$= \prod_{j=s}^{t} (1 - p_{cj})$. Simplifications possible, e.g., $Q_{1:\tau} = (1 - p_c)^\tau$ for $\mathcal{M}_{bh}$

The first observation has a 0 response, therefore

```
> i <- 1
> (fitted(zipfit, type.fitted = "pstr0") / fitted(zipfit, type.fitted = "pobs0"))[i]

 [1] 0.75523
```

This suggests that there is about a 75.5 percent chance that the 0 response is due to a structural zero as opposed to both sources.

17.2 The Positive-Bernoulli Distribution

The positive-binomial can be considered a special case of a 'positive-Bernoulli' distribution. The former makes the same assumptions as the binomial (fixed number and independence of trials, two outcomes, unchanging probabilities of success between trials), whereas the latter is less restrictive. In some applications, the response is a sequence of Bernoulli trials which permits an examination of some of the binomial assumptions. In this section, we describe the positive-Bernoulli distribution in the context of capture–recapture experiments for wildlife surveys, so that animals (more generally, units or individuals) are used for members of the population. Capture–recapture surveys have found numerous applications in ecology and epidemiology. Table 17.2 summarizes most of the terminology and notation used in this section.

Suppose we have a *closed* population of N animals of a certain type of species. By 'closed', we mean that there are no births or deaths, and no emigration or immigration. Such an assumption might be reasonable if the overall time period is short enough. Further to the binomial assumptions, the following are also assumed: animals do not lose their tags, and tags are recorded correctly. Animals are sampled at τ occasions, e.g., trapping a nocturnal mammal species on seven consecutive nights. If an animal is captured for the first time, then it is marked or tagged so that it can be identified individually later and it is immediately returned to the population to remix. Thus each animal caught has a capture history: a τ-vector of 1s and 0s denoting capture/recapture and noncapture, respectively. In total, we have a sample consisting of n animals, and the $n \times \tau$ response matrix $\mathbf{Y}$ has 0/1 values with rows containing at least one 1. The aim of the experiment is to estimate the unknown population size N using the observed capture history data $\mathbf{Y}$ and any other additional information collected on captured individuals such as weight or sex, or environmental information such as rainfall or temperature.

Enumerate the individuals in the sample by $i = 1, \ldots, n$ and those never captured by $i = n + 1, \ldots, N$. The general form of the full likelihood function is

$$L^\dagger \;=\; K \cdot \prod_{i=1}^{N} \prod_{j=1}^{\tau} p_{ij}^{y_{ij}} (1 - p_{ij})^{1-y_{ij}} \tag{17.10}$$

where $y_{ij} = 0$ or 1, and p_{ij} is the probability that animal i is captured at sampling occasion j. Here, K is independent of the p_{ij} but may depend on N. One can write (17.10) as

$$L^\dagger \;=\; K \cdot \left\{ \prod_{i=1}^{n} \prod_{j=1}^{\tau} p_{ij}^{y_{ij}} (1 - p_{ij})^{1-y_{ij}} \right\} \cdot \left\{ \prod_{i=n+1}^{N} \prod_{j=1}^{\tau} (1 - p_{ij}) \right\}. \tag{17.11}$$

The RHS of (17.11) is unknown, therefore cannot be used because it refers to data that has not been collected. Consequently, no MLE of N will be available unless some homogeneity assumption is made about the uncaptured animals.

In practice, a *conditional* likelihood function of the form

$$L \;=\; \prod_{i=1}^{n} \frac{\displaystyle\prod_{j=1}^{\tau} p_{ij}^{y_{ij}} (1 - p_{ij})^{1-y_{ij}}}{1 - \displaystyle\prod_{s=1}^{\tau} (1 - p_{is}^*)} \tag{17.12}$$

can be used (Huggins, 1989). The denominator is the probability that an animal is captured at least once. It is a conditional likelihood because it involves only the captured animals. The quantities p_{is}^* are the p_{ij} adjusted for possible behavioural effects—see $\mathcal{M}_b$ and $\mathcal{M}_{tb}$ below. Equation (17.12) differs from the positive binomial likelihood by the normalizing constants $\log\left(\frac{\tau}{\tau \bar{y}_i}\right)$, where $\bar{y}_i$ is the sample proportion of captures. This difference must be removed if comparisons between models are to be made using criteria such as AIC and BIC. Further details are given in Sect. 17.2.1.

Otis et al. (1978) describe 8 variant models which apply to the positive-Bernoulli model (17.12) under various assumptions, i.e., capture/recapture probabilities can depend on time (t), individual heterogeneity (h) and behavioural response (b), which are described below. Models which depend on one or a combination of these effects are defined using subscripts, e.g., $\mathcal{M}_{th}$ depends on time and heterogeneity. Half of the variants are prefixed with "h". Since the conditional likelihood (17.12) belongs to the exponential family, all 8 models are GLMs (Huggins and Hwang, 2011).

Heterogeneity models ($\mathcal{M}_h$, $\mathcal{M}_{th}$, $\mathcal{M}_{bh}$ and $\mathcal{M}_{tbh}$) allow for each individual animal to have its own probability of capture/recapture, independent of other an-

Table 17.3   Upper table gives the relationships between the 8 Otis et al. (1978) models and function calls. Note: see Table 17.5, and Sect. 3.3.1.2 details the `parallel` argument. Lower table gives the $\boldsymbol{\eta}$ for the 8 models. The g = logit link is the default for all.

Model	family =
$\mathcal{M}_0/\mathcal{M}_h$	`posbinomial(omit.constant = TRUE)`
	`posbernoulli.b(drop.b = FALSE ~ 0)`
	`posbernoulli.t(parallel.t = FALSE ~ 0)`
	`posbernoulli.tb(drop.b = FALSE ~ 0, parallel.t = FALSE ~ 0)`
$\mathcal{M}_b/\mathcal{M}_{bh}$	`posbernoulli.b()`
	`posbernoulli.tb(drop.b = FALSE ~ 1, parallel.t = FALSE ~ 0)`
$\mathcal{M}_t/\mathcal{M}_{th}$	`posbernoulli.t()`
	`posbernoulli.tb(drop.b = FALSE ~ 0, parallel.t = FALSE ~ 1)`
$\mathcal{M}_{tb}/\mathcal{M}_{tbh}$	`posbernoulli.tb()`

Model	$\boldsymbol{\eta}^T$
$\mathcal{M}_0/\mathcal{M}_h$	$g(p)$
$\mathcal{M}_b/\mathcal{M}_{bh}$	$(g(p_c), g(p_r))$
$\mathcal{M}_t/\mathcal{M}_{th}$	$(g(p_1), \ldots, g(p_\tau))$
$\mathcal{M}_{tb}/\mathcal{M}_{tbh}$	$(g(p_{c1}), \ldots, g(p_{c\tau}), g(p_{r2}), \ldots, g(p_{r\tau}))$

imals. Hence, individual covariates such as weight and gender can be included in the linear/additive predictors. Non-heterogeneity models ($\mathcal{M}_0$, $\mathcal{M}_b$, $\mathcal{M}_t$ and $\mathcal{M}_{tb}$) are simply intercept-only models, i.e., $\sim$ 1. Consequently, the description of the models below can be grouped into pairs, e.g., $\mathcal{M}_h$ is $\mathcal{M}_0$ with covariates. Both $\mathcal{M}_0$ and $\mathcal{M}_h$ are serviced by `family = posbinomial(omit.constant = TRUE)` since $p_i = p_{ij} = p_{is}^*$ in (17.12), and the others by a `posbernoulli.`-type family function. They have the probabilities in (17.12) as their default fitted values. All but $\mathcal{M}_0/\mathcal{M}_h$ are amenable to reduced rank regression, e.g., Sect. 17.4.

The 8 models have a nested structure of which $\mathcal{M}_{tbh}$ is the most general. The simpler models can be fitted as special cases of more complex models with appropriate arguments set. Table 17.3 summarizes these. It is noted that, although `posbernoulli.tb()` may be used to fit the simpler models, its computation is less efficient in terms of memory and speed.

It is useful to consider the special case of $\tau = 2$ sampling occasions given in Table 17.4 to illustrate the interrelationships between the models. A short description of the variants are as follows.

$\boxed{\mathcal{M}_0/\mathcal{M}_h}$ The null model $\mathcal{M}_0$ is the simplest and has homogeneous capture probabilities $H_0 : p_{ij} = p$. It indicates that all animals have the same probability of capture regardless of the sampling occasion. In its original formulation, the catchability p in $\mathcal{M}_0$ did not depend on any covariates (intercept-only). Since VGAM allows for general η, the model is better described as $\mathcal{M}_h$. Both models may be fitted by `posbinomial()` by summing the response as counts over the sampling occasions. The default link is $\eta = g(p)$ for g = logit link.

Note that Otis et al. (1978) described the extreme case for $\mathcal{M}_h$ where $p_{ij} = p_i$ with p_i being parameters in their own right.

Table 17.4 Capture history sample space and corresponding probabilities for various models of a $\tau = 2$ sample capture–recapture experiment of a closed population. Here, p_j refers to sampling period j. The "00" row is never realized in sample data. The **VGAM** family function for fitting the model is given in the last row. Models without the "h" prefix are intercept-only models.

Capture	Joint probability			
history	$\mathcal{M}_0/\mathcal{M}_h$	$\mathcal{M}_b/\mathcal{M}_{bh}$	$\mathcal{M}_t/\mathcal{M}_{th}$	$\mathcal{M}_{tb}/\mathcal{M}_{tbh}$
01	$(1-p)p$	$(1-p_c)\,p_c$	$(1-p_1)p_2$	$(1-p_{c1})\,p_{c2}$
10	$p(1-p)$	$p_c(1-p_r)$	$p_1(1-p_2)$	$p_{c1}(1-p_{r2})$
11	p^2	$p_c\,p_r$	$p_1\,p_2$	$p_{c1}\,p_{r2}$
00	$(1-p)^2$	$(1-p_c)^2$	$(1-p_1)(1-p_2)$	$(1-p_{c1})(1-p_{c2})$
$M = \dim(\boldsymbol{\eta})$	1	2	$2\ (=\tau)$	$3\ (=2\tau-1)$
Family	posbinomial()	posbernoulli.b()	posbernoulli.t()	posbernoulli.tb()

While this could possibly be fitted by creating a covariate of the form `factor(1:n)` there would be far too many parameters for comfort. Such an extreme case is not recommended, to avoid over-parameterization.

$\boxed{\mathcal{M}_t/\mathcal{M}_{th}}$ For $\mathcal{M}_t$, the probabilities of capture are the same for each animal but may vary with time, i.e., $H_0 : p_{ij} = p_j$. In **VGAM**, the default links and constraints are

$$\boldsymbol{\eta} \;=\; (\mathrm{logit}(p_1), \ldots, \mathrm{logit}(p_\tau))^T \quad \text{and} \quad \mathbf{H}_1 = \mathbf{I}_\tau, \quad \mathbf{H}_k = \mathbf{1}_\tau, \quad (17.13)$$

and these can be seen in

```
> args(posbernoulli.t)

  function (link = "logit", parallel.t = FALSE ~ 1, iprob = NULL,
      p.small = 1e-04, no.warning = FALSE)
  NULL
```

Of course, the number of parameters grows proportionally with τ.

$\boxed{\mathcal{M}_b/\mathcal{M}_{bh}}$ It is well-known that some species of animals acquire a behavioural change from their first capture. Let p_c and p_r be the probability of capture and recapture, respectively. If $p_c < p_r$ or $p_r < p_c$, then these are known as being trap-happy and trap-shy, respectively. Trap-happiness may be due to the consumption of food during capture. Trap-shyness may be due to the overall negative experience of the animal due to injury, handling, fear, etc. Let $(\eta_1, \eta_2) = (\mathrm{logit}(p_c), \mathrm{logit}(p_r))$ be the default link functions for `posbernoulli.b()`. Then the denominator of (17.12) becomes $1 - (1 - p_c)^\tau$ and, because argument I2 = FALSE,

$$g(p_c) \;=\; \eta_1 \;=\; \eta_c, \qquad\qquad (17.14)$$
$$g(p_r) \;=\; \eta_2 \;=\; \eta_b + \eta_c, \quad \text{say}, \qquad (17.15)$$

where η_c and η_b model the capture and behavioural effects. That is, the difference in parallelism $\eta_b(\boldsymbol{x}) = \eta_2(\boldsymbol{x}) - \eta_1(\boldsymbol{x})$ can be ascribed as the behavioural effect on the g-scale. For example, variables whose coefficients in η_b are positive correspond to trap-happiness.

Unless the data set is large, one will want to keep η_b as simple as possible, such as by using a dummy variable—then it is convenient to absorb it into the intercept term. This is achieved by the default

$$\mathbf{H}_1 = \begin{pmatrix} 0 & 1 \\ 1 & 1 \end{pmatrix}, \quad \mathbf{H}_2 = \mathbf{H}_3 = \cdots = \begin{pmatrix} 1 \\ 1 \end{pmatrix}. \quad (17.16)$$

Hence the first coefficient $\beta^*_{(1)1} = \eta_b$. If $\mathbf{H}_k \neq (1,1)^T$ (i.e., not parallel), then argument I2 specifies whether $\mathbf{H}_k$ is as $\mathbf{H}_1$ above or $\mathbf{I}_M$, the usual trivial constraint matrix. We have

```
> args(posbernoulli.b)

  function (link = "logit", drop.b = FALSE ~ 1,
      type.fitted = c("likelihood.cond","mean.uncond"), I2 = FALSE,
      ipcapture = NULL, iprecapture = NULL, p.small = 1e-04,
      no.warning = FALSE)
  NULL
```

To remove any behavioural effect, set `drop.b = FALSE ~ 0` to remove the first column from the constraint matrix $\mathbf{H}_1$. Then this reduces to $\mathcal{M}_0/\mathcal{M}_h$.

$\boxed{\mathcal{M}_{tb}/\mathcal{M}_{tbh}}$ For these, the $(2\tau - 1)$-vector $\boldsymbol{\eta}$ is defined in Table 17.3. There are three arguments which determine whether there are behavioural effects and/or time effects: `parallel.b`, `parallel.t` and `drop.b`. The last two are as above. Their defaults can be seen

```
> args(posbernoulli.tb)

  function (link = "logit", parallel.t = FALSE ~ 1,
      parallel.b = FALSE ~ 0, drop.b = FALSE ~ 1,
      type.fitted = c("likelihood.cond", "mean.uncond"), imethod = 1,
      iprob = NULL, p.small = 1e-04, no.warning = FALSE,
      ridge.constant = 0.01, ridge.power = -4)
  NULL
```

and their effect on the $\mathbf{H}_k$ can be seen in Table 17.5. One would usually want to keep the behavioural effect to be equal over different sampling occasions, therefore `parallel.b` should be normally left to its default. Allowing it to be `FALSE` for a covariate x_k means an additional $\tau - 1$ parameters—something that is not warranted unless the data set is very large and/or the behavioural effect varies a lot over time.

Given the above models, how can N be estimated? Let

$$p_i(\boldsymbol{\beta}^*) = 1 - \prod_{s=1}^{\tau} (1 - p^*_{is}) \quad (17.17)$$

be the probability that animal i is captured at least once in the course of the study. It is the denominator of (17.12). Here, $\boldsymbol{\beta}^*$ are all the regression coefficients of the model related to the capture (but not recapture) probabilities. Then

$$\widehat{N} = \sum_{i=1}^{n} p_i(\boldsymbol{\beta}^*)^{-1} \quad (17.18)$$

Table 17.5 For the general $\mathcal{M}_{tbh}(\tau)$ family `posbernoulli.tb()`, the constraint matrices corresponding to the arguments `parallel.t`, `parallel.b` and `drop.b`. In each cell, the LHS matrix is $\mathbf{H}_k$ when `drop.b` is `FALSE` for x_k. The RHS matrix is when `drop.b` is `TRUE` for x_k; it simply deletes the left submatrix of $\mathbf{H}_k$. See also Table 17.3. Notes: (i) the default for `posbernoulli.tb()` is $\mathbf{H}_1$ = the LHS matrix of the top-right cell, and $\mathbf{H}_k$ = the RHS matrix of the top-left cell. (ii) $\mathbf{I}_{\tau[-1,]} = (\mathbf{0}_{\tau-1}|\mathbf{I}_{\tau-1})$.

	parallel.t	!parallel.t
parallel.b	$\begin{pmatrix} \mathbf{0}_\tau & \mathbf{1}_\tau \\ \mathbf{1}_{\tau-1} & \mathbf{1}_{\tau-1} \end{pmatrix}, \begin{pmatrix} \mathbf{1}_\tau \\ \mathbf{1}_{\tau-1} \end{pmatrix}$	$\begin{pmatrix} \mathbf{0}_\tau & \mathbf{I}_\tau \\ \mathbf{1}_{\tau-1} & \mathbf{I}_{\tau[-1,]} \end{pmatrix}, \begin{pmatrix} \mathbf{I}_\tau \\ \mathbf{I}_{\tau[-1,]} \end{pmatrix}$
!parallel.b	$\begin{pmatrix} \mathbf{0}_{\tau\times(\tau-1)} & \mathbf{1}_\tau \\ \mathbf{I}_{\tau-1} & \mathbf{1}_{\tau-1} \end{pmatrix}, \begin{pmatrix} \mathbf{1}_\tau \\ \mathbf{1}_{\tau-1} \end{pmatrix}$	$\begin{pmatrix} \mathbf{0}_{\tau\times(\tau-1)} & \mathbf{I}_\tau \\ \mathbf{I}_{\tau-1} & \mathbf{I}_{\tau[-1,]} \end{pmatrix}, \begin{pmatrix} \mathbf{I}_\tau \\ \mathbf{I}_{\tau[-1,]} \end{pmatrix}$

is unbiased (and the Horvitz and Thompson (1952) estimator) for the population size N. It can be shown that an associated estimate of the variance of $\widehat{N}(\boldsymbol{\beta}^*)$ is

$$s^2(\boldsymbol{\beta}^*) = \sum_{i=1}^n p_i(\boldsymbol{\beta}^*)^{-2} \left[1 - p_i(\boldsymbol{\beta}^*)\right] \tag{17.19}$$

if $\boldsymbol{\beta}^*$ is known. If $\boldsymbol{\beta}^*$ is to be estimated, then one can use

$$\mathrm{Var}\left(\widehat{N}(\widehat{\boldsymbol{\beta}}^*)\right) \approx s^2(\widehat{\boldsymbol{\beta}}^*) + \widehat{\boldsymbol{d}}^T \widehat{\mathrm{Var}}(\widehat{\boldsymbol{\beta}}^*)\,\widehat{\boldsymbol{d}}, \tag{17.20}$$

where

$$\boldsymbol{d} = \frac{dN(\boldsymbol{\beta}^*)}{d\boldsymbol{\beta}^*} = \sum_{i=1}^n p_i(\boldsymbol{\beta}^*)^{-2}\,\frac{dp_i(\boldsymbol{\beta}^*)}{d\boldsymbol{\beta}^*}$$

$$= \sum_{i=1}^n \frac{-1}{p_i(\boldsymbol{\beta}^*)^2} \sum_{s=1}^\tau \left[\prod_{t=1,\,t\neq s}^\tau (1-p_{it}^*)\right] \frac{\partial p_{is}^*}{\partial\boldsymbol{\beta}^*}. \tag{17.21}$$

This follows from a Taylor series expansion of $\widehat{N}(\widehat{\boldsymbol{\beta}}^*)$ about $\widehat{N}(\boldsymbol{\beta}^*)$.

In closing, we note in passing that several capture–recapture methods can be fitted as loglinear GLMs, e.g., as implemented by Rcapture. From Baillargeon and Rivest (2007), let $\boldsymbol{y} = (y_1,\ldots,y_\tau)^T$ be the capture history for a particular animal. For simplicity, assume variant $\mathcal{M}_0$ so that $P(\boldsymbol{y}) = (1-p)^{\tau-y^*} p^{y^*}$ where $y^* = \boldsymbol{y}^T\mathbf{1}$ is the number of times the animal is caught. The expected number of animals in the population having capture history $\boldsymbol{y}$ is

$$\mu_{\boldsymbol{y}} = N \times P(\boldsymbol{y}) = \exp\left\{\log\left(N(1-p)^\tau\right) + y^*\,\mathrm{logit}\,p\right\}. \tag{17.22}$$

Stacking all $2^\tau - 1$ possibilities of $\boldsymbol{y}$ together suggests fitting $\boldsymbol{\mu} = \exp\{\mathbf{X}\boldsymbol{\beta}\}$, with $\mathbf{X}$ being a $(2^\tau - 1) \times 2$ model matrix with rows of the form $(1, y_j^*)$. Let the coefficients be $\boldsymbol{\beta} = (\beta_1, \beta_2)^T$, say. Then one can use $\widehat{N} = n + \exp(\widehat{\beta_1})$. This is justified because $\exp(\beta_1) = \exp\{\log(N(1-p)^\tau)\} = N \times P(\boldsymbol{y_0}) = \mu_0$ where $\boldsymbol{y} = \boldsymbol{0}$ is unobservable, and μ_0 is the expected number of animals never captured. Of course, the most straightforward approach to (17.22) is to assume a Poisson model.

17.2.1 Further Software Details

All the family functions of Table 17.3 except `posbinomial()` should have an $n \times \tau$ capture history matrix as the response, preferably with column names. With z_{ij} defined (Table 17.2) as indicators of the past capture of animal i, these are stored on **VGAM** objects as the `cap.hist1` component in the `extra` slot. Also, there is a component called `cap1` which indicates on which sampling occasion the first capture occurred.

All family functions return a point estimate $\widehat{N}$ (17.18) in the `extra` slot with component name `N.hat`. Likewise, its standard error based on the square root of (17.20) has component name `SE.N.hat`. The `posbinomial()` family function differs from the others in that the number of trials can differ from observation to observation; if so then the two values are not computed.

The AIC, BIC and other information criteria are widely used by biologists especially (Burnham and Anderson, 2002) to choose between the 8 models. For fitting $\mathcal{M}_0/\mathcal{M}_h$ by `posbinomial()`, one must set argument `omit.constant = TRUE` if model comparisons are to be made with `AIC()` or `BIC()`, since one needs to omit the log-normalizing constant $\log \binom{\tau}{\tau \bar{y}_i}$ from its log-likelihood so that it is comparable with the logarithm of (17.12).

By default, all the family functions have fitted values corresponding to the probabilities in the conditional likelihood function (17.12), viz.

$$\widehat{p}_{ij}^{\,y_{ij}} \left(1 - \widehat{p}_{ij}\right)^{1-y_{ij}} \cdot \left[1 - \prod_{s=1}^{\tau} \left(1 - \widehat{p}_{i,cs}\right)\right]^{-1}.$$

Alternatively, the unconditional means of the Y_j can be returned as the fitted values upon selecting `type.fitted = "mean"` argument. They are $\mu_1 = E(Y_1) = p_{c1}/(1 - Q_{1:\tau})$, $\mu_2 = [(1 - p_{c1}) \, p_{c2} + p_{c1} \, p_{r2}]/(1 - Q_{1:\tau})$, and for $j = 3, 4, \ldots, \tau$,

$$\mu_j = (1 - Q_{1:\tau})^{-1} \left\{ p_{cj} \, Q_{1:(j-1)} + p_{rj} \left[p_{c1} + \sum_{s=2}^{j-1} p_{cs} \, Q_{1:(s-1)} \right] \right\}. \quad (17.23)$$

Estimator $\widehat{N}(\boldsymbol{\beta}^*)$ in (17.18) may be unstable when any of the $p_i(\boldsymbol{\beta}^*)$ are very close to 0, and by default, a warning is issued if this occurs. The family functions also have arguments which specify exactly what is meant by being close to 0, and to suppress the warning if so desired.

Yee et al. (2015) give the first and expected second derivatives for the $\mathcal{M}_{tbh}$, $\mathcal{M}_{bh}$ and $\mathcal{M}_{th}$ models. In the implementation of `posbernoulli.tb()`, arguments `ridge.constant` and `ridge.power`, add a ridge parameter to the some of the diagonal EIM elements to ensures that the working weight matrices are positive-definite. The ridge factor decays to zero as iterations proceed, and it plays a negligible role upon convergence. See Sect. 9.2.1 for details.

17.2.2 Deermice Example

Huggins (1991) reports an analysis involving fitting all 8 variations of the positive-Bernoulli model to a small deer mouse (*Peromyscus maniculatus*) data set. A deer mouse is a small rodent native to North America, and about 8 to 10 cm long, not counting the length of the tail. There were 38 individual mice caught over 6 trapping occasions. Individual body weight, sex and age were also recorded, which we use as covariates to model heterogeneity. The data are given in the data frame `deermice`:

```
> cbind(head(deermice), '    ' = tail(rownames(deermice)), tail(deermice))

   y1 y2 y3 y4 y5 y6 sex age weight      y1 y2 y3 y4 y5 y6 sex age weight
1   1  1  1  1  1  1   1   0      y   33   0  0  0  0  1  0   0   y      14
2   1  0  0  1  1  1   1   1      y   34   0  0  0  0  1  0   1   y      11
3   1  1  0  0  1  1   1   0      y   35   0  0  0  0  1  0   0   a      24
4   1  1  0  1  1  1   1   0      y   36   0  0  0  0  0  1   0   y       9
5   1  1  1  1  1  1   1   0      y   37   0  0  0  0  0  1   0   a      16
6   1  1  0  1  1  1   1   0      a   38   0  0  0  0  0  1   1   a      19
```

We mimic part of the analysis by fitting the $\mathcal{M}_{bh}$, $\mathcal{M}_{tb}$ and $\mathcal{M}_{tbh}$ models as follows.

```
> deermice$Age <- 2 - as.numeric(deermice$age)    # 0 == young,  1 == adult
> deermice$Sex <- 1 - as.numeric(deermice$sex)    # 0 == female, 1 == male
> M.bh  <- vglm(cbind(y1, y2, y3, y4, y5, y6) ~ weight + Sex + Age,
                posbernoulli.b, data = deermice)
> M.tb  <- vglm(cbind(y1, y2, y3, y4, y5, y6) ~ 1,
                posbernoulli.tb, data = deermice)
> M.tbh <- vglm(cbind(y1, y2, y3, y4, y5, y6) ~ weight + Sex + Age,
                posbernoulli.tb, data = deermice)
```

Based on AIC values, the analysis concluded that $\mathcal{M}_{bh}$ was superior. Then

```
> coef(M.bh)

(Intercept):1 (Intercept):2        weight           Sex           Age
      1.17596      -2.90702       0.15917       0.91629      -1.88451

> sqrt(diag(vcov(M.bh)))

(Intercept):1 (Intercept):2        weight           Sex           Age
      0.40981       0.90493       0.06432       0.35024       0.63498
```

which agree with the results of Huggins (1991). The first coefficient, 1.18, is positive and hence implies a trap-happy effect. The Wald statistic for the behavioural effect, being 2.87, suggests the effect is real. Lastly,

```
> c(M.bh@extra$N.hat, M.bh@extra$SE.N.hat)

[1] 47.1442  7.3219
```

suggests there are $\widehat{N} = 47$ mice in the population of interest, with an associated standard error of 7.32.

To perform some model checking, we now confirm that the component function of `weight` is indeed linear. To do this we apply some smoothing, but do not allow the smooth to be too flexible because of the size of the data set.

```
> deermice.bh <- vgam(cbind(y1, y2, y3, y4, y5, y6) ~ s(weight, df = 3) + Sex + Age,
                posbernoulli.b, data = deermice)
> plot(deermice.bh, se = TRUE, las = 1, lcol = "blue", scol = "orange",
       rcol = "purple", scale = 5)
```

Plots of the component functions against each covariate are given in Fig. 17.3. In general, `weight` does seem to have a (positive) linear effect on the logit scale. Young deer mice appear more easily caught compared to adults, and gender seems to have a smaller effect than weight. A more formal test of linearity is

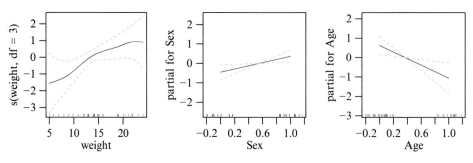

Fig. 17.3 Estimated component functions for the `deermice` VGAM.

```
> summary(deermice.bh)


Call:
vgam(formula = cbind(y1, y2, y3, y4, y5, y6) ~ s(weight, df = 3) +
    Sex + Age, family = posbernoulli.b, data = deermice)

Number of linear predictors:    2

Names of linear predictors: logit(pcapture), logit(precapture)

Dispersion Parameter for posbernoulli.b family:    1

Log-likelihood: -137.93 on 69.04 degrees of freedom

Number of iterations:   12

DF for Terms and Approximate Chi-squares for Nonparametric Effects

                     Df Npar Df Npar Chisq P(Chi)
(Intercept):1         1
(Intercept):2         1
s(weight, df = 3)     1        2        3.22 0.1942
Sex                   1
Age                   1
```

and this suggests there is no significant nonlinearity. This is in agreement
with Hwang and Huggins (2011) who used kernel smoothing.

17.2.3 Prinia Example

This example looks at the Yellow-bellied Prinia (*Prinia flaviventris*), a common
bird species located in Southeast Asia. A capture–recapture experiment was conducted at the Mai Po Nature Reserve in Hong Kong during 1991, where captured individuals had their wing lengths measured and fat index recorded. A total
of $\tau = 19$ weekly capture occasions were considered, where $n = 151$ distinct birds
were captured. Here, we use (standardized) wing length and fat index (0 or 1) as
covariates, in conjunction with `posbinomial()`. Part of the relevant data are

```
> head(prinia)[, 1:4]

    length fat cap noncap
1  1.006504   1   5     14
2  1.264626   1   3     16
3 -0.025983   1   6     13
4  3.071478   0   1     18
5  0.438636   1   5     14
6  0.748382   0   1     18
```

We allow smoothing the wing length variable as follows.

```
> M.h.GAM <-
    vgam(cbind(cap, noncap) ~ s(length, df = 3) + fat, data = prinia,
        posbinomial(omit.const = TRUE, parallel = TRUE ~ s(length, df = 3) + fat))
> c(M.h.GAM@extra$N.hat, M.h.GAM@extra$SE.N.hat)

[1] 447.63 108.89

> plot.info <- plot(M.h.GAM, se = TRUE)
```

The fitted capture probabilities, with and without fat content, are plotted against wing length in Fig. 17.4. The code to do this is as follows.

```
> fit2.info <- plot.info@preplot[[1]]
> fat.effect <- coef(M.h.GAM)["fat"]
> intercept <- coef(M.h.GAM)["(Intercept)"]
> ooo <- order(fit2.info$x)
> centring.const <- mean(prinia$length) - coef(M.h.GAM)["s(length, df = 3)"]
> plotframe <- data.frame(lin.pred.b = intercept + fat.effect * 1 +
                            centring.const + fit2.info$y[ooo],
                          lin.pred.0 = intercept + fat.effect * 0 +
                            centring.const + fit2.info$y[ooo],
                          x2 = fit2.info$x[ooo])
> plotframe <- transform(plotframe,
                    up.lin.pred.b = lin.pred.b + 2 * fit2.info$se.y[ooo],
                    lo.lin.pred.b = lin.pred.b - 2 * fit2.info$se.y[ooo],
                    up.lin.pred.0 = lin.pred.0 + 2 * fit2.info$se.y[ooo],
                    lo.lin.pred.0 = lin.pred.0 - 2 * fit2.info$se.y[ooo])
> plotframe <- transform(plotframe,
                       fv.b = logit(   lin.pred.b, inverse = TRUE),
                    up.fv.b = logit(up.lin.pred.b, inverse = TRUE),
                    lo.fv.b = logit(lo.lin.pred.b, inverse = TRUE),
                       fv.0 = logit(   lin.pred.0, inverse = TRUE),
                    up.fv.0 = logit(up.lin.pred.0, inverse = TRUE),
                    lo.fv.0 = logit(lo.lin.pred.0, inverse = TRUE))
> with(plotframe,
     matplot(x2, cbind(up.fv.b, fv.b, lo.fv.b), type = "l", col = "blue",
           lty = c(2, 1, 2), las = 1, cex.lab = 1.5, lwd = 1.5, cex.axis = 1.5,
           main = "", ylab = ~ hat(p), xlab = "Wing length (standardized)"))
> with(plotframe, matlines(x2, cbind(up.fv.0, fv.0, lo.fv.0), col = "darkorange",
                  lty = c(2, 1, 2), lwd = 1.5))
> legend("topleft", legend = c("Fat present", "Fat not present"), bty = "n",
       lwd = 1.5, col = c("blue", "darkorange"), merge = TRUE, cex = 1.5)
```

The capture probabilities appear larger for individuals with fat content present. As is usually the case, the approximate ±2 pointwise SEs become wider at the boundaries.

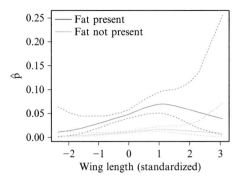

Fig. 17.4 Capture probability estimates with approximate ±2 pointwise SEs, versus wing length with (*blue*) and without (*orange*) fat content present fitting a $\mathcal{M}_h$-VGAM, using the `prinia` data.

17.2.4 A $\mathcal{M}_{tbh}$ Example

Here, we will attempt to mimic the results of Huggins (1989), who fitted a $\mathcal{M}_{tbh}$ to a tiny data set of 18 individuals and 10 trapping occasions. For the ith individual, the model will be written as $(i = 1, \ldots, 18, j = 1, \ldots, 10 = \tau)$

$$\operatorname{logit} p_{ij} = \beta^*_{(1)1} z_{ij} + \beta^*_{(2)1} + \beta^*_{(1)2} \cdot \mathbf{x}2_i + \beta^*_{(1)3} \cdot \mathbf{x}3_j, \qquad (17.24)$$

where $\beta^*_{(1)1}$ is the behavioural effect, and z_{ij} is defined in Table 17.2. Variable x2 is an ordinary individual covariate such as weight, but x3 is a time-varying or occasion-specific covariate such as temperature or daily rainfall that is handled using the `xij` argument described in Sect. 3.4. Variable x3 takes on variable values z1–z10 in the data frame depending on the linear predictor.

Firstly, let's look at some of the data and, after renaming the variables, delete two observations which were never caught (so that $n = 18$).

```
> head(Huggins89table1[, -c(5, 10, 15)], 4)

   x2 y01 y02 y03 y05 y06 y07 y08 y10 t01 t02 t03 t05 t06 t07 t08 t09 t10
1 6.3   0   1   0   1   0   0   1   0  5.8 2.9 3.7 3.5 3.6 1.9   3 4.8 4.7
2 4.0   0   1   1   1   0   0   1   0  5.8 2.9 3.7 3.5 3.6 1.9   3 4.8 4.7
3 4.7   0   1   0   1   1   1   1   0  5.8 2.9 3.7 3.5 3.6 1.9   3 4.8 4.7
4 4.8   0   0   1   1   0   1   1   0  5.8 2.9 3.7 3.5 3.6 1.9   3 4.8 4.7

> small.table1 <- transform(Huggins89table1, x3.tij = t01,
                    T02 = t02, T03 = t03, T04 = t04, T05 = t05, T06 = t06,
                    T07 = t07, T08 = t08, T09 = t09, T10 = t10)
> small.table1 <- subset(small.table1,
                  y01 + y02 + y03 + y04 + y05 + y06 + y07 + y08 + y09 + y10 > 0)
```

Here, the Z02–Z10 are for the recapture η_j and there is no Z01 because recapture is not possible at the first time occasion. Then (17.24) can be fitted by

```
> fit.tbh <-
    vglm(cbind(y01, y02, y03, y04, y05, y06, y07, y08, y09, y10) ~ x2 + x3.tij,
         xij = list(x3.tij ~ t01 + t02 + t03 + t04 + t05 +
                             t06 + t07 + t08 + t09 + t10 +
                             T02 + T03 + T04 + T05 +
                     T06 + T07 + T08 + T09 + T10 - 1),
         posbernoulli.tb(parallel.t = TRUE ~ x2 + x3.tij),
         data = small.table1, trace = FALSE,
         form2 = ~ x2 + x3.tij +
                   t01 + t02 + t03 + t04 + t05 + t06 + t07 + t08 + t09 + t10 +
                   T02 + T03 + T04 + T05 + T06 + T07 + T08 + T09 + T10)
```

The **form2** argument is required if **xij** is used, and it needs to include all the variables in the model. It is from this formula that $\mathbf{X}_{\mathrm{VLM}}$ is constructed; the relevant columns are extracted to construct the diagonal matrix in (3.40) in the specified order of diagonal elements given by **xij**. Their names need to be uniquely specified.

To check, the constraint matrices are

```
> head(constraints(fit.tbh, matrix = TRUE), 4)

      (Intercept):1 (Intercept):2 x2 x3.tij
[1,]              0             1  1       1
[2,]              0             1  1       1
[3,]              0             1  1       1
[4,]              0             1  1       1

> tail(constraints(fit.tbh, matrix = TRUE), 4)

       (Intercept):1 (Intercept):2 x2 x3.tij
[16,]              1             1  1       1
[17,]              1             1  1       1
[18,]              1             1  1       1
[19,]              1             1  1       1
```

Then the $\widehat{\beta}^{*}_{(j)1}$ and their standard errors are

```
> coef(fit.tbh)

(Intercept):1 (Intercept):2           x2     x3.tij
      1.09376      -0.63056      0.38449   -0.83692

> sqrt(diag(vcov(fit.tbh)))    # SEs

(Intercept):1 (Intercept):2           x2     x3.tij
      0.53648       1.36080      0.22291    0.18567
```

These results largely agree with Huggins (1989). The first coefficient, 1.09, is positive and hence implies a trap-happy effect. The Wald statistic for the behavioural effect is 2.04, which suggests that the effect is real.

Estimates of the population size can be obtained from

```
> fit.tbh@extra$N.hat        # Estimate of the population size //

  [1] 20.935

> fit.tbh@extra$SE.N.hat     # The standard error

  [1] 3.4559
```

17.2.5 Using Select()

We now refit `fit.tbh` using `Select()` to illustrate the avoidance of manual manipulation of cumbersome formulas and response matrices with many columns, especially as related to arguments `form2` and `xij`. For example, if `pdata` is a data frame with columns y01, y02, ..., y30, then `Select(pdata, "y")` will return the matrix `cbind(y01, y02, ..., y30)`, provided that there are no other variables beginning with `"y"`. Another example is

```
Select(pdata, "t", as.formula = TRUE, lhs = "ymatrix", rhs = "0")
```

which will return the formula `ymatrix ~ t01 + t02 + ⋯ + t30 + 0`, provided that there are no other variables beginning with `"t"`. At present, the arguments of `Select()` are

```
> args(Select)

  function (data = list(), prefix = "y", lhs = NULL, rhs = NULL,
      rhs2 = NULL, rhs3 = NULL, as.character = FALSE, as.formula.arg = FALSE,
      tilde = TRUE, exclude = NULL, sort.arg = TRUE)
  NULL
```

Starting with `Huggins89table1`, the following code works quite generally provided that the original variables are labelled as y01, y02, ..., and t01, t02, ...

```
> Hdata <- subset(Huggins89table1, rowSums(Select(Huggins89table1, "y")) > 0)
> Hdata.T <- Select(Hdata, "t")        # A 10-column submatrix copy
> colnames(Hdata.T) <- gsub("t", "T", colnames(Hdata.T))   # Rename colnames
> Hdata <- data.frame(Hdata, Hdata.T)
> Hdata <- transform(Hdata, x3.tij = y01)
> Form2 <- Select(Hdata, prefix = TRUE, as.formula = TRUE)
> Xij   <- Select(Hdata, c("t", "T"), as.formula = TRUE,
                  sort = FALSE, rhs = "0", lhs = "x3.tij", exclude = "T01")
> fit.tbh <- vglm(Select(Hdata, "y") ~ x2 + x3.tij, form2 = Form2, xij = list(Xij),
                  posbernoulli.tb(parallel.t = TRUE ~ x2 + x3.tij),
                  data = Hdata, trace = FALSE)  # Setting trace = TRUE is a good idea
> coef(fit.tbh)  # Same as before

    (Intercept):1 (Intercept):2           x2        x3.tij
          1.09376      -0.63056      0.38449      -0.83692
```

The argument `Form2` might be described as greedy here because it contains more terms than is strictly needed. In fact, it contains all the columns of `Hdata` actually.

Note that this demonstrates the ability to enter a matrix response without an explicit `cbind()`. Also, if `Y <- Select(Hdata, "y")`, then `vglm(Y ~ ⋯)` can work as well—however, this can be argued as having bad style because the response is detached from the data frame.

In closing, it is noted that an alternative of `Select()` is to use the R function `subset()`, for example

```
subset(pdata, TRUE, select = y01:y10)   # Same as with(pdata, cbind(y01, ..., y10))
```

is equivalent to `with(pdata, cbind(y01, ..., y10))`. More generally,

```
subset(pdata, TRUE, select = grepl("^y", colnames(pdata)))
```

selects all variables beginning with "y". However, `subset()` does not generate formulas.

17.2.6 Ephemeral and Enduring Memory Example

Yang and Chao (2005) consider modelling the behavioural effect with both enduring (long-term) and ephemeral (short-term) memory components. For example, the short-term component depends on whether or not the animal was caught on the most recent sampling occasion. We call this a lag-1 effect. In the example of this section, which combines aspects of the previous 3 examples, we illustrate how this may be easily achieved within the VGLM framework; it is another case of using the `xij` argument. We retain the enduring component as with the $\mathcal{M}_{tbh}$: the first column of $\mathbf{H}_1$ is $(\mathbf{0}_\tau^T, \mathbf{1}_{\tau-1}^T)^T$ applies to all the recapture probabilities. For simplicity, we first consider a lag-1 effect only for the short-term component.

In the following, we fit a $\mathcal{M}_{tbh}$ model to `deermice` with both long-term and short-term effects:

$$\operatorname{logit} p_{cs} = \beta_{(2)1}^* + \beta_{(1)2}^* \, \texttt{sex} + \beta_{(1)3}^* \, \texttt{weight},$$
$$\operatorname{logit} p_{rt} = \beta_{(1)1}^* + \beta_{(2)1}^* + \beta_{(1)2}^* \, \texttt{sex} + \beta_{(1)3}^* \, \texttt{weight} + \beta_{(1)4}^* \, y_{t-1},$$

where $s = 2, \ldots, \tau$, $t = 1, \ldots, \tau$ and $\tau = 6$. The inclusion of y_{t-1} is to model the short-term effect, which is only activated upon a capture or recapture, and lasts to the next sampling occasion.

```
> deermice <- transform(deermice, Lag1 = y1)
> M.tbh.lag1 <-
    vglm(cbind(y1, y2, y3, y4, y5, y6) ~ sex + weight + Lag1, data = deermice,
         posbernoulli.tb(parallel.t = FALSE ~ 0,
                         parallel.b = FALSE ~ 0,
                         drop.b = FALSE ~ 1),
         xij = list(Lag1 ~ fill(y1) + fill(y2) + fill(y3) + fill(y4) +
                      fill(y5) + fill(y6) +
                      y1 + y2 + y3 + y4 + y5),
         form2 = ~ sex + weight + Lag1 +
                    fill(y1) + fill(y2) + fill(y3) + fill(y4) +
                    fill(y5) + fill(y6) +
                    y1 + y2 + y3 + y4 + y5 + y6)
> coef(M.tbh.lag1)

 (Intercept):1 (Intercept):2            sex        weight           Lag1
     1.2776930     -0.4268384     -1.1513794     0.0058372      0.0230575
```

The coefficient of `Lag1`, 0.0231, is the estimated ephemeral effect $\widehat{\beta}^*_{(1)4}$. The estimated enduring effect $\widehat{\beta}^*_{(1)1}$ has value 1.2777. Note that the `fill()` function is used to create 6 variables having 0 values, i.e., $\mathbf{0}_n$.

There is an alternative method to fit the above model; here, we set $\mathbf{H}_{\texttt{Lag1}} = (\mathbf{0}^T_\tau, \mathbf{1}^T_{\tau-1})^T$, and the variables `fill(y1)`,...,`fill(y6)` can be replaced by variables that do not need to be 0. Importantly, the two methods have $\mathbf{X}^{\#}_{(ik)}\mathbf{H}_k$ in (3.40) being the same regardless. The second alternative method requires constraint matrices to be inputted using the `constraints` argument. For example,

```
> deermice <- transform(deermice, Lag1 = y1)
> deermice <- transform(deermice, f1 = y1, f2 = y1, f3 = y1, f4 = y1,
                                   f5 = y1, f6 = y1)
> tau <- 6
> H2 <- H3 <- cbind(rep(1, 2*tau-1))
> H4 <- cbind(c(rep(0, tau), rep(1, tau-1)))
> M.tbh.lag1.method2 <-
    vglm(cbind(y1, y2, y3, y4, y5, y6) ~ sex + weight + Lag1, data = deermice,
         posbernoulli.tb(parallel.b = TRUE ~ 0, parallel.t = TRUE ~ 0),
         constraints = list("(Intercept)" = cbind(H4, 1), sex = H2, weight= H3,
                            Lag1 = H4),
         xij = list(Lag1 ~ f1 + f2 + f3 + f4 + f5 + f6 +
                           y1 + y2 + y3 + y4 + y5),
         form2 = Select(deermice, prefix = TRUE, as.formula = TRUE))
> coef(M.tbh.lag1.method2)

(Intercept):1 (Intercept):2           sex         weight           Lag1
    1.2776930     -0.4268384    -1.1513794      0.0058372      0.0230575
```

is identical. In closing, it can be noted that more complicated models can be handled. For example, the use of `pmax()` to handle lag-2 effects as follows.

```
> deermice <- transform(deermice, Lag2 = y1)
> M.tbh.lag2 <-
    vglm(cbind(y1, y2, y3, y4, y5, y6) ~ sex + weight + Lag2, data = deermice,
         posbernoulli.tb(parallel.t = FALSE ~ 0,
                         parallel.b = FALSE ~ 0,
                         drop.b = FALSE ~ 1),
         xij = list(Lag2 ~ fill(y1) + fill(y2) + fill(y3) + fill(y4) +
                           fill(y5) + fill(y6) +
                           y1 + pmax(y1, y2) + pmax(y2, y3) + pmax(y3, y4) +
                           pmax(y4, y5)),
         form2 = ~ sex + weight + Lag2 +
                   fill(y1) + fill(y2) + fill(y3) + fill(y4) +
                   fill(y5) + fill(y6) +
                   y1 + pmax(y1, y2) + pmax(y2, y3) + pmax(y3, y4) +
                   pmax(y4, y5) + y6)
> coef(M.tbh.lag2)

(Intercept):1 (Intercept):2           sex         weight           Lag2
    1.9676972     -0.4417297    -1.2172470      0.0074673     -0.7222657
```

It is left as an exercise to the reader (Ex. 17.11) to show how both lag-1 and lag-2 effects may be estimated from the same model.

17.3 The Zero-Inflated Binomial Distribution

One application area of the zero-inflated binomial (ZIB) distribution is occupancy survey modelling. Here, data is collected by repeatedly sampling n sites and counting the total number of times a species is present at each site. For example, Webb et al. (2014) consider a migratory and endangered swift parrot called *Lathamus discolor* which nests in tree hollows in Tasmania, Australia. Suppose that all sites are either continuously occupied or unoccupied during the survey period. Detection is imperfect so that there is a detection probability p_{ij} of sighting the species at the jth of N_i visits to site i, given that it is truly occupied there. The occupancy probability ϕ_i^* is the chance the species is really present at site i. If it is assumed that detections occur independently, and are constant across the survey period $(p_{i1} = \cdots = p_{i,N_i} (= p_i, \text{say}))$, then $Y_i^* \sim \text{Binomial}(N_i, z_i p_i)$ where $Y_i^* = N_i Y_i$ is the total number of times the species is detected there. The latent binary variable z_i indicates whether site i is truly occupied. The use of Y_i^* is because Y_i is retained as a sample proportion. Of course, $y_i^* > 0$ implies $z_i = 1$, and $z_i = 0$ implies $y_i^* = 0$. The model can be fitted with `zibinomialff()`.

If $N_i = 1$, then it is not possible to separate out the effects of p_i and ϕ_i^*. Indeed, imperfect detection makes estimation and modelling difficult in general. Another major weakness is that the p_{ij} may depend on the abundance of the species occupying the site—an unknown quantity. A recent article highlighting the subtleties and difficulties of occupancy survey modelling is Welsh et al. (2013); see also McCrea and Morgan (2015, Chap. 6).

17.4 RR-ZIPs and RR-ZA Models

In Chap. 5 we saw reduced-rank regression (RRR) was applied to some of the models described in this chapter, to good effect. In particular, the ZIP becomes a (rank-1) *reduced-rank zero-inflated Poisson* distribution, or RR-ZIP. In a nutshell, this RR-VGLM variant is motivated by applications where the probability of a non-zero value may have a monotonic relationship with the Poisson mean.

'Constrained zero-inflated generalized additive models', or COZIGAMs, were defined by Liu and Chan (2010) as (5.19)–(5.20) but with η_j as additive predictors. One could use regression splines to estimate the component functions, therefore a COZIGAM might be fitted using something akin to

```
cozivgam.zip <- rrvglm(y ~ bs(x2, 3) + ns(x3, 3), zipoissonff(zero = NULL),
                       data = zdata)
```

In the case of (5.19)–(5.20) remaining as η_j = linear predictors, the RR-ZIP might be better called a COZIGLM or COZIVGLM rather than a COZIGAM. The coupling can be seen, for example, if μ increases then η_1 increases, and then η_2 and ϕ increase provided that a_{21} is positive.

The statistical and software infrastructure allows natural extensions to other models, e.g., a COZIVGLM applied to a NB-2 model might be fitted by

```
rrzinb <-
cozivglm.nb2 <-
  rrvglm(y ~ x2 + x3, zinegbinomial(zero = NULL), data = zdata, str0 = 3)
```

Here, str0 stands for a structural zero, and rrzinb stands for an RR-ZINB. Equations (5.19)–(5.20) hold for this model too, and since k is estimated as an intercept-only, the variance function is that of NB-2. The ability to fit COZIVGLMs and COZIVGAMs based on Poisson and NB distributions suggests a more accurate naming system (e.g., COZIVGLM-ZIP and COZIVGAM-ZINB-2), however there is good reason to loosely refer them simply as RR-ZIPs and RR-ZINBs, etc.

17.4.1 RR-ZAPs and RR-ZABs and Other Variants

Of course, the RR-VGLM idea can be conveyed to ZAPs and ZABs too, etc. What does one obtain? The RR-zero-altered Poisson distribution, or RR-ZAP, yields the coupling

$$\frac{\phi(\boldsymbol{x})}{1 - \phi(\boldsymbol{x})} \propto [\mu(\boldsymbol{x})]^{a_{21}^{-1}} . \tag{17.25}$$

That is, the odds of observing a zero is proportional to the Poisson mean raised to some power—this is highly interpretable.

The RR-zero-altered binomial distribution, or RR-ZAB, with a logit link applied to both $\phi = P(Y = 0)$ and the binomial probability of success p, results in the coupling

$$\frac{\phi(\boldsymbol{x})}{1 - \phi(\boldsymbol{x})} = K \cdot \left(\frac{p(\boldsymbol{x})}{1 - p(\boldsymbol{x})}\right)^{\alpha} \tag{17.26}$$

for parameters α and K to be estimated. That is, the odds of an observed zero is proportional to the odds of success in an ordinary binomial distribution raised to some power—this is highly interpretable too. Note that α is unconstrained, e.g., it is unnecessary for $\alpha > 0$.

As a final example here, RRR can be applied to the $\mathcal{M}_{bh}$ model to yield the relationship

$$\frac{p_r(\boldsymbol{x})}{1 - p_r(\boldsymbol{x})} = K_1 \cdot \left(\frac{p_c(\boldsymbol{x})}{1 - p_c(\boldsymbol{x})}\right)^{a_{21}} \tag{17.27}$$

for $p_c(\boldsymbol{x})$, $p_r(\boldsymbol{x})$, K_1 and a_{21} to be estimated. A typical call is of the form

```
rrvglm(cbind(y1, y2, y3, y4, y5) ~ x2 + x3 + x4, posbernoulli.b, data = pdata)
```

which may generate some warnings that may be ignored. Of course, an RR-$\mathcal{M}_{th}$ might be a good idea, especially with large τ and/or many covariates. But a difficulty with an RR-$\mathcal{M}_{tbh}$ is that the RRR cannot model p_{cj} and p_{rj} in a symmetric manner, e.g., for a rank-1 model, by constraining two elements of $\mathbf{A}$ to be equal.

Bibliographic Notes

General references to the types of models covered in this chapter include Winkelmann and Boes (2006), Winkelmann (2008), Cameron and Trivedi (2013). Other references are Welsh et al. (1996), Kleiber and Zeileis (2008), Tutz (2012). Yee (2014) develops RRR for coupling two η_js, and mentions some other R packages for fitting ZI- and ZA-distributions.

Some references on capture–recapture methods related to the positive-Bernoulli distribution: Huggins (1991) who justified the use of (17.12) for the purpose of asymptotic theory, Williams et al. (2002), Amstrup et al. (2005), McCrea and Morgan (2015).

Exercises

Ex. 17.1.   Logarithmic and Positive-NB Distributions
Show that the logarithmic distribution (`logff()` in Table 11.10) is the limiting distribution of the positive (0-truncated) negative binomial distribution (Table 17.6) as $k \to 0^+$. Hint: a series expansion of the reciprocal gamma function is $1/\Gamma(k) = k + \gamma k^2 + O(k^3)$.   [Johnson et al. (2005)]

Ex. 17.2.   Explain why Table 17.6 has no `posgeometric()`.

Ex. 17.3.   If $\boldsymbol{\theta} = (\phi, \lambda)^T$ in the ZIP distribution (17.9), determine the score vector, and show that the EIM is given by

$$\begin{pmatrix} \dfrac{1 - e^{-\lambda}}{(1 - \phi)(\phi + (1 - \phi)\,e^{-\lambda})} & \dfrac{-e^{-\lambda}}{\phi + (1 - \phi)\,e^{-\lambda}} \\[2ex] \dfrac{-e^{-\lambda}}{\phi + (1 - \phi)\,e^{-\lambda}} & \dfrac{1 - \phi}{\lambda} - \dfrac{\phi(1 - \phi)\,e^{-\lambda}}{\phi + (1 - \phi)\,e^{-\lambda}} \end{pmatrix}. \tag{17.28}$$

Ex. 17.4.   Derive the score vector and EIM of the ZIB detailed in Table 17.7.

Ex. 17.5.   Family function `posbernoulli.tb()` has $\boldsymbol{\eta}$ defined as in Table 17.3 rather than as $\boldsymbol{\eta} = (g(p_{c1}), g(p_{c2}), g(p_{r2}), \ldots, g(p_{c\tau}), g(p_{r\tau}))^T$. Using the matrix-band format described in Sect. 18.3.5, show that the chosen enumeration requires less storage for the working weight matrices. Approximately what percentage is the savings when τ is large?

Ex. 17.6.   CDFs for ZA- and ZI-Distributions
Let F be the CDF corresponding to the parent distribution in Fig. 17.1.

(a) Write down the CDF of the ZA-distribution as a function of ω and F.
(b) Do the same for the ZI-distribution as a function of ϕ and F.
(c) Invert your answer in (a) so as to obtain an expression for the p-quantile, where $0 < p < 1$.
(d) Invert your answer in (b) so as to obtain an expression for the p-quantile, where $0 < p < 1$.

Ex. 17.7.   Capture–recapture Models Fitted to `hare` in Rcapture
Consider the `hare` data frame in package Rcapture.

(a) Fit models $\mathcal{M}_0$, $\mathcal{M}_b$, $\mathcal{M}_t$, $\mathcal{M}_{tb}$ models to the data.
(b) Obtain the estimates of N from the models. Do they differ much?
(c) Compute the AICs of the models and suggest which fit is to be preferred. For that one, obtain an approximate 95% confidence interval for N.
(d) Repeat (c) using BICs. Is there much of a difference in the answers?

Ex. 17.8. $\mathcal{M}_{tbh}(\tau = 3)$
Consider the $\mathcal{M}_{tbh}$ model with $\tau = 3$ sampling occasions.

(a) Write down the equivalent of Table 17.4 for this model.
(b) With covariates x2 and x3 plus the intercept, and with $\tau = 3$, write down $\boldsymbol{\eta}$, and the constraint matrices corresponding to `parallel.b = TRUE ~ x2 - 1` and `parallel.t = FALSE ~ x3`.

Ex. 17.9.   Positive-Bernoulli EIMs for $\mathcal{M}_{tbh}(\tau = 2)$
Consider sample data generated by $\mathcal{M}_{tb}(\tau = 2)$ in Table 17.4.

(a) Derive the marginal distributions $f_1(y_1)$ and $f_2(y_2)$. Derive their means μ_1 and μ_2.
(b) Derive the score vector and EIM.

Ex. 17.10.   Positive-Bernoulli EIMs for $\mathcal{M}_{bh}(\tau = 2)$ and $\mathcal{M}_{bh}(\tau = 3)$

(a) Consider the $\mathcal{M}_{bh}$ model (17.14)–(17.15) with $\tau = 2$ sampling occasions. This is also illustrated in Table 17.4. Show that, for one observation, $-E(\partial^2 \ell/\partial p_r^2) = p_c/\{p_r(1 - p_r)(1 - (1 - p_c)^2)\}$. Explain why the EIM is diagonal, and derive its other diagonal element.
(b) Repeat (a) with the $\tau = 3$ case. Show that, for one observation, $-E(\partial^2 \ell/\partial p_c^2) = (3 - 6p_c + 7p_c^2 - 3p_c^3)/\{p_c(1 - p_c)(1 - (1 - p_c)^3)\} + \partial A/\partial p_c$ where $A = 3(1 - p_c)^2/(1 - (1 - p_c)^3)$.
(c) Explain why an RR-$\mathcal{M}_{bh}$ model results in (17.27).
(d) Write down an expression for η_j produced by

```
rrvglm(cbind(y1, y2, y3, y4, y5) ~ x2 + x3 + x4, posbernoulli.t, data = pdata)
```

**Ex. 17.11.   **Adapt the short-term memory model `M.tbh.lag2` from Sect. 17.2.6 so that it has a parameter for a lag-1 effect and another parameter for a lag-2 effect. Of course, there will be the usual enduring effect $\widehat{\beta}^*_{(1)1}$.

Ex. 17.12.   `posbernoulli.t()` with the `xij` Argument
In Sect. 17.2.4, a $\mathcal{M}_{tbh}$ model was fitted to simulated data.

(a) Fit a $\mathcal{M}_{th}$ model using `posbernoulli.t()`.
(b) Fit the equivalent model as (a) but using `posbernoulli.tb()`.

Ex. 17.13.   Mean and Variances

(a) From (17.2)–(17.3), derive the mean and variance for the positive-Poisson, positive-binomial and positive-geometric distributions.
(b) From (17.8), do the same for these ZI-distributions.
(c) From (17.5)–(17.6), do the same for these ZA-distributions.

Ex. 17.14.   Occupancy Modelling
A biologist decides to visit n sites to see whether a tree species grows there. At each site visit, he has a probability p_i of detecting the species ($i = 1, \ldots, n$).

He will go to each site a maximum of K times but will stop going there once he detects the species there. Here, K is fixed. Each site is occupied by the species with probability ψ_i. Let Y_i^* be the number of visits to site i.

(a) Write down the probability function and log-likelihood.
(b) Derive the first derivatives with respect to (p_i, ψ_i). Derive the EIM too.
(c) † Write a VGAM family function to fit this model. The response will be an $n \times K$ matrix of 0s and 1s. Reading along a row, once a 1 is encountered, what goes after that is ignored. A row made up entirely of 0s means that species was never seen there.

Ex. 17.15.   Explain why a zero-inflated binomial distribution fitted to Bernoulli responses will fail.

Ex. 17.16.   EIMs—ZI-Distributions
Consider (17.7) and let $A = \phi + (1-\phi)f(0) = P(Y = 0)$ where $f(0) = P(Y^* = 0)$ where Y^* denotes the random variable from the parent distribution.

(a) Let ℓ_{i+} and ℓ_{i0} be parts of the log-likelihood corresponding to positive y_i and $y_i = 0$, respectively. Show that

   (i) $\partial \ell_{i0}/\partial \phi = [1 - f(0)]/A$ and $\partial \ell_{i0}/\partial \theta = [(1-\phi)/A]\,\partial f(0)/\partial \theta$,
   (ii) $\partial \ell_{i+}/\partial \phi = -1/(1-\phi)$ and $\partial \ell_{i+}/\partial \theta = \partial \ell_i^*/\partial \theta$.

(b) Show that the score component $\partial \ell/\partial \phi$ has zero mean.
(c) Show that the EIM for a zero-inflated 1-parameter discrete distribution, that is parameterized in terms of (ϕ, θ), is

$$
\begin{pmatrix}
\dfrac{1}{A}\dfrac{(1-f(0))}{(1-\phi)} & \dfrac{1}{A}\cdot\dfrac{\partial f(0)}{\partial \theta} \\[2ex]
\dfrac{1}{A}\cdot\dfrac{\partial f(0)}{\partial \theta} & J
\end{pmatrix}
\tag{17.29}
$$

where a conditional expectation is used for $J =$

$$
-E\left[\frac{\partial^2 \ell_i^*}{\partial \theta^2}\right]\cdot(1-A) - \left[\frac{1-\phi}{A}\right]\cdot\left\{A\,\frac{\partial^2 f(0)}{\partial \theta^2} - (1-\phi)\left(\frac{\partial f(0)}{\partial \theta}\right)^2\right\}.
\tag{17.30}
$$

   Hint: $E(Y) = E(Y^*)/(1 - f(0))$ for ℓ_{i+}.
(d) Use (17.29) to show the EIM for the ZIP distribution is (17.28). Show it is positive-definite for $(\phi, \lambda) \in (0, 1) \times (0, \infty)$.
(e) Use (17.29) to derive the EIM for the ZI-binomial and ZI-geometric.

Ex. 17.17.   Show that, for one observation from a positive normal distribution,

$$
-E\left[\frac{\partial^2 \ell}{\partial \mu^2}\right] = \frac{1}{\sigma^2}\left\{1 - \frac{\mu\phi}{\sigma(1-\Phi)} - \frac{\phi^2}{(1-\Phi)^2}\right\},
$$

$$
-E\left[\frac{\partial^2 \ell}{\partial \sigma^2}\right] = \frac{2}{\sigma^2} - \frac{\mu\phi}{\sigma^3(1-\Phi)}\left\{1 + \frac{\mu^2}{\sigma^2} + \frac{\phi\mu}{\sigma(1-\Phi)}\right\},
$$

$$
-E\left[\frac{\partial^2 \ell}{\partial \mu\,\partial \sigma}\right] = \frac{\phi}{\sigma^2(1-\Phi)}\left\{1 + \frac{\mu^2}{\sigma^2} + \frac{\phi\mu}{\sigma(1-\Phi)}\right\},
$$

where $\phi = \phi(-\mu/\sigma)$ and $\Phi = \Phi(-\mu/\sigma)$.

Ex. 17.18. Try fitting positive-Poisson and positive-NB distributions to the `corbet` butterfly data. Assess your fits using some goodness-of-fit test.

Ex. 17.19. Suppose the response Y = number of babies born is regressed against x_2 = `age` and x_3 = `sex`, in a large data set collected from a human cross-sectional study, as follows:

```
vglm(nbabies ~ age + sex, zipoisson, data = zdata, crit = "coeff", trace = TRUE)
```

(a) Explain why this regression fails by using a relevant selection from the following terms: (i) interaction, (ii) quasi-complete separation, (iii) orthogonal parameters, (iv) canonical link, (v) structural 0, (vi) Hauck-Donner effect, (vii) idempotent, (viii) seemingly unrelated regression.
(b) How might one proceed with a regression analysis?
(c) Why is `crit = "coeff"`, `trace = TRUE` above a good idea?

Ex. 17.20. The reciprocal of a positive normal random variable is said to have an *alpha distribution*. Show that

$$F(y; \alpha, \beta) = \frac{\Phi(\alpha - \beta/y)}{\Phi(\alpha)}$$

for some $\alpha > 0$ and $\beta > 0$ (express these as functions of μ and σ), and that the resulting density is

$$f(y; \alpha, \beta) = \frac{\beta}{\sqrt{2\pi} \, y^2 \, \Phi(\alpha)} \cdot \exp\left\{ -\frac{1}{2} \left(\alpha - \frac{\beta}{y} \right)^2 \right\}.$$

Ex. 17.21. Consider the variable `pub1stAuthor` (the number of first-author publications, from the MathSciNet database) of the `profs.nz` data frame.

(a) Fit a zero-inflated Poisson model to these data with `I(2014 - firstyear)` as explanatory. Model the probability of a structural 0 as intercept-only.
(b) Do the same but with `log(2014 - firstyear)` as explanatory.
(c) Which of (a) and (b) seems most satisfactory? Why?

Ex. 17.22. Consider the `ugss` data frame.

(a) Choose a response from the set {`movies`, `piercings`, `receivetxt`, `sendtxt`, `tattoos`}. Then, using explanatory variables `age` and `sex`, fit the following models: (i) quasi-Poisson, (ii) NB-1, (iii) NB (NB-2), (iv) NB-P, (v) ZIP, (vi) ZINB, (vii) ZAP, (viii) ZANB.
(b) Perform some goodness-of-fit test such as AIC or BIC to suggest which model appears best.
(c) For the 'best' model:

   (i) Interpret the estimated regression coefficients.
   (ii) Try estimating the effect of `age` using smoothing. Is there much nonlinearity?
   (iii) Explore interactions between `age` and `sex`. Are there significant interactions?

Ex. 17.23.   Consider the `belcap` dental data frame.

(a) For the response `dmfte`, and using explanatory variables `log(0.5 + dmftb)`, `gender`, `ethnic` and `school`, fit a ZIP.
(b) Is the model really linear with respect to `log(0.5 + dmftb)`? Try smoothing.
(c) For whatever model deemed best, interpret the estimated regression coefficients.

Ex. 17.24.   Consider all the females in the `xs.nz` data frame, and the number of `babies` born to them.

(a) Fit a simple zero-inflated NB and zero-inflated Poisson regression. Use an intercept-only term.
(b) Add the variables `age` and `ethnicity` to your ZINB model. Try to allow for some smoothing.
(c) Add the variables `age` and `ethnicity` to your ZIP model. Allow for some smoothing.
(d) Comment on the four models. Is any one of them to be preferred?

We begin to die the moment we are born, and the end is linked to the beginning.
—Manilius

Table 17.6 Summary of positive and ZA discrete distributions currently in VGAM. For ZA-types, $\omega = \Pr(\text{an observed } 0)$ (argument pobs0). Argument onempobs0 corresponds to $\omega^* = 1 - \omega$. Notes: (i) Those η_j which are underlined are intercept-only, by default. (ii) The mean is returned as the fitted value, by default.

Distribution	Probability function $f(y; \boldsymbol{\theta})$	Support	Default $\boldsymbol{\eta}^T$	Mean	VGAM family
Positive Bernoulli	See (17.12)	$(\{0,1\})^\tau \setminus \{\mathbf{0}_\tau\}$	See Sect. 17.2	See Sect. 17.2.1	posbernoulli.[tb]()
Positive binomial	$f_{\mathrm{B}}(y) = \dfrac{1}{1 - (1-p)^N}\dbinom{N}{Ny} p^{Ny}(1-p)^{N(1-y)}$	$\dfrac{1}{N}\left(\dfrac{1}{N}\right)1$	(logit p)	$\dfrac{p}{1 - (1-p)^N}$	posbinomial()
Positive NB	$f_{\mathrm{NB}}(y) = \dfrac{\dbinom{y+k-1}{y}\left(\dfrac{\mu}{\mu+k}\right)^y\left(\dfrac{k}{k+\mu}\right)^k}{1 - [k/(k+\mu)]^k}$	$1(1)\infty$	(log μ, log k)	$\dfrac{\mu}{1 - [k/(k+\mu)]^k}$	posnegbinomial()
Positive Poisson	$f_{\mathrm{P}}(y) = \dfrac{1}{1 - e^{-\lambda}}\dfrac{e^{-\lambda}\lambda^y}{y!}$	$1(1)\infty$	(log λ)	$\dfrac{\lambda}{1 - e^{-\lambda}}$	pospoisson()
ZA binomial	$I[y=0]\,\omega + I[y>0]\,(1-\omega)\,f_{\mathrm{B}}(y)$ $I[y=0]\,(1-\omega^*) + I[y>0]\,\omega^*\,f_{\mathrm{B}}(y)$	$0\left(\dfrac{1}{N}\right)1$	(logit ω, logit p) $\left(\text{logit } p,\ \underline{\text{logit } \omega^*}\right)$	$\dfrac{(1-\omega)\,p}{1 - (1-p)^N}$ $\dfrac{\omega^*\,p}{1 - (1-p)^N}$	zabinomial() zabinomialff()
ZA geometric	$I[y=0]\,\omega + I[y>0]\,(1-\omega)\,p(1-p)^{y-1}$ $I[y=0]\,(1-\omega^*) + I[y>0]\,\omega^*\,p(1-p)^{y-1}$	$0(1)\infty$	(logit ω, logit p) $\left(\text{logit } p,\ \underline{\text{logit } \omega^*}\right)$	$(1-\omega)/p$ ω^*/p	zageometric() zageometricff()
ZA NB	$I[y=0]\,\omega + I[y>0]\,(1-\omega)\,f_{\mathrm{NB}}(y)$ $I[y=0]\,(1-\omega^*) + I[y>0]\,\omega^*\,f_{\mathrm{NB}}(y)$	$0(1)\infty$	(logit ω, log λ, log k) $\left(\text{log } \lambda,\ \underline{\text{log } k},\ \text{logit } \omega^*\right)$	$\dfrac{(1-\omega)\,\mu}{1 - [k/(k+\mu)]^k}$ $\dfrac{\omega^*\,\mu}{1 - [k/(k+\mu)]^k}$	zanegbinomial() zanegbinomialff()
ZA Poisson	$I[y=0]\,\omega + I[y>0]\,(1-\omega)\,f_{\mathrm{P}}(y)$ $I[y=0]\,(1-\omega^*) + I[y>0]\,\omega^*\,f_{\mathrm{P}}(y)$	$0(1)\infty$	(logit ω, log λ) $\left(\text{log } \lambda,\ \underline{\text{logit } \omega^*}\right)$	$\dfrac{(1-\omega)\,\lambda}{1 - e^{-\lambda}}$ $\dfrac{\omega^*\,\lambda}{1 - e^{-\lambda}}$	zapoisson() zapoissonff()

Table 17.7 Summary of zero-inflated (ZI) discrete distributions currently supported by VGAM. Here, ϕ is the probability of a structural zero (argument `pstr0`). Argument `onempstr0` corresponds to $\phi^* = 1 - \phi$. Notes: (i) Those η_j which are underlined are intercept-only, by default. (ii) The mean is returned as the fitted value, by default. (iii) All family functions have associated **dpqr**-type functions which permit zero–deflation.

Distribution	Probability function $f(y; \boldsymbol{\theta})$	Support	Default $\boldsymbol{\eta}^T$	Mean	VGAM family
ZI binomial	$I[y=0]\,\phi + (1-\phi)\binom{N}{Ny}p^{Ny}(1-p)^{N(1-y)}$	$0\left(\frac{1}{N}\right)1$	$(\mathrm{logit}\,\phi,\ \mathrm{logit}\,p)$	$(1-\phi)\,p$	`zibinomial()`
	$I[y=0]\,(1-\phi^*) + \phi^*\binom{N}{Ny}p^{Ny}(1-p)^{N(1-y)}$		$(\mathrm{logit}\,p,\ \underline{\mathrm{logit}\,\phi^*})$	$\phi^*\,p$	`zibinomialff()`
ZI geometric	$I[y=0]\,\phi + (1-\phi)\,p\,(1-p)^y$	$0(1)\infty$	$(\mathrm{logit}\,\phi,\ \mathrm{logit}\,p)$	$(1-\phi)\dfrac{1-p}{p}$	`zigeometric()`
	$I[y=0]\,(1-\phi^*) + \phi^*\,p\,(1-p)^y$		$(\mathrm{logit}\,p,\ \underline{\mathrm{logit}\,\phi^*})$	$\phi^*\dfrac{1-p}{p}$	`zigeometricff()`
ZI NB	$I[y=0]\,\phi + (1-\phi)\binom{y+k-1}{y}\dfrac{\mu^y\,k^k}{(\mu+k)^{y+k}}$	$0(1)\infty$	$(\mathrm{logit}\,\phi,\ \log\mu,\ \underline{\log k})$	$(1-\phi)\,\mu$	`zinegbinomial()`
	$I[y=0]\,(1-\phi^*) + \phi^*\binom{y+k-1}{y}\dfrac{\mu^y\,k^k}{(\mu+k)^{y+k}}$		$(\log\mu,\ \underline{\log k},\ \underline{\mathrm{logit}\,\phi^*})$	$\phi^*\,\mu$	`zinegbinomialff()`
ZI Poisson	$I[y=0]\,\phi + (1-\phi)\dfrac{e^{-\lambda}\lambda^y}{y!}$	$0(1)\infty$	$(\mathrm{logit}\,\phi,\ \log\lambda)$	$(1-\phi)\,\lambda$	`zipoisson()`
	$I[y=0]\,(1-\phi^*) + \phi^*\dfrac{e^{-\lambda}\lambda^y}{y!}$		$(\log\lambda,\ \underline{\mathrm{logit}\,\phi^*})$	$\phi^*\,\lambda$	`zipoissonff()`

Chapter 18
On VGAM Family Functions

Our goal in programming with data is to turn ideas into software, quickly and faithfully.
—Chambers (1998)

18.1 Introduction

This chapter gives some details about what a **VGAM** family function consists of. The casual reader need not delve into the details, however, developers and serious users should benefit. Such readers are assumed to be familiar with aspects of the R language, including the object-oriented programming features of the S4 language. **VGAM** has a reasonably flexible design that, theoretically, allows users to solve for the MLEs of straightforward problems. At its simplest, one just creates a basic **VGAM** family function that has the appropriate derivatives and working weights and initial values, etc. Due to space limitations, only skeletal details are provided.

An outline of this chapter is as follows. We begin by looking at the associated topic of **VGAM** link functions. Then Sect. 18.3 describes the basics of simple family functions. Section 18.4 provides details on more advanced features. Section 18.5 gives further examples. Section 18.6 provides the technical details behind the smart prediction facility of Sect. 8.2.5. For those wanting just to solve a simple one-off problem, reading up to Sect. 18.3.1 might suffice. For **VGAM** family functions that allow a suite of arguments, Sect. 18.3.2 onwards will be necessary.

Here is a checklist for those contemplating writing a **VGAM** family function.

(i) Do the usual regularity conditions hold for the model? Is scoring a suitable algorithm?
(ii) Do closed-form expressions for the EIM exist, and if so, can it be computed? If not, can it be approximated by generating random score vectors?—this will mean the first derivatives are needed and an **r**-type function written. Details of SFS are given in Sect. 9.2.2.

T.W. Yee, *Vector Generalized Linear and Additive Models*,
Springer Series in Statistics, DOI 10.1007/978-1-4939-2818-7_18

Table 18.1 Compulsory arguments in a **VGAM** link function for $g_j(\theta_j) = \eta_j$, and their defaults. The first argument should always be named `theta`.

Argument + default	Function
`theta`	The default is θ_j, but is interpreted as η_j if `inverse = TRUE`. May be numerical or character. If character then `short` and `tag` are accessed
`inverse = FALSE`	Logical. If `TRUE` then return the inverse link function $\theta_j = g_j^{-1}(\eta_j)$
`deriv = 0`	Returns $d^k\eta_j/d\theta_j^k$ for $k = 0, 1, 2$. If `inverse = TRUE` then return $d^k\theta_j/d\eta_j^k$
`short = TRUE`	Logical. If `TRUE` then return a short (character) label, else a long label, e.g., `"logit(theta)"` and `"log(theta/(1-theta))"`
`tag = FALSE`	If `TRUE` then add a descriptor to the front of the label, e.g., `"Logit: logit(theta)"`

(iii) Does the writer have lots of time, skill and patience? More often than not, getting even the simplest function bug-free requires much more work than first anticipated. There is a steep learning curve, and a reasonable requisite of theory too. Hacking is usually needed, and this is partly a result of the limited documentation available and the absence of support.

(iv) Is the problem a one-off exercise that could be easier solved using a general purpose function, such as `optim()` or `nlm()`? Is it really worthwhile writing and documenting a function? Who else will use it, and how will it be disseminated to the community of users? As an R package?

(v) Optionally, can the log-likelihood be computed? If so, then it is preferably done in the form of a `d`-type function with a `log` argument.

A warning to the reader: writing even a simple **VGAM** family function is not to be considered an easy task! Another warning is that some **VGAM** internals may change in the future, hence software maintenance is definitely required.

Something to be mindful of is that there are four stages of complexity regarding writing **VGAM** family functions. These are the (recall that $M_1 \equiv \dim(\boldsymbol{\eta})$ for one response)

(i) $M_1 = 1$ case with 1 response,
(ii) $M_1 = 1$ case with multiple responses,
(iii) $M_1 > 1$ case with 1 response,
(iv) $M_1 > 1$ case with multiple responses.

All good **VGAM** family functions ought to handle multiple responses, e.g., of the form `cbind(y1, y2, y3)`, where possible. However, this is an optional feature to have. If it does, then it potentially can be modelled as an RCIM (Sect. 5.7). Over time, more case (i) functions may be upgraded to case (ii), as well as case (iii) to case (iv). Programming for (i) and (ii) is relatively easy, but cases (iii)–(iv) generally need support in the form the functions listed in Table 18.5. For example, case (iii) may require `iam()` to map the (s, t) element of the EIM to `wz`. Case (iv) is the most difficult, and it involves interleaving the columns of matrices, and consequently the code becomes more complicated. Functions such as `arwz2wz()` and `w.wz.merge()` can facilitate this latter case.

Where the EIM is programmable, some suggested families to scrutinize, respectively, for the above cases, are:

(i) `borel.tanner`, `fgm()`, `posbinomial()`, `simple.exponential()`,
(ii) `biclaytoncop()`, `lindley()`, `maxwell()`, `rayleigh()`, `zetaff()`,

(iii) `binom2.or()`, `multinomial()`, `posbernoulli.b()`,
(iv) `gamma2()`, `kumar()`, `uninormal()`, `weibullR()`, `zipoisson()`.

These examples of (i) and (iii) hold *at the time of writing*, but may be upgraded in the future to (ii) and (iv), respectively. Likewise, where the EIM is unavailable but random variates are available, some suggested families to examine, respectively, for the above cases, are:

(i) `hzeta()`, `skewnormal()`,
(ii) `yulesimon()`,
(iii) `fff()`, `riceff()`, `slash()`,
(iv) `negbinomial()`, `perks()`, `posnegbinomial()`, `zinegbinomial()` (Sect. 18.5.2).

Should the OIM and EIM coincide, `deriv3()` might be of help, to avoid manual programming of certain simple derivatives (Table 18.5).

18.2 Link Functions

Table 1.2 gives a summary of some **VGAM** link functions currently available. These are the g_j in the basic $\eta_j = g_j(\theta_j)$ formula. The functions are incompatible with families associated with `glm()` such as `binomial()` and `poisson()`. Table 18.1 gives details on what the functions are expected to return. As a specific example, let's consider the logit link. Here are a few basic calls.

```
> pvec <- seq(0.1, 0.8, by = 0.1)
> eta <- 0:4
> logit(pvec)

   [1] -2.19722 -1.38629 -0.84730 -0.40547  0.00000  0.40547  0.84730  1.38629

> logit(eta, inverse = TRUE)   # Also known as the antilogit

   [1] 0.50000 0.73106 0.88080 0.95257 0.98201

> logit("prob")

   [1] "logit(prob)"

> logit("prob", tag = TRUE, short = FALSE)

   [1] "Logit: log(prob/(1-prob))"

> logit(pvec, deriv = 1)

   [1] 11.1111  6.2500  4.7619  4.1667  4.0000  4.1667  4.7619  6.2500

> logit(pvec, deriv = 1, inverse = TRUE)

   [1] 0.09 0.16 0.21 0.24 0.25 0.24 0.21 0.16

> logit(pvec, deriv = 2)

   [1] -98.7654 -23.4375  -9.0703  -3.4722   0.0000   3.4722   9.0703  23.4375

> logit(pvec, deriv = 2, inverse = TRUE)

   [1]  0.072  0.096  0.084  0.048  0.000 -0.048 -0.084 -0.096
```

The `logit()` function is easily understood given that $\eta = \log\{\theta/(1-\theta)\}$ implies $\theta = e^\eta/(1+e^\eta)$, $d\theta/d\eta = \theta(1-\theta)$ and $d^2\theta/d\eta^2 = \theta(1-\theta)(1-2\theta)$.

Link functions are invoked within **VGAM** family functions using `eta2theta()` and `theta2eta()`, as well as `dtheta.deta()` and `d2theta.deta2()` (Table 18.2). This is because the link chosen by the user is passed into the appropriate slot using `substitute()`. Actually, "passed into" might be more accurately replaced by "embedded" or "infixed". Here are some simple calls to compute some of the above.

```
> theta2eta(pvec, "logit")

  [1] -2.19722 -1.38629 -0.84730 -0.40547  0.00000  0.40547  0.84730  1.38629

> eta2theta(eta, "logit")

  [1] 0.50000 0.73106 0.88080 0.95257 0.98201

> dtheta.deta(pvec, "logit")

  [1] 0.09 0.16 0.21 0.24 0.25 0.24 0.21 0.16

> d2theta.deta2(pvec, "logit")

  [1]  0.072  0.096  0.084  0.048  0.000 -0.048 -0.084 -0.096
```

Actually, these four functions also have a third argument called `earg`, which is an extra `arg`ument to allow other parameters to be passed in. This will be seen in Sect. 18.3.4.

18.2.1 Chain Rule Formulas

Most computations involving link functions and IRLS make use of the chain rule in some form. In order to handle $\eta_j = g_j(\theta_j)$, the following formulas are useful:

$$\frac{\partial \ell_i}{\partial \eta_j} = \frac{\partial \ell_i}{\partial \theta_j} \frac{\partial \theta_j}{\partial \eta_j}, \tag{18.1}$$

$$\frac{\partial^2 \ell_i}{\partial \eta_j^2} = \frac{\partial^2 \ell_i}{\partial \theta_j^2} \left(\frac{\partial \theta_j}{\partial \eta_j}\right)^2 + \left\{\frac{\partial \ell_i}{\partial \theta_j} \frac{\partial^2 \theta_j}{\partial \eta_j^2}\right\}, \tag{18.2}$$

$$\frac{\partial^2 \ell_i}{\partial \eta_j \, \partial \eta_k} = \frac{\partial^2 \ell_i}{\partial \theta_j \, \partial \theta_k} \frac{\partial \theta_j}{\partial \eta_j} \frac{\partial \theta_k}{\partial \eta_k}, \quad j \neq k, \tag{18.3}$$

$$E\left(\frac{\partial \ell_i}{\partial \boldsymbol{\theta}}\right) = \mathbf{0}. \tag{18.4}$$

Sometimes a η_j involves more than one parameter, e.g., the NB-C_2-2 model which is a negative binomial regression with its canonical link $\eta_1 = \log(\mu/(\mu+k))$ (Sect. 11.3.3) involving both its parameters. Then, more generally for vector $\boldsymbol{\eta}$, the following formulas are required.

Table 18.2 Functions that make use of a **VGAM** link function. The argument `earg` is a list, which often defaults to `list(inverse = FALSE, deriv = 0, short = TRUE, tag = FALSE)`.

Function	Value and comments
`dtheta.deta(theta, link, earg)`	$d\theta/d\eta$
`d2theta.deta2(theta, link, earg)`	$d^2\theta/d\eta^2$
`eta2theta(eta, link, earg)`	$\theta = g^{-1}(\eta)$
`theta2eta(theta, link, earg)`	$\eta = g(\theta)$
`link2list()`	Processes a **VGAM** family function link argument (Sect. 18.3.3)
`namesof(theta, link, earg, tag = FALSE, short = TRUE)`	A description of the link function. Useful in `@blurb`, and for assigning `predictors.names` in `@initialize`

$$\frac{\partial \ell_i}{\partial \boldsymbol{\eta}} = \frac{\partial \boldsymbol{\theta}^T}{\partial \boldsymbol{\eta}} \frac{\partial \ell_i}{\partial \boldsymbol{\theta}}, \tag{18.5}$$

$$\frac{\partial^2 \ell_i}{\partial \boldsymbol{\eta} \, \partial \boldsymbol{\eta}^T} = \frac{\partial \boldsymbol{\theta}^T}{\partial \boldsymbol{\eta}} \frac{\partial^2 \ell_i}{\partial \boldsymbol{\theta} \, \partial \boldsymbol{\theta}^T} \frac{\partial \boldsymbol{\theta}}{\partial \boldsymbol{\eta}^T} + \left\{ \sum_j \frac{\partial \ell_i}{\partial \theta_j} \frac{\partial^2 \theta_j}{\partial \boldsymbol{\eta} \, \partial \boldsymbol{\eta}^T} \right\}. \tag{18.6}$$

An example of (18.5)–(18.6) is when η_j and η_k are both functions of θ_s and θ_t, e.g., the NB-C$_2$-2 model. Then

$$\frac{\partial \ell_i}{\partial \eta_j} = \frac{\partial \ell_i}{\partial \theta_s} \frac{\partial \theta_s}{\partial \eta_j} + \left\{ \frac{\partial \ell_i}{\partial \theta_t} \frac{\partial \theta_t}{\partial \eta_j} \right\}, \tag{18.7}$$

$$\frac{\partial^2 \ell_i}{\partial \eta_j^2} = \frac{\partial^2 \ell_i}{\partial \theta_s^2} \left(\frac{\partial \theta_s}{\partial \eta_j} \right)^2 + 2 \frac{\partial^2 \ell_i}{\partial \theta_s \, \partial \theta_t} \frac{\partial \theta_s}{\partial \eta_j} \frac{\partial \theta_t}{\partial \eta_j} + \frac{\partial^2 \ell_i}{\partial \theta_t^2} \left(\frac{\partial \theta_t}{\partial \eta_j} \right)^2 +$$

$$\left\{ \frac{\partial \ell_i}{\partial \theta_s} \frac{\partial^2 \theta_s}{\partial \eta_j^2} + \frac{\partial \ell_i}{\partial \theta_t} \frac{\partial^2 \theta_t}{\partial \eta_j^2} \right\}, \tag{18.8}$$

$$\frac{\partial^2 \ell_i}{\partial \eta_j \, \partial \eta_k} = \frac{\partial^2 \ell_i}{\partial \theta_s^2} \frac{\partial \theta_s}{\partial \eta_j} \frac{\partial \theta_s}{\partial \eta_k} + \frac{\partial^2 \ell_i}{\partial \theta_s \, \partial \theta_t} \left(\frac{\partial \theta_s}{\partial \eta_j} \frac{\partial \theta_t}{\partial \eta_k} + \frac{\partial \theta_s}{\partial \eta_k} \frac{\partial \theta_t}{\partial \eta_j} \right) +$$

$$\frac{\partial^2 \ell_i}{\partial \theta_t^2} \frac{\partial \theta_t}{\partial \eta_j} \frac{\partial \theta_t}{\partial \eta_k} + \left\{ \frac{\partial \ell_i}{\partial \theta_s} \frac{\partial^2 \theta_s}{\partial \eta_j \, \partial \eta_k} + \frac{\partial \ell_i}{\partial \theta_t} \frac{\partial^2 \theta_t}{\partial \eta_j \, \partial \eta_k} \right\}, \quad j \neq k. \tag{18.9}$$

The implication of (18.4) is that all the above terms in braces $\{\cdots\}$ vanish if one uses the EIM instead of the OIM. Most **VGAM** family functions have $\eta_j = g_j(\theta_j)$, therefore should Fisher scoring be implemented, (18.2) and (18.3) reduce to

$$-E\left[\frac{\partial^2 \ell_i}{\partial \eta_j^2} \right] = -E\left[\frac{\partial^2 \ell_i}{\partial \theta_j^2} \right] \cdot \left(\frac{\partial \theta_j}{\partial \eta_j} \right)^2, \tag{18.10}$$

$$-E\left[\frac{\partial^2 \ell_i}{\partial \eta_j \, \partial \eta_k} \right] = -E\left[\frac{\partial^2 \ell_i}{\partial \theta_j \, \partial \theta_k} \right] \cdot \frac{\partial \theta_j}{\partial \eta_j} \frac{\partial \theta_k}{\partial \eta_k}, \quad j \neq k. \tag{18.11}$$

18.3 Family Function Basics

The essential requirement to create a **VGAM** family function is a model that can be estimated by IRLS. This usually means a model whose log-likelihood (3.7), score vector (3.14), and EIM (3.11) can be computed. If EIMs are unavailable, then the ability to generate random variates from the model is the second-best choice. An optional requirement is some measure of the fit, e.g., a deviance function or, more commonly, just the log-likelihood ℓ. If there is no objective function to be minimized or maximized, then iterations will continue until the changes in the regression coefficients (3.9) are sufficiently small.

In terms of writing code, Table 18.3 enumerates the most important slots of the typical **VGAM** family function. For example, the variable `wz` computed in `@weight` typically is $w_i \mathbf{W}_i$ where the elements of $\mathbf{W}_i$ are defined by (3.11), whereas `@deriv` typically returns $w_i \boldsymbol{u}_i$ where the elements of $\boldsymbol{u}_i$ are defined by (3.14). In general, Fisher scoring is much preferred over Newton-Raphson. This is mainly because all the $\mathbf{W}_i$ are required to be positive-definite over a larger parameter space.

18.3.1 A Simple VGAM Family Function

As a first example, here's a simple family function for the exponential distribution $f(y; \lambda) = \lambda e^{-\lambda y}$ where $y > 0$, and $\lambda > 0$ is the rate parameter. It has $\eta = \log \lambda$ embedded in it, therefore it is unnecessarily too rigid. And like most **VGAM** family functions for continuous distributions, it does not handle data on the boundaries, e.g., $y_i = 0$ will cause a failure.

```
> print(simple.exponential)

  function() {
    new("vglmff",
      blurb = c("Simple exponential distribution\n",
                "Link:     log(rate)\n"),
      deviance = function(mu, y, w, residuals = FALSE, eta, extra = NULL,
                          summation = TRUE) {
        devy <- -log(y) - 1
        devmu <- -log(mu) - y / mu
        devi <- 2 * (devy - devmu)
        if (residuals) {
          sign(y - mu) * sqrt(abs(devi) * c(w))
        } else {
          dev.elts <- c(w) * devi
          if (summation) sum(dev.elts) else dev.elts
        }
      },
```

Table 18.3 The most important slots of a typical **VGAM** family object (of S4 class `"vglmff"`), and their purposes. Those in the bottom half of the table are optional.

Slot	Type	Purpose
`@blurb`	character string	Descriptive. Usually calls `namesof()`
`@infos`	`function(...)`	Returns a list with, e.g., M_1 (`M1`), Q_1 (`Q1`) and logical `multipleResponses`, etc. These are data-independent 'constants'
`@linkinv`	`function(eta, extra = NULL)`	Returns fitted values, e.g., a matrix with rows $\boldsymbol{\mu}_i^T$
`@constraints`	expression	Processes the constraint matrices. This comprises the (i) `constraints` argument, and (ii) arguments such as `parallel` and `zero`. Calls `cm.VGAM()`, `cm.zero.VGAM()`, `negzero.expression.VGAM`, etc.
`@initialize`	expression	Allows error checking for x and y, etc.; preprocesses `w` and `y` if necessary; computes initial `mustart` and/or `etastart`; assigns `predictors.names`
`@last`	expression	Assigns `misc$link`, `misc$earg` and other information; evaluated after final IRLS iteration
`@loglikelihood`	`function(mu, y, w, resid = FALSE, eta, extra = NULL, summation = TRUE)`	Returns ℓ, or $[(\ell)_{is}]$ if `summation = FALSE`
`@vfamily`	character string	For identification, and may be used for S4 classes in the future. Assigned the family function name
`@deriv`	expression	Returns an $n \times M$ matrix of score vectors; has rows $w_i \, \partial \ell_i / \partial \boldsymbol{\eta}^T$, cf. (3.7) and (3.14)
`@weight`	expression	Computes working weights `wz`, $w_i \mathbf{W}_i$, in matrix-band format ((3.11) and Sect. 18.3.5). Evaluated immediately after `@deriv` so that assigned variables can be used without interruption
`@deviance`	`function(mu, y, w, resid = FALSE, eta, extra = NULL, summation = TRUE)`	This slot is optional. Returns the deviance, or deviance contributions if `summation = FALSE`, or deviance residuals (Sects. 2.3.2, 3.7.4) if `resid = TRUE` (and `NULL` if they do not exist)
`@first`	expression	Optional; evaluated at the beginning of IRLS iterations
`@linkfun`	`function(mu, extra = NULL)`	Returns $n \times M$ eta $(\boldsymbol{\eta}_1, \dots, \boldsymbol{\eta}_n)^T$. Sometimes not needed because it may not be possible to compute $\boldsymbol{\eta}_i$, given $\boldsymbol{\mu}_i$. Indeed, for some models, $\boldsymbol{\mu}_i$ does not even exist! Usually this slot is only applicable for $M_1 = 1$ families
`@simslot`	`function(object, nsim)`	Returns `nsim` random variates $\widehat{\boldsymbol{y}}_i$ (Sect. 8.4.3)

```
    loglikelihood = function(mu, y, w, residuals = FALSE, eta, extra = NULL,
                             summation = TRUE) {
      if (residuals) return(NULL)
      if (summation) sum(c(w) * dexp(y, rate  = 1 / mu, log = TRUE)) else
        c(w) * dexp(y, rate  = 1 / mu, log = TRUE)
    },
    initialize = expression({
      predictors.names <- "loge(rate)"
      mustart <- y + (y == 0) / 8
    }),
    linkinv = function(eta, extra = NULL) exp(-eta),
    linkfun = function(mu,  extra = NULL) -log(mu),
    vfamily = "simple.exponential",
    deriv = expression({
      rate <- 1 / mu
      dl.drate <- mu - y
      drate.deta <- dtheta.deta(rate, "loge")
      c(w) * dl.drate * drate.deta
    }),
    weight = expression({
      ned21.drate2 <- 1 / rate^2   # EIM
      wz <- c(w) * drate.deta^2 * ned21.drate2
      wz
    }))
  }
<environment: namespace:VGAM>
```

Family functions in **VGAM** have the form

```
my.family.function <- function(<argument list>) {
  new("vglmff", ...)
}
```

where the slots of the `"vglmff"` object are described in Table 18.3. This creates an object of S4 class `"vglmff"` (the name might be changed in the future) which is fed into modelling functions such as `vglm()` and `vgam()` via their `family` argument.

What is the statistical derivation behind this function? Recall that $\mu = \lambda^{-1}$ and $\mathrm{Var}(Y) = \lambda^{-2}$. Because $\lambda > 0$, it is natural for a log link on the parameter to be taken. Thus $\log \lambda = \eta$, which equals $-\log \mu$. Now $w_i \, \ell_i(\lambda_i; y_i) = w_i \log f(y_i; \lambda_i) = w_i \left(\log \lambda_i - \lambda_i \, y_i \right)$ so that

$$
w_i \frac{\partial \ell_i}{\partial \lambda_i} \;=\; w_i \left(\frac{1}{\lambda_i} - y_i \right) \;=\; w_i \left(\mu_i - y_i \right) \quad \text{and}
$$

$$
w_i \frac{\partial^2 \ell_i}{\partial \lambda_i^2} \;=\; -w_i \frac{1}{\lambda_i^2} \;=\; w_i \, E\!\left(\frac{\partial^2 \ell_i}{\partial \lambda_i^2} \right).
$$

These relationships coupled with

$$
\frac{\partial \ell_i}{\partial \eta_i} \;=\; \frac{\partial \ell_i}{\partial \lambda_i} \frac{\partial \lambda_i}{\partial \eta_i} \quad \text{and} \quad -E\!\left(\frac{\partial^2 \ell_i}{\partial \eta_i^2} \right) \;=\; -E\!\left(\frac{\partial^2 \ell_i}{\partial \lambda_i^2} \right) \cdot \left(\frac{\partial \lambda_i}{\partial \eta_i} \right)^2,
$$

Table 18.4 Some variables in `vglm.fit()` and `vgam.fit()`, and their properties. Daggered (†) variables should be assigned values. See also Table 8.5.

Variable	Purpose
`coefstart`	To be assigned a vector of coefficients in `@initialize`
`control`	List, the result of `vglm.control()`, etc. Has components `maxits`, etc.
`deriv.mu` †	$n \times M$ matrix of score vectors (3.14) returned by `@deriv`
`eta`	$n \times M$ matrix of linear predictors. May be a vector if $M = 1$
`etastart` †	To be assigned a matrix of $\boldsymbol{\eta}_i^T$ in `@initialize`
`extra` †	List containing useful data and/or intermediate computations
`Hlist`	A p_{VLM}-list containing named constraint matrices $\mathbf{H}_k$
`intercept.only`	Logical: `TRUE` if the only explanatory variable is an intercept, otherwise `FALSE`
`M`	M, the total number of linear/additive predictors η_j
`misc` †	List containing useful data and/or intermediate computations
`mu`	n-row matrix or n-vector of fitted values. Not necessarily a mean
`mustart`	To be assigned a matrix of fitted values in `@initialize`
`n`	Number of observations in the data set; is $n = n_{\mathrm{LM}} = \text{nrow}(\mathbf{x})$
`ncolHlist`	A p_{VLM}-vector with elements $\text{ncol}(\mathbf{H}_k)$
`ncol.X.vlm`	Number of variables in $\mathbf{X}_{\mathrm{VLM}} = \text{ncol}(\text{X.vlm.save})$; is p_{VLM}
`p`	Number of variables in $\mathbf{X}_{\mathrm{LM}} = \text{ncol}(\mathbf{x})$; is $p = p_{\mathrm{LM}}$
`predictors.names` †	To be assigned a vector of M names for the η_j
`w`	Prior weights w_i; the `weights` argument. Default: `rep(1, n)`, but may be a $n \times S$ matrix
`wz` †	The working weight matrices (3.11) returned by `@weight`. Format is in matrix-band representation (Sect. 18.3.5)
`x`	LM matrix, $\mathbf{X}_{\mathrm{LM}}$ $(n \times p)$. Column 1 is $\mathbf{1}_n$ for optional intercept
`X.vlm.save`	VLM matrix, $\mathbf{X}_{\mathrm{VLM}}$ is $nM \times p_{\mathrm{VLM}}$
`Xm2`	LM matrix for argument `form2`
`y`	Response. May be a n-vector or a $n \times Q$ matrix

explain `@deriv` and `@weight`, respectively (the EIM was used rather than the OIM). The MLE is given by setting $\partial \ell_i / \partial \lambda_i = 0$ giving $\widehat{\lambda}_i = y_i^{-1}$ so that $\ell_{i,\max} = -\log y_i - 1$.

Unfortunately, `simple.exponential()` has some crippling deficiencies, e.g., there is no choice of a link function, and it does not handle multiple responses. Section 18.3.4 remedies these shortcomings. In the meanwhile, here are some notes about **VGAM** family functions.

1. Notationally, recall that $\mathbf{Y} = [(y_{iq})]$ for $q = 1, \ldots, Q$ is the 'total' response allowing for multiple responses. Let $Q_1 = \dim(\boldsymbol{y}_i)$ for one response so that $Q = Q_1 \cdot S$ for S responses. This matches $M = M_1 \cdot S$. For models with fixed Q_1, the corresponding family function has the slot `@infos` having a list component `Q1` and `M1` to store Q_1 and M_1. If these are not set, then it is assumed that $M_1 = Q_1 = 1$.
   Let M_1 be defined as in Table A.4, viz. as the number of η_j for a single response. Many **VGAM** family functions can handle multiple responses (e.g., `cbind(y1, y2, y3)`), and so it is necessary internally to know the value of M_1. This is returned by the `infos` slot. For example,

Table 18.5 Some internal utility functions useful for writing **VGAM** family functions.

Function	Purpose and comments
arwz2wz()	May be useful in @weight when $M > 2$ and if multiple responses are supported. It receives as input the working weights in an **array** of dimension $n \times S \times t$, say, where $t \in \{1, 2, \ldots, M(M+1)/2\}$, and then copies the elements into the appropriate **wz** format such as (18.14). An example can be seen in gpd()@weight
deriv3()	Can avoid laborious differentiation. Used by, e.g., bistudentt() foldnormal(), hzeta(), mix2normal(), slash()
dimm(M, hbw = M)	Returns M^* in (18.12), given M. The name **dimm** stands for **dim**ension-of-**m**atrix. Argument **hbw** is the half-bandwidth
grid.search()	Performs a grid search to help locate the maximum or minimum of a function (usually ℓ). Useful in @initialize
iam(j, k, M, both = TRUE, diag = TRUE)	Either maps (j, k) onto $\{1, 2, \ldots, M^*\}$ as in (18.13), or supplies the indices of the array. The name **iam** stands for index-of-array-to-**m**atrix
interleave.VGAM(L, M)	Interleaves columns with rows, e.g., interleave.VGAM(10, 2) gives 1, 6, 2, 7, 3, 8, 4, 9, 5, 10. An example based on kumar() is in Sect. 18.3.5
w.wz.merge()	May be used in @weight for merging w with wz, e.g., (18.14). If $S = 1$ then c(w) * wz is returned. This function is particularly useful for models with $M > 1$. Examples include negbinomial() and uninormal()
w.y.check()	May be used in @initialize for checking the w and y input, e.g., that they are conformable, and it allows some basic range checking. Only one response, if stipulated, can be policed too. It may return w and y as matrices of the required dimension

```
> appletree <- data.frame(y = 0:7, w = c(70, 38, 17, 10, 9, 3, 2, 1))
> apple.nbfit <- vglm(cbind(y, y) ~ 1, negbinomial, data = appletree, weights = w)
> (keep <- unlist(apple.nbfit@family@infos()))

                M1                  Q1 multipleResponses              lmu
               "2"                  "1"            "TRUE"            "loge"
             lsize                zero
            "loge"                "-2"
```

This shows that, for negative binomial regression, there are $M_1 = 2$ parameters per response, that each response takes up $Q_1 = 1$ column of the response matrix, and that every abs(as.numeric(keep["zero"])) = 2nd η_j is intercept-only (i.e., the k parameters). In general, $M = SM_1$ where S is the number of responses, hence $Q = Q_1 S$ is the total number of columns of the LHS of the formula.

2. The computations for vglm() and vgam() are actually performed by vglm.fit() and vgam.fit() respectively. It is here that the slots of **VGAM** family functions are utilized. Table 18.4 lists the main variables involved. Some of these need to be assigned values; others such as w and y can be modified with care. The variables in @initialize, @deriv and @weight are global because these slots are evaluated as expression()s.

18.3.2 Initial Values

The main purpose of `@initialize` is to compute starting values. All **VGAM** family functions should be *self-starting*, and it suffices to assigning to the variable `etastart` an $n \times M$ matrix of linear predictors $(\boldsymbol{\eta}_1, \ldots, \boldsymbol{\eta}_n)^T$. Less generally, assigning `mustart` is only applicable if there is a slot `linkfun` which can convert this to `etastart`.

Choosing good starting values is an art and is not always easy, but it is often crucial to the success of the scoring algorithm. For simple univariate models, choosing $\widehat{\eta}_i^{(0)} = g(y_i)$ often works. This follows, as it can be shown for GLMs that

$$g(y_i) \approx g(\mu_i) + g'(\mu_i)(y_i - \mu_i) = \eta_i + \left(\frac{\partial \mu_i}{\partial \eta}\right)^{-1}(y_i - \mu_i) = z_i,$$

which tells us that the adjusted dependent variable z_i is a local approximation to $g(y_i)$. However, sometimes $g(y_i)$ is undefined for certain values of y_i, and this must be taken care of, e.g., $\log(0)$ is undefined for the Poisson distribution.

A second strategy is to choose $\boldsymbol{\mu}_i^{(0)}$ equalling its MLE, if tractable. For this, an improvement involves perturbing the initial values towards the individual data values. In particular, the argument `ishrinkage` that appears in a number of family functions such as `negbinomial()` denotes the parameter s in (8.5). The resulting initial fitted value (usually for a 'mean'-like parameter) is perturbed slightly towards each individual response value, such as illustrated in Fig. 8.1. In particular, using $s = 0.5$ and $\widetilde{\mu} = \overline{y}$ is akin to using `(weighted.mean(y, w) + y) / 2`, and has the advantage that the initial values are in the interval $(\min(y_i), \max(y_i))$, and consequently links such as `logit()` and `loge()` should work with binary and count data, etc.

18.3.3 Arguments in VGAM Family Functions

Although there are no arguments in `simple.exponential()`, almost all **VGAM** family functions offer a variety, such as a

- link, e.g., `cumulative(link = "logit")`,
- initial value, e.g.,

```
> args(betabinomialff)

  function (lshape1 = "loge", lshape2 = "loge", ishape1 = 1, ishape2 = NULL,
      imethod = 1, ishrinkage = 0.95, nsimEIM = NULL, zero = NULL)
  NULL
```

Here, `ishape1` and `ishape2` are optional initial values. Those with a default value of `NULL`, as in `ishape2` here, are usually based on other initial values (such as `ishape1` here). The argument `imethod` is described in Sect. 8.3.1.

- constraint, e.g.,

```
> args(cratio)

  function (link = "logit", parallel = FALSE, reverse = FALSE,
      zero = NULL, whitespace = FALSE)
  NULL
```

The arguments `parallel` and `zero` refer to linear constraints on the functions (Sects. 3.3 and 18.3.6).

This richer set of options contrasts sharply with the paucity of arguments of `glm()` family functions, which is usually just `link`.

Table 18.2 mentions `link2list()` for processing a VGAM family function link argument such as `lshape`. As an example, here are the first few lines of `gpd()`.

```
> head(gpd, 10)


 1  function (threshold = 0, lscale = "loge", lshape = logoff(offset = 0.5),
 2      percentiles = c(90, 95), iscale = NULL, ishape = NULL, tolshape0 = 0.001,
 3      type.fitted = c("percentiles", "mean"), giveWarning = TRUE,
 4      imethod = 1, zero = -2)
 5  {
 6      type.fitted <- match.arg(type.fitted, c("percentiles", "mean"))[1]
 7      lscale <- as.list(substitute(lscale))
 8      escale <- link2list(lscale)
 9      lscale <- attr(escale, "function.name")
10      lshape <- as.list(substitute(lshape))
```

Presently, the call to `link2list()` for each link argument, such as `lshape` here, is the standard way such are processed. Extra parameters end up being stored in variables beginning with "e", e.g., `eshape` is a list with a component called `offset` having the value `0.5`. Then `eshape` is passed into the `earg` argument of `eta2theta()`, etc.

18.3.4 Extending the Exponential Distribution Family

We now extend `simple.exponential()` in 5 ways. Firstly, we allow for multiple responses $y^{(s)}$ for $s = 1, \ldots, S$. Secondly, suppose one wishes to introduce a known[1] location parameter a_s, i.e., $f(y^{(s)}; \lambda_s) = \lambda_s \exp\{-\lambda_s (y^{(s)} - a_s)\}$ for $y^{(s)} > a_s$. One then has

$$E(Y^{(s)}) = \mu_s = a_s + \lambda_s^{-1} \quad \text{and} \quad \text{Var}(Y^{(s)}) = \lambda_s^{-2} = (\mu_s - a_s)^2 .$$

The default value of a_s is zero. Thirdly, we allow the user to input a link function applied to λ_s so that $\eta_j = g(\lambda_j)$ for g not necessarily `loge()`. Fourthly, we will allow a choice between the EIM and the OIM in computing the working weights. Fifthly, we allow parallelism constraints, intercept-only constraints, and some control over initial values. The VGAM family function, which is considerably longer, is

---

[1] The location parameter of an exponential distribution must be treated as known. Due to the memoryless property, it would be unestimable if unknown. And if unknown, the regularity conditions would be violated as the support depends on it.

```
> print(better.exponential)

  function(link = "loge", location = 0, expected = TRUE,
            ishrinkage = 0.95, parallel = FALSE, zero = NULL) {
    link <- as.list(substitute(link))
    earg <- link2list(link)
    link <- attr(earg, "function.name")

    new("vglmff",
    blurb = c("Exponential distribution\n\n",
              "Link:     ", namesof("rate", link, earg, tag = TRUE), "\n",
              "Mean:     ", "mu = ", if (all(location == 0)) "1 / rate" else
              if (length(unique(location)) == 1)
              paste(location[1], "+ 1 / rate") else "location + 1 / rate"),
    constraints = eval(substitute(expression({
      constraints <- cm.VGAM(matrix(1, M, 1), x = x, bool = .parallel ,
                          constraints = constraints, apply.int = TRUE)
      constraints <- cm.zero.VGAM(constraints, x, .zero , M)
    }), list( .parallel = parallel, .zero = zero ))),
    infos = eval(substitute(function(...) {
      list(M1 = 1, Q1 = 1, multipleResponses = TRUE, zero = .zero )
    }, list( .zero = zero ))),
    deviance = function(mu, y, w, residuals = FALSE, eta,
                        extra = NULL, summation = TRUE) {
      location <- extra$location
      devy <- -log(y - location) - 1
      devmu <- -log(mu - location) - (y - location ) / (mu - location)
      devi <- 2 * (devy - devmu)
      if (residuals) sign(y - mu) * sqrt(abs(devi) * w) else {
        dev.elts <- c(w) * devi
        if (summation) sum(dev.elts) else dev.elts
      }
    },
    initialize = eval(substitute(expression({
    checklist <- w.y.check(w = w, y = y, ncol.w.max = Inf, ncol.y.max = Inf,
                        out.wy = TRUE, colsperw = 1, maximize = TRUE)
    w <- checklist$w  # So ncol(w) == ncol(y)
    y <- checklist$y

    extra$ncoly <- ncoly <- ncol(y)
    extra$M1 <- M1 <- 1
    M <- M1 * ncoly

    extra$location <- matrix( .location , n, ncoly, byrow = TRUE)  # By row!
    if (any(y <= extra$location))
      stop("all responses must be greater than argument 'location'")

    mynames1 <- if (M == 1) "rate" else paste("rate", 1:M, sep = "")
    predictors.names <-
      namesof(mynames1, .link , earg = .earg , short = TRUE)
```

```
        if (length(mustart) + length(etastart) == 0)
            mustart <- matrix(colSums(y * w) / colSums(w), n, M, byrow = TRUE) *
                    .ishrinkage + (1 - .ishrinkage ) * y + 1 / 8
        if (!length(etastart))
            etastart <- theta2eta(1 / (mustart - extra$location), .link , .earg )
    }), list( .location = location, .link = link, .earg = earg,
                .ishrinkage = ishrinkage ))),
    linkinv = eval(substitute(function(eta, extra = NULL)
        extra$location + 1 / eta2theta(eta, .link , earg = .earg ),
    list( .link = link, .earg = earg ))),
    last = eval(substitute(expression({
        misc$link <- rep( .link , length = M)
        misc$earg <- vector("list", M)
        names(misc$link) <- names(misc$earg) <- mynames1
        for (ii in 1:M)
            misc$earg[[ii]] <- .earg
        misc$location <- .location
        misc$expected <- .expected
    }), list( .link = link, .earg = earg,
                .expected = expected, .location = location ))),
    linkfun = eval(substitute(function(mu, extra = NULL)
        theta2eta(1 / (mu - extra$location), .link , earg = .earg ),
    list( .link = link, .earg = earg ))),
    loglikelihood =
    function(mu, y, w, residuals = FALSE, eta, extra = NULL, summation = TRUE)
        if (residuals) stop("loglikelihood residuals not implemented yet") else {
            rate <- 1 / (mu - extra$location)
            ll.elts <- c(w) * dexp(y - extra$location, rate = rate, log = TRUE)
            if (summation) sum(ll.elts) else ll.elts
        },
    vfamily = c("better.exponential"),
    simslot = eval(substitute(function(object, nsim) {
        pwts <- if (length(pwts <- object@prior.weights) > 0)
                    pwts else weights(object, type = "prior")
        if (any(pwts != 1)) warning("ignoring prior weights")
        mu <- fitted(object)
        rate <- 1 / (mu - object@extra$location)
        rexp(nsim * length(rate), rate = rate)
    }, list( .link = link, .earg = earg ))),
    deriv = eval(substitute(expression({
        rate <- 1 / (mu - extra$location)
        dl.drate <- mu - y
        drate.deta <- dtheta.deta(rate, .link , earg = .earg )
        c(w) * dl.drate * drate.deta
    }), list( .link = link, .earg = earg ))),
    weight = eval(substitute(expression({
        ned2l.drate2 <- (mu - extra$location)^2
        wz <- ned2l.drate2 * drate.deta^2  # EIM
        if (! .expected ) {  # Use the OIM, not the EIM
            d2rate.deta2 <- d2theta.deta2(rate, .link , earg = .earg )
            wz <- wz - dl.drate * d2rate.deta2
        }
        c(w) * wz
    }), list( .link = link, .expected = expected, .earg = earg ))))
}
<environment: namespace:VGAM>
```

Here are some notes.

1. The function `substitute()` is the workhorse for substituting argument values of a function into an S expression or function that will be evaluated later. Here, `better.exponential()` repeatedly substitutes the argument `link` as `.link`— it is conventional in VGAM for the substituted variable name to begin with a ".". to aid in identification. Rather than substitute the location parameter into *all* the components, the above code has the value placed into the `extra` list in `@initialize`, and the remainder of the function use it through `extra`. This technique is particularly beneficial if the size of the argument is large because substituting a large structure can be inefficient, e.g., for

```
vglm(y ~ x2, better.exponential(location = 1:length(y)), data = edata)
```

where `y` is a large vector.

2. Ideally, at least one of `@deviance` and `@loglikelihood` should exist. If it does, then VGAM uses `@deviance` to test convergence at each IRLS step, otherwise `@loglikelihood` is used, else the regression coefficients as a final resort. It is sometimes a good idea to program both where possible. Having either enables half-stepping to occur—each IRLS iteration is guaranteed to be an improvement if the next step happens to 'overshoot'. Half-stepping means that a half-step will be taken until the change in deviance or loglikelihood is an improvement over the current iteration (Table 8.2). Half-stepping is not possible with `criterion = "coef"` since there is no objective function.

3. The argument `earg` is a list that is eventually inputted into `do.call()`. It contains several components, including `inverse`, `deriv`, `short` and `tag`. For example, for `link = logoff(offset = 0.5)`, `earg` has a component called `offset` with the value 0.5.

4. Families which handle multiple responses and/or have $M > 1$ should possess a `zero` argument, and if they make sense, `parallel`, `exchangeable` and `nointercept`. All these are handled in the `constraints` slot. Table 18.6 is a list of common arguments that set up constraint matrices conveniently, and this is the subject of Sect. 18.3.6. Stylistically, the conventions described in Sect. 8.1 applying to the naming of functions and arguments are worth adhering to.

5. For VGAM family functions which operate on a common type of data, it is convenient for users to have a uniform type of processing of the response `y`. For example, `binom2.or()` and `binom2.rho()` both pertain to a bivariate binary response. Another example are the group of categorical models in Table 14.1. Then it makes sense for each member of that group of family functions to call a common R expression or function in `@initialize`. Developers therefore may need to examine the `@initialize` slot of any similar family functions to check whether this is so.

6. As well as writing a VGAM function, it is customary to try write matching `dpqr`-type functions where possible. If so, then the details in Sect. 11.1 are useful.

7. The `loglikelihood` and `deviance` slots have the form

```
> args(acat()@loglikelihood)

  function (mu, y, w, residuals = FALSE, eta, extra = NULL, summation = TRUE)
  NULL
```

The `summation` argument means that the log-likelihood ℓ is returned. If it is set to `FALSE`, then it returns a vector or matrix with elements $w_{is}\ell_{is}$, whose sum is ℓ. Currently the `residuals` argument is not used much at all, except in the case of GLM families where they are well-defined. Nevertheless, there is general provision for deviance residuals as well as log-likelihood residuals to be returned in the future.

8. Yee and Stephenson (2007) propose applying quasi-Newton updates to the working weight matrices within the IRLS algorithm. More experience since then has shown that this technique does not work as well as using simulated Fisher scoring (Sect. 9.2.2). This inferiority applies to both the estimation process and the resulting standard errors of the estimates.

9. Although not used in this example, functions such as `expm1()` and `log1p()` should be used where possible to avoid catastrophic cancellation when programming certain formulas. For example, the EIM for the zero-inflated Poisson (17.28) involves a term $1 - e^{-\lambda}$ which is better coded as `-expm1(-lambda)`. This remains positive for positive `lambda`, whereas naïve coding will return a 0 for `lambda` less than about 10^{-16} and will result in an EIM that is not positive-definite. Another example is to use `log1p(-rho^2)` for the log-N_2 density (13.9).

18.3.5 The *wz* Data Structure

The variable `wz` stores the working weight matrices $w_i\mathbf{W}_i$ in a special format named the *matrix-band* format. This format comprises an $n \times M^*$ matrix where

$$
M^* = \sum_{j=1}^{hbw} (M - j + 1) = \frac{1}{2} hbw (2M - hbw + 1) \tag{18.12}
$$

is the number of columns. Here, hbw refers to the *half-bandwidth* of the matrix, which is an integer between 1 and M inclusive: a diagonal matrix has unit half-bandwidth, a tridiagonal matrix has half-bandwidth 2, etc. For a general matrix $M^* = M(M+1)/2$. As an example, if $M = 4$ then `wz` will have up to $M^* = 10$ columns enumerating the unique elements of (symmetric) $\mathbf{W}_i$ as follows:

$$
\mathbf{W}_i = \begin{pmatrix} 1 & 5 & 8 & 10 \\ & 2 & 6 & 9 \\ & & 3 & 7 \\ & & & 4 \end{pmatrix}. \tag{18.13}
$$

The order is firstly the diagonal, then the band above that, and so on until the $(1, M)$ element. The function `iam()` described below helps manipulate the elements more conveniently and assists with the overall bookkeeping. If $\mathbf{W}_i$ *is* banded, then `wz` needs not have $\frac{1}{2}M(M+1)$ columns; only M^* columns suffice, and the rest of the elements are implicitly zero—this conserves storage.

The matrix-band format is adopted for parsimony. As well as reducing the size of `wz` itself in most cases, the matrix-band format often makes the computation of `wz` more simple and efficient. A final reason is that we sometimes need to input $\mathbf{W}_i$ into **VGAM**. If `wz` is $M \times M \times n$, then `vglm(..., weights = wz)` will result in an error, whereas `wz` being an $n \times M^*$ matrix will work.

With S responses, the combined working weights $\mathbf{W}_i$ have the form of a block-diagonal matrix $\mathrm{diag}(w_{i1}\mathbf{W}_{i1}, \ldots, w_{iS}\mathbf{W}_{iS})$ because one can have different prior weights w_{is} for the sth response. For example, for $S = 3$ responses of an $M_1 = 2$ model, the overall enumeration is

$$\mathbf{W}_i \;=\; \begin{pmatrix} 1 & 7 & & & & \\ & 2 & & & & \\ & & 3 & 9 & & \\ & & & 4 & & \\ & & & & 5 & 11 \\ & & & & & 6 \end{pmatrix} \tag{18.14}$$

(cf. (18.13)). This tridiagonal matrix requires at least 11 columns of which columns 8 and 10 are all 0s. Sometimes elements 1, 3 and 5 are computed in some matrix for $E(-\partial^2 \ell_{is}/\partial \eta_{1s}^2)$, and elements 2, 4 and 6 in another matrix for $E(-\partial^2 \ell_{is}/\partial \eta_{2s}^2)$, and elements 7, 9 and 11 in a third matrix for $E(-\partial^2 \ell_{is}/(\partial \eta_{1s}\partial \eta_{2s}))$; then piecing them together as in (18.14) can be finicky, however this may be facilitated by `w.wz.merge()` and/or `arwz2wz()` (Table 18.5). These functions may also be useful for multiplying each $\mathbf{W}_{is}$ by its prior weight w_{is}, e.g., columns 1, 2 and 7 need to be multiplied by w_{i1}, etc. If $S = 1$, then `w.wz.merge()` returns `c(w) * wz`.

With S responses, here is an example of the use of `interleave.VGAM()`. It applies to `kumar()` (Sect. 18.5.1) and we let $S = 5$, say. The variables `mynames1` and `mynames2` are used separately for the two shape parameters, and these are interleaved to obtain `predictors.names`.

```
>    ncoly <- 5  # Suppose there are 5 responses
>    lshape1 <- lshape2 <- "loge"  # Defaults
>    eshape1 <- eshape2 <- NULL  # No extra arguments
>    M1 <- 2  # 2 parameters per response
>
>    M <- M1 * ncoly
>    mynames1 <- paste("shape1", if (ncoly > 1) 1:ncoly else "", sep = "")
>    mynames2 <- paste("shape2", if (ncoly > 1) 1:ncoly else "", sep = "")
>    predictors.names <-
         c(namesof(mynames1, lshape1 , earg = eshape1 , tag = FALSE),
           namesof(mynames2, lshape2 , earg = eshape2 , tag = FALSE))[
             interleave.VGAM(M, M = M1)]
>    mynames1

 [1] "shape11" "shape12" "shape13" "shape14" "shape15"

>    mynames2

 [1] "shape21" "shape22" "shape23" "shape24" "shape25"

>    predictors.names

 [1] "loge(shape11)" "loge(shape21)" "loge(shape12)" "loge(shape22)"
 [5] "loge(shape13)" "loge(shape23)" "loge(shape14)" "loge(shape24)"
 [9] "loge(shape15)" "loge(shape25)"
```

18.3.5.1 Auxiliary Functions

The functions listed in Table 18.5 are available to facilitate the use of the matrix-band format. Here are some common examples of usage.

1. For the 4×4 example (18.13), the call

```
> iam(2, 3, M = 4)

[1] 6
```

returns the position of the $(2, 3)$ element. A specific example of its use is `betaR()` which currently only handles a single response. Its `weight` slot is

```
> betaR()@weight

expression({
    trig.sum <- trigamma(shapes[, 1] + shapes[, 2])
    ned2l.dshape12 <- trigamma(shapes[, 1]) - trig.sum
    ned2l.dshape22 <- trigamma(shapes[, 2]) - trig.sum
    ned2l.dshape1shape2 <- -trig.sum
    wz <- matrix(as.numeric(NA), n, dimm(M))
    wz[, iam(1, 1, M)] <- ned2l.dshape12 * dshapes.deta[, 1]^2
    wz[, iam(2, 2, M)] <- ned2l.dshape22 * dshapes.deta[, 2]^2
    wz[, iam(1, 2, M)] <- ned2l.dshape1shape2 * dshapes.deta[,
        1] * dshapes.deta[, 2]
    c(w) * wz
})
```

A second type of use of `iam()` takes the form of, e.g.,

```
> iam(NA, NA, M = 4, both = TRUE, diag = TRUE)

$row.index
 [1] 1 2 3 4 1 2 3 1 2 1

$col.index
 [1] 1 2 3 4 2 3 4 3 4 4
```

which returns the indices for the respective array coordinates for successive columns of matrix-band format, as in (18.13). If `diag = FALSE`, then the first 4 elements in each vector are omitted. Note that the first two arguments of `iam()` are not used here, and they have been assigned NAs for simplicity. Here is an example of its use to compute the working weights of the multinomial logit model. The diagonal elements are $w_i \, \mu_{ij}(1 - \mu_{ij})$ for $j = 1, \ldots, M$, and the off-diagonal elements are $-w_i \, \mu_{ij}\mu_{ik}$ for $j \neq k$. This is programmed in `multinomial()@weight` and appears as something like

```
wz <- mu[, 1:M] * (1 - mu[, 1:M])
if (M > 1) {
    index <- iam(NA, NA, M = M, both = TRUE, diag = FALSE)
    wz <- cbind(wz, -mu[, index$row] * mu[, index$col])
}
```

2. If $M > 1$, then a typical use is as follows. Consider the **VGAM** family function `uninormal()` for the univariate normal distribution. Its default is $\boldsymbol{\eta} = (\mu, \log(\sigma))^T$, and for $\boldsymbol{\theta} = (\mu, \sigma)^T$, it has ith EIM equal to $\sigma_i^{-2} \cdot \mathrm{diag}(1, 2)$. Then

```
> uninormal()@weight

  expression({
      wz <- matrix(as.numeric(NA), n, M)
      ned2l.dmu2 <- 1/sdev^2
      if (FALSE) {
          ned2l.dva2 <- 0.5/Varm^2
      }
      else {
          ned2l.dsd2 <- 2/sdev^2
      }
      wz[, M1 * (1:ncoly) - 1] <- ned2l.dmu2 * dmu.deta^2
      wz[, M1 * (1:ncoly)] <- if (FALSE) {
          ned2l.dva2 * dva.deta^2
      }
      else {
          ned2l.dsd2 * dsd.deta^2
      }
      w.wz.merge(w = w, wz = wz, n = n, M = M, ndepy = ncoly)
  })
```

The code block corresponding to if (FALSE) is for $\boldsymbol{\eta} = (\mu, \log(\sigma^2))^T$ corresponding to argument var.arg = FALSE. Note that M1 is already set to 2, and wz has M columns. An alternative is to have wz <- matrix(0, n, dimm(M)) in the top line but this is wasteful in terms of storage, especially for multiple responses. Incidentally, one might instead use c(TRUE, FALSE) and c(FALSE, TRUE) in the indexing of the columns of wz.

18.3.6 Implementing Constraints Within Family Functions

For most models, certain types of linear constraints on the functions are common, and thus they should be made convenient for the user to choose. In particular, these include the parallel, exchangeable and zero arguments (Table 18.6 and Sect. 3.3). Writers of VGAM family functions should implement any such arguments anticipated for that model.

Table 18.6 shows that there are two groups of arguments: the first group is assigned a logical or a formula with a logical response, and the second group is assigned a numerical vector. VGAM has the function cm.VGAM() to process the first group. The second group has a separate function for each argument. Here are some brief details.

- cm.VGAM() applies to arguments such as parallel and exchangeable, viz. having the format described in Sect. 3.3.1 such as TRUE, TRUE ~ x2 + x3 + x4, FALSE ~ 0.
- cm.zero.VGAM() processes a numerical vector, specifying those values of j for which η_j are intercept-only. Similarly, cm.nointercept.VGAM() processes a numerical vector, specifying those values of j for which η_j have no intercept. The role of these two functions is to delete certain columns off relevant constraint matrices.

Table 18.6 Common arguments that set up constraint matrices. The syntax for the upper table are, e.g., TRUE, FALSE, TRUE ~ x2 + x3 - 1, FALSE ~ 0. The syntax for the lower table are, e.g., NULL (none), 2:3, -c(1, 3), c(1, -3). The upper table arguments are processed by cm.VGAM().

Argument	Comments
drop.b	e.g., for posbernoulli.b(), posbernoulli.tb(), Table 17.3, drop the behavioural effect?
eq.sd	e.g., for mix2normal(), are the standard deviation parameters equal?
exchangeable	e.g., for binom2.or(), binom2.rho(), loglinb2()
parallel	Sets $\mathbf{H}_k = \mathbf{1}_M$, i.e., $\beta_{(s)k} = \beta_{(t)k}$ for all $s, t \in \{1, 2, \ldots, M\}$, for selected k, e.g., for cumulative(), negbinomial(), posbernoulli.b()
zero	Sets $\beta_{(j)k} = 0$ for selected j (for $k = 2, 3, \ldots$), e.g., for almost all VGAM family functions with $M > 1$. Processed by cm.zero.VGAM()
nointercept	Sets $\beta_{(j)1} = 0$ for selected j. Processed by cm.nointercept.VGAM()

All functions should be invoked in @constraints. Here is an example.

```
> args(multinomial)

  function (zero = NULL, parallel = FALSE, nointercept = NULL,
      refLevel = "last", whitespace = FALSE)
  NULL

> print(multinomial()@constraints)

  expression({
      constraints <- cm.VGAM(matrix(1, M, 1), x = x, bool = FALSE,
          apply.int = TRUE, constraints = constraints)
      constraints <- cm.zero.VGAM(constraints, x, NULL, M)
      constraints <- cm.nointercept.VGAM(constraints, x, NULL,
          M)
  })
```

The first argument of cm.VGAM() is the constraint matrix: $\mathbf{1}_M$ for the parallelism constraint. The second argument is always the LM matrix x. The third argument is the argument itself, which is substitute()d in. The fourth argument should always be the variable constraints. The argument apply.int = TRUE means that, by default, the constraint *is* applied to the intercept if that argument has the value TRUE. A contrasting example is cumulative(parallel = TRUE), which will not apply the parallelism to the intercept. The programmer needs to decide whether the value TRUE applies the constraint to the intercept or not, and program it accordingly.

The function cm.zero.VGAM() is even simpler. The line containing that function should be pasted in its entirety as above, except that the argument's value is substitute()d into the third argument.

The function cm.nointercept.VGAM() is programmed in a very similar manner to cm.zero.VGAM(). The argument's value is substitute()d into the third argument.

18.4 Some Other Topics

18.4.1 Writing R Packages and Documentation

R package creation is documented in the "Writing R Extensions" document. Details are given regarding a host of topics, including the NAMESPACE file for declaring which variables, functions and S4-style classes to export and import, and the DESCRIPTION file that includes specifying package dependencies. In VGAM, currently not all the contents of NAMESPACE are documented in an online help file.

To create the online help file, the `prompt()` function may be used to create the corresponding .Rd file upon which to edit.

18.4.2 Some S4 Issues

VGAM operates using the S4 object-oriented programming system. The S4-style classes reflect the structure apparent from Fig. 1.2, and Table 18.7 is a summary of these for the major modelling functions. Most modelling functions return an object having a class equal to the name of the modelling function, e.g., a `vglm()` fit has class `"vglm"`. Lines of inheritance also follow Fig. 1.2 to a large degree, e.g., `"vgam"` extend `"vglm"` by the addition of more slots such as `@Bspline`. We say `"vgam"` contains `"vglm"`.

All the classes in VGAM can be seen by

```
> getClasses("package:VGAM")

    [1] "Coef.qrrvglm"        "Coef.rrvgam"         "Coef.rrvglm"
    [4] "grc"                 "qrrvglm"             "rcim"
    [7] "rrvgam"              "rrvglm"              "summary.qrrvglm"
   [10] "summary.rrvgam"      "summary.rrvglm"      "summary.vgam"
   [13] "summary.vglm"        "summary.vlm"         "SurvS4"
   [16] "vcov.qrrvglm"        "vgam"                "vglm"
   [19] "vglmff"              "vlm"                 "vlmsmall"
   [22] "vsmooth.spline"      "vsmooth.spline.fit"
```

For more information about one class, try, for example,

```
> getClass("vglm")
```

which produces much output including the 'distance' between known subclasses and superclasses, as well as

```
> extends("rrvglm")   # "vlmsmall" is experimental & may be deprecated in the future

    [1] "rrvglm"    "vglm"      "vlm"       "vlmsmall"
```

Methods functions can be determined by `showMethods()`, e.g.,

```
> showMethods(classes = "vglm")
```

Writers of VGAM family functions need not become overly involved with S4 object-oriented programming, because their function is largely fed into the `family` argument of the modelling function. However, new methods *do* require S4. Suppose that

Table 18.7 Classes of fitted objects in **VGAM** and inheritance relationships (using "$\subset$"). These are subject to future change.

Modelling function	S4 class	Comments
`vlm()`	`"vlm"`	Not really useful in practice, and might be written for completeness
`vglm()`	`"vglm"`	`"vlm"` $\subset$ `"vglm"`, i.e., `"vlm"` is a subclass of `"vglm"`, or `"vglm"` is a superclass of `"vlm"`, or `"vglm"` extends `"vlm"`
`vgam()`	`"vgam"`	`"vglm"` $\subset$ `"vgam"`
`rrvglm()`	`"rrvglm"`	`"vglm"` $\subset$ `"rrvglm"`
`cqo()`	`"qrrvglm"`	`"rrvglm"` $\subset$ `"qrrvglm"`. Function `cqo()` might be better (or later) called `qrrvglm()`
`cao()`	`"rrvgam"`	`"qrrvglm"` $\subset$ `"rrvgam"`. Function `cao()` might be better (or later) called `rrvgam()`
`rcim()`	`"rcim"`	`"rcim"` $\subset$ `"rrvglm"`
`grc()`	`"grc"`	`"grc"` $\subset$ `"rcim"`

one wanted to write an accessor function called `Depvar()` to return the response or dependent variable as held in the `@y` slot. Then

```
Depvar.vlm <- function(object, ...) object@y

if (!isGeneric("Depvar"))
  setGeneric("Depvar", function(object, ...) standardGeneric("Depvar"))

setMethod("Depvar", "vlm", function(object, ...) Depvar.vlm(object, ...))
```

is a very simple implementation: `Depvar(fit)` should work for any `fit` which inherits from `"vlm"`. The choice here of dispatching with respect to `"vlm"` objects means that all the major classes of models (Table 1.1) should be handled by this one function. That is, `"vlm"` is the fundamental subclass, as seen in Fig. 1.2.

Other particularly useful S4 functions to the programmer include `as()`, `is()`, `new()`, `setClass()`, `slotNames()`.

18.5 Examples

This section looks at two more examples. Prospective writers of family functions will probably need to examine these in detail and invest substantial time hacking into the code. The source code is available in the `.tar.gz` form and not in the `.zip` file for Windows machines.

18.5.1 The Kumaraswamy Distribution Family

This distribution, which is summarized in Table 12.11, has a **VGAM** family function that is more representative of most other family functions. It handles multiple responses, has $M_1 = 2$ and its EIM is tractable. Here it is.

```
> print(kumar)

  function(lshape1 = "loge", lshape2 = "loge",
           ishape1 = NULL,   ishape2 = NULL,
           grid.shape1 = c(0.4, 6.0), tol12 = 1.0e-4, zero = NULL) {
    lshape1 <- as.list(substitute(lshape1))
    eshape1 <- link2list(lshape1)
    lshape1 <- attr(eshape1, "function.name")
    lshape2 <- as.list(substitute(lshape2))
    eshape2 <- link2list(lshape2)
    lshape2 <- attr(eshape2, "function.name")

    if (length(ishape1) &&
        (!is.Numeric(ishape1, length.arg = 1, positive = TRUE)))
      stop("bad input for argument 'ishape1'")
    if (length(ishape2) && !is.Numeric(ishape2))
      stop("bad input for argument 'ishape2'")

    if (!is.Numeric(tol12, length.arg = 1, positive = TRUE))
      stop("bad input for argument 'tol12'")
    if (!is.Numeric(grid.shape1, length.arg = 2, positive = TRUE))
      stop("bad input for argument 'grid.shape1'")

    if (length(zero) &&
        !is.Numeric(zero, integer.valued = TRUE))
      stop("bad input for argument 'zero'")

    new("vglmff",
    blurb = c("Kumaraswamy distribution\n\n",
              "Links:    ", namesof("shape1", lshape1, eshape1, tag = FALSE), ", ",
                            namesof("shape2", lshape2, eshape2, tag = FALSE), "\n",
              "Mean:     shape2 * beta(1 + 1 / shape1, shape2)"),
    constraints = eval(substitute(expression({
      dotzero <- .zero
      M1 <- 2
      eval(negzero.expression.VGAM)
    }), list( .zero = zero ))),
    infos = eval(substitute(function(...) {
      list(M1 = 2, Q1 = 1, expected = TRUE, multipleResponses = TRUE,
           lshape1 = .lshape1 , lshape2 = .lshape2 , zero = .zero )
    }, list( .zero = zero, .lshape1 = lshape1, .lshape2 = lshape2 ))),
    initialize = eval(substitute(expression({
      checklist <- w.y.check(w = w, y = y, Is.positive.y = TRUE,
                             ncol.w.max = Inf, ncol.y.max = Inf,
                             out.wy = TRUE, colsyperw = 1, maximize = TRUE)
      w <- checklist$w
      y <- checklist$y # Now 'w' and 'y' have the same dimension.
      if (any((y <= 0) | (y >= 1)))
        stop("the response must be in (0, 1)")
```

```
    extra$ncoly <- ncoly <- ncol(y)
    extra$M1 <- M1 <- 2
    M <- M1 * ncoly
    mynames1 <- paste("shape1", if (ncoly > 1) 1:ncoly else "", sep = "")
    mynames2 <- paste("shape2", if (ncoly > 1) 1:ncoly else "", sep = "")
    predictors.names <-
        c(namesof(mynames1, .lshape1 , earg = .eshape1 , tag = FALSE),
          namesof(mynames2, .lshape2 , earg = .eshape2 , tag = FALSE))[
          interleave.VGAM(M, M = M1)]

    if (!length(etastart)) {
      kumar.Loglikfun <- function(shape1, y, x, w, extraargs) {
        mediany <- colSums(y * w) / colSums(w)
        shape2 <- log(0.5) / log1p(-(mediany^shape1))
        sum(c(w) * dkumar(y, shape1 = shape1, shape2 = shape2, log = TRUE))
      }

      shape1.grid <- seq( .grid.shape1[1], .grid.shape1[2], len = 19)
      shape1.init <- if (length( .ishape1 )) .ishape1 else
        grid.search(shape1.grid, objfun = kumar.Loglikfun,
                    y = y,  x = x, w = w)
      shape1.init <- matrix(shape1.init, n, ncoly, byrow = TRUE)

      mediany <- colSums(y * w) / colSums(w)
      shape2.init <- if (length( .ishape2 )) .ishape2 else
        log(0.5) / log1p(-(mediany^shape1.init))
      shape2.init <- matrix(shape2.init, n, ncoly, byrow = TRUE)

      etastart <- cbind(theta2eta(shape1.init, .lshape1 , earg = .eshape1 ),
                        theta2eta(shape2.init, .lshape2 , earg = .eshape2 ))[,
                   interleave.VGAM(M, M = M1)]
    }
}), list( .lshape1 = lshape1, .lshape2 = lshape2,
          .ishape1 = ishape1, .ishape2 = ishape2,
          .eshape1 = eshape1, .eshape2 = eshape2,
          .grid.shape1 = grid.shape1 ))),
linkinv = eval(substitute(function(eta, extra = NULL) {
  shape1 <- eta2theta(eta[, c(TRUE, FALSE)], .lshape1 , earg = .eshape1 )
  shape2 <- eta2theta(eta[, c(FALSE, TRUE)], .lshape2 , earg = .eshape2 )
  shape2 * (base::beta(1 + 1/shape1, shape2))
}, list( .lshape1 = lshape1, .lshape2 = lshape2,
         .eshape1 = eshape1, .eshape2 = eshape2 ))),
last = eval(substitute(expression({
  misc$link <- c(rep( .lshape1 , length = ncoly),
                 rep( .lshape2 , length = ncoly))[interleave.VGAM(M, M = M1)]
  temp.names <- c(mynames1, mynames2)[interleave.VGAM(M, M = M1)]
  names(misc$link) <- temp.names
```

```
   misc$earg <- vector("list", M)
   names(misc$earg) <- temp.names
   for (ii in 1:ncoly) {
     misc$earg[[M1*ii-1]] <- .eshape1
     misc$earg[[M1*ii  ]] <- .eshape2
   }
}), list( .lshape1 = lshape1, .lshape2 = lshape2,
          .eshape1 = eshape1, .eshape2 = eshape2 ))),
loglikelihood = eval(substitute(
function(mu, y, w, residuals = FALSE, eta, extra = NULL, summation = TRUE) {
   shape1 <- eta2theta(eta[, c(TRUE, FALSE)], .lshape1 , earg = .eshape1 )
   shape2 <- eta2theta(eta[, c(FALSE, TRUE)], .lshape2 , earg = .eshape2 )
   if (residuals) {
     stop("loglikelihood residuals not implemented yet")
   } else {
     ll.elts <- c(w) * dkumar(x = y, shape1, shape2, log = TRUE)
     if (summation) sum(ll.elts) else ll.elts
   }
}, list( .lshape1 = lshape1, .lshape2 = lshape2,
         .eshape1 = eshape1, .eshape2 = eshape2 ))),
vfamily = c("kumar"),
simslot = eval(substitute(
function(object, nsim) {
  eta <- predict(object)
  shape1 <- eta2theta(eta[, c(TRUE, FALSE)], .lshape1 , earg = .eshape1 )
  shape2 <- eta2theta(eta[, c(FALSE, TRUE)], .lshape2 , earg = .eshape2 )
  rkumar(nsim * length(shape1), shape1 = shape1, shape2 = shape2)
}, list( .lshape1 = lshape1, .lshape2 = lshape2,
         .eshape1 = eshape1, .eshape2 = eshape2 ))),
deriv = eval(substitute(expression({
   shape1 <- eta2theta(eta[, c(TRUE, FALSE)], .lshape1 , earg = .eshape1 )
   shape2 <- eta2theta(eta[, c(FALSE, TRUE)], .lshape2 , earg = .eshape2 )
   dshape1.deta <- dtheta.deta(shape1, link = .lshape1 , earg = .eshape1 )
   dshape2.deta <- dtheta.deta(shape2, link = .lshape2 , earg = .eshape2 )
   dl.dshape1 <- 1 / shape1 + log(y) - (shape2 - 1) * log(y) *
                   (y^shape1) / (1 - y^shape1)
   dl.dshape2 <- 1 / shape2 + log1p(-y^shape1)
   dl.deta <- c(w) * cbind(dl.dshape1 * dshape1.deta,
                           dl.dshape2 * dshape2.deta)
   dl.deta[, interleave.VGAM(M, M = M1)]
}), list( .lshape1 = lshape1, .lshape2 = lshape2,
          .eshape1 = eshape1, .eshape2 = eshape2 ))),
weight = eval(substitute(expression({
  ned2l.dshape11 <- (1 + (shape2 / (shape2 - 2)) *
    ((digamma(shape2) - digamma(2))^2 -
    (trigamma(shape2) - trigamma(2)))) / shape1^2
  ned2l.dshape22 <- 1 / shape2^2
  ned2l.dshape12 <-
     (digamma(2) - digamma(1 + shape2)) / ((shape2 - 1) * shape1)
```

```
      index1 <- (abs(shape2 - 1) < .tol12 )   # Fix up singular point at shape2 == 1
      ned21.dshape12[index1] <- -trigamma(2) / shape1[index1]
      index2 <- (abs(shape2 - 2) < .tol12 )   # Fix up singular point at shape2 == 2
      ned21.dshape11[index2] <- (1 - 2 * psigamma(2, deriv = 2)) / shape1[index2]^2

      wz <- array(c(c(w) * ned21.dshape11 * dshape1.deta^2,
                    c(w) * ned21.dshape22 * dshape2.deta^2,
                    c(w) * ned21.dshape12 * dshape1.deta * dshape2.deta),
                  dim = c(n, M / M1, 3))
      wz <- arwz2wz(wz, M = M, M1 = M1)
      wz
   }), list( .lshape1 = lshape1, .lshape2 = lshape2,
             .eshape1 = eshape1, .eshape2 = eshape2, .tol12 = tol12 ))))
}
<environment: namespace:VGAM>
```

Here are some short notes.

1. As $\ell_i = \log \alpha_i + \log \beta_i + (\alpha_i - 1) \log y_i + (\beta_i - 1) \log(1 - y_i^{\alpha_i})$, the score vector comprises $\alpha_i^{-1} + \log y_i - (\beta_i - 1)(\log y_i) y_i^\alpha / (1 - y_i^{\alpha_i})$ and $\beta_i^{-1} + \log(1 - y_i^{\alpha_i})$. The EIM requires digamma and trigamma evaluations (Sect. A.4.1). A small complication is that singularities for two EIM elements exist at $\beta = 1$ and $\beta = 2$. These are easily dealt with an application of the L'Hospital rule:

$$\lim_{\beta \to 1} \frac{\psi(2) - \psi(1 + \beta)}{\beta - 1} = -\psi'(2),$$

$$\lim_{\beta \to 2} \frac{\beta}{\beta - 2} \left\{ [\psi(\beta) - \psi(2)]^2 - [\psi'(\beta) - \psi'(2)] \right\} = -2\,\psi''(2).$$

2. A grid search is conducted in @initialize as a function of shape1. Given a value of shape1, a suitable value of shape2 can be obtained from its CDF $F(y) = 1 - (1 - y^\alpha)^\beta$.
3. The first subscript of array() variables varies the fastest, followed by the second, etc. This property is exploited by wz when it is inputted into arwz2wz().

18.5.2 Simulated Fisher Scoring

Working weights approximated by simulated Fisher scoring (Sect. 9.2.2) have a @weight slot comprising of a for() loop running over the nsimEIM simulations. Each of these simulations is vectorized. Here is an example involving a (currently) $S = 1$ response family called slash().

```
> VGAM:::slash.control

function(save.weights = TRUE, ...) {
    list(save.weights = save.weights)
}
<environment: namespace:VGAM>
```

```
> slash()@deriv

  expression({
      mu <- eta2theta(eta[, 1], link = "identitylink", earg = list(
          theta = , inverse = FALSE, deriv = 0, short = TRUE, tag = FALSE))
      sigma <- eta2theta(eta[, 2], link = "loge", earg = list(theta = ,
          bvalue = NULL, inverse = FALSE, deriv = 0, short = TRUE,
          tag = FALSE))
      dmu.deta <- dtheta.deta(mu, link = "identitylink", earg = list(
          theta = , inverse = FALSE, deriv = 0, short = TRUE, tag = FALSE))
      dsigma.deta <- dtheta.deta(sigma, link = "loge", earg = list(
          theta = , bvalue = NULL, inverse = FALSE, deriv = 0,
          short = TRUE, tag = FALSE))
      zedd <- (y - mu)/sigma
      d3 <- deriv3(~w * log(1 - exp(-(((y - mu)/sigma)^2)/2)) -
          log(sqrt(2 * pi) * sigma * ((y - mu)/sigma)^2), c("mu",
          "sigma"))
      eval.d3 <- eval(d3)
      dl.dthetas <- attr(eval.d3, "gradient")
      dl.dmu <- dl.dthetas[, 1]
      dl.dsigma <- dl.dthetas[, 2]
      ind0 <- (abs(zedd) < 2.22044604925031e-13)
      dl.dmu[ind0] <- 0
      dl.dsigma[ind0] <- -1/sigma[ind0]
      c(w) * cbind(dl.dmu * dmu.deta, dl.dsigma * dsigma.deta)
  })

> slash()@weight

  expression({
      run.varcov <- 0
      ind1 <- iam(NA, NA, M = M, both = TRUE, diag = TRUE)
      sd3 <- deriv3(~w * log(1 - exp(-(((ysim - mu)/sigma)^2)/2)) -
          log(sqrt(2 * pi) * sigma * ((ysim - mu)/sigma)^2), c("mu",
          "sigma"))
      for (ii in 1:(250)) {
          ysim <- rslash(n, mu = mu, sigma = sigma)
          seval.d3 <- eval(sd3)
          dl.dthetas <- attr(seval.d3, "gradient")
          dl.dmu <- dl.dthetas[, 1]
          dl.dsigma <- dl.dthetas[, 2]
          temp3 <- cbind(dl.dmu, dl.dsigma)
          run.varcov <- run.varcov + temp3[, ind1$row] * temp3[,
              ind1$col]
      }
      run.varcov <- run.varcov/250
      wz <- if (intercept.only)
          matrix(colMeans(run.varcov, na.rm = FALSE), n, ncol(run.varcov),
              byrow = TRUE)
      else run.varcov
      dthetas.detas <- cbind(dmu.deta, dsigma.deta)
      wz <- wz * dthetas.detas[, ind1$row] * dthetas.detas[, ind1$col]
      c(w) * wz
  })
```

Here are some notes.

1. The family function's `.control()` function sets the value of `save.weights` to `TRUE`. This means that the working weights are attached to the object after convergence. For families having 'proper' EIMs programmed in, `save.weights = FALSE` and they are recomputed when needed, e.g., to obtain standard errors. This eliminates the need to attach a large data structure on the object. It is a good idea to set `save.weights = TRUE` for families implementing simulated Fisher scoring because recomputing them later tends to be too expensive, and there are reproducibility issues.

2. The R function `deriv3()` is used to differentiate the loglikelihood with respect to the parameters. While this saves the programmer's time, the execution time is usually longer than manually programming. Manually programming also allows safer evaluation if the L'Hospital rule is needed, e.g., Ex. 16.13.

3. The code in the `@weight` slot is based on the code from `@deriv`. One simply replaces `y` by `ysim`, where the latter are random variates generated at the current parameter values. In particular, the ith row of `run.varcov` stores

$$\mathrm{Var}\left(\frac{\partial \ell_i}{\partial \boldsymbol{\theta}}\right) \;\approx\; N^{-1} \sum_{s=1}^{N} \frac{\partial \ell_i}{\partial \boldsymbol{\theta}} \frac{\partial \ell_i}{\partial \boldsymbol{\theta}^T} \tag{18.15}$$

where $N = $ `nsimEIM`, and the partial derivatives are random score vectors.

4. For models where the only explanatory variable is an intercept, one can average the working weights over the n observations as well. Usually, one can only overage over the `nsimEIM` simulations for each i. Should the grand mean be used, this should result in more accurate working weight matrices.

18.6 Writing Smart Functions †

In this section we look at the more technical aspects of smart prediction (Sect. 8.2.5) and how to write smart functions.

Smart prediction is based on the fundamental property that the evaluation of the model frame during prediction is done in precisely the same order as the original fit. Consequently we can use a data structure known in computer science as a first-in first-out (FIFO) *queue* (cf. a last-in first-out (LIFO) stack) held temporarily in a pre-ordained place to store the data-dependent parameters while the original model frame is constructed from the formula. Upon the model's convergence the contents of the queue are stored on the fitted object in the `smart.prediction` slot. Upon prediction, the queue contents are copied back into the hidden location and reused during the construction of the second model frame. Each function that has data-dependent parameters need to know how to write to and read from the hidden location—they are consequently called smart. Table 18.8 summarizes the functions and variables in **VGAM** implementing smart prediction.

The first example is to write `sm.min1()`, say, to help fit the equivalent of `fit2` in Sect. 8.2.5. Then

```
> print(sm.min1)

function(x) {
  x <- x  # Evaluate x; needed for nested calls, e.g., sm.bs(sm.scale(x)).
  minx <- min(x)
  if (smart.mode.is("read")) {
    smart  <- get.smart()
    minx <- smart$minx  # Overwrite its value
  } else if (smart.mode.is("write"))
    put.smart(list(minx = minx))
  minx
}
<environment: namespace:VGAM>
attr(,"smart")
[1] TRUE
```

Incidentally, this function lacks the generality of **min()**, e.g., it has no **na.rm** argument and it should be only used on a vector **x**.

The functions **put.smart()** and **get.smart()** are opposites—the former writes a list to a specified location, and the latter retrieves it. The function **smart.mode.is()** returns **TRUE** or **FALSE** depending on its argument.

Here is a second example. The function **sm.scale()** was written so that models such as **fit1** could be predicted from. For simplicity, we have written a similar function to **sm.scale()** called **sm.scale1()** which only standardizes a numerical vector (**sm.scale()** handles matrices). It performs no error checking and here is what it looks like:

```
> print(sm.scale1)

function(x, center = TRUE, scale = TRUE) {
  x <- x  # Evaluate x; needed for nested calls, e.g., sm.bs(sm.scale(x)).
  if (!is.vector(x))
    stop("argument 'x' must be a vector")
  if (smart.mode.is("read")) {
    smart  <- get.smart()
    return((x - smart$Center) / smart$Scale)
  }
  if (is.logical(center))
    center <- if (center) mean(x) else 0
  if (is.logical(scale))
    scale <- if (scale) sqrt(var(x)) else 1
  if (smart.mode.is("write"))
    put.smart(list(Center = center,
                   Scale  = scale))
  (x - center) / scale
}
<environment: namespace:VGAM>
attr(,"smart")
[1] TRUE
```

Here are some technical details needed for writing a smart function. They operates in three modes: "neutral", "write" and "read".

- In "neutral" mode (assumed so unless "write" or "read") it operates like an ordinary function and simply returns the result, e.g., **min(x)**.
- In "write" mode (at fitting time) it writes out the data-dependent parameters that need saving into a special data structure called **.smart.prediction** in an R environment called **VGAM:::smartpredenv** and then returns the result as well.

When the regression modelling function is finishing, `.smart.prediction` is attached to the object's `smart.prediction` slot.

- In "read" mode (at prediction time) the `smart.prediction` slot is copied back into `smartpredenv` (and called `.smart.prediction`) by the prediction methods function. It is now available for reading by the smart function, which is invoked the second time when the model frame is computed. If the original parameters are needed, then the smart function will access them from `.smart.prediction`.

In "read" mode, smart functions can be programmed in two ways. The first way is without recursion, e.g., `sm.scale1()`. The second way by recursion operates by evaluating a `do.call()` once the original parameters have been reinstated. The cleanest way is if the information originally written is in a list, with components whose names match the function's arguments—see `sm.scale.default()`, `sm.scale2()` below, `sm.bs()` and `sm.ns()` as examples. If there are only one or two parameters, or if the expression is simple, then the first option is usually the best. A disadvantage of the recursive call is the possible need of extra nuisance arguments.

A line such as `x <- x` (where `x` is the primary argument of the smart function) is needed at the very beginning of the function to cause lazy evaluation to work immediately. This is necessary because of terms such as `sm.bs(sm.scale1(x))`, where a smart function calls another smart function. The inner `sm.scale1()` needs to be evaluated using its smart parameters before the outer `bs` is evaluated. The statement `x <- x` ensures that. In general, writing smart function poses potential pitfalls for inexperienced programmers and it requires careful testing. Upon writing a smart function, a logical attribute `"smart"` should be assigned `TRUE` so that `is.smart()` can return true, e.g.,

```
sm.myfunction <- function(x, ...) {
...
}
attr(sm.myfunction, "smart") <- TRUE
```

Here is an equivalent function of `sm.scale1()`, called `sm.scale2()`:

```
> print(sm.scale2)

function(x, center = TRUE, scale = TRUE) {
  x <- x  # Evaluate x; needed for nested calls, e.g., sm.bs(sm.scale(x)).
  if (!is.vector(x))
    stop("argument 'x' must be a vector")
  if (smart.mode.is("read")) {
    return(eval(smart.expression))  # Recursion used
  }
  if (is.logical(center))
    center <- if (center) mean(x) else 0
  if (is.logical(scale))
    scale <- if (scale) sqrt(var(x)) else 1
  if (smart.mode.is("write"))
    put.smart(list(center = center,
                   scale  = scale,
                   match.call = match.call()))
  (x - center) / scale
}
<environment: namespace:VGAM>
attr(,"smart")
[1] TRUE
```

It runs the second way because the expression `smart.expression` invokes the recursion. In order for `smart.expression` to work it is crucial that:

(a) the list in `put.smart()` to have exactly the same names as the arguments of the smart function (here, they are `center` and `scale`), and

(b) the primary argument of the smart function is called `x`.

As another recursive example,

```
> print(sm.min2)

  function(x, .minx = min(x)) {
    x <- x  # Evaluate x; needed for nested calls, e.g., sm.bs(sm.scale(x)).
    if (smart.mode.is("read")) {  # Use recursion
      return(eval(smart.expression))
    } else
    if (smart.mode.is("write"))
      put.smart(list( .minx = .minx , match.call = match.call()))
    .minx
  }
  <environment: namespace:VGAM>
  attr(,"smart")
  [1] TRUE
```

implements another simple `sm.min()`-type function. Outwardly, it differs from `sm.min1()` in that `.minx` is a nuisance argument. That it begins with a period is to signify that it should not be used by the user.

VGAM's smart prediction writes three data structures (`.smart.prediction`, `.smart.prediction.counter`, and `.smart.prediction.mode`) to smartpredenv. They are deleted after both fitting and prediction is complete using `wrapup.smart()`. The user should be oblivious to their existence.

The character variable `.smart.prediction.mode` equals `"neutral"`, `"read"` or `"write"`. It is in `"write"` mode when the smart functions write their arguments out (i.e., when model is originally fitted), and in `"read"` mode while predicting, and in `"neutral"` otherwise, e.g., at the command line, or while predicting but needing to be momentarily out of `"read"` and `"write"` mode so that it acts like an ordinary function.

On set up, `.smart.prediction` is a list with `max.smart` empty components and the variable `.smart.prediction.counter` is assigned 0. When a smart function writes out its data-dependent parameters, `.smart.prediction.counter` is incremented and the argument of `put.smart()` is written to that component of `.smart.prediction`. If more than `max.smart` components are used up, then `.smart.prediction` is lengthened automatically. The list `.smart.prediction` is trimmed of any unused components just prior to being attached to the object.

Similarly, upon prediction, `.smart.prediction` (`object@smart.prediction`) is placed back in `smartpredenv`, and `.smart.prediction.counter` is assigned 0. When in read mode, `.smart.prediction.counter` is incremented and `get.smart()` returns the `.smart.prediction.counter`th component of the list `.smart.prediction`. As mentioned above, this data structure is known as a queue.

Table 18.8 Functions and variables for smart prediction, supplied in **VGAM** for programmers.

Function or variable	Comments
`is.smart()`	Returns a logical. Is a function smart?
`VGAM:::smartpredenv`	Pre-ordained hidden environment where the queue `.smart.prediction` resides
`get.smart()`	Reads the next list component from `.smart.prediction`
`put.smart()`	Writes the next list component to `.smart.prediction`
`.smartprediction`	List, each component contains a term's data-dependent parameters
`.smart.prediction.counter`	Non-negative integer, points to the last element in the queue when writing, and the first element when reading
`max.smart`	An argument of `setup.smart()`, the length of `.smartprediction`, e.g., 30. Automatically expands to a larger value if needed
`smart.expression`	Expression, useful for a smart function that uses recursion. Has some limitations, e.g., first argument name `x`
`smart.mode.is()`	Returns a character string of the smart prediction mode: `"read"`, `"write"` or `"neutral"`
`.smart.prediction.mode`	Character, the smart prediction mode, has the value `"read"`, `"write"` or `"neutral"`. This variable is best accessed by `smart.mode.is()`
`setup.smart()`	Sets up the data structures ready for smart prediction
`wrapup.smart()`	Wraps up smart prediction by deleting the data structures

Bibliographic Notes

Chambers and Hastie (1991), Chambers (1998), and Chambers (2008) give authoritative accounts of the S3 and S4 languages, although the online help and documentation are most recent. The latter two references deal with the S4 language.

Many books and articles have now been written on **R** programming. Some more recent ones include Braun and Murdoch (2008), Gentleman (2009), Adler (2010), Jones et al. (2014), Wickham (2015). Altman and Jackman (2011) has some reflective thoughts on writing software. Fox and Weisberg (2011, Chap.8) is a brief description of a few selected **R** programming topics. The CRAN website lists a lot of contributed documentation and includes resources in a variety of languages.

There are many potential areas where **VGAM** family functions may be written, because IRLS is a suitable estimation algorithm, for example, generalized estimating equations (Wild and Yee, 1996) and robust regression (Huber and Ronchetti, 2009).

Exercises

Ex. 18.1. Modify `better.exponential()` so that it has an additional default argument called `irate = NULL`, which is an optional initial value for the rate parameter. If inputted by the user, then it should override all other self-starting initial values.

Ex. 18.2. Consider the NB-C, viz. negative binomial regression with the canonical link $\eta_1 = \log(\mu/(\mu+k))$. Derive the first and expected second derivatives based on (18.7)–(18.9). Use (11.2) and $\eta_2 = \log k$.

Ex. 18.3. Suppose x, $\mathbf{A}$ (symmetric) and $\mathbf{X}$ consist of constants, and $\ell(\boldsymbol{\eta})$ is a log-likelihood with $\boldsymbol{\eta} = \mathbf{X}\boldsymbol{\beta}$. Show the following:

$$\frac{\partial\, \boldsymbol{x}^T \boldsymbol{\beta}}{\partial \boldsymbol{\beta}} = \boldsymbol{x}, \tag{18.16}$$

$$\frac{\partial\, \boldsymbol{\beta}^T \mathbf{A} \boldsymbol{\beta}}{\partial \boldsymbol{\beta}} = 2\mathbf{A}\boldsymbol{\beta}, \tag{18.17}$$

$$\frac{\partial^2 \ell}{\partial \boldsymbol{\beta}\, \partial \boldsymbol{\beta}^T} = \mathbf{X}^T \frac{\partial^2 \ell}{\partial \boldsymbol{\eta}\, \partial \boldsymbol{\eta}^T} \mathbf{X}. \tag{18.18}$$

Ex. 18.4. † Find some univariate discrete or continuous distribution currently unimplemented in VGAM, whose EIM can be derived or accessed from the literature.

(a) Write a (simpler) VGAM family function to fit this model—it should handle one such response.
(b) If practical, write the dpqr-type functions associated with the distribution. If there is an r-type function, use it to test your answer to (a).
(c) Extend your answer to (a) by allowing for multiple responses.
(d) Use prompt() to write a suitable .Rd file for the family function and, if relevant, another for the dpqr-type functions.

Ex. 18.5. † Consider a zero-inflated beta-binomial distribution based on $\phi =$ probability of a structural zero, and positive shape parameters a and b. Complete Ex. 18.4 specifically for this distribution. Hint: Table 11.10, Sect. 11.4 and Ex. 17.16 may be useful. Call the family function zibetabinomialff().

Ex. 18.6. † Wong (1989) [*Biometrika* 76(1), 55–60] describes a simple linear regression model with measurement error where the joint distribution of (x, y) satisfies $x = \xi + \varepsilon_x$, $y = \eta + \varepsilon_y$, $\xi \sim N(\mu, \tau^2)$, and the errors are uncorrelated. Then the joint distribution is subsequently bivariate normal

$$N_2\left(\begin{pmatrix} \mu \\ \alpha + \beta\mu \end{pmatrix}, \begin{pmatrix} \tau^2 + \sigma_0^2 & \beta\tau^2 \\ \beta\tau^2 & \beta^2\tau^2 + \sigma_0^2 \end{pmatrix}\right),$$

and the EIM is tractable when the common error variance $\mathrm{Var}(\varepsilon_x) = \mathrm{Var}(\varepsilon_y) = \sigma_0^2$ is known. Write a VGAM family function called simple.lm.eiv(sigma0) to implement this model. Use the orthogonal parameterization detailed in the paper involving parameters $\boldsymbol{\theta} = (\lambda_0, \lambda_1, \lambda_2, \beta)^T$. Test out your function using some simulated data.

Ex. 18.7.   New Link Functions

(a) For the links of Table 18.9 derive $d^k\eta/d\theta^k$ for $k = 1$ and 2, as well as $\theta = g^{-1}(\eta)$.
(b) † Write VGAM-compatible link functions for some self-selected entries from the table. Choose suitable default values for a and b if needed. Test them out on some data sets. Write the .Rd files too.

Table 18.9 Link functions for Ex. 18.7. Here, a and b are prespecified constants, and T_ν is the CDF of a Student-t distribution with $\mathtt{df} = \nu$ degrees of freedom.

Link	$g(\theta)$	Range of θ
qtlink(df)	$T_\nu^{-1}(\theta)$	$(0, 1)$
pregibonlink()	$\dfrac{\theta^{a-b} - 1}{a - b} - \dfrac{(1 - \theta)^{a+b} - 1}{a + b}$	$(0, 1)$
logitpowerlink()	$\log\left(\theta^a / (1 - \theta^a)\right)$	$(0, 1)$
oddspowerlink()	$a^{-1}\left([\theta/(1-\theta)^a] - 1\right)$	$(0, 1)$
aolink()	$\log\left(\left[(1-\theta)^{-a} - 1\right]/a\right)$	$(0, 1)$
neglognegloglink()	$-\log(-\log\theta)$	$(0, 1)$
logit10link()	$\log_{10}(\theta/(1 - \theta))$	$(0, 1)$
loglog1plink()	$\log(\log(1 + \theta))$	$(0, \infty)$
log10link()	$\log_{10}\theta$	$(0, \infty)$

Ex. 18.8.   † Smart Functions (Table 18.8)
Write the following smart functions and their corresponding .Rd files: $\mathtt{sm.cut()}$, $\mathtt{sm.max()}$, $\mathtt{sm.mean()}$, $\mathtt{sm.min()}$, $\mathtt{sm.sd()}$, $\mathtt{sm.var()}$. Try to implement the full capabilities that their unsmart functions offer. Test out your functions well.

Correct with care, if you expect to write anything which shall be worthy of a second perusal.
—Quintus Horatius Flaccus

Appendix A
Background Material

All who debate on matters of uncertainity, should be free from prejudice, partiality, anger, or compassion.
—Caius Sallustius Crispus

A.1 Some Classical Likelihood Theory

Most of the VGLM/VGAM framework is infrastructure directed towards maximizing a full-likelihood model, therefore it is useful to summarize some supporting results from classical likelihood theory. The following is a short summary of a few selected topics serving the purposes of this book. The focus is on aspects of direct relevance to the practitioners and users of the software. The presentation is informal and nonrigorous; rigorous treatments, including justification and proofs, can be found in the texts listed in the bibliographic notes. The foundation of this subject was developed by Fisher a century ago (around the decade of WW1), and he is regarded today as the father of modern statistics.

A.1.1 Likelihood Functions

The usual starting point is to let Y be a random variable with density function $f(y; \boldsymbol{\theta})$ depending on $\boldsymbol{\theta} = (\theta_1, \ldots, \theta_p)^T$, a multidimensional unknown parameter. Values that Y can take are denoted in lower-case, i.e., y. By 'density function', here is meant a probability (mass) function for a discrete-valued Y, and probability density function for continuous Y. We shall refer to f as simply the density function, and use integration rather than summation to denote quantities such as expected values, e.g., $E(Y) = \int f(y) \, dy$ where the range of integration is over the *support* of the distribution, i.e., those values of y where $f(y) > 0$ (called $\mathcal{Y}$).

A lot of statistical practice centres upon making inference about $\boldsymbol{\theta}$, having observed $Y = y$. As well as obtaining an estimate $\widehat{\boldsymbol{\theta}}$, it is customary to cite some measure of accuracy or plausibility of the estimate, usually in the form of its

© Thomas Yee 2015

T.W. Yee, *Vector Generalized Linear and Additive Models*,
Springer Series in Statistics, DOI 10.1007/978-1-4939-2818-7

standard error, SE($\widehat{\boldsymbol{\theta}}$). It is also common to conduct hypothesis tests, e.g., for a one-parameter model, test the null hypothesis $H_0 : \theta = \theta_0$ for some known and fixed value θ_0.

Let $\boldsymbol{\Omega}$ be the parameter space, which is the set of possible values that $\boldsymbol{\theta}$ can take. For example, if $Y \sim N(\mu, \sigma^2)$ where $\boldsymbol{\theta} = (\mu, \sigma)^T$, then $\boldsymbol{\Omega} = \mathbb{R} \times (0, \infty) = \mathbb{R} \times \mathbb{R}_+$. Another simple example is the beta distribution having positive shape parameters $\boldsymbol{\theta} = (s_1, s_2)^T$, therefore $\boldsymbol{\Omega} = \mathbb{R}_+^2$. Clearly, $\boldsymbol{\Omega} \subseteq \mathbb{R}^p$.

In the wider VGLM/VGAM framework, some of our responses $\boldsymbol{y}_i$ may be multivariate, therefore let $\boldsymbol{Y} = (\boldsymbol{Y}_1^T, \ldots, \boldsymbol{Y}_n^T)^T$ be a random vector of n observations, each $\boldsymbol{Y}_i$ being a random vector. We observe $\boldsymbol{y} = (\boldsymbol{y}_1^T, \ldots, \boldsymbol{y}_n^T)^T$ in totality.

Each $\boldsymbol{y}_i$ can be thought of as being a realization from some statistical distribution with joint density function $f(\boldsymbol{y}_i; \boldsymbol{\theta})$. With n observations, the joint density function can be written $f(\boldsymbol{y}; \boldsymbol{\theta})$. We say that a (parametric) statistical model is a set of possible density functions indexed by $\boldsymbol{\theta} \in \boldsymbol{\Omega}$, i.e.,

$$\mathcal{M}_{\boldsymbol{\theta}} \;=\; \{f(\cdot; \boldsymbol{\theta}) : \boldsymbol{\theta} \in \boldsymbol{\Omega}\},$$

which may be simplified to just $\mathcal{M}$.

The approach considered in this book is to assume that the user knows such a family of distributions. Often this strong assumption is groundless, and therefore parametric models may give misleading results. A method that lies in between the fully-parametric method adopted in this book and nonparametric methods is based on using an empirical likelihood (Owen, 2001), which gives the best of both worlds. The empirical likelihood supplies information at a sufficient rate that reliable confidence intervals/regions and hypothesis tests can be constructed.

Of course, parameterizations are not unique, e.g., for many distributions in Chap. 12, the scale parameter b is used so that the form y/b appears in the density, whereas some practitioners prefer to use its reciprocal, called the rate, and then the densities have the term λy. Two other examples, from Table 12.11, are the beta and beta-binomial distributions which are commonly parameterized in terms of the shape parameters, otherwise the mean and a dispersion parameter.

Regardless of the parameterization chosen, the parameter must be *identifiable*. This means that each element of $\mathcal{M}_{\boldsymbol{\theta}}$ corresponds to exactly one value of $\boldsymbol{\theta}$. Stated another way, if $\boldsymbol{\theta}_1$ and $\boldsymbol{\theta}_2 \in \boldsymbol{\Omega}$ with $\boldsymbol{\theta}_1 \neq \boldsymbol{\theta}_2$ then the densities $\mathcal{M}_{\boldsymbol{\theta}_1}(\boldsymbol{y}) \neq \mathcal{M}_{\boldsymbol{\theta}_2}(\boldsymbol{y})$. As an example, consider the multinomial logit model (1.25) described in Sect. 14.2. We can have

$$P(Y = j | \boldsymbol{x}) = \frac{e^c}{e^c} \frac{\exp\{\eta_j\}}{\sum_{k=1}^{M+1} \exp\{\eta_k\}} = \frac{\exp\{\eta_j + c\}}{\sum_{k=1}^{M+1} \exp\{\eta_k + c\}},$$

for any constant c, hence the $M+1$ η_js are non-identifiable. In practice, we choose $c = -\eta_t$ for some t, and then redefine the η_j. The family function `multinomial()` chooses $t = M + 1$, by default, as the reference group so that $\eta_{M+1} \equiv 0$, but $t = 1$ is another popular software default.

A.1.2 Maximum Likelihood Estimation

Maximum likelihood estimation is the most widely used general-purpose estimation procedure in statistics. It centres on the *likelihood function* for $\boldsymbol{\theta}$, based on the observation of $\boldsymbol{Y} = \boldsymbol{y}$:

$$L(\boldsymbol{\theta}; \boldsymbol{y}) \;=\; f(\boldsymbol{y}; \boldsymbol{\theta}), \qquad \boldsymbol{\theta} \in \boldsymbol{\Omega}. \tag{A.1}$$

With a philosophical twist, two quantities can be seen contrasted here: the likelihood function that is a function of the parameter $\boldsymbol{\theta}$, given the data $\boldsymbol{y}$, cf. the density that is a function of the data $\boldsymbol{y}$, given the parameter $\boldsymbol{\theta}$. The likelihood function is thus the probability of observing what we got $(\boldsymbol{y})$ as a function of $\boldsymbol{\theta}$ based on our model. Clearly, this holds for discrete responses, but it can be easily justified for continuous responses too (see below). Thus maximum likelihood estimation treats the data as being fixed and given, and it determines $\boldsymbol{\theta}$ which makes our observed data most probable.

It is much more convenient to work on a log-scale. One major reason for this monotone transformation is that data is very commonly assumed to be independent, so we can obtain additivity of log-likelihood contributions. Also, rather than having a single observation $Y = y$, it is more general to have $\boldsymbol{Y}_i = \boldsymbol{y}_i$ for $i = 1, \ldots, n$, where n is the sample size. Putting these two properties together,

$$L(\boldsymbol{\theta}; \boldsymbol{y}) \;=\; f(\boldsymbol{y}; \boldsymbol{\theta}) \;=\; \prod_{i=1}^{n} f(\boldsymbol{y}_i; \boldsymbol{\theta}) \;=\; \prod_{i=1}^{n} L_i, \tag{A.2}$$

where the data is $\boldsymbol{y} = (\boldsymbol{y}_1, \ldots, \boldsymbol{y}_n)^T$.

Now, taking the logarithm of this joint distribution gives the *log-likelihood* function

$$\ell(\boldsymbol{\theta}; \boldsymbol{y}) \;=\; \sum_{i=1}^{n} \log f(\boldsymbol{y}_i; \boldsymbol{\theta}) \;=\; \sum_{i=1}^{n} \ell_i. \tag{A.3}$$

The fact that this is a sum will enable us later to state large sample properties of ML estimators by application of the law of large numbers.

Maximum likelihood estimation involves maximizing L, or equivalently, ℓ. We can write

$$\widehat{\boldsymbol{\theta}} \;=\; \underset{\boldsymbol{\theta} \in \Omega}{\arg\max}\, \ell(\boldsymbol{\theta}; \boldsymbol{y}),$$

and the solution need not be unique or even exist. Unless $\widehat{\boldsymbol{\theta}}$ is on the boundary, we obtain $\widehat{\boldsymbol{\theta}}$ by solving $\partial \ell(\boldsymbol{\theta})/\partial \boldsymbol{\theta} = \boldsymbol{0}$. Iterative methods (Sect. A.1.2.4) are commonly employed to obtain the *maximum likelihood estimate* $\widehat{\boldsymbol{\theta}}$, because no closed-form expression can be obtained.

In maximizing ℓ, it is the relative values of $\ell(\boldsymbol{\theta})$ that matter, not their values in absolute terms. Hence, some authors omit any additive constants not involving $\boldsymbol{\theta}$ from ℓ but still use "=" in (A.3). This actually holds implicitly for continuous responses $\boldsymbol{Y}$ because the probability that $\boldsymbol{Y} = \boldsymbol{y}$ is actually 0, hence fundamentally, (A.1) is actually of the form $f(\boldsymbol{y}; \boldsymbol{\theta}) \cdot \varepsilon$ which is a 'real' probability—it represents the chances of observing a value in a small set of volume centred at $\boldsymbol{y}$. Then (A.2) involves a $\propto$ because the width of the volume, as measured by ε, does not depend on $\boldsymbol{\theta}$, and therefore (A.3) is equality up to a constant. For families such as `posbinomial()`, it is necessary to set the argument `omit.constant` to `TRUE` when comparing nested models that have different normalizations (Sect. 17.2.1).

ML estimators are functions of quantities known as *sufficient statistics*. A statistic is simply a function of the sample space $\mathcal{S}$, and it will be denoted here by T. Sufficient statistics are statistics that reduce the data into two parts: a useful part and an irrelevant part. The sufficient statistic contains all the information about $\boldsymbol{\theta}$ that is contained in $\boldsymbol{Y}$, and it is not unique. By considering only the useful part, sufficient statistics allow for a form of data reduction. The usual definition of a statistic T that is sufficient for $\mathcal{M}_{\boldsymbol{\theta}}$ of $\boldsymbol{Y}$ is that the conditional distribution $f(\boldsymbol{Y}|T = t)$ does not depend on $\boldsymbol{\theta}$, for all values of t. However, this definition is not as useful as one would like. Fortunately, there is a famous result called the factorization theorem that is more useful than the original definition, because it provides a method for testing whether a statistic T is sufficient, as well as obtaining T in the first place. It can be stated as follows.

Factorization Theorem   A statistic $T(\boldsymbol{Y})$ is sufficient for $\mathcal{M}_{\boldsymbol{\theta}}$ iff there exist non-negative functions $g(\cdot; \boldsymbol{\theta})$ and h such that

$$f(\boldsymbol{y}; \boldsymbol{\theta}) = g(T(\boldsymbol{y}); \boldsymbol{\theta}) \cdot h(\boldsymbol{y}). \tag{A.4}$$

Then clearly maximizing a likelihood via f is equivalent to maximizing g only, because h is independent of $\boldsymbol{\theta}$.

Some well-known examples of sufficient statistics are as follows.

(i) If $Y_i \sim \text{Poisson}(\mu)$ independently, then $\sum_i Y_i$ is sufficient for $\theta = \mu$. Similarly, if $Y_i \sim \text{Binomial}(n = 1, \mu)$ is a sequence of independent Bernoulli random variables, then $\sum_i Y_i$ is also sufficient for $\theta = \mu$. In both cases, there is a reduction of n values down to one value.

(ii) If the Y_i are a random sample from an $N(\mu, \sigma^2)$ distribution, then $(\bar{y}, s^2)$ are sufficient for $\boldsymbol{\theta} = (\mu, \sigma)^T$. This is reduction of an n-vector down to 2 values.

A.1.2.1 Notation

The standard notation

$$\frac{\partial \ell(\boldsymbol{\theta})}{\partial \boldsymbol{\theta}} = \left(\frac{\partial \ell(\boldsymbol{\theta})}{\partial \theta_1}, \cdots, \frac{\partial \ell(\boldsymbol{\theta})}{\partial \theta_p} \right)^T = \left(\frac{\partial \ell(\boldsymbol{\theta})}{\partial \boldsymbol{\theta}^T} \right)^T,$$

$$\frac{\partial \boldsymbol{b}(\boldsymbol{\theta})}{\partial \boldsymbol{\theta}^T} = \left[\left(\frac{\partial b_j(\boldsymbol{\theta})}{\partial \theta_k} \right) \right] = \left(\frac{\partial \boldsymbol{b}^T(\boldsymbol{\theta})}{\partial \boldsymbol{\theta}} \right)^T, \quad \text{and} \quad \frac{\partial^2 \ell(\boldsymbol{\theta})}{\partial \boldsymbol{\theta} \, \partial \boldsymbol{\theta}^T} = \left[\left(\frac{\partial^2 \ell(\boldsymbol{\theta})}{\partial \theta_j \, \partial \theta_k} \right) \right]$$

is adopted.

Before describing the Fisher scoring algorithm which is central to this book, it is necessary to define some standard quantities first. Let the *score* (or *gradient*) vector be defined as

$$\boldsymbol{U}(\boldsymbol{\theta}) = \frac{\partial \ell(\boldsymbol{\theta})}{\partial \boldsymbol{\theta}}, \tag{A.5}$$

the *Hessian* as

$$\boldsymbol{\mathcal{H}}(\boldsymbol{\theta}) = \frac{\partial \boldsymbol{U}(\boldsymbol{\theta})}{\partial \boldsymbol{\theta}^T} = \frac{\partial^2 \ell}{\partial \boldsymbol{\theta} \, \partial \boldsymbol{\theta}^T}, \tag{A.6}$$

and the *(observed) information matrix* as

$$\boldsymbol{\mathcal{I}}_O(\boldsymbol{\theta}) = -\boldsymbol{\mathcal{H}}(\boldsymbol{\theta}) = -\frac{\partial^2 \ell}{\partial \boldsymbol{\theta} \, \partial \boldsymbol{\theta}^T}. \tag{A.7}$$

Sometimes it is necessary to distinguish between the true value of $\boldsymbol{\theta}$ (called $\boldsymbol{\theta}_*$) and $\boldsymbol{\theta}$ itself. If not, then $\boldsymbol{\theta}$ is used for both meanings.

The acronym "MLE" is used loosely to stand for: maximum likelihood estimation, maximum likelihood estimator, and maximum likelihood estimate.

A.1.2.2 Regularity Conditions

To formalize the method of MLE more adequately, some mathematical properties required of $\mathcal{M}_{\boldsymbol{\theta}}$ must be established. These are called regularity conditions. A distribution satisfying them is called *regular*, otherwise it is nonregular.

Regularity Condition I   The dimension of $\boldsymbol{\theta}$ is fixed. A counterexample is a problem used commonly to motivate James-Stein estimation, which is that $Y_i \sim N(\mu_i, \sigma^2 = 1)$ independently. Then $\boldsymbol{\theta} = (\mu_1, \ldots, \mu_n)^T$ grows with increasing n. Neyman and Scott (1948) showed that MLEs could be inconsistent when the number of parameters increased with n. In such cases, a method to eliminate unnecessary parameters is often sought, e.g., by integrating or conditioning them out.

Regularity Condition II   The parameter $\boldsymbol{\theta}$ is identifiable.

Regularity Condition III   The distributions $\mathcal{M}_{\boldsymbol{\theta}}$ have a common support, i.e., are independent of $\boldsymbol{\theta}$. Here are some counterexamples.

(i) The simplest is $Y_i \sim \text{Unif}(0, \theta)$, so that its support is a function of θ.
(ii) Another common type of example is a 3-parameter density parameterized by a location (a), scale (b) and shape (s) parameter, and whose support is defined on (a, ∞). A specific example of this that has received considerable attention is the 3-parameter Weibull distribution, whose CDF can be written as $1 - \exp\{-[(y-a)/b]^s\}$ for $y > a$, and 0 otherwise. Another example of this sort is the 3-parameter lognormal distribution where $\log(Y - a) \sim N(\mu, \sigma^2)$ so that $a < Y < \infty$.
(iii) The generalized extreme value distribution (GEV; Sect. 16.2) depends on the unknown parameter values. This problem is studied in depth in Smith (1985), who also considered the 3-parameter Weibull distribution.

Regularity Condition IV $\boldsymbol{\Omega}$ is an open set (of $\mathbb{R}^p$).

Regularity Condition V   The true value $\boldsymbol{\theta}_*$ lies in the interior of $\boldsymbol{\Omega}$.

Regularity Condition VI   The first three derivatives of ℓ exist on an open set containing $\boldsymbol{\theta}_*$ (call it $\mathcal{A}$, say), and $\partial^3 \log f(y; \boldsymbol{\theta})/(\partial\theta_s \, \partial\theta_t \, \partial\theta_u) \leq M(y)$ uniformly for $\boldsymbol{\theta} \in \mathcal{A}$, where $0 < E(M(y)) < \infty$.

The next condition addresses the interchange of the order of double differentiation with respect to $\boldsymbol{\theta}$ and integration over $\mathcal{S}$.

Regularity Condition VII   For all $\boldsymbol{y} \in \mathcal{Y}$ and $\boldsymbol{\theta} \in \boldsymbol{\Omega}$, ℓ is twice-differentiable with

$$\frac{\partial}{\partial\boldsymbol{\theta}} \int_{\mathcal{Y}} f(\boldsymbol{y}; \boldsymbol{\theta}) \, d\boldsymbol{y} = \int_{\mathcal{Y}} \frac{\partial}{\partial\boldsymbol{\theta}} f(\boldsymbol{y}; \boldsymbol{\theta}) \, d\boldsymbol{y},$$

and

$$\frac{\partial^2}{\partial\boldsymbol{\theta} \, \partial\boldsymbol{\theta}^T} \int_{\mathcal{Y}} f(\boldsymbol{y}; \boldsymbol{\theta}) \, d\boldsymbol{y} = \int_{\mathcal{Y}} \frac{\partial^2}{\partial\boldsymbol{\theta} \, \partial\boldsymbol{\theta}^T} f(\boldsymbol{y}; \boldsymbol{\theta}) \, d\boldsymbol{y}.$$

A commonly used counterexample of regularity conditions VI–VII is the double exponential (Laplace) distribution (Sect. 15.3.2), whose derivative does not exist at the location parameter.

A.1.2.3 Fisher Information

A very important quantity in MLE theory is the *Fisher information*, which can manifest itself in the form of the *Fisher information matrix*, or *expected information matrix* (EIM). This measures the average amount of information about the parameter $\boldsymbol{\theta}$ over all possible observations, not just those actually observed. Intuitively, it measures the average amount of curvature of ℓ at the MLE $\widehat{\boldsymbol{\theta}}$. If the data provides a lot of information about $\boldsymbol{\theta}$, then the peak at the MLE will be sharp, not flat, because the parameter has a large effect on the likelihood function. Flatness, or a lack of steepness, denotes a lot of uncertainty in the estimated parameter. The EIM can be defined as

$$\boldsymbol{\mathcal{I}}_E(\boldsymbol{\theta}) \;=\; \mathrm{Var}\left(\frac{\partial \ell}{\partial \boldsymbol{\theta}}\right). \tag{A.8}$$

The Fisher information has some basic properties:

1. For independent observations, it is *additive*; and for i.i.d. random variables, this can be written as

$$\boldsymbol{\mathcal{I}}_E(\boldsymbol{\theta}) \;=\; n\boldsymbol{\mathcal{I}}_{E1}(\boldsymbol{\theta}),$$

where $\boldsymbol{\mathcal{I}}_{E1}(\boldsymbol{\theta})$ is the EIM for the first observation. This makes intuitive sense, because increasing n ought to increase the amount of information there is about $\boldsymbol{\theta}$. That the total Fisher information is the sum of each observation's Fisher information will be shown later to imply that the amount of uncertainty in $\widehat{\boldsymbol{\theta}}$ should decrease with increasing n, i.e., $\mathrm{Var}(\widehat{\boldsymbol{\theta}})$ should decrease in a matrix sense.
2. It is *positive-semidefinite*. Practically, for us it is positive-definite over a large part of $\boldsymbol{\Omega}$, though singular EIMs can occur as extreme cases in likelihood theory.
3. It changes under transformations, and the EIM under monotonic transformations is readily available, as follows. Let $g_j(\boldsymbol{\theta})$ be a set of p invertible functions that are differentiable. Then

$$\boldsymbol{\mathcal{I}}_E(\boldsymbol{g}) \;=\; \frac{\partial \boldsymbol{\theta}^T}{\partial \boldsymbol{g}} \, \boldsymbol{\mathcal{I}}_E(\boldsymbol{\theta}) \, \frac{\partial \boldsymbol{\theta}}{\partial \boldsymbol{g}^T} \tag{A.9}$$

where $\boldsymbol{g} = (g_1(\boldsymbol{\theta}), \ldots, g_p(\boldsymbol{\theta}))^T$. This result is used much in this book, both directly and indirectly, e.g., the variance-covariance matrix (A.27) for the delta method, and it lurks in the background of (18.6), (18.9) and (18.11).
   As a simple example, if $\tau = g(\theta)$ where g is smooth and $g'(\theta) \neq 0$, then $\boldsymbol{\mathcal{I}}_{E1}(\tau) = \boldsymbol{\mathcal{I}}_{E1}(\theta)/[g'(\theta)]^2$. Applied specifically to $Y \sim \mathrm{Poisson}(\lambda)$, then $\boldsymbol{\mathcal{I}}_{E1}(\lambda) = 1/\lambda$, and for $\tau = \sqrt{\lambda}$, $\boldsymbol{\mathcal{I}}_{E1}(\tau) = (4\lambda)/4 = 4$, which is independent of λ (this is known as the Poisson variance-stabilizing transformation).
4. For some models with $p > 1$, it is possible for the (j, k) EIM element to be equal to 0 ($j \neq k$). If so, then θ_j and θ_k are said to be *orthogonal*, and this implies *asymptotic independence* between them. An important consequence of two parameters being orthogonal is that the MLE of one parameter varies only slowly with the other parameter. Indeed, for some models where several

parameterizations have been proposed, it is not uncommon to prefer ones with orthogonal parameters because of the stability they produce. Computationally, it can lead to faster convergence and be numerically well-conditioned. And in the case of **VGAM**, less storage may arise because of the matrix-band format used to represent EIMs (Sect. 18.3.5), e.g., for the bivariate odds ratio model it has the form (McCullagh and Nelder, 1989, p.228)

$$\begin{pmatrix} \times & \times & 0 \\ \times & \times & 0 \\ 0 & 0 & \times \end{pmatrix}$$

so that the working weights can be stored in an $n \times 4$ matrix, which is a saving of $2n$ doubles compared to n general 3×3 working weight matrices. If necessary, one might reorder the θ_j so that the non-zero values cluster about the diagonal band; this idea holds for family function `posbernoulli.tb()` (Ex. 17.5). For more details, see Cox and Reid (1987) and Young and Smith (2005).

Some Examples of EIMs

The **VGAM** package implements Fisher scoring on most parts, therefore each model must have EIMs that are tractable or can be approximated. In the latter case, Sect. 9.2 describes some methods. We now illustrate the former case by considering simple distributions that have closed-form expressions for the EIM elements. These examples come from the **VGAM** package.

1. `betaR()`   The standard beta density, as implemented by [dpqr]beta(), parameterizes the density in terms of the two positive shape parameters, and it is

$$f(y; s_1, s_2) = \frac{y^{s_1-1} (1-y)^{s_2-1}}{Be(s_1, s_2)} = \frac{y^{s_1-1} (1-y)^{s_2-1} \Gamma(s_1 + s_2)}{\Gamma(s_1) \Gamma(s_2)}$$

for $y \in (0,1)$. For one observation, $\ell = (s_1 - 1)\log y + (s_2 - 1)\log(1-y) + \log \Gamma(s_1 + s_2) - \log \Gamma(s_1) - \log \Gamma(s_2)$, from which the derivatives are

$$\frac{\partial \ell}{\partial s_1} = \log y + \psi(s_1 + s_2) - \psi(s_1),$$

$$\frac{\partial \ell}{\partial s_2} = \log(1 - y) + \psi(s_1 + s_2) - \psi(s_2),$$

$$-\frac{\partial^2 \ell}{\partial s_j^2} = \psi'(s_j) - \psi'(s_1 + s_2), \quad j = 1, 2,$$

$$-\frac{\partial^2 \ell}{\partial s_1 \partial s_2} = -\psi'(s_1 + s_2).$$

The second derivatives are not functions of y, and therefore the OIM and EIM coincide, both being

$$\begin{pmatrix} \psi'(s_1) - \psi'(s_1 + s_2) & -\psi'(s_1 + s_2) \\ -\psi'(s_1 + s_2) & \psi'(s_2) - \psi'(s_1 + s_2) \end{pmatrix}.$$

2. `rayleigh()`   Sometimes the property $E[\partial \ell / \partial \theta_j] = 0$ can be used to good effect when working out elements of the EIM, as the following simple

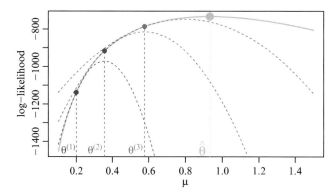

Fig. A.1 The first few Newton-like iterations for a Poisson regression fitted to the V1 data set. The *solid orange curve* is $\ell(\theta)$ with $\theta = \mu$. The initial value is $\theta^{(1)} = 0.2$. Each iteration $\theta^{(a)}$ corresponds to the maximum of the quadratic (*dashed curves*) from the previous iteration.

example shows. From Table 12.8, the density of the Rayleigh distribution is $y \cdot \exp\{-2^{-1}(y/b)^2\}/b^2$ for positive y and positive scale parameter b. Then, for one observation, $\ell = \log y - 2^{-1}(y/b)^2 - 2 \log b$ so that $\ell' = ([y/b]^2 - 2)/b$. Equating this to 0 implies that $E(Y^2) = 2b^2$. Then $-\ell'' = (3y^2 - 2b^2)/b^4$ so that the EIM is $(3 \times 2b^2 - 2b^2)/b^4 = 4/b^2$.

A.1.2.4 Newton-Like Algorithms

Given that an iterative method will be used to solve for the MLE, let's expand ℓ in a first-order Taylor series about the current estimate at iteration $a - 1$:

$$\ell(\boldsymbol{\theta}^{(a)}) \approx \ell(\boldsymbol{\theta}^{(a-1)}) + (\boldsymbol{\theta}^{(a)} - \boldsymbol{\theta}^{(a-1)})^T \frac{\partial \ell(\boldsymbol{\theta}^{(a-1)})}{\partial \boldsymbol{\theta}}.$$

Now take the first derivatives: $\partial \ell / \partial \boldsymbol{\theta}$ evaluated at $\boldsymbol{\theta}^{(a)}$ is equal to

$$\begin{aligned}
\frac{\partial \ell(\boldsymbol{\theta}^{(a)})}{\partial \boldsymbol{\theta}} &= \frac{\partial \ell(\boldsymbol{\theta}^{(a-1)})}{\partial \boldsymbol{\theta}} + \frac{\partial^2 \ell(\boldsymbol{\theta}^{(a-1)})}{\partial \boldsymbol{\theta} \, \partial \boldsymbol{\theta}^T} \left(\boldsymbol{\theta}^{(a)} - \boldsymbol{\theta}^{(a-1)} \right) \qquad (A.10) \\
&= \boldsymbol{U}(\boldsymbol{\theta}^{(a-1)}) + \boldsymbol{\mathcal{H}}(\boldsymbol{\theta}^{(a-1)}) \left(\boldsymbol{\theta}^{(a)} - \boldsymbol{\theta}^{(a-1)} \right).
\end{aligned}$$

Ideally, the next iteration will be very good, or even better, it will be optimal. If so, then $\boldsymbol{\theta}^{(a)}$ will have the value $\widehat{\boldsymbol{\theta}}$, which is the MLE—and then its score vector will be $\boldsymbol{0}$. Thus we will be totally optimistic and set the LHS of (A.10) to $\boldsymbol{0}$. Upon rearrangement, this leads to the Newton-Raphson step

$$\boldsymbol{\theta}^{(a)} = \boldsymbol{\theta}^{(a-1)} - \boldsymbol{\mathcal{H}}(\boldsymbol{\theta}^{(a-1)})^{-1} \, \boldsymbol{U}(\boldsymbol{\theta}^{(a-1)}). \qquad (A.11)$$

The algorithm converges quickly at a quadratic convergence rate, provided that ℓ is well-behaved (close to a quadratic) in a neighbourhood of the maximum, and if the starting value is close enough to the solution. By a 'quadratic convergence rate', it is meant that

$$\lim_{a\to\infty} \frac{\|\boldsymbol{\theta}^{(a)} - \widehat{\boldsymbol{\theta}}\|}{\|\boldsymbol{\theta}^{(a-1)} - \widehat{\boldsymbol{\theta}}\|^2} = c$$

for some positive c. What this means in practice is that the number of correct decimal places doubles at each iteration near the solution.

Figure A.1 illustrates the idea behind Newton-like algorithms for a simple one-parameter problem involving a Poisson regression fitted to the V1 data set. Starting at $\theta^{(0)} = 0.2$, successive quadratics are fitted to approximate ℓ and obtain the next iteration $\theta^{(a)}$. These quadratics match the derivatives $\ell^{(\nu)}(\theta^{(a-1)})$ for $\nu = 0, 1, 2$.

The Newton-Raphson algorithm requires the inversion of an order-p matrix, which is $O(p^3)$ and therefore expensive for very large p, and it does require the programming of the $p(p+1)/2$ unique elements of $\boldsymbol{\mathcal{H}}$. And a Newton-Raphson step is not guaranteed to be an improvement: $\ell(\boldsymbol{\theta}^{(a)}) < \ell(\boldsymbol{\theta}^{(a-1)})$ is a possibility. There have been many modifications proposed to the plain Newton-Raphson algorithm, but that is beyond the scope of this book; for more details see, e.g., Dennis and Schnabel (1996), Nocedal and Wright (2006), Weihs et al. (2014).

An alternative procedure proposed by Fisher is to replace the OIM by the EIM. The result is

$$\boldsymbol{\theta}^{(a)} = \boldsymbol{\theta}^{(a-1)} + \boldsymbol{\mathcal{I}}_E^{-1}(\boldsymbol{\theta}^{(a-1)})\,\boldsymbol{U}(\boldsymbol{\theta}^{(a-1)}), \tag{A.12}$$

which is known as *Fisher's method of scoring*, or just *Fisher scoring*. This method usually possesses only a linear convergence rate, meaning

$$\lim_{a\to\infty} \frac{\|\boldsymbol{\theta}^{(a)} - \widehat{\boldsymbol{\theta}}\|}{\|\boldsymbol{\theta}^{(a-1)} - \widehat{\boldsymbol{\theta}}\|} = c,$$

for some $0 < c < 1$, however typically $c \approx 0$ so that the convergence rate is quite acceptable. As the n EIMs are usually positive-definite, this means that each step is in an ascent direction, and half-stepping can be used to guarantee an improvement at each step (Sect. 3.5.4).

Fisher scoring is implemented by VGAM mainly for two reasons. The first is that, for most models, the EIMs are positive-definite over a large portion of the parameter space $\boldsymbol{\Omega}$, in contrast to OIMs which tend to be positive-definite in a smaller subset. As an example, consider the Rayleigh distribution above. Clearly, $-\ell''$ is positive for $y > \sqrt{2/3}\,b$, whereas the EIM is positive for all b. As mentioned elsewhere, IRLS requires *each* of the n EIMs to be positive-definite, not just their sum. The second reason is that EIMs are often simpler than the OIM. Fisher scoring may be performed by using the iteratively reweighted (generalized) least squares algorithm—see Sect. 3.2 for details. For GLMs with a canonical link, the OIM equals the EIM, therefore Newton-Raphson and Fisher scoring coincide.

How can one know whether one has reached the true solution? We say that $\widehat{\boldsymbol{\theta}}$ is a *stationary point* if $\boldsymbol{U}(\widehat{\boldsymbol{\theta}}) = \boldsymbol{0}$. Iterative numerical methods may converge to a stationary point called a *local* maximum, e.g., when ℓ is multimodal such as Fig. 12.1 and the initial values are not very good. Also, if $\boldsymbol{\mathcal{I}}_O(\widehat{\boldsymbol{\theta}})$ is positive-definite, then $\widehat{\boldsymbol{\theta}}$ is a relative maximum. Equivalently, all its eigenvalues are positive, but if $\widehat{\boldsymbol{\mathcal{H}}}$ has positive and negative eigenvalues, then $\widehat{\boldsymbol{\theta}}$ is known as a *saddle point*. For some models, it can be proven that ℓ is concave in $\boldsymbol{\theta}$. If so, then the MLE is unique, and any local solution is the global solution. For example, for several categorical regression models, see Pratt (1981).

Incidentally, another common Newton-like method known as the Gauss-Newton method is used, particularly in nonlinear regression. This approximates the Hessian by $\sum_i \boldsymbol{u}(\boldsymbol{\theta}^{(a-1)})\,\boldsymbol{u}(\boldsymbol{\theta}^{(a-1)})^T$. It has the advantage that only first derivatives are needed, however it can suffer from the so-called *large residual problem* that causes its convergence to be very slow.

A.1.3 Properties of Maximum Likelihood Estimators

Under regularity conditions, MLEs have many good properties. They are described as *asymptotic* because $n \to \infty$. We write $\widehat{\boldsymbol{\theta}}_n$ to emphasize the MLE is based on a sample of size n, because this is enlightening in the case of i.i.d. observations. Recall here that $\boldsymbol{\theta}_*$ is the true value of $\boldsymbol{\theta}$. The properties of MLEs include the following.

1. *Asymptotic consistency*:   for all $\varepsilon > 0$ and $\boldsymbol{\theta}_* \in \Omega$,

$$\lim_{n\to\infty} P[\|\widehat{\boldsymbol{\theta}}_n - \boldsymbol{\theta}_*\|_\infty > \varepsilon] \;=\; 0. \tag{A.13}$$

That is, the distribution of $\widehat{\boldsymbol{\theta}}_n$ collapses around $\boldsymbol{\theta}_*$. Here, the maximum (infinity) norm is used to show that the usual plim definition (A.32) is applied element-by-element to $\boldsymbol{\theta}_n$. It is common to write $\widehat{\boldsymbol{\theta}}_n \xrightarrow{\mathcal{P}} \boldsymbol{\theta}_*$ (convergence in probability). This is called weak consistency; a stronger form based on almost sure convergence in probability can be defined.

2. *Asymptotic normality*: $\widehat{\boldsymbol{\theta}}_n$ is asymptotically $N_p(\boldsymbol{\theta}_*, \boldsymbol{\mathcal{I}}_E^{-1}(\boldsymbol{\theta}_*))$ as $n \to \infty$, i.e.,

$$\widehat{\boldsymbol{\theta}}_n \xrightarrow{\mathcal{D}} N_p(\boldsymbol{\theta}_*, \boldsymbol{\mathcal{I}}_E^{-1}(\boldsymbol{\theta}_*)) \tag{A.14}$$

(convergence in distribution). For i.i.d. data, this can be stated as

$$\sqrt{n}\left(\widehat{\boldsymbol{\theta}}_n - \boldsymbol{\theta}_*\right) \xrightarrow{\mathcal{D}} N_p(\boldsymbol{0}, \boldsymbol{\mathcal{I}}_{E1}^{-1}(\boldsymbol{\theta}_*)). \tag{A.15}$$

Thus under i.i.d. conditions, $\widehat{\boldsymbol{\theta}}_n$ converges to $\boldsymbol{\theta}_*$ in distribution at a $\sqrt{n}$-rate. In consequence of the above,

$$(\widehat{\boldsymbol{\theta}}_n - \boldsymbol{\theta}_*)^T \, \boldsymbol{\mathcal{I}}_E(\boldsymbol{\theta}_*) \, (\widehat{\boldsymbol{\theta}}_n - \boldsymbol{\theta}_*) \;\sim\; \chi_p^2 \tag{A.16}$$

as $n \to \infty$.

3. *Asymptotically unbiasedness*: $E(\widehat{\boldsymbol{\theta}}_n) \to \boldsymbol{\theta}_*$ as $n \to \infty$, for all $\boldsymbol{\theta}_* \in \Omega$.

4. *Asymptotically efficiency*:   If a most-efficient (unbiased) estimator exists, then it will be the MLE. See the Cramér-Rao inequality of Sect. A.1.3.1.

5. *Invariance*:   Another fundamental property is that if $\widehat{\boldsymbol{\theta}}$ is the MLE, then under a different parameterization $g(\boldsymbol{\theta})$ (where g is some monotone function of $\boldsymbol{\theta}$), the MLE of $g(\boldsymbol{\theta})$ is $g(\widehat{\boldsymbol{\theta}})$. This means we can choose the most convenient parameterization, or one having superior properties. Maximum likelihood estimation is also invariant under transformation of the observations. This can be seen from (A.30): the LHS density is $f_Y(y; \boldsymbol{\theta})$ and the RHS is $f_X(x(y); \boldsymbol{\theta}) \cdot |dx/dy|$ where dx/dy is independent of $\boldsymbol{\theta}$.

6. Under mild regularity conditions,

$$E\left(\frac{\partial \ell}{\partial \boldsymbol{\theta}}\right) = \mathbf{0}, \tag{A.17}$$

$$\boldsymbol{\mathcal{I}}_E(\boldsymbol{\theta}) = E\left(\frac{\partial \ell}{\partial \boldsymbol{\theta}}\frac{\partial \ell}{\partial \boldsymbol{\theta}^T}\right) = -E\left(\frac{\partial^2 \ell}{\partial \boldsymbol{\theta}\,\partial \boldsymbol{\theta}^T}\right) = -E\left(\frac{\partial}{\partial \boldsymbol{\theta}^T}\,\boldsymbol{U}\right). \tag{A.18}$$

7. Under regularity conditions, the score itself is asymptotically normal. In particular,

$$\boldsymbol{U}(\boldsymbol{\theta}_*) \sim N_p(\mathbf{0}, \boldsymbol{\mathcal{I}}_E(\boldsymbol{\theta}_*)) \tag{A.19}$$

as $n \to \infty$.

A.1.3.1 The Cramér-Rao Inequality

A simplified version of the famous Cramér-Rao inequality is stated as follows. Under regularity conditions and i.i.d. conditions, for all n and unbiased estimators $\widehat{\boldsymbol{\theta}}_n$,

$$\mathrm{Var}(\widehat{\boldsymbol{\theta}}_n) - \boldsymbol{\mathcal{I}}_E^{-1}(\boldsymbol{\theta}) \tag{A.20}$$

is positive-semidefinite. It is usually stated for the one-parameter case only, in which case

$$\frac{1}{n\,\boldsymbol{\mathcal{I}}_{E1}(\theta)} = \boldsymbol{\mathcal{I}}_E^{-1}(\theta) \leq \mathrm{Var}(\widehat{\theta}_n). \tag{A.21}$$

That is, the inverse of the EIM (known as the Cramér-Rao lower bound; CRLB) is a lower bound for the variance of an unbiased estimator; it is used as a benchmark to compare the performance of any unbiased estimator. An approximation to the multiparameter case (A.20) is to apply (A.21) to each diagonal element of $\boldsymbol{\mathcal{I}}_E^{-1}(\boldsymbol{\theta})$.

For some models, equality in (A.21) can be attained, therefore that estimator is (fully) efficient, or *best*, or a *minimum variance unbiased estimator* (MVUE). For other models, there exists no unbiased estimator that achieves the lower bound. Typically, the MLE achieves the CRLB.

While unbiasedness of an estimator is considered a good thing for many people, a viable option is to consider biased estimators which have a lower mean-squared error

$$\mathsf{MSE} = E\left[\sum_{j=1}^{p}(\widehat{\theta}_j - \theta_{*j})^2\right] = E[\|\widehat{\boldsymbol{\theta}} - \boldsymbol{\theta}_*\|^2] = \mathrm{trace}\{E[(\widehat{\boldsymbol{\theta}} - \boldsymbol{\theta}_*)(\widehat{\boldsymbol{\theta}} - \boldsymbol{\theta}_*)^T]\}.$$

The decomposition

$$\mathsf{MSE}(\widehat{\boldsymbol{\theta}}) = \mathrm{trace}\{\mathrm{Var}(\widehat{\boldsymbol{\theta}})\} + \|\mathrm{Bias}(\widehat{\boldsymbol{\theta}})\|^2 \tag{A.22}$$

is in contrast to the variance of the estimator with its bias $E(\widehat{\boldsymbol{\theta}}) - \boldsymbol{\theta}_*$.

A.1.4 Inference

Based on the above properties, MLE provides confidence intervals/regions for estimated quantities, tests of goodness-of-fit, and tests for the comparison of models. Loosely, one can view confidence intervals/regions and hypothesis testing as two sides of the same coin. Our summary here will separate out the two. Sometimes we partition $\boldsymbol{\theta} = (\boldsymbol{\theta}_1^T, \boldsymbol{\theta}_2^T)^T$ where $p_j = \dim(\boldsymbol{\theta}_j)$, and treat $\boldsymbol{\theta}_2$ as a nuisance parameter. Let the true value of $\boldsymbol{\theta}_1$ be $\boldsymbol{\theta}_{*1}$.

A.1.4.1 Confidence Intervals and Regions

There are two common methods, although three are listed here to parallel the hypothesis testing case.

1. **Wald Test**   Based on (A.16),

$$\left(\widehat{\boldsymbol{\theta}} - \boldsymbol{\theta}_*\right)^T \mathbf{V}^{-1} \left(\widehat{\boldsymbol{\theta}} - \boldsymbol{\theta}_*\right) \;\dot\sim\; \chi_p^2$$

in large samples. Here, $\mathbf{V}^{-1}$ is commonly chosen to be one of the following: (a) $\boldsymbol{\mathcal{I}}_E(\widehat{\boldsymbol{\theta}})$, (b) $\boldsymbol{\mathcal{I}}_O(\widehat{\boldsymbol{\theta}})$. The idea behind these is to use any consistent estimator, and both choices are equivalent to 1st-order approximation. Based on 2nd-order approximations and conditional arguments, Efron and Hinkley (1978) argued that the OIM is superior as an estimator of variance. As VGAM implements Fisher scoring, type (a) serves as the basis for the estimated variance-covariance matrix.

Based on the above, an approximate normal-theory $100(1 - \alpha)\%$ confidence region for $\boldsymbol{\theta}_1$ is the ellipsoid defined as the set of all $\boldsymbol{\theta}_{1*}$ satisfying

$$(\widehat{\boldsymbol{\theta}}_1 - \boldsymbol{\theta}_{1*})^T \boldsymbol{\mathcal{I}}_E(\widehat{\boldsymbol{\theta}}_1) (\widehat{\boldsymbol{\theta}}_1 - \boldsymbol{\theta}_{1*}) \;\leq\; \chi_{p_1}^2(\alpha).$$

For VGAM, an approximate $100(1 - \alpha)\%$ confidence interval for θ_j is given by

$$\widehat{\theta}_j \;\pm\; z(\alpha/2) \, \mathrm{SE}(\widehat{\theta}_j), \tag{A.23}$$

where the SE derives from the EIM, which is of the form $(\mathbf{X}_{\mathrm{VLM}}^T \mathbf{W} \mathbf{X}_{\mathrm{VLM}})^{-1}$ (Eq. (3.21); see Sect. 3.2 for details).

2. **Score Test**   Like the Wald test, confidence regions may be proposed which are based on a quadratic approximation to ℓ. Consequently, parameterizations which improve this approximation will give more accurate results, e.g., with the aid of parameter link functions. However, since the score test method is the least common of the three, no details are given here apart from a small mention in the hypothesis testing situation below.

3. **Likelihood Ratio Test (LRT)**   Let the *profile likelihood* for $\boldsymbol{\theta}_1$ be

$$R(\boldsymbol{\theta}_1) \;=\; \max_{\boldsymbol{\theta}_2} \; L(\boldsymbol{\theta}_1, \boldsymbol{\theta}_2)/L(\widehat{\boldsymbol{\theta}}).$$

Then the LR subset statistic $-2 \log R(\boldsymbol{\theta}_{*1}) \sim \chi_{p_1}^2$ asymptotically, therefore an approximate $100(1 - \alpha)\%$ confidence region for $\boldsymbol{\theta}_1$ is the set of all $\boldsymbol{\theta}_{1*}$ such that

$$-2 \log R(\boldsymbol{\theta}_{1*}) \ < \ \chi^2_{p_1}(\alpha).$$

For a simple 1-parameter model, this reduces to the set of all θ values satisfying

$$2 \left[\ell(\widehat{\theta}; \boldsymbol{y}) - \ell(\theta; \boldsymbol{y}) \right] \ \leq \ \chi^2_1(\alpha). \tag{A.24}$$

The methods function `confint.glm()` in MASS computes confidence intervals for each coefficient of a fitted GLM, based on the method of profile likelihoods. More generally, we can write the profile log-likelihood of $\boldsymbol{\theta}_1$ as $\ell_P(\boldsymbol{\theta}_1, \widehat{\boldsymbol{\theta}}_2(\boldsymbol{\theta}_1))$, where $\widehat{\boldsymbol{\theta}}_2(\boldsymbol{\theta}_1)$ is the MLE of $\boldsymbol{\theta}_2$ given $\boldsymbol{\theta}_1$. Being of lower dimension, ℓ_P is often used for inference, e.g., if $\widehat{\boldsymbol{\theta}}_2(\boldsymbol{\theta}_1)$ is easy.

A.1.4.2 Hypothesis Testing

For hypothesis testing, there are three well-known ways for tests of $H_0 : \boldsymbol{\theta} = \boldsymbol{\theta}_0$ where $\boldsymbol{\theta}_0$ is known and fixed. None of the tests are uniformly better, although the LRT is considered superior in many problems. Another advantage of the LRT is that it is invariant under nonlinear reparameterizations—this is not so for the Wald test, and for the score test, invariance depends on the choice of $\mathbf{V}$.

1. **Wald Test**   Based on (A.16) and under the null hypothesis $H_0 : \boldsymbol{\theta} = \boldsymbol{\theta}_0$,

$$\left(\widehat{\boldsymbol{\theta}} - \boldsymbol{\theta}_0 \right)^T \mathbf{V}^{-1} \left(\widehat{\boldsymbol{\theta}} - \boldsymbol{\theta}_0 \right) \ \dot{\sim} \ \chi^2_p \tag{A.25}$$

in large samples. Here, $\mathbf{V}^{-1}$ is commonly chosen to be one of the following: (a) $\mathcal{I}_E(\widehat{\boldsymbol{\theta}})$, (b) $\mathcal{I}_O(\widehat{\boldsymbol{\theta}})$, (c) $\mathcal{I}_E(\boldsymbol{\theta}_0)$, (d) $\mathcal{I}_O(\boldsymbol{\theta}_0)$. The idea behind (a)–(b) is to use any consistent estimator.

This result can be extended to arbitrary linear combinations of $\boldsymbol{\theta}$. In particular, for the linear combination $\boldsymbol{e}_j^T \boldsymbol{\theta} = \theta_j$, and $\boldsymbol{\theta}_0 = \mathbf{0}$, we usually take the square root and obtain the Wald statistic for $H_0 : \theta_j = 0$

$$z_0 \ = \ \frac{\widehat{\theta}_j - 0}{\sqrt{\widehat{\mathrm{Var}}(\widehat{\theta}_j)}} \ = \ \frac{\widehat{\theta}_j}{\mathrm{SE}(\widehat{\theta}_j)},$$

which is treated as a Z-statistic (or a t-ratio for LMs). One-sided tests are then accommodated, e.g., $H_1 : \theta_j < 0$ or $H_1 : \theta_j > 0$, in which case the p-values are $\Phi(z_0)$ and $\Phi(-z_0)$ provided $\widehat{\theta}_j < 0$ and $\widehat{\theta}_j > 0$, respectively [and $2\Phi(-|z_0|)$ for the 2-sided alternative $H_1 : \theta_j \neq 0$]. Alternatively, Z^2 may be treated as having an approximate χ^2_1 distribution. For VGLMs, VGAM prints out Wald statistics (usually type (a)) with the methods function `summary()`. The 4-column table of estimates, SEs, Wald statistics and p-values can be obtained by, e.g.,

```
> coef(summary(vglmObject))    # Entire table
> coef(summary(vglmObject))[, "Pr(>|z|)"] # p-values
```

Given a fitted model (including an LM or GLM) that has $\widehat{\boldsymbol{\theta}}$ and some estimate $\widehat{\mathrm{Var}}(\widehat{\boldsymbol{\theta}})$ obtainable by `coef()` and `vcov()`, the function `linearHypothesis()` in car can test a system of linear hypotheses based on the Wald test.

2. **Score Test**   Using the result (A.19) and under $H_0 : \boldsymbol{\theta} = \boldsymbol{\theta}_0$,

$$U(\boldsymbol{\theta}_0)^T \, \boldsymbol{\mathcal{I}}_E^{-1}(\boldsymbol{\theta}_0) \, U(\boldsymbol{\theta}_0) \; \sim \; \chi_p^2 \qquad\qquad (A.26)$$

asymptotically (Rao, 1948). The EIM is evaluated at the hypothesized value $\boldsymbol{\theta}_0$, but at the MLE $\widehat{\boldsymbol{\theta}}$ is an alternative. Both versions of the test are valid; in fact, they are asymptotically equivalent. One advantage of using $\boldsymbol{\theta}_0$ is that calculation of the MLE may be bypassed. One disadvantage is that the test can be inconsistent (Freedman, 2007). In spite of their simplicity, score tests are not as commonly used as Wald and LR tests. Further information about score tests is at, e.g., Rao (1973). The package mdscore implements a modified score test for GLMs that offers improvements in accuracy when n is small.

3. **Likelihood Ratio Test (LRT)**   This test is based on a comparison of maximized likelihoods for nested models. Suppose we are considering two models, $\mathcal{M}_1$ and $\mathcal{M}_2$ say, such that $\mathcal{M}_1 \subseteq \mathcal{M}_2$. That is, $\mathcal{M}_1$ is a subset or special case of $\mathcal{M}_2$. For example, one may obtain a simpler model $\mathcal{M}_1$ by setting some of the θ_j in $\mathcal{M}_2$ to zero, and we want to test the hypothesis that those elements are indeed zero.

The basic idea is to compare the maximized likelihoods of the two models. The maximized likelihood under the smaller model $\mathcal{M}_1$ is

$$\sup_{\boldsymbol{\theta} \in \mathcal{M}_1} L(\boldsymbol{\theta}; \boldsymbol{y}) \; = \; L(\widehat{\boldsymbol{\theta}}_{\mathcal{M}_1}; \boldsymbol{y}),$$

where $\widehat{\boldsymbol{\theta}}_{\mathcal{M}_1}$ is the MLE of $\boldsymbol{\theta}$ under model $\mathcal{M}_1$. Likewise, the maximized likelihood under the larger model $\mathcal{M}_2$ has the same form

$$\sup_{\boldsymbol{\theta} \in \mathcal{M}_2} L(\boldsymbol{\theta}; \boldsymbol{y}) \; = \; L(\widehat{\boldsymbol{\theta}}_{\mathcal{M}_2}; \boldsymbol{y}),$$

where $\widehat{\boldsymbol{\theta}}_{\mathcal{M}_2}$ is the MLE of $\boldsymbol{\theta}$ under model $\mathcal{M}_2$. The ratio of these quantities,

$$\lambda \; = \; \frac{L(\widehat{\boldsymbol{\theta}}_{\mathcal{M}_1}; \boldsymbol{y})}{L(\widehat{\boldsymbol{\theta}}_{\mathcal{M}_2}; \boldsymbol{y})},$$

lies in $[0, 1]$. Values close to 0 indicate that the smaller model is not acceptable compared to the larger model, while values close to unity indicate that the smaller model is almost as good as the large model.

Under regularity conditions, the *likelihood ratio test statistic*

$$-2 \log \lambda \; = \; 2 \log L(\widehat{\boldsymbol{\theta}}_{\mathcal{M}_2}; \boldsymbol{y}) - 2 \log L(\widehat{\boldsymbol{\theta}}_{\mathcal{M}_1}; \boldsymbol{y}) \; \rightarrow \; \chi_\nu^2$$

where $\nu = \dim(\mathcal{M}_2) - \dim(\mathcal{M}_1)$, the difference in the number of parameters in the two models. When applied to GLMs, the LRT is also known as the deviance test.

LRTs may be performed using lrtest(), e.g., for the following two vglm() objects where the simpler model is a special case of the more complex model,

```
> # Models must be nested
> lrtest(Complex.model, Simpler.model)
```

returns the LRT statistic and p-value.

In the above, the Wald and score tests were for $H_0 : \boldsymbol{\theta} = \boldsymbol{\theta}_0$, however, hypothesis tests involving only a subset of $\dim(\boldsymbol{\theta}_0)$ parameters are easily handled: replace p in (A.25) and (A.26) by p_0 and choose the relevant submatrix of $\mathbf{V}^{-1}$.

All three tests are asymptotically equivalent, and therefore can be expected to give similar results in large samples. In small samples, simulation studies have suggested that LRTs are generally the best. Note that the calculation of a LRT requires fitting two models ($\mathcal{M}_1$ and $\mathcal{M}_2$), compared to only one model for the Wald test ($\mathcal{M}_2$), and sometimes no model at all for the score test. However, note that the Hauck-Donner phenomenon (Sect. 2.3.6.2) may affect the Wald test but not the LRT.

The three test statistics have an elegant geometric interpretation that is illustrated in Fig. A.2a,b for the single-parameter case. In a nutshell, the pertinent features are the horizontal and vertical distances between $\ell(\theta_0)$ and $\ell(\widehat{\theta})$, and the slope $\ell'(\theta_0)$. This example comes from a negative binomial $\mathrm{NB}(\mu, k)$ distribution fitted to the `machinists` data set. The two plots are for $\theta = k$ and $\theta = \log k$. Here, $H_0 : k = \frac{1}{3}$, chosen for illustrative purposes only.

- The Wald test statistic is a function of $|\widehat{\theta} - \theta_0|$. Heuristically, the justification is to expand $\ell(\theta_0)$ about $\widehat{\theta}$ in a Taylor series under the assumption that the null hypothesis is true:

$$\ell(\theta_0) \approx \ell(\widehat{\theta}) + \frac{1}{2} \ell''(\widehat{\theta})(\theta_0 - \widehat{\theta})^2$$

  because $\ell'(\widehat{\theta}) = 0$ and $H_0 : \theta_* = \theta_0$. Then the Wald test statistic

$$(\theta_0 - \widehat{\theta}) \left[-\ell''(\widehat{\theta}) \right] (\theta_0 - \widehat{\theta}) \approx 2\{\ell(\widehat{\theta}) - \ell(\theta_0)\}$$

  i.e., approximates the LRT statistic. Here, choice (b) in (A.25) provides the metric. Expanded the way it appears here, the Wald test statistic is the squared *horizontal* distance after some standardization.
- The score test is a function of $\ell'(\theta_0)$, i.e., its *slope*. If $\widehat{\theta}$ approaches θ_0, then this derivative gets closer to 0, hence we would tend to reject the null hypothesis if the slope is very different from zero. Heuristically, it can be justified by expanding $\ell'(\widehat{\theta})$ about θ_0 in a Taylor series under the assumption that the null hypothesis is true:

$$\ell'(\widehat{\theta}) = 0 \approx \ell'(\theta_0) + \ell''(\theta_0)(\widehat{\theta} - \theta_0) + \frac{1}{2} \ell'''(\theta_0)(\widehat{\theta} - \theta_0)^2$$

  so that $(\widehat{\theta} - \theta_0) \approx \ell'(\theta_0)/\{-\ell''(\theta_0)\}$. Choosing choice (d) in (A.25), we can write $\sqrt{\boldsymbol{\mathcal{I}}_O(\theta_0)}\, (\widehat{\theta} - \theta_0) \approx \ell'(\theta_0)/\sqrt{\boldsymbol{\mathcal{I}}_O(\theta_0)}$. Both sides are approximately standard normally distributed. Upon squaring both sides,

$$(\widehat{\theta} - \theta_0)\, \boldsymbol{\mathcal{I}}_O(\theta_0)\, (\widehat{\theta} - \theta_0) = U(\theta_0)\, \boldsymbol{\mathcal{I}}_O(\theta_0)\, U(\theta_0)$$

  which is a Wald test statistic expressed in terms of the gradient at the hypothesized value.
- The LRT statistic is a function of $\ell(\widehat{\theta}) - \ell(\theta_0)$, in fact, it is simply twice that. This corresponds to the labelled *vertical* distance.

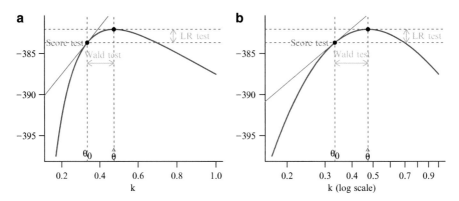

Fig. A.2 Negative binomial $NB(\mu, k)$ distribution fitted to the `machinists` data set. The y-axis is ℓ. Let $\theta = k$ and $\theta^* = \log k$. **(a)** $\ell(\theta)$ is the *solid blue curve*. **(b)** $\ell(\theta^*)$ is the *solid blue curve*. Note: for $H_0 : \theta = \theta_0$ (where $\theta_0 = \frac{1}{3}$), the likelihood-ratio test, score test and Wald test statistics are based on quantities highlighted with respect to ℓ. In particular, the score statistic is based on the tangent $\ell'(\theta_0)$.

Figure A.2b shows the same problem but under the reparameterization $\theta = \log k$. The log-likelihood is now more symmetric about $\widehat{\theta}$, i.e., its quadratic approximation is improved, therefore we would expect inferences to be more accurate compared to the first parameterization.

A.1.4.3 Delta Method

The *delta method* is a general method for obtaining approximate standard errors of functions of the parameter. Its basic idea is local linearization via derivatives. Let $\phi = g(\boldsymbol{\theta})$ be some function of the parameter. Apply a Taylor-series expansion about the true value:

$$\widehat{\phi} \;=\; g(\widehat{\boldsymbol{\theta}}) \;=\; g(\boldsymbol{\theta}_*) + (\widehat{\boldsymbol{\theta}} - \boldsymbol{\theta}_*)^T \frac{\partial g(\boldsymbol{\theta}_*)}{\partial \boldsymbol{\theta}} + \frac{1}{2}(\widehat{\boldsymbol{\theta}} - \boldsymbol{\theta}_*)^T \frac{\partial^2 g(\boldsymbol{\theta}_*)}{\partial \boldsymbol{\theta}\, \partial \boldsymbol{\theta}^T} (\widehat{\boldsymbol{\theta}} - \boldsymbol{\theta}_*) + \cdots,$$

hence

$$\sqrt{n}\left(g(\widehat{\boldsymbol{\theta}}) - g(\boldsymbol{\theta}_*)\right) \;\approx\; \sqrt{n}\,(\widehat{\boldsymbol{\theta}} - \boldsymbol{\theta}_*)^T (\partial g(\boldsymbol{\theta}_*)/\partial \boldsymbol{\theta}).$$

Consequently, from (A.14),

$$g(\widehat{\boldsymbol{\theta}}_n) - g(\boldsymbol{\theta}_*) \;\xrightarrow{\mathcal{D}}\; N_p\!\left(\mathbf{0},\; (\partial g(\boldsymbol{\theta}_*)/\partial \boldsymbol{\theta}^T)\, \boldsymbol{\mathcal{I}}_E^{-1}(\boldsymbol{\theta}_*)\, (\partial g(\boldsymbol{\theta}_*)/\partial \boldsymbol{\theta})\right). \quad \text{(A.27)}$$

To make use of this result, all quantities are computed at the MLE: for large n,

$$\mathrm{SE}(\widehat{\phi}) \;\approx\; \left\{\sum_{j=1}^{p}\sum_{k=1}^{p} \frac{\partial g}{\partial \theta_j} \frac{\partial g}{\partial \theta_k}\, \widehat{v}_{jk}\right\}^{\frac{1}{2}} \;=\; \left\{\frac{\partial g(\widehat{\boldsymbol{\theta}})}{\partial \boldsymbol{\theta}^T}\, \widehat{\mathrm{Var}}(\widehat{\boldsymbol{\theta}})\, \frac{\partial g(\widehat{\boldsymbol{\theta}})}{\partial \boldsymbol{\theta}}\right\}^{\frac{1}{2}}, \quad \text{(A.28)}$$

i.e., all the partial derivatives are evaluated at $\widehat{\boldsymbol{\theta}}$. In the case of $p = 1$ parameter, (A.28) reduces to

$$\mathrm{SE}(\widehat{\phi}) \;\approx\; \left| \frac{\mathrm{d}g}{\mathrm{d}\theta} \right| \sqrt{\widehat{v}_{11}}, \tag{A.29}$$

where $\mathrm{d}g/\mathrm{d}\theta$ is evaluated at $\widehat{\theta}$.

For simple intercept-only models, VGAM uses the delta method in calls of the form `vcov(vglmObject, untransform = TRUE)`. This is possible because (A.29) is readily computed for models having the form $\eta_j = g_j(\theta_j) = \beta_{(j)1}$ for simple links. The accuracy of the method depends on the functional form of g_j and the precision of $\widehat{\theta}_j$.

A.2 Some Useful Formulas

A.2.1 Change of Variable Technique

Suppose that a random variable X has a known PDF $f_X(x)$, and $Y = g(X)$ is some transformation of X, where $g : \mathbb{R} \to \mathbb{R}$ is any differentiable monotonic function. That is, g is increasing or decreasing, therefore is invertible (one-to-one). Then the PDF of Y, by the change-of-variable formula, is

$$f_Y(y) \;=\; f_X\big(g^{-1}(y)\big) \cdot \left| \frac{d}{dy}\, g^{-1}(y) \right| \;=\; f_X(x(y)) \cdot \left| \frac{dx}{dy} \right|. \tag{A.30}$$

A.2.2 Series Expansions

The following series expansions are useful, e.g., to work out the first and expected second derivatives of the GEV and GPD, as $\xi \to 0$:

$$\log(1 + z) \;=\; z - \frac{z^2}{2} + \frac{z^3}{3} - \frac{z^4}{4} + \cdots \quad \text{for } |z| \le 1 \text{ and } z \ne -1,$$

$$e^z \;=\; \lim_{n \to \infty} \left(1 + \frac{z}{n}\right)^n,$$

$$(1 + z)^\alpha \;=\; 1 + \alpha z + \frac{\alpha(\alpha - 1)}{2!} z^2 + \frac{\alpha(\alpha - 1)(\alpha - 2)}{3!} z^3 + \cdots, \quad \text{for } |z| \le 1,$$

$$(1 + x)^{-1} \;=\; 1 - x + x^2 - x^3 + \cdots \quad \text{for } -1 < x < 1.$$

A.2.3 Order Notation

There are two types of Landau's O-notation which are convenient abbreviations for us.

A.2.3.1 For Algorithms

Here, the $O(\cdot)$ notation is mainly used to measure the approximate computational expense of algorithms, especially in terms of time and memory. For functions $f(n)$ and $g(n)$, we say $f(n) = O(g(n))$ if and only if there exists two (positive and finite) constants c and n_0 such that

$$|f(n)| \leq c|g(n)| \tag{A.31}$$

for all $n \geq n_0$. For us, f and g are positive-valued, therefore (A.31) states that f does not increase faster than g. Saying that the computing time of an algorithm is $O(g(n))$ implies that its execution time takes no more than some constant multiplied by $g(n)$.

It can be shown that, e.g., $O(1) < O(\log n) < O(n) < O(n \log n) < O(n^2) < O(n^3) < O(2^n) < O(n!) < O(n^n)$. In any pairwise comparison, these inequalities usually do not hold in practice unless n is sufficiently large. As an example, the fastest known sorting algorithms for elements of a general n-vector cost $O(n \log n)$ whereas simpler algorithms such as bubble sort cost $O(n^2)$. Some people have suggested that usually an algorithm should be no more than $O(n \log n)$ to be practically manageable for very large data sets.

The so-called big-O notation, described above implicitly for integer n, is also useful and similarly defined for a real argument. For example, an estimator with an asymptotic bias of $O(h^2)$ has less asymptotic bias than another estimator whose asymptotic bias is $O(h)$, because $h \to 0^+$ as $n \to \infty$. Such considerations are made in, e.g., Sect. 2.4.6.2.

A.2.3.2 For Probabilities

In direct parallel with the above, the *order in probability* notation deals with convergence in probability of sets of random variables. A sequence of random variables $X_1, X_2, \ldots$ is said to *converge in probability* to the random variable X if, for all $\varepsilon > 0$,

$$\lim_{n \to \infty} P[|X_n - X| > \varepsilon] = 0. \tag{A.32}$$

The random variable X is called the *probability limit* of X_n, and it is written plim $X_n = X$, or alternatively, as $X_n \xrightarrow{\mathcal{P}} X$.

Now if $\{X_n\}$ is a set of random variables and $\{a_n\}$ is a set of constants, then $X_n = O_p(a_n)$ if for all $\varepsilon > 0$, there exists a finite $N > 0$ such that

$$P\left[\left|\frac{X_n}{a_n}\right| > N\right] < \varepsilon, \tag{A.33}$$

for all n. If $X_n = O_p(a_n)$, then we say that X_n/a_n is *stochastically bounded*. As an example, we say that $\{X_n\}$ is at most of order in probability n^k if, for every $\varepsilon > 0$, there exists a real N so that $P[n^{-k}|X_n| > N] < \varepsilon$ for all n.

A.2.4 Conditional Expectations

Provided that all the expectations are finite, for random variables X and Y,

$$
\begin{aligned}
E(Y) &= E_X\{E(Y|X)\}, & \text{(A.34)} \\
E[g(Y)] &= E_X\{E[g(Y)|X]\} \text{ (iterated expectation)}, & \text{(A.35)} \\
\text{Var}(Y) &= E_X\{\text{Var}(Y|X)\} + \text{Var}_X\{E(Y|X)\} \text{ (conditional variance).} & \text{(A.36)}
\end{aligned}
$$

One application of these formulas is the beta-binomial distribution (Sect. 11.4).

A.2.5 Random Vectors

Here are some basic results regarding random vectors $\boldsymbol{X} = (X_1, \dots, X_n)^T$ and $\boldsymbol{Y} = (Y_1, \dots, Y_n)^T$, i.e., vectors of random variables.

1. $E(\boldsymbol{X}) = \boldsymbol{\mu_X}$, where the ith element of $\boldsymbol{\mu_X}$ is $E(X_i)$. Similarly, $E(\boldsymbol{Y}) = \boldsymbol{\mu_Y}$.
2. $\text{Cov}(\boldsymbol{X}, \boldsymbol{Y}) = E[(\boldsymbol{X} - \boldsymbol{\mu_X})(\boldsymbol{Y} - \boldsymbol{\mu_Y})^T]$, with $\text{Var}(\boldsymbol{X}) = \text{Cov}(\boldsymbol{X}, \boldsymbol{X})$ ($= \boldsymbol{\Sigma_X}$, say). We write $\boldsymbol{X} \sim (\boldsymbol{\mu_X}, \boldsymbol{\Sigma_X})$.
3. $\text{Cov}(\mathbf{A}\boldsymbol{X}, \mathbf{B}\boldsymbol{Y}) = \mathbf{A}\,\text{Cov}(\boldsymbol{X}, \boldsymbol{Y})\,\mathbf{B}^T$ for conformable matrices $\mathbf{A}$ and $\mathbf{B}$ of constants.
4. $E[\boldsymbol{X}^T \mathbf{A} \boldsymbol{X}] = \boldsymbol{\mu_X}^T \mathbf{A} \boldsymbol{\mu_X} + \text{trace}(\mathbf{A}\boldsymbol{\Sigma_X})$.
5. $\text{trace}(\mathbf{AB}) = \text{trace}(\mathbf{BA})$ for conformable matrices $\mathbf{A}$ and $\mathbf{B}$.
6. $\text{rank}(\mathbf{A}) = \text{rank}(\mathbf{A}^T) = \text{rank}(\mathbf{A}^T \mathbf{A}) = \text{rank}(\mathbf{AA}^T)$.
7. If $\mathbf{A}$ is $n \times n$ with eigenvalues $\lambda_1, \dots, \lambda_n$, then

$$
\text{trace}(\mathbf{A}) = \sum_{i=1}^{n} \lambda_i, \qquad \det(\mathbf{A}) = \prod_{i=1}^{n} \lambda_i.
$$

8. A symmetric matrix $\mathbf{A}$ is positive-definite if $\boldsymbol{x}^T \mathbf{A} \boldsymbol{x} > 0$ for all $\boldsymbol{x} \neq \boldsymbol{0}$. Such matrices have positive eigenvalues, are invertible, and have a Cholesky decomposition that exists and is unique.

Some proofs for these can be found in, e.g., Seber and Lee (2003).

A.3 Some Linear Algebra

Least squares computations are usually based on orthogonal methods such as the QR factorization and singular value decomposition, because they are numerically more stable than naïve methods. They almost always give more accurate answers. A few details are given below.

A.3.1 Cholesky Decomposition

Given an $n \times n$ symmetric positive-definite matrix $\mathbf{A}$, its Cholesky decomposition $\mathbf{A} = \mathbf{U}^T \mathbf{U}$ where $\mathbf{U}$ is an upper-triangular matrix (i.e., $(\mathbf{U})_{ij} \equiv U_{ij} = 0$ for $i > j$) with positive diagonal elements. When $\mathbf{A}$ is 1×1, then $\mathbf{U}$ is just the square root of the element A_{11}. The computation of $\mathbf{U}$ might be written:

Iterate: For $i = 1, \ldots, n$

(i) $U_{ii} = \sqrt{A_{ii} - \sum_{k=1}^{i-1} U_{ki}^2}$

(ii) Iterate: For $j = i+1, \ldots, n$

$U_{ij} = (A_{ij} - \sum_{k=1}^{i-1} U_{ki} U_{kj})/U_{ii}$

The first operation is to compute $U_{11} = \sqrt{A_{11}}$. The algorithm requires $\frac{1}{3}n^3 + O(n^2)$ flops, which is about half the cost of the more general $\mathbf{LU}$ decomposition (Gaussian elimination).

Solving the linear system of equations $\mathbf{Ax} = \mathbf{y}$ can be achieved by first solving $\mathbf{U}^T \mathbf{z} = \mathbf{y}$ by forward substitution, and then solving $\mathbf{Ux} = \mathbf{z}$ by backward substitution. Each of these steps requires $n^2 + O(n)$ flops. Forward substitution here might be written as

---

Iterate: For $i = 1, \ldots, n$

$z_i = (y_i - \sum_{k=1}^{i-1} U_{ki} z_k)/U_{ii}$

The first operation is to compute $z_1 = y_1/U_{11}$. Likewise, backward substitution here might be written as

---

Iterate: For $i = n, \ldots, 1$

$x_i = (z_i - \sum_{k=i+1}^{n} U_{ik} x_k)/U_{ii}$

The first operation is to compute $x_n = z_n/U_{nn}$.

A variant of the above is the *rational* Cholesky decomposition, which can be written $\mathbf{A} = \mathbf{LDL}^T$, where $\mathbf{L}$ is a *unit* lower-triangular matrix, and $\mathbf{D}$ is a diagonal matrix with positive diagonal elements. By 'unit', we mean that the diagonal elements of $\mathbf{L}$ are all unity. This variant avoids computing n square roots in the usual algorithm, and should be used if $\mathbf{A}$ is banded with only a few bands, e.g., tridiagonal. (A matrix $\mathbf{T}$ is tridiagonal if $(\mathbf{T})_{ij} = 0$ for $|i - j| > 1$).

If $\mathbf{A}$ is a band matrix, with $(2m + 1)$ elements in its central band, then the Hutchinson and de Hoog (1985) algorithm is a method for computing the $2m + 1$ central bands of its inverse. The rational Cholesky decomposition of $\mathbf{A}$ has an $\mathbf{L}$ which is $(m + 1)$-banded, and the approximate cost is $\frac{1}{3}m^3 + nm^2 + O(m^2)$ flops. For cubic smoothing splines, $m = 2$ and the algorithm can be applied to compute the GCV.

Incidentally, a common method of measuring the width of a symmetric band matrix is by its half-bandwidth, e.g., $c = (2m+1)$ elements in its central band corresponds to a half-bandwidth of $(c+1)/2 = m+1$. Hence diagonal and tridiagonal matrices have half-bandwidths 1 and 2, etc.

A.3.2 Sherman-Morrison Formulas

If $\mathbf{A}$ is invertible, and $\boldsymbol{u}$ and $\boldsymbol{v}$ are vectors with $1 + \boldsymbol{v}^T \mathbf{A}^{-1} \boldsymbol{u} \neq 0$, then the Sherman-Morrison formula is

$$\left(\mathbf{A} + \boldsymbol{u}\boldsymbol{v}^T\right)^{-1} = \mathbf{A}^{-1} - \frac{\mathbf{A}^{-1}\boldsymbol{u}\boldsymbol{v}^T\mathbf{A}^{-1}}{1 + \boldsymbol{v}^T \mathbf{A}^{-1} \boldsymbol{u}}. \tag{A.37}$$

If $\mathbf{A}$ is invertible, then the Sherman-Morrison-Woodbury formula is

$$\left(\mathbf{A} + \mathbf{U}\mathbf{V}\right)^{-1} = \mathbf{A}^{-1} - \mathbf{A}^{-1}\mathbf{U}\left(\mathbf{I} + \mathbf{V}\mathbf{A}^{-1}\mathbf{U}\right)^{-1}\mathbf{V}\mathbf{A}^{-1}. \tag{A.38}$$

Incidentally, provided all inverses exist,

$$\begin{pmatrix} \mathbf{A}_{11} & \mathbf{A}_{12} \\ \mathbf{A}_{21} & \mathbf{A}_{22} \end{pmatrix}^{-1} = \begin{pmatrix} \mathbf{A}^{11} & \mathbf{A}^{12} \\ \mathbf{A}^{21} & \mathbf{A}^{22} \end{pmatrix} \tag{A.39}$$

where $\mathbf{A}^{11} = \mathbf{A}_{11}^{-1} + \mathbf{A}_{11}^{-1}\mathbf{A}_{12}\left(\mathbf{A}_{22} - \mathbf{A}_{21}\mathbf{A}_{11}^{-1}\mathbf{A}_{12}\right)^{-1}\mathbf{A}_{21}\mathbf{A}_{11}^{-1}$ or equivalently, $\mathbf{A}^{11} = \left(\mathbf{A}_{11} - \mathbf{A}_{12}\mathbf{A}_{22}^{-1}\mathbf{A}_{21}\right)^{-1}$.

A.3.3 QR Method

The QR decomposition of an $n \times p$ matrix $\mathbf{X}$ with $n > p$ is

$$\mathbf{X} = \mathbf{Q}\mathbf{R} = (\mathbf{Q}_1 \; \mathbf{Q}_2)\begin{pmatrix} \mathbf{R}_1 \\ \mathbf{O} \end{pmatrix} = \mathbf{Q}_1\mathbf{R}_1, \tag{A.40}$$

where $\mathbf{Q}$ ($n \times n$) is orthogonal (i.e., $\mathbf{Q}^T\mathbf{Q} = \mathbf{Q}\mathbf{Q}^T = \mathbf{I}_n$, or equivalently, $\mathbf{Q}^{-1} = \mathbf{Q}^T$) and $\mathbf{R}_1$ ($p \times p$) is upper triangular.

In R, the function qr() computes the QR factorization, and there are associated functions such as qr.coef(), qr.qty(), qr.Q() and qr.R(). These functions are based on LINPACK by default, but there is a logical argument for qr() in the form of LAPACK = FALSE that can be set to TRUE to call LAPACK instead. One can think of LAPACK (Anderson et al., 1999) as a more modern version of LINPACK (Dongarra et al., 1979).

Given a rank-p model matrix $\mathbf{X}$, solving the normal equations (2.6) by the QR method means that the OLS estimate $\widehat{\boldsymbol{\beta}} = \mathbf{R}_1^{-1}\mathbf{Q}_1^T\boldsymbol{y}$ is easily computed because qr.qty() returns $\mathbf{Q}_1^T\boldsymbol{y}$, and back substitution can be then used. As $\mathbf{X}$ is of full column-rank, all the diagonal elements of $\mathbf{R}_1$ are nonzero (positive by convention, actually).

It is easily verified that if the diagonal elements of $\mathbf{R}_1$ are positive (trivially achieved by negating certain columns of $\mathbf{Q}_1$ if necessary) then $\mathbf{R}_1$ corresponds to the Cholesky decomposition of $\mathbf{X}^T\mathbf{X}$, i.e., $\mathbf{X}^T\mathbf{X} = \mathbf{R}_1^T\mathbf{R}_1$. But the QR decomposition is the preferred method for computing $\widehat{\boldsymbol{\beta}}$ because there is no need to compute the sum-of-squares and cross-products matrix $\mathbf{X}^T\mathbf{X}$—doing so squares the condition number, so that if the columns of $\mathbf{X}$ are almost linearly dependent, then there will be a loss of accuracy. In general, orthogonal methods do not exacerbate ill-conditioned matrices.

For large n and p, the cost of performing a QR decomposition on $\mathbf{X}$ using Householder reflections[1] is approximately $2np^2$ floating point operations. This is about twice the cost of solving the normal equations by Cholesky when $n \gg p$.

A.3.4 Singular Value Decomposition

The singular value decomposition (SVD) of $\mathbf{X}$ as above is

$$\mathbf{X} \; = \; \mathbf{U}\mathbf{D}\mathbf{V}^T, \tag{A.41}$$

where $\mathbf{U}$ $(n \times p)$ is such that $\mathbf{U}^T\mathbf{U} = \mathbf{I}_p$, and $\mathbf{V}$ $(p \times p)$ is orthogonal, and $\mathbf{D}$ is a $p \times p$ diagonal matrix with non-negative elements d_{ii} (called the *singular values*). The matrix $\mathbf{U}$ here comprises the first p columns of an orthogonal matrix, much like $\mathbf{Q}_1$ does to $\mathbf{Q}$ in (A.40).

It is easy to show that the eigenvalues of $\mathbf{X}^T\mathbf{X}$ are d_{ii}^2, and it is usual to sort the singular values so that $d_{11} \geq d_{22} \geq \cdots \geq d_{pp} \geq 0$. With this enumeration, the eigenvectors of $\mathbf{X}^T\mathbf{X}$ make up the columns of $\mathbf{V}$, and the first p eigenvectors of $\mathbf{X}\mathbf{X}^T$ make up the columns of $\mathbf{U}$. A common method for determining the rank of $\mathbf{X}$ is to count the number of nonzero singular values, however, comparisons with 0 are made in light of the machine precision, i.e, `.Machine$double.eps`. In R, `svd()` computes the SVD by LAPACK, and the cost is approximately $6np^2 + 11p^3$ flops—which can be substantially more expensive than the QR decomposition.

A special case of the SVD is when $\mathbf{X}$ is square, symmetric and positive-definite. Then its SVD can be written as

$$\mathbf{X} \; = \; \mathbf{P}\boldsymbol{\Lambda}\mathbf{P}^T, \tag{A.42}$$

where $\boldsymbol{\Lambda}$ has the sorted eigenvalues of $\mathbf{X}$ along its diagonal, and $\mathbf{P}$ is orthogonal with the respective eigenvectors of $\mathbf{X}$ defining its columns. Equation (A.42) is known as the *spectral decomposition* or eigendecomposition of $\mathbf{X}$, and a useful consequence is that powers of $\mathbf{X}$ have the simple form

$$\mathbf{X}^s \; = \; \mathbf{P}\boldsymbol{\Lambda}^s\mathbf{P}^T, \tag{A.43}$$

e.g., $s = \pm\frac{1}{2}$ especially.

A.4 Some Special Functions

Many densities or their log-likelihoods are expressed in terms of special functions. A few of the more common ones are mentioned here.

---

[1] Ex. A.6; another common algorithm by Givens rotations entails an extra cost of about 50%

A.4.1 Gamma, Digamma and Trigamma Functions

The gamma function is defined for $x > 0$ as

$$\Gamma(x) \;=\; \int_0^\infty t^{x-1}\, e^{-t}\, dt \tag{A.44}$$

and can be computed by `gamma(x)`, and its logarithm by `lgamma(x)`. For positive integer a,

$$\Gamma(a+1) \;=\; a\,\Gamma(a) \;=\; a! \tag{A.45}$$

and Stirling's approximation for large x is

$$\Gamma(x+1) \;\sim\; \sqrt{2\pi x}\, x^x\, e^{-x}. \tag{A.46}$$

A useful limit is

$$\lim_{n\to\infty} \frac{\Gamma(n+\alpha)}{\Gamma(n)\, n^\alpha} \;=\; 1 \quad \forall \alpha \in \mathbb{R}. \tag{A.47}$$

The incomplete gamma function

$$P(a,x) \;=\; \frac{1}{\Gamma(a)} \int_0^x t^{a-1}\, e^{-t}\, dt \tag{A.48}$$

may be evaluated by `pgamma(x, a)`.

Derivatives of the log-gamma function are often encountered in discrete and continuous distributions. For such, define $\psi(x) = \Gamma'(x)/\Gamma(x)$ as the digamma function, and $\psi'(x)$ as the trigamma function.

For the digamma function, since $\psi(x+1) = \psi(x) + x^{-1}$, it follows that for integer $a \geq 2$,

$$\psi(a) \;=\; -\gamma + \sum_{i=1}^{a-1} i^{-1} \quad \text{where} \quad -\psi(1) \;=\; \gamma \approx 0.5772 \tag{A.49}$$

is the Euler–Mascheroni constant. For large x, a series expansion for the digamma function is

$$\psi(x) \;=\; \log x - \frac{1}{2x} + \sum_{k=1}^{\infty} \frac{B_{2k}}{2k\, x^{2k}} \;=\; \log x - \frac{1}{2x} - \frac{1}{12 x^2} + \cdots, \tag{A.50}$$

where B_k is the kth Bernoulli number.

For the trigamma function, since $\psi'(x+1) = \psi'(x) - x^{-2}$, it follows that for integer $a \geq 2$, $\psi'(a) = \pi^2/6 - \sum_{i=1}^{a-1} i^{-2}$ because $\psi'(1) = \pi^2/6$. For large x, a series expansion for the trigamma function is

$$\psi'(x) \;=\; \frac{1}{x} + \frac{1}{2 x^2} + \sum_{k=1}^{\infty} \frac{B_{2k}}{x^{2k+1}} \;=\; \frac{1}{x} + \frac{1}{2 x^2} + \frac{1}{6 x^3} - \frac{1}{30 x^5} + \cdots. \tag{A.51}$$

Higher-order derivatives of $\psi(x)$ may be computed by `psigamma()`.

A.4.2 Beta Function

The beta function is defined as

$$Be(a,b) \;=\; \int_0^1 t^{a-1}\,(1-t)^{b-1}\,dt, \quad 0 < a,\; 0 < b. \tag{A.52}$$

Then

$$Be(a,b) \;=\; \frac{\Gamma(a)\,\Gamma(b)}{\Gamma(a+b)}. \tag{A.53}$$

The incomplete beta function is

$$I_x(a,b) \;=\; \frac{Be_x(a,b)}{Be(a,b)}, \tag{A.54}$$

where

$$Be_x(a,b) \;=\; \int_0^x t^{a-1}\,(1-t)^{b-1}\,dt. \tag{A.55}$$

The function $I_x(a,b)$ can be evaluated by `pbeta(x, a, b)`.

A.4.3 The Riemann Zeta Function

The Riemann zeta function is defined by

$$\zeta(s) \;=\; \sum_{n=1}^{\infty} n^{-s}, \quad \Re(s) > 1. \tag{A.56}$$

Analytic continuation via

$$\zeta(s) \;=\; 2^s\,\pi^{s-1}\,\sin(\pi s/2)\,\Gamma(1-s)\,\zeta(1-s)$$

implies that it can be defined for all $\Re(s)$, with $\zeta(1) = \infty$. Some special values are $\zeta(2) = \pi^2/6$, and $\zeta(4) = \pi^4/90$. Euler found that for integer $n \geq 2$, $\zeta(2n) = A_{2n}$ where A_{2n} is rational. Indeed, $A_{2n} = \frac{1}{2}(-1)^{n+1} B_{2n}\,(2\pi)^{2n}/(2n)!$ in terms of the Bernoulli numbers.

A.4.4 Erf and Erfc

The *error function*, `erf(x)`, is defined for all x as

$$\frac{2}{\sqrt{\pi}} \int_0^x \exp(-t^2)\,dt, \tag{A.57}$$

therefore is closely related to the CDF $\Phi(\cdot)$ of the standard normal distribution. The inverse function is defined for $x \in [-1,1]$, i.e., `erf(x, inverse = TRUE)`.

The *complementary error function*, `erfc(x)`, is defined as `1-erf(x)`. Its inverse function is defined for $x \in [0, 2]$.

A.4.5 The Marcum Q-Function

The (generalized) Marcum Q-function is defined as

$$Q_m(a, b) = \int_b^\infty x \left(\frac{x}{a}\right)^{m-1} \exp\left\{-\frac{x^2 + a^2}{2}\right\} I_{m-1}(ax)\, dx \qquad \text{(A.58)}$$

$$= \exp\left\{-\frac{a^2 + b^2}{2}\right\} \sum_{k=1-m}^\infty \left(\frac{a}{b}\right)^k I_k(ab)$$

where $a \geq 0$, $b \geq 0$ and m is a positive integer. Here, I_{m-1} is a modified Bessel function of the first kind of order $m-1$ (as in Table A.1). The Marcum Q-function is used, e.g., as a CDF for noncentral chi-squared and Rice distributions, i.e., `price()`.

The case $m = 1$ is known as the ordinary Marcum Q-function.

A.4.6 Exponential Integral, Debye Function

The exponential integral, which is defined for real x, can be computed by `expint()` and is

$$Ei(x) = \int_{-\infty}^x t^{-1} e^t\, dt, \quad x \neq 0. \qquad \text{(A.59)}$$

The function `expexpint()` computes $e^{-x} Ei(x)$, and `expint.E1()` computes

$$E_1(x) = \int_x^\infty t^{-1} e^{-t}\, dt, \quad x \geq 0. \qquad \text{(A.60)}$$

The Debye function $D_n(x)$ is defined as

$$D_n(x) = \frac{n}{x^n} \int_0^x \frac{t^n}{e^t - 1}\, dt \qquad \text{(A.61)}$$

for $x \geq 0$ and $n = 0, 1, 2, 3, \ldots$.

A.4.7 Bessel Functions

Bessel functions appear widely in probability and statistics, e.g., distributions for directional data such as those defined on circles and spheres, Poisson processes and distributions (the most notable being the difference of two Poisson distributions, called the Skellam distribution). Of the various kinds, Table A.1 lists the most relevant ones relating to Chaps. 11–12.

Table A.1 Bessel functions (modified and unmodified) of order ν. The order nu may be fractional.

Function	Formula	R function	Name
$I_\nu(x)$	$\sum_{m=0}^{\infty} \frac{1}{m!\,\Gamma(m+\nu+1)} \left(\frac{x}{2}\right)^{2m+\nu}$	`besselI(x, nu)`	Modified Bessel function of the first kind
$K_\nu(x)$	$\lim_{\lambda\to\nu} \frac{\pi}{2} \frac{I_{-\lambda}(x) - I_\lambda(x)}{\sin(\lambda\pi)}$	`besselK(x, nu)`	Modified Bessel function of the third kind
$J_\nu(x)$	$\sum_{m=0}^{\infty} \frac{(-1)^m}{m!\,\Gamma(m+\nu+1)} \left(\frac{x}{2}\right)^{2m+\nu}$	`besselJ(x, nu)`	Bessel function of the first kind
$Y_\nu(x)$	$\lim_{\lambda\to\nu} \frac{J_\lambda(x)\cos(\lambda\pi) - J_{-\lambda}(x)}{\sin(\lambda\pi)}$	`besselY(x, nu)`	Bessel function of the second kind (Weber's function)

Bibliographic Notes

There are multitudes of books covering statistical inference and likelihood theory in detail, e.g., Edwards (1972), Rao (1973), Cox and Hinkley (1974), Silvey (1975), Barndorff-Nielsen and Cox (1994), Lindsey (1996), Welsh (1996), Severini (2000), Owen (2001), Casella and Berger (2002), Young and Smith (2005), Boos and Stefanski (2013). Most texts on mathematical statistics include at least a chapter on MLE, e.g., Knight (2000), Bickel and Doksum (2001), Shao (2003). Another book on statistical inference, which is compact and is concentrated on concepts, is Cox (2006). Hypothesis testing is treated in detail in Lehmann and Romano (2005).

A readable and applied account of models based on ML estimation is Azzalini (1996). Another applied book based on likelihood is Clayton and Hills (1993). GLMs are covered in detail in McCullagh and Nelder (1989); see also Lindsey (1997), Dobson and Barnett (2008). There have been a number of extensions of GLMs proposed. One of them, called "multivariate GLMs" by Fahrmeir and Tutz (2001, Sect.3.1.4). Another is the idea of composite link functions (Thompson and Baker, 1981). Standard texts for GAMs are Hastie and Tibshirani (1990) and Wood (2006).

A comprehensive account on many aspects of linear algebra, both theoretically and numerically, is Hogben (2014). Another, Golub and Van Loan (2013), remains an authoritative reference on matrix computations.

Detailed treatments of many special functions can be found in, e.g., Abramowitz and Stegun (1964), Gil et al. (2007), Olver et al. (2010).

Exercises

Ex. A.1. Let $\mathbf{A}$ and $\mathbf{B}$ be general $n \times n$ matrices, and $\boldsymbol{x}$ and $\boldsymbol{y}$ be general n-vectors. Work out the cost (expressed in $O(\cdot)$ complexity) of computing the following quantities in terms of the number of multiplications and additions, e.g., $n(n-1) = n^2 + O(n)$ multiplications, $n - 1 = n + O(1)$ additions.

(a) $\mathbf{A} + \mathbf{B}$,
(b) $5\,\mathbf{A}$,
(c) $\boldsymbol{x}^T \boldsymbol{y}$,
(d) $\mathbf{A}\,\boldsymbol{x}$,
(e) $\boldsymbol{x}^T \mathbf{A}\,\boldsymbol{x}$,
(f) $\mathbf{AB}$,
(g) $\mathrm{trace}(\mathbf{A})$,
(h) $\mathrm{trace}(\mathbf{A}^T \mathbf{A})$.
(i) Which is cheaper for computing $\mathbf{AB}\boldsymbol{x}$: $\mathbf{A}(\mathbf{B}\boldsymbol{x})$ or $(\mathbf{AB})\boldsymbol{x}$? By how much?

Ex. A.2. Prove that if $f_1 = O(g_1)$ and $f_2 = O(g_2)$ then $f_1 \cdot f_2 = O(g_1 \cdot g_2)$.

Ex. A.3. The R function `sort()`, by default, uses an algorithm called Shellsort. There are variants of this algorithm, but suppose the running time is $O(n^{4/3})$. Suppose it takes 2.4 seconds to sort 2 million (random) observations on a certain machine. Very crudely, how long might it be expected to sort 11 million (random) observations on that machine?

Ex. A.4. Use the results of Sect. A.2.4 to derive the mean and variance of Y_i^* for the beta-binomial distribution, i.e., (11.13).

Ex. A.5. From Sect. A.3.2, if $\mathbf{K}$ is a positive-definite matrix, show that

$$\left(\mathbf{I} + \mathbf{TKT}^T \right)^{-1} = \mathbf{I} - \mathbf{T} \left(\mathbf{K}^{-1} + \mathbf{T}^T \mathbf{T} \right)^{-1} \mathbf{T}^T. \qquad (A.62)$$

Ex. A.6. QR Factorization by the Householder Reflections
Suppose $\mathbf{X}$ is $n \times p$ with $n > p$ and of rank p. A Householder matrix is of the form

$$\boldsymbol{P} = \mathbf{I}_n - \frac{2\boldsymbol{v}\boldsymbol{v}^T}{\boldsymbol{v}^T \boldsymbol{v}} \qquad (A.63)$$

for some n-vector $\boldsymbol{v} \neq \mathbf{0}$.

(a) Show that $\boldsymbol{P}$ is symmetric and orthogonal.
(b) If $\boldsymbol{v} = \boldsymbol{x} - \boldsymbol{y}$ with $\|\boldsymbol{x}\|_2 = \|\boldsymbol{y}\|_2$, show that $\boldsymbol{P}\boldsymbol{x} = \boldsymbol{y}$.
(c) Let $\boldsymbol{x}_{(1)}$ be the first column of $\mathbf{X}$. Suppose we want to choose $\boldsymbol{v}$ so that $\boldsymbol{P}\boldsymbol{x}_{(1)} = c\,\boldsymbol{e}_1$ for some $c \neq 0$. Show that selecting $\boldsymbol{v} = \boldsymbol{x}_{(1)} + \alpha\boldsymbol{e}_1$ with $\alpha = \pm\|\boldsymbol{x}_{(1)}\|_2$ will achieve this. Given the choice of the sign of α, why is $\alpha = \mathrm{sign}(x_{11}) \cdot \|\boldsymbol{x}_{(1)}\|_2$ the better choice?
(d) Now for the kth column of $\mathbf{X}$, suppose we want to annihilate elements below the kth diagonal, leaving elements above the kth diagonal unchanged. Let $\boldsymbol{x}_{(k)} = (\boldsymbol{x}_{(k)}^{*T}, \boldsymbol{x}_{(k)}^{**T})^T$ be the kth column of $\mathbf{X}$, for $k = 2, \ldots, p$, where the first element of $\boldsymbol{x}_{(k)}^{**}$ is the diagonal element x_{kk}. We want to choose $\boldsymbol{v}_k$

so that $\boldsymbol{\mathcal{P}}_k\, \boldsymbol{x}_{(k)} = (\boldsymbol{x}_{(k)}^{*T}, c_k, \boldsymbol{0}_{n-k}^T)^T$ for some $c_k \neq 0$. Show that selecting $\boldsymbol{v}_k = (\boldsymbol{0}_{k-1}^T, x_{kk} + \alpha_k, \boldsymbol{x}_{(k)[-1]}^{**T})^T$ with $\alpha_k = \pm \|\boldsymbol{x}_{(k)}^{**}\|_2$ achieves this.

(e) Show that the product of two orthogonal matrices is orthogonal.

(f) Deduce that $\mathbf{Q}_1$ comprises the first p columns of the product $\boldsymbol{\mathcal{P}}_1 \boldsymbol{\mathcal{P}}_2 \cdots \boldsymbol{\mathcal{P}}_p$, and that $\mathbf{R} = \boldsymbol{\mathcal{P}}_p \cdots \boldsymbol{\mathcal{P}}_2 \boldsymbol{\mathcal{P}}_1 \mathbf{X}$, in the QR factorization (A.40) of $\mathbf{X}$.

Ex. A.7.   QR Factorization and Hilbert Matrices
Hilbert matrices, which are defined by $(\mathbf{X})_{ij} = (i + j - 1)^{-1}$ for $i, j = 1, \ldots, n$, are notorious for being ill-conditioned for n as little as 8 or 9. Compute the QR decomposition of the 8×4 left submatrix of an order-8 Hilbert matrix by explicitly computing the Householder matrices $\boldsymbol{\mathcal{P}}_1, \boldsymbol{\mathcal{P}}_2, \boldsymbol{\mathcal{P}}_3, \boldsymbol{\mathcal{P}}_4$ described in the previous exercise. Then check your answer with `qr()`.

Ex. A.8.   QR Method and Weighted Least Squares

(a) Extend the algorithm for estimating the OLS $\widehat{\boldsymbol{\beta}}$ by the QR method to handle WLS.

(b) For (a), how can $\widehat{\mathrm{Var}(\boldsymbol{\beta})}$ be computed?

Ex. A.9.   Show that

(a) the inverse of a nonsingular upper triangular matrix is also upper triangular,

(b) the product of two upper triangular matrices is upper triangular.

Ex. A.10.   Express the error function (A.57), and its inverse, in terms of $\Phi(\cdot)$ or $\Phi^{-1}(\cdot)$.

Ex. A.11.   Consider the log-gamma function. Show that $\log \Gamma(y+a) - \log \Gamma(y) \sim a \log y$ as $y \to \infty$, where $0 < a \ll y$.

Ex. A.12.   Digamma Function

(a) Verify the recurrence formula $\psi(z + 1) = \psi(z) + z^{-1}$.

(b) The digamma function has a single root on the positive real line. Apply the Newton-Raphson algorithm (A.11) to compute this root to at least 10 decimal places.

Ex. A.13.   Derive the score vector and EIM for the following distributions, to show that they involve digamma and trigamma functions.

(a) The log-F distribution (`logF()`).

(b) The Dirichlet distribution (`dirichlet()`).

Everything comes to an end which has a beginning.
—Marcus Fabius Quintilianus

Glossary

See Tables A.2, A.3, A.4, A.5.

© Thomas Yee 2015
T.W. Yee, *Vector Generalized Linear and Additive Models*,
Springer Series in Statistics, DOI 10.1007/978-1-4939-2818-7

Table A.2 Summary of some notation used throughout the book. Some R commands are given.

Notation	Comments
μ	Mean
$\widetilde{\mu}$	Median
$u_+ = \max(u, 0)$	Positive part of u, with $u_+^p = (u_+)^p$ and not $(u^p)_+$, `pmax(u, 0)`
$u_- = -\min(u, 0)$	Negative part of u, so that $u = u_+ - u_-$ & $\|u\| = u_+ + u_-$, `-pmin(u, 0)`
$\lfloor u \rfloor$	Floor of u, the largest integer not greater than u, e.g., $\lfloor 28.1 \rfloor = 28$, `floor(u)`
$\lceil u \rceil$	Ceiling of u, the smallest integer not less than u, e.g., $\lceil 28.1 \rceil = 29$, `ceiling(u)`
$\text{sign}(u)$	Sign of u, -1 if $u < 0$, $+1$ if $u > 0$, 0 if $u = 0$, `sign(u)`
$I(\text{statement})$	Indicator function, $1/0$ if statement is true/false, `as.numeric(statement)`
$\mathbb{C}$	Complex plane (excluding infinity), with $\Re(z) =$ the real part of z
$\mathbb{N}^0$	Set of all nonnegative integers, $0(1)\infty$
$\mathbb{N}^+$	Set of all positive integers, $1(1)\infty$
$\mathbb{R}$	Real line (excluding infinity), i.e., $(-\infty, \infty)$
$\mathbb{Z}$	Set of all integers
$a(b)c$	$\{a, a+b, a+2b, \ldots, c\}$; `seq(a, c, by = b)`
$\|\boldsymbol{x}\|_p$	$(\sum_i \|x_i\|^p)^{1/p}$, the p-norm of $\boldsymbol{x}$, so that $\|\boldsymbol{x}\|_\infty = \max(\|x_1\|, \|x_2\|, \ldots)$. By default, $p = 2$ so that $\|\boldsymbol{x}\|$ is the length of $\boldsymbol{x}$
$\|\boldsymbol{x} - \boldsymbol{y}\|$	Euclidean distance between two vectors $\boldsymbol{x}$ and $\boldsymbol{y}$, i.e., $\sqrt{(\boldsymbol{x} - \boldsymbol{y})^T (\boldsymbol{x} - \boldsymbol{y})}$, `norm(x - y, "2")`
$\mathbf{1}_M$	M-vector of 1s, `rep(1, M)`
$\mathbf{0}_n$	n-vector of 0s, `rep(0, n)`
$\boldsymbol{e}_i$	$(0, \ldots, 0, 1, 0, \ldots, 0)^T$, a vector of zeros, but with a one in the ith position, `diag(n)[, i, drop = FALSE]`
$\text{ncol}(\mathbf{A})$	Number of columns of matrix $\mathbf{A}$, `ncol(A)`. And $\mathcal{R}_k = \text{ncol}(\mathbf{H}_k)$
$\text{vec}(\mathbf{A})$	Vectorization of matrix $\mathbf{A}$ by columns, $(\boldsymbol{a}_1^T, \ldots, \boldsymbol{a}_n^T)^T$, `c(A)`
$\boldsymbol{x}_{[-1]i}$	The vector $\boldsymbol{x}_i$ with the first element deleted, `x[-1]`
$\mathbf{B}_{[-1,]}$	The matrix $\mathbf{B}$ with the first row deleted, `B[-1, ]`
$\mathbf{B}_{[,-1]}$	The matrix $\mathbf{B}$ with the first column deleted, `B[, -1]`
$\otimes$	Kronecker product, $\mathbf{A} \otimes \mathbf{B} = [(a_{ij}\mathbf{B})]$, `kronecker(A, B)`
$\circ$	Hadamard (element-by-element) product, $(\mathbf{A} \circ \mathbf{B})_{ij} = \mathbf{A}_{ij}\mathbf{B}_{ij}$, `A * B`

Table A.3 Summary of further notation used throughout the book.

Notation	Comments
$\sim$	Is distributed as
$\sim$	Is asymptotically equivalent to, or converges to (e.g., (2.75), (2.79))
$\dot\sim$	Is approximately distributed as
$\xrightarrow{\mathcal{D}}$	Convergence in distribution, i.e., $\{Y_i\} \xrightarrow{\mathcal{D}} Y$ if $\lim_{n\to\infty} F_n(y) = F_Y(y)$ for all y where F_Y is continuous (Y_i has CDF F_i)
$\xrightarrow{\mathcal{P}}$	Convergence in probability, (A.32)
$\phi(z)$	PDF of a standard normal, $N(\mu = 0, \sigma^2 = 1)$, $(2\pi)^{-\frac{1}{2}} e^{-\frac{1}{2}z^2}$ for $z \in \mathbb{R}$, `dnorm(z)`
$\Phi(z)$	CDF of a standard normal, `pnorm(z)`
$z(\alpha)$	$(1-\alpha)$-quantile of $N(0,1)$, i.e., `qnorm(1-alpha)`, `qnorm(alpha, lower.tail = FALSE)`
$\chi^2_\nu(\alpha)$	$(1-\alpha)$-quantile of a chi-square distribution with ν degrees of freedom, i.e., `qchisq(1-alpha, df = nu)`, `qchisq(alpha, df = nu, lower.tail = FALSE)`
$t_\nu(\alpha)$	$(1-\alpha)$-quantile of a Student t distribution with ν degrees of freedom, i.e., `qt(1-alpha, df = nu)`, `qt(alpha, df = nu, lower.tail = FALSE)`
iff	If and only if, i.e., a necessary and sufficient condition, $\iff$
$\log x$	Natural logarithm, $\log_e$, ln, `log(x)`
$\Gamma(x)$	Gamma function $\int_0^\infty t^{x-1} e^{-t}\, dt$ for $x > 0$, Sect. A.4.1, `gamma(x)`
$\psi(x) = \Gamma'(x)/\Gamma(x)$	Digamma function, $d\log\Gamma(x)/dx$, `digamma(x)`
$\psi'(x)$	Trigamma function, `trigamma(x)`
$\gamma = -\psi(1)$	Euler–Mascheroni constant, ≈ 0.57722, `-digamma(1)`
Cauchy sequence	A sequence $\{\boldsymbol{x}_n\}$ in a vector space $\mathcal{V}$ satisfying: given any $\varepsilon > 0$, $\exists N \in \mathbb{N}^+$ such that $\|\boldsymbol{x}_m - \boldsymbol{x}_n\| \le \varepsilon$ whenever $m, n \ge N$
$\mathcal{L}_2(a,b)$	$\{f : f$ is a Lebesgue square integrable function on $(a,b)\}$, i.e., $\int_a^b |f(t)|^2\, dt < \infty$. For $(a,b) = \mathbb{R}$, we write $\mathcal{L}_2$
$\mathcal{C}^k[a,b]$	$\{f : f', f'', \ldots, f^{(k)}$ all exist and are continuous on $[a,b]\}$. Note that $f \in \mathcal{C}^k[a,b]$ implies that $f \in \mathcal{C}^{k-1}[a,b]$. Also, $\mathcal{C}[a,b] \equiv \mathcal{C}^0[a,b] = \{f(t) : f(t)$ continuous and real valued for $a \le t \le b\}$
$\mathcal{W}_2^m[a,b]$	A *Sobolev space* of order m is $\{f : f^{(j)}, j = 0, \ldots, m-1$, are absolutely continuous on $[a,b]$, and $f^{(m)} \in \mathcal{L}_2[a,b]\}$
f absolutely continuous on $[a,b]$	$\forall \varepsilon > 0$, $\exists \delta > 0$ such that $\sum_{i=1}^n |f(x_i') - f(x_i)| < \varepsilon$ whenever $\{[x_i, x_i'] : i = 1, \ldots, n\}$ is a finite collection of mutually disjoint subintervals of $[a,b]$ with $\sum_{i=1}^n |x_i - x_i'| < \delta$. That is, f is differentiable almost everywhere and equals the integral of its derivative
$l_p(\mathbb{R}^n)$	$\{\boldsymbol{x} = (x_1, \ldots, x_n)^T : (\sum_{i=1}^n |x_i|^p)^{1/p} < \infty$ for $1 \le p < \infty\}$
Convex function $f : \mathcal{X} \to \mathbb{R}$	$f(tx_1 + (1-t)x_2) \le t f(x_1) + (1-t) f(x_2)\ \forall t \in [0,1]$ and $x_1, x_2 \in \mathcal{X}$, e.g., x^2 and e^x on $\mathbb{R}$. A sufficient condition is that $f''(x) > 0\ \forall x \in \mathcal{X}$
Concave function $f : \mathcal{X} \to \mathbb{R}$	$f(tx_1 + (1-t)x_2) \ge t f(x_1) + (1-t) f(x_2)\ \forall t \in [0,1]$ and $x_1, x_2 \in \mathcal{X}$, e.g., $\sqrt{x}$ and $\log x$ on $(0, \infty)$. A sufficient condition is that $f''(x) < 0\ \forall x \in \mathcal{X}$

Table A.4 Summary of some quantities. Data is $(y_i, \boldsymbol{x}_i)$ for $i = 1, \ldots, n$. See also Table 8.5. The indices $i = 1, \ldots, n$, $j = 1, \ldots, M$, $k = 1, \ldots, p$, $s = 1, \ldots, S$, $q = 1, \ldots, Q$. Starred quantities are estimated, as well as $\mathbf{C}$ and $\mathbf{A}$.

Notation	Comments
S	Number of responses. If $S > 1$ then these are "multiple responses"
M_1	Number of η_j for a single response
M	Number of η_j (summed over all S responses), e.g., $M = M_1 S$
Q_1	$\dim(\boldsymbol{y}_i)$ for a single response, hence $Q = Q_1 S$
$\mathbf{Y} = (\boldsymbol{y}_1, \ldots, \boldsymbol{y}_n)^T = (\boldsymbol{y}^{(1)}, \ldots, \boldsymbol{y}^{(Q)})$	Response matrix, is $n \times Q$
$\mathbf{X} = \mathbf{X}_{\text{LM}} = (\boldsymbol{x}_1, \ldots, \boldsymbol{x}_n)^T = (\boldsymbol{x}^{(1)}, \ldots, \boldsymbol{x}^{(p)})$	LM (model) matrix $[(x_{ik})]$, is $n \times p$ ($= n_{\text{LM}} \times p_{\text{LM}}$); $\boldsymbol{x}^{(1)} = \mathbf{1}_n$ if there is an intercept term
$\mathbf{X}_{\text{VLM}}$	VLM (model) matrix, $(nM) \times p_{\text{VLM}}$ ($= n_{\text{VLM}} \times p_{\text{VLM}}$), (3.18), (3.20)
$\boldsymbol{x} = (x_1, \ldots, x_p)^T = (\boldsymbol{x}_1^T, \boldsymbol{x}_2^T)^T$	Vector of explanatory variables, with $x_1 = 1$ if there is an intercept term, $\boldsymbol{x}_1$ is $p_1 \times 1$, and $\boldsymbol{x}_2$ is $p_2 \times 1$. Sometimes $\boldsymbol{x} = (x_1, \ldots, x_d)^T$, especially when referring to additive models
$\boldsymbol{x}_i^T = (x_{i1}, \ldots, x_{ip}) = (\boldsymbol{x}_{1i}^T, \boldsymbol{x}_{2i}^T)$	ith row of $\mathbf{X}$
$\boldsymbol{x}_{ij} = (x_{i1j}, \ldots, x_{ipj})^T$	Vector of explanatory variables for $\eta_j(\boldsymbol{x}_{ij})$. Explanatory variables specific to η_j (see `xij` argument). Partitioned into $\boldsymbol{x}_i^*$ and $\boldsymbol{x}_{ij}^*$ as in (3.35)
$\mathbf{X}_{\texttt{form2}}$	LM (model) matrix for argument `form2`. Has n rows
$\boldsymbol{\eta} = (\eta_1, \ldots, \eta_M)^T$	Vector of linear/additive predictors, with $\boldsymbol{\eta}_i = (\eta_{1i}, \ldots, \eta_{Mi})^T$
$\mathbf{H}_k = \left(\boldsymbol{h}_k^{(1)}, \ldots, \boldsymbol{h}_k^{(\mathcal{R}_k)}\right) = (\boldsymbol{h}_{1k}, \ldots, \boldsymbol{h}_{Mk})^T$	Constraint matrix ($M \times \mathcal{R}_k$) for x_k. Known, fixed and of full column-rank, (3.25)
$\boldsymbol{\eta}_i = \sum_{k=1}^{p} \mathbf{H}_k \boldsymbol{\beta}_{(k)}^* x_{ik}$	Vector of linear predictors, (3.27)
$\boldsymbol{\eta}_i = \sum_{k=1}^{d} \mathbf{H}_k \boldsymbol{f}_k^*(x_{ik})$	Vector of additive predictors, (3.25)
$\eta_j(\boldsymbol{x}_i) = \sum_{k=1}^{p} \beta_{(j)k} x_{ik}$	jth linear predictor (without constraints), (1.1)
$\eta_j(\boldsymbol{x}_i) = \sum_{k=1}^{d} f_{(j)k}(x_{ik})$	jth additive predictor (without constraints), (1.2)
$\boldsymbol{f}_k^*(x_k) = \left(f_{(1)k}^*(x_k), \ldots, f_{(\mathcal{R}_k)k}^*(x_k)\right)^T$	A $\mathcal{R}_k$-vector of smooth functions of x_k
$\mathbf{C} = (\boldsymbol{c}_{(1)}, \ldots, \boldsymbol{c}_{(R)}) = (\boldsymbol{c}_1, \ldots, \boldsymbol{c}_{p_2})^T$	Matrix of constrained coefficients, (5.3)
$\mathbf{A} = (\boldsymbol{a}_{(1)}, \ldots, \boldsymbol{a}_{(R)}) = (\boldsymbol{a}_1, \ldots, \boldsymbol{a}_M)^T$	Matrix of regression coefficients, (5.4)
$\boldsymbol{\nu} = (\nu_1, \ldots, \nu_R)^T = \mathbf{C}^T \boldsymbol{x}_2$	Vector of R latent variables or gradients, (5.1)
$\boldsymbol{\nu}_i = (\nu_{i1}, \ldots, \nu_{iR})^T = \mathbf{C}^T \boldsymbol{x}_{2i}$	ith site score
$\boldsymbol{\beta}_{(k)}^* = (\beta_{(1)k}^*, \ldots, \beta_{(\mathcal{R}_k)k}^*)^T$	Coefficients for x_k to be estimated, (3.28)
$\mathbf{B} = (\boldsymbol{\beta}_1 \quad \boldsymbol{\beta}_2 \quad \cdots \quad \boldsymbol{\beta}_M) = \left(\mathbf{H}_1 \boldsymbol{\beta}_{(1)}^* \mid \cdots \mid \mathbf{H}_p \boldsymbol{\beta}_{(p)}^*\right)^T$	Matrix of VLM/VGLM regression coefficients, $p \times M$, (1.32), (3.29)

Table A.5 Summary of some further quantities. See also Table 8.5.

Notation	Comments
$\mathbf{A}^T$	Transpose of $\mathbf{A}$, $(\mathbf{A}^T)_{ij} = (\mathbf{A})_{ji}$
$\boldsymbol{\beta}^\dagger$	$\text{vec}(\mathbf{B}) = (\boldsymbol{\beta}_1^T, \ldots, \boldsymbol{\beta}_M^T)^T$, (3.8)
$\boldsymbol{\theta}$	A generic vector of parameters to be estimated, often $(\theta_1, \ldots, \theta_p)^T$, can denote its true value
$\boldsymbol{\theta}_*$	The true value of $\boldsymbol{\theta}$, used occasionally when needed, p.536
$\mathcal{H}$	Hat matrix, (2.10)
$\mathcal{H}$	Hessian matrix, $[(\partial^2 \ell/(\partial \boldsymbol{\theta} \, \partial \boldsymbol{\theta}^T))]$, (A.6)
$\boldsymbol{\mathcal{I}}_E$	Expected (Fisher) information matrix (EIM), (A.8)
$\boldsymbol{\mathcal{I}}_{E1}$	EIM for one observation
$\boldsymbol{\mathcal{I}}_O$	Observed information matrix, $-\mathcal{H}$ (OIM), (A.7)
$\mathcal{P}$	Householder matrix, (Ex. A.6)
$Y_{(i)}$	ith order statistic of $Y_1, Y_2, \ldots, Y_n$, so that $Y_{(1)} \leq Y_{(2)} \leq \cdots \leq Y_{(n)}$
$\overline{y}_{i\bullet}$	Mean of y_{ij} over all j, $\sum_{j=1}^{n_i} y_{ij}/n_i$, (Sect. 1.5.2.4)
D	Deviance, e.g., (3.53)

References

Abramowitz, M. and I. A. Stegun (Eds.) 1964. *Handbook of Mathematical Functions*. New York: Dover.

Adams, N., M. Crowder, D. J. Hand, and D. Stephens (Eds.) 2004. *Methods and Models in Statistics*. London: Imperial College Press.

Adler, J. 2010. *R in a Nutshell*. Sebastopol: O'Reilly.

Agresti, A. 2010. *Analysis of Ordinal Categorical Data* (2nd ed.). Hoboken: Wiley.

Agresti, A. 2013. *Categorical Data Analysis* (Third ed.). Hoboken: Wiley.

Agresti, A. 2015. *Foundations of Linear and Generalized Linear Models*. Hoboken: Wiley.

Ahn, S. K. and G. C. Reinsel 1988. Nested reduced-rank autoregressive models for multiple time series. *Journal of the American Statistical Association* 83(403):849–856.

Ahsanullah, M. H. and G. G. Hamedani 2010. *Exponential Distribution: Theory and Methods*. New York: Nova Science.

Aigner, D. J., T. Amemiya, and D. Poirer 1976. On the estimation of production frontiers: Maximum likelihood estimation of the parameters of a discontinuous density function. *International Economic Review* 17(2):377–396.

Aitkin, M., B. Francis, J. Hinde, and R. Darnell 2009. *Statistical Modelling in R*. Oxford: Oxford University Press.

Akaike, H. 1973. Information theory and an extension of the maximum likelihood principle. In B. N. Petrov and F. Csáki (Eds.), *Second International Symposium on Information Theory*, pp. 267–281. Budapest: Akadémiai Kaidó.

Albert, A. and J. A. Anderson 1984. On the existence of maximum likelihood estimates in logistic regression models. *Biometrika* 71(1):1–10.

Allison, P. 2004. Convergence problems in logistic regression. See Altman et al. (2004), pp. 238–252.

Altman, M., J. Gill, and M. P. McDonald 2004. *Numerical Issues in Statistical Computing for the Social Scientist*. Hoboken: Wiley-Interscience.

Altman, M. and S. Jackman 2011. Nineteen ways of looking at statistical software. *Journal of Statistical Software* 42(2), 1–12.

Amemiya, T. 1984. Tobit models: a survey. *Journal of Econometrics* 24(1–2):3–61.

Amemiya, T. 1985. *Advanced Econometrics*. Oxford: Blackwell.

Amodei, L. and M. N. Benbourhim 1991. A vector spline approximation with application to meteorology. In P. J. Laurent, A. Le Méhauté, and L. L. Schumaker (Eds.), *Curves and Surfaces*, pp. 5–10. Boston: Academic Press.

© Thomas Yee 2015

T.W. Yee, *Vector Generalized Linear and Additive Models*,
Springer Series in Statistics, DOI 10.1007/978-1-4939-2818-7

Amstrup, S. C., T. L. McDonald, and B. F. J. Manly 2005. *Handbook of Capture–Recapture Analysis*. Princeton: Princeton University Press.

Anderson, E., Z. Bai, C. Bischof, S. Blackford, J. Demmel, J. Dongarra, J. Du Croz, A. Greenbaum, S. Hammarling, A. McKenney, and D. Sorensen 1999. *LAPACK Users' Guide* (Third ed.). Philadelphia: SIAM Publications.

Anderson, J. A. 1984. Regression and ordered categorical variables. *Journal of the Royal Statistical Society, Series B* 46(1):1–30. With discussion.

Anderson, T. W. 1951. Estimating linear restrictions on regression coefficients for multivariate normal distributions. *Annals of Mathematical Statistics* 22(3):327–351.

Andrews, H. P., R. D. Snee, and M. H. Sarner 1980. Graphical display of means. *American Statistician* 34(4):195–199.

Arnold, B. C. 2015. *Pareto Distributions* (Second ed.). Boca Raton: Chapman & Hall/CRC.

Aronszajn, N. 1950. Theory of reproducing kernels. *Transactions of the American Mathematical Society* 68(3):337–404.

Ashford, J. R. and R. R. Sowden 1970. Multi-variate probit analysis. *Biometrics* 26(3):535–546.

Azzalini, A. 1996. *Statistical Inference: Based on the Likelihood*. London: Chapman & Hall.

Azzalini, A. 2014. *The Skew-normal and Related Families*. Cambridge: Cambridge University Press.

Baillargeon, S. and L.-P. Rivest 2007. Rcapture: Loglinear models for capture–recapture in R. *Journal of Statistical Software* 19(5):1–31.

Baker, F. B. and S.-H. Kim 2004. *Item Response Theory: Parameter Estimation Techniques* (Second ed.). New York: Marcel Dekker.

Balakrishnan, N. and A. P. Basu (Eds.) 1995. *The Exponential Distribution: Theory, Methods, and Applications*. Amsterdam: Gordon and Breach.

Balakrishnan, N. and C.-D. Lai 2009. *Continuous Bivariate Distributions* (Second ed.). New York: Springer.

Balakrishnan, N. and V. B. Nevzorov 2003. *A Primer on Statistical Distributions*. New York: Wiley-Interscience.

Banerjee, S. and A. Roy 2014. *Linear Algebra and Matrix Analysis for Statistics*. Boca Raton: CRC Press.

Barndorff-Nielsen, O. E. and D. R. Cox 1994. *Inference and Asymptotics*. London: Chapman & Hall.

Barrodale, I. and F. D. K. Roberts 1974. Solution of an overdetermined system of equations in the ℓ_1 norm. *Communications of the ACM* 17(6):319–320.

Beaton, A. E. and J. W. Tukey 1974. The fitting of power series, meaning polynomials, illustrated on band-spectroscopic data. *Technometrics* 16(2):147–185.

Beirlant, J., Y. Goegebeur, J. Segers, J. Teugels, D. De Waal, and C. Ferro 2004. *Statistics of Extremes: Theory and Applications*. Hoboken: Wiley.

Bellman, R. E. 1961. *Adaptive Control Processes*. Princeton: Princeton University Press.

Belsley, D. A., E. Kuh, and R. E. Welsch 1980. *Regression Diagnostics: Identifying Influential Data and Sources of Collinearity*. New York: John Wiley & Sons.

Berlinet, A. and C. Thomas-Agnan 2004. *Reproducing Kernel Hilbert Spaces in Probability and Statistics*. Boston: Kluwer Academic Publishers.

Berndt, E. K., B. H. Hall, R. E. Hall, and J. A. Hausman 1974. Estimation and inference in nonlinear structural models. *Ann. Econ. and Soc. Measur.* 3–4: 653–665.

Bickel, P. J. and K. A. Doksum 2001. *Mathematical Statistics: Basic Ideas and Selected Topics* (Second ed.). Upper Saddle River: Prentice Hall.

Bilder, C. M. and T. M. Loughin 2015. *Analysis of Categorical Data with R.* Boca Raton: CRC Press.

Birch, J. B. 1980. Some convergence properties of iterated least squares in the location model. *Communications in Statistics B* 9(4):359–369.

Bock, R. D. and M. Leiberman 1970. Fitting a response model for n dichotomously scored items. *Psychometrika* 35(2):179–197.

Boos, D. D. and L. A. Stefanski 2013. *Essential Statistical Inference.* New York: Springer.

Bowman, K. O. and L. R. Shenton 1988. *Properties of Estimators for the Gamma Distribution.* New York: Marcel Dekker.

Braun, W. J. and D. J. Murdoch 2008. *A First Course in Statistical Programming with R.* Cambridge: Cambridge University Press.

Buja, A., T. Hastie, and R. Tibshirani 1989. Linear smoothers and additive models. *The Annals of Statistics* 17(2):453–510. With discussion.

Burnham, K. P. and D. R. Anderson 2002. *Model Selection and Multi-Model Inference: A Practical Information-Theoretic Approach* (Second ed.). New York: Springer.

Byrd, R. H. and D. A. Pyne 1979. Some results on the convergence of the iteratively reweighted least squares. In *ASA Proc. Statist. Computat. Section*, pp. 87–90.

Cameron, A. C. and P. K. Trivedi 2013. *Regression Analysis of Count Data* (Second ed.). Cambridge: Cambridge University Press.

Cantoni, E. and T. Hastie 2002. Degrees-of-freedom tests for smoothing splines. *Biometrika* 89(2):251–263.

Carroll, R. J. and D. Ruppert 1988. *Transformation and Weighting in Regression.* New York: Chapman and Hall.

Casella, G. and R. L. Berger 2002. *Statistical Inference* (Second ed.). Pacific Grove: Thomson Learning.

Castillo, E., A. S. Hadi, N. Balakrishnan, and J. M. Sarabia 2005. *Extreme Value and Related Models with Applications in Engineering and Science.* Hoboken: Wiley.

Chambers, J. M. 1998. *Programming with Data: A Guide to the S Language.* New York: Springer.

Chambers, J. M. 2008. *Software for Data Analysis: Programming with R.* Statistics and Computing. New York: Springer.

Chambers, J. M. and T. J. Hastie (Eds.) 1991. *Statistical Models in S.* Pacific Grove: Wadsworth/Brooks Cole.

Cheney, W. and D. Kincaid 2012. *Numerical Mathematics and Computing* (Seventh ed.). Boston: Brooks/Cole.

Chotikapanich, D. (Ed.) 2008. *Modeling Income Distributions and Lorenz Curves.* New York: Springer.

Christensen, R. 1997. *Log-linear Models and Logistic Regression* (Second ed.). New York: Springer-Verlag.

Christensen, R. 2011. *Plane Answers to Complex Questions: The Theory of Linear Models* (4th ed.). New York: Springer-Verlag.

Christensen, R. H. B. 2013. *Analysis of ordinal data with cumulative link models—estimation with the R-package ordinal*. R package version 2013.9–30.

Claeskens, G. and N. L. Hjort 2008. *Model Selection and Model Averaging*. Cambridge Series in Statistical and Probabilistic Mathematics. Cambridge: Cambridge University Press.

Clayton, D. and M. Hills 1993. *Statistical Models in Epidemiology*. Oxford: Oxford University Press.

Cleveland, W. S. 1979. Robust locally weighted regression and smoothing scatterplots. *Journal of the American Statistical Association* 74(368):829–836.

Cleveland, W. S. and S. J. Devlin 1988. Locally weighted regression: An approach to regression analysis by local fitting. *Journal of the American Statistical Association* 83(403):596–610.

Cleveland, W. S., E. Grosse, and W. M. Shyu 1991. Local regression models. See Chambers and Hastie (1991), pp. 309–376.

Cohen, Y. and J. Cohen 2008. *Statistics and Data with R: An Applied Approach Through Examples*. Chichester: John Wiley & Sons.

Coles, S. 2001. *An Introduction to Statistical Modeling of Extreme Values*. London: Springer-Verlag.

Consul, P. C. and F. Famoye 2006. *Lagrangian Probability Distributions*. Boston: Birkhäuser.

Cook, R. D. and S. Weisberg 1982. *Residuals and Influence in Regression*. Monographs on Statistics and Applied Probability. London: Chapman & Hall.

Cox, D. R. 2006. *Principles of Statistical Inference*. Cambridge: Cambridge University Press.

Cox, D. R. and D. V. Hinkley 1974. *Theoretical Statistics*. London: Chapman & Hall.

Cox, D. R. and N. Reid 1987. Parameter orthogonality and approximate conditional inference. *Journal of the Royal Statistical Society, Series B* 49(1):1–39. With discussion.

Crawley, M. J. 2005. *Statistics: An Introduction using R*. Chichester: John Wiley & Sons.

Crowder, M. and T. Sweeting 1989. Bayesian inference for a bivariate binomial distribution. *Biometrika* 76(3):599–603.

Dalgaard, P. 2008. *Introductory Statistics with R* (Second ed.). New York: Springer.

Davino, C., C. Furno, and D. Vistocco 2014. *Quantile Regression: Theory and Applications*. Chichester: Wiley.

Davison, A. C. 2003. *Statistical Models*. Cambridge: Cambridge University Press.

Davison, A. C. and E. J. Snell 1991. Residuals and diagnostics. See Hinkley et al. (1991), pp. 83–106.

de Boor, C. 2001. *A Practical Guide to Splines (Revised Edition)*. New York: Springer.

de Gruijter, D. N. M. and L. J. T. Van der Kamp 2008. *Statistical Test Theory for the Behavioral Sciences*. Boca Raton, FL, USA: Chapman & Hall/CRC.

de Haan, L. and A. Ferreira 2006. *Extreme Value Theory*. New York: Springer.

de Vries, A. and J. Meys 2012. *R for Dummies*. Chichester: Wiley.

De'ath, G. 1999. Principal curves: a new technique for indirect and direct gradient analysis. *Ecology* 80(7):2237–2253.

del Pino, G. 1989. The unifying role of iterative generalized least squares in statistical algorithms. *Statistical Science* 4(4):394–403.

Dempster, A. P., N. M. Laird, and D. B. Rubin 1977. Maximum likelihood from incomplete data via the *EM* algorithm. *Journal of the Royal Statistical Society, Series B* 39(1):1–38. With discussion.

Dempster, A. P., N. M. Laird, and D. B. Rubin 1980. Iteratively reweighted least squares for linear regression when errors are normal/independent distributed. In P. R. Krishnaiah (Ed.), *Multivariate Analysis–V: Proceedings of the Fifth International Symposium on Multivariate Analysis*, pp. 35–57. Amsterdam: North-Holland Publishing Company.

Dennis, J. E. and R. B. Schnabel 1996. *Numerical Methods for Unconstrained Optimization and Nonlinear Equations*. Philadelphia: Society for Industrial and Applied Mathematics.

Devroye, L. 1986. *Non-Uniform Random Variate Generation*. New York: Springer-Verlag.

Dobson, A. J. and A. Barnett 2008. *An Introduction to Generalized Linear Models* (Third ed.). Boca Raton: Chapman & Hall/CRC Press.

Dongarra, J. J., J. R. Bunch, C. B. Moler, and G. W. Stewart 1979. *LINPACK User's Guide*. Philadelphia: SIAM Publications.

Edwards, A. W. F. 1972. *Likelihood*. Cambridge: Cambridge University Press.

Efron, B. 1986. Double exponential families and their use in generalized linear regression. *Journal of the American Statistical Association* 81(395):709–721.

Efron, B. 1991. Regression percentiles using asymmetric squared error loss. *Statistica Sinica* 1(1):93–125.

Efron, B. 1992. Poisson overdispersion estimates based on the method of asymmetric maximum likelihood. *Journal of the American Statistical Association* 87(417):98–107.

Efron, B. and D. V. Hinkley 1978. Assessing the accuracy of the maximum likelihood estimator: observed versus expected Fisher information. *Biometrika* 65(3):457–487. With discussion.

Eilers, P. H. C. and B. D. Marx 1996. Flexible smoothing with *B*-splines and penalties. *Statistical Science* 11(2):89–121.

Elandt-Johnson, R. C. 1971. *Probability Models and Statistical Methods in Genetics*. New York: Wiley.

Embrechts, P., C. Klüppelberg, and T. Mikosch 1997. *Modelling Extremal Events for Insurance and Finance*. New York: Springer-Verlag.

Eubank, R. L. 1999. *Spline Smoothing and Nonparametric Regression* (Second ed.). New York: Marcel-Dekker.

Everitt, B. S. and D. J. Hand 1981. *Finite Mixture Distributions*. London: Chapman & Hall.

Fahrmeir, L., T. Kneib, S. Lang, and B. Marx 2011. *Regression: Models, Methods and Applications*. Berlin: Springer.

Fahrmeir, L. and G. Tutz 2001. *Multivariate Statistical Modelling Based on Generalized Linear Models* (Second ed.). New York: Springer-Verlag.

Fan, J. and I. Gijbels 1996. *Local Polynomial Modelling and Its Applications*. London: Chapman & Hall.

Fan, J. and J. Jiang 2005. Nonparametric inferences for additive models. *Journal of the American Statistical Association* 100(471):890–907.

Fan, J. and Q. Yao 2003. *Nonlinear Time Series: Nonparametric and Parametric Methods*. New York: Springer.

Faraway, J. J. 2006. *Extending the Linear Model with R: Generalized Linear, Mixed Effects and Nonparametric Regression Models.* Boca Raton: Chapman and Hall/CRC.

Faraway, J. J. 2015. *Linear Models with R* (Second ed.). Boca Raton: Chapman & Hall/CRC.

Fessler, J. A. 1991. Nonparametric fixed-interval smoothing with vector splines. *IEEE Transactions on Signal Processing* 39(4):852–859.

Finkenstadt, B. and H. Rootzén (Eds.) 2003. *Extreme Values in Finance, Telecommunications and the Environment.* Boca Raton: Chapman & Hall/CRC.

Firth, D. 1991. Generalized linear models. See Hinkley et al. (1991), pp. 55–82.

Firth, D. 1993. Bias reduction of maximum likelihood estimates. *Biometrika* 80(1):27–38.

Firth, D. 2003. Overcoming the reference category problem in the presentation of statistical models. *Sociological Methodology* 33(1):1–18.

Firth, D. and R. X. de Menezes 2004. Quasi-variances. *Biometrika* 91(1):65–80.

Fishman, G. S. 1996. *Monte Carlo: Concepts, Algorithms, and Applications.* New York: Springer-Verlag.

Fitzenberger, B., R. Koenker, and J. A. F. Machado (Eds.) 2002. *Economic Applications of Quantile Regression.* Berlin: Springer-Verlag.

Forbes, C., M. Evans, N. Hastings, and B. Peacock 2011. *Statistical Distributions* (fouth ed.). Hoboken: John Wiley & Sons.

Fox, J. and S. Weisberg 2011. *An R Companion to Applied Regression* (Second ed.). Thousand Oaks: Sage Publications.

Freedman, D. A. 2007. How can the score test be inconsistent? *American Statistician* 61(4):291–295.

Freedman, D. A. and J. S. Sekhon 2010. Endogeneity in probit response models. *Political Analysis* 18(2):138–150.

Freund, J. E. 1961. A bivariate extension of the exponential distribution. *Journal of the American Statistical Association* 56(296):971–977.

Friedman, J. H. and W. Stuetzle 1981. Projection pursuit regression. *Journal of the American Statistical Association* 76(376):817–823.

Frühwirth-Schnatter, S. 2006. *Finite Mixture and Markov Switching Models.* New York: Springer.

Gabriel, K. R. and S. Zamir 1979. Lower rank approximation of matrices by least squares with any choice of weights. *Technometrics* 21(4):489–498.

Gauch, Hugh G., J., G. B. Chase, and R. H. Whittaker 1974. Ordinations of vegetation samples by Gaussian species distributions. *Ecology* 55(6):1382–1390.

Gentle, J. E., W. K. Härdle, and Y. Mori 2012. *Handbook of Computational Statistics: Concepts and Methods* (Second ed.). Berlin: Springer.

Gentleman, R. 2009. *R Programming for Bioinformatics.* Boca Raton: Chapman & Hall/CRC.

Geraci, M. and M. Bottai 2007. Quantile regression for longitudinal data using the asymmetric Laplace distribution. *Biostatistics* 8(1):140–154.

Gil, A., J. Segura, and N. M. Temme 2007. *Numerical Methods for Special Functions.* Philadelphia: Society for Industrial and Applied Mathematics.

Gill, J. and G. King 2004. What to do when your Hessian is not invertible: Alternatives to model respecification in nonlinear estimation. *Sociological Methods & Research* 33(1):54–87.

Gilleland, E., M. Ribatet, and A. G. Stephenson 2013. A software review for extreme value analysis. *Extremes* 16(1):103–119.

Goldberger, A. S. 1964. *Econometric Theory.* New York: Wiley.

Golub, G. H. and C. F. Van Loan 2013. *Matrix Computations* (Fourth ed.). Baltimore: Johns Hopkins University Press.

Gomes, M.I., and A. Guillou. 2015. Extreme value theory and statistics of univariate extremes: a review. *International Statistical Review* 83(2):263–292.

Goodman, L. A. 1981. Association models and canonical correlation in the analysis of cross-classifications having ordered categories. *Journal of the American Statistical Association* 76(374):320–334.

Gower, J. C. 1987. Introduction to ordination techniques. In P. Legendre and L. Legendre (Eds.), *Developments in Numerical Ecology*, pp. 3–64. Berlin: Springer-Verlag.

Green, P. J. 1984. Iteratively reweighted least squares for maximum likelihood estimation, and some robust and resistant alternatives. *Journal of the Royal Statistical Society, Series B* 46(2):149–192. With discussion.

Green, P. J. and B. W. Silverman 1994. *Nonparametric Regression and Generalized Linear Models: A Roughness Penalty Approach.* London: Chapman & Hall.

Greene, W. H. 2012. *Econometric Analysis* (Seventh ed.). Upper Saddle River: Prentice Hall.

Greene, W. H. and D. A. Hensher 2010. *Modeling Ordered Choices: A Primer.* Cambridge: Cambridge University Press.

Gu, C. 2013. *Smoothing Spline ANOVA Models* (Second ed.). New York, USA: Springer.

Gumbel, E. J. 1958. *Statistics of Extremes.* New York, USA: Columbia University Press.

Gupta, A. K. and S. Nadarajah (Eds.) 2004. *Handbook of Beta Distribution and Its Applications.* New York, USA: Marcel Dekker.

Hao, L. and D. Q. Naiman 2007. *Quantile Regression.* Thousand Oaks, CA, USA: Sage Publications.

Härdle, W. 1987. *Smoothing Techniques With Implementation in S.* New York, USA: Springer-Verlag.

Härdle, W. 1990. *Applied Nonparametric Regression.* Cambridge: Cambridge University Press.

Härdle, W., H. Liang, and J. Gao 2000. *Partially Linear Models.* New York, USA: Springer.

Härdle, W., M. Müller, S. Sperlich, and A. Werwatz 2004. *Nonparametric and Semiparametric Models.* Berlin: Springer.

Harezlak, J., D. Ruppert, and M.P. Wand. 2016. *Semiparametric regression in R.* New York: Springer.

Harper, W. V., T. G. Eschenbach, and T. R. James 2011. Concerns about maximum likelihood estimation for the three-parameter Weibull distribution: Case study of statistical software. *American Statistician* 65(1):44–54.

Harrell, F. E. 2001. *Regression Modeling Strategies: With Applications to Linear Models, Logistic Regression, and Survival Analysis.* New York, USA: Springer.

Harville, D. A. 1997. *Matrix Algebra From a Statistician's Perspective.* New York, USA: Springer-Verlag.

Hastie, T. 1996. Pseudosplines. *Journal of the Royal Statistical Society, Series B* 58(2):379–396.

Hastie, T. and W. Stuetzle 1989. Principal curves. *Journal of the American Statistical Association* 84(406):502–516.

Hastie, T. and R. Tibshirani 1993. Varying-coefficient models. *Journal of the Royal Statistical Society, Series B* 55(4):757–796.

Hastie, T. J. and D. Pregibon 1991. Generalized linear models. See Chambers and Hastie (1991), pp. 195–247.

Hastie, T. J. and R. J. Tibshirani 1990. *Generalized Additive Models.* London: Chapman & Hall.

Hastie, T. J., R. J. Tibshirani, and J. H. Friedman 2009. *The Elements of Statistical Learning: Data Mining, Inference and Prediction* (Second ed.). New York, USA: Springer-Verlag.

Hauck, J. W. W. and A. Donner 1977. Wald's test as applied to hypotheses in logit analysis. *Journal of the American Statistical Association* 72(360):851–853.

He, X. 1997. Quantile curves without crossing. *American Statistician* 51(2):186–192.

Heinze, G. and M. Schemper 2002. A solution to the problem of separation in logistic regression. *Statistics in Medicine* 21(16):2409–2419.

Hensher, D. A., J. M. Rose, and W. H. Greene 2014. *Applied Choice Analysis* (Second ed.). Cambridge: Cambridge University Press.

Hilbe, J. M. 2009. *Logistic Regression Models.* Boca Raton, FL, USA: Chapman & Hall/CRC.

Hilbe, J. M. 2011. *Negative Binomial Regression* (Second ed.). Cambridge, UK; New York, USA: Cambridge University Press.

Hinkley, D. V., N. Reid, and E. J. Snell (Eds.) 1991. *Statistical Theory and Modelling. In Honour of Sir David Cox, FRS*, London. Chapman & Hall.

Hogben, L. (Ed.) 2014. *Handbook of Linear Algebra* (Second ed.). Boca Raton, FL, USA: Chapman & Hall/CRC.

Hörmann, W., J. Leydold, and G. Derflinger 2004. *Automatic Nonuniform Random Variate Generation.* Berlin: Springer.

Horvitz, D. G. and D. J. Thompson 1952. A generalization of sampling without replacement from a finite universe. *Journal of the American Statistical Association* 47(260):663–685.

Huber, P. J. 2011. *Data Analysis: What Can Be Learned From the Past 50 Years.* Hoboken, NJ, USA: Wiley.

Huber, P. J. and E. M. Ronchetti 2009. *Robust Statistics* (second ed.). New York, USA: Wiley.

Huggins, R. and W.-H. Hwang 2011. A review of the use of conditional likelihood in capture–recapture experiments. *International Statistical Review* 79(3):385–400.

Huggins, R. M. 1989. On the statistical analysis of capture experiments. *Biometrika* 76(1):133–140.

Huggins, R. M. 1991. Some practical aspects of a conditional likelihood approach to capture experiments. *Biometrics* 47(2):725–732.

Hui, F. K. C., S. Taskinen, S. Pledger, S. D. Foster, and D. I. Warton 2015. Model-based approaches to unconstrained ordination. *Methods in Ecology and Evolution* 6(4):399–411.

Hurvich, C. M. and C.-L. Tsai 1989. Regression and time series model selection in small samples. *Biometrika* 76(2):297–307.

Hutchinson, M. F. and F. R. de Hoog 1985. Smoothing noisy data with spline functions. *Numerische Mathematik* 47(1):99–106.

Hwang, W.-H. and R. Huggins 2011. A semiparametric model for a functional behavioural response to capture in capture–recapture experiments. *Australian & New Zealand Journal of Statistics* 53(4):403–421.

Ichimura, H. 1993. Semiparametric least squares (SLS) and weighted SLS estimation of single-index models. *J. Econometrics* 58(1–2):71–120.

Ihaka, R. and R. Gentleman 1996. R: A language for data analysis and graphics. *Journal of Computational and Graphical Statistics* 5(3):299–314.

Imai, K., G. King, and O. Lau 2008. Toward a common framework for statistical analysis and development. *Journal of Computational and Graphical Statistics* 17(4):892–913.

Izenman, A. J. 1975. Reduced-rank regression for the multivariate linear model. *Journal of Multivariate Analysis* 5(2):248–264.

Izenman, A. J. 2008. *Modern Multivariate Statistical Techniques: Regression, Classification, and Manifold Learning.* New York, USA: Springer.

James, G., D. Witten, T. Hastie, and R. Tibshirani 2013. *An Introduction to Statistical Learning with Applications in R.* New York, USA: Springer.

Joe, H. 2014. *Dependence Modeling with Copulas.* Boca Raton, FL, USA: Chapman & Hall/CRC.

Johnson, N. L., A. W. Kemp, and S. Kotz 2005. *Univariate Discrete Distributions* (Third ed.). Hoboken, NJ, USA: John Wiley & Sons.

Johnson, N. L., S. Kotz, and N. Balakrishnan 1994. *Continuous Univariate Distributions* (Second ed.), Volume 1. New York, USA: Wiley.

Johnson, N. L., S. Kotz, and N. Balakrishnan 1995. *Continuous Univariate Distributions* (Second ed.), Volume 2. New York, USA: Wiley.

Johnson, N. L., S. Kotz, and N. Balakrishnan 1997. *Discrete Multivariate Distributions.* New York, USA: John Wiley & Sons.

Jones, M. C. 1994. Expectiles and M-quantiles are quantiles. *Statistics & Probability Letters* 20(2):149–153.

Jones, M. C. 2002. Student's simplest distribution. *The Statistician* 51(1):41–49.

Jones, M. C. 2009. Kumaraswamy's distribution: A beta-type distribution with some tractability advantages. *Statistical Methodology* 6(1):70–81.

Jones, O., R. Maillardet, and A. Robinson 2014. *Introduction to Scientific Programming and Simulation Using R* (Second ed.). Boca Raton, FL, USA: Chapman and Hall/CRC.

Jongman, R. H. G., C. J. F. ter Braak, and O. F. R. van Tongeren (Eds.) 1995. *Data Analysis in Community and Landscape Ecology.* Cambridge: Cambridge University Press.

Jørgensen, B. 1984. The delta algorithm and GLIM. *International Statistical Review* 52(3):283–300.

Jørgensen, B. 1997. *The Theory of Dispersion Models.* London: Chapman & Hall.

Jorgensen, M. 2001. Iteratively reweighted least squares. In A. H. El-Shaarawi and W. W. Piegorsch (Eds.), *Encyclopedia of Environmetrics*, Volume 2, pp. 1084–1088. Chichester, New York, USA: Wiley.

Kaas, R., M. Goovaerts, J. Dhaene, and M. Denuit 2008. *Modern Actuarial Risk Theory Using R* (Second ed.). Berlin: Springer.

Kateri, M. 2014. *Contingency Table Analysis. Methods and Implementation Using R.* New York, USA: Birkhäuser/Springer.

Kennedy, William J., J. and J. E. Gentle 1980. *Statistical Computing.* New York, USA: Marcel Dekker.

Keogh, R. H. and D. R. Cox 2014. *Case-Control Studies.* New York, USA: Cambridge University Press.

Kleiber, C. and S. Kotz 2003. *Statistical Size Distributions in Economics and Actuarial Sciences.* Hoboken, NJ, USA: Wiley-Interscience.

Kleiber, C. and A. Zeileis 2008. *Applied Econometrics with R*. New York, USA: Springer.

Klugman, S. A., H. H. Panjer, and G. E. Willmot 2012. *Loss Models: From Data to Decisions* (4th ed.). Hoboken, NJ, USA: Wiley.

Klugman, S. A., H. H. Panjer, and G. E. Willmot 2013. *Loss Models: Further Topics*. Hoboken, NJ, USA: Wiley.

Knight, K. 2000. *Mathematical Statistics*. Boca Raton, FL, USA: Chapman & Hall/CRC.

Kocherlakota, S. and K. Kocherlakota 1992. *Bivariate Discrete Distributions*. New York, USA: Marcel Dekker.

Koenker, R. 1992. When are expectiles percentiles? (problem). *Econometric Theory* 8(3):423–424.

Koenker, R. 2005. *Quantile Regression*. Cambridge: Cambridge University Press.

Koenker, R. 2013. Discussion: Living beyond our means. *Statistical Modelling* 13(4):323–333.

Koenker, R. and G. Bassett 1978. Regression quantiles. *Econometrica* 46(1):33–50.

Kohn, R. and C. F. Ansley 1987. A new algorithm for spline smoothing based on smoothing a stochastic process. *SIAM Journal on Scientific and Statistical Computing* 8(1):33–48.

Konishi, S. and G. Kitagawa 2008. *Information Criteria and Statistical Modeling*. Springer Series in Statistics. New York, USA: Springer.

Kooijman, S. A. L. M. 1977. Species abundance with optimum relations to environmental factors. *Annals of Systems Research* 6:123–138.

Kosmidis, I. 2014a. Bias in parametric estimation: reduction and useful side-effects. *WIREs Computational Statistics* 6:185–196.

Kosmidis, I. 2014b. Improved estimation in cumulative link models. *Journal of the Royal Statistical Society, Series B* 76(1):169–196.

Kosmidis, I. and D. Firth 2009. Bias reduction in exponential family nonlinear models. *Biometrika* 96(4):793–804.

Kosmidis, I. and D. Firth 2010. A generic algorithm for reducing bias in parametric estimation. *Electronic Journal of Statistics* 4:1097–1112.

Kotz, S., T. J. Kozubowski, and K. Podgórski 2001. *The Laplace Distribution and Generalizations: a Revisit with Applications to Communications, Economics, Engineering, and Finance*. Boston, MA, USA: Birkhäuser.

Kotz, S. and S. Nadarajah 2000. *Extreme Value Distributions: Theory and Applications*. London: Imperial College Press.

Kotz, S. and J. R. van Dorp 2004. *Beyond Beta: Other Continuous Families of Distributions with Bounded Support and Applications*. Singapore: World Scientific.

Kozubowski, T. J. and S. Nadarajah 2010. Multitude of Laplace distributions. *Statistical Papers* 51(1):127–148.

Lange, K. 2002. *Mathematical and Statistical Methods for Genetic Analysis* (Second ed.). New York, USA: Springer-Verlag.

Lange, K. 2010. *Numerical Analysis for Statisticians* (Second ed.). New York, USA: Springer.

Lange, K. 2013. *Optimization* (Second ed.). New York, USA: Springer.

Lawless, J. F. 1987. Negative binomial and mixed Poisson regression. *The Canadian Journal of Statistics* 15(3):209–225.

Lawless, J. F. 2003. *Statistical Models and Methods for Lifetime Data* (Second ed.). Hoboken, NJ, USA: John Wiley & Sons.

Leadbetter, M. R., G. Lindgren, and H. Rootzén 1983. *Extremes and Related Properties of Random Sequences and Processes*. New York, USA: Springer-Verlag.

Leemis, L. M. and J. T. McQueston 2008. Univariate distribution relationships. *American Statistician* 62(1):45–53.

Lehmann, E. L. and G. Casella 1998. *Theory of Point Estimation* (Second ed.). New York, USA: Springer.

Lehmann, E. L. and J. P. Romano 2005. *Testing Statistical Hypotheses* (3rd ed.). New York, USA: Springer.

Lesaffre, E. and A. Albert 1989. Partial separation in logistic discrimination. *Journal of the Royal Statistical Society, Series B* 51(1):109–116.

Libby, D. L. and M. R. Novick 1982. Multivariate generalized beta distributions with applications to utility assessment. *Journal of Educational and Statistics* 7(4):271–294.

Lindsay, B. G. 1995. *Mixture Models: Theory, Geometry and Applications*, Volume 5. Hayward CA, USA: NSF-CBMS Regional Conference Series in Probability and Statistics, IMS.

Lindsey, J. K. 1996. *Parametric Statistical Inference*. Oxford: Clarendon Press.

Lindsey, J. K. 1997. *Applying Generalized Linear Models*. New York, USA: Springer-Verlag.

Liu, H. and K. S. Chan 2010. Introducing COZIGAM: An R package for unconstrained and constrained zero-inflated generalized additive model analysis. *Journal of Statistical Software* 35(11):1–26.

Liu, I. and A. Agresti 2005. The analysis of ordered categorical data: An overview and a survey of recent developments. *Test* 14(1):1–73.

Lloyd, C. J. 1999. *Statistical Analysis of Categorical Data*. New York, USA: Wiley.

Loader, C. 1999. *Local Regression and Likelihood*. New York, USA: Springer.

Lopatatzidis, A. and P. J. Green 1998. Semiparametric quantile regression using the gamma distribution. *Unpublished manuscript*.

Maddala, G. S. 1983. *Limited Dependent and Qualitative Variables in Econometrics*. Cambridge: Cambridge University Press.

Mai, J.-F. and M. Scherer 2012. *Simulating Copulas: Stochastic Models, Sampling Algorithms, and Applications*. London: Imperial College Press.

Maindonald, J. H. and W. J. Braun 2010. *Data Analysis and Graphics Using R: An Example-Based Approach* (Third ed.). Cambridge: Cambridge University Press.

Marra, G. and R. Radice 2010. Penalised regression splines: theory and application to medical research. *Statistical Methods in Medical Research* 19(2):107–125.

Marshall, A. W. and I. Olkin 2007. *Life Distributions: Structure of Nonparametric, Semiparametric, and Parametric Families*. New York, USA: Springer.

McCrea, R. S. and B. J. T. Morgan 2015. *Analysis of Capture–Recapture Data*. Boca Raton, FL, USA: Chapman & Hall/CRC.

McCullagh, P. 1980. Regression models for ordinal data. *Journal of the Royal Statistical Society, Series B* 42(2):109–142. With discussion.

McCullagh, P. 1989. Some statistical properties of a family of continuous univariate distributions. *Journal of the American Statistical Association* 84(405):125–129.

McCullagh, P. and J. A. Nelder 1989. *Generalized Linear Models* (Second ed.). London: Chapman & Hall.

McFadden, D. 1974. Conditional logit analysis of qualitative choice behavior. In P. Zarembka (Ed.), *Conditional Logit Analysis of Qualitative Choice Behavior*, pp. 105–142. New York, USA: Academic Press.

McLachlan, G. J. and D. Peel 2000. *Finite Mixture Models*. New York, USA: Wiley.

Mikosch, T. 2006. Copulas: tales and facts (with rejoinder). *Extremes* 9(1): 3–20,55–62.

Miller, A. 2002. *Subset Selection in Regression* (Second ed.). Boca Raton, FL, USA: Chapman & Hall/CRC.

Miller, J. J. and E. J. Wegman 1987. Vector function estimation using splines. *Journal of Statistical Planning and Inference* 17:173–180.

Morris, C. N. 1982. Natural exponential families with quadratic variance functions. *The Annals of Statistics* 10(1):65–80.

Mosteller, F. and J. W. Tukey 1977. *Data Analysis and Regression*. Reading, MA, USA: Addison-Wesley.

Murthy, D. N. P., M. Xie, and R. Jiang 2004. *Weibull Models*. Hoboken, NJ, USA: Wiley.

Myers, R. H., D. C. Montgomery, G. G. Vining, and T. J. Robinson 2010. *Generalized Linear Models With Applications in Engineering and the Sciences* (Second ed.). Hoboken, NJ, USA: Wiley.

Nadarajah, S. and S. A. A. Bakar 2013. A new R package for actuarial survival models. *Computational Statistics* 28(5):2139–2160.

Nelder, J. A. and R. W. M. Wedderburn 1972. Generalized linear models. *Journal of the Royal Statistical Society, Series A* 135(3):370–384.

Nelsen, R. B. 2006. *An Introduction to Copulas* (Second ed.). New York, USA: Springer.

Newey, W. K. and J. L. Powell 1987. Asymmetric least squares estimation and testing. *Econometrica* 55(4):819–847.

Neyman, J. and E. L. Scott 1948. Consistent estimates based on partially consistent observations. *Econometrica* 16(1):1–32.

Nocedal, J. and S. J. Wright 2006. *Numerical Optimization* (Second ed.). New York, USA: Springer.

Nosedal-Sanchez, A., C. B. Storlie, T. C. M. Lee, and R. Christensen 2012. Reproducing kernel Hilbert spaces for penalized regression: A tutorial. *American Statistician* 66(1):50–60.

Novak, S. Y. 2012. *Extreme Value Methods with Applications to Finance*. Boca Raton, FL, USA: CRC Press.

Olver, F. W. J., D. W. Lozier, R. F. Boisvert, and C. W. Clark (Eds.) 2010. *NIST Handbook of Mathematical Functions*. New York, USA: National Institute of Standards and Technology, and Cambridge University Press.

Osborne, M. R. 1992. Fisher's method of scoring. *International Statistical Review* 60(1):99–117.

Osborne, M. R. 2006. Least squares methods in maximum likelihood problems. *Optimization Methods and Software* 21(6):943–959.

Otis, D. L., K. P. Burnham, G. C. White, and D. R. Anderson 1978. Statistical inference from capture data on closed animal populations. *Wildlife Monographs* 62:3–135.

Owen, A. B. 2001. *Empirical Likelihood*. Boca Raton, FL, USA: Chapman & Hall/CRC.

Page, L. A., S. Hajat, and R. S. Kovats 2007. Relationship between daily suicide counts and temperature in England and Wales. *British Journal of Psychiatry* 191(2):106–112.

Pal, N., C. Jin, and W. K. Lim 2006. *Handbook of Exponential and Related Distributions for Engineers and Scientists*. Boca Raton, FL, USA: Chapman & Hall/CRC.

Palmer, M. 1993. Putting things in even better order: the advantages of canonical correspondence analysis. *Ecology* 74(8):2215–2230.

Palmgren, J. 1989. Regression models for bivariate binary responses. Technical Report 101, Biostatistics Dept, University of Washington, Seattle, USA.

Park, B. U., E. Mammen, Y. K. Lee, and E. R. Lee 2015. Varying coefficient regression models: a review and new developments. *International Statistical Review* 83(1):36–64.

Pickands, J. 1975. Statistical inference using extreme order statistics. *The Annals of Statistics* 3(1):119–131.

Plackett, R. L. 1965. A class of bivariate distributions. *Journal of the American Statistical Association* 60(310):516–522.

Poiraud-Casanova, S. and C. Thomas-Agnan 2000. About monotone regression quantiles. *Statistics & Probability Letters* 48(1):101–104.

Powers, D. A. and Y. Xie 2008. *Statistical Methods for Categorical Data Analysis* (Second ed.). Bingley, UK: Emerald.

Pratt, J. W. 1981. Concavity of the log likelihood. *Journal of the American Statistical Association* 76(373):103–106. Correction p.954, Vol 77.

Prentice, R. L. 1974. A log gamma model and its maximum likelihood estimation. *Biometrika* 61(3):539–544.

Prentice, R. L. 1986. Binary regression using an extended beta-binomial distribution, with discussion of correlation induced by covariate measurement errors. *Journal of the American Statistical Association* 81(394):321–327.

Prescott, P. and A. T. Walden 1980. Maximum likelihood estimation of the parameters of the generalized extreme-value distribution. *Biometrika* 67(3):723–724.

Randall, J. H. 1989. The analysis of sensory data by generalized linear model. *Biometrics Journal* 31(7):781–793.

Rao, C. R. 1948. Large sample tests of statistical hypotheses concerning several parameters with applications to problems of estimation. *Mathematical Proceedings of the Cambridge Philosophical Society* 44(1):50–57.

Rao, C. R. 1973. *Linear Statistical Inference and its Applications* (Second ed.). New York, USA: Wiley.

Rasch, G. 1961. On general laws and the meaning of measurement in psychology. *Proceedings of the Fourth Berkeley Symposium on Mathematical Statistics and Probability* 4:321–333.

Reinsch, C. H. 1967. Smoothing by spline functions. *Numerische Mathematik* 10(3):177–183.

Reinsel, G. C. and R. P. Velu 1998. *Multivariate Reduced-Rank Regression: Theory and Applications*. New York, USA: Springer-Verlag.

Reinsel, G. C. and R. P. Velu 2006. Partially reduced-rank multivariate regression models. *Statistica Sinica* 16(3):899–917.

Reiss, R.-D. and M. Thomas 2007. *Statistical Analysis of Extreme Values: with Applications to Insurance, Finance, Hydrology and Other Fields* (Third ed.). Basel, Switzerland: Birkhäuser.

Rencher, A. C. and G. B. Schaalje 2008. *Linear Models in Statistics* (second ed.). New York, USA: John Wiley & Sons.

Richards, F. S. G. 1961. A method of maximum-likelihood estimation. *Journal of the Royal Statistical Society, Series B* 23(2):469–475.

Richards, S. J. 2012. A handbook of parametric survival models for actuarial use. *Scandinavian Actuarial Journal* 2012(4):233–257.

Ridout, M. S. 1990. Non-convergence of Fisher's method of scoring—a simple example. *GLIM Newsletter* 20(6).

Rinne, H. 2009. *The Weibull Distribution.* Boca Raton, FL, USA: CRC Press.

Ripley, B. D. 1996. *Pattern Recognition and Neural Networks.* Cambridge: Cambridge University Press.

Ripley, B. D. 2004. Selecting amongst large classes of models. See Adams et al. (2004), pp. 155–170.

Rose, C. and M. D. Smith 2002. *Mathematical Statistics with* Mathematica. New York, USA: Springer.

Rose, C. and M. D. Smith 2013. *Mathematical Statistics with* Mathematica. eBook.

Rubin, D. B. 2006. Iteratively reweighted least squares. In *Encyclopedia of Statistical Sciences,* Volume 6. Wiley.

Ruppert, D., M. P. Wand, and R. J. Carroll 2003. *Semiparametric Regression.* Cambridge: Cambridge University Press.

Ruppert, D., M. P. Wand, and R. J. Carroll 2009. Semiparametric regression during 2003–2007. *Electronic Journal of Statistics* 3(1):1193–1256.

Sakamoto, Y., M. Ishiguro, and G. Kitagawa 1986. *Akaike Information Criterion Statistics.* Dordrecht, Netherlands: D. Reidel Publishing Company.

Schenker, N. and J. F. Gentleman 2001. On judging the significance of differences by examining the overlap between confidence intervals. *American Statistician* 55(3):182–186.

Schepsmeier, U. and J. Stöber 2014. Derivatives and Fisher information of bivariate copulas. *Statistical Papers* 55(2):525–542.

Schimek, M. G. (Ed.) 2000. *Smoothing and Regression: Approaches, Computation, and Application.* New York, USA: Wiley.

Schnabel, S. K. and P. H. C. Eilers 2009. Optimal expectile smoothing. *Computational Statistics & Data Analysis* 53(12):4168–4177.

Schumaker, L. L. 2007. *Spline Functions: Basic Theory* (Third ed.). Cambridge: Cambridge University Press.

Schwarz, G. 1978. Estimating the dimension of a model. *The Annals of Statistics* 6(2):461–464.

Seber, G. A. F. 2008. *A Matrix Handbook for Statisticians.* Hoboken, NJ, USA: Wiley.

Seber, G. A. F. and A. J. Lee 2003. *Linear Regression Analysis* (Second ed.). New York, USA: Wiley.

Seber, G. A. F. and C. J. Wild 1989. *Nonlinear Regression.* New York, USA: Wiley.

Self, S. G. and K.-Y. Liang 1987. Asymptotic properties of maximum likelihood estimators and likelihood ratio tests under nonstandard conditions. *Journal of the American Statistical Association* 82(398):605–610.

Senn, S. 2004. John Nelder: From general balance to generalised models (both linear and hierarchical). See Adams et al. (2004), pp. 1–12.

Severini, T. A. 2000. *Likelihood Methods in Statistics.* New York, USA: Oxford University Press.

Shao, J. 2003. *Mathematical Statistics* (Second ed.). New York, USA: Springer.

Shao, J. 2005. *Mathematical Statistics: Exercises and Solutions.* New York, USA: Springer.

Silverman, B. W. 1984. Spline smoothing: The equivalent variable kernel method. *The Annals of Statistics* 12(3):898–916.

Silverman, B. W. 1985. Some aspects of the spline smoothing approach to non-parametric regression curve fitting. *Journal of the Royal Statistical Society, Series B* 47(1):1–21. With discussion.

Silvey, S. D. 1975. *Statistical Inference*. London: Chapman & Hall.

Simonoff, J. S. 2003. *Analyzing Categorical Data*. New York, USA: Springer-Verlag.

Sklar, A. 1959. Fonctions de répartition à n dimensions et leurs marges. *Publications de l'Institut de Statistique de L'Université de Paris* 8:229–231.

Small, C. G. and D. L. McLeish 1994. *Hilbert Space Methods in Probability and Statistical Inference*. New York, USA: Wiley.

Smith, M. and R. Kohn 2000. Nonparametric seemingly unrelated regression. *Journal of Econometrics* 98(2):257–281.

Smith, R. L. 1985. Maximum likelihood estimation in a class of nonregular cases. *Biometrika* 72(1):67–90.

Smith, R. L. 1986. Extreme value theory based on the r largest annual events. *Journal of Hydrology* 86(1–2):27–43.

Smith, R. L. 2003. Statistics of extremes, with applications in environment, insurance and finance. See Finkenstadt and Rootzén (2003), pp. 1–78.

Smithson, M. and E. C. Merkle 2013. *Generalized Linear Models for Categorical and Continuous Limited Dependent Variables*. London: Chapman & Hall/CRC.

Smyth, G. K. 1989. Generalized linear models with varying dispersion. *Journal of the Royal Statistical Society, Series B* 51(1):47–60.

Smyth, G. K. 1996. Partitioned algorithms for maximum likelihood and other nonlinear estimation. *Statistics and Computing* 6(3):201–216.

Smyth, G. K., A. F. Huele, and A. P. Verbyla 2001. Exact and approximate REML for heteroscedastic regression. *Statistical Modelling* 1(3):161–175.

Spector, P. 2008. *Data Manipulation with R*. New York, USA: Springer Verlag.

Srivastava, V. K. and T. D. Dwivedi 1979. Estimation of seemingly unrelated regression equations: A brief survey. *Journal of Econometrics* 10(1):15–32.

Srivastava, V. K. and D. E. A. Giles 1987. *Seemingly Unrelated Regression Equations Models: Estimation and Inference*. New York, USA: Marcel Dekker.

Stacy, E. W. 1962. A generalization of the gamma distribution. *Annals of Mathematical Statistics* 33(3):1187–1192.

Takane, Y., H. Yanai, and S. Mayekawa 1991. Relationships among several methods of linearly constrained correspondence analysis. *Psychometrika* 56(4):667–684.

Tawn, J. A. 1988. An extreme-value theory model for dependent observations. *Journal of Hydrology* 101(1–4):227–250.

Taylor, J. W. 2008. Estimating value at risk and expected shortfall using expectiles. *Journal of Financial Econometrics* 6(2):231–252.

Taylor, L. R. 1961. Aggregation, variance and the mean. *Nature* 189(4766):732–735.

ter Braak, C. J. F. 1986. Canonical correspondence analysis: A new eigenvector technique for multivariate direct gradient analysis. *Ecology* 67(5):1167–1179.

ter Braak, C. J. F. 1995. Calibration. See Jongman et al. (1995), pp. 78–90.

ter Braak, C. J. F. and I. C. Prentice 1988. A theory of gradient analysis. In *Advances in Ecological Research*, Volume 18, pp. 271–317. London: Academic Press.

ter Braak, C. J. F. and P. F. M. Verdonschot 1995. Canonical correspondence analysis and related multivariate methods in aquatic ecology. *Aquatic Sciences* 57(3):255–289.

ter Braak, C. J. F., and P. Šmilauer 2015. Topics in constrained and unconstrained ordination. *Plant Ecology* 216(5):683–696.

Thompson, R. and R. J. Baker 1981. Composite link functions in generalized linear models. *Journal of the Royal Statistical Society, Series C* 30(2):125–131.

Tibshirani, R. 1996. Regression shrinkage and selection via the lasso. *Journal of the Royal Statistical Society, Series B* 58(1):267–288.

Titterington, D. M., A. F. M. Smith, and U. E. Makov 1985. *Statistical Analysis of Finite Mixture Distributions.* New York, USA: Wiley.

Tobin, J. 1958. Estimation of relationships for limited dependent variables. *Econometrica* 26(1):24–36.

Trivedi, P. K. and D. M. Zimmer 2005. Copula modeling: An introduction for practitioners. *Foundations and Trends in Econometrics* 1(1):1–111.

Tutz, G. 2012. *Regression for Categorical Data.* Cambridge: Cambridge University Press.

van den Boogaart, K. G. and R. Tolosana-Delgado 2013. *Analyzing Compositional Data with R.* Berlin: Springer.

Venables, W. N. and B. D. Ripley 2002. *Modern Applied Statistics With S* (4th ed.). New York, USA: Springer-Verlag.

von Eye, A. and E.-E. Mun 2013. *Log-linear Modeling: Concepts, Interpretation, and Application.* Hoboken, NJ, USA: Wiley.

Vuong, Q. H. 1989. Likelihood ratio tests for model selection and nonnested hypotheses. *Econometrica* 57(2):307–333.

Wahba, G. 1982. Vector splines on the sphere, with application to the estimation of vorticity and divergence from discrete, noisy data. In W. S. K. Zeller (Ed.), *Multivarate Approximation Theory*, Volume 2, pp. 407–429. Birkhäuser: Verlag.

Wahba, G. 1990. *Spline models for observational data*, Volume 59 of *CBMS-NSF Regional Conference Series in Applied Mathematics.* Philadelphia, PA, USA: Society for Industrial and Applied Mathematics (SIAM).

Wand, M. P. and M. C. Jones 1995. *Kernel Smoothing.* London: Chapman & Hall.

Wand, M. P. and J. T. Ormerod 2008. On semiparametric regression with O'Sullivan penalized splines. *Australian & New Zealand Journal of Statistics* 50(2):179–198.

Wang, Y. 2011. *Smoothing Splines: Methods and Applications.* Boca Raton, FL, USA: Chapman & Hall/CRC.

Webb, M. H., S. Wotherspoon, D. Stojanovic, R. Heinsohn, R. Cunningham, P. Bell, and A. Terauds 2014. Location matters: Using spatially explicit occupancy models to predict the distribution of the highly mobile, endangered swift parrot. *Biological Conservation* 176:99–108.

Wecker, W. E. and C. F. Ansley 1983. The signal extraction approach to nonlinear regression and spline smoothing. *Journal of the American Statistical Association* 78(381):81–89.

Wedderburn, R. W. M. 1974. Quasi-likelihood functions, generalized linear models, and the Gauss-Newton method. *Biometrika* 61(3):439–447.

Wegman, E. J. 1981. Vector splines and the estimation of filter functions. *Technometrics* 23(1):83–89.

Weihs, C., O. Mersmann, and U. Ligges 2014. *Foundations of Statistical Algorithms: With References to R Packages.* Boca Raton, FL, USA: CRC Press.

Weir, B. S. 1996. *Genetic Data Analysis II.* Sunderland, MA, USA: Sinauer.

Welsh, A. H. 1996. Robust estimation of smooth regression and spread functions and their derivatives. *Statistica Sinica* 6:347–366.

Welsh, A. H., R. B. Cunningham, C. F. Donnelly, and D. B. Lindenmayer 1996. Modelling the abundances of rare species: statistical models for counts with extra zeros. *Ecological Modelling* 88(1–3):297–308.

Welsh, A. H., D. B. Lindenmayer, and C. F. Donnelly 2013. Fitting and interpreting occupancy models. *PLOS One* 8(1):1–21.

Welsh, A. H. and T. W. Yee 2006. Local regression for vector responses. *Journal of Statistical Planning and Inference* 136(9):3007–3031.

Wickham, H. 2015. *Advanced R*. Boca Raton, FL, USA: Chapman & Hall/CRC.

Wild, C. J. and T. W. Yee 1996. Additive extensions to generalized estimating equation methods. *Journal of the Royal Statistical Society, Series B* 58(4):711–725.

Wilkinson, G. N. and C. E. Rogers 1973. Symbolic description of factorial models for analysis of variance. *Journal of the Royal Statistical Society, Series C* 22(3):392–399.

Williams, B. K., J. D. Nichols, and M. J. Conroy 2002. *Analysis and Management of Animal Populations*. London: Academic Press.

Williams, D. A. 1975. The analysis of binary responses from toxicological experiments involving reproduction and teratogenicity. *Biometrics* 31(4):949–952.

Winkelmann, R. 2008. *Econometric Analysis of Count Data* (5th ed.). Berlin: Springer.

Winkelmann, R. and S. Boes 2006. *Analysis of Microdata*. Berlin: Springer.

Withers, C. S. and S. Nadarajah 2009. The asymptotic behaviour of the maximum of a random sample subject to trends in location and scale. *Random Operators and Stochastic Equations* 17(1):55–60.

Wold, S. 1974. Spline functions in data analysis. *Technometrics* 16(1):1–11.

Wood, S. N. 2006. *Generalized Additive Models: An Introduction with R*. London: Chapman and Hall.

Wooldridge, J. M. 2006. *Introductory Econometrics: A Modern Approach* (5th ed.). Mason, OH, USA: South-Western.

Yanai, H., K. Takeuchi, and Y. Takane 2011. *Projection Matrices, Generalized Inverse Matrices, and Singular Value Decomposition*. New York, USA: Springer.

Yang, H.-C. and A. Chao 2005. Modeling animals' behavioral response by Markov chain models for capture–recapture experiments. *Biometrics* 61(4):1010–1017.

Yasuda, N. 1968. Estimation of the interbreeding coefficient from phenotype frequencies by a method of maximum likelihood scoring. *Biometrics* 24(4):915–934.

Yatchew, A. 2003. *Semiparametric Regression for the Applied Econometrician*. Cambridge: Cambridge University Press.

Yee, T. W. 1998. On an alternative solution to the vector spline problem. *Journal of the Royal Statistical Society, Series B* 60(1):183–188.

Yee, T. W. 2000. Vector splines and other vector smoothers. In J. G. Bethlehem and P. G. M. van der Heijden (Eds.), *Proceedings in Computational Statistics COMPSTAT 2000*, pp. 529–534. Heidelberg: Physica-Verlag.

Yee, T. W. 2004a. A new technique for maximum-likelihood canonical Gaussian ordination. *Ecological Monographs* 74(4):685–701.

Yee, T. W. 2004b. Quantile regression via vector generalized additive models. *Statistics in Medicine* 23(14):2295–2315.

Yee, T. W. 2006. Constrained additive ordination. *Ecology* 87(1):203–213.

Yee, T. W. 2010a. The **VGAM** package for categorical data analysis. *Journal of Statistical Software* 32(10):1–34.

Yee, T. W. 2010b. VGLMs and VGAMs: an overview for applications in fisheries research. *Fisheries Research* 101(1–2):116–126.

Yee, T. W. 2014. Reduced-rank vector generalized linear models with two linear predictors. *Computational Statistics & Data Analysis* 71:889–902.

Yee, T. W. and A. F. Hadi 2014. Row-column interaction models, with an R implementation. *Computational Statistics* 29(6):1427–1445.

Yee, T. W. and T. J. Hastie 2003. Reduced-rank vector generalized linear models. *Statistical Modelling* 3(1):15–41.

Yee, T. W. and N. D. Mitchell 1991. Generalized additive models in plant ecology. *Journal of Vegetation Science* 2(5):587–602.

Yee, T. W. and A. G. Stephenson 2007. Vector generalized linear and additive extreme value models. *Extremes* 10(1–2):1–19.

Yee, T. W., J. Stoklosa, and R. M. Huggins 2015. The VGAM package for capture–recapture data using the conditional likelihood. *Journal of Statistical Software* 65(5):1–33.

Yee, T. W. and C. J. Wild 1996. Vector generalized additive models. *Journal of the Royal Statistical Society, Series B* 58(3):481–493.

Yeo, I.-K. and R. A. Johnson 2000. A new family of power transformations to improve normality or symmetry. *Biometrika* 87(4):954–959.

Young, G. A. and R. L. Smith 2005. *Essentials of Statistical Inference.* Cambridge: Cambridge University Press.

Yu, K. and J. Zhang 2005. A three-parameter asymmetric Laplace distribution and its extension. *Communications in Statistics - Theory and Methods* 34(9–10):1867–1879.

Yu, P. and C. A. Shaw 2014. An efficient algorithm for accurate computation of the Dirichlet-multinomial log-likelihood function. *Bioinformatics* 30(11):1547–54.

Zellner, A. 1962. An efficient method of estimating seemingly unrelated regressions and tests for aggregation bias. *Journal of the American Statistical Association* 57(298):348–368.

Zhang, C. 2003. Calibrating the degrees of freedom for automatic data smoothing and effective curve checking. *Journal of the American Statistical Association* 98(463):609–628.

Zhang, Y. and O. Thas 2012. Constrained ordination analysis in the presence of zero inflation. *Statistical Modelling* 12(6):463–485.

Zhu, M., T. J. Hastie, and G. Walther 2005. Constrained ordination analysis with flexible response functions. *Ecological Modelling* 187(4):524–536.

Zuur, A. F. 2012. *A Beginner's Guide to Generalized Additive Models with R.* Newburgh, UK: Highland Statistics Ltd.

Zuur, A. F., E. N. Ieno, and E. H. Meesters 2009. *A Beginner's Guide to R.* New York, USA: Springer.

Zuur, A. F., A. A. Saveliev, and E. N. Ieno 2012. *Zero Inflated Models and Generalized Linear Mixed Models with R.* Newburgh, UK: Highland Statistics Ltd.

Index

© Thomas Yee 2015
T.W. Yee, *Vector Generalized Linear and Additive Models*,
Springer Series in Statistics, DOI 10.1007/978-1-4939-2818-7

Printed by Printforce, the Netherlands